Late-Quaternary Environments
of the United States
Volume 1
The Late Pleistocene

Late-Quaternary Environments of the United States

H. E. Wright, Jr., Editor

Volume 1

The Late Pleistocene

Stephen C. Porter, Editor

University of Minnesota Press, Minneapolis

Copyright © 1983 by the University of Minnesota.
All rights reserved.
Published by the University of Minnesota Press.
2037 University Avenue Southeast, Minneapolis, MN 55414
Printed in the United States of America.

Library of Congress Cataloging in Publication Data

Main entry under title:
Late-Quaternary environments of the United States.
  Includes bibliographies.
  Contents: v. 1. The late Pleistocene/Stephen C.
Porter, editor—v. 2. The Holocene/H. E.
Wright, Jr., editor.
  1. Geology, Stratigraphic—Quaternary—Congresses.
2. Geology—United States—Congresses. I. Wright,
H. E. II. Porter, Stephen C.
QE696.L29   1983   551.7'9'0973   83-5804

ISBN 0-8166-1252-8 (set)
ISBN 0-8166-1169-6 (v. 1)
ISBN 0-8166-1171-8 (v. 2)

Contents

Contributors to This Volume   vii

Preface   ix

Introduction—Stephen C. Porter   xi

Glaciation

1. The Late Wisconsin Glacial Record of the Laurentide Ice Sheet in the United States
    D. M. Mickelson, Lee Clayton, D. S. Fullerton, and H. W. Borns, Jr.   3

2. The Cordilleran Ice Sheet in Alaska
    Thomas D. Hamilton and Robert M. Thorson   38

3. The Cordilleran Ice Sheet in Washington, Idaho, and Montana
    Richard B. Waitt, Jr., and Robert M. Thorson   53

4. Late Wisconsin Mountain Glaciation in the Western United States
    Stephen C. Porter, Kenneth L. Pierce, and Thomas D. Hamilton   71

Nonglacial Environments

5. Late-Pleistocene Fluvial Systems
    Victor R. Baker   115

6. Depositional Environment of Late Wisconsin Loess in the Midcontinental United States
    Robert V. Ruhe   130

7. Sangamon and Wisconsinan Pedogenesis in the Midwestern United States
    Leon R. Follmer   138

8. Trends in Late-Quaternary Soil Development in the Rocky Mountains and Sierra Nevada
    of the Western United States
    Ralph R. Shroba and Peter W. Birkeland   145

9. The Periglacial Environment in North America during Wisconsin Time
    Troy L. Péwé   157

10. Pluvial Lakes of the Western United States
    George I. Smith and F. Alayne Street-Perrott   190

Coastal and Marine Environments

11. Sea Level and Coastal Morphology of the United States through the Late Wisconsin Glacial Maximum
    Arthur L. Bloom   215

12. The Ocean around North America at the Last Glacial Maximum
    John Imbrie, Andrew McIntyre, and T. C. Moore, Jr.   230

Pleistocene Biota

13. Vegetational History of the Northwestern United States Including Alaska
    Calvin J. Heusser   239

14. Late Wisconsin Paleoecology of the American Southwest
    W. Geoffrey Spaulding, Estella B. Leopold, and Thomas R. Van Devender   259

15. Vegetational History of the Eastern United States 25,000 to 10,000 Years Ago
    W. A. Watts   294

16. Terrestrial Vertebrate Faunas
    Ernest L. Lundelius, Jr., Russell W. Graham, Elaine Anderson, John Guilday, J. Alan Holman, David W. Steadman, and S. David Webb   311

17. Late Wisconsin Fossil Beetles in North America
    Alan V. Morgan, Anne Morgan, Allan C. Ashworth, and John V. Matthews, Jr.   354

18. The Antiquity of Man in America
    Frederick Hadleigh West   364

Climatology

19. Paleoclimatic Evidence from Stable Isotopes
    Irving Friedman   385

20. Late-Pleistocene Climatology
    R. G. Barry   390

Contributors to This Volume

Elaine Anderson, *730 Magnolia Street, Denver, Colo. 80220.*

Allan C. Ashworth, *Department of Geology, North Dakota State University, Fargo, N.D. 58105.*

Victor R. Baker, *Department of Geosciences, University of Arizona, Tucson, Ariz. 85721.*

R. G. Barry, *Cooperative Institute for Research in Environmental Sciences, University of Colorado, Boulder, Colo. 80309.*

Peter W. Birkeland, *Department of Geological Sciences, University of Colorado, Boulder, Colo. 80309.*

Arthur L. Bloom, *Department of Geological Sciences, Cornell University, Ithaca, N.Y. 14853.*

H. W. Borns, Jr., *Department of Geological Sciences, University of Maine, Orono, Maine 04473.*

L. Clayton,*Wisconsin Geological Survey, Madison, Wis. 53706.*

Leon R. Follmer, *Illinois State Geological Survey, Urbana, Il. 61801.*

Irving Friedman, *U.S. Geological Survey, Federal Center, MS 913, Denver, Colo. 80225.*

Russell W. Graham, *Illinois State Museum, Springfield, Il. 62706.*

John Guilday, *deceased.*

Thomas D. Hamilton, *U.S. Geological Survey, Gould Hall, University of Alaska, Anchorage, Alaska 99504.*

Calvin J. Heusser, *Department of Biology, New York University, Box 608, Tuxedo, N.Y. 10987.*

J. Alan Holman, *Department of Geology, Michigan State University, East Lansing, Mich. 48823.*

John Imbrie, *Department of Geological Sciences, Brown University, Providence, R.I. 02912.*

Estella B. Leopold, *Quaternary Research Center, University of Washington, Wash. 98195.*

Ernest L. Lundelius, Jr., *Department of Geology, University of Texas, Austin, Tx. 78712.*

Andre McIntyre, *Lamont-Doherty Geological Observatory, Palisades, N.Y. 10964.*

John V. Matthews, Jr., *Geological Survey of Canada, Ottawa, Ontario KIA 0E8, Canada.*

D. M. Mickelson, *Department of Geology and Geophysics, University of Wisconsin, Madison, Wis. 53706.*

T. C. Moore, Jr., *Exxon Production Research Co., Houston, Tx. 77001.*

Anne Morgan, *Department of Biology, University of Waterloo, Ontario N2L 3G1.*

Alan V. Morgan, *Department of Earth Sciences, University of Waterloo, Ontario N2L 3G1.*

Troy L. Péwé, *Department of Geology, Arizona State University, Tempe, Ariz. 85281.*

Kenneth L. Pierce, *U.S. Geological Survey, Federal Center, Denver, Colo. 80225.*

Stephen C. Porter, *Quaternary Research Center, University of Washington, Seattle, Wash. 98195.*

Robert V. Ruhe, *Water Resources Research Center, Indiana University, Bloomington, In. 47401.*

Ralph R. Shroba, *U.S. Geological Survey, MS 954, Box 25046, Federal Center, Denver, Colo. 80225.*

George I. Smith, *U.S. Geological Survey, Menlo Park, Calif. 94025.*

W. Geoffrey Spaulding, *Quaternary Research Center, University of Washington, Seattle, Wash. 98195.*

David W. Steadman, *National Museum of Natural History, Washington, D.C. 20515.*

F. Alayne Street-Perrott, *School of Geography, Oxford OX1 3TB, England.*

Robert M. Thorson, *Museum of Natural History, University of Alaska, College, Alaska 99701.*

Thomas R. Van Devender, *Department of Geosciences, University of Arizona, Tucson, Ariz. 95721.*

Richard B. Waitt, Jr., *U.S. Geological Survey, Vancouver, Wash. 98660.*

W. A. Watts, *Provost's House, Trinity College, Dublin 2, Ireland.*

S. David Webb, *Florida State Museum, University of Florida, Gainesville, Fla. 32601.*

Frederick Hadleigh West, *Laboratory of Prehistoric Archaeology, Williams College, Williamstown, Mass. 01267.*

Preface

To many, the Pleistocene Epoch, which constitutes the earlier and major part of the Quaternary Period, is synonymous with the glacial ages and is viewed as a time of dramatic environmental change. Although there now is abundant evidence that glaciations occurred not only during the Pleistocene but through much of the Tertiary Period as well, and therefore are more appropriately considered as late-Cenozoic phenomena, the Pleistocene was a time when successive glacial intervals were most extreme and when their effect on the world's landscapes was most pronounced. The dynamic nature of Pleistocene environmental changes is most obvious in the case of the last glaciation, in part because of its proximity to our own time and in part because the geologic record of the changes is more extensive than that of earlier glacial episodes. For these reasons, the opportunities for assessing ice-age environmental conditions reliably are more numerous for late-Pleistocene time, and the resulting reconstructions generally are regarded as reasonably representative of conditions during earlier Pleistocene glaciations.

Two decades ago, work was begun on a major synthesis of the Quaternary history of the United States, the final product of which appeared in monographic form at the time of the Seventh Congress of the International Association for Quaternary Research (now the International Quaternary Union, or INQUA) in Boulder, Colorado (Wright and Frey, 1965). In that volume, the entire Quaternary record was reviewed by representatives from the major scientific disciplines engaged in Quaternary research. The past decade and a half has seen a remarkable expansion of research in this field, in part because of the intrinsic interest in the subject but also because of the increasing need for information about surface processes and the environmental history of our planet. An underlying theme for much of this research involves the question of the magnitude, character, and rates of climatic change, about which significant insights are only likely to be provided by the Quaternary statigraphic record. The flowering of activity in Quaternary-related studies has led to the establishment of national organizations such as the American Quaternary Association (AMQUA), the organization of specialized research centers and institutes in more than half a dozen major American universities, and the ap-

pearance of new journals dealing exclusively with interdisciplinary Quaternary research.

The present volume grew out of a series of conferences held by Soviet and American scientists under the auspices of the U.S./U.S.S.R. Cooperative Agreement in the field of Environmental Protection. The meetings, at which physical and biological scientists from the two countries exchanged information and ideas, developed around the theme of environmental changes in the Northern Hemisphere during the last glaciation and the Holocene. Accordingly, it was agreed that monographs encompassing these intervals of time would be published in both english- and russian-language editions, with publication to coincide with the Eleventh Congress of the International Quaternary Union in Moscow in 1982. The American contribution was designed to appear in two complementary volumes, one dealing with the late Pleistocene and one with the Holocene, both emphasizing environmental reconstruction insofar as possible. In the present volume, emphasis is placed on the interval 25,000 to 10,000 yr. B.P., and, although most contributors to the volume deal exclusively with this time period, in some chapters both earlier and later events are also considered. To some extent, these monographs serve to update *The Quaternary of the United States*, published in 1965, although they are more restricted in temporal scope. They emphasize important new data and interpretations as well as new techniques and approaches that have led to significant advances in Quaternary science.

Because of the increasing number of scientists working on Quaternary problems and the corresponding increase in new information resulting from field and laboratory studies of Quaternary phenomena, some aspects of these volumes undoubtedly will be superseded within a few years. But this is an encouraging sign, for it speaks well for the vigor of this research field and for the importance Quaternary environmental science has achieved within the broader framework of the natural sciences.

Reference

Wright, H. E., Jr., and Frey, D. G. (eds.) (1965). "The Quaternary of the United States." Princeton University Press, Princeton, N.J.

Introduction
Stephen C. Porter

Environments and Environmental Change during the Late Pleistocene

The varied natural landscapes of the United States owe much of their present character to geologic and biologic events of the last 25,000 years, the interval encompassing the most recent glacial-interglacial cycle. Many of the forces that shaped the land surface in the past can be seen at work today, but others no longer operate in the same areas. To a large extent, the scenic grandeur of the American West, the agriculturally productive soils of the Midwest, and the country's abundant water and timber resources and important economic mineral deposits are traceable to events of the past 25 millennia. Over a substantial part of the country, cities and towns have been constructed and roads built on sediments that were deposited within this most recent interval of geologic time. Thus, there are both intellectual and practical motivations for pursuing studies of the late-Quaternary environmental history of the United States, for they not only provide important insights into the natural history of the landscape but can also produce the kinds of information needed to solve problems posed by urban development, natural hazards, and management of natural resources.

Also of increasing concern is the question of climatic change on short time scales, for the progressive modification of atmospheric chemistry brought on by industrial burning of fossil fuels may lead to a substantial shift in global climate detectable by the middle of the next century. Studies of natural environments of the Quaternary provide important data bearing on the magnitude, character, and rates of climatic change and the manner in which surface environments respond to such changes. Such information should permit us to anticipate some of the biologic and physical changes that may occur over the coming centuries.

The preserved geologic record of the United States for that part of the late Pleistocene between about 25,000 and 10,000 years ago presents a picture of dynamic environmental change resulting from the shift from full-glacial to postglacial conditions. The range of latitudes encompassed (19 °N to 71 °N), the breadth of the continent (4000 km), and the extremely varied surface topography (sea level to 6190 m) mean that the diversity of natural environments, both present and past, is unmatched in any other country. For this reason, the late-Pleistocene record of the United States is an extremely varied one reflecting the complex interplay of many geologic and biologic processes in a wide range of environmental settings.

The Last Glaciation

At the culmination of the last (Wisconsin) glacial age about 20,000 to 18,000 years ago, the region was strikingly different from what it is today. Much of the northern third of the contiguous states was buried beneath the Laurentide and Cordilleran ice sheets, which to the north covered nearly all of Canada under ice as much as 3500 m thick. The southernmost point reached by the Laurentide glacier was at 39 °N latitude in the Mississippi River drainage, where major lobes occupied the basins of the Great Lakes. The two ice sheets constituted the single greatest influence on the development of the present topography in the northern part of the country, for they eroded the landscape, left extensive morainal deposits in their terminal zones, disrupted drainage systems, locally dammed extensive proglacial lakes, produced voluminous meltwater sediments that were deposited as outwash valley trains and reworked by the wind into sand dunes and widespread blankets of loess, and produced a major effect on the atmospheric circulation of the North American continent. The ice sheets also markedly reduced air and ground temperatures near their margins and contributed to the development of a rigorous periglacial zone, which through some sectors was characterized by perennially frozen ground and tundra vegetation. Along its eastern margin, the Laurentide glacier extended beyond the present shoreline onto the now-submerged continental shelf in the Gulf of Maine. The Cordilleran ice sheet in the northwestern states, which reached its southernmost limit in the Puget lowland of western Washington about 14,000 years ago, encountered more rugged terrain and possessed an irregular lobate margin. It overwhelmed the northern Cascade Range and buried all but the highest summits, which rose above its surface as sharp-crested nunataks. At its seaward limit, the glacier extended beyond the present coastline of Washington and British Columbia and is presumed to have terminated near the edge of the continental shelf.

Mountain glaciers formed in highlands south of the ice sheets in the western United States as well as in Alaska. Reconstructions indicate that at the glacial maximum the snow line was lowered about 1000 m below its present level in the middle latitudes, but probably less in the Arctic, in response to changes of temperature and precipitation. Although the timing of ice advances in the mountains is generally less well known than the timing of the fluctuations of the southern margins of the continental ice sheets, limited dates and stratigraphic relationships suggest that the culminating advances may have been broadly in phase.

Pluvial lakes that formed in intermontane basins of the American West apparently reached their largest dimensions at about the same time the mountain glaciers did, in contrast to low-latitude lakes of central Africa, which were low during glacial times and high during the early Holocene (Street and Grove, 1979). The magnitude of the change of specific climatic factors responsible for the rise of the Great Basin lakes is not yet known with certainty; conflicting hypotheses argue for wet and cool or, alternatively, dry and cold full-glacial conditions in that region.

Although attempts have been made to construct maps depicting vegetational conditions in the United States under full-glacial conditions on the basis of fossil-pollen data, well-dated sites of appropriate age are few in number and are limited mainly to the eastern half of the country. Except for a few scattered localities, sites in the montane West that could be used to develop such maps are not available. In the desert Southwest, important new data have come to light from studies of plant macrofossil remains in ancient packrat (*Neotoma*) middens that span the full-glacial interval and provide critical evidence, at the species level, of local vegetational environments and altitudinal shifts of ecotone boundaries. The general picture of full-glacial conditions that emerges from vegetational studies is often difficult to interpret with confidence, for some species assemblages have no exact modern counterparts. It is clear, however, that distributional patterns at the glacial maximum did not always reflect a simple latitudinal or altitudinal displacement of modern ecotones.

Reconstructions of faunal distributions also suffer from a paucity of well-dated sites having an adequate geographic spread. Local faunas provide evidence of mammalian populations that included many now-extinct species. Conflicting hypotheses regarding the nature of the full-glacial vegetation in Alaska hinge partially on vertebrate fossil evidence and inferences regarding the environmental conditions and food resources to which large-mammal faunas were adapted.

The extent to which global sea level fell during the glacial maximum is not known with certainty (estimates range from less than 100 m to more than 150 m), and shoreline reconstructions for full-glacial conditions reflect this uncertainty. Regardless of the exact amount of the marine regression, it is clear that broad expanses of now-submerged continental margins were exposed and in places the coastline configuration was substantially different. The most significant geographic changes occurred off the eastern seaboard of the United States, where the 100-m isobath lies some 120 km beyond the present shoreline, and adjacent to Alaska, where the shallow Bering Shelf became an emergent land mass (Beringia) more than 1000 km wide connecting Asia and North America. Fossil plants and animals dredged and cored from the continental shelves provide evidence of the habitats that existed there during the glacial maximum and during the early part of the subsequent marine transgressions.

The emergence of the Bering Shelf at times of glacially lowered sea level bears importantly on the controversial question of when human beings first entered the New World. Characteristics of ancient bones and bone fragments recovered in the Old Crow Basin of the Yukon Territory have led some investigators to conclude that humans of Asian origin were present in northwestern North America during the last glaciation and, in fact, that they may have arrived

there before the Sangamon Interglaciation (Jopling et al., 1981). Others are skeptical of such an interpretation and maintain that the earliest unequivocal, stratigraphically controlled cultural evidence for humans in the New World dates to late-glacial times, perhaps no earlier than 15,000 to 14,000 years ago.

Sea-surface temperatures of the full-glacial oceans off North American have recently been reconstructed from microfossil data obtained from deep-sea cores, which permit definition of the distributional patterns of planktonic assemblages (CLIMAP Project Members, 1976). The results show that in both the Atlantic and Pacific Oceans the polar water mass expanded, thereby conpressing the belts of subpolar and transitional waters, shifting middle-latitude zonal currents, and steepening thermal gradients. Ocean waters off the United States were markedly cooler north of latitude 42 °N at the glacial maximum, especially in winter.

A new and promising approach to paleclimatic reconstructions has appeared with the development of global circulation models of the atmosphere that can be investigated with modern high-speed computers. Through the use of ice-age sea-surface temperatures derived by the CLIMAP project, mapped distribution of glacier ice, inferred distribution of sea ice, and estimated albedo for land surfaces as inputs to the models, simulated ice-age climates have been generated (e.g. Gates, 1976; Manabe and Hahn, 1977). The results of several model experiments differ in detail (Figure I-1), but the patterns of ice-age temperature and precipitation departures across the United States are reasonably consistent with what is inferred about full-glacial conditions based on terrestrial paleoclimatic data. Through assessment of the magnitude of ice-age climatic change in continental regions from physical and biologic evidence and through refinement of knowledge about the basic input parameters, the validity of the model results can be evaluated and improved models can be developed. Such modeling of surface environmental conditions for different intervals in the past should provide new insights into unresolved problems of Quaternary paleoclimatology.

Late-Glacial Environmental Changes

A variety of physical and biologic evidence suggests that environmental patterns of the full-glacial generally persisted until about 14,000 years ago, with only minor reversals. Thereafter, however, conditions changed dramatically as the Earth withdrew from the glacial age and moved toward full-interglacial conditions. For many natural systems the change was not complete until well into the Holocene, but for some systems interglacial patterns were already in existence by the end of the Pleistocene.

The dating of late-glacial events is often sufficiently well controlled that response lags of different systems to climatic change can be evaluated. Although final major recession of the southern margin of the Laurentide ice sheet began shortly after 14,000 years ago and the ice had largely retreated from the United States by the beginning of the Holocene, the process of deglaciation continued in Canada for another four to five millennia. By contrast, the southern margin of the Cordilleran ice sheet had retreated to, or north of, the Canadian border by about 12,000 years ago, and by the beginning of the Holocene it probably had receded to, or within, present glacier limits in source areas.

Radiocarbon dates from areas of mountain glaciation in the American West indicate that both the San Juan Mountains and the Yellowstone Plateau became largely deglaciated before about 14,000

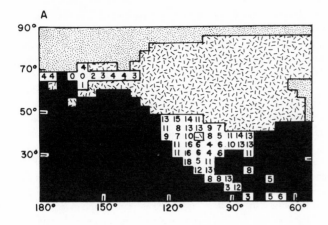

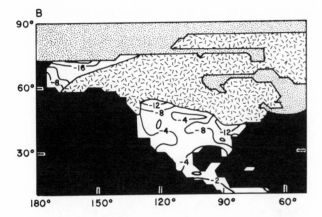

Figure I-1. Simulated differences of summer air temperatures (in degrees Celsius) over ice-free land areas generated by global circulation model experiments. Both experiments illustrated used CLIMAP-generated boundary conditions, including distribution of glacier ice (coarse stippled pattern) and inferred distribution of sea ice (fine stippled pattern). A. Present minus ice-age July surface temperatures based on RAND model. (Modified from Gates, 1976: Figure 7, from *Science*, volume 191, pp. 1138-44), copyright 1976 by the American Association for the Advancement of Science.) B. Ice-age minus present July-August surface air temperatures derived from Geophysical Fluid Dynamics Laboratory model experiment. (Modified from Manake and Hahn, 1977: Figure 20.)

years ago in response to a rising snow line that left most of the high and extensive source areas of the glacier systems below the snow line. A similar early deglaciation is inferred for the small ice cap on Mauna Kea in Hawaii, which formed under full-glacial climate as the snow line was lowered to intersect the upper slopes of the volcano. Although in most mountain ranges radiocarbon dating control is not adequate to determine the exact timing of deglaciation, by the end of the Pleistocene most highlands had apparently lost the bulk of their ice-age glacier cover. The distribution of late-glacial Glacier Peak tephra in the North Cascades, for example, indicates that the heads of at least some major valleys were free of ice before about 12,000 years ago. Similarly early dates for deglaciation have been obtained from Alaska, where some large valley systems were free or nearly free of glacier ice by 12,000 or 11,000 years ago.

With the onset of deglaciation, certain large ice-dammed lakes were

suddenly emptied, and the resulting floods dramatically affected the landscape over which they passed. Among the most impressive examples was the release of waters from glacial Lake Missoula in Montana, which inundated a vast tract of land in the Columbia River Basin of Washington. The Channeled Scabland, a product of a succession of such floods during the Pleistocene, represents only a part of the affected drainage system, for the floodwaters left a record of their passage all along the Columbia Valley, as well as in deposits of the deep-sea fan at the base of the continental slope beyond the mouth of the river. The lengthy and often lively controversy surrounding the origin of the Channeled Scabland and its related deposits provides an example of the difficulties sometimes encountered when attempting to interpret late-glacial environments that frequently contrasted dramatically with those of the present.

The change to postglacial climatic conditions also affected pluvial lakes in closed basins of the western United States. Although lake levels remained generally high until about 14,000 years ago, after that date all lakes receded systematically, and most became dry by the middle Holocene. However, the response was not synchronous regionally, for lakes first began to wane in the southern Basin and Range region about 13,000 years ago, whereas lakes farther north remained high or at intermediate levels until the end of the Pleistocene.

As continental ice sheets wasted, meltwaters returning to the world oceans caused sea level to rise eustatically, inundating the exposed margins of the continental shelf. The steepest part of reconstructed sea-level curves for the deglacial hemicycle lies between about 14,000 and 10,000 years ago, an interval during which the major Northern Hemisphere ice sheets of North America and Europe were greatly reduced in volume and most smaller glaciers largely or entirely wasted away. In coastal regions formerly covered by Laurentide and Cordilleran ice, rising sea level inundated isostatically depressed land areas and left a marine record that now, after completion of isostatic recovery, lies above present sea level. In Washington State such deposits are found to an altitude of at least 200 m, while in Maine the marine limit rises to a maximum of about 150 m. In Beringia the transgression led to widespread submergence of the Bering Shelf and severance of the land connection between Siberia and Alaska.

Beetles are among the most sensitive indicators of environmental change because of their mobility, their specific ecologic tolerances, and the often excellent preservation of their hard parts. During full-glacial times, insects that formerly occupied regions covered by the continental ice sheets were displaced to refugia in Beringia, in a belt across the northern United States south of the ice limit, and possibly in the now-submerged continental shelves. The environmental changes accompanying deglaciation led to rapid dispersal of beetle populations. In the eastern United States, beetle remains record the transition from open to forested conditions and appear simultaneously in the fossil record with the first spruce. Considerable range changes occurred as recently as 12,000 to 10,000 years ago; for some species postglacial migration distances were as much as 2000 to 3000 km. As more data are assembled, it becomes apparent that fossil beetles hold the promise of providing important information about postglacial migration routes that will improve environmental reconstructions.

Disharmonious floras and faunas containing species that appear ecologically incompatible in modern terms are characteristic of the late Pleistocene in the United States. The lack of modern analogues suggests either that various species responded at different times or migrated at different rates in response to the change from a glacial to an interglacial climatic regime, thereby producing ever-changing mixes of species, or that fossil populations reflect changing climatic regimes that were different from those of the present world. Pollen data clearly indicate the progressive migration of forest species from refugia of full-glacial times into deglaciated terrain, a process for which there are many modern analogues, although on much more limited geographic scales. The distances through which individual species migrated are often measured in many hundreds or even thousands of kilometers, and the paths they took from widely separated full-glacial refugia imply that the late-glacial interval was characterized by an ever-changing mosaic of vegetation. Such changes further suggest that the reconstruction of late-glacial environments on the basis of biotic information can be perilous unless the ecologic tolerances of species are well understood.

Major modifications in vertebrate faunas also occurred during late-glacial times. Many late-Pleistocene faunas appear to record climates with smaller seasonal extremes than today. Assemblages that are generally regarded as being boreal, temperate, or subtropical in the contemporary world have been found in associations having no modern analogues.

The transition from late-Pleistocene to modern vertebrate faunas was relatively abrupt. It occurred near the end of the Pleistocene about 11,000 to 10,000 years ago and was marked by the extinction of many species of large vertebrates, the cause of which is a much debated question. One hypothesis focuses on the arguments that the extinctions coincided with a time of major environmental change and that the climatic stresses of the late-glacial world may have played a major role both in causing the wave of extinctions and in restricting the ranges of extant species. Others have emphasized the association of the remains of extinct species with paleo-indian kill sites and have argued that the appearance of hunters in central North America at the end of the last glacial age led to a widespread overkill of the populous late-Pleistocene large-mammal fauna. Although some smaller vertebrate species also became extinct (e.g., reptiles, amphibians, and birds), about a third of the extinct forms were predators or scavengers, and nearly all extinct forms were large, with an average body weight of more than 60 kg. A large percentage of the extinctions apparently occurred at the end of the Pleistocene, but some now-extinct species disappeared as early as 18,000 years ago; others persisted well into the Holocene. In view of the many dramatic environmental changes that were occurring at the end of the glacial age, an unequivocal cause-and-effect relationship is difficult to prove. This clearly is one area of research in which data from a number of related Quaternary disciplines are needed to help solve an intriguing scientific problem.

The end of the Pleistocene, arbitrarily placed at 10,000 yr B.P., heralded no cessation of environmental change. There remained a vast ice sheet centered over eastern Canada, much of the land in the northern part of the United States was experiencing isostatic uplift as the crust continued to recover from the recently vanished glacier load, rising sea level still stood some 20 m below its present position, and shifts of biota were dramatically changing the character of the landscape. The environmental changes of the latest Pleistocene and the Holocene can, therefore, be viewed as a continuum in part reflecting the ongoing recovery from the glacial age but also reflecting secondary climatic variations superimposed on the general first-order trend. The complex dynamic adjustments of landscape elements to a new interglacial state, set in motion as the glacial age terminated, continued into the Holocene and in some cases persist to the present day. Among the significant advances in Quaternary research during the past two decades has been a general recognition of the importance these dynamic interactions have played in shaping the modern landscape. As a result, new questions are posed and new avenues of research are suggested that should ensure the generation of fresh hypotheses and a continued lively debate about the Quaternary evolution of the United States.

References

CLIMAP Project Members (1976). The surface of the ice-age Earth. *Science* 191, 1131-38.

Gates, W. L. (1976). Modeling the ice-age climate. *Science* 191, 1138-44.

Jopling, A. V., Irving, W. N., and Beebe, B. F. (1981). Stratigraphic, sedimentological, and faunal evidence for the occurrence of pre-Sangamonian artefacts in northern Yukon. *Arctic* 34, 3-33.

Manabe, S., and Hahn, D. G. (1977). Simulation of the tropical climate of an ice age. *Journal of Geophysical Research* 82, 3889-3911.

Street, F. A., and Grove, A. T. (1979). Global maps of lake-level fluctuations since 30,000 yr B.P. *Quaternary Research* 12, 83-118.

Glaciation

CHAPTER 1

The Late Wisconsin Glacial Record
of the Laurentide Ice Sheet in the United States

D. M. Mickelson, Lee Clayton, D. S. Fullerton, and H. W. Borns, Jr.

Introduction

The Laurentide ice sheet covered most of north-central and north-eastern North America during late Wisconsin time. Recent reconstructions of the ice sheet by Bryson and others (1969), Prest (1969), Ives and others (1975), and Denton and Hughes (1980) provide a good deal of information about the behavior of the ice sheet through time and about the nature of its margins. However, many questions remain unanswered. The extent and nature of the marine-based part of the ice sheet along its eastern edge are still debated. The amount of interaction with the Cordilleran ice sheet and the possible existence of an ice-free corridor are still in question. These issues are discussed by Mayewski and others (1980) and are not pursued here. Instead, we review the record of Late Wisconsin Laurentide glaciation in the United States.

Since the publication of ''The Quaternary of the United States'' (Wright and Frey, 1965) numerous individual papers (e.g., Fullerton, 1980) and several compilations of papers (e.g., Black, et al., 1973; Mahaney, 1976; Denton and Hughes, 1980) have added to our understanding of the glacial history of the Laurentide ice sheet in the north-central and northeastern United States. Although new radiocarbon dates and the reevaluation of earlier ones in the light of new geomorphic and stratigraphic interpretations have led to a further refinement of the chronology of glacial and related events, we still do not understand details of the chronology of glacial events. In this chapter, time lines for ice-margin positions, based on radiocarbon dates, are tentative and will, no doubt, be reinterpreted in the future. Events are assumed to be synchronous, but supporting radiocarbon dates are not adequate to prove this concept. The chronology of events in several areas has been reinterpreted by using only stratigraphically controlled wood dates in or beneath till when they contradict dates on dispersed organic matter.

The morphometry of landforms has been studied in local areas, but major summaries of the physiography of glacial deposits have been undertaken only in a few areas. In fact, the realization that many landforms now found on younger surfaces are inherited features has been a major development in the study of the Great Lakes area during the last 20 years. Parts of many ''moraines'' do not mark stillstands of the margin of the last ice. In this chapter's section on the southern

Great Lakes, we attempt to define types of moraines and their significance as ice-margin indicators.

It has also become clear that markedly different landforms have been produced along different parts of the margin of the Laurentide ice sheet in the United States. Traditional concepts developed in ''active-ice'' terrain cannot be used to interpret the genesis of the thick supraglacial deposits of the western region (Figure 1-1).

Although most recent studies of glacial process have been done on modern glaciers, studies along various parts of the former Laurentide ice sheet have led to postulated modes of deposition of materials and formation of landforms in a few areas. Much more remains to be done in the application of observations on modern glaciers to the genesis of glacial landforms of Pleistocene age, and we feel that future attention should turn more to this field of study.

During the last 5 years, attempts have been made to formulate models of the Laurentide ice sheet as it was at or near its maximum 20,000 to 18,000 years ago. Although some observations in the western region have produced input parameters for glaciologic models (e.g., Mathews, 1974), most models have used numerous assumptions based on glaciologic theory.

In this chapter, the glacial history of the southern border of the Laurentide ice sheet is summarized; correlations are based on the assumption of synchrony of ice advances and a reinterpretation of radiocarbon dates. The number of local stratigraphic and geographic terms is kept to a minimum so that the discussion can be easily read. (We have attempted to follow the terminology used by the original authors for informal and formal stratigraphic names. We use upper-case letters for formal names and lowercase letters for informal names. Our use of stratigraphic names does not imply formal approval of various stratigraphic names and does not necessarily conform to the U.S. Geological Survey's usage.) However, we have used many of these terms in the figures and tables.

Many people contributed ideas and comments during the preparation of this manuscript, and we appreciate their assistance. We especially appreciate the review of a draft manuscript done by Ned K. Bleuer, Parker E. Calkin, Donald F. Eschman, George R. Hallberg, W. Hilton Johnson, S. C. Porter, and H. E. Wright, Jr. Their comments are greatly appreciated and we apologize if not all of our correlations concur with those they suggested.

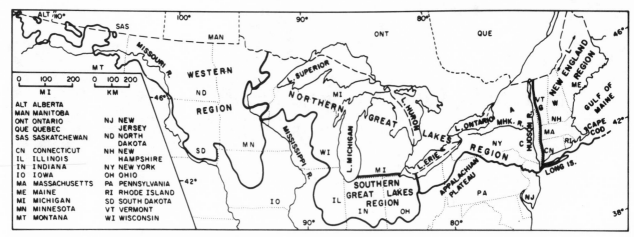

Figure 1-1. Locations of regions and places discussed in the text. The state and province abbreviations are given in the illustration. Other abbreviations are: (A) Adirondack Mountains, (C) Catskill Mountains, (G) Green Mountains, (L) Longfellow Mountains, (MHK. R.) Mohawk River, (W) White Mountains.

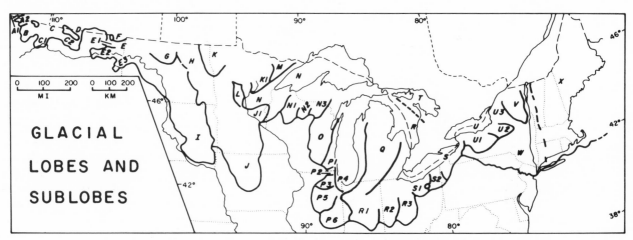

Figure 1-2. Lobes and sublobes of the Late Wisconsin Laurentide ice sheet in the United States. Lobes are designated by letter and sublobes by number. Listed from west to east:

A, Alberta	E2, Glasgow	L, Wadena	P3, Princeton	S1, Killbuck
A1, St. Mary	E3, Yellowstone	M, Rainy	P4, Joliet	S2, Grand River
A2, Milk River	F, Plentywood-Scobey	N, Lake Superior	P5, Peoria	T, Georgian Bay
B, Shelby	G, Souris	N1, Chippewa	P6, Decatur	U, Lake Ontario-St. Lawrence
C, Havre	H, Leeds	N2, Wisconsin Valley	Q, Saginaw	U1, Finger Lakes
C1, Loma	I, James	N3, Langlade	R, Lake Huron	U2, Oneida
C2, Malta	J, Des Moines	O, Green Bay	R1, East White	U3, Black River
D, Whitewater	J1, Grantsburg	P, Lake Michigan	R2, Miami	V, Adirondack Mountains
E, Missouri Valley	K, Red River	P1, Delavan	R3, Scioto	W, Lake Champlain-Hudson River
E1, Poplar	K1, St. Louis	P2, Harvard	S, Lake Erie	X, New England

Physical Setting of Ice Advances

BED SHAPE AND LOBATION OF ICE

The shape of the underlying land surface created major ice lobes and sublobes along the southern border of the Laurentide ice sheet (Figure 1-2). Because the direction of the ice flow changed markedly with advance and retreat of the ice during the Late Wisconsin, any attempt to portray the direction is deceptive. The directions of the ice flow are diagrammed in Figure 1-3. At no time, however, was ice actually moving along all the flow lines shown.

A map of the Late Wisconsin glacier bed would resemble a map of the present land surface (Figure 1-4), except that the land near Hudson Bay is higher today because of glacial rebound. The amount of crustal depression that had already taken place when ice first advanced during Late Wisconsin time is not known; neither is the amount of crustal depression during the maximum extent of ice because the ice thickness is not known. Any reasonable reconstruction of ice-surface profiles, however, shows that much of the land surface in the northern Great Lakes area (Figure 1-1) and eastward was at or below present sea level. Thus, ice flowed up a steeper slope than is

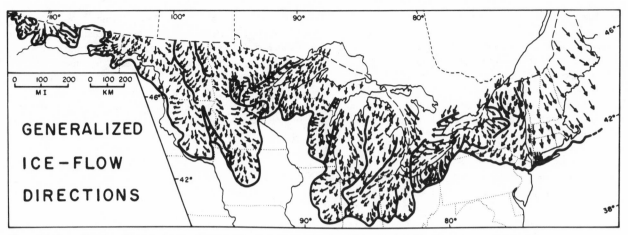

Figure 1-3. Generalized Late Wisconsin ice-flow directions indicating only flow directions during latest glacial advance at any specific locality.

present today. Because different lobes advanced and retreated at slightly different times and because the distance of advance varied, deposits along many lobe junctions overlap.

In the western region, the land surface slopes eastward from an altitude of more than 1000 m near the Rocky Mountains in Montana and Alberta. In Alberta, ice flowed westward, directly up the slope of the Great Plains. In western North Dakota, ice flow was controlled by broad, shallow lowlands as it moved southeastward, nearly perpendicular (Figure 1-3) to the regional slope (Figure 1-4) of the Great Plains; farther east, in eastern North Dakota and northern Minnesota, the ice sheet flowed southward. The Des Moines and James lobes (Figure 1-2) advanced across the Continental Divide and flowed southward down the regional slope into the Mississippi River drainage.

In the Great Lakes region, ice flow was profoundly influenced by the basins that now hold the Great Lakes. Ice was directed toward the southwest in the Lake Superior, Wadena, and Rainy lobes (Figure 1-2) and extended considerably farther west into Minnesota than to the south into Wisconsin, where it crossed Precambrian uplands. Ice flowing onto these uplands climbed a steep slope out of the lake basin, and, as a result, glacial deposits there have a high proportion of

supraglacial till. To the east, ice funneled southward down the Green Bay basin and still farther down the deeper Lake Michigan basin (Figures 1-2 and 1-3). An upland in central Michigan retarded ice flow, and at times the Huron lobe extended down its southwest-trending lowland as far west as Illinois. Abutting the ice margin in Ohio, Pennsylvania, and New York is the Appalachian Plateau (Figure 1-1), which rose to the south and east in front of the ice margin (Figure 1-4). East of Lake Ontario, the St. Lawrence lowland (Figure 1-1) extends northeastward to the sea. Ice flowing from the north and northeast was diverted along this lowland toward the Great Lakes (Figure 1-3).

The subglacial topography also influenced the ice sheet in a number of ways other than by simply directing ice flow. In the western region, extensive proglacial lakes formed in many valleys, and each drained southeastward toward eastern South Dakota and southern Minnesota, where the water entered the Mississippi River drainage. The ponded meltwater probably hastened glacial disintegration. Similar lakes formed in the basins of the Great Lakes and no doubt influenced the shape of the ice margin during deglaciation.

The lack of a ready outlet for meltwater in many areas may also have influenced conditions at the glacier bed. A buildup of subglacial

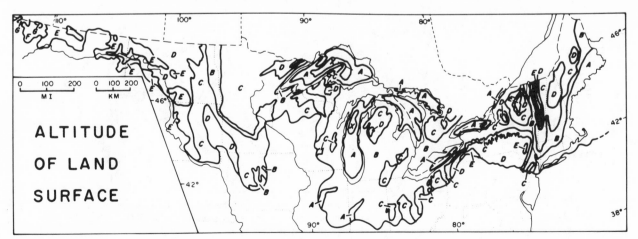

Figure 1-4. Generalized present land-surface altitude of area covered by the Late Wisconsin Laurentide ice sheet. (A) Sea level to 150 m, (B) 150 to 300 m, (C) 300 to 350 m, (D) 450 to 600 m, (E) 600 to 900 m, (F) 900 to 1200 m, (G) more than 1200 m, (X) below sea level.

pore pressure may have decreased shear strength at the bed and allowed the ice to move rapidly with a gentle surface slope. In the western area, lakes were probably too shallow to allow the formation of true ice shelves, but it is possible that the Great Lakes did support small ice shelves during the glacier's recession.

Another influence of bed shape on glacial process can be seen in Figure 1-10. Thin supraglacial till occurs near the centerline of lobes with beds that were flat or sloping downhill, including the lobes of north-central Montana as well as the Souris, Leeds, James, and Des Moines lobes. However, along the sides of the Souris, James, and Des Moines lobes, where the ice was flowing up against steep slopes, thick deposits of supraglacial till occur. The uphill flow resulted in decelerating ice movement and intensified shearing of subglacial material upward into the ice. A similar relationship occurred in the northern Great Lakes region, but in the southern part the uphill flow of the ice did not produce intense shearing on a regional scale, perhaps because of temperature differences.

Finally, the bed shape also determined the location of the lakes that collected fine-grained sediment, which later was reworked to form clayey or silty till throughout the region. Most clayey till found in the southern part of the area covered by the Laurentide ice sheet was probably derived from lake sediment.

BEDROCK LITHOLOGY

Most till was probably derived from preexisting sediment rather than directly from bedrock. This appears to be true across the United States, especially in the western and Great Lakes regions, where bedrock outcrops are few and the surface relief is low. However, on a regional scale, till texture and lithology is determined largely by the lithology of bedrock lying up-ice (Anderson, 1957; Dreimanis and Vagners, 1969). Noncalcerous till is confined to areas of crystalline bedrock where loess or calcareous lake sediment was not a significant contributor to the till.

Most till in the United States deposited by the Laurentide ice sheet contains granite, gneiss, schist, basalt, and a variety of other kinds of Precambrian igneous and metamorphic rock derived from the Cana-

dian Shield. In New York and New England (Figure 1-1), Paleozoic and Mesozoic terranes also contributed metamorphic and igneous rock types (Figure 1-5). Bordering the shield in the northern Great Lakes and western areas is a band of basal Paleozoic sandstone that in places is over 100 km wide (Figure 1-5, number 5). From eastern North Dakota eastward through western New York, Ordovician, Silurian, and Devonian rocks are primarily carbonate (Figure 1-5, number 3), although some sandstone (e.g., St. Peter sandstone) and shale (Figure 1-5, number 4) are present. In New York State and New England, carbonate rock alternates with the more common shale and sandstone units (Figure 1-5, number 4). The upper Paleozoic and Mesozoic are dominated by shale and sandstone (Figure 1-5, number 6), but in places coal and other continental deposits are present.

Nature of the Glacial Record

Major differences in the glacial record exist along the southern portion of the former ice sheet. In New England, part of the ice sheet was probably marine based, and its behavior was influenced by changes in sea level and the presence of a bounding ice shelf. Although the extent of Late Wisconsin ice has been debated (Grant, 1977; Mayewski et al., 1980; Hughes, 1981), it seems likely that the rapid deglaciation of the St. Laurence lowland as a calving bay produced a large, stagnating ice mass over northern and eastern New England relatively early during deglaciation. Because of the fairly high relief in this area, mountain summits soon protruded above the ice surface, and glacial deposits formed along ice margins in mountain valleys.

In the Great Lakes region, relief is lower than in New England (Figure 1-4). This area has traditionally been known for its end moraines and other "active-ice" features. In most places in the region, relatively clean ice deposited little supraglacial drift; such ice in some areas became nearly stagnant, but it left little evidence of this fact because the ice was clean.

In the northern Great Lakes and western areas (Figure 1-1), supraglacial debris is more abundant. Although subglacial till predominates, locally large areas of thick, hummocky supraglacial till

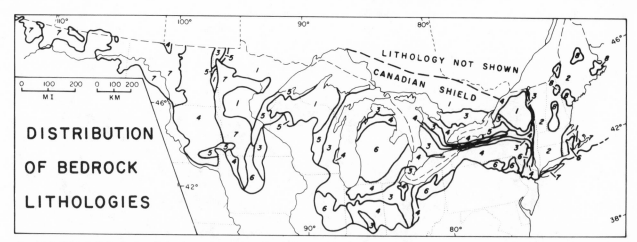

Figure 1-5. Generalized distribution of bedrock types in area covered by the Laurentide ice sheet in Late Wisconsin time. (1) Precambrian igneous and metamorphic rocks. (2) Paleozoic metamorphic and basic igneous rocks (various grades of metamorphism). (3) Paleozoic limestone and dolomite. (4) Older and younger Paleozoic and Mesozoic shale with siltstone, limestone, and dolomite; locally common sandstone. (5) Late Precambrian and older Paleozoic sandstone with minor shale, limestone, and dolomite. (6) Younger Paleozoic shale and sandstone with minor limestone, coal, and conglomerate. (7) Mesozoic and Tertiary sand, silt, clay, and coal. (8) Paleozoic intrusive granites.

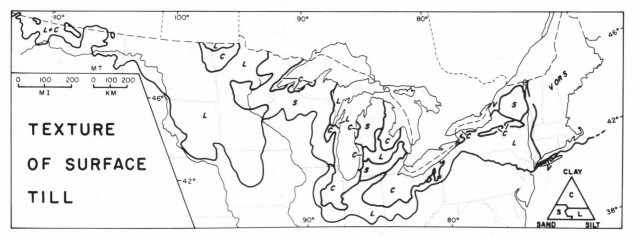

Figure 1-6. Generalized map of the texture of surface tills. The symbols in triangular diagram at the lower right mean the following. Corners of the triangle represent 100% clay, 100% silt, 100% sand. Clayey (C) includes clay, sandy clay, clay loam, silty clay loam, and silty clay. Sandy (S) includes sand, loamy sand, sandy clay loam, and sandy loam. Loamy (L) includes loam, silt loam, and silt. Variable textures (labeled V) include clay to sand textures too complex to map at this scale.

take the place of the end moraines and flat till plains of the southern Great Lakes region as the major landform element. In addition, drumlins are an important glacial landform.

The major differences among these four areas can probably be explained by the nature of the underlying material, the shape of the bed, the proximity to large water bodies, and the physical conditions of the ice sheet. The last factor differed considerably between 25,000 and 10,000 years ago.

The nature of the glacial record in various areas has made it necessary to use a different approach for each of the areas discussed in the text. Only in the southern Great Lakes area is stratigraphic control on Early and mid-Late Wisconsin deposits sufficient even to attempt a discussion of readvances as separate, distinct, area-wide events. Similar advances probably took place in the northern Great Lakes and New England, but they have not been documented.

WESTERN REGION

Till Character and Thickness

Much of the till in the western region is derived from shale and siltstone or preexisting lake sediment and glacial deposits. Because of its composition, most till has a loamy or clayey texture, although in places sandstone has contributed to a sandier till. (The term "loamy" describes a grain-size distribution with approximately equal amounts of silt and sand and 10% to 40% clay. See Figure 1-6.) The till is calcareous and generally contains a relatively large amount of smectite. The high clay content of the till influenced landform genesis in two ways. It reduced the permeability of the glacial deposits and thereby retarded the escape of subglacial water. This resulted in increased subglacial pore pressure and reduced basal shear stress. The high clay content also increased the mobility of the supraglacial till; this resulted in the flow-till hummocks characteristic of the region.

The approximate thickness of Late Wisconsin glacial and nonglacial sediment is shown in Figure 1-7. The lithostratigraphy is not well known in most areas, and where it has been studied the age of only the upper few units is known. Nevertheless, some general trends can be shown. Late Wisconsin drift has an average thickness of approximately 10 m in the region.

In much of the Red River valley (Figure 1-2), Late Wisconsin sediment is about 50 m thick. Lake Agassiz deposits are 40 m thick in

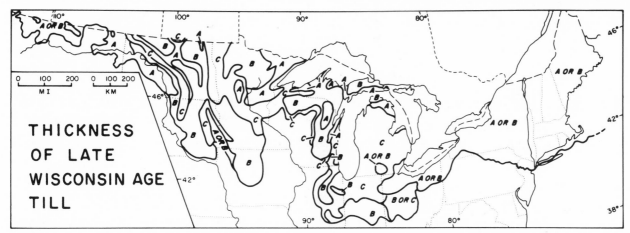

Figure 1-7. Generalized thickness of Late Wisconsin till. Thicknesses are estimated in most places, and boundaries are arbitrary. (A) 0 to 5 m, (B) 5 to 20 m, (C) more than 20 m.

some areas. One Late Wisconsin till sheet is 30 m thick in the middle of the valley, and probably at least two other Late Wisconsin till sheets are present (Arndt, 1977). Surface relief is generally less than 1 m.

In most areas of thick sediment shown in Figure 1-7, the stratigraphy is not known. The upland areas have hummocky topography that resulted from the collapse of supraglacial sediment. The thickness of the supraglacial till of the last glacial advance is approximately equal to the local relief, which is typically about 10 to 20 m (Clayton and Moran, 1974; Clayton et al., 1980). At one site in northwestern North Dakota, the supraglacial till is more than 40 m thick; wood buried under 40 m of flow till has been dated 10,350 ± 300 yr B.P. (W-1817) (Moran et al., 1973), and the region was free of active ice long before the wood was buried.

Thin till is shown in Figure 1-7 where the total Pleistocene sequence is known to be thin (pre-Pleistocene outcrops are common), where the till stratigraphy is well enough known to suggest thin till, and where the presence of numerous drumlins with sand cores indicates Late Wisconsin erosion rather than deposition.

Chronology

Western Montana

The positions of ice margin in western Montana cannot be correlated with those in northeastern Montana on a lithostratigraphic basis owing to the geographic separation of the deposits and the absence of radiocarbon control. Available stratigraphic and morphological data from Montana and published data from Alberta and Saskatchewan are consistent with a Late Wisconsin age for the surface drift shown in Figure 1-8. Although some workers in Alberta have concluded that Late Wisconsin ice may not have extended as far south as the Montana boundary (e.g., Reeves, 1973; Stalker, 1977; Stalker and Harrison, 1977), we feel that this interpretation is not supported by firm chronologic or stratigraphic evidence.

In the St. Mary sublobe (Figure 1-2), till of the Laurentide ice sheet overlies "early Pinedale" alpine glacial deposits (Calhoun, 1906; Alden, 1932; Richmond, 1965; G.M. Richmond, unpublished map). After the "early Pinedale" deglaciation in the St. Mary River valley, the continental ice merged with "middle Pinedale" alpine ice in the Belly River valley, just north of the Montana-Alberta boundary; and

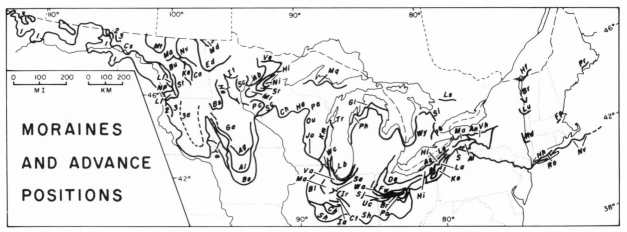

Figure 1-8. Selected moraines, moraine systems, and advance positions (often including end moraine) listed alphabetically by state.

Illinois (IL): Bl, Bloomington moraine system; Ct, Chatsworth moraine, Ellis moraine, Paxton moraine; Ia, Illiana moraine systems; Ir, Iroquois moraine system; Lb, Lake Border moraine system; Ma, Marseilles moraine system; Sa, St. Anne moraine; Sh, Shelbyville moraine system; Va, Valparaiso moraine system; Wc, West Chicago moraine system.

Indiana (IN): Fw, Fort Wayne moraine; Sj, St. Johns moraine; Uc, Union City moraine; Wa, Wabash moraine.

Iowa (IO): Ag, Algona moraine; Al, Altamont advance; Be, Bemis advance; Ta, Tazewell advance.

Maine (ME): Pr, Pineo Ridge advance.

Massachusetts (MA): Fp, Fresh Pond advance; Nv, Nantucket-Vineyard moraine.

Michigan (MI): Gl, Greatlakean advance; Mq, Marquette advance; Ph, Port Huron moraine.

Minnesota (MN): Ab, Alborn advance; Bs, Big Stone advance; Ga, Gary advance; He, Hewitt advance; Hi, Highland moraine; It, Itasca moraine; Ml, Mille Lacs moraine; Ni, Nickerson moraine; Pc, Pine City advance; Sc, St. Croix moraine; Sr, Split Rock advance; Ve, Vermillion moraine.

Montana (MT): Advances 1, 2, 3.

New York (NY): Al, Almond-Kent moraine system; An, Alden moraine; Ma, Marilla moraine, Hamburg moraine, Gowanda moraine; Br, Bridport advance; Hf, Highland Front moraine; Hh, Harbor Hill moraine; Le, Lake Escarpment moraine; Lu, Luzerne advance; Rd, Rosendale advance; Ro, Ronkonkoma moraine; S, Salamanca Reentrant; Vh, Valley Heads moraine system and Finger Lakes area.

North Dakota (ND): Bu, Burnstad advance; Co, Cooperstown advance; Cs, Charlson advance; Ed, Edinburg advance; Ke, Kensal advance; Ll, Long Lake advance; Ma, Martin advance; Md, Marchand advance; Mt, Minot advance; Np, Napolean advance; Nv, North Viking advance; St, Streeter advance.

Ohio (OH): As, Astubula moraine; Br, Broadway moraine; De, Defiance moraine; Hi, Hiram moraine; Ke, Kent moraine system; La, Lavery moraine system; Po, Powell moraine.

Ontario (ONT): Ls, Lake Simcoe moraine; Pa, Paris moraine; Si, Singhampton moraine; Wy, Wyoming moraine.

South Dakota (SD): Advances 2, 3, 3a.

Wisconsin (WI): Ch, Chippewa moraine; Ha, Harrison moraine; Jo, Johnstown moraine; Km, Kettle moraine; Ou, Outer moraine; Pa, Parrish moraine; Sc, St. Croix moraine; Tr, Two Rivers moraine.

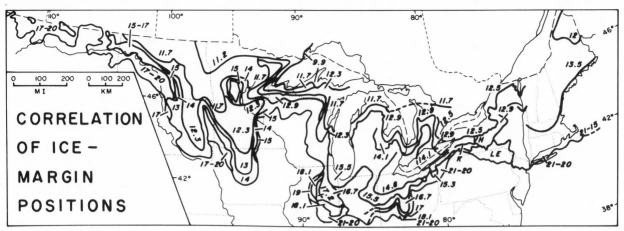

Figure 1-9. Tentative correlation of ice-margin positions or readvance limits. Areas where no ice-front positions are shown are poorly known. Ages in thousands of years before present.

in the St. Mary River valley a glacial lake was dammed between "middle Pinedale" ice and the Laurentide ice sheet (G.M. Richmond, unpublished map). Elsewhere in the Rocky Mountains, the "early Pinedale" maximum ice position was reached after 22,400 ± 1230 yr B.P. (DIC-870), 22,040 +740/−820 yr B.P. (DIC-871), and 21,730 +610/−670 yr B.P. (DIC-869) (Madole, 1980; R.F. Madole, personal communication, 1981). If correlations of alpine drifts in the Rocky Mountains are valid, the expansion of the "early Pinedale" ice lobe into southern Alberta may have culminated at around 21,000 yr B.P. The synchronous advances of the continental ice and the "middle Pinedale" alpine ice in the St. Mary River valley occurred after very extensive "early Pinedale" deglaciation; the till of the continental ice sheet is inferred to have been deposited no earlier than 20,000 yr B.P. and perhaps as late as 18,000 yr B.P. (Figure 1-9).

The physical properties and surface morphological characteristics of the till in the St. Mary and Milk River sublobes and the Shelby lobe (Figure 1-2) are similar. The surface till was assigned a Late Wisconsin age by Richmond (1960), Colton and others (1961), Richmond and Lemke (1965), Lemke and others (1965), and Westgate (1965, 1968). The circular disintegration ridges and other collapsed features that are common in these three areas have been modified only slightly and are as "youthful" as those of the Charlson drift in North Dakota (Figure 1-8), which may have an approximate radiocarbon age of 14,000 yr B.P. (Clayton et al., 1980: Figure 35) on the basis of correlation with events far to the east. The surface till in the Havre lobe is the same lithostratigraphic unit as that in the Shelby lobe. The ice-margin position of the Shelby lobe (shown in Figure 1-8) is based on morphological evidence, but the precise date of this advance has not been fixed.

Eastern Montana to Minnesota

The ice-margin correlations suggested in Figures 1-8 and 1-9 are in part from Clayton and others (1980) and Clayton and Moran (1982). Correlation from area to area has been based in part on direct matching of ice-marginal features at area boundaries. However, positions identified in one area, in some cases, are obscure in adjacent areas because of confusing evidence or lack of study or because they have been overlapped by younger drift. In such cases, correlation has sometimes been based on lithostratigraphy, radiocarbon chronology,

or similarities in suites of landforms. Some correlations have been based on the assumed shape of a glacial lobe, but detailed glacial modeling generally has not yet been done. Arriving at a precise model seems difficult because knowledge of the variations in basal shear stress and marginal accumulations is lacking.

Most earlier chronologies (before Clayton et al., 1980), such as those of Moran and Clayton (1972), Klassen (1972, 1975), Teller (1976), Moran and others (1976), Stalker (1977), Christiansen (1979), and Teller and Fenton (1980), suggested earlier dates for many of the advances than those indicated here because they made use of radiocarbon dates from sediment probably contaminated by older carbon. Clayton and others (1980), Teller and others (1980), and Nambudiri and others (1980) show that a significant proportion of the organic material in fine-grained Pleistocene sediment in the region is derived from early Tertiary coal or Cretaceous black shale. Marl and shell dates present similar problems because of potential contamination from Paleozoic limestone. Because there is generally no way to distinguish the dates for the contaminated from those for the uncontaminated marl, shell, charcoal, peat, and organic clay, the following chronology is based on wood dates, which are less likely to be contaminated (Table 1-1).

The oldest Late Wisconsin wood date that we know of in the region is 20,670 ± 1500 yr B.P. (GX-2741) (Matsch et al., 1972) from the till in eastern South Dakota. A date of 20,000 ± 800 yr B.P. (0-1325) (Ruhe, 1969: 201) is from wood at the base of the till of an advance (designated Ta in Figure 1-8) in Iowa, which Ruhe (1969) correlates with the Tazewell advance of Illinois.

There is no radiocarbon control for the westward correlatives of the Tazewell, but margin 2 in South Dakota and the Long Lake advance (Ll, Figure 1-8) or Napoleon advance (Np, Figure 1-8) in North Dakota are possible correlatives. The Napoleon has generally been guessed (Clayton et al., 1980) to be of Early Wisconsin age, but a Late Wisconsin age seems possible. The most likely equivalent in Montana (margin 1) is thought to be Late Wisconsin.

The Burnstad advance (Bu, Figure 1-8) followed the Long Lake advance in North Dakota. Largely on the basis of the similar morphology of the terminal deposits, it is tentatively correlated westward to the Charlson advance (Cs, Figure 1-8) and to ice margin 2 (Figure 1-8) in eastern Montana and eastward to the St. Croix ice margin (Sc, Figure 1-8) in Minnesota.

Table 1-1. Preliminary Correlation of Selected Tills, Moraines, and Ice-Front Positions in the Northern Great Lakes Region

Years B.P.	James and Souris Lobes		Red River and Des Moines Lobes		Lake Superior, Rainy, and Wadena Lobes		Green Bay Lobe		Lake Michigan Lobe		Lake Huron and Saginaw Lobes	
	Unit	Event	Unit	Event	Unit	Event	Unit	Event	Unit	Event	Unit	Event
10,000				Douglas T, Hanson Creek ④		Marquette A ④						
11,000			Huot, Falconer Fm	Edinburg A								
12,000	Minot A, Martin A, Cooperstown A Kensal A, Big Stone A ③		Dahlen Fm	Alborn A		Nickerson M	Middle Inlet T, Glenmore T	Athelstane M, Denmark M ⑤	Two Rivers T	Two Rivers M ⑤	Valders T (Michigan)	Greatlakean A
				Pine City, Algona M ③		Split Rock M	Chilton T, Kirby Lake T		Valders T, Haven T, Manitowoc T	Inner Port Huron M		Bay City M, Inner Port Huron M
13,000							Silver Cliff T, Branch River T	Mountain M	Shorewood T, Ozaukee T	Outer Port Huron M	St. Joseph T, Port Huron T	Wyoming M, Outer Port Huron M ⑥
14,000	Advance 3, Streeter M		New Ulm Fm, Upper Red Lake Falls Fm	Altamont M (Iowa) ② Bemis M (Iowa) ①	Independence, Cromwell Fm	Mille Lacs M						
15,000												
16,000	Advance 2, Burnstad A		Lower Red Lake Falls Fm			St. Croix M, Itasca M	Mapleview T, Wayside T	Outer M, Johnstown M				
17,000												
20,000	Long Lake A		Marcoux Fm	"Tazewell" A (Iowa)		Hewitt A (?)						
21,000												

Table 1-1 *continued.*

Central Lake Erie Lobe		Eastern Lake Erie and Western Lake Ontario Lobes		Central and Eastern Lake Ontario Lobe		Hudson-Champlain Lobe, St. Lawrence Lowland		Eastern New England		
Unit	Event	Unit	Event	Unit	Event	Unit	Event	Unit	Event	Years B.P.
										10,000
										11,000
			Lake Simcoe M (?)				St. Narcisse-St. Faustin M (?)		Grand Falls M (?)	
										12,000
Wildfield T, Halton T	Trafalgar M, Palgrave M, Waterdown M	Halton T, Upper Leaside T	Alden M							
Wentworth T	Singhamton M, Paris M	Wentworth T	Marilla M, Hamburg M, Girard M		Cazenovia Ms, Tulley M, Hinckley M		Bridport A		Pineo Ridge A	
										13,000
									Pond Ridge M	
							Luzerne A			
Ashtabula T	Ashtabula M		Lake Escarpment Ms	Valley Heads D			Rosendale A		Fresh Pond M	14,000
Hiram T	Hiram M	Port Stanley T							Buzzards Bay M. Sandwich M	
										15,000
Lavery T, Hayesville T	Lavery M			Almond T	Almond M					
									Nantucket-Vineyard M	16,000
										17,000
										20,000
Navarre T, Kent T	Kent M	Kent T	Kent M				Harbor Hill A			
										21,000

The first well-dated event is the Bemis advance (Be, Figure 1-8) of the Des Moines lobe in central Iowa, which culminated at about 14,000 yr B.P. Six dates between 14,700 ± 400 and 13,820 ± 400 yr B.P. (W-153, W-152, I-1402, W-517, I-1268, W-513) (Ruhe, 1969) are from trees that were pushed over by ice or were buried in growth position by loess and till.

Clayton and others (1980: Figure 34) tentatively correlate the Bemis of Iowa with the type Bemis margin of eastern South Dakota, in agreement with Ruhe (1969). However, Matsch (1972) indicates that the supposed Bemis equivalent of southeastern Minnesota may be older. Farther west the Bemis may correlate with the Burnstad or Streeter margin (St, Figure 1-8) of North Dakota and margin 3 of northeastern Montana.

Another advance occurred a few hundred years later; the ice margin in most areas then was within a few to tens of kilometers of the 14,000 yr B.P. margin. In South Dakota and Iowa, it is called the Altamont advance (Al, Figure 1-8) (ice margin 3, South Dakota Geological Survey, 1971; Ruhe, 1969) and may correlate with the Streeter ice margin of North Dakota (Clayton et al., 1980).

A date of 12,760 ± 120 yr B.P. (Y-595) (Lemke et al., 1965) from wood in outwash beyond margin 3a (Figure 1-8) of the James lobe in South Dakota gives the latest possible date for retreat from margin 3. A date of 12,970 ± 250 yr B.P. (W-626) (Ruhe, 1969: 62-65 and 220) from wood under outwash of the Algona advance (Ag, Figure 1-8) of the Des Moines lobe gives the latest possible date for the retreat from the Altamont limit.

Clayton and others (1980) tentatively correlate margin 3a of the James lobe with the Gary (Ga, Figure 1-8) and Algona ice margins of the Des Moines lobe. It may correlate with margin 3 in eastern Montana. Seven wood dates from till behind margin 3a in South Dakota were previously interpreted as either postdating or predating advance 3a. Those dates cluster around about 12,400 or 12,300 yr B.P. (W-801, W-987, W-1189, W-1372, W-1756, Y-452, Y-925) (Lemke et al., 1965: 22; Levin et al., 1965: 378; Ives et al., 1967: 510; Stuiver, 1969: 578). Geologists from the South Dakota Geological Survey generally have assumed that these dates predate the surface till (Steece and Howells, 1965; Christensen and Stephens, 1967; Hedges, 1968, 1972, 1975; Christensen, 1974, 1977; Koch, 1975), because the wood was thought to be from the base of the surface till sheet or under it. Other geologists (e.g., Clayton et al., 1980) working to the northwest have either ignored the dates or assumed that the wood is younger than the surface till and that it was buried by postglacial mass movement. Lake Agassiz was thought to be in existence then or shortly after, leaving too little time for the James and Red River lobes to melt back and readvance to the Kensal (Ke, Figure 1-8) and later ice margins and then melt back again to open the Lake Agassiz basin. However, strictly according to wood dates, Lake Agassiz came into existence shortly before 11,000 yr B.P., and about 1000 years may have elapsed between the retreat from margin 3a and the formation of Lake Agassiz. We have assumed that advance 3a occurred after, rather than before, the 12,400 to 12,300 yr B.P. dates.

There are similar dates from the Des Moines lobe. Several wood dates from till of the Algona (Ag, Figure 1-8) advance are in the range of 11,700 to 12,600 yr B.P. (11,740 ± 100, DIC-1361; 12,020 ± 170, F-8768; 12,610 ± 250, ISGS-641) (G.R. Hallberg, personal communication, 1981).

The Kensal margin of eastern North Dakota represents a readvance of at least several tens of kilometers. To the northwest, it apparently was overlapped by younger advances; and, to the southeast,

it may correlate with, or be slightly older than, the Big Stone margin (Bs, Figure 1-8) of the Des Moines lobe.

The Martin (Ma, Figure 1-8) and Cooperstown (Co, Figure 1-8) ice margins in North Dakota are slightly older than the Minot (Mt, Figure 1-8) and North Viking (Nv, Figure 1-8) ice margins. The Edinburg margin (Ed, Figure 1-8) of the Red River lobe marks a readvance of at least 200 km (Teller and Fenton, 1980). In southern Manitoba, it correlates with the Darlingford margin (Halstead, 1959). The Alborn (Ab, Figure 1-8) ice margin of the St. Louis sublobe could correlate with any of the North Dakota margins from the North Viking to the Burnstad.

None of the later glacial advances in the region has been directly dated, but dates related to Lake Agassiz provide minimum ages for retreat of the Red River lobe from the United States. Lake Agassiz occupied the Red River valley of eastern North Dakota, northwestern Minnesota, and southern Manitoba as the Red River lobe retreated. During the early part of the Cass and Lockhart phases, the outlet to the south through the Minnesota River valley was being cut down. The lake level then stabilized until the beginning of the Moorhead phase, when the Lake Superior lobe retreated far enough to open a lower eastern outlet into northern Lake Superior in Ontario. The Marquette advance (Mq, Figure 1-8) of the Lake Superior lobe closed the eastern outlet, causing Lake Agassiz to rise again to the southern outlet at the beginning of the Emerson phase. The beginning of the Emerson phase, at about 9900 yr B.P., is the best-dated Agassiz event. Seven dates from wood buried on the unconformity below sediment of the Emerson phase range from 10,080 ± 280 to 9810 ± 300 yr B.P. (W-900, W-1005, GSC-391, I-3880, W-993, W-1361, W-1360) (Lowdon et al., 1967: 165; Ashworth et al., 1972; Moran et al., 1973). Dates for the Moorhead phase include 10,960 ± 300, 10,820 ± 190, and 10,340 ± 170 yr B.P. (W-723, TAM-1, and I-5213) (Moran et al., 1973). We know of no wood dates directly related to the Lockhart phase, but enough time is required between the Edinburg advance and the Moorhead phase to deposit the offshore sediment of the Brenna Formation, which is 10 to 40 m thick in much of the Red River valley (Arndt, 1977: plates 1 and 2). Therefore, the Edinburg advance must have occurred sometime before about 11,000 yr B.P.

Three wood dates seem to conflict with this chronology. A date of 12,680 ± 300 yr B.P. (W-1757) (Ives et al., 1967: 511) is from fluvial sediment with no overlying till behind ice margin 3a of the James lobe. A date of 12,025 ± 205 yr B.P. (S-553) (Christiansen, 1979: Table 1) is from wood in fluvial sediment behind the equivalent of the Minot margin in southeastern Saskatchewan. A date of 11,600 ± 430 yr B.P. (GSC-1081) (Klassen, 1972: 550) comes from wood in fluvial sediment behind the Minot margin in southeastern Manitoba. The last two occur in meltwater channels that were draining into Lake Agassiz. Each of these dates is older than what would seem appropriate for the chronology shown in Table 1-1. Each sample was located in and overlain by fluvial sediment; this fact suggests that the samples may once have been covered by till that was later eroded by fluvial activity. Each of the dates is only slightly more than one standard deviation too old, and so the anomaly might be attributable to statistical variation.

We know of several hundred radiocarbon dates from the area and time range being considered here, but only 27 have been used in our chronology. All but wood dates have been rejected because of unresolvable contamination problems, and only wood dates that seem to be from stratigraphically significant materials have been used.

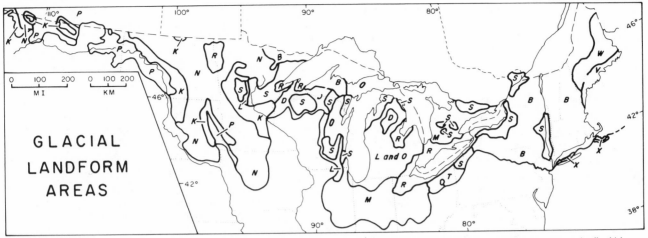

Figure 1-10. Glacial landform areas of the portion of the United States covered by the Late Wisconsin Laurentide ice sheet. (B) Generally thin but locally thick till over bedrock topography and older drift. (D) Extensive pitted outwash separated by low till uplands. (K) Thick, clayey supraglacial till forming high-relief hummocky topography. (L) Thick, sandy, and loamy supraglacial till forming high-relief hummocky topography. (M) Active-ice moraine topography with moraines of low local relief and nearly flat till plain between composed mostly of subglacial till with thin superglacial till. (N) Thin supraglacial clayey or loamy till forming very low-relief hummocky topography, washboard or minor moraines and drumlins. (O) Low-relief subglacial till with some drumlins and extensive pitted outwash. (P) Drumlinized thin till draped over older topography washboard moraines but with few end moraines. (R) Plains formed by glaciolacustrine sediment or with thin till on surface. (S) Drumlins and streamlined forms of subglacial till with only thin superglacial deposits. (T) Thin till over thick pre-Late Wisconsin drift. (V) Glaciomarine sediment in lowlands surrounding till and bedrock hills and washboard moraines. (W) Low-relief till-and-outwash surface with very long esker systems. (X) Thin till over glaciotechtonically thrust and stacked stratified sediments.

Other dates could be used to construct other chronologies. For example, Christiansen (1979: Figure 13) dates the next margin north of the Minot margin in Saskatchewan as 14,000 yr B.P. (3000 years earlier than we have dated it) on the basis of a bone date (S-685) and two organic-sediment dates (GSC-1369 and a date extrapolated from dates S-522 through S-526). Teller and others (1980) suggest that these dates and most of the other dates Christiansen uses are from contaminated samples (only one was wood) because his chronology requires that the terminus of the Red River lobe be in southern Manitoba at the same time that the Des Moines lobe was advancing into central Iowa (for which there are numerous wood dates). We know of no evidence for a separate Des Moines ice cap or James ice cap and of no logical reason to assume one.

Landform Assemblages

The western region, from Montana through central Minnesota, has four distinct suites of landforms. The outermost suite (P, Figure 1-10) is characterized by thin till draped over the preglacial landscape. Most of the rest of the region is characterized by hummocky supraglacial till, thin with low relief in the lowlands (N, Figure 1-10) and thick with high relief in the uplands (K, Figure 1-10). A fourth suite (R, Figure 1-10) is characterized by low-relief, clayey till deposited in a glacial lake basin; it is like suite R in the Great Lakes region. (It is not discussed further here.)

Suite P

The till of suite P is generally less than 10 m thick, and that of the last advance is generally less than 3 m thick. Because of the till's thinness, the topography in much of the area consists of slightly modified preglacial topography, with a largely integrated drainage network. The thin till typically has a subdued collapse topography with flow-till landforms no higher than 3 m (much like suite N, which is described

below). In many areas, erosion has exposed pre-Pleistocene sediment or rock. In addition, the following landforms are present in some areas. Thick till with collapse topography is present along the margins of the St. Mary and Milk River sublobes and the Shelby and Whitewater lobes. Thin till with long, narrow drumlins occurs in the middle of the Shelby lobe and Malta sublobe (Colton et al., 1961). Low ridges like the "cross-valley moraines" of Andrews (1963) and the "sublacustrine moraines" of Barnett and Holdsworth (1974) occur where the Shelby and Missouri Valley lobes were fronted by lakes.

Evidence is scanty for interpreting the style of deglaciation in suite P. Ice margins were only slightly deflected around interfluvial uplands (Colton et al., 1961; Clayton et al., 1980), indicating the presence of thick ice. Around the outer edge of the Shelby and Whitewater lobes, marginal ridges have radii of curvature of more than 13 km, indicating that subglacial topography exerted little influence on ice flow during maximum glaciation and that the ice was thick. Farther up-glacier, the radii diminish and older arcuate patterns are truncated progressively by younger patterns with contrasting orientations. In some areas the truncations may record significant regional readvances, but in most areas they appear to record minor fluctuations of ice discharge during net deglaciation. The pattern of deglaciation associated with each readvance is similar to that associated with maximum glaciation. Marginal compressive shearing near the limits of the advances produced narrow, arcuate belts of thicker supraglacial till with collapse topography; farther up-glacier, less-intense shearing resulted in thinner till, and erosion features characterize the center of lobe. This pattern is similar to the one presented by Clayton and Moran (1974).

The pattern of deglaciation in the Havre lobe differed greatly from that in the Shelby and Whitewater lobes (Figure 1-2). Deglaciation in the Havre lobe occurred primarily through the rapid retreat of the ice margin. Collapse topography and ice-contact deposits occur locally, primarily along the axes of the sublobes. Although readvances are recorded by multiple till units in the area of the Havre lobe, the limits of readvance are not generally marked by constructional topography or unusually thick till.

Suites N and K

Most of the rest of the western region (Figure 1-10) consists of suites N and K. Suite N has low relief and thin supraglacial till, and it occurs in lowland areas; suite K has high relief and thick supraglacial till, and it occurs in upland areas (Clayton et al., 1980). These suites resemble suite P, but suites N and K have abundant glacial thrust masses, which are apparently lacking in suite P; and suite P has "ice-crack moraines," which are lacking in N and K. The chief difference, however, is that suite P is dominated by nonglacial morphology with just a thin veneer of glacial sediment, whereas most of the topography of suites N and K is the result of glacial processes.

Much of suite N, especially in the north, consists of thin flow till with subdued topography resulting from supraglacial collapse processes. That is, it is dead-ice moraine similar in origin to the topography of suite K but with lower relief. In about a tenth of the area in suite N, the flow till is so thin that the dominant landform is long, narrow drumlins that resulted from the subglacial molding of lodgment till or older Pleistocene or pre-Pleistocene material. The minor moraines (washboard moraine, type B; Elson, 1968) characteristic of this region were apparently composed of meltout or flow till in the northern part of the region (Clayton and Moran, 1974) and lodgment till in the southern part (Kemmis et al., 1981). Also characteristic of suite N are glacial thrust masses, eskers, meltwater channels, outwash plains, and proglacial lake plains.

Most of suite K consists of thick flow till with high-relief collapse topography. End moraines are rare in both suites K and N. Associated features include glacial thrust masses, collapsed supraglacial fluvial and lacustrine deposits, and ice-walled-lake plains. Drumlins, eskers, outwash plains, and meltwater channels are generally absent from suite K.

Glacial thrust masses are characteristic of both suites K and N (Clayton et al., 1980: 44-48). Many individual thrust masses are roughly equidimensional and about a kilometer across, but some transversely elongated ones are tens of kilometers long. They are composed of material with shear strength much greater than the 100 kN per square meter of shear stress that can be developed at a glacier bed. Some special circumstances were needed to reduce shear strength so that the glacier could move these masses of sediment (Clayton and Moran, 1974: 93 and 103-4). They occur over aquifers, and their shape is controlled by the shape of the aquifer. This indicates that the shear strength was reduced by elevated pore pressure transmitted by the aquifer from farther upglacier. Some occur above aquifers that today continue out beyond the ice-margin position and would seem to provide an easy route for the dissipation of pore pressure. At the time of thrusting, the ends of these aquifers must have been sealed, apparently by subglacial permafrost. Judging from the relationship of thrust masses to ice-margin positions, the permafrost must have extended back at least 2 km under the terminus.

In contrast to the glacier of suite P, the glacier of suites N and K was strongly influenced by the subglacial topography. This fact indicates that it was flatter and thinner than the glacier in suite P. The lateral margins of the James and Des Moines lobes sloped southward at about 0.5 m per kilometer. Although the slope of the lateral margin and the slope of the centerline of the lobe are not identical, a gentle slope is indicated, as Mathews (1974) points out by suggesting a basal shear stress roughly a tenth as great as in "normal" glaciers.

A key factor in the region of suites N and K is the presence of smectite-rich Cretaceous shale. The unique landforms of this region are restricted to the area of Cretaceous shale or to the area of Pleistocene sediment derived from the shale. The shale provided a base of very low permeability to retard the escape of subglacial water, and it resulted in slightly permeable till with few widely continuous aquifers. As the ice moved southwestward off the Canadian Shield in Keewatin and northern Saskatchewan, it had a steep profile and few terminal lobes (Prest et al., 1968; Prest, 1969). However, where it crossed onto Cretaceous shale, reduced underdrainage resulted in increased subglacial pore pressure, which reduced the basal shear resistance and allowed the ice profile to flatten and the ice velocity to increase. Here the margin became more highly lobate, and flow lines turned to the southeast down the Prairie ice stream in the low area between the Missouri Plateau and the main Laurentide ice sheet. The rapidly moving ice permitted the rapid advance of the terminus, and the gentle surface profile permitted the ice's rapid retreat. The radiocarbon chronology of the region (Clayton et al., 1980: 58-76) indicates terminus fluctuations on the order of 1 km per year. The gentle profile, the rapid ice movement, and the upward slope of the bed resulted in rapidly decelerating flow in the terminal areas and the intense shearing that produced the widespread supraglacial till. The elevated pore pressure facilitated glacial thrusting.

Boulton and Jones (1979) suggest that the reduced ice-surface slope was a result of the weakness of the subglacial material of the region and that most of the slippage occurred below the ice-sediment interface. This seems highly unlikely because of the intact stratigraphic framework in much of the area. Subglacial shearing of the magnitude suggested by Boulton and Jones should have destroyed any stratigraphic unity in the region.

SOUTHERN GREAT LAKES REGION

Till Character and Thickness

Throughout the southern Great Lakes region (Figure 1-1) till is calcareous, the carbonate having been derived primarily from Paleozoic sedimentary rock (Figure 1-5) and Quaternary lake sediment. The till units in different lobes and within each lobe have been distinguished primarily by texture and color but also by characteristics such as pebble lithology (e.g., Anderson, 1957), carbonate and clay-mineral content (e.g. Willman and Frye, 1970; Steiger and Holowaychuk, 1971; Gooding, 1973), heavy minerals (Dreimanis et al., 1957), and elemental content (Wilding et al., 1971).

In general, older Late Wisconsin till has a loamy texture (Figure 1-6). Much of the till matrix presumably was derived from older glacial deposits, weathered regolith, and bedrock eroded far up-ice (Wayne, 1963; Goldthwait et al., 1965; Willman and Frye, 1970; Indiana Geological Survey, 1979). Till of later readvances, which contains incorporated lake sediment, is more clayey (Figure 1-6).

The total thickness of Pleistocene sediment in much of the region exceeds 50 m, but much of the sediment is older than Late Wisconsin. Individual till units of specific readvances generally are less than 5 m thick, except in end moraines, where the thickness may exceed 75 m (ver Steeg, 1933, 1938; Horberg, 1950; Wayne, 1956b; Piskin and Bergstrom, 1967). The thickness of the Late Wisconsin part of the sequence is generalized in Figure 1-7.

Significance of End Moraines

Many narrow, sharply defined ridges that have been mapped as end moraines in the southern Great Lakes region are records of glacial readvances (terminal moraines) or stillstands of ice margins (recessional moraines). However, the "moraine" ridges are not all products of single episodes of till deposition at the margin of an active ice sheet. The complex moraine systems (e.g., the Shelbyville, Bloomington, Illiana, Marseilles, and Valparaiso systems in Illinois and the Knightstown system in Indiana) (Figure 1-8) and many of the broader ridges with multiple crests or smooth or gently undulating surfaces (e.g, the Powell, Broadway, St. Johns, and Defiance moraines in Ohio) are composite landforms. In some places they are composed of several imbricated till ridges, and in other places they are overridden moraines. It still other places they are segments of true end moraine (as the term is defined later in this section).

The existence of overridden end moraines has produced speculation about the significance of end moraines in reconstructions of the history of Late Wisconsin glacial fluctuations in the southern Great Lakes region (e.g., Wright, 1971, 1976; Bleuer, 1974a). Buried or overridden moraines in Illinois were noted, described, or illustrated by Powers and Ekblaw (1940), Leighton and others (1948), Horberg (1953), Bretz (1955), Kempton (1963), Frye and others (1965), Willman and Frye (1970), Kempton and others (1971), Johnson (1976), J. T. Wickham (1979), Lineback (1979), and others. Overridden moraines in Indiana were mapped by Wayne (1958, 1965a, 1965b) and Wayne and others (1966) and were discussed by Bleuer (1974a, 1975). Specific aspects of Totten's (1969, 1973) analysis of overridden end moraines in the Scioto sublobe (Figure 1-1) in Ohio have been questioned (Goldthwait, 1971).

In general, at any particular place a ridge is either a true end moraine or a palimpsest moraine. True end moraines are either simple, superposed, or rock cored (Figure 1-11). Palimpsest moraines and superposed moraines are both overridden moraines, but their interpretation differs considerably.

A *true end moraine* (normally called end moraine) is a ridge composed predominantly or entirely of till and formed at an ice margin during the last episode of till deposition. The thickness of the youngest till in the ridge is significantly greater than the thickness behind it. All types of true end moraines can be either terminal or recessional. A *simple end moraine* is a ridge produced by a thickening of the till of the youngest advance. The ridge is not underlain by a bedrock ridge or a buried end moraine but owes its topographic form to the till within it (Figure 1-11A).

A *rock-cored end moraine* is a type of true end moraine that is cored by bedrock. The thickness of the till of the latest glacial advance is significantly greater in than behind the moraine (Figure 1-11C). The existence of a bedrock core does not alter the significance of the ridge—the ice margin stabilized temporarily in response to topographic control exerted by a bedrock ridge, and an unusually thick accumulation of till was deposited on the crest of the ridge.

A *superposed end moraine* is a till ridge that consists of an overridden end moraine on which is superposed a younger end moraine (Figure 1-11B). Part of the relief and form of the ridge is inherited from the buried moraine, but much of the relief and form (particularly the width) is related to the formation of the younger moraine. The thickness of the younger till on the crest of the ridge commonly is two to five or more times as great as that on adjacent till surfaces, and many superposed moraines have multiple ridge crests.

Superposition of end moraines was a normal occurrence in the southern Great Lakes region and probably was caused by the inherent instability of the thin ice in the marginal zone of the ice sheet. An older end moraine 1 to 5 km wide, with relief of 10 to 30 m, might have served as a significant obstacle to ice flow in the thin marginal zone of the ice sheet. The relief was not enough to prevent the overriding of the moraine by readvancing ice, but the effect of the intersection of the broad ridge with the low-gradient parabolic profile of the ice surface, both during advance and recession, caused the ice margin to remain in the position of the older moraine for a significantly longer time there than in adjacent areas. Consequently, a much greater thickness of younger till was deposited on the older moraine than on adjacent till-plain surfaces. When the culmination of a readvance occurred when the ice margin was a short distance (4 km or less) beyond an older moraine, the subglacial relief of the older moraine beneath the thin ice of the marginal zone caused the ice to retreat rapidly to the position of the older moraine, where an equilibrium profile was reestablished, and a "near-terminal" moraine of younger till was formed on the older end moraine. The configurations of superposed terminal moraines, superposed "near-terminal" moraines, and superposed recessional moraines reflect the shape of the ice margin during the latest interval of deglaciation, in spite of the fact that the moraines are superposed.

A *palimpsest moraine* (Figure 1-11D) is a till ridge that is composed of a thin veneer of younger till draped over an older end moraine. The relief and surface form of the ridge were inherited from the buried moraine; the younger till is not thicker (and commonly it is thinner) on the crest of the ridge than on the ground moraine surfaces on both sides of the ridge.

Readvances in the southern Great Lakes region can be recognized by (1) the truncation or overlapping of older ridges by younger moraines, (2) the abrupt changes in the alignment of concentric ridges in series, (3) the presence of subdued, overridden moraine segments upglacier from sharply defined ridges, (4) the abrupt changes in the composition of tills coincidental with the distal margins of moraines, (5) the presence of large masses of incorporated lake sediment in the till of an end moraine and the general absence of incorporated sediments in till beyond the moraine, and (6) the occurrence of heads of outwash valley trains and meltwater sluiceways

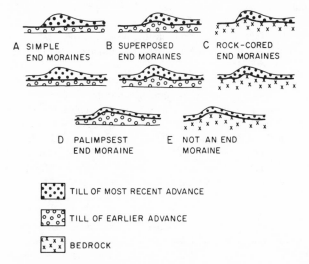

Figure 1-11. Cross sections of moraines and morainelike ridges showing internal composition.

coincidental with end moraines. Determining the distance of re-advance is difficult because the distance of ice-margin recession before the readvance seldom can be ascertained.

Synchrony of Major Glacial Advances

Several authors have speculated that Late Wisconsin readvances in adjacent lobes and sublobes in Illinois, Indiana, and Ohio were not synchronous (e.g., Wright, 1971b; Bleuer, 1980). Because of apparent asynchrony and other phenomena, some readvances have been attributed to glacial surges (Goldthwait, 1968; Goldthwait and Rosengreen, 1969; Wright, 1971; Quinn and Goldthwait, 1979; Bleuer, 1980). However, published stratigraphic and chronologic controls are compatible with the assumption that nearly all (if not all) major glacial readvances were synchronous in Illinois, Indiana, and western Ohio. An absence of conclusive evidence for synchrony in many instances reflects only a paucity of stratigraphic and chronologic controls. Minor fluctuations in individual sublobes may sometimes have been asynchronous. During net ice retreat within the basins of the present Great Lakes, differences in the ablation of calving margins or rapid changes in the levels of proglacial lakes may have caused asynchronous readvances in portions of some ice masses.

An apparent asynchrony in adjacent sublobes reflects several factors. First, the distance of synchronous readvances and ice-margin recessions in adjacent sublobes differed. Consequently, in one sublobe the till of an earlier advance occurs at the surface south of the limit of a younger advance; in an adjacent sublobe the older till occurs only in the subsurface as a consequence of overridding during a more extensive younger advance. Some workers have assumed that the presence of an older till at the surface in one sublobe and its absence at the surface in an adjacent sublobe indicate that the earlier advance occurred in only one sublobe.

Second, efforts to define the limits of readvances have focused on the identification and correlation of end moraines and not on the identification of pinch-outs of thin till units that do not have distinctive topographic expression. The limits of readvances in many places are not marked by end moraines.

Third, stratigraphic studies in Illinois and Indiana since 1958 have focused on defining and tracing formal lithostratigraphic units (Frye and Willman, 1960; Wayne, 1963; Frye et al., 1965; Wayne and Zumberge, 1965; Willman and Frye, 1970; Johnson, 1976). The reconstruction of ice-margin fluctuations on the basis of criteria other than the boundaries between formally named lithostratigraphic units has received little attention. Late Wisconsin tills in central Indiana are lithologically and texturally similar (Harrison, 1959, 1963; Gooding, 1973), and multiple advances generally are recognized only when proglacial or nonglacial sediment occurs between till units. Because of compositional similarity, till units representing two or more glacial readvances in studied sections have been relegated (as they should be) to a single lithostratigraphic unit (e.g., the Cartersburg Till Member of the Trafalgar Formation [Table 1-2]). Although it is certainly basic to the resolution of Quaternary history, emphasis on the definition of formal lithostratigraphic units in some instances may have hampered or obscured the resolution of the history of minor glacial readvances and recessions.

Fourth, few stratigraphic studies of central Indiana and western Ohio have been published since 1965; and, although many sections described in earlier studies have been reexamined in light of more recent stratigraphic concepts, techniques, and information, the information is incomplete and unpublished (N. K. Bleuer, personal com-munication, 1981). Multiple till sections in the East White sublobe were described by Thornbury (1937), Wayne and Thornbury (1951, 1955), Thornbury and Wayne (1953, 1957), Thornbury and Deane (1955), Harrison (1963), Wayne (1965b, 1973), Burger and others (1966), Schneider and Wayne (1967), and others, but the till units described were not traced and correlated regionally on the basis of stratigraphic criteria. Very detailed sections in the Miami sublobe were described and correlated by Gooding (1957, 1963, 1965, 1975), Gamble (1958), Durrell (1961), Brace (1968), Wright (1970), Blackman (1970), and others, but the till units were not traced as stratigraphic units into the East White sublobe. Very little is known about the vertical succession of till units north of the 16,700 yr B.P. advance limit (Figure 1-9) in central and eastern Indiana and western Ohio.

We have decided, therefore, that unless asynchrony can be demonstrated, we will assume synchrony of events for purposes of correlation. The question of synchrony is an important facet of the history of the Late Wisconsin Laurentide ice sheet, and perhaps our decision will encourage more critical research on this topic.

Chronology

Although end moraines in the southern Great Lakes region were mapped and correlated more than 80 years ago (Leverett, 1896, 1897, 1899, 1902), disagreement still exists about the precise timing of advances, the synchrony of advances in adjacent lobes and sublobes, and the relationship between end moraines and specific lithostratigraphic units. In Table 1-2 we show tentative correlations of till units in the Lake Michigan lobe and the East White, Miami, and Scioto sublobes (Figure 1-2). The major glacial advances discussed in this chapter often include several minor advances. Because the magnitude of synchronous advances often differed in adjacent sublobes, advances often appear to have culminated at slightly different times. Thus, it is not uncommon to have an overlapping of lithostratigraphic units within the major advances listed in Table 1-2.

Reasonably firm radiocarbon control in Illinois and Indiana is restricted to the advances around 21,000, 20,000, 19,000, and 18,100 yr B.P. Reliable radiocarbon controls for the Erie interstade and other interstades younger than 18,000 yr B.P. are not available in the southern Great Lakes region. (Significant radiocarbon ages younger than 15,500 yr B.P. from materials in and between tills north of this area are discussed in the section on the northern Great Lakes region.)

Because of limited radiocarbon control and also miscorrelation of stratigraphic units in the literature, we have chosen not to adopt all aspects of the chronologic frameworks outlined in earlier summaries (Frye et al., 1965; Goldthwait et al., 1965; Wayne and Zumberge, 1965; Goldthwait and Rosengreen, 1969; Willman and Frye, 1970; Dreimanis and Goldthwait, 1973; Frye and Willman, 1973; Morner and Dreimanis, 1973; Gooding, 1975; Goldthwait, 1976; Johnson, 1976; Dreimanis, 1977b).

The distribution of tills of significant glacial readvances in surface exposures (as correlated in Table 1-2) is illustrated in Figure 1-9. The correlations of these ice-margin positions are very tentative. They are based on (1) published correlations of lithostratigraphic units in adjacent lobes and sublobes, (2) published and unpublished time-distance diagrams of ice-margin fluctuations in sublobes, and (3) fence diagrams constructed from several dozen published stratigraphic sections in conjunction with radiocarbon ages from many of the sections, and (4) the assumption that major advances in adjacent

Note: Ages listed here for major glacial events are from wood in till; ages for "interstades" are from wood or other materials from intertill or subtill stratigraphic horizons. Brackets enclose duplicate or replicate measurements of the same sample or age measurements of different samples from the same stratigraphic unit or horizon at a single locality. Radiocarbon ages marked with ⊕ from Illinois are not tabulated. Ages and stratigraphic relations of samples are presented in Willman and Frye (1970), Kempton et al. (1971), Johnson, Gross, and Moran (1971), Johnson et al. (1971, 1972), Johnson (1976), Follmer et al. (1979), and other publications of the Illinois State Geological Survey. Unpublished age (ISGS-767) from ponded sediment above Batestown Till and beneath loess from W. H. Johnson. Names in quotation marks are informal.

Table 1-2. Tentative Correlation of Late Wisconsin Tills and Selected End Moraines and Moraine Systems, Illinois, Indiana, and Western Ohio

Ice-Margin Position	Lake Michigan Lobe (Till)	Lake Michigan Lobe (Moraine)	East White Sublobe (and Northwestern Indiana) — Till	East White — Moraine	Miami Sublobe (and Northeastern Indiana) — Till	Miami — Moraine	Scioto Sublobe (and Northwestern Ohio) — Till	Scioto — Moraine	Scioto — Dates
				Lake Huron Lobe					
14,800 yr B.P. Advance	Part of Wadsworth Till Member	Lake Border Moraines	Part of New Holland Till Member	Tinley Moraine	Part of New Holland Till Member	Defiance Moraine	Lake Till	Defiance Moraine	
	Early Lake Chicago		Early Lake Chicago		Lake Maumee I		Lake Maumee I		
	Part of Wadsworth Till Member	Part of Valparaiso Moraine System	Part of New Holland Till Member	Part of Valparaiso Moraine System	Part of New Holland Till Member	Fort Wayne Moraine (Superposed) Wabash Moraine	Tymochtee Till	Fort Wayne Moraine Wabash Moraine	
					Significant Ice-Margin Recession		Significant Ice-Margin Recession		
15,500 yr B.P. Advance	Part of Wadsworth Till Member	Most of Valparaiso Moraine System	Part of New Holland Till Member	Part of Valparaiso Moraine System	Part of New Holland Till Member (Indiana); Unnamed Till (Ohio)	Salamonie Moraine; Mississinewa Moraine; Union City Moraine	Unnamed Till	St. Johns Moraine; Broadway Moraine; Powell Moraine	>14,780 ± 192 (OWU-83)
	"Kankakee Flood"		"Kankakee Torrent"		"Erie Interstade"		"Erie Interstade"		
16,100 yr B.P. and 16,700 yr B.P. Advances	Haeger and Yorkville Till Members	Rockdale Moraine; Minooka and St. Anne Moraines; Much of Marseilles Moraine System; Gilman and Cullom Moraines; Paxton, Ellis, and Chatsworth Moraines	Part of "Cartersburg" Till Member	Iroquois Moraine; Type Crawfordsville Moraine; Part of Knightstown Moraine System	Part of "Bloomington" Till	Eldorado Moraine	Upper Darby (Darby II) Till	London Moraine; Bloomingburg Moraine	
	Minor "Interstade"		Minor "Interstade"		Minor "Interstade"		Minor "Interstade"		
17,200 yr B.P. Advance	Snider Till Member	Most of Illiana Moraine System	Part of "Cartersburg" Till Member	Part of Knightstown Moraine System	Most of "Bloomington" Till	Hamburg Moraine; Farmersville Moraine	Lower Darby (Darby I) Till	Esboro Moraine; Glendon Moraine; Reesville and Logan Elm Moraines	
	Major "Interstade" 17,690 ± 270 (ISGS-767)		Major "Interstade"		Major "Interstade"		Major "Interstade"		
18,100 yr B.P. Advance	Malden and Batestown Till Members	Champaign and Urbana Moraines; Parts of Arcola, Normal, and Eureka Moraines; Part of Bloomington Moraine System	Part of "Cartersburg" Till Member	Part of Knightstown Moraine System	"Champaign" Till	Inner Camden Moraine; Outer Camden and Springfield Moraines — 18,400 ± 300 (OWU-286)	Upper Caesar (Caesar II) Till	Inner Cuba Moraine; Outer Cuba Moraine	17,290 ± 436 (OWU-76); 18,000 ± 400 (W-331); 18,500 ± 420 (Y-448); [17,880 ± 224 (OWU-52); 18,050 ± 400 (W-91)]
	Major "Interstade"		Major "Interstade"		Major "Interstade"		Major "Interstade"		
19,000 yr B.P. Advance	Part of Tiskilwa Till Member	Much of Bloomington Moraine System	Part of "Cartersburg" Till Member		Part of "Shelbyville" Till (Subsurface)		Unnamed Till (Subsurface)		
	Minor "Interstade"		"Interstadial" Silt		"Interstadial" Silt		"Interstadial" Silt		19,735 ± 475 (OWU-257); 19,500 ± 500 (DAL-77); 19,850 ± 765 (OWU-452)
20,000 yr B.P. Advance	Delavan and Piatt Till Members (and Part of Tiskilwa Till Member)	Parts of Leroy, Shirley, and Cerro Gordo Moraines; Segments of Shelbyville Moraine System	Part of "Center Grove" Till Member	Most of "Shelbyville" Moraine 20,110 ± 360 (ISGS-475)	Most of "Shelbyville" Till	Most of Hartwell Moraine — 19,100 ± 300 (W-724); 19,620 ± 470 (I-6182); 19,800 ± 300 (L-467); 19,800 ± 500 (I-1776); 19,900 ± 500 (I-1778); 19,980 ± 500 (W-92); [19,620 ± 372 (OWU-102); 20,275 ± 620 (OWU-102bis)]; 20,500 ± 800 (W-304)	Lower Caesar (Caesar I) Till	Buried Wilmington Moraine	19,800 ± 400 (I-4795); 20,400 ± 700 (W-2459); 20,820 ± 600 (W-2465)
	Major "Interstade"		"Interstadial" Silt — 19,930 ± 900 (PIC-31); 19,930 ± 990 (PIC-17); 20,100 ± 900 (W-598); 20,300 ± 800 (W-597); 20,660 ± 180 (ISGS-541); 20,800 ± 800 (W-579); 20,900 ± 500 (I-5216)			"Connorsville" Interstadial Deposits — 18,750 ± 300 (W-738); 20,000 ± 500 (I-610)		"Interstadial" Silt		17,340 ± 390 (OWU-256); 19,303 ± 1080 (OWU-488); 20,700 ± 1000 (DAL-14)
21,000 yr B.P. Advance	Glenburn (Subsurface) and Fairgrange Till Members (and Part of Tiskilwa Till Member)	Part of Arcola Moraine; Segments of Shelbyville Moraine System	Part of "Center Grove" Till Member	Segments of "Shelbyville Moraine" — 20,900 ± 800 (W-580); 21,400 ± 650 (W-668); 22,300 ± 800 (W-595)	Fayette Till	Overridden Parts of Hartwell Moraine (?) — 20,290 ± 800 (I-1007); [20,000 ± 500 (L-397-C); 20,700 ± 600 (W-37)]; 21,150 ± 450 (I-4345); 21,340 ± 125 (GrN-4514); 21,600 ± 400 (W-648); 22,000 ± 1000 (W-414); [21,940 ± 130 (ISGS-116); 22,230 +415/−430 (DIC-47)]; [19,535 ± 655 (OWU-490); 25,100 ± 1600 (DAL-5); 27,650 ± 1000 (I-6183)]	Boston Till	Vandervoort and Mt. Olivine Moraines	20,910 ± 240 (ISGS-44); 21,080 ± 200 (ISGS-42); 21,400 ± 600 (W-88); 21,600 ± 1000 (W-127); 23,000 ± 850 (Y-449)
	"Farmdalian" Deposits and Soil		"Interstadial" Deposits and Soil — 16,555 ± 250 (SI-4); 17,145 ± 210 (SI-4); 18,899 ± 270 (OWU-42); 18,911 ± 407 (OWU-119); 19,906 ± 691 (OWU-8); 20,230 ± 200 (Y-1248); 21,400 ± 700 (DAL-10-A); 21,700 ± 1200 (DAL-10-B); 19,500 ± 800 (W-165); 20,500 ± 800 (W-577); 21,010 ± 350 (ISGS-477); 21,340 ± 1860 (PIC-18); 21,400 ± 860 (PIC-32); 23,300 ± 600 (W-663)			"New Paris Interstade" and "Sidney Interstade" Deposits and Soil — 22,400 ± 260 (GrN-4513); 22,430 ± 140 (GrN-1761); 22,800 ± 200 (GrN-4512); 22,480 ± 800 (W-356); 23,000 ± 800 (W-188); 23,300 ± 500 (I-7345); 23,400 ± 600 (W-1688); 24,440 +560/−590 (DIC-47); 24,790 ± 780 (OWU-140-B)		"Interstadial" Deposits		21,140 ± 1435 (OWU-159); 22,225 ± 1650 (OWU-160); 24,800 +495/−525 (DIC-82); 26,220 +390/−395 (DIC-57); 32,090 +1375/−1660 (DIC-70); 32,420 +1405/−1700 (DIC-52)

Principal Sources of Stratigraphic and Chronologic Control (Lake Michigan Lobe): Kempton, 1963; Frye et al., 1965; Kempton and Hackett, 1968; Willman and Frye, 1970; Willman, 1971; Johnson, Gross, and Moran, 1971; Johnson et al., 1971, 1972; Kempton et al., 1971; Frye and Willman, 1973; Johnson, 1976; Follmer et al., 1979; Wickham, 1979; Lineback, 1979

Principal Sources of Stratigraphic and Chronologic Control (East White Sublobe): Thornbury, 1937, 1958; Wayne and Thornbury, 1951, 1955; Thornbury and Wayne, 1953, 1957; Thornbury and Deane, 1955; Wayne, 1958, 1963, 1965a, 1965b, 1966, 1968, 1973; Zumberge, 1960; Wier and Gray, 1961; Harrison, 1963; Wayne and Zumberge, 1965; Burger et al., 1966; Wayne et al., 1966; Schneider and Wayne, 1967; Schneider and Keller, 1970; Bleuer and Moore, 1972; Bleuer, 1974b, 1975

Principal Sources of Stratigraphic and Chronologic Control (Miami Sublobe): Gooding, 1957, 1961, 1963, 1965, 1975; Gamble, 1958; Goldthwait, 1958, 1976; Durrell, 1961; Goldthwait et al., 1961, 1965; Forsyth, 1965a, 1965b; Wayne and Zumberge, 1965; Wayne, 1958, 1966, 1968; Burger et al., 1971; Wilding et al., 1971; Gray et al., 1972; Johnson and Keller, 1972; Dreimanis and Goldthwait, 1973

Principal Sources of Stratigraphic and Chronologic Control (Scioto Sublobe): Goldthwait and Forsyth, 1955; Goldthwait, 1958, 1968, 1976; Goldthwait et al., 1961, 1965; Forsyth, 1965b; Teller, 1967; Goldthwait and Rosengren, 1969; Wilding et al., 1971; Totten, 1973; Dreimanis and Goldthwait, 1973; Rosengren, 1974; Gooding, 1975

sublobes were synchronous. Important glacial events (advances and recessions of ice margins) are referred to informally as "advances" in order to permit the comparison of events from one lobe or sublobe to another with a minimum use of stratigraphic and geographic names. Within each of these major advances there were often minor advances that were not synchronous.

The error ranges of most radiocarbon ages in Table 1-2 span longer time intervals than the duration of either glacial advances or "interstades" within 50 km of the Late Wisconsin glacial limit. The statistical errors assigned to the measured ages are one standard deviation, and it is expected that the true radiocarbon ages of one-third of the samples listed in Table 1-2 may be older or younger than the stated error ranges. The measured ages also may be in error due to contamination or analytical error, and the statistical errors cannot be ignored even when the measured ages are reliable. The stratigraphic position of an interstadial deposit is not necessarily determined by the radiocarbon age of a sample from the deposit, and the stratigraphic position of a till unit is not necessarily determined by the age of a sample from the till or from interstadial deposits beneath or above the till unit. From a practical standpoint, till units must be identified and traced as lithostratigraphic units on the basis of their inherent physical and chemical properties, and the chronology of events should be established only in relation to the framework of known stratigraphic units. Unfortunately, the till of a minor readvance might not be distinguished from an earlier or later advance because of the similarity of units. Likewise, different lithostratigraphic units might represent different facies of a single advance.

In principle, the chronologic framework of glacial advances and recessions in each sublobe should be based on multiple age measurements from each "interstadial" deposit. Because of the time-transgressive nature of the till units and "interstadial" deposits, the ages of culminations of readvances can be determined accurately only from local time-distance diagrams that are controlled by a number of radiocarbon ages. Table 1-2 demonstrates that the present correlations of glacial advances in the southern Great Lakes region (as well as other published correlations) have relatively little firm chronologic control. Correlations of advances are speculative, and the assigned ages are used only for purposes of discussion.

The Advance about 21,000 Yr B.P.

The advance about 21,000 yr B.P. may be represented by the Fairgrange Till Member and the basal part of Tiskilwa Till Member of the Wedron Formation in Illinois and extreme western Indiana, part of the Center Grove Till Member of the Trafalgar Formation in central Indiana, the Fayette till in the Miami sublobe, and the Boston till in the Scioto sublobe (Table 1-2). The age of the outermost surface drift of Late Wisconsin age on the east side of the Scioto sublobe (Figure 1-2) is uncertain, because till units in that region have not been traced as stratigraphic units. The assignment of the drift to the advance about 21,000 yr B.P. is based solely on an early solid-carbon date of 21,400 ± 600 B.P. (W-88) from wood in till near Newark, Ohio (Goldthwait, 1958). The limits of the advances around 20,000 and 18,100 yr B.P. also may be present in the same area.

The limits of the advances around 21,000 and 20,000 yr B.P. were nearly identical from central Illinois to central Ohio. Approximately half of the Late Wisconsin glacial limit in that region is represented by drift of the advance about 21,000 yr B.P., and half by drift of the advance about 20,000 yr B.P. (Figure 1-9).

The distance of ice-margin retreat following the advance about 21,000 yr B.P. is uncertain. Frye and Willman (1973) infer a retreat of approximately 96 km in Illinois. Johnson (1976) infers that the ice margin in the Decatur sublobe (Figure 1-2) in Illinois may have retreated to, or north of, the position of the present Valparaiso moraine system (Figure 1-8). In the East White and Miami sublobes (Figure 1-2), the ice margin retreated at least 55 km, and probably much farther, as is indicated by the distribution of sites from which dated wood has been recovered from "interstadial" silt or from the base of the till of the readvance about 20,000 yr B.P. The ice margin of the Scioto sublobe also retreated at least 55 km. Dreimanis (1977b) infers a recession of 40 to 100 km in Indiana and Ohio. In the Miami sublobe, the interval of deglaciation and subsequent time-transgressive readvance is referred to as the Connorsville interstade (Gooding, 1963, 1975).

The Advance about 20,000 Yr B.P.

The advance about 20,000 yr B.P. (Figure 1-9, Table 1-2) apparently is represented by the Delavan Till Member, the Piatt Till Member and part of the Tiskilwa Till Member of the Wedron Formation in Illinois, part of the Center Grove Till Member of the Trafalgar Formation in the East White sublobe, much of the Shelbyville till in the Miami sublobe, and the subsurface lower Caesar (Caesar I) till in the Scioto sublobe.

Where the limit of the advance about 21,000 yr B.P. was over-ridden by drift of the advance about 20,000 yr B.P. the buried earlier limit probably is no more than a few kilometers north of the younger limit; where the older drift is exposed as a fringe south of the younger limit, the fringe is also narrow. In view of the observation that the ice margin retreated 55 km to perhaps more than 100 km before the advance about 20,000 yr B.P., the close coincidence of the two drift limits in four sublobes, from west-central Illinois to central Ohio, is remarkable.

The distance of ice-margin retreat from the limit of the advance about 20,000 yr B.P. is not known. A major retreat (approximately 150 km) in Illinois before the deposition of the surface Tiskilwa Till Member (Table 1-2) in the Peoria sublobe (Figure 1-2) is inferred by Frye and Willman (1973). The Tiskilwa Till Member (or a till of equivalent age in the Decatur sublobe) is not included in the time-distance diagram by Johnson (1976). The stratigraphic successions in the East White and Miami sublobes are too poorly documented to provide a basis for estimates of the magnitude of retreat. If radiocarbon ages of 19,500 ± 500 (DAL-77) and 19,850 ± 765 yr B.P. (OWU-452) from intertill silt of Columbus, Ohio (Table 1-2), are reliable, the margin of the Scioto sublobe may have retreated at least 100 km.

The Advance about 19,000 Yr B.P.

The advance about 19,000 yr B.P. apparently is represented by the surface Tiskilwa Till Member of the Wedron Formation in the Peoria and Princeton sublobes in Illinois, part of the Cartersburg Till Member of the Trafalgar Formation in the East White sublobe, part of the Shelbyville till in the Miami sublobe, and an unnamed subsurface till in the Scioto sublobe (Table 1-2).

The advance is of major significance in west-central Illinois, where much of the massive Bloomington moraine system (Bl, Figure 1-8) was formed as a result of it (S. S. Wickham, 1979). A single lithostratigraphic unit (the Tiskilwa Till Member) forms the basal part of the Wedron Formation in central northern Illinois from the outer limit of the Bloomington moraine system (the limit of Late Wisconsin glaciation) to or nearly to the Lake Michigan basin (Willman, 1971). In parts of the region, this single lithostratigraphic unit consists of compositionally similar (and therefore not distinguishable) subunits that were deposited as a result of three or more significant glacial readvances. These subunits of the single Tiskilwa Till Member recorded advances that deposited the Fairgrange, Delavan, and Tiskilwa Till in the southern part of the Peoria sublobe. The surface Tiskilwa Till Member of the advance about 19,000 yr B.P. in the south (Figure 1-9) is younger than much of the till that constitutes the complex single lithostratigraphic unit (Tiskilwa Till Member) in north-central Illinois.

The overlap of Tiskilwa Till Member over Delavan Till Member (Willman and Frye, 1970; Frye and Willman, 1973; Johnson, 1976; Follmer et al., 1979; Lineback, 1979) does not necessarily indicate that a major interval of deglaciation occurred prior to the "Tiskilwa advance"; the two tills are similar in composition, and the geographic distribution of sites where the tills are superposed has not been determined. However, Frye and Willman (1973) infer that the ice margin may have retreated at least 150 km from the limit of the Delavan Till Member before the "Tiskilwa advance."

The till of the advance about 19,000 yr B.P. has not been identified with certainty in the Decatur sublobe in Illinois; it may be a subunit of the Piatt Till Member. In Indiana and western Ohio, apparently, the till units deposited as a result of the advance occur only in the subsurface.

The advance about 19,000 yr B.P. was followed by extensive deglaciation in Illinois. Frye and Willman (1973) estimate that the ice margin retreated at least 210 km. A tentative correlation of the Lemont drift made by Bretz (1955) and Horberg and Potter (1955) in extreme northeastern Illinois with the Malden Till Member (Bogner, 1974; Johnson, 1976) supports Willman's (1971) conclusion that, following deposition of the Tiskilwa Till Member, the ice margin retreated nearly to, or into, the Lake Michigan basin. The distances of ice-margin recession in Indiana and western Ohio are not known.

The Advance about 18,100 Yr B.P.

The advance about 18,100 yr B.P. in Illinois is represented by the Malden and Batestown Till Members of the Wedron Formation (Table 1-2). The stratigraphic break between the Malden-Batestown Till Members and older tills is marked by a change in till composition (Willman and Frye, 1970; Johnson, 1976). The tentative correlation of the Lemont drift with the Malden Till Member, as noted above, leads to an interpretation that the ice margin may have readvanced from a position in the present Lake Michigan basin. Consequently, the deposition of the Malden and Batestown Till Members may represent a maximum readvance of more than 200 km.

In the East White sublobe (Figure 1-2), the advance about 18,100 yr B.P. is represented by part of the Cartersburg Till Member of the Trafalgar Formation. In the Miami sublobe, it is represented by the Champaign till (Table 1-2). The limit of the advance about 19,000 yr B.P. was overridden by the advance about 18,100 yr B.P. in both sublobes, and the older till occurs only in the subsurface. In the Scioto sublobe, the advance is represented by the upper Caesar (Caesar II) till. The limits of the advances about 20,000 and 19,000 yr B.P. also were overridden by the advance about 18,100 in the Scioto sublobe.

The distance of ice-margin retreat from the limit of the advance about 18,100 yr B.P. in Illinois is uncertain. Frye and Willman (1973) estimate a retreat of approximately 175 km before the deposition of

the Yorkville Till Member, but the occurrence of the Snider Till Member between the Malden and Yorkville Till Members was not recognized as a separate unit at that time. Johnson (1976) infers that the ice margin may have retreated to, or north of, the limit of the advance about 15,500 yr B.P. (Figure 1-9). Distances of ice-margin retreat in the East White and Miami sublobes are not known due to the lack of subsurface stratigraphic and chronologic information. A retreat of at least 55 km may have occurred in the Scioto sublobe.

The Advance about 17,200 Yr B.P.

The advance about 17,200 yr B.P. appears to be represented by the Snider Till Member of the Wedron Formation in Illinois, part of the Cartersburg Till Member of the Trafalgar Formation in the East White sublobe, the Bloomington till in the Miami sublobe, and the lower Darby (Darby I) till in the Scioto sublobe (Table 1-2). The advance is of major significance in western and central Ohio (Goldthwait et al., 1961, 1965; Forsyth, 1965b; Goldthwait, 1968, 1976; Forsyth and Goldthwait, 1980).

The interpreted age of the culmination of the advance is based on time-distance diagrams constructed on the assumption that the rates of advance and retreat associated with the advances about 17,200 and 16,700 yr B.P. were approximately the same as those of the radiocarbon-controlled earlier advances. Reliable radiocarbon ages from interstadial silt beneath the till and in the till of the advance have not been published. An age of 17,290 ± 346 yr B.P. (OWU-76) has been attributed to the advance (Dreimanis and Goldthwait, 1973), but the sample was from the upper Caesar till (Table 1-2) and not the lower Darby till. An age of 17,340 ± 390 yr B.P. (OWU-256) from plant fragments in intertill silt north of the Reesville moraine has been attributed to the interstade between the advances about 18,100 and 17,200 yr B.P. (Goldthwait, 1968; Goldthwait and Rosengreen, 1969; Mörner and Dreimanis, 1973; Forsyth and Goldthwait, 1980). Charcoal from the same stratigraphic horizon (Moos, 1970) yielded ages of 19,303 ± 1080 yr B.P. (OWU-488) (Ogden and Hay, 1973) and 20,700 ± 1000 yr B.P. (DAL-14) (Ogden and Hart, 1976). The underlying and overlying tills were not identified on the basis of objective compositional properties or stratigraphic tracing of till units, and the silt was correlated primarily on the basis of the 17,340 yr B.P. age of sample OWU-256. The older ages are considered more reliable; the silt tentatively is considered to occur between the Boston and lower Caesar tills, rather than between the upper Caesar and lower Darby tills (Table 1-2).

The distance of ice-margin retreat from the limits of the advance about 17,200 yr B.P. is not known, but it probably was not greater than 30 km in any sublobe. Detailed stratigraphic information is lacking north of the limit of the younger limit of the advance about 16,700 yr B.P. throughout the southern Great Lakes region. The till sheets of the two advances in each sublobe have similar compositional properties, and the geographic distribution of sites where the two till sheets are superposed is not known. In the Scioto sublobe, the ice margin apparently retreated approximately 15 km, although a retreat of more than 30 km is possible.

The Advances about 16,700 and 16,100 Yr B.P.

The advances about 16,700 and 16,100 yr B.P. appear to be represented by the Yorkville Till Member of the Wedron Formation in Illinois, part of the Cartersburg Till Member of the Trafalgar Formation in the East White sublobe, part of the Bloomington till in the

Miami sublobe, and the upper Darby (Darby II) till in the Scioto sublobe (Table 1-2).

The Snider and Yorkville Till Members (Table 1-2) in northeastern Illinois cannot be differentiated in surface exposures on the basis of texture or carbonate content in some places (Moore and Johnson, 1981). Distinguishing the two till sheets as separate formal lithostratigraphic units does not appear to be warranted in some parts of Illinois. Nevertheless, the till units seem to represent distinct, significant readvances of the ice margin. At least two significant readvances are recorded by multiple units of Yorkville Till Member (Killey, 1980), but the advances are included in a single glacial event in Table 1-2).

In west-central Indiana, the East White sublobe's ice margin advanced at least 40 km westward over older Decatur sublobe Snider Till (Bleuer, 1975; Moore, 1978), which indicates a major change in ice dispersal paths in the Lake Michigan and Lake Huron lobes or a change in the dominance of flow in the two lobes.

Bleuer (1974b, 1975) considers the surface till of the type Crawfordsville moraine to be of the same age as the surface till of the Iroquois moraine and also to be younger than the Yorkville Till Member of the Paxton, Ellis, Chatsworth, and St. Anne moraines (Ct, Sa, Figure 1-8) in Illinois. The Lake Huron lobe till of the Crawfordsville moraine apparently represents the readvance about 16,100 yr B.P. that locally overrode the limit of the advance about 16,700 yr B.P. in western Indiana (Figure 1-9 and Table 1-2). The ice margin of the Lake Michigan lobe may have formed the Rockdale moraine in Illinois, which is composed of Yorkville Till Member, at approximately the same time. Thus, the "Crawfordsville" advance in Indiana is inferred to have culminated approximately 16,100 yr B.P. (Figure 1-9 and Table 1-2).

The age of the culmination of the advance about 16,700 yr B.P. is based on time-distance diagrams constructed in the manner outlined in the discussion of the advance about 17,200 yr B.P. The location of the advance limit in central Indiana and the Miami sublobe has not been established on stratigraphic grounds. In the Miami sublobe, the advance limit may coincide with the distal margin of the Eldorado moraine; in the Scioto sublobe, apparently, it coincides with the distal margin of the Bloomingburg moraine or the boulder belt associated with the moraine (Goldthwait, 1952; Goldthwait et al., 1961; Goldthwait and Rosengreen, 1969).

The time interval represented by the time-transgressive proglacial and nonglacial deposits that occur between the till sheets of the advances about 16,700 or 16,100 and 15,500 yr B.P. in the Lake Huron and Lake Erie lobes and the East White, Miami, and Scioto sublobes is referred to as the Erie interstade (Dreimanis and Goldthwait, 1973; Mörner and Dreimanis, 1973; Dreimanis, 1977b). No radiocarbon ages have been obtained from Erie interstadial deposits in the southern Great Lakes region.

The Haeger Till Member of the Wedron Formation (Willman and Frye, 1970) is exposed in a relatively small, wedge-shaped area south of the Wisconsin-Illinois state line. The till is overlapped by the Wadsworth Till Member (Lineback, 1979) and, apparently, it is the same age as the youngest part of the Yorkville Till Member (Table 1-2). Thick outwash overlies and is interbedded with the sandy and gravelly till, and ice-contact deposits are widespread (Powers and Ekblaw, 1940; Willman and Frye, 1970; McComas et al., 1972; Johnson, 1976). The supraglacial deposits and glaciofluvial sediment associated with the Haeger Till Member apparently were deposited during the early part of the Erie Interstade.

Deglaciation in the East White, Miami, and Scioto sublobes was marked by widespread downwasting. Hummocky, stratified deposits and eskers on till of the advance about 16,700 yr B.P. are overlain by thin, clayey till of the advance about 15,500 yr B.P. in Indiana (Wayne, 1968) and Ohio (D.S. Fullerton, unpublished data).

Much of the collapse topography in the Saginaw lobe (Figure 1-2) was formed during the early part of the Erie interstade. The "Kankakee torrent" (Ekblaw and Athy, 1925; Zumberge, 1960; Wayne and Zumberge, 1965), or "Kankakee flood" (Willman and Frye, 1970), was not a truly catastropic event. Meltwater released by large-scale stagnation in the Saginaw lobe was confined on all sides by active ice in the East White sublobe and in the Lake Erie, Lake Huron, and Lake Michigan basins (Figure 1-2). Consequently, the discharge was funneled southwestward into northeastern Illinois. In Michigan and extreme northern Indiana, the Kankakee deposits are collapsed and reworked glacial sediment, boundary ice-contact and glaciofluvial deposits, and collapsed lake deposits. In Illinois and extreme western Indiana, they are chiefly fluvial and offshore lake deposits underlain by Yorkville Till Member or Cartersburg Till and overlain by thick eolian sand.

The distance of ice-margin retreat from the limit of the advance about 16,700 yr B.P. in the Michigan lobe is not known. Frye and Willman (1973) estimate that the ice margin retreated at least 145 km from the limit of the Yorkville Till Member in Illinois. However, the Yorkville Till Member, as currently mapped (Lineback, 1979), extends south of the limit recognized by Frye and Willman, and the distance of ice-margin retreat may have been more than 150 km. The Lake Huron lobe retreated into the Lake Huron basin at least as far north as Goderich, Ontario, and the Lake Erie lobe margin retreated at least as far northeast as the Niagara Escarpment at the southwest end of the Lake Ontario basin (Dreimanis, 1969; Dreimanis and Goldthwait, 1973; Mörner and Dreimanis, 1973). A series of ice-dammed lakes occupied the Lake Erie basin, part of the Lake Huron basin, and perhaps part of the Lake Ontario basin during the Erie interstade.

The Advance about 15,500 Yr B.P.

The advance about 15,500 yr B.P. is represented by part of the Wadsworth Till Member of the Wedron Formation in Illinois, part of the New Holland Till Member of the Lagro Formation in Indiana, and an unnamed till in western Ohio (Table 1-2). In Illinois and extreme northwestern Indiana, the limit of the advance coincides with the distal margin of the Valparaiso moraine system (Va, Figure 1-8). In northeastern Indiana and extreme western Ohio, it coincides with the distal margin of the Union City moraine (Uc, Figure 1-8); in central western Ohio, it coincides with the distal margin of the Powell moraine (Po, Figure 1-8) (Fullerton, 1980). The New Holland Till Member overlies sediment of the "Kankakee torrent" (Hartke et al., 1975).

The Lake Huron lobe's ice margin readvanced more than 600 km to the position of the Union City and Powell moraines (Mörner and Dreimanis, 1973; Evenson and Dreimanis, 1976). The limit of the advance in eastern Indiana and western Ohio is marked by a change in till texture (clay content), elemental composition, magnetic susceptibility, depth of leaching, and soil (Forsyth, 1956, 1965b; Goldthwait, 1959; Wayne, 1968; Steiger and Holowaychuk, 1971; Wilding et al., 1971; Gooding, 1973; Forsyth and Goldthwait, 1980).

Wood from till north of the Powell moraine has yielded an age of 14,780 ± 192 yr B.P. (OWU-83) (Ogden and Hay, 1965) (Table 1-2);

however, the location and stratigraphic significance of the sample are uncertain. The sample is considered as providing only a minimum age of ice-margin recession from the position of the Powell moraine.

The distance of ice-margin retreat from the limit of the advance about 15,500 yr B.P. in the southern Great Lakes region is not known. Extensive ice-marginal lakes formed in northeastern Indiana and northwestern Ohio (Forsyth, 1973), and the ice margin may have retreated nearly to, or north of, the present southern shore of Lake Erie.

The Advance about 14,800 Yr B.P.

The advance about 14,800 yr B.P. is represented by part of the Wadsworth Till Member of the Wedron Formation in Illinois, part of the New Holland Till Member of the Lagro Formation in Indiana, and the Tymochtee till in northwestern Ohio (Table 1-2).

In northeastern Indiana and northwestern Ohio, the limit of the advance coincides approximately with the distal margin of the Wabash moraine (Wa, Figure 1-8), and the limit is marked by a change in texture (clay content), calcium carbonate equivalent, and calcium-magnesium ratio of the till. The compositional change has been attributed to the incorporation of large quantities of proglacial lake sediments during the readvance (Goldthwait, 1959; Goldthwait et al., 1961, 1965; Steiger and Holowaychuk, 1971; Gooding, 1973; Forsyth and Goldthwait, 1980).

In some places, the Wabash moraine and younger Fort Wayne moraine (Fw, Figure 1-8) are overridden (Bleuer, 1974a); in other places, they are true end moraines. The overridden segments apparently are superposed moraines, and both moraines record stillstands of the ice margin during the glacial event about 14,800 yr B.P.

After the formation of the Fort Wayne moraine, lakes formed in the southwestern part of the Lake Erie basin and in the southern part of the Lake Michigan basin (Table 1-2). A subsequent glacial readvance resulted in the extinction of the lake in the Lake Michigan basin and the formation of the Tinley moraine and younger Lake Border moraines (Lb, Figure 1-8) (Bretz, 1955; Schneider, 1968; Willman, 1971). An advance of the ice margin in the Lake Erie basin to the position of the Defiance moraine (De, Figure 1-8) was synchronous with the "Tinley advance" (Fullerton, 1980). Although parts of the Defiance moraine are overridden (Goldthwait, 1959), other parts of the moraine are true end moraine of the readvance, and the overridden parts apparently are superposed moraine. The very clayey till of the readvance to the position of the Defiance moraine in Ohio is referred to as the Lake till.

Minimum ages for the culmination of the advance about 14,800 yr B.P. are provided by four radiocarbon dates from depressions on, or north of, the till limit in Ohio: $14,500 \pm 150$ yr B.P. (ISGS-402) (S. N. Totten in Crowl, 1980), $14,300 \pm 450$ yr B.P. (W-198) (Rubin and Suess, 1955; Goldthwait, 1958), $14,290 \pm 130$ yr B.P. (ISGS-72) (Coleman, 1973; Totten, 1973), and $14,050 \pm 75$ yr B.P. (ISGS-348) (Totten, 1976).

Landform Assemblages

Most of the glaciated terrain south of the limit of the advance about 15,500 yr B.P. in Illinois (Figure 1-9) is charactized by broad, low "moraine" ridges and moraine systems with smooth, gentle slopes separated by flat, gently undulating or rolling till plains (M, Figure 1-10). Many of the ridges have a relief of 15 to 30 m, are 1 to 3 km wide, and can be traced for distances of 80 to 160 km (Leighton et al.,

1948; Willman and Frye, 1970). Where several parallel or concentric ridges are closely spaced or where ridges merge or overlap, they are grouped into moraine systems (the Shelbyville, Bloomington, Iliana, and Marseilles moraine systems). Individual moraine ridges generally are not bedrock controlled; nevertheless, the overall pattern of ridges and moraine systems within each sublobe reflects the bedrock's physiography (Horberg and Anderson, 1956; Willman and Frye, 1970). The low relief of the till plains is due chiefly to the thick mantle of subsurface till; the underlying bedrock surface commonly has considerable relief (Leighton et al., 1948; Horberg, 1950; Piskin and Bergstrom, 1967). The bedrock is exposed only very locally. Consequently, the texture and lithology of Late Wisconsin till in most of Illinois and Indiana (Figure 1-6) to a great extent reflect the incorporation of older Quaternary deposits and proglacial sediment. However, clasts usually are derived from outcrops far to the north.

The glaciated terrain south of the limit of the about 15,500 yr B.P. advance in the East White sublobe (Figures 1-2 and 1-9) is predominantly a gently rolling to nearly flat till plain. Segments of end moraines occur chiefly at or near the limits of glacial advances and in the broad interlobate areas between the East White and Miami sublobes (Wier and Gray, 1961; Wayne, 1965a; Wayne et al., 1966).

South of the limits of the advance about 15,500 yr B.P. in the Miami sublobe, the limits of the several advances are marked by segments of end moraine or belts of overlapping or superposed end moraines (Goldthwait et al., 1961, 1965; Gray et al., 1972). The moraines and moraine belts are 1 to 10 km wide, with a general local relief of 10 to 15 m. The till surfaces between moraines or belts of moraines are rolling to hilly, and the bedrock is exposed locally on valley walls and ridges.

The Scioto sublobe (Figures 1-2 and 1-9) south of the limit of the advance about 15,500 yr B.P. is a region of topographic contrasts. The central part of the sublobe is a gently undulating to rolling till plain. On the west side of the sublobe, a series of splayed, arcuate till ridges or belts of hummocky topography with relief as great as 90 m interrupts the rolling to hilly till plain (Goldthwait et al., 1961, 1965). The ridges merge northwestward into the suture between the Miami and Scioto sublobes. On the east side of the sublobe, ice movement was obstructed by bedrock hills and ridges. Consequently, the limits of successive readvances probably are superposed in places. Discrete end moraines that can be traced great distances or that can be correlated confidently with the splayed moraines on the west side of the sublobe are lacking. Rather, bedrock-controlled belts of hummocky till without definite linear trends and complicated belts of overlapping or merged till ridges are separated by areas of rolling or hilly till plain (Goldthwait, 1952; Conley, 1956; Goldthwait et al., 1961; Forsyth, 1966).

Low-relief (less than 3 m) hummocks are common on most of the moraines south of the limit of the advance about 15,500 yr B.P. in Illinois and the East White sublobe in Indiana (Wayne, 1956a, 1964, 1967, 1968; Harrison, 1963; Schneider et al., 1963; Willman and Frye, 1970; Hooten, 1973; Bleuer, 1974b; Bleuer and Moore, 1975; Lineback, 1975; Johnson, 1978; J.T. Wickham, 1979). Wayne (1968) proposes that the low-relief features originated by downwasting of relatively clean ice, with hummocks being formed as a result of the melting of debris-rich basal ice. This seems to be a likely explanation, and it implies that much less supraglacial debris was produced during deglaciation here than in areas to the north.

Belts of Precambrian boulders along the crests of several moraines and between moraines in Ohio have been described by Goldthwait

(1952, 1969), Goldthwait and Forsyth (1955), Forsyth and Goldthwait (1962), Goldthwait and Rosengreen (1969), and Quinn and Goldthwait (1979). Similar concentrations of boulders occur in Indiana. Presumably, the boulders, which occur primarily on the surface, were transported englacially and were deposited on subglacial till. Thus, it appears that, in most of the southern Great Lakes region south of the limit of the advance around 15,500 yr B.P., the ice was relatively clean and only far-traveled erratics occurred above the bed of the ice.

North of the limit of the advance about 15,500 yr B.P. in Illinois and northwestern Indiana and throughout the Saginaw lobe, the moraines have higher local relief and in many places are composed of ice-contact stratified deposits (Leighton et al., 1948; Bretz, 1955; Zumberge, 1960; Johnson et al., 1965; Wayne and Zumberge, 1965; Schneider and Johnson, 1967; Willman, 1971). These areas of high-relief moraines appear to occur where ice climbed out of a basin or abutted adjacent lobes — both situations in which compressive flow was likely.

Northeastern Indiana and western Ohio north of the limit of the advance about 15,500 yr B.P. are characterized by a rolling to very flat till plain and lake plain that is crossed by a series of low-relief, subdued, and generally concentric till ridges (Wayne, 1958, 1968; Goldthwait et al., 1961, 1965; Wayne and Zumberge, 1965; Burger et al., 1971; Johnson and Keller, 1972). The surfaces of the ridges generally are smooth or undulating, with local areas of subdued hummocks and closed depressions; rugged hummocky topography is rare or absent. Individual ridges generally are 0.5 to 3 km wide; where ridges merge, overlap, or are superposed, the belts are 2 to 8 km wide. In general, the relief on the till of the end-moraine surfaces is considerably lower than the relief on the end moraines in the northern Great Lakes and western regions.

Many areas between end moraines in the southern Great Lakes region are flat to gently rolling till plain. Although small-scale linear features are present locally (Schneider et al., 1963), drumlin fields were not formed. With the exception of the Saginaw lobe, evidence of glacial thrusting and intense compressive flow is almost nonexistent in the southern Great Lakes region, although it is relatively common in the western and northern Great Lakes regions. Bedrock exposures are uncommon; however, clear evidence of subglacial abrasion and subglacial stream erosion, presumably during Late Wisconsin time, has been documented on islands in Lake Erie and along the southern shore of Lake Erie (Carney, 1910; Goldthwait, 1973).

The relative scarcity of thick supraglacial drift indicating intense compressive flow, the absence of proved ice-thrust features, the absence of true drumlins, and the lack of documented evidence of even sporadic or discontinous permafrost suggest that the glacier bed in the southern Great Lakes region was at the pressure melting point throughout most of the Late Wisconsin glaciation.

NORTHERN GREAT LAKES REGION

Till Character and Thickness

Till of the Lake Superior lobe (Figure 1-2) is derived primarily from the Precambrian shield, and clasts are chiefly igneous and metamorphic rocks and upper-Precambrian sandstone (Figure 1-5). Older Late Wisconsin till units have a sandy loam matrix (40% to 70% sand, 25% to 60% silt, and 5% to 25% clay) and are noncalcareous. Younger till units are much more clayey (55% to 75% clay) (Johnson, 1980; Need, 1980). The thickness of the till varies considerably (Figure 1-7).

In the Green Bay and Lake Michigan lobes to the east, a source of carbonate rocks is available (Figure 1-5), and the till is calcareous, with carbonate clasts averaging about 25% in the northwest to 80% locally along the southern margin of Lake Michigan. Till deposited before about 13,000 yr B.P. has a sandy texture (Figure 1-6). Younger tills are reddish brown, have clay-loam and loam textures, and are highly calcareous (Lineback et al., 1974; Taylor, 1979; McCartney and Mickelson, 1982; Acomb et al., 1982). Tills of the Lake Michigan and Saginaw lobes in Michigan generally have loam and sandy loam textures and have a lithologic composition similar to tills of the Michigan and Green Bay lobes in Wisconsin.

Southeast of Lake Erie, in northeastern Ohio, northwestern Pennsylvania, and western New York, the ice of the Erie lobe and Killbuck and Grand River sublobes (Figure 1-2) advanced up the regional slope (Figure 1-4) onto dissected Paleozoic sedimentary rocks of the Appalachian Plateau (Figure 1-5). Older Late Wisconsin tills are yellowish brown when oxidized and have loam and fine-sandy loam textures; tills younger than the Erie interstade have darker colors and are more clayey (Shepps et al., 1959; White, 1960, 1961, 1969; Muller, 1963; White et al., 1969; Gross and Moran, 1971). Late Wisconsin tills in northeastern Ohio and northwestern Pennsylvania generally are thin (White, 1971).

Late Wisconsin tills in northwestern Pennsylvania were traced into southwestern New York as lithostratigraphic units by Muller (1956, 1960, 1963). The oldest Late Wisconsin tills have a silt loam and loam matrix, and younger ones are more clayey.

Late Wisconsin tills in west-central and east-central New York and the Finger Lakes region generally have a loam texture (Figures 1-6 and 1-9), although clayey till is present locally on the lake plain south of Lake Ontario. The thickness of till units varies. In parts of the Valley Heads moraine system, it may exceed 50 m. Late Wisconsin tills south, southeast, and east of the Finger Lakes region generally have a silt loam or loam matrix. The regional characteristics of till have been discussed by Coates (1963, 1974, 1976), Moss and Ritter (1962), Denny and Lyford (1963), Muller (1965c), Cadwell (1973a, 1973b), Kirkland (1973a, 1973b), and others.

In the Adirondack Mountains, till derived from complex Precambrian igneous and metamorphic rocks is thin, sandy, and gravelly (commonly more than 60% sand and less than 20% clay) (Buddington and Leonard, 1962; Denny, 1974; Craft, 1976, 1979). To the north in the St. Lawrence lowland (Figures 1-1 and 1-6), till is chiefly derived from Paleozoic sedimentary rocks and generally has sandy or sandy loam texture.

Till in the Hudson lowland (Figure 1-1) has a variable texture (Figure 1-6) and a complex lithology derived from sedimentary, metamorphic, and igneous rocks (Figure 1-5). Its thickness is variable and may exceed 30 m in the terminal moraine area of Long Island and New Jersey.

Chronology

Events before 13,500 Yr B.P. in Minnesota,
Wisconsin, and the Northern Peninsula of Michigan

The Wadena lobe (Figure 1-2) has been considered a sublobe of the Red River lobe derived from the Keewatin ice center, but its landforms and deposits are more similar to those of the Superior lobe than to those of lobes to the west. Ice of the Rainy lobe flowed across Ontario and northeastern Minnesota to central Minnesota, and ice of the Lake Superior lobe flowed from the James Bay lowland, through the deep Lake Superior basin, and into central Minnesota.

The sequence of events in the Lake Superior, Rainy, and Wadena lobes presented here is largely that developed by Wright (1972) and Wright and others (1973). However, we have reinterpreted the chronology because some of the published ages were derived from fine-grained organic material that may have been contaminated by older carbon and because the chronology of those summaries is incompatible with more firmly controlled chronologies in adjacent areas. As a result, we suggest here that some of the events occurred later than Wright (1972) and Wright and others (1973) propose.

The oldest known Late Wisconsin radiocarbon date from the region is 20,670 + 1500/− 1000 yr B.P. (GX-2741) (Matsch et al., 1972) from wood in till of the Lake Superior lobe in eastern South Dakota. The first Late Wisconsin (possibly Early Wisconsin) glacial event in the region was the advance during the Hewitt phase, when the Wadena lobe formed drumlins in west-central Minnesota (He, Figure 1-8). Wright (1972) suggests that the Hewitt phase occurred in Early Wisconsin or pre-Wisconsin time on the basis of a date earlier than 40,000 yr B.P. (W-1232) from ''plant litter' (Levin et al., 1965) above till of the Hewitt phase in central Minnesota. However, this date may be too old because of contamination with older carbon. The occurrence of an ice-free area in central Minnesota throughout Late Wisconsin time seems unlikely when the known extent of the ice in adjacent areas is considered.

The Hewitt phase was followed by the St. Croix phase, during which the Wadena, Rainy, and Superior lobes advanced to the Itasca (It, Figure 1-8) and St. Croix (Sc, Figure 1-8) ice-margin positions. Outwash fans on the distal side of the St. Croix moraine in central Minnesota are pitted with ice-block depressions (Wright, 1972). If the moraine marks the limit of the advance, then the Hewitt and St. Croix phases were separated by a time interval no longer than that needed for all buried Hewitt ice to melt, and both phases may have been Late Wisconsin in age. Wright (1972) and Wright and others (1973) suggest that the St. Croix phase occurred earlier than the time we show in Table 1-1. Their suggestion is based on radiocarbon dates from postglacial lake deposits overlying the St. Croix till: 20,500 ± 400 yr B.P. (I-5443) and 15,850 ± 240 yr B.P. (I-5048). These dates are from fine-grained organic material that may have been contaminated by older carbon. Wright (1972) and Wright and others (1973) correlate the St. Croix margin of the Lake Superior and Rainy lobes with the Itasca margin of the Wadena lobe because of the presence of coalescing outwash fans in front of the two moraines. By the same reasoning, we believe that the Itasca margin correlates with the outer limit of the Red River lobe because of coalescing outwash fans at the interlobe area, which indicate that the St. Croix margin and the eastern Des Moines margin were in contact in south-central Minnesota. This contrasts with the interpretation of Wright (1972) and Wright and others (1973) that the lobes were separate and that the St. Croix margin is older than the eastern Des Moines margin in central Minnesota. Just to the south, this interlobe margin is covered by deposits of a later advance of the Grantsburg sublobe (Figure 1-2). The outermost margin of the Des Moines lobe in central Minnesota may correlate with the outermost margin of the Des Moines lobe in Iowa, where wood dates indicate that it is no older than 14,000 yr B.P. It could be somewhat older (as indicated in Table 1-1).

The St. Croix moraine has been traced eastward into Wisconsin, where it forms an interlobe junction with the Chippewa moraine (Ch, Figure 1-8). No dates of wood under or within this till have been obtained from the marginal zones of the Chippewa, Wisconsin Valley,

Langlade sublobes or the Green Bay lobe (Figure 1-2) (Black, 1976b). Mickelson and others (1974) suggest that, although some cross-cutting of moraines occurs in Wisconsin, the Harrison, Parish, and Outer moraines (Ha, Pa, Ou, Figure 1-8) are products of a single glacial event.

In the Lake Superior lobe of Minnesota, deposits of the Automba phase are referred to as the Cromwell Formation (Wright, 1972; Wright et al., 1973). This lobe advanced to the Mille Lacs (Ml, Figure 1-8) margin, and the Rainy lobe ice advanced to the Vermillion margin (Ve, Figure 1-8). Although radiocarbon control is lacking, this phase preceded (Wright, 1972; Wright et al., 1973) the Split Rock phase (Sr, Figure 1-8, Table 1-1) of the Lake Superior lobe, which we suggest occurred at approximately 12,300 yr B.P.

Several radiocarbon dates provide minimum ages for deglaciation in the Green Bay lobe. Black (1976b) lists ages of 12,800 ± 400 (UCLA-632), 10,420 ± 300 (W-820), 12,220 ± 250 (W-762), 12,000 ± 500 (W-641), 15,560 ± 150 (WIS-442), 11,560 ± 350 (W-2015), 13,120 ± 130 (WIS-431), and 12,870 ± 125 (WIS-388) yr B.P. All are from fine-grained organic material or wood in calcareous lake sediments, and they do not fix the time of initial deglaciation accurately. Maher (1981) has obtained dates of 12,880 ± 125 (WIS-1004), 12,520 ± 160 (WIS-1075), and 12,260 ± 115 (WIS-1073) yr B.P. for noncalcareous basal organic sediment in Devils Lake, at the distal edge of the Johnstown moraine of the Green Bay lobe (Figure 1-2; Jo, Figure 1-8). Devils Lake is not an ice-block lake, and the dispersed organic matter is noncalcareous. Maher suggests that perhaps ice remained at, or close to, its maximum extent until approximately 13,000 yr B.P. in the Green Bay lobe.

The outermost (Johnstown) moraine in the southern part of the Green Bay lobe traditionally has been correlated across the Kettle Interlobate moraine (Chamberlin, 1878; Alden, 1918) with the West Chicago moraine (Wc, Figure 1-8) in Illinois (Frye et al., 1965). The Haeger Till Member of the Wedron Formation of Illinois, which comprises the West Chicago moraine, is older than the Wadsworth Till Member (Table 1-2). Although numerous small end-moraine segments are present in the Green Bay lobe, no major readvances before the 13,000 yr B.P. readvance are recognized.

Events before 13,500 yr B.P. in Eastern Ohio, Pennsylvania, New York, and New Jersey

The glacial record in northeastern Ohio and northwestern Pennsylvania differs from that in the southern Great Lakes region to the west and southwest. Late Wisconsin ice of the Lake Erie lobe (Figure 1-2) moved by compressive flow out of the Lake Erie basin and onto the Appalachian Plateau (Figure 1-1). A major bedrock high caused the development of the Killbuck and Grand River sublobes in Ohio; part of the physiographic high was a nunatak during maximum Late Wisconsin glaciation.

Continuous Late Wisconsin end moraines and moraine systems were not formed in northeastern Ohio and northwestern Pennsylvania before the advance about 15,500 yr B.P. Constructional topography on the older Late Wisconsin tills nearly everywhere is hummocky collapse topography without pronounced linear trends. Linear ridges generally are overridden ridges of pre-Late Wisconsin till, ice-contact deposits, or bedrock with thin veneers of Late Wisconsin till. Both continuous end moraines and belts of overridden end moraines are present north of the limit of the advance about 15,500 yr B.P., however, from northeastern Ohio through central New

York. Broader belts of end moraine on younger till surfaces in the eastern Killbuck and Grand River sublobes commonly are areas of hummocky collapsed topography developed on older Late Wisconsin till, pre-Late Wisconsin till, and/or ice contact deposits with thin veneers of younger till. Few of the "end moraines" in the northern parts of the Killbuck and Grand River sublobes are true end moraines; most are palimpsest moraines or superposed moraines.

In eastern Ohio, western Pennsylvania, and western New York, four major Late Wisconsin advances are recorded by named lithostratigraphic units. Only the earliest (represented by the Navarre and Kent Tills in the Killbuck and Grand River sublobes, respectively) is controlled by radiocarbon ages. Six dates from wood and twigs in proglacial lake sediment beneath Kent Till (White, 1968; Fullerton and Groenewold, 1974) indicate that the Lake Erie lobe ice margin reached Cleveland, Ohio, at 23,500 to 23,000 yr B.P. In many exposures, the Navarre and Kent Tills are composed of multiple units. The number of readvances and the magnitude of ice-margin recessions and readvances represented by the till units are not known. In most places, the broad Kent moraine apparently is a belt of pre-Late Wisconsin hummocky topography overridden by Late Wisconsin ice. The Kent Till on the end-moraine surface and on flatter surfaces to the north and south is generally less than 3 m thick, and the topography of the belt is palimpsest (White, 1969, 1971, 1974; White et al., 1969).

Subsequent advances of the Killbuck sublobe resulted in the deposition of the Hayesville Till (probably about 15,500 yr B.P.) and Hiram Till (probably about 14,800 yr B.P.) (Table 1-1) (White, 1961; Fullerton, 1980). In the Grand River sublobe, the same advances resulted in the deposition of the Lavery Till and Hiram Till (White, 1960; Fullerton, 1980).

A date on wood in marl of 14,000 ± 350 yr B.P. (W-365) (Droste et al., 1959) in northwestern Pennsylvania has been considered a minimum age for the Lavery Till (White et al., 1969; Crowl, 1980; Crowl and Sevon, 1980). However, this site is more than 20 km beyond the limit of Lavery Till as mapped by Shepps (1959a, 1959b) and Shepps and others (1959), and the supposedly thin Lavery Till in the vicinity of the site could be older than Lavery Till where defined.

The minimum ages of the Hiram Till are 14,500 ± 150 yr B.P. (ISGS-40) (S. N. Totten in Crowl, 1980; Crowl and Sevon, 1980), 14,290 ± 130 yr B.P. (ISGS-72) (Coleman, 1973; Totten, 1973), and 14,050 ± 75 yr B.P. (ISGS-348) (Totten, 1976).

The younger Ashtabula Till (Table 1-1) (White, 1960) was deposited as a result of a significant glacial readvance during the existence of glacial Lake Maumee in the Lake Erie basin approximately 14,100 yr B.P. (Fullerton, 1980). The ridges of the Ashtabula moraine system, however, are superposed moraines, with cores of older till. The minimum ages of the Ashtabula Till are discussed by Fullerton (1980).

In western New York, Kent till (Table 1-1) forms the Late Wisconsin drift border as far as the Salamanca reentrant (S, Figure 1-8) (Muller, 1963). Narrow belts of Lavery and Hiram Tills project south of the Lake Escarpment moraine system (Le, Figure 1-8) in New York (Muller, 1963: Plate 1).

The Ashtabula moraine system in Ohio and Pennsylvania, the Lake Escarpment moraine system in western New York, and the Valley Heads moraine system in west-central New York constitute a single belt of splayed, imbricated, overlapping, and overridden till ridges. The three continuous systems are topographically controlled by escarpments and drainage divides. The different names applied to the belt of moraines reflect the precedence of terminology in different

regions, rather than distinct genesis or age. The systems everywhere are controlled by bedrock topography, and they differ in that respect from moraine systems in the southern Great Lakes region. The three systems have been inferred to be synchronous and to record a single glacial event (Muller, 1965c, 1977b; Coates, 1974; Calkin and Miller, 1977). Some ridges in each system were probably formed during the advances about 14,800 and 14,100 yr B.P. However, the youngest moraine ridges of the Lake Escarpment system in New York are younger than those in the Ashtabula system in Pennsylvania and Ohio, and the youngest ridges in parts of the Valley Heads system are younger than any ridges in the Ashtabula and Lake Escarpment systems.

The location of the Late Wisconsin drift border south of the Finger Lakes (Figures 1-8 and 1-9) is uncertain. The Kent moraine (Ke, Figure 1-8) continues eastward from the reentrant as a morphological feature (Muller, 1960, 1963, 1965a, 1965b). The Almond moraine (Al, Figure 1-8) (Connally, 1961, 1964) in the western Finger Lakes region of central New York apparently is the morphological continuation of the Kent moraine, and it has been considered the outer limit of Late Wisconsin glaciation by Muller (1965a, 1965b, 1965c, 1977a, 1977b) and others. The Olean drift, south of the Almond-Kent moraine, therefore, has been considered to be of Early Wisconsin age.

Crowl and Sevon (1980) and Crowl (1980) conclude that radiocarbon dates from bogs and ponds in eastern Pennsylvania prove that all the Olean drift in Pennsylvania and New York is Late Wisconsin in age. However, the Olean drift may consist of two or more till units deposited during different glaciations (Coates, 1976), and the surface drift in eastern Pennsylvania is not necessarily the same age as the type surface drift in the Salamanca reentrant in western New York. The significance of the bog ages in eastern Pennsylvania is uncertain. However, the assignment of a Late Wisconsin age to the drift in Pennsylvania east of the Lehigh valley is supported by soil development, landforms, and postdepositional modification of the drift (Berg, 1975; Sevon, 1975a, 1975b, 1975c; Berg, et al., 1977; Sevon and Berg, 1978).

The morphological feature referred to as the Kent moraine east of the Salamanca reentrant may be younger than the Kent moraine in Ohio, western Pennsylvania, and extreme southwestern New York, and it may be a morphological continuation of the Lavery moraine. (Compare Plate 1 in Muller, 1963; Figure 1 in Calkin and Miller, 1977; and Muller, 1977b). If this is true, the limit of the advance of about 15,500 yr B.P. could be extended eastward, as the Lavery, Kent, and Almond moraines, to the western Finger Lakes region, where the Almond drift apparently is overlapped by, or merges with, segments of the Valley Heads moraine system (Vh, Figure 1-8) (Connally, 1961, 1964; Muller, 1965a, 1965c). Existing data from New York and Pennsylvania are compatible with the assignment of a Late Wisconsin age to the type Olean drift.

The Valley Heads moraine system probably represents a series of ice-margin fluctuations from about 15,500 B.P. until approximately 12,900 yr B.P. At least five lithologically and texturally similar till units are present locally in the system, but the stratigraphy and sequence of glacial events have not been determined. As ice margins fluctuated at and north of the moraine system, lakes were dammed in central and western New York between the ice margins and the Appalachian Plateau. The controversial history of the lakes is outlined by Calkin (1970), Muller (1977a), Calkin and Miller (1977), and Fullerton (1980).

The Adirondack Mountains (Figure 1-2) apparently were covered by continental ice during maximum Late Wisconsin glaciation (Craft, 1976, 1979), although cirques were occupied by local glaciers during deglaciation. Chronologic control of events in the Adirondacks before 13,000 yr B.P. is lacking.

Ice of the Champlain-Hudson River lobe (Figure 1-2) flowed southward and southwestward (Figure 1-3) to New Jersey and eastern Pennsylvania. To the east, the ice stream abutted the New England lobe, and the two lobes jointly formed moraines on Long Island.

The age of the Ronkonkoma moraine (Ro, Figure 1-8), the outermost end moraine on Long Island, is uncertain. Connally and Sirkin (1973), Sirkin and Mills (1975), Gustavson (1976), Sirkin (1980), and Sirkin and Stuckenrath (1980) infer that all the surface drift on Long Island is Late Wisconsin in age. However, Mills and Wells (1974), Coates (1976), and others infer that at least part of the Ronkonkoma moraine is pre-Late Wisconsin in age; Belknap (1980) draws a similar conclusion on the basis of amino acid studies.

The chronology of Late Wisconsin events in the Champlain-Hudson River lobe is also uncertain because firm radiocarbon control is not available. Connally and Sirkin (1970, 1973) infer that the Harbor Hill moraine (Hh, Figure 1-8) north of the Ronkonkoma moraine on Long Island was formed before 17,000 yr B.P. and that the ice margin retreated with a series of pauses or readvances in the Wallkill Valley and on the west side of the Hudson lowland. A younger documented (but not dated) readvance of about 25 km to the Rosendale ice-margin position (Rd, Figure 1-8) is inferred to have culminated about 14,000 yr B.P. (Connally and Sirkin, 1973), 16,000 yr B.P. (Connally, 1980), or 14,300 yr B.P. (Fullerton, 1980). The younger Luzerne advance (Lu, Figure 1-8) (Connally and Sirkin, 1971, 1973) culminated about 13,200 yr B.P. (Connally and Sirkin, 1973) or 14,100 yr B.P. (Fullerton, 1980). Younger ice-margin positions in the Hudson and Champlain lowlands are discussed in the next section.

Advances between 13,000 and 12,000 Yr B.P.

In Minnesota, the Automba phase of the Lake Superior lobe was followed by the Split Rock phase. Three dates from postglacial sediment overlying Split Rock deposits (Cromwell Formation) suggest (Wright, 1972: 530; Wright et al., 1973: 164) that the Split Rock phase occurred earlier than the date given in Table 1-1: 16,150 ± 550 yr B.P. (W-1973), 15,250 ± 220 yr B.P. (I-5051), and 13,480 ± 350 yr B.P. (W-1762). However, the dates are from fine-grained organic material that may have been contaminated with older carbon. Wright (1972: 533-35) correlates the Split Rock phase with the Pine City phase (Pc, Figure 1-8) of the Grantsburg sublobe of the Des Moines lobe in eastern Minnesota because Split Rock outwash formed a delta in a lake dammed by the Grantsburg sublobe. Both phases may include multiple ice advances.

The Split Rock margin has not been traced eastward across Wisconsin, but the inferred age of the advance suggests a correlation with the well-documented early Port Huron advance (Ph, Figure 1-8) of the Green Bay, Lake Michigan, and Lake Huron lobes (Table 1-1) at approximately 12,900 yr B.P.

In the Green Bay and Lake Michigan lobes of Wisconsin, no dates are available from wood in or beneath the several tills of the advance at 12,900 yr B.P. (Figure 1-9 and Table 1-1). At least two Port Huron advances have been documented (Acomb et al., 1982; McCartney and Mickelson, 1982) that postdate the withdrawal of the ice from the margin of the advance about 14,000 yr B.P. and antedate the development of the Two Creeks forest bed at approx-

imately 11,800 yr B.P. Several papers (Black, 1966, 1974, 1976b, 1978, 1980; Evenson et al., 1974, 1976, 1978; Farrand, 1976) include discussions of Late Wisconsin events in this area, and so we do not review the controversy here. Port Huron ice-margin positions can be correlated across Lake Michigan with the Whitehall and Manistee moraines in western Michigan and their correlatives in eastern Michigan, the outer and inner Port Huron moraines. The maximum age of the outer Port Huron moraine is dated by wood in St. Joseph till in Ontario (13,100 ± 110 yr B.P. [GSC-2313]) and by several dates (12,920 ± 400 [W-430], 12,900 ± 200 [I-3175], and 12,800 ± 250 [Y-240]) associated with the rise and stabilization of Lake Whittlesey in the Huron-Erie basins (Fullerton, 1980). This lake was dammed by the early Port Huron advance. Radiocarbon dates from the Cheboygan bryophyte bed north of the outer and inner Port Huron moraines in northern Michigan have been assumed to date the first Port Huron advance in the central Great Lakes area (Farrand et al., 1969). These authors and others (Farrand and Eschman, 1974; Evenson et al., 1976; Dreimanis, 1977a, 1977b) assume that the dates on bryophyte fragments beneath till date the advance of ice to the Port Huron moraine at approximately 13,000 yr B.P. Fullerton (1980) suggests that the bryophyte ages may be unreliable and that the till beneath the bryophytes may be younger than the Port Huron moraines. Correlation of the till at this site with till to the south and west in the Lake Michigan basin is tenuous.

The Port Huron moraine in the southern part of the Lake Huron basin in Michigan is correlated with the Wyoming moraine (Wy, Figure 1-8) in Ontario (Evenson and Dreimanis, 1976; Dreimanis, 1977a, 1977b) In both areas the moraines have double crests. In the Georgian Bay lobe (Figure 1-2), ice advanced at this time to the position of the Singhampton moraine (Si, Figure 1-8) (Chapman and Putnam, 1966; Dreimanis, 1977a, 1977b). In the Lake Erie and Lake Ontario lobes, the Paris moraine (Pa, Figure 1-8) locally represents the limit of the advance about 12,900 yr B.P.; elsewhere, the limit of the readvance is beyond the Paris moraine (Karrow, 1963). Muller (1977b) infers that the early Port Huron equivalent in New York is either the Hamburg or Marilla moraine (Ma, Figure 1-8). Calkin and Miller (1977) conclude that at least locally the limit of the advance is south of both of these moraines, and Fullerton (1980) argues that it is much farther south, at the position of the Girard moraine.

The advance about 12,900 yr B.P. is represented by part of the Valley Heads moraine system in the western and central Finger Lakes region in New York and by the Cazenovia and Tully moraine systems in the eastern Finger Lakes region and Mohawk lowland. It is represented by the Hinckley moraine system along the southwest margin of the Adirondack Mountains (Fullerton, 1980), by the Bridport advance limit in the Champlain lowland (Connally and Sirkin, 1973), and possibly by the Loon Lake advance limit (Denny, 1974) in the northeastern Adirondacks (Fullerton, 1980).

A later readvance in western and central New York caused waters in the Lake Erie basin to rise from post-Lake Wayne low levels to the Lake Warren III level and waters in the Finger Lakes region to rise to the Lake Warren level (Fullerton, 1980). The Alden moraine in the Lake Erie basin, the Geneva, Union Springs, Auburn, Stanwix, and Camden moraines in the Finger Lakes region, and the Ellenburg Depot moraine (Denny, 1974) in the St. Lawrence lowland northeast of the Adirondack Mountains are considered to be the tentative limits of the advance (Fullerton, 1980). The culmination of the advance occurred between 12,600 and 12,400 yr B.P. The minimum ages of the advance (Fullerton, 1980) do not closely restrict the age of the advance.

Advances after 12,000 Yr B.P.

Several sites along the west side of the Lake Michigan lobe and in the Green Bay lobe (Figure 1-2) have yielded apparently reliable dates from wood under till. The Two Creeks forest bed dates average about 11,850 yr B.P. (Black and Rubin, 1967-1968; Black, 1976b). The till over the bed (Two Rivers and Glenmore tills) records an advance of about 125 km to the Two Rivers moraine (Tr, Figure 1-8) (Evenson et al., 1974). If the Cheboygan bryophyte ages in Michigan are unreliable, the retreat before the advance may have been much more extensive (Fullerton, 1980). Black (1966, 1969b) and McCartney and Mickelson (1982) trace the boundary of the post-Two Creeks advance northward, but it has not been traced westward to Minnesota.

The equivalent of this advance in the Lake Superior lobe of Minnesota is probably the Nickerson advance, which deposited the Barnum Formation. The Nickerson phase may be represented by two advances, one before 12,000 yr B.P. and one later. According to Wright (1972; Wright et al., 1973), the Nickerson phase occurred at about the same time as the Alborn phase of the St. Louis sublobe because drainage from the latter ice by way of the St. Louis River was diverted around the terminus of the Lake Superior lobe during the Nickerson phase (Ni, Figure 1-8). The Alborn phase has been estimated to date to about 12,000 yr B.P. because the post-Alborn stage of Lake Aitken has been dated at $11,710 \pm 325$ and $11,560 \pm 400$ yr B.P. (W-502 and W-1141), but a more exact date cannot be determined. We have elected to show (in Table 1-1) both the Alborn and Nickerson phases culminating at 11,700 yr B.P., synchronous with the post Two Creeks advance to the east. The Nickerson phase probably occurred no later than about 11,500 yr B.P. The St. Louis sublobe could not have supplied meltwater to the St. Louis River after that time, because the Red River lobe could not have supplied ice to the St. Louis sublobe while Lake Agassiz was in existence.

The limit of the advance about 11,700 yr B.P. has been traced eastward in the northern part of the southern Michigan (Gl, Figure 1-8) (Burgis, 1977; Burgis and Eschman, 1981) to the Huron basin, but a correlative farther east has not been defined. The Lake Simcoe moraine in Ontario and the St. Narcisse and St. Faustin moraines in the St. Lawrence lowland may be equivalents (Fullerton, 1980).

The Lake Superior lobe produced several moraines, including the Sands-Sturgeon moraine (Saarnisto, 1974) at approximately 11,000 yr B.P. By that time, active ice had retreated entirely out of the Great Lakes region into Canada everywhere east of Lake Superior.

The Nemadji phase of Lake Superior began after the retreat from the Nickerson margin at approximately 11,500 yr B.P. (Farrand, 1969). Ice continued to occupy at least part of the Lake Superior basin until about 11,000 yr B.P., when ice retreated far enough to open the eastern spillway of Lake Agassiz into northern Lake Superior. This spillway was blocked by a readvance of the Lake Superior lobe at the start of the Emerson phase of Lake Agassiz (Elson, 1967; Clayton and Moran, 1982. This probably was the Marquette advance (Mq, Figure 1-8), which occurred about 9900 yr B.P. (Figure 1-9), as indicated by seven wood dates in till and outwash from northern Wisconsin and northern Michigan: $10,230 \pm 300$ yr B.P. (W-3896) (Hughes, 1978), $10,230 \pm 500$ yr B.P. (W-1414) (Black, 1976b), $10,220 \pm 500$ yr B.P. (M-359) (Saarnisto, 1974), $10,100 \pm 100$ yr B.P. (WIS-409) (Black, 1976b), 9850 ± 300 yr B.P. (W-3866) (Hughes, 1978), 9780 ± 250 yr B.P. (W-3904) (Hughes, 1978), and 9730 ± 140 (I-5082) (Black, 1976b).

The Superior lobe continued to block the eastern outlet of Lake Agassiz until the end of the Emerson phase at about 9300 yr B.P. (Elson, 1967; Clayton and Moran, 1982). At that time, the ice margin retreated from the Superior basin for the last time, and Lake Agassiz again spilled into northern Lake Superior.

Landform Assemblages

Several landforms and assemblages of landforms are more characteristic of the northern Great Lakes region than other areas. Among these are drumlins, tunnel channels, and large amounts of supraglacial debris in end moraines. (The distribution of glacial landscapes is illustrated in Figure 1-10.) In this discussion, we concentrate on interpreting these differences.

Large drumlin fields (S, Figure 1-10), absent on the Late Wisconsin surface in the southern Great Lakes area, are common in Minnesota, Wisconsin, northern Michigan, and New York. A comprehensive study of all drumlin fields has not been undertaken, but individual drumlin fields have been described by Alden (1918) and Whittecar and Mickelson (1976, 1979) in Wisconsin, Fairchild (1907, 1929) and Miller (1972) in New York, and Wright (1957) in Minnesota. Muller (1974) reviews the formation of drumlins and points out the characteristics of drumlins in New York and other regions, and King (1974) examines the morphometry of the drumlins in New York and Wisconsin.

Drumlins throughout the northern Great Lakes region have diverse compositions, ranging from all till, to till and stratified drift, to bedrock with a thin till cover. The tills in the drumlins also have diverse textures (Muller, 1974). Till of similar texture occurs to the south (Figure 1-6), where no drumlins occur, so it seems unlikely that till texture is a major factor in determining whether or not drumlins form.

Most drumlins occur in a spreading pattern; this suggests lateral extension and longitudinal compressive flow. This could be, in part, a result of an upward-sloping bed, as in the case of the drumlins of southern Wisconsin and New York; but in other situations drumlins formed as ice flowed down a regional slope (e.g., in northern Wisconsin). Drumlin fields tend to occur in a zone behind the end moraine or limit of readvance, and evidently most drumlins formed relatively close to the ice margin. It seems likely, therefore, that drumlins formed in or near the ablation zone, a fact that could account for the compressive flow in the drumlin areas.

Areas in the southern Great Lakes region have bed topography similar to that in drumlinized areas (eg., south of the Lake Michigan and Lake Erie basins). However, no drumlins were formed there even though compressive flow must have existed in a broad zone behind the ice margin. Therefore, the nature of the ice sheet itself must have been different in the two areas; the most likely difference was the existence of a frozen-bed zone along the ice margin in the northern Great Lakes during much of Late Wisconsin time. A number of other landforms support this hypothesis.

Cushing (*in* Wright et al., 1964) and Wright (1973) have described subglacial tunnel channels in the Lake Superior lobe behind the St. Croix moraine (SC, Figure 1-8). These are clearly associated with drumlins (Wright, 1973: Figure 2), and it is suggested that they were formed by catastrophic drainage of subglacial water from beneath an ice sheet that was frozen to the bed near the margin. Similar features are present in the Green Bay lobe of Wisconsin, together with smaller tunnel channels that extend less than 20 km back from the former ice margin. These channels lead to large outwash fans at the moraine front and, because of their size and abundance, must have carried water at different times. These features are present in northern Illinois but cannot be documented to the south in Illinois, Indiana, and Ohio (N. K. Bleuer, W. H. Johnson, and S. N. Totten, personal communication, 1980).

End moraines in northern Wisconsin, Minnesota, and Michigan generally are narrower and have higher relief than the broad, low-relief end moraines of the southern Great Lakes region (except in interlobate areas). Although no one has quantitatively analyzed the relative amounts of supraglacial and subglacial drift in the moraines of the two areas, the amount of supraglacial drift apparently increases northward in the moraines from central Illinois to northern Wisconsin. This may also be due to the presence of a frozen-bed zone near the margin, with a freezing-on zone behind the margin in the area where drumlins formed. Apparent thrust masses similar to those in the western area are present in the Green Bay lobe of Wisconsin.

Independent evidence also suggests that conditions were conducive to having a frozen-bed zone along the ice margin in Wisconsin and Minnesota. Wright (1973) infers from pollen evidence that mean annual temperatures along the ice margin may have been -8°C or colder. Black (1965, 1969a, 1976a) reports numerous ice-wedge casts west and south of the Late Wisconsin border in Wisconsin. (See also Péwé, 1982.) Although these casts have not been dated, it seems likely that they are of Late Wisconsin age and indicate permafrost conditions.

There is not yet enough available information to estimate how long a frozen-bed remained along the perimeter of the ice sheet during Late Wisconsin time. Wright (1973) suggests that by the time of the Split Rock and Nickerson advances the frozen toe was narrower than it had been, but perhaps it still existed. In the Lake Michigan lobe area of Wisconsin, changes in consolidation properties of tills suggest a transition from cold- to warm-based ice some time between 13,000 and 12,000 yr B.P. (Mickelson et al., 1979). Drumlins continued to form, however, in post-Two Creeks time in the upper peninsula of Michigan and in southern Michigan.

NEW ENGLAND REGION

Till Character and Thickness

Most of New England (Figure 1-1) has fairly high relief; however, southeastern Massachusetts, Long Island, and New York are parts of a low-relief coastal plain. The character of the drift in the Coastal Plain is quite similar over a distance of about 400 km along the southern limit of the ice and it differs from that of the drift in northern New England both in composition and in morphology (Flint, 1971). New England, in contrast to the middle of the continent, is a region of high relief, steep slopes, and fine topographic texture. Bedrock lithologies are predominantly metamorphic and igneous rocks rich in quartz; few rocks are rich in primary carbonate (Figure 1-5).

In general, the color of New England tills derived from crystalline bedrock varies from light to dark gray, depending on the amount of mafic minerals in the source terrane. However, tills are red or black when derived from areally limited exposures of Triassic or Pennsylvanian sedimentary rocks. The matrix in most tills ranges from sandy to silty with less than 10% clay-sized particles. However, the

"drumlin tills" of the Boston area and some of the tills of coastal Maine, for example, contain as much as 20% to 30% clay-size particles, but little of the clay consists of clay minerals (Figure 1-6).

Unfortunately, very little regional research has been done to characterize the tills of New England beyond that summarized by Schafer and Hartshorn (1965). Two types of till, identified primarily by texture and occasionally by composition, have been recognized over broad areas of southern New England (Flint, 1961). Bedrock lithologies vary greatly over short distances, especially within crystalline terranes. Representative sampling of these highly variable tills would be a prodigious task of questionable value in view of current research priorities for the region.

Schafer and Hartshorn (1965) estimate that the average thickness of drift in New England is probably less than 10 m and perhaps closer to 5 m (Figure 1-7). Flint (1930) estimates an average thickness of 1.5 to 3 m for till in Connecticut, and Goldthwait and Kruger (1938) estimate 3 to 5 m for till in New Hampshire. Such estimates are not available for the other states in New England, but they probably would be similar.

Chronology

In the midcontinental region, till stratigraphy can be used to document the advance and retreat of the ice margin during Late Wisconsin time. However, readvances in New England did not produce tills with significantly different compositions over wide areas.

Several sections displaying multiple tills have been reported in recent years from New England (e.g., White, 1947; Caldwell, 1959; Flint, 1961; Kaye, 1961; Stewart and MacClintock, 1964; Pessl, 1966; Pessl and Koteff, 1970; Borns and Calkin, 1977). Some suggest at least two major phases of glaciation for the region. Wood in the base of a complex drift section near Wallingford, Connecticut, was dated as older than 40,000 yr B.P. (Y-451) (Schafer and Hartshorn, 1965), and peat overlain by till near Worcester, Massachusetts, was dated as older than 38,000 yr B.P. (W-647; L-380) (Schafer and Hartshorn, 1965). A dated section at New Sharon, Maine (Caldwell, 1959; Borns and Calkin, 1977), is the only one that clearly demonstrates at least two different glaciations separated by a nonglacial interval. There, a subaerial weathering zone containing organic debris dated as older than 52,000 yr B.P. (Y-2683) is found between two basal tills. The presence of multiple till sections at widely distributed localities in New England suggests the possibility of one or more pre-Late Wisconsin ice advances in the region.

The Late Wisconsin terminal position of the Laurentide ice sheet east of Long Island generally has been assumed to be marked by the Vineyard-Nantucket moraine line (Schafer, 1961, 1980; Kaye, 1964; Schafer and Hartshorn, 1965) and perhaps by the distribution of coarse gravel on the continental shelf (Schlee and Pratt, 1970). However, this view has been questioned recently; Grant (1977) suggests that the maximum position of the Late Wisconsin ice was far north of the edge of the continental shelf and that it more closely approximated the present coasts of Maine and perhaps New Hampshire. Three radiocarbon dates from basal till in the Great South Channel southeast of Cape Cod — 26,600 ± 500 yr B.P. (W-3953), 20,350 ± 450 (W-3968) yr B.P. (corrected ages 24,290 and 19,500 yr B.P., respectively) (Bothner and Spiker, 1980), and 20,330 ± 600 yr B.P. (W-3901) (Oldale, 1980) — indicate that the till was deposited during Late Wisconsin time at a location just north of the terminal position of the ice as indicated by Schlee and Pratt (1970). These dates, as well as a model for reconstructing the extent and disintegration of the Late Wisconsin ice sheet in the Gulf of Maine (Fastook et al., 1980) support Schlee and Pratt's (1970) estimation of the extent of the ice.

Stratigraphic evidence from Zack's Cliff on Martha's Vineyard south of Cape Cod (Figure 1-1) (Kaye, 1964, 1980) indicates that the ice margin was at that location as late as approximately 15,300 yr B.P. Evidence presented by Connally and Sirkin (1973) and Sirkin and Stuckenrath (1980) from western Long Island, New York, suggests that the ice margin farther west had reached its maximum position by about 21,700 yr B.P. and had receded by approximately 18,000 yr B.P. Apparently, the Late Wisconsin Laurentide ice sheet reached its southernmost position at various times between 21,000 and 15,000 yr B.P. in different places in southern New England.

The recession of the ice margin on the central and eastern sections of the continental shelf was accompanied by a marine transgression across the Gulf of Maine and into eastern coastal Massachusetts, coastal New Hampshire, and coastal and central Maine. The receding ice margin reached the position of the present New England coast north and east of Boston between 14,000 and 13,000 yr B.P. (Borns, 1973). At about that time, the ice margin in southern New England probably trended approximately west from southern New Hampshire through central Connecticut (Flint, 1953) or Massachusetts and then extended northeastward, forming a reentrant in the ice perhaps as far north as Bridport (Figure 1-8) south of Lake Champlain in the Hudson-Champlain lowland (Connally and Sirkin, 1973).

The ice margin retreated a distance of approximately 300 km across the Gulf of Maine while retreating only about 75 km into central Connecticut or Massachusetts. This rapid recession in the Gulf of Maine is best explained by a rapidly calving marine margin in contrast to a grounded southern margin in southern New England. The reentrant in the Hudson-Champlain lowland could be explained partially by the presence of the various ice-marginal lakes (Connally and Sirkin, 1973), which would probably have enhanced the rate of marginal recession by calving.

Before about 1960, the deglaciation of New England was thought to have involved progressively marginal recession to the north or northwest toward the St. Lawrence lowland in Canada, with the recession giving rise to local areas of separation, stagnation, and disintegration of the ice as topographic thresholds hindered the flow and finally emerged from the ice. According to this model, all of New England would have been deglaciated by about 12,000 to 11,000 yr B.P. In 1963, Borns suggested that a residual ice cap may have persisted between the Maine coast and the St. Lawrence lowland by approximately 12,800 yr B.P., perhaps caused by a marine transgression up the St. Lawrence lowland from the Gulf of St. Lawrence. This transgression is recorded by deposits of the Goldthwait and Champlain seas. Such a possibility has been recognized by others (e.g., Shilts, 1976; LaSalle et al., 1977; Thomas, 1977; Lowell, 1980, 1981; Genes et al., 1981). Richard (1978) demonstrates that the transgressing sea reached the Ottawa area by at least 12,700 yr B.P., and Stuiver and Borns (1975) show that the receding ice margin stood along the coast of Maine between 13,300 and 13,000 yr B.P. These data clearly indicate that an ice cap existed between the Maine coast and the St. Lawrence lowland no later than 13,000 yr B.P. Progressive separation of the ice cap from the main mass of Laurentide ice, caused by the calving embayment that moved southwest up the St. Lawrence lowland, possibly resulted in sections of the ice cap being separated as early as 14,000 yr B.P. (Mayewski, et al., 1980).

The development of rapidly calving embayments both in the St. Lawrence lowland and in coastal Maine and New Hampshire would have promoted a rapid drawdown of the surface of the ice cap (Fastook et al., 1980) that exposed the highlands of New Hampshire and Maine as nunataks between 14,000 and 13,000 yr B.P. (Borns and Calkin, 1977; Borns, 1980; Davis and Davis, 1980; Hughes, 1980; Mayewski et al., 1980). The geography and chronology of the dissipation of this ice cap are the subject of current research. However, the geographically limited data available suggests that the final dissipation of the ice was accomplished after 11,500 yr B.P. (e.g., Davis and Davis, 1980; Kite and Stuckenrath, 1980).

Deglaciation in New England while the ice margin was close to the present position of the coastline appears to have involved the marginal retreat of an internally active ice sheet (Shafer and Hartshorn, 1965; Borns, 1973; Stuiver and Borns, 1975; Goldsmith and Schafer, 1980; Koteff, 1980; Larsen and Hartshorn, 1980; Schafer, 1980; Stone and Peper, 1980). The glacier's terminus in northern coastal Massachusetts, coastal New Hampshire, and central coastal Maine receded as a calving marine front, whereas in southern New England the grounded margin receded subaerially.

Landform Assemblages

As the ice margin retreated across the Gulf of Maine and the present coastal zones of New Hampshire and Maine, it was accompanied by a marine transgression. The recession of this marine-based portion of the ice sheet is documented by hundreds of emerged submarine end moraines and ice-contact marine deltas and by a discontinuous blanket of glaciomarine silt and clay (V, Figure 1-10) (Leavitt and Perkins, 1935; Borns, 1973; Smith, 1981).

Two types of end moraines are recognized. The larger type occurs in segments up to 15 km long and 20 m high. The smaller washboard moraines commonly occur in clusters 2 to 5 m high with segments up to a few kilometers long.

The moraines in general are composed of basal till, flow till, and submarine outwash and sand and gravel. The relative amounts of these sediment types vary considerably among individual moraines, as well as within an end moraine. The large moraines are commonly referred to as "stratified moraines" because they are composed primarily of stratified submarine outwash with lesser amounts of till (Leavitt and Perkins, 1935; Borns, 1981; Smith, 1981).

The washboard moraines have not been studied as well as the large moraines but appear also to be composed of mixtures of till and submarine sand-and-gravel outwash. In general, the lithologic composition of these coastal deposits reflects the crystalline terrane from which they were derived. On the other hand, the texture largely reflects the effect of large volumes of meltwater reaching the sea from the base of the glacier margin as well as the hydrodynamics of the sea into which the meltwater and sediments were discharged.

Inland from the coastal zone, the ice appears to have dissipated by marginal recession and thinning (Goldthwait, 1938; Goldsmith and Schafer, 1980; Koteff, 1980; Schafer, 1980; Stone and Peper, 1980; Larsen, 1981). Much of the initial lowering of the surface of the glacier was probably caused by drawdown into calving marine embayments in the Gulf of Maine and the St. Lawrence lowland (Thomas, 1977; Fastook et al., 1980; Mayewski et al., 1980; Hughes, 1981). In southern New England, the nonmarine portion of the ice margin dissipated because of a negative mass balance. Ultimately, as the ice margin receded, perhaps as internally active ice with a stagnant fringe (Schafer and Hartshorn, 1965; Koteff, 1980; Stone and Peper, 1980;

Caldwell et al., 1981), the thinning glacier encountered underlying topographic highs, especially in the Green, White, and Longfellow mountains (Figure 1-1) (Goldthwait, 1938; Borns and Calkin, 1977). The ice sheet then progressively separated and finally dissipated as stagnant residual masses in low areas. This general style of dissipation resulted in a variety of ice-contact stratified deposits, glaciolacustrine sediments, and outwash, but with a general lack of recognizable end moraines throughout most of north-central New England.

Regional Synthesis

We have attempted in earlier parts of this chapter not only to describe the sequence of glacial events along the southern margin of the Laurentide ice sheet but also to provide information about the character of glacial sediments and landforms in different areas. It was our intention that this information would be used to interpret temperature conditions, regions of compressive and extending flow in the ice sheet, and ice-surface profiles. It is clear, however, given our current state of knowledge concerning the conditions of formation for various landforms, that only generalizations can be made. Recently, attention has turned to this area of research, and we hope this trend will continue.

In the western region in Montana, North Dakota, South Dakota, and parts of Minnesota, ice near the margin of the glacier clearly was frozen to its bed. Only rare wood dates between 21,000 and 12,000 yr B.P. have been obtained north of the Des Moines lobe in Iowa and the James lobe in South Dakota. Large, intact thrust masses and drumlins are restricted to the northern part of the western region as well. Thick supraglacial till is abundant, perhaps in part because of upward movement of debris behind a margin frozen at the bed but also because of the compressive flow caused by underlying topography.

Clayton and others (1980) suggest that a frozen terminal zone about 2 km wide existed in North Dakota between 14,000 and 12,000 yr B.P. We suggest that the zone may have been somewhat wider and that the drumlin areas represent a partially frozen bed where freezing-on at the base was taking place. Evidence of former permafrost is common in fluvial sediments outside the Late Wisconsin drift border in southwestern North Dakota, but precise dates are not available.

In the southern James and Des Moines lobes, drumlins and large, intact thrust masses are absent. Thick supraglacial till is present, but only where ice moved by compressive flow up the regional slope. The Des Moines lobe advanced over *Picea* forest between 15,000 and 14,000 yr B.P.; this suggests that permafrost may not have been present that far south at that time. We suggest that in this area the glacier bed was wet from the margin to the central part of the lobe.

In the northern Great Lakes region, there are permafrost indicators of Late Wisconsin age (Péwé, 1982). Birks (1976) reports that the transition from tundra to forest vegetation occurred at approximately 14,700 to 13,600 yr B.P. (on the basis of nonwood radiocarbon dates) in central Minnesota, and Wright (1980) suggests that about 14,000 yr B.P., at the time of the maximum extent of the Des Moines lobe, shrub-tundra bordered the northeast edge of the Des Moines lobe while spruce forest was present to the south. The Cheboygan bryophyte bed of northern Michigan (Farrand et al., 1969) contains *Dryas* dated between 13,000 and 10,000 yr B.P. In eastern Wisconsin, no Late Wisconsin wood dates older than about 13,100 yr B.P. have been obtained. This evidence, as well as the presence of

drumlins, the occurrence of fairly thick supraglacial deposits even where ice was flowing down the regional slope, and the abundance of subglacial (catastrophic?) tunnel channels, supports the existence of an ice margin frozen to its bed. The drumlins may represent an area of partially frozen bed where freezing-on was taking place. In New York and parts of Ontario, reported evidence of intact thrust masses and the presence of drumlins and permafrost (Morgan, 1972) are also consistent with this model.

In the southern Great Lakes region, forest vegetation was present along the ice margin throughout Late Wisconsin time. Studies by Gruger (1972a, 1972b) indicate that *Picea* grew in southern Illinois as the Late Wisconsin ice advanced. To the east, in Indiana and Ohio, *Picea* was evidently dominant (Burns, 1958; Goldthwait, 1958, 1959; Gooding, 1963; Kapp and Gooding, 1964). Although Whitehead (1973) suggests a zone of tundra vegetation along the perimeter of the ice sheet in this area, evidence of long-term tundra conditions is lacking.

Periglacial frost phenomena have been described from several localities in Illinois (Sharp, 1942; Horberg, 1949; Ekblaw and Willman, 1955; De Heinzelin, 1957; Frye and Willman, 1958; Wayne, 1967). However, precise dates on most of these features have not been determined, and none of the features reported before 1970 can definitely be attributed to permafrost (Black, 1976a). The Dekalb mounds in northern Illinois (Flemal et al., 1970; Flemal, 1972) cannot be considered proof of permafrost, because similar features have been reported elsewhere as collapse features. Black (1976a), in a review of permafrost features in this area, concludes that the evidence for permafrost in Illinois, Indiana, and Ohio is not convincing.

In the southern Great Lakes region, supraglacial till is relatively thin, except in areas of intense compressive flow caused by the underlying topography or by the junction of ice lobes. Drumlins, intact thrust masses, and subglacial tunnel channels, so common in the northern Great Lakes and western regions, are absent here. This area, like the Des Moines lobe in Iowa, was wet based at its margin and for a considerable distance upglacier.

In New England, the ice sheet was different from that to the west. Evidence of conditions along the ice margin during the glacial maximum is generally lacking because much of the area is now submerged. Certainly during ice retreat, the glacier margin was greatly affected by the higher relief in this area and by a wide calving margin in the Gulf of Maine and the St. Lawrence lowland. Proglacial and subglacial outwash is abundant on the present land surface, and it seems unlikely that the bed was frozen after 15,000 yr B.P. No evidence for or against a frozen bed before that time is available.

By about 13,000 yr B.P., ice in southern and eastern New England had been cut off from the main part of the Laurentide ice sheet by the opening of the St. Lawrence River and the emergence of highlands in northern New England. At that time, the whole ice mass probably was below the equilibrium line and extensive marginal and subglacial melting was taking place.

Summary and Conclusions

Late Wisconsin glacial ice entered the United States some time before 23,000 yr B.P. In New England, the ice sheet evidently was marine based along its eastern margin. There, the ice margin had evidently reached its maximum position by about 21,000 yr B.P. and was retreating by about 18,000 yr B.P. The receding ice margin reached the present coast of Maine, New Hampshire, and Massachusetts by about 13,500 yr B.P. Shortly after that time, the St. Lawrence valley was deglaciated, leaving a residual ice cap in northern New England. The final dissipation of the ice did not take place until after 11,500 yr B.P.

The retreat of the ice left numerous drumlins (many of them rock cored), long esker systems, and, in coastal areas, washboard moraines. Long, continuous end moraines like those in the southern Great Lakes area are absent.

In the southern Great Lakes region, major readvances of the ice margin took place at about 21,000, 20,000, 19,000, 18,100, 17,200, 16,700, 15,500, and 14,800 yr B.P. before ice retreated into the northern Great Lakes region. The ice sheet had its greatest extent between 21,000 and 19,000 yr B.P. throughout the southern Great Lakes area.

Most of the advance positions are marked by end moraines, end moraine systems, or changes in the texture and lithology of tills in stratigraphic succession. Moraines are commonly separated by low-relief till plains. Drumlins are absent from the region, and tunnel channels, eskers, and ice-thrust masses are rare.

In the northern Great Lakes area, early advances are not well dated. By 14,500 yr B.P., ice had retreated into the basins of the Great Lakes. Subsequent readvances took place at approximately 12,900, 12,300, 11,700, and 9900 yr B.P. Glacial ice was gone from the United States by shortly after 9500 yr B.P. In the northern Great Lakes region, end moraines typically have high-internal relief, and drumlins, eskers, and tunnel channels are common in many areas. Till texture and lithology vary according to their source.

In the western region, the ice sheet reached its maximum position in much of South Dakota, North Dakota, and Montana between 20,000 and 17,000 yr B.P. In parts of southern Minnesota, central Iowa, and southern South Dakota, deposits of an advance at about 14,000 yr B.P. mark the outer limit of Late Wisconsin glaciation. Till in this region typically has fine texture and is calcareous. In most of this region, large areas of thick supraglacial deposits are present, as are drumlins, tunnel channels, and ice-thrust masses. In the southern portion of the area, deposits resemble those of the southern Great Lakes area.

References

Acomb, L. J., Mickelson, D. M., and Evenson, E. B. (1982). Till stratigraphy and late glacial events in the Lake Michigan lobe of eastern Wisconsin. *Geological Society of America Bulletin* 93, 289-96.

Alden, W. C. (1918). "Quaternary Geology of Southeastern Wisconsin." U.S. Geological Survey Professional Paper 106.

Alden, W. C. (1932). "Physiography and Glacial Geology of Eastern Montana and Adjacent Areas." U.S. Geological Survey Professional Paper 174.

Anderson, R. C. (1957). Pebble and sand lithology of the major Wisconsin glacial lobes of the central lowland. *Geological Society of America Bulletin* 68, 1415-49.

Andrews, J.T. (1963). Cross-valley moraines of the Rimrock and Isortoq river valleys, Baffin Island, Northwest Territories: A descriptive analysis. *Geography Bulletin* 19, 49-77.

Arndt, B. M. (1977). "Stratigraphy of Offshore Sediment, Lake Agassiz, North Dakota." North Dakota Geological Survey Report of Investigation 60.

Ashworth, A. C., Clayton, L., and Bickley, W. B. (1972). The Mosbeck site: A paleoenvironmental interpretation of the late Quaternary history of Lake Agassiz based on fossil insect and mollusk remains. *Quaternary Research* 2, 176-88.

Barnett, D. M., and Holdsworth, G. (1974). Origin, morphology, and chronology of sublacrustrine moraines, Generator Lake, Baffin Island, Northwest Territories, Canada. *Canadian Journal of Earth Sciences* 11, 380-408.

Belknap, D. F. (1980). Amino acid geochronology and the Quaternary of New England and Long Island. *In* "American Quaternary Association, Sixth Biennial Meeting, Abstracts and Program, 18-20 August 1980," pp. 16-17. Institute for Quaternary Studies, University of Maine, Orono.

Berg, T. M. (1975). "Geology and Mineral Resources of the Brodheadsville Quadrangle, Monroe and Carbon Counties, Pennsylvania." Pennsylvania Geological Survey, 4th Series, Atlas 205, a.

Berg, T. M., Sevon, W. D., and Bucek, M. F. (1977). "Geological and Mineral Resources of the Pocono Pines and Mount Pocono Quadrangles, Monroe County, Pennsylvania." Pennsylvania Geological Survey, 4th Series, Atlas 204, c and d.

Birks, H. J. B. (1976). Late-Wisconsinan vegetational history at Wolf Creek, central Minnesota. *Ecological Monographs* 46, 395-429.

Black, R. F. (1965). Ice-wedge casts of Wisconsin. *Wisconsin Academy of Science, Arts and Letters* 54, 187-222.

Black, R. F. (1966). Valders glaciation in Wisconsin and upper Michigan: A progress report. *University of Michigan Great Lakes Research Division Publication* 15, 169-75.

Black, R. F. (1969a). Slopes in southwestern Wisconsin, U.S.A., periglacial or temperate? *Biuletyn Peryglacjalny* 18, 69-81.

Black, R. F. (1969b). Valderan glaciation in western upper Michigan. *In* "Proceedings of the 12th Conference on Great Lakes Research," pp. 116-23. International Association of Great Lakes Research.

Black, R. F. (1974). Late Pleistocene shorelines and stratigraphic relations in the Late Michigan basin: Discussion. *Geological Society of America Bulletin* 85, 659-60.

Black, R. F. (1976a). Periglacial features indicative of permafrost: Ice and soil wedges. *Quaternary Research* 6, 3-26.

Black, R. F. (1976b). Quaternary geology of Wisconsin and contiguous upper Michigan. *In* "Quaternary Stratigraphy of North America" (W. C. Mahaney, ed.), pp. 93-117. Dowden, Hutchinson, and Ross, Stroudsburg, Pa.

Black, R. F. (1978). Comment on "Greatlakean substage: A replacement for Valderan substage in the Lake Michigan basin." *Quaternary Research* 9, 119-23.

Black, R. F. (1980). Valders-Two Creeks, Wisconsin, revisited: The Valders Till is most likely post-Twocreekan. *Geological Society of America Bulletin* 91 (part 1), 713-23.

Black, R. F. Goldthwait, R. F., and Willman, H. B. (eds.) (1973). "The Wisconsinan Stage." Geological Society of America Memoir 136.

Black, R. F., and Rubin, M. (1967-1968). Radiocarbon dates of Wisconsin. *Wisconsin Academy of Science, Arts and Letters, Transactions* 56, 99-115.

Blackman, M. J. (1970). A detailed study of the Pleistocene history of a portion of Preble County, Ohio. M.S. thesis, Miami University of Ohio, Miami.

Bleuer, N. K. (1974a). Buried till ridges in the Fort Wayne area, Indiana, and their regional significance. *Geological Society of America Bulletin* 85, 917-20.

Bleuer, N. K. (1974b). "Distribution and Significance of Some Ice-Disintegration Features in West-Central Indiana." Indiana Geological Survey Occasional Paper 8.

Bleuer, N. K. (1975). "The Stone Creek Section: A Historical Key to the Glacial Stratigraphy of West-Central Indiana." Indiana Geological Survey Occasional Paper 11.

Bleuer, N. K. (1980). Recurrent surging (?) of the Wisconsinan Huron-Erie lobe in Indiana. *In* "American Quaternary Association, Sixth Biennial Meeting, Abstracts and Program, 18-20 August 1980," p. 28. Institute for Quaternary Studies, University of Maine, Orono.

Bleuer, N. K., and Moore, M. C. (1975). Buried pinchout of Saginaw lobe drift in northeastern Indiana. *Indiana Academy of Science Proceedings* 84, 362-72.

Bleuer, N. K., and Moore, M. C. (1975). Glacial stratigraphy of the Fort Wayne, Indiana, area and the drainage of glacial Lake Maumee. *Indiana Academy of Science Proceedings* 81, 195-209.

Bogner, J. (1974). The Lemont drift of northeastern Illinois: A regional approach. *Geological Society of America, Abstracts with Programs* 6, 493.

Borns, H. W., Jr. (1963). Preliminary report on the age and distribution of the late Pleistocene ice in north-central Maine. *American Journal of Science* 261, 738-40.

Borns, H. W., Jr. (1973). Late Wisconsin fluctuations of the Laurentide ice sheet in southern and eastern New England. *In* "The Wisconsinan Stage" (R. F. Black, R. P. Goldthwait, and H. B. Willman, eds.), pp. 37-45. Geological Society of America Memoir 136.

Borns, H. W., Jr. (1980). Separation of the late Wisconsin ice sheet over the Border and Longfellow Mountains, Maine-Quebec, and its implications. *Geological Society of America, Abstracts with Programs* 12, 25.

Borns, H. W., Jr. (1981). Mode of recession of the Late Wisconsin Laurentide ice sheet in coastal Maine. *Geological Society of America, Abstracts with Programs* 13, 123.

Borns, H. W., Jr., and Calkin, P. E. (1977). Quaternary glaciation, west-central Maine. *Geological Society of America Bulletin* 88, 1773-84.

Bothner, M. H., and Spiker, E. C. (1980). Upper Wisconsinan till recovered on the continental shelf southeast of New England. *Science* 210, 423-25.

Boulton, G. S., and Jones, A. S. (1979). Stability of temperate ice caps and ice sheets resting on beds of deformable sediment. *Journal of Glaciology* 24, 29-43.

Brace, B. R. (1968). Pleistocene stratigraphy of the Hamilton quadrangle. M.S. thesis, Miami University of Ohio, Miami.

Bretz J H. (1955). "Geology of the Chicago Region: Part 2. The Pleistocene." Illinois Geological Survey Bulletin 65.

Bryson, R. A., Wendland, M. W., Ives, J. D., and Andrews, J. T. (1969). Radiocarbon isochrones on the disintegration of the Laurentide ice sheet. *Arctic and Alpine Research* 1, 1-14.

Buddington, A. F., and Leonard, B. F. (1962). "Regional Geology of the St. Lawrence County Magnetic District, Northwest Adirondacks, New York." U.S. Geological Survey Professional Paper 376.

Burger, A. M. Forsyth, J. L., Nicoll, R. S., and Wayne, W. S. (1971). Geologic map of the 1° × 2° Muncie quadrangle, Indiana and Ohio, showing bedrock and unconsolidated deposits. Indiana Geological Survey Regional Map 5, part B.

Burger, A. M., Rexroad, C. B., Schneider, A. F., and Shaver, R. H. (1966). Excursions in Indiana geology. *Indiana Geology Survey Guidebook* 12, 1-61.

Burgis, W. A. (1977). Late Wisconsinan history of northeastern lower Michigan. Ph.D. dissertation, University of Michigan, Ann Arbor.

Burgis, W. A., and Escman, D. F. (1981). Late-Wisconsinan history of northeastern lower Michigan. *In* "Friends of the Pleistocene, Midwest Section, Guidebook." Department of Geology, University of Michigan, Ann Arbor.

Burns, G. W. (1958). Wisconsin age forests in western Ohio: II. Vegetation and burial conditions. *Ohio Journal of Science* 58, 220-30.

Cadwell, D. H. (1973a). Glacial geology of the Chanango River valley near Binghamton, New York. *In* "Glacial Geology of the Binghamton: Western Catskill Region" (D. R. Coates, ed.), pp. 31-39. State University of New York Publications in Geomorphology 3.

Cadwell, D. H. (1973b). Late Wisconsinan chronology of the Chanango River valley and vicinity, New York. Ph.D. dissertation, State University of New York at Binghamton, Binghamton.

Caldwell, D. W. (1959). "Glacial Lake and Glacial Marine Clays of the Farmington Area, Maine.' Maine Geological Survey Special Geological Study 3.

Caldwell, D. W., Thompson, W., and Hanson, L. S. (1981). Styles of deglaciation above the marine limit in central and western Maine. *Geological Society of America, Abstracts with Programs* 13, 124-25.

Calhoun, F. H. H. (1906). "The Montana Lobe of the Keewatin Ice Sheet." Geological Survey Professional Paper 50.

Calkin, P. E. (1970). Strand lines and chronology of the glacial Great Lakes in northwestern New York. *Ohio Journal of Science* 70, 78-96.

Calkin, P. E., and Miller, K. E. (1977). Late Quaternary environment and man in western New York. *Annals of the New York Academy Sciences* 288, 297-315.

Carney, F. (1910). Glacial erosion on Kelley's Island, Ohio. *Geological Society of America Bulletin* 20, 640-45.

Chamberlin, T. C. (1878). On the extent and significance of the Wisconsin kettle moraine. *Wisconsin Academy of Science, Arts and Letters* 4, 201-34.

Chapman, L. J., and D. F. Putnam, 1966. "The Physiography of Southern Ontario." 2nd ed. University of Toronto Press, Toronto, Ontario.

Christensen, C. M. (1974). "Geology and Water Resources of Bon Homme County, South Dakota." South Dakota Geological Survey Bulletin 21-I.

Christensen, C. M. (1977). "Geology and Water Resources of McPherson, Edmunds, and Faulk Counties, South Dakota." South Dakota Geological Survey Bulletin 26-I.

Christensen, C. M., and Stephens, J. C. (1967). "Geology and Hydrology of Clay County, South Dakota." South Dakota Geological Survey Bulletin 19-I.

Christiansen, E. A. (1979). The Wisconsinan deglaciation of southern Saskatchewan and adjacent areas. *Canadian Journal of Earth Sciences* 16, 913-38.

Clayton, L., and Moran, S. R. (1974). A glacial process-form model. *In* "Glacial Geomorphology" (D. R. Coates, ed.), pp. 89-119.State University of New York at Binghamton Publications in Geomorphology.

Clayton, L., and Moran, S. R. (1982). Chronology of Late Wisconsinan glaciation in middle North America. *Quaternary Science Reviews* 1, 55-82.

Clayton, L., Moran, S. R., and Bluemle, J. P. (1980). "Explanatory Text to Accompany the Geologic Map of North Dakota." North Dakota Geological Survey Report of Investigation 69.

Coates, D. R. (1963). Geomorphology of the Binghamton area. *In* "Geology of South-Central New York" (D. R. Coates, ed.), pp. 97-105. New York State Geological Association 35th Annual Meeting Guidebook. State University of New York at Binghamton, Binghamton.

Coates, D. R. (1974). Reappraisal of the glaciated Appalachian Plateau. *In* "Glacial Geomorphology" (D. R. Coates, ed.), pp. 205-43. State University of New York at Binghamton Publications in Geomorphology.

Coates, D. R. (1976). Quaternary stratigraphy of New York and Pennsylvania. *In* "Quaternary Stratigraphy of North America" (W. C. Mahaney, ed.), pp. 65-90. Dowden, Hutchinson, and Ross, Stroudsberg, Pa.

Coleman, D. D. (1973). Illinois State Geological Survey radiocarbon dates IV. *Radiocarbon* 15, 75-85.

Colton, R. B., Lemke, R. W., and Lindvall, R. M. (1961). Glacial map of Montana east of the Rocky Mountains. U.S. Geological Survey Miscellaneous Geologic Investigations Map I-327.

Conley, J. F. (1956). The glacial geology of Fairfield County, Ohio. M.S. thesis, Ohio State University, Columbus.

Connally, G. G. (1961). "The Glacial Geology of the Western Finger Lakes Region, New York; Progress Report II." New York State Museum and Science Service, Albany.

Connally, G. G. (1964). The Almond moraine of the western Finger Lakes region, New York. Ph.D. dissertation, Michigan State University, East Lansing.

Connally, G. G. (1980). Late Wisconsinan deglaciation of western Vermont: Readvances, sequences, and proglacial lakes. *Geological Society of America, Abstracts with Programs*, 12, 29.

Connally, G. G., and Sirkin, L. A. (1970). Late glacial history of the upper Wallkill Valley, New York. *Geological Society of America Bulletin* 81, 3297-3305.

Connally, G. G., and Sirkin, L. A. (1971). The Luzerne readvance near Glen Falls, New York. *Geological Society of America Bulletin* 82, 989-1008.

Connally, G. G., and Sirkin, L. A. (1973) Wisconsinan history of the Hudson-Champlain lobe. *In* "The Wisconsinan Stage" (R. F. Black, R. P. Goldthwait, and H. B. Willman, eds.), pp. 47-69. Geological Society of America Memoir 136.

Craft, J. L. (1976). Pleistocene local glaciation in the Adirondack Mountains, New York. Ph.D. dissertation, University of Western Ontario, London.

Craft, J. L. (1979). "Evidence of Local Glaciation, Adirondack Mountains, New York." Friends of the Pleistocene, Eastern Section, Guidebook.

Crowl, G. H. (1980). Woodfordian age of the Wisconsin glacial border in northeastern Pennsylvania. *Geology* 8, 51-55.

Crowl, G. H., and Sevon, W. D. (1980). "Glacial Border Deposits of Late Wisconsinan Age in Northeastern Pennsylvania." Pennsylvania Geological Survey General Geology Report 71.

Davis, P. T., and Davis, R. B. (1980). Interpretation of minimum-limiting radiocarbon dates for deglaciation of Mount Katahdin area, Maine. *Geology* 8, 396-400.

De Heinzelin, J. (1957). "Pleistocene Geology in the Middle West: A Final Report of a Study Travel." Institut Royal des Sciences Naturelles de Belgique, Brussels.

Denny, C. S. (1974). "Pleistocene Geology of the Northeast Adirondack Region, New York." U.S. Geological Survey Professional Paper 786.

Denny, C. S., and Lyford, W. H. (1963). "Surficial Geology and Soils of the Elmira-Williamsport Region, New York and Pennsylvania." U.S. Geological Survey Professional Paper 379.

Denton, G. H., and Hughes, T. J. (1980). "The Last Great Ice Sheets." John Wiley and Sons, New York.

Driemanis, A. (1969). Late-Pleistocene lakes in the Ontario and Erie basins. *In* "Proceedings of the 12th Conference on Great Lakes Research," pp. 170-80. International Association of Great Lakes Research.

Dreimanis, A. (1977a). Correlation of Wisconsin glacial events between the eastern Great Lakes and the St. Lawrence lowlands. *Geographie Physique et Quaternaire*, 31, 37-51.

Dreimanis, A. (1977b). Late Wisconsin glacial retreat in the Great Lakes region, North America. *New York Academy of Science Annals* 288, 70-89.

Dreimanis, A., and Goldthwait, R. P. (1973). Wisconsinan glaciation in the Huron, Erie, and Ontario lobes. *In* "The Wisconsinan Stage" (R. F. Black, R. P. Goldthwait, and H. B. Willman, eds.), pp. 71-106. Geological Society of America Memoir 136.

Dreimanis, A., Reaveley, G. H., Cook, R. J. B., Knox, K. S., and Moretti, F. J. (1957). Heavy mineral studies in tills of Ontario and adjacent areas. *Journal of Sedimentary Petrology* 27, 148-61.

Dreimanis, A., and Vagners, U. (1969). Lithologic relation of till to bedrock. *In* "Quaternary Geology and Climate" (H. E. Wright, Jr. ed.), pp. 93-98. National Academy of Science Publication 1701.

Droste, J., Rubin, M., and White, G. W. (1959). Age of marginal Wisconsin drift at Corry, northwestern Pennsylvania. *Science* 130, 160.

Durrell, R. H. (1961). Roadlog. *In* "Pleistocene Geology of the Cincinnati Region (Kentucky, Ohio, and Indiana)" (R. H. Durrell, J. L. Forsyth, and R. P. Goldthwait, eds.), pp. 65-98. Geological Society of America, North-Central Section, Guidebook, New York.

Ekblaw, G. E., and Athy, L. F. (1925). Glacial Kankakee torrent in northeastern Illinois. *Geological Society of America Bulletin* 36, 417-28.

Ekblaw, G. E., and Willman, H. B. (1955). Farmdale drift near Danville, Illinois. *Illinois Academy of Science Transactions* 47, 129-38.

Elson, J. A. (1967). Geology of glacial Lake Agassiz. *In* "Life, Land, and Water" (W. S. Mayer-Oakes, ed.), pp. 36-95. University of Manitoba Press, Winnipeg.

Elson, J. A. (1968). Washboard moraines and other minor moraine types. *In* "The Encyclopedia of Geomorphology" (R. W. Fairbridge, ed.), pp. 1213-18. Reinhold, New York.

Evenson, E. B. (1973). Late Pleistocene shorelines and stratigraphic relations in the Lake Michigan basin. *Geological Society of America Bulletin* 84, 2281-98.

Evenson, E. B., and Dreimanis, A. (1976). Late glacial (14,000-10,000 years B.P.) history of the Great Lakes region and possible correlations. *In* "Quaternary Glaciations in the Northern Hemisphere," Report 3, pp. 217-38. International Geological Correlation Program Project 73/1/24.

Evenson, E. B., Farrand, W. R., and Eschman, D. F. (1974). Late Pleistocene shorelines and stratigraphic relations in the Lake Michigan basin: Reply. *Geological Society of America Bulletin* 85, 661-64.

Evenson, E. B., Farrand, W. R., Eschman, D. F., Mickelson, D. M., and Maher, L. J. (1976). Greatlakean substage: A replacement for Valderan substage in the Lake Michigan basin. *Quaternary Research* 6, 411-24.

Evenson, E. B., Farrand, W. R., Eschman, D. F., Mickelson, D. M., and Maher, L. J. (1978). Reply to comments by P. F. Karrow and R. F. Black. *Quaternary Research* 9, 123-29.

Fairchild, H. L. (1907). Drumlins of central western New York. *New York State Museum Bulletin* 111, 391-443.

Fairchild, H. L. (1929). New York drumlins. *Rochester (N.Y.) Academy of Science Proceedings* 7, 1-37.

Farrand, W. R. (1969). The Quaternary history of Lake Superior. *In* "Proceedings of the 12th Conference of Great Lakes Research," pp. 181-97. International Association of Great Lakes Research.

Farrand, W. R. (1976). Was there really a Valders? *The Michigan Academician* 8, 477-86.

Farrand, W. R., and Eschman, D. F. (1974). Glaciation of the southern Peninsula of Michigan: A review. *The Michigan Academician* 7, 31-56.

Farrand, W. R., Zahner, R., and Benninghoff, W. S. (1969). Cary-Port Huron interstade: Evidence from a buried bryophyte bed, Cheboygan County, Michigan. *In* "United States Contributions to Quaternary Research" (S. A. Schumm and W. C. Bradley, eds.), pp. 249-62. Geological Society of America Special Paper 123.

Fastook, J., Hyland, M., and Hughes, T. (1980). A numerical model for reconstruction and disintegration of the late Wisconsin glaciation in the Gulf of Maine. *Geological Society of America, Abstracts with Programs* 12, 34.

Flemal, R. C. (1972). Ice injection origin of the DeKalb mounds, north-central Illinois,

U.S.A. *In* "Proceedings of the International Geological Congress, 24th Session, Montreal, Section 12, Quaternary Geology," pp. 130-35.

Flemal, R. C., Hester, J. L., and Hinckley, K. C. (1970). The De Kalb mounds: Possible remnants of pingos. *In* "Guidebook for Thirty-Fourth Annual Tri-State Field Conference" (I. E. Odum and M. Weiss, eds.), pp. 65-72. Department of Geology, Northern Illinois University, De Kalb.

Flint, R. F. (1930). "The Glacial Geology of Connecticut." Connecticut State Geological and Natural History Survey Bulletin 47.

Flint, R. F. (1953). Probable Wisconsin substages and Late Wisconsin events in northeastern United States and southeastern Canada. *Geological Society of America Bulletin* 64, 897-919.

Flint, R. F. (1961). Two tills in southern Connecticut. *Geological Society of America Bulletin* 72, 1687-91.

Flint, R. F. (1971). "Glacial and Quaternary Geology." John Wiley and Sons, New York.

Follmer, L. R., McKay, E. D., Lineback, J. A., Gross, D. L., and others (1979). "Wisconsinan, Sangamonian, and Ilinoian Stratigraphy in Central Illinois." Illinois Geological Survey Guidebook 13.

Forsyth, J. L. (1956). Glacial geology of Logan and Shelby Counties, Ohio. Ph.D. dissertation, Ohio State University, Columbus.

Forsyth, J. L. (1961). Pleistocene geology. *In* "Geology of Knox County" (S. I. Root, J. Ridriguez, and J. L. Forsyth, eds.), pp. 107-38. Ohio Geological Survey Bulletin 59.

Forsyth, J. L. (1962). Glacial geology. *In* "Geology of Fairfield County" (E. W. Wolfe, J. L. Forsyth, and G. D. Dove, eds.), pp. 116-49. Ohio Geological Survey Bulletin 60.

Forsyth, J. L. (1965a). Age of buried soil in the Sidney, Ohio, area. *American Journal of Science* 263, 571-97.

Forsyth, J. L. (1965b). Contribution of soils to the mapping and interpretation of Wisconsin tills in western Ohio. *Ohio Journal of Science* 65, 220-27.

Forsyth, J. L. (1966). "Glacial Map of Licking County, Ohio." Ohio Geological Survey Report of Investigations 59.

Forsyth, J. L. (1973). Late-glacial and postglacial history of western Lake Erie. *The Compass of Sigma Gamma Epsilon* 51, 16-26.

Forsyth, J. L. and Goldthwait, R. P. (1962). "Midwest Friends of the Pleistocene Field Guide." Friends of the Pleistocene, Midwest Section, Guidebook.

Forsyth, J. L., and Goldthwait, R. P. (1980). Rapid Late Wisconsin deglaciation of western Ohio. *In* "American Quaternary Association, Sixth Biennial Meeting, Abstracts and Program, 18-20 August 1980," pp. 80-81. Institute for Quaternary Studies, University of Maine, Orono.

Frye, J. C., and Willman, H. B. (1958). Permafrost features near the Wisconsin glacial margin in Illinois. *American Journal of Science* 256, 518-24.

Frye, J. C., and Willman, H. B. (1960). "Classification of the Wisconsinan Stage in the Lake Michigan Lobe." Illinois Geological Survey Circular 285.

Frye, J. C., and Willman, H. B. (1973). Wisconsinan climatic history interpreted from the Lake Michigan lobe deposits and soils. *In* "The Wisconsinan Stage" (R. F. Black, R. P. Goldthwait, and H. B. Willman, eds.), pp. 135-52. Geological Society of America Memoir 136.

Frye, J. C., Willman, H. B., and Black, R. P. (1965). Outline of glacial geology of Illinois and Wisconsin. *In* "The Quaternary of the United States" (H. E. Wright, Jr., and D. G. Frey, eds.), pp. 43-61. Princeton University Press, Princeton, N.J.

Fullerton, D. S. (1980). "Preliminary Correlation of Post-Erie Interstadial Events (16,000-10,000 Radiocarbon Years before Present), Central and Eastern Great Lakes Region, and Hudson, Champlain, and St. Lawrence Lowlands, United States and Canada." U.S. Geological Survey Professional Paper 1089.

Fullerton, D. S., and Groenewold, G. H. (1974). Quaternary stratigraphy of Garfield Heights (Cleveland) Ohio: Additional observations. *Geological Society of America, Abstracts with Programs* 6, 509-10.

Gamble, E. E. (1958). "Descriptions and Interpretations of Some Pleistocene Sections in Wayne County, Indiana." Earlham College Science Bulletin 3.

Genes, A. N., Newman, W. A., and Brewer, T. (1981). Late Wisconsin deglaciation of northern Maine. *Geological Society of America, Abstracts with Programs* 13, 135.

Goldsmith, R. and Schafer, J. P. (1980). Deglaciation of eastern Connecticut. *Geological Society of America, Abstracts with Programs* 12, 38.

Goldthwait, J. W. (1938). The uncovering of New Hampshire by the last ice sheet. *American Journal of Science* 236, 345-72.

Goldthwait, J. W., and Kruger, F. C. (1938). Weathered rock in and under the drift in New Hampshire. *Geological Society of America Bulletin* 49, 1183-97.

Goldthwait, R. P. (1952). "The 1952 Field Conference of Friends of the Pleistocene." Friends of the Pleistocene, Midwest Section, Guidebook.

Goldthwait, R. P. (1958). Wisconsin age forests in western Ohio: I. Age and glacial events. *Ohio Journal of Science* 58, 209-19.

Goldthwait, R. P. (1959). Scenes in Ohio during the last ice age. *Ohio Journal of Science* 59, 193-216.

Goldthwait, R. P. (1968). Two loesses in central southwest Ohio. *In* "The Quaternary of Illinois" (R. E. Bergstrom, ed.), pp. 41-47. University of Illinois College of Agriculture Special Publication 14.

Goldthwait, R. P. (1969). Boulder belt moraines. Paper presented at the 78th Annual Meeting of the Ohio Academy of Science.

Goldthwait, R. P. (1971). Introduction to till today. *In* "Till: A Symposium" (R. P. Goldthwait et al., eds), pp. 3-26. Ohio State University Press, Columbus.

Goldthwait, R. P. (1973). Till deposition versus glacial erosion. *In* "Research in Polar and Alpine Geomorphology" (B. D. Fahey and R. D. Thompson, eds.), pp. 159-66. Proceedings of the Third Guelph Symposium on Geomorphology. Geo Abstracts Limited, Norwich, England.

Goldthwait, R. P. (1976). Paleosols and Quaternary history of western Ohio (abstract). *In* "Quaternary Stratigraphy of North America" (W. C. Mahaney, ed.), pp. 159-60. Dowden, Hutchinson, and Ross, Stroudsburg, Pa.

Goldthwait, R. P., Dreimanis, A., Forsyth, J. L., Karrow, P. F., and White G. W. (1965). Pleistocene deposits of the Erie lobe. *In* "The Quaternary of the United States" (H. E. Wright, Jr., and D. Frey, eds.), pp. 85-97. Princeton University Press, Princeton, N.J.

Goldthwait, R. P., and Forsyth, J. L. (1955). Pleistocene chronology of southwestern Ohio. *In* Indiana and Ohio Geological Surveys Fifth Biennial Field Conference, Guidebook, pp. 35-72.

Goldthwait, R. P., and Rosengreen, T. (1969). Till stratigraphy from Columbus southwest to Highland County, Ohio. *In* Geological Society of America, North-Central Section, Third Annual Meeting, Field Trip Guidebook," Trip 2, pp. 2-1 to 2-17.

Goldthwait, R. P., White, G. W., and Forsyth, J. L. (1961, revised 1967). Glacial map of Ohio. U.S. Geological Survey Miscellaneous Investigations Map I-316.

Gooding, A. M. (1957). "Pleistocene Terraces in the Upper Whitewater Drainage Basin, Southeastern Indiana." Earlham College Science Bulletin 2.

Gooding, A. M. (1961). Illinoian and Wisconsin history in southeastern Indiana. *In* "Pleistocene Geology of the Cincinnati Region (Kentucky, Ohio, and Indiana)," pp.99-128. Geological Society of America, North-Central Section, Guidebook, New York.

Gooding, A. M. (1963). Illinoian and Wisconsin glaciations in the Whitewater basin, southeastern Indiana, and adjacent areas. *Journal of Geology* 71, 665-82.

Gooding, A. M. (1965). Southeastern Indiana. *In* "Guidebook for Field Conference G, Great Lakes-Ohio River Valley," pp. 43-53. Seventh Congress of the International Association for Quaternary Research, Boulder, Colo.

Gooding, A. M. (1973). "Characteristics of Late Wisconsinan Tills in Eastern Indiana." Indiana Geological Survey Bulletin 49.

Gooding, A. M. (1975). The Sidney interstadial and Late Wisconsin history in Indiana and Ohio. *American Journal of Science* 275, 993-1011.

Grant, D. R. (1977). Glacial style and ice limits, the Quaternary stratigraphic record, and changes of land and ocean level in the Atlantic Provinces, Canada. *Geographie Physique et Quaternaire* 31, 247-60.

Gray, H. H., Forsyth, J. L., Schneider, A. F., and Gooding, A. M. (1972). Geologic map of the 1° × 2° Cincinnati quadrangle, Indiana and Ohio, showing bedrock and unconsolidated deposits. Indiana Geological Survey Regional Geologic Map 7, part B.

Gross, D. L., and Moran, S. R. (1971). Grain-size and mineralogical gradations within tills of the Allegheny Plateau. *In* "Till: A Symposium" (R. P. Goldthwait et al., eds.), pp 251-74. Ohio State University Press, Columbus.

Grüger, E. (1972a). Late Quaternary vegetational development in south-central Illinois. *Quaternary Research* 2, 217-31.

Grüger, E. (1972b). Pollen and seed studies of Wisconsinan vegetation in Illinois, U.S.A. *Geological Society of America Bulletin* 83, 2715-34.

Gustavson, T. C. (1976). Paleotemperature analysis of the marine Pleistocene of Long Island, New York, and Nantucket Island, Massachusetts. *Geological Society of America Bulletin* 87, 1-8.

Halstead, E. C. (1959). "Ground-Water Resources of the Brandon Map-Area, Manitoba." Geological Survey of Canada Memoir 300.

Harrison, W. (1959). "Petrographic Similarity of Wisconsin Tills in Marion County, Indiana." Indiana Geological Survey Report of Progress 15.

Harrison, W. (1963). "Geology of Marion County, Indiana." Indiana Geological Survey Bulletin 28.

Hartke, E. J., Hill, J. R., and Reshkin, M. (1975). "Environmental Geology of Lake and Porter Counties, Indiana: An Aid to Planning." Indiana Geological Survey Special Report 11.

Hedges, L. S. (1968). "Geology and Water Resources of Beadle County, South Dakota." South Dakota Geological Survey Bulletin 18-I.

Hedges, L. S. (1972). Geology and Water Resources of Campbell County, South Dakota." South Dakota Geological Survey Bulletin 20-I.

Hedges, L. S. (1975). "Geology and Water Resources of Charles Mix and Douglas Counties, South Dakota." South Dakota Geological Survey Bulletin 22-I.

Hooten, J. E. (1973). Glacial geology of eastern Champaign County, Illinois. M.S. thesis, University of Illinois, Urbana.

Horberg, L. (1949). A possible fossil ice wedge in Bureau County, Illinois. *Journal of Geology* 57, 132-36.

Horberg, L. (1950). "Bedrock Topography of Illinois." Illinois Geologic Survey Bulletin 73.

Horberg, L. (1953). "Pleistocene Deposits below the Wisconsin Drift in Northeastern Illinois." Illinois Geological Survey Report of Investigations 165.

Horberg, L., and Anderson, R. C. (1956). Bedrock topography and Pleistocene glacial lobes in central United States. *Journal of Geology* 64, 101-16.

Horberg, L., and Potter, P. E., (1955). "Stratigraphic and Sedimentologic Aspects of the Lemont Drift of Northeastern Illinois." Illinois Geological Survey Report of Investigations 185.

Hughes, J. D. (1978). A post-Two Creeks buried forest in Michigan's northern peninsula. *In* "Proceedings of the 24th Meeting of the Institute of Lake Superior Geology," p. 16.

Hughes, T. J. (1980). Numerical reconstruction of paleo-ice sheets. *In* "The Last Great Ice Sheets" (G.H. Denton and T.J. Hughes, eds.), pp. 222-61. John Wiley and Sons, New York.

Hughes, T. J. (1981). Models of glacial reconstruction and deglaciation applied to maritime Canada and New England. *Geological Society of America, Abstracts with Programs* 13, 138.

Indiana Geological Survey (1979). Map of Indiana showing unconsolidated deposits. Miscellaneous Map 26.

Ives, J. D., Andrews, J. T. and Barry, R. G. (1975). Growth and decay of the Laurentide ice sheet and comparisons with Fenno-Scandinavia. *Die Naturwissenschaften* 62, 118-25.

Ives, P. C., Levin, B., Oman, C. L., and Rubin, M. (1967). U.S. Geological Survey radiocarbon dates IX. *Radiocarbon* 9, 505-29.

Johnson, G. H., and Keller, S. J. (1972). Geologic map of the 1° × 2° Fort Wayne quadrangle, Indiana, Michigan, and Ohio, showing bedrock and unconsolidated deposits. Indiana Geologic Survey Regional Geologic Map 8, part B.

Johnson, G.H., Schneider, A.F., and Ulrich, H.P. (1965). "Glacial Geology and Soils of the Area around Lake Maxinkuckee." Indiana Academy of Science, Field Trip Guidebook, Bloomington.

Johnson, M. D. (1980). Origin of the Lake Superior red clay and glacial history of Wisconsin's Lake Superior shoreline west of the Bayfield peninsula. M.S. thesis, University of Wisconsin, Madison.

Johnson, W.H. (1976). Quaternary stratigraphy in Illinois: Status and problems. *In* "Quaternary Stratigraphy of North America" (W.C. Mahaney, ed.), pp. 161-96. Dowden, Hutchinson, and Ross, Stroudsburg, Pa.

Johnson, W.H. (1978). Patterned ground, wedge-shaped bodies, and circular landforms in central and eastern Illinois. *Geological Society of America, Abstracts with Programs* 10, 257.

Johnson, W.H., Follmer, L.R., Gross, D.L., and Jacobs, A.M. (1972). "Pleistocene Stratigraphy of East-Central Illinois." Illinois Geological Survey Guidebook Series 9.

Johnson, W.H., Glass, H.D., Gross, D.L., and Moran, S.R. (1971). "Glacial Drift of the Shelbyville Moraine at Shelbyville, Illinois." Illinois Geological Survey Circular 459.

Johnson, W. H., Gross, D. L., and Moran, S. R. (1971). Till stratigraphy of the Danville region, east-central Illinois. In "Till: A Symposium" (R.P. Goldthwait et al., eds.), pp. 184-210. Ohio State University Press, Columbus.

Kapp, R. O., and Gooding, A. M. (1964). Pleistocene vegetational studies in the Whitewater basin, southeastern Indiana. Journal of Geology 72, 307-26.

Karrow, P. F. (1963). "Pleistocene Geology of the Hamilton-Galt Area." Ontario Department of Mines Geological Report 16.

Kaye, C. A. (1961). "Pleistocene Stratigraphy of Boston, Massachusetts." U.S. Geological Survey Professional Paper 424-B, pp. B73-B76.

Kaye, C. A. (1964). "Outline of Pleistocene Geology of Martha's Vineyard, Massachusetts." U.S. Geological Survey Professional Paper 501-C, pp. C134-139.

Kaye, C. A. (1980). Late Wisconsinan glaciation of Martha's Vineyard and a peripheral ice cap for coastal New England. Geological Society of America, Abstracts with Programs 12, 44.

Kemmis, T. J., Hallberg, G. R., and Lutenegger, A. S. (1981). "Depositional Environments of Glacial Sediments and Landforms on the Des Moines Lobe, Iowa." Iowa Geological Survey Guidebook 6, Iowa City.

Kempton, J. P. (1963). "Subsurface Stratigraphy of the Pleistocene Deposits of Central Northern Illinois." Illinois Geological Survey Circular 356.

Kempton, J. P., DuMontelle, P. B., and Glass, H. D. (1971). Subsurface stratigraphy of the Woodfordian tills in the McLean County region, Illinois. In "Till: A Symposium" (R. P. Goldthwait et al., eds.), pp. 217-33. Ohio State University Press, Columbus.

Kempton, J. P., and Hackett, J. E. (1968). Stratigraphy of the Woodfordian and Altonian drifts of central northern Illinois. In "The Quaternary of Illinois" (R. E. Bergstrom, ed.), pp. 27-34. University of Illinois College of Agriculture Special Publication 14.

Killey, M. M. (1980). Physical and mineralogical variations in the Yorkville Till Member, Grundy and adjacent counties, Illinois. Geological Society of America, Abstracts with Programs 12, 231.

King, C. A. M. (1974). Morphometry in glacial geomorphology. In "Glacial Geomorphology" (D. R. Coates, ed.), pp. 147-62. State University of New York at Binghamton Publications in Geomorphology.

Kirkland, J. T. (1973a). Glacial geology of the western Catskills. Ph.D. dissertation, State University of New York at Binghamton, Binghamton.

Kirkland, J. T. (1973b). Glaciation of the western Catskill Mountains. In "Glacial Geology of the Binghamton-Western Catskill Region" (D. R. Coates, ed.), pp. 57-68. State University of New York at Binghamton Publications in Geomorphology.

Kite, J. S., and Stuckenrath, R. (1980). Late-glacial and Holocene geologic history of the middle St. John River valley: Northern Maine and northwestern New Brunswick. Geological Society of America, Abstracts with Programs 12, 46.

Klassen, R. W. (1972). Wisconsin events and the Assiniboine and Qu'Appelle Valleys of Manitoba and Saskatchewan. Canadian Journal of Earth Sciences 9, 544-60.

Klassen, R. W. (1975). "Quaternary Geology and Geomorphology of Assiniboine and Qu'Appelle Valleys of Manitoba and Saskatchewan." Geological Survey of Canada Bulletin 228.

Koch, N. C. (1975). "Geology and Water Resources of Marshall County, South Dakota." South Dakota Geological Survey Bulletin 23-I.

Koteff, C. (1980). Patterns of Late Wisconsinan deglaciation in New England. Geological Society of America, Abstracts with Programs 12, 67.

Larsen, F. D. (1981). Deglaciation of central Vermont. Geological Society of America, Abstracts with Programs 13, 142.

Larsen, F. D., and Harshorn, J. H. (1980). Deglaciation of the Connecticut Valley of Massachusetts by an active ice lobe. Geological Society of America, Abstracts with Programs 12, 68.

LaSalle, P., Martineau, G., and Chauvin, L. (1977). "Morphology, Stratigraphy and Deglaciation, Beauce-Notre-Dame-Mountains-Laurentide Park Area." Ministry of Natural Resources, Quebec, report DPV-516, pp. 1-74.

Leavitt, H. W., and Perkins, E. H. (1935). "Glacial Geology of Maine." Maine Technology Experimental Station Bulletin 30 (2).

Leighton, M. M., Ekblaw, G. E., and Horberg, L. (1948). Physiographic divisions of Illinois. Journal of Geology 56, 16-33.

Lemke, R. W., Laird, W. M., Tipton M. J., and Lindvall, R. M. (1965). Quaternary geology of the northern Great Plains. In "The Quaternary of the United States" (H. E. Wright, Jr., and D. G. Frey, eds.), pp. 15-27. Princeton University Press, Princeton, N.J.

Leverett, F. (1896). The water resources of Illinois. In "United States Geological Survey, 17th Annual Report, 1895-1896," part II, pp. 695-849.

Leverett, F. (1897). The water resources of Indiana and Ohio. In "United States Geological Survey, 18th Annual Report," part IV, pp. 419-560.

Leverett, F. (1899). "The Illinois Glacial Lobe." U.S. Geological Survey Monograph 38.

Leverett, F. (1902). "Glacial Formations and Drainage Features of the Erie and Ohio Basins." U.S. Geological Survey Monograph 41.

Levin, B., Ives, P. C., Oman, C. L., and Rubin, M. (1965). U.S. Geological Survey radiocarbon measurements VIII. Radiocarbon 7, 372-98.

Lineback, J. A. (1975). Glacial landforms on Wisconsinan and Illinoian drift in east-central Ilinois mapped from Skylab photographs. Geological Society of America, Abstracts with Programs 7, 809.

Lineback, J. A. (1979). Quaternary deposits of Illinois. Illinois Geological Survey Map.

Lineback, J. A., Gross, D. L., and Meyer, R. P. (1974). "Glacial Tills under Lake Michigan." Illinois State Geological Survey Environmental Geology Notes 69.

Lowdon, J. A., Fyles, J. E., and Blake, W., Jr. (1967). Geological Survey of Canada radiocarbon dates VI. Radiocarbon 9, 156-97.

Lowell, T. V. (1980). Late Wisconsin ice extent in Maine: Evidence from Mt. Desert Island and the St. John River area. M.S. thesis, University of Maine, Orono.

Lowell, T. V. (1981). Late Wisconsin ice-flow reversal and deglaciation, northwestern Maine. Geological Society of America, Abstracts with Programs 13, 143.

McCartney, M. C., and Mickelson, D. M. (1982). Late Woodfordian and Greatlakean history of the Green Bay lobe, Wisconsin. Geological Society of America Bulletin 93, 297-302.

McComas, M. R., Kempton, J. P., and Hinkley, K. C. (1972). "Geology, Soils, and Hydrogeology of Volo Bog and Vicinity, Lake County, Illinois." Illinois Geological Survey Environmental Geology Notes 57.

Madole, R. F. (1980). "Glacial Lake Devlin and the Chronology of Pinedale Glaciation on the East Slope of the Front Range, Colorado." U.S. Geological Survey Open-File Report 80-725.

Mahaney, W. C. (Ed.) (1976). "Quaternary Stratigraphy of North America." Dowden, Hutchinson, and Ross, Stroudsburg, Pa.

Maher, L. J., Jr. (1981). The Green Bay sublobe began to retreat 12,500 B.P.: Total pollen influx during the early Greatlakean substage (11,900 to 10,900 B.P.) was but half the influx during the Twocreekan. Geological Society of America, Abstracts with Programs 13, 288.

Mathews, W. H. (1974). Surface profiles of the Laurentide ice sheet in its marginal areas. Journal of Glaciology 13, 37-43.

Matsch, C. L. (1972). Quaternary geology of southwestern Minnesota. In "Geology of Minnesota: A Centennial Volume" (P.K. Sims and G.B. Morey, eds.), pp. 548-60. Minnesota Geological Survey, St. Paul.

Matsch, C. L., Rutford, R. H., and Tipton, M. J. (1972). "Quaternary Geology of Northeastern South Dakota and Southwestern Minnesota." Minnesota Geological Survey Guidebook Series 7.

Mayewski, P., Denton, G. H., and Hughes, T. J. (1980). Late Wisconsin ice sheets in North America. In "The Last Great Ice Sheets" (G.H. Denton and T.J. Hughes, eds.), pp. 67-178. John Wiley and Sons, New York.

Mickelson, D. M., Acomb, L. J., and Edil, T. B. (1979). The origin of preconsolidated and normally consolidated tills in eastern Wisconsin, USA. In "Moraines and Varves" (C. Schlüchter, ed.), pp. 179-88. Balkema, Rotterdam.

Mickelson, D. M., Nelson, A. R., and Stewart, M. (1974). Glacial events in north-central Wisconsin. In "Late Quaternary Environments of Wisconsin" (J. C. Knox and D. M. Mickelson, eds.), pp. 163-81. American Quaternary Association Third Biennial Meeting, Madison.

Miller, J. W., Jr. (1972). Variations in New York drumlins. Annals of the Association of American Geographers 62, 418-23.

Mills, H. C., and Wells, P. D. (1974). Ice-shove deformation and glacial stratigraphy of Port Washington, Long Island, New York. *Geological Society of America Bulletin* 85, 357-64.

Moore, D. W., and Johnson, W. H. (1981). Stratigraphy of Woodfordian till and lake beds in Iroquois and neighboring counties, Illinois. *Geological Society of America, Abstracts with Programs* 13, 310.

Moore, M. C. (1978). Lake Michigan lobe origin of Woodfordian-age till in north-central Indiana. *Geological Society of America, Abstracts with Programs* 10, 279.

Moos, M. H. (1970). The age and significance of a paleosol in Fayette County, Ohio. M.S. thesis, Ohio State University, Columbus.

Moran, S. R., and Clayton, L. (1972). Lake Agassiz and the history of the Des Moines lobe. *Geological Society of America, Abstracts with Programs* 5, 602-3.

Moran, S. R., Clayton, L., Scott, M. W., and Brophy, J. A. (1973). "Catalog of North Dakota Radiocarbon Dates." North Dakota Geological Survey Miscellaneous Series 53.

Moran, S. R., Arndt, M., Bluemle, J. P., Camara, M., Clayton, L., Fenton, M. M., Harris, K. L., Hobbs, H. C., Keatinge, R., Sackreiter, D. K., Salomon, N. L., and Teller, J. (1976). Quaternary stratigraphy and history of North Dakota, southern Manitoba, and northwestern Minnesota. *In* "Quaternary Stratigraphy of North America" (W. C. Mahaney, ed.), pp. 133-58. Dowden, Hutchinson, and Ross, Stroudsburg, Pa.

Morgan, A. V. (1972). Late Wisconsin ice-wedge polygons near Kitchner, Ontario, Canada. *Canadian Journal of Earth Sciences* 9, 607-17.

Mörner, N.-A., and Dreimanis, A. (1973). The Erie interstate. *In* "The Wisconsinan Stage" (R. F. Black, R. P. Goldthwait, and H. B. Willman, eds.), pp. 107-34. Geological Society of America Memoir 136.

Moss, J. H., and Ritter, D. F. (1962). New evidence regarding the Binghamton substage in the region between the Finger Lakes and Catskills, New York. *American Journal of Science* 260, 81-106.

Muller, E. H. (1956). Texture as a basis for correlation of till sheets in Chautauqua County, western New York (abstract). *Geological Society of America Bulletin* 67, 1819.

Muller, E. H. (1960). "Glacial geology of Cattaraugus County, New York." Friends of the Pleistocene, Eastern Section, 23rd Reunion, Guidebook.

Muller, E. H. (1963). "Geology of Chautauqua County, New York: Part II. Pleistocene Geology." New York State Museum of Science Survey Bulletin, 392.

Muller, E. H. (1965a). Genesee Valley and the Binghamton problem. *In* "Guidebook for Field Conference A, New England-New York State," 71-79. Seventh Congress of the International Association for Quaternary Research, Boulder, Colo.

Muller, E. H. (1965b). Olean to Buffalo. *In* "Guidebook for Field Conference A, New England-New York State," 79-84. Seventh Congress of the International Association for Quaternary Research, Boulder, Colo.

Muller, E. H. (1965c). The Quaternary geology of New York. *In* "The Quaternary of the United States" (H. E. Wright Jr., and D. G. Frey, eds.), pp. 99-112. Princeton University Press, Princeton, N.J.

Muller, E. H. (1974). Origins of drumlins. *In* "Glacial Geomorphology" (D. R. Coates, ed.), pp. 187-204. State University of New York at Binghamton Publications in Geomorphology.

Muller, E. H. (1977a). Late glacial and early post-glacial environments in western New York. *New York Academy of Sciences Annals* 288, 223-33.

Muller, E. H. (1977b). Quaternary geology of New York, Niagara sheet. New York State Museum and Science Service, Map and Chart Series 28.

Nambudiri, E. M. V., Teller, J. T., and Last, W. M. (1980). Pre-Quaternary microfossils: A guide to errors in radiocarbon dating. *Geology* 8, 123-26.

Need, E. A. (1980). Till stratigraphy and glacial history of Wisconsin's Lake Superior shoreline: Wisconsin Point to Bark River. M.S. thesis, University of Wisconsin, Madison.

Ogden, J. G., III, and Hart, W. C. (1976). Dalhousie University natural radiocarbon measurements I. *Radiocarbon* 18, 43-49.

Ogden, J. G., III, and Hay, R. J. (1965). Ohio Wesleyan University natural radiocarbon measurements II. *Radiocarbon* 7, 166-73.

Ogden, J. G., III, and Hay, R. J. (1973). Ohio Wesleyan University natural radiocarbon measurements V. *Radiocarbon* 15, 350-66.

Oldale, R. N. (1980). The Sankaty Head section of Nantucket Island, Massachusetts,

and its correlation with the drift of coastal southern New England and Long Island, New York. *In* "American Quaternary Association, Sixth Biennial Meeting, Abstracts and Program, 18-20 August 1980," pp. 151-2, Institute for Quaternary Studies, University of Maine, Orono.

Pessl, F., Jr. (1966). "A Two-Till Locality in Northeastern Connecticut." U.S. Geological Survey Professional Paper 550-D, pp. D89-D93.

Pessl, F., Jr., and Koteff, C. (1970). Glacial and postglacial stratigraphy along Nash Stream, northern New Hampshire. *In* "Guidebook for Field Trips in the Rangeley Lakes-Dead River Basin Region, Western Maine" (G. M. Boone, ed.), pp. G1-G15. Proceedings of the New England Intercollegiate Geological Conference, 62nd Annual Meeting.

Péwé, T. L. (1983). The periglacial environment in North America in Wisconsin time. *In* "Late-Quaternary Environments of the United States," Vol. 1, "The Late Pleistocene" (S. C. Porter, ed.), pp. 157-89. University of Minnesota Press, Minneapolis.

Piskin, K., and Bergstrom, R. E. (1967). "Glacial Drift in Illinois: Thickness and Character." Illinois Geological Survey Circular 416.

Powers, W. E., and Ekblaw, G. E. (1940). Glaciation of the Grays Lake, Illinois, quadrangle. *Geologic Society of America Bulletin* 51, 1329-35.

Prest, V. K. (1969). Retreat of Wisconsin and recent ice in North America. Geological Survey of Canada Map 1257A.

Quinn, M. J., and Goldthwait, R. P. (1979). "Glacial Geology of Champaign County, Ohio." Ohio Geological Survey Report of Investigations 3.

Reeves, B. O. K. (1973). The nature and age of the contact between the Laurentide and Cordilleran ice sheets in the western interior of North America. *Arctic and Alpine Research* 5, 1-16.

Richard, S. H. (1978). "Age of the Champlain Sea and 'Lampsilis Lake' Episode in the Ottawa-St. Lawrence Lowlands." Geological Survey of Canada Paper 78-1C, pp. 23-28.

Richmond, G. M. (1960). "Correlation of Alpine and Continental Glacial Deposits of Glacier National Park and Adjacent High Plains, Montana." U.S. Geological Survey Professional Paper 400-B, pp. B223-B224.

Richmond, G. M. (1965). Relations of alpine and continental deposits in the St. Mary Valley and adjacent High Plains. *In* "Guidebook for Field Conference E, Northern and Middle Rocky Mountains," pp. 58-62. Seventh Congress of the International Association for Quaternary Research, Boulder, Colo.

Richmond, G. M., and Lemke, R. W. (1965). Glacial Lake Cut Bank and the margin of the continental drift. *In* "Guidebook for Field Conference E, Northern and Middle Rocky Mountains," pp. 56-57. Seventh Congress of the International Association for Quaternary Research, Boulder, Colo.

Rubin, M., and Suess, H. E. (1955). U.S. Geological Survey radiocarbon dates II. *Science* 121, 481-88.

Ruhe, R.V. (1969). "Quaternary Landscapes in Iowa." Iowa State University Press, Ames.

Saarnisto, M. (1974). The deglaciation history of the Lake Superior region and its climatic implications. *Quaternary Research* 4, 316-39.

Schafer, J. P. (1961). "Correlation of the End Moraines in Southern Rhode Island." U.S. Geological Survey Professional Paper 424-D, pp. D68-D70.

Schafer, J. P. (1980). The last ice sheet in Rhode Island. *Geological Society of America, Abstracts with Programs* 12, 80.

Schafer, J. P., and Hartshorn, J. H. (1965). The Quaternary of New England. *In* "The Quaternary of the United States" (H. E. Wright, Jr., and D. G. Frey, eds.), pp. 113-28. Princeton University Press, Princeton, N.J.

Schlee, J., and Pratt, R. M. (1970). "Atlantic Continental Shelf and Slope of the United States." U.S. Geological Survey Professional Paper 529-4.

Schneider, A. F. (1968). The Tinley moraine in Indiana. *Indiana Academy of Science Proceedings* 77, 271-78.

Schneider, A. F., and Johnson, G. H. (1967). Late Wisconsin glacial history of the area around Lake Maxinkuckee. *Indiana Academy of Science Proceedings* 76, 328-34.

Schneider, A. F., Johnson, G. H., and Wayne, W. J., (1963). Some linear features in west-central Indiana. *Indiana Academy of Science Proceedings* 72, 172-73.

Schneider, A. F., and Keller, S. J. (1970). Geologic map of the 1° × 2° Chicago quadrangle, Indiana, Illinois, and Michigan, showing bedrock and unconsolidated deposits. Indiana Geological Survey Regional Geologic Map 4, part B.

Schneider, A. F., and Wayne, W. J. (1967). Pleistocene stratigraphy of west-central

Indiana. *In* "Geologic Society of America, North-Central Section, First Annual Meeting, Guidebook," Trip 3, pp. 75-103.

Sevon, W. D. (1975a). "Geology and Mineral Resources of the Christmas and Pohopoco Mountain Quadrangles, Carbon and Monroe Counties, Pennsylvania." Pennsylvania Geological Survey, 4th Series, Atlas 195, a and b.

Sevon, W. D. (1975b). "Geology and Mineral Resources of the Hickory Run and Blakeslee Quadrangles, Carbon and Monroe Counties, Pennsylvania." Pennsylvania Geological Survey, 4th Series, Atlas 194, c and d.

Sevon, W. D. (1975c). "Geology and Mineral Resources of the Tobyhanna and Buck Hill Falls Quadrangles, Monroe County, Pennsylvania." Pennsylvania Geological Survey, 4th Series, Atlas 204, a and b.

Sevon, W. D., and Berg, T. M. (1978). "Geology and Mineral Resources of the Skytop Quadrangle, Monroe and Pike Counties, Pennsylvania." Pennsylvania Geological Survey, 4th Series, Atlas 214, a.

Sharp, R. P. (1942). Periglacial involutions in northeastern Illinois. *Journal of Geology* 50, 113-33.

Shepps, V. C. (1959a). Glacial geology of northwestern Pennsylvania. *In* "Geological Society of America, North-Central Section, Guidebook for Field Trips," pp. 167-88.

Shepps, V. C. (1959b). "Introduction to the Glacial Geology of Crawford and Erie Counties, Pennsylvania." Field Conference of Pennsylvania Geologists, 24th Annual Meeting, Guidebook.

Shepps, V. C., White, G.W., Droste, J. B., and Sitler, R. F. (1959). "Glacial Geology of Northwestern Pennsylvania." Pennsylvania Geological Survey Bulletin G32.

Shilts, W. W. (1976). Glacial events in southern Quebec-northern New England: A reappraisal. *Geological Society of America, Abstracts with Programs* 8, 267.

Sirkin, L. (1980). Late Wisconsinan glaciation of Long Island, New York, to Block Island, Rhode Island. *Geological Society of America, Abstracts with Programs* 12, 83.

Sirkin, L. A., and Mills, H. (1975). Wisconsinan glacial stratigraphy and structure of northwestern Long Island. *In* "Guidebook to Field Excursions" (M.P. Wolff, ed.), pp. 299-324. New York State Geological Association, 47th Annual Meeting, Hampstead, N.Y.

Sirkin, L. A., and Stuckenrath, R. (1980). The Port Washingtonian warm interval in the northern Atlantic Coastal Plain. *Geological Society of America Bulletin* 91, (part 1), 332-36.

Smith, G. W. (1981). Chronology of Late Wisconsin deglaciation of coastal Maine. *Geological Society of America, Abstracts with Programs* 13, 178.

South Dakota Geological Survey (1971). Generalized glacial map of South Dakota. South Dakota Geological Survey Educational Series Map 2.

Stalker, A. MacS. (1977). The probable extent of classical Wisconsin ice in southern and central Alberta. *Canadian Journal of Earth Sciences* 14, 2614-19.

Stalker, A. MacS., and Harrison, J. E. (1977). Quaternary glaciation of the Waterton-Castle River region of Alberta. *Bulletin of Canadian Petroleum Geology* 25, 882-906.

Steece, F. V., and Howells, L. W. (1965). "Geology and Ground Water Supplies in Sanborn County, South Dakota." South Dakota Geological Survey Bulletin 17-I.

Steiger, J. R., and Holowaychuk, N. (1971). Particle-size and carbonate analysis of glacial till and lacustrine deposits in western Ohio. *In* "Till: A Symposium" (R. P. Goldthwait et al., eds.), pp. 275-89. Ohio State University Press, Columbus.

Stewart, D. P., and MacClintock, P. (1964). The Wisconsin stratigraphy of northern Vermont. *American Journal of Science* 262, 1089-97.

Stone, B. D., and Peper, J. D. (1980). Topographic control of the deglaciation of eastern Massachusetts: Ice lobation and marine incursion. *Geological Society of America, Abstracts with Programs* 12, 85.

Stuiver, M. (1969). Yale natural radiocarbon measurements IX. *Radiocarbon* 11, 545-658.

Stuiver, M., and Borns, H. W., Jr. (1975). Late Quaternary marine invasion in Maine: Its chronology and associated crustal movement. *Geological Society of America Bulletin* 86, 99-104.

Taylor, L. D. (1979). Preliminary analysis of glacial stratigraphy in the vicinity of Port Huron and Two Rivers till sheets, Manistee, Michigan. *Geological Society of America, Abstracts with Programs* 9, 658.

Teller, J. T. (1967). The glacial geology of Clinton County, Ohio. Ohio Geological Survey Report of Investigations 67 (map).

Teller, J. T. (1976). Lake Agassiz deposits in the main offshore basin of southern Manitoba. *Canadian Journal of Earth Sciences* 13, 27-43.

Teller, J. T., and Fenton, M. M. (1980). Late Wisconsinan glacial stratigraphy and history of southeastern Manitoba. *Canadian Journal of Earth Sciences* 17, 19-35.

Teller, J. T., Moran, S. R., and Clayton, L. (1980). The Wisconsinan deglaciation of southern Saskatchewan and adjacent areas: Discussion. *Canadian Journal of Earth Sciences* 17, 539-41.

Thomas, R. H. (1977). Calving bay dynamics and ice sheet retreat up the St. Lawrence Valley system. *Geographie Physique et Quaternaire* 31, 347-56.

Thornbury, W. D. (1937). "Glacial Geology of Southern and South-Central Indiana." Indiana Geologic Survey publication.

Thornbury, W. D. (1958). The geomorphic history of the upper Wabash Valley. *American Journal of Science* 256, 449-69.

Thornbury, W. D., and Deane, H. L. (1955). "The Geology of Miami County, Indiana." Indiana Geologic Survey Bulletin 8.

Thornbury, W. D., and Wayne, W. J. (1953). Wisconsin stratigraphy of the Wabash Valley and west-central Indiana. *In* Illinois and Indiana Geological surveys, Fourth Biennial Pleistocene Field Conference, Itinerary," pp. 74-98.

Thornbury, W. D., and Wayne, W. J. (1957). "Field Guide and Road Log for Study of Kansan, Illinoian, and Early Tazewell Tills, Loesses, and Associated Formations in South-Central Indiana." Friends of the Pleistocene, Midwest Section, Eighth Annual Field Conference, Guidebook.

Totten, S. N. (1969). Overridden recessional moraines of north-central Ohio. *Geological Society of America Bulletin* 80, 1931-45.

Totten, S. N. (1973). "Glacial Geology of Richland County, Ohio." Ohio Geological Survey Report of Investigations 88.

Totten, S. N. (1976). The "up-in-the-air" late Pleistocene beaver pond near Lodi, Medina County, northern Ohio. *Geological Society of America, Abstracts with Programs* 8, 514.

ver Steeg, K. (1933). The thickness of the glacial deposits in Ohio. *Science* (new series) 78, 459.

ver Steeg, K. (1938). Thickness of the glacial drift in western Ohio. *Journal of Geology* 46, 654-59.

Wayne, W. J. (1956a). Pleistocene periglacial environment in Indiana (abstract). *Indiana Academy of Science Proceedings* 65, 164.

Wayne, W. J. (1956b). "Thickness of drift and Bedrock Physiography of Indiana North of the Wisconsin Glacial Boundary." Indiana Geological Survey Report of Progress 7.

Wayne, W. J. (1958). Glacial geology of Indiana. Indiana Geological Survey, Atlas of Mineral Resources, Map 10.

Wayne, W. J. (1963). "Pleistocene Formations of Indiana." Indiana Geologic Survey Bulletin 25.

Wayne, W. J. (1964). "Pleistocene Patterned Ground and Periglacial Temperatures in Indiana" (abstract). Geological Society of America Special Paper 76, pp. 176-77.

Wayne, W. J. (1965a). "The Crawfordsville and Knightstown Moraines in Indiana." Indiana Geologic Survey Report of Progress 28.

Wayne, W. J. (1965b). Western and central Indiana. *In* "Guidebook for Field Conference G, Great Lakes-Ohio River Valley," pp. 27-42. Seventh Congress of the International Association for Quaternary Research, Boulder, Colo.

Wayne, W. J. (1966). Ice and land—A review of the Tertiary and Pleistocene history of Indiana. *In* "Natural Features of Indiana" (A. A. Lindsey, Ed.), pp. 21-39. Indiana Academy of Sciences Sesquicentennial volume, Bloomington.

Wayne, W. J. (1967). Periglacial features and climatic gradient in Illinois, Indiana, and western Ohio, east-central United States. *In* "Quaternary Paleoecology" (E. J. Cushing and H. E. Wright, Jr., eds.), pp. 393-414. Yale University Press, New Haven, Conn.

Wayne, W. J. (1968). The Erie lobe margin in east-central Indiana during the Wisconsin glaciation. *Indiana Academy of Science Proceedings* 77, 279-91.

Wayne, W. J. (1973). Multiple tills at Wabash, Indiana (abstract). *Indiana Academy of Science Proceedings* 83, 242-43.

Wayne, W. J., Johnson, G. H., and Keller, S. J. (1966). Geologic map of the 1° × 2° Danville quadrangle, Indiana and Illinois, showing bedrock and unconsolidated deposits. Indiana Geological Survey Map 2, part B.

Wayne, W. J., and Thornbury, W. D. (1951). "Glacial Geology of Wabash County, Indiana." Indiana Geological Survey Bulletin 5.

Wayne, W. J., and Thornbury, W. D. (1955). Wisconsin stratigraphy of northern and eastern Indiana. *In* "Indiana and Ohio Geologic Surveys, Fifth Biennial Pleistocene Field Conference," pp. 1-34.

Wayne, W. J., and Zumberge, J. H. (1965). Pleistocene geology of Indiana and Michigan. *In* "The Quaternary of the United States" (H. E. Wright, Jr., and D. G. Frey, eds.), pp. 63-84. Princeton University Press, Princeton, N.J.

Westgate, J. A. (1965). The Pleistocene stratigraphy of the Foremost-Cypress Hills area, Alberta. *In* "Alberta Society of Petroleum Geologists, 15th Annual Field Conference, Guidebook: Part 1. Cypress Hills Plateau," pp. 85-111.

Westgate, J. A. (1968). "Surficial Geology of the Foremost-Cypress Hills Area, Alberta." Alberta Research Council Bulletin 22.

White, G. W. (1960). "Classification of Wisconsin Glacial deposits in Northeastern Ohio." U.S. Geological Survey Bulletin 1121-A, pp. A1-A12.

White, G. W. (1961). "Classification of Glacial Deposits in the Killbuck Lobe, Northeast-Central Ohio." U.S. Geological Survey Professional Paper 424-C, pp. C71-C73.

White, G. W. (1968). Age and correlation of Pleistocene deposits at Garfield Heights (Cleveland), Ohio. *Geological Society of America Bulletin* 79, 749-55.

White, G. W. (1969). Pleistocene deposits of the northwestern Allegheny Plateau, U.S.A. *Quarterly Journal of the Geological Society of London* 124, 131-49.

White, G. W. (1971). Thickness of Wisconsinan tills in Grand River and Killbuck lobes, northeastern Ohio and northwestern Pennsylvania. *In* "Till: A Symposium" (R. P. Goldthwait et al., eds.), pp. 149-63. Ohio State University Press, Columbus.

White, G. W. (1974). Buried glacial geomorphology. *In* "Glacial Geomorphology" (D. R. Coates, ed.), pp. 331-49. State University of New York at Binghamton Publications in Geomorphology.

White, G. W., and Totten, S. M. (1979). "Glacial Geology of Ashtabula County, Ohio." Ohio Geological Survey Report of Investigation 112.

White, G. W., Totten, S. M., and Gross, D. L. (1969). "Pleistocene Stratigraphy of Northwestern Pennsylvania." Pennsylvania Geological Survey Bulletin G55.

White, S. E. (1947). Two tills and the development of glacial drainage in the vicinity of Stafford Springs, Connecticut. *American Journal of Science* 245, 754-78.

Whitehead, D. R. (1973). Late-Wisconsin vegetational changes in unglaciated eastern North America. *Quaternary Research* 3, 621-31.

Whittecar, G. R., and Mickelson, D. M. (1976). Sequence of till deposition and erosion in drumlins. *Boreas* 6, 213-17.

Whittecar, G. R., and Mickelson, D. M. (1979). Composition, internal structures, and an hypothesis of formation for drumlins, Waukesha County, Wisconsin, U.S.A. *Journal of Glaciology* 22, 357-71.

Wickham, J. T. (1979). "Glacial Geology of North-Central and Western Champaign County, Illinois." Illinois Geologic Survey Circular 506.

Wickham, S. S. (1979). The Tiskilwa Till Member, Wedron Formation: A regional study. *Geological Society of America, Abstracts with Programs* 11, 260.

Wier, C. E., and Gray, H. H. (1961). Geologic map of the Indianapolis 1° × 2° quadrangle, Indiana and Illinois, showing bedrock and unconsolidated deposits. Indiana Geologic Survey Regional Geologic Map 1.

Wilding, L. P., Drees, L. R., Smeck, N. E., and Hall, G. F. (1971). Mineral and elemental composition of Wisconsin-age till deposits in west-central Ohio. *In* "Till: A Symposium" (R. P. Goldthwait et al., eds.), pp. 290-317. Ohio State University Press, Columbus.

Willman, H. B. (1971). "Summary of the geology of the Chicago Area." Illinois Geologic Survey Bulletin 94.

Wright, F. M. (1970). Pleistocene stratigraphy of the Farmersville and the northern part of the Middletown quadrangles, southwestern Ohio. M.A. thesis, Miami University of Ohio, Miami.

Wright, H. E., Jr. (1957). Wadena drumlin field, Minnesota. *Geografiska Annaler* 39, 19-31.

Wright, H. E., Jr. (1971). Retreat of the Laurentide ice sheet from 14,000 to 9,000 years ago. *Quaternary Research* 1, 316-30.

Wright, H. E., Jr. (1972). Quaternary history of Minnesota. *In* "Geology of Minnesota: A Centennial Volume (P.K. Sims and G.B. Morey, eds.), pp. 515-47, Minnesota Geological Survey, St. Paul.

Wright, H. E., Jr. (1973). Tunnel valleys, glacial surges, and subglacial hydrology of the Superior lobe, Minnesota. *In* "The Wisconsinan Stage" (R. F. Black, R. P. Goldthwait, and H.B. Willman, eds.), pp. 251-76. Geological Society of America Memoir 136.

Wright, H. E., Jr. (1976). Ice retreat and revegetation in the western Great Lakes area. *In* "Quaternary Stratigraphy of North America" (W. C. Mahaney, ed.), pp. 119-32. Dowden, Hutchinson, and Ross, Stroudsburg, Pa.

Wright, H. E., Jr. (1980). Surge Moraines of the Klutan Glacier, Yukon Territory, Canada: Origin, wastage, vegetation succession, lake development, and application to the late-glacial of Minnesota. *Quaternary Research* 14, 2-18.

Wright, H. E., Jr., Cushing, E. J., and Baker, R. G. (1964). "Midwest Friends of the Pleistocene, Guidebook, Eastern Minnesota."

Wright, H. E., Jr. and Frey, D. G. (eds.) (1965). "The Quaternary of the United States." Princeton University Press, Princeton, N.J.

Wright, H. E., Jr., Matsch, C. L., and Cushing, E. J. (1973). Superior and Des Moines lobes. *In* "The Wisconsinan Stage" (R. F. Black, R. P. Goldthwait, and H. B. Willman, eds.), pp. 153-85. Geological Society of America Memoir 136.

Wright, H. E., Jr., Mattson, L. A., and Thomas, J. A. (1970). Geology of the Cloquet quadrangle, Carlton County, Minnesota. Minnesota Geological Survey Geological Map series GM-3.

Zumberge, J. H. (1960). Correlation of the Wisconsin drifts in Illinois, Indiana, Michigan, and Ohio. *Geological Society of America Bulletin* 71, 1177-88.

CHAPTER 2

The Cordilleran Ice Sheet in Alaska

Thomas D. Hamilton and Robert M. Thorson

Introduction

Glaciation in Alaska has been primarily alpine in character. The mountains of the northern and central parts of the state, which generally are low (1000 to 3000 m altitude) and remote from major sources of precipitation, generated glaciers that were small compared to those farther south (Porter et al., 1983). Modern glaciers and Pleistocene ice complexes have been much more extensive throughout southern Alaska, where mountains are higher (up to 4000 to 6000 m) and are close to moisture sources in the North Pacific and the Gulf of Alaska (Figure 2-1). Individual ice tongues occupied most northern valleys of the Alaska Range, but glaciers south of the range crest coalesced into the vast ice caps, intermontane glaciers, and piedmont glaciers that constituted the northwestern part of the Cordilleran ice sheet. This glacier complex was fed by ice streams that originated primarily in two arcuate mountain chains. The Aleutian Range and Alaska Range, which confined the ice sheet on the north and west, together are more than 2000 km long. A second chain, around the Gulf of Alaska coast, extends 1900 km from Kodiak Island through the Kenai and Chugach Mountains into southeastern Alaska. The structural depression between the two mountain chains is interrupted by the Talkeetna Mountains, which separate the Cook Inlet-Susitna lowland region from the Copper River basin, and by the Wrangell Mountains, which bound the Copper River basin to the east (Figure 2-2).

The glacial record of southern Alaska has been summarized in several compilations that were published around the time of the Seventh Conference of the International Association for Quaternary Research (Karlstrom, 1964; Karlstrom et al., 1964; Coulter et al., 1965; Péwé et al., 1965; Péwé, Hopkins, and Giddings, 1965). This chapter focuses on subsequent studies and emphasizes particularly areas around the Copper River basin and the complex glacial and marine record of the Cook Inlet-Susitna lowland region. The alpine glaciers of the northern Alaska Range, which were largely separate from the coalescent glaciers of the ice sheet, are reviewed elsewhere (Porter et al., 1983).

Modern Setting

Most of south-central Alaska has a subarctic maritime to transitional climate characterized by annual precipitation of as much as 300 cm, moderate to heavy (150 to 500 cm) winter snowfall (Johnson and Hartman, 1969: 60-69), cool summers, and long but moderate winters. Tectonism is active today, and the Holocene sea-level record shows complex interactions of eustatic, glacial-isostatic, and orogenic changes (Miller, 1973; Williams and Coulter, 1981). Abrupt crustal displacements have accompanied strong earthquakes (Plafker, 1969), and more gradual isostatic movements are associated with crustal unloading by retreating coastal glaciers (Sirkin and Tuthill, 1971; Clark, 1978). Significant crustal displacements during the Holocene are recorded along strike-slip faults (Plafker et al., 1978), and uplift along parts of the Gulf of Alaska coast has averaged as much as 1 cm per year (Hudson et al.,1976). Volcanism is active throughout much of the region (Johnston, 1979; Benson and Motyka, 1979).

The Copper River basin (Figure 2-2) is a broad, poorly drained interior lowland, generally 500 to 850 m in altitude, surrounded by glacierized mountains (Wahrhaftig, 1965: 38-39). The Alaska Range, at the northern margin of the basin, is highest in the Mount Hayes area and declines eastward toward the Canadian border. The Wrangell Mountains, which indent the eastern flank of the basin, consist of shield and composite volcanoes 2500 to 5000 m high that bear summit ice caps and radiating valley glaciers. The Chugach Mountains, which rise to 2500 to 3300 m along the southern flank of the basin, support larger ice caps and ice fields. The Talkeetna Mountains, which are lower than the coastal ranges and lie in their precipitation shadow, lack summit ice caps and support relatively short (5 to 12 km) valley glaciers.

The Cook Inlet-Susitna lowland region (Wahrhaftig, 1965: 36) is a broad northeast-trending trough about 50 to 60 km wide at altitudes below 200 m. It is bordered to the north and west by the eastward-curving Aleutian and Alaska Ranges; its eastern margin is formed by the Kenai, Chugach, and Talkeetna Mountains. The Aleutian Range generally is low, but Quaternary volcanoes rise to 3000 m and support ice caps with radiating valley glaciers. The Alaska Range also is low (less than 2000 m) west of Cook Inlet, but it rises northward to the high and extensively glacierized peaks around Mount McKinley. Generally short (10 to 20 km) valley glaciers flow north in the precipitation shadow of the McKinley massif, but glaciers south of the drainage divide are as long as 40 to 70 km. The Kenai Mountains rise to only 1200 to 1800 m, but they receive more snowfall than the

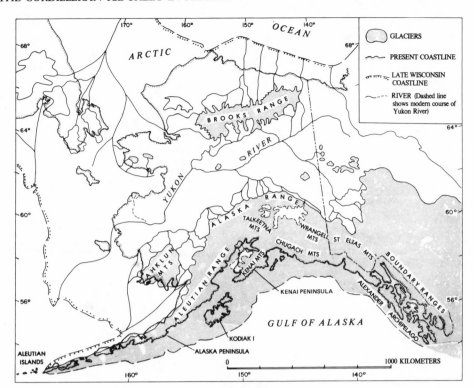

Figure 2-1. Alaska during Late Wisconsin glaciation, including Cordilleran ice sheet, smaller mountain glaciers, emergent Bering platform, and major routes of proglacial drainage across continental shelves.

Alaska Range and support large ice fields with outlet glaciers that extend to the sea.

The modern glaciation threshold in south-central Alaska is controlled mainly by moisture transported northward from the Gulf of Alaska and northeast up Cook Inlet from the North Pacific (Figure 2-2). (The "glaciation threshold," called "glaciation limit" or "glaciation level" by other authors, is the lowest altitude in a given locality at which glaciers can develop. See the discussion in Porter et al., 1983.) Some additional precipitation from the Bering Sea reaches the western flank of the Alaska Range but has little effect on the area farther east. The glaciation threshold is lowest on the seaward flank of mountains that border the Gulf of Alaska. It rises steeply northward and is as high as 2000 m on the inland flanks of high peaks at the head of Prince William Sound and near the Canadian border. The threshold also is low (1000 to 1200 m) in the Cook Inlet-Susitna lowland area, which serves as a conduit for moist air masses that originate over the North Pacific. Progressively drier conditions prevail farther to the north and east, and the threshold is generally high thoughout the eastern Talkeetna Mountains, the Copper River basin, and the eastern Alaska Range. It is highest in the precipitation shadows of the Wrangell Mountains and the Mount McKinley and Mount Hayes areas of the Alaska Range.

Late Wisconsin Glaciation

In much of Alaska, Late Wisconsin glaciation has not been differentiated from other glacial events of late-Pleistocene age (Coulter et al., 1965) or earlier Wisconsin age (Péwé, 1975, 1976). Recent field studies supported by extensive radiocarbon dating have enabled the delineation of Late Wisconsin drift through much of the Brooks Range and through northern valleys of the Alaska Range (Porter et al., 1983). Similar separation and mapping of the Late Wisconsin

limits of the Cordilleran ice sheet are less readily accomplished because the glacial record of this vast region is poorly known in some areas and controversial in others and because much of the ice sheet terminated in proglacial lakes, in marine embayments, or on the presently submerged continental shelf. In the following sections, we attempt to define the position of the Late Wisconsin succession within the late-Pleistocene glacial records of south-central Alaska, and then we attempt to reconstruct the glaciation threshold, ice dynamics, chronology, and limits of this final fluctuation of the Cordilleran ice sheet.

LATE-PLEISTOCENE GLACIAL RECORD

The mountains of southern Alaska have undergone repeated episodes of glaciation that began in Miocene time (Denton and Armstrong, 1969; Plafker and Addicott, 1976) and have continued to the present. Studies of ice-rafted detritus in sediment cores from the deep-sea floor in the Gulf of Alaska (Kent et al., 1971; von Huene et al., 1976) show sequences of particularly intense glaciation during the middle and late Pleistocene. According to the expanded time scale and correlations of von Huene and others (1976: 417-19), extensive ice advances some 90,000 to 50,000 and 25,000 to 10,000 years ago around the Gulf of Alaska followed a major interglacial recession (Figure 2-3A). The younger advance corresponds to Late Wisconsin glaciation as dated elsewhere in Alaska (Porter et al., 1983). We infer that the late-Pleistocene history of the Cordilleran ice sheet throughout south-central Alaska must correspond closely to the marine record in the Gulf of Alaska, because ice that filled the lowlands north of the coastal mountains was part of a dynamic system that was continuous with south-flowing tidewater glaciers.

The glacial history of Cook Inlet, as described by Karlstrom (1964), consists of five Pleistocene ice advances that were separated by intervals of soil formation and weathering. The youngest three advances

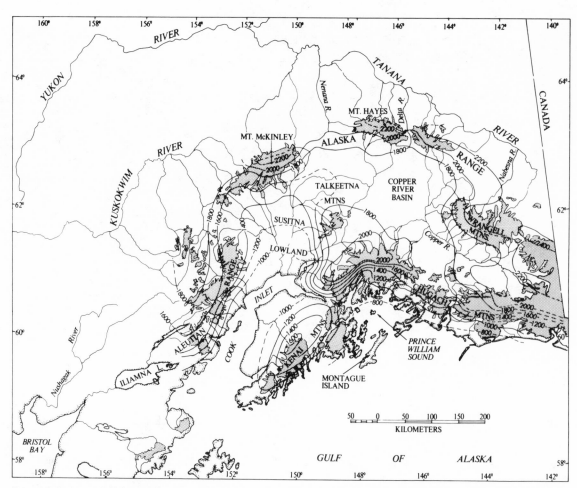

Figure 2-2. South-central Alaska, including principal mountain ranges, existing glaciers (shaded), and modern glaciation threshold (contours at 200-m intervals; dashed where speculative).

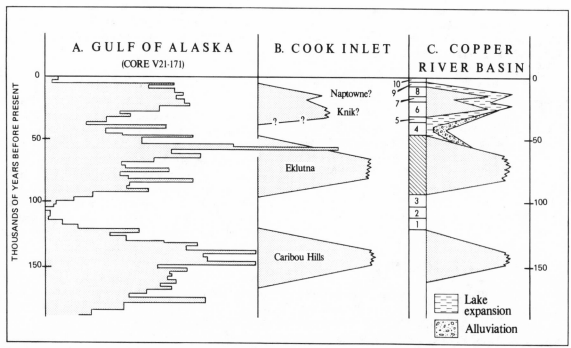

Figure 2-3. Graphs showing (A) abundance of ice-rafted sand in Gulf of Alaska core and inferred correlations with (B) Cook Inlet glacial sequence and with (C) stratigraphic and glacial records from Copper River basin (A is from von Huene et al., 1976: 416; and B and C are modified from Péwé et al., 1965).

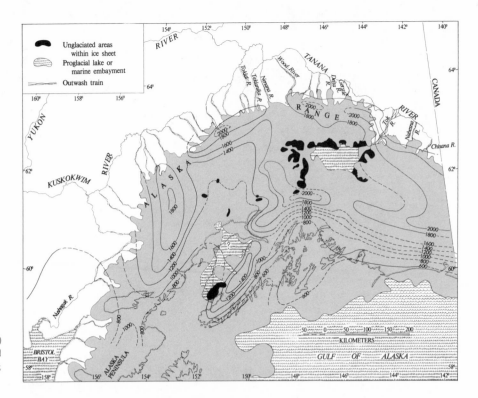

Figure 2-4. Glaciers of south-central Alaska (shaded) during Late Wisconsin glaciation, including inferred glaciation threshold (contours at 200-m intervals; dashed where speculative).

(Eklutna, Knik, and Naptowne Glaciations) were separated by intervals that were brief compared to the preceding nonglacial episode (Figure 2-3B). On the basis of inferred weathering rates, Karlstrom believes that this entire sequence may have taken place within the past 120,000 years (1964: 63; 1976). The floor of Cook Inlet was completely filled by ice during the Eklutna phase, but the subsequent advances involved less extensive piedmont glaciers, and parts of Cook Inlet remained free of ice during Knik and Naptowne time. Although Karlstrom believes that the Eklutna ice advance is pre-Wisconsin in age, stratigraphic evidence from Bristol Bay (D. M. Hopkins, personal communication, 1981) and the Alaska Range (Weber et al., 1981) suggests that the youngest expansion of the Cordilleran ice sheet equivalent in magnitude to the Eklutna advance followed rather than preceded the last interglaciation. We speculate that the disintegration of coastal glaciers at the close of the Eklutna advance could have caused the pronounced increase in the amount of ice-rafted detritus that is evident at about 90,000 to 50,000 yr B.P. in the sediment record of the Gulf of Alaska (Figure 2-3A). We provisionally equate the Knik and Naptowne ice advances with the Late Wisconsin ice-rafting episode in the Gulf of Alaska.

Bluff exposures along the Copper River record multiple ice advances during the late Pleistocene. Ice extended across the basin floor during major glacial intervals. During times of less extensive glaciation, ice streams terminated in large proglacial lakes that occupied the basin when expanding glaciers in the Chugach Mountains blocked the Copper River. A bluff exposure 80 m high described by Ferrians and Nichols (Péwé et al., 1965: 102-5) records four glacial advances that alternated with episodes of fluvial and lacustrine sedimentation (Table 2-1 and Figure 2-3C). The oldest glaciation is indicated only by till blocks incorporated in younger fluvial sediments (1) that evidently are part of a continuous sequence of interglacial deposits (1-3) formed when the Copper River was not impeded by glaciers in the Chugach Mountains. (The numbers in parentheses refer to the stratigraphic units in Table 2-1 and Figure 2-3B.) A younger glaciation is indicated

by blocks of till and varved lakebeds redeposited in younger alluvium (4) and also by deformed sedimentary deposits (3) beneath a prominent disconformity. An expanding lake in the Copper River basin (5) preceded a later glacial advance (6). The glacier subsequently thinned or retreated but evidently did not recede from the lower course of the Copper River, and a lake (7) continued to occupy the basin. The deformed pebbly beds near the top of unit 7 mark the final advance of the glacier (8), which was followed by a late-glacial lacustrine episode (9). The lake then drained, and the river incised to its present level. Our correlation of this section with the Gulf of Alaska and Cook Inlet records is based on the similarity of the interglacial episodes that preceded a long interval characterized by generally expanded glaciers. Wisconsin glaciation was marked by an initial ice advance that was more erosive than subsequent advances and may have been more extensive. The two younger advances were separated by an interval of lacustrine deposition that was minor compared to the longer and more complex interstade that preceded them. They could represent a sequence of two glacial advances during Late Wisconsin time.

LATE WISCONSIN GLACIATION THRESHOLD

The glaciation threshold reconstructed in Figure 2-4 is based on the lowest altitudes of highlands associated with undissected cirques that probably supported glaciers during Late Wisconsin time. This surface may have been deformed by orogeny and by postglacial isostatic readjustments in some areas, and it may be inaccurate where cirques were buried beneath thickening ice streams and spreading ice caps. In addition, the values were taken from small-scale topographic maps on which cirques that contained small or short-lived glaciers would be poorly portrayed. Nevertheless, the reconstruction shows general trends that are compatible with the late-Quaternary topography and moisture sources in south-central Alaska, and it should suggest a useful approximation of air-mass movements during the Late Wisconsin.

Table 2-1.
Stratigraphic Section of Copper River Bluff (62°05′N, 145°25′W)

Unit	Thickness (m)	Description
10	0.9	Modern vegetation mat above eolian silt and sand
9	2.0	Dark gray clayey silt with lenses of silt and volcanic ash; late-glacial lacustrine deposit; gradational contact with unit 8
*8	4.5-7.5	Till; subrounded pebbles, cobbles, and boulders in dark gray silty sand matrix; thin sandy lenses in upper 0.7 m
7	6.0-7.5	Laminated to massive blue gray clayey silt to silty clay, containing several 2 to 10 cm tephra beds; upper 1.5 m pebbly and locally deformed
*6	16-21	Clayey silt with dropstones; grades downward into till that contains stones up to boulder size and blocks of redeposited varved clay and silt
5	7.6	Laminated light gray to dark bluish gray silty clay containing scattered pebbles and cobbles; overlies medium to coarse sand with steep, south-dipping foreset beds that contain scattered pebbles and cobbles
4	8.0	Fine gravel above horizontally bedded medium to coarse fluvial sand with thin pumiceous pebbly zones and scattered blocks of till and varved silt and clay
*		———————— Disconformity ————————
3	10.8	Horizontally bedded gray to brown silt and fine sand, upper 1 m highly contorted; overlies fluvial sand and gravel with local oxide staining
2	0.9	Angular pebbles and cobbles, generally andesite, in matrix of silty sand; color varies from pink, green, and brick red to gray; volcanic mud flow deposits from Mount Sanford
1	18.3	Interbedded fluvial sand and coarse gravel containing detrital blocks of till and lacustrine sediment; locally iron stained; cross bedding dips steeply southward

Source: Modified from Ferrians and Nichols (Péwé et al., 1965: 102-5).
Note: Three episodes of inferred Wisconsin glaciation are marked by asterisks.

According to our reconstruction, the Late Wisconsin glaciation threshold generally parallels the modern threshold (Figure 2-2) at lower altitudes and resembles Péwé's (1975: 31) more-generalized Wisconsin snow-line surface in Alaska. During Late Wisconsin time, the glaciation threshold was lowest along the Gulf of Alaska coast, and it was also low in the lower Cook Inlet-Alaska Peninsula region. It rose abruptly across the Kenai and Chugach Mountains and attained high values in the precipitation shadows inland from the highest peaks. Low threshold values indicate continued northeastward transport of moisture through Cook Inlet and the Susitna lowland. Glaciation should have been particularly extensive near the head of this trough along the southern flank of the Alaska Range near Mount McKinley and along the western flank of the Talkeetna Mountains. Both of these mountain masses served as additional barriers to moisture, with pronounced precipitation shadows on the northern sides of the Mount McKinley and Mount Hayes massifs and a high glaciation threshold east of the Talkeetna Mountains throughout the Copper River basin.

Apparent depressions of the Late Wisconsin glaciation threshold below modern values (compare Figures 2-2 and 2-4) are minimum values because of the error factors discussed above. Apparently, the

depression averaged about 200 m along the northern flank of the Alaska Range east of the Nenana River, where precipitation was most strongly inhibited by the mountains and ice fields to the south. Uplift of this tectonically active area (Porter et al., 1983) might also have reduced the difference between Late Wisconsin and modern values. Farther west, the threshold was depressed as much as 300 to 350 m, probably because more moisture was available for glacier expansion, and, consequently, the lowering of temperatures during Late Wisconsin time had a greater effect. South of the Alaska Range, the threshold was depressed at least 200 m, but most of the lower cirques in this region were covered by glaciers that originated in the higher source areas. Low threshold depressions around the Gulf of Alaska also are controlled in part by late-Quaternary orogenic and isostatic uplift, and the deflection of the 600-m threshold contour around Montague Island confirms Plafker's assertion (1969: 120-123) that much of this area has been a major center of uplift in south-central Alaska.

GLACIAL DYNAMICS

Late Wisconsin glacier advances in south-central Alaska created a nearly continuous ice complex, the form and flow pattern of which (Figure 2-5) were controlled largely by the region's mountainous terrain and by differing rates of snow accumulation in the individual source areas. The boundaries shown in Figure 2-5 were compiled from published sources (Karlstrom, 1964; Karlstrom et al., 1964; Coulter et al., 1965; Péwé et al., 1965; Péwé, 1975:16, 72) and from our own unpublished field mapping, analyses of radar and Landsat imagery, and interpretation of stratigraphic records and radiocarbon dates. Late Wisconsin glacier limits are drawn at the edge of the continental shelf around the Gulf of Alaska because seismic-reflection profiles, bottom morphology, and sediment cores demonstrate the widespread presence of slightly modified glacial features and overconsolidated subglacial deposits across much of this submerged platform (von Huene et al., 1967; Molnia and Carlson, 1975; Molnia et al., 1980; P. R. Carlson and B. F. Molnia, personal communication, 1981).

The basal flow lines of former ice streams (Figure 2-5) were reconstructed from the orientations of major glacial valleys; the former glacier lobes as defined by end moraines; and the trends of lateral moraines, rock basins, and streamlined bodies of drift and bedrock. These deposits and erosional forms are morphologically similar to dated glacial features of Late Wisconsin age in the Alaska Range.

The reconstructed flow lines and ice limits define major ice streams of the Cordilleran ice sheet and their source areas in the Alaska Range, the Kenai and Chugach Mountains, and parts of the Aleutian Range. Smaller glaciers radiated from the Wrangell Mountains and originated in the Talkeetna Mountains. The ice streams formed piedmont lobes around the margins of the Copper River basin and Cook Inlet; they also nourished glaciers that extended north across the Alaska Range, west around the Talkeetna Mountains, and south across the Gulf of Alaska shelf. Separate ice caps evidently developed on the continental shelf east of the Aleutian Range opposite Iliamna Lake and southwest of Kodiak Island (Karlstrom and Ball, 1969).

Glacier advances into the Copper River basin were inhibited by generally high glaciation thresholds in surrounding source areas. The principal direction of ice flow was northwest down the Chitina Valley from sources in the Chugach, Wrangell, and St. Elias Mountains. Smaller ice streams, fed from ice fields in the Chugach Mountains,

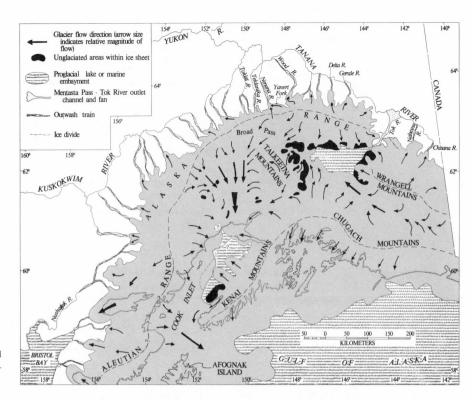

Figure 2-5. Late Wisconsin glaciation of south-central Alaska, including ice distribution (shaded), flow patterns, and divides.

entered the basin from the south. Glaciers originating in the Alaska Range, the Talkeetna Mountains, and the western Wrangell Mountains were much smaller and generally reached only the margin of the basin. Glaciers from the north terminated against low foothills, as mapped by Ferrians and Nichols (Péwé et al., 1965: 94); glaciers from the south evidently terminated in a deep lake (Nichols and Yehle, 1969; Péwé, 1975: 72-73; Williams and Johnson, 1980).

Glacier limits and flow patterns around Cook Inlet are poorly defined because much of this area is presently submerged or was inundated during Late Wisconsin time. The influx of glaciers into the upper inlet was dominated by large ice streams from the southern flank of the McKinley massif, which evidently received large amounts of snowfall during Late Wisconsin time, as it does today. Somewhat smaller ice streams from parts of the Alaska Range farther to the west (Nelson and Reed, 1978) and from western valleys of the Talkeetna Mountains coalesced with glaciers from the Mount McKinley region to form a great compound ice tongue some 200 km long that probably terminated somewhere in the upper inlet. Large piedmont lobes extended east from the Aleutian Range, terminating within the inlet, locally crossing to its eastern shore, and probably crossing its mouth. Lesser ice tongues from the east, where precipitation was impeded by topographic barriers, terminated on the Kenai Peninsula.

Glacier ice and meltwater crossed the Alaska Range in three places. The largest ice stream flowed down the Delta Valley to build the Donnelly moraine; a smaller glacier extended into the Nenana Valley, where it joined the larger and more vigorous Yanert Fork glacier to build the Riley Creek moraine (Porter et al., 1982). The proglacial lake in the Copper River basin may have discharged northeastward through Mentasta Pass, perhaps episodically. The outlet stream extended down the Tok River valley to its mouth, where it formed an alluvial fan that partly blocked the upper Tanana River.

Glaciers radiated outward through western valleys of the Talkeetna Mountains, but the eastern part of this range was par-

tially unglaciated and elsewhere supported ice bodies that generally were smaller than those farther west. Many northern valleys were overwhelmed by southwest-flowing glaciers that were generated principally in the Alaska Range. A larger ice stream originating mainly in the Alaska Range flowed around the north end of the Talkeetna Mountains, where it scoured a deep glacial trough in the Broad Pass area.

The diverse ice streams and outlet glaciers of the Cordilleran ice sheet could have experienced different glacial histories and regimens. Glaciers in the Copper River basin were influenced by a relatively continental climate and by fluctuations of the lake that filled the basin's center. Ice streams in the Susitna lowland were fed from source areas with large amounts of snowfall; their discharge was relatively great, and subglacial erosion was more vigorous. Glaciers around Cook Inlet were influenced by precipitation shadows on the inland flank of the Kenai Mountains, by additional blockage of moist air masses by glaciers at the inlet's mouth, by marine invasion caused by isostatic depression, and possibly by local climatic effects resulting from marine transgressions. Glaciers that crossed the Alaska Range and the Talkeetna Mountains would have been controlled by ice levels in nearby parts of the Cordilleran ice sheet, whereas neighboring cirque-fed mountain glaciers would have fluctuated in response to local climatic controls. For these reasons, fluctuations of glaciers in south-central Alaska may not have always been in phase. Advances and recessions may be related to regional ice dynamics, changes in lake levels, sea-level fluctuations, and erosion rates of outlet glaciers, in addition to climatic change. Regional correlations, therefore, need to be made with great caution.

RADIOCARBON CHRONOLOGY

Many stratigraphic sections in the Copper River region have yielded organic samples that are older than the time range of conventional radiocarbon dating (Table 2-2). These samples are from fluvial and

Table 2-2.

Late-Pleistocene and Early-Holocene Radiocarbon Dates for Copper River Basin and Vicinity

Locality	Date (yr B.P.) (and Lab Number)	Location, Material, and Stratigraphic Position	Reference
33	9510 ± 475 (I-AGS-9)	Bering Lake; sedge peat, 3.3 m depth in muskeg	Heusser, 1960: Figure 21 and p. 223
32	9520 ± 350 (W-1654)	Valdez; freshwater peat above till and below marine delta deposits	Williams and Coulter, 1981
31	10,250 ± 250 (W-767)	Little Nelchina River; peat from paleosol developed on terrace gravel	Péwé et al., 1965: 109
30	10,390 ± 350 (I-AGS-5)	Copper River delta; basal peat	R 1, p. 31; Heusser, 1960: Figure 20 and p. 222
29	10,565 ± 200 (GX-249)	Maclaren River; lowest peat exposed in core of ''Pingo'' (probably a palsa)	Péwé et al., 1965: 91
28	10,900 ± 160 (Y-2301) 10,980 ± 150 (I-6092) 11,100 ± 120 (SI-1103)	White River; basal peat and organic silt on Macauley till	Denton, 1974
27	11,270 ± 200 (Y-2306)	White River; basal peat on Macauley till	Denton, 1974
26	11,300 ± 200 (W-1524)	Mentasta Valley; organic silt below peat and above alluvium that postdates local glaciation	R 9, p. 522
25	11,390 ± 300 (W-848)	Gulkana River; peat below silt and clay and above bedded sand, silt and clay	R 2, p. 172
24	9,100 ± 80 (UCLA-1858) 11,800 ± 750 (UCLA-1859)	Tangle Lakes; organic matter at base of postglacial section (UCLA-1859) and wood fragments at horizon of spruce influx (UCLA-1858)	Schweger, in press
23	13,280 ± 400 (W-583)	East of Tyone Creek; basal peat above outwash gravel; outwash is graded to delta at 750 m lake level	Péwé et al., 1965: 111; Karlstrom et al., 1964 (section M2)
22	24,900 ± 325 (BETA-1822)	Oshetna bluff; wood from gravel deposit that interfingers with sandy drift	Thorson et al., 1981
21	30,670 ± 1,050 (I-11,380)	Dadina River; silty humus beneath till, sand, and gravel and diamicton	J. R. Williams, personal communication, 1980
20	28,300 ± 1,000 (W-1343) 31,300 ± 1,000 (W-843)	Sanford River; peat from top (W-1343) and bottom (W-843) of 2-m silty sand unit beneath sand and gravel; dates low stand of proglacial lake in Copper River basin	R 7, p. 391 R 2, p. 171
19	11,535 ± 140 (BETA-1821) 21,730 ± 390 (DIC-1861) 29,450 ± 610 (BETA-1819) 31,070 ± 860 (DIC-1862) ± 960 (DIC-1862) 32,000 ± 2,735 (BETA-1820)	Susitna River near Tyone River; BETA-1821 is peaty silt above sand deposit with basal lag boulders; BETA-1820 is wood fragments below the lag boulders; DIC-1861, from second bluff, is woody peat from cross-bedded sand below till; BETA-1819 is bone from gravel below the sand; DIC-1862 is wood from interstadial below the gravel	Thorson et al., 1981
18	> 29,000 (I-271)	Copper Center; wood within oxidized till	R 5, p. 63
17	> 35,000 (W-357)	Tyone Creek; peat beneath till and gravel (section M1)	Karlstrom et al., 1964
16	> 35,000 (W-373)	Tonsina River; wood fragments from sand and gravel beneath till, diamicton, and lacustrine beds	Rubin and Alexander, 1958
15	> 35,000 (W-377)	Gakona River; forest bed beneath glaciofluvial and glaciolustrine beds of last glaciation	Karlstrom et al., 1964 (section M5)
14	> 37,000 (W-307)	Copper River; peat and wood in paleosol above till and beneath 30 m of sand and outwash gravel; top of section eroded	Rubin and Alexander, 1958
13	> 37,000 (GSC-1576)	White River; wood nodules and plant stems from silt beds above till and below till and outwash	Denton, 1974

Table 2-2.
continued

Locality	Date (yr. B.P.) (and Lab Number)	Location, Material, and Stratigraphic Position	Reference
12	> 38,000 (W-295)	Nelchina River; peat and woody debris beneath till and gravel	Karlstrom et al., 1964 (section M3)
11	> 38,000 (W-842)	Nelchina River; organic silty sand beneath till, glaciofluvial, and glaciolacustrine deposits	R 2, p. 168
10	> 38,000 (W-969)	Copper Center; log in terrace gravel overlain by boulder bed	R 6, p. 67
9	> 38,000 (W-1337)	Gulkana River; wood from sand overlain by sand, gravel, and diamicton	R 7, p. 391
8	9,650 ± 370 (W-1161) 10,730 ± 300 (W-2171) > 32,000 (I-364) > 38,000 (W-1379)	Slana River; W-1161 is peat and silt on fine sand over gravel; W-2171 is organic silt at base of alluvium overlying lacustrine beds; I-364 and W-1379 are wood and peat from silt and fine sand overlain by thick (11 m) sand deposit	R 6, p. 63 R 12, p. 331 R 5, p. 63 R 7, p. 392
7	> 38,000 (W-1390)	Chitina River; wood below clayey silt and above gravel and diamicton	R 7, p. 392
6	> 38,000 (W-1424)	Gakona River; peat overlain and underlain by sand and gravel	R 7, p. 393
5	> 40,000 (W-977)	Lower Chitina River; wood fragments from till	R 6, p. 62
4	9,240 ± 300 (W-487) 11,440 ± 400 (W-429)	Slana River area; top (W-487) and bottom (W-429) of peat bed overlain by coarse gravel and underlain by coarse gravel and diamicton	Karlstrom et al., 1964 (section M6)
	10,300 ± 200 (W-1526)	Organic silt from sand and gravel	R 9, pp. 525-26
	11,190 ± 300 (W-1377)	Wood and peat (W-1377) from sand below Ahtell Creek peat bed	R 7, p. 392
	17,600 ± 400 (W-1134) > 42,000 (W-1162)	Wood fragments (W-1134) near base of thick (15 m) unit of sand grading upward into gravel; peat (W-1162) below laminated silt that underlies the 15-m sand unit; peat probably redeposited	R 6, pp. 64-65
3	> 47,000 (Y-2305) > 47,000 (Y-2389)	White River; organic matter in gravel overlain by lacustrine deposits containing dropstones and underlain by till	Denton, 1974
2	> 47,000 (Y-2308)	White River; organic matter in gravel overlain by till and underlain by two tills	Denton, 1974
1	9,400 ± 300 (W-714) > 46,000 (GrN-4165) > 49,000 (GrN-4448) 43,440 ± 250 (GrN-4086) 58,600 ± 1100 (GrN-4798)	Gakona River basal peat (W-714) above glaciolacustrine deposits; all remaining dates are on wood below the glaciolacustrine beds; GrN-4165 and GrN-4448 are conventional dates; GrN-4086 and GrN-4798 are enrichment dates (GrN-4798 is the preferred date for this sample	Péwé et al., 1965; Karlstrom et al., 1964 (section M4); R 14, p. 17

Note: Localities are shown by number in Figure 2-6.

Abbreviations: R 1, R 2, etc. = volume 1, volume 2, etc., of *American Journal of Science Radiocarbon Supplement*.

lacustrine deposits that generally lie above till and underlie glacial, lacustrine, and fluvial sediments assigned to the youngest Pleistocene glaciation. Some of the "infinite" radiocarbon dates are from logs; this indicates that the basin was at least partially forested before the last major ice advance. Organic sediments dated as older than 38,000 to 35,000 yr B.P. occur beneath till or relict glacial boulders throughout the southern part of the basin (10, 11, 12, 16, 17); they also occur east of the basin in the White River valley (2 and 13). (The numbers in parentheses refer to localities designated in Table 2-2 and Figure 2-6.) Comparable sediments dated as older than 47,000 to 35,000 yr B.P. underlie proglacial lake deposits along the lower

Gakona River farther north in the basin (1, 6, 15) and in northern parts of the White River valley (3), areas that evidently lay beyond the outer limits of glacier advances of Late Wisconsin age. Other organic sediments dated as older than 38,000 to 32,000 yr B.P. are preserved beneath thick sand deposits along the Slana River (8) near the Mentasta Pass outlet from the Copper River basin. Several other samples of "infinite" age may predate a Middle Wisconsin interstade, as is apparently indicated by rounded clasts of redeposited peat (3) and by wood from weathered till (18). Several samples from bluffs along the deeply incised lower courses of the Chitina and Copper Rivers (5, 7, 14) may also be older than Middle Wisconsin. At locality 14, for ex-

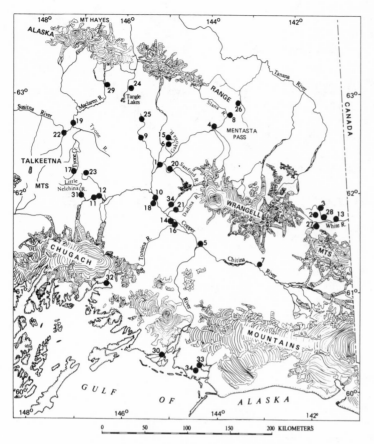

Figure 2-6. Copper River basin, including major geographic features and locations of radiocarbon samples.

ample, as D. R. Nichols (*in* Rubin and Alexander, 1958: 1481) states, two tills with thick intervening outwash and lacustrine deposits have been eroded from the top of the measured section (described in Table 2-2).

Several finite radiocarbon dates have been obtained from interstadial deposits of Middle Wisconsin age in the Copper River basin. The oldest, an enrichment date of 58,600 ± 1100 yr B.P. (1), could be spurious, because trace amounts of contamination can produce an apparent age in this time range for samples that actually are much older and because this date is inconsistent with another enrichment date from the same horizon. However, an age of about 60,000 yr B.P. is not unreasonable for a major Wisconsin interstade if the time scale on the Gulf of Alaska ice-rafting record (Figure 2-3A) is more or less correct. A more reliably dated locality (19) is along a series of bluffs cut by the Susitna River near the mouth of Tyone River (Figure 2-7), where wood, peat, and bone dated at 32,000 to 29,000 yr B.P. lie below drift assigned to the last glaciation and above glaciolacustrine deposits that accumulated during an earlier interval of ice expansion. Fluvial sand, which resembles the distal outwash described from the Brooks Range (Porter et al., 1983), began accumulating some time after 29,000 yr B.P. and was overridden by glaciers or overlapped by a proglacial lake some time after about 22,000 yr B.P. The deposition of recessional fluvial deposits associated with this advance had ceased by about 11,500 yr B.P.

Organic sediments that accumulated along the Sanford River about 31,000 to 28,000 yr B.P. (20) represent a low lake stand in the Cop-

per River basin that was preceded and followed by higher levels of the glacier-dammed lake. A comparable date from the Dadina River (21) shows that the formation of silty humus about 30,700 yr B.P. was followed by renewed glacier advances and by deposition of multiple diamictons (tills?) (J. R. Williams, personal communication, 1981). The dates from these localities demonstrate that glaciers retreated sufficiently to cause lower lake levels or possibly the drainage of the basin during an interval that perhaps lasted from about 60,000 to 25,000 yr B.P. A date of 24,000 yr B.P. for the onset of the last glaciation was obtained from ice-contact stratified drift at Oshetna Bluff (22). Deposits in the Slana River area (4) that may have similar origin have been dated at 17,600 yr B.P. near their base.

Limiting dates on deglaciation are consistent throughout the Copper River region. Basal peat above outwash gravel east of Tyone Creek (23) has been dated at 13,300 yr B.P. and evidently provides a minimum limiting age for a lake at about 750 m altitude (Péwé et al., 1965) that formed after the full-glacial lake stand that occurs at higher altitudes around the basin (Nichols and Yehle, 1969; D. R. Nichols, personal communication, 1981). A major peat-forming interval in the Slana River area (4) lasted from about 11,500 to 9200 yr B.P., and correlative wood, peat, and organic silt occur above Late Wisconsin outwash throughout that area (4, 8, 26). Basal organic matter near the Tangle Lakes (24) is about 11,800 years old; basal organic matter above till and alluvium in the White River valley, near Mentasta Pass, and along the Gulkana and Susitna Rivers (19, 25, 26, 27, 28) all date in the range of about 11,500 to 11,000 yr B.P. A younger basal date of about 10,500 yr B.P. from the Maclaren River area (29) probably reflects only the limited depth of organic deposits accessible at that exposure, but another young date, 10,250 yr B.P., on a terrace of the Little Nelchina River (31) might provide a closer limiting date for the recession of the Nelchina Glacier. Basal dates of 10,400 to 9500 yr B.P. at Valdez (32) and along the Gulf of Alaska coast (30, 33) probably reflect regional sea-level history, which evidently was dominated by postglacial coastal uplift (Plafker, 1969) and by eustatic sea-level changes in the Valdez area (J. R. Williams, personal communication, 1981). The dates indirectly reflect ice wastage in the Chugach Mountains and the Copper River basin, which controlled the isostatic component of the uplift record, but they do not necessarily provide close limiting ages for deglaciation.

The radiocarbon record of Late Wisconsin events in the Cook Inlet region is more obscure than that for the Copper River basin. Problems include the widespread presence of Tertiary coal fragments and lignified wood redeposited in Quaternary sedimentary deposits, scarcity of dates and detailed stratigraphic studies west of Cook Inlet and the Susitna Valley, and possible truncation of some stratigraphic sections in exposed coastal bluffs. Eleven samples (localities 1-10) from the east side of Cook Inlet and two from the west side have "infinite" ages, most of which are greater than 30,000 yr B.P. (The numbers in parentheses refer to localities designated in Table 2-3 and Figure 2-8.) These samples must be of Middle Wisconsin age or older, and most lie stratigraphically below glacial deposits variously assigned to the Knik and Naptowne ice advances. Two additional dates (2 and 11), reported as about 34,000 and 39,000 yr B.P., are close to the limit of conventional radiocarbon dating and have very large counting errors; they probably also provide minimum ages for the samples. A date of about 30,700 yr B.P. farther along the Susitna River (12) may represent a more reliable maximum age for the last glaciation west of the Talkeetna Mountains. This date is significantly younger than most maximum limiting dates on the last glaciation of the Cook

Inlet area; this suggests that glaciers may have occupied Cook Inlet before the upper Susitna Valley was glaciated, that Late Wisconsin glaciers may not have been extensive around Cook Inlet, or that the Cook Inlet samples may be contaminated with redeposited wood or coal fragments.

Four concordant dates of marine shells from the glaciomarine Bootlegger Cove Clay near Anchorage (13-15) range from 15,000 to 13,500 years (Schmoll et al., 1972). These sediments overlie diamicton of Knik age and may interfinger with ice-contact deposits formed during glacier recession; they are overlapped by till and outwash of the Elmendorf Moraine, which is correlated with the Naptowne advance (Karlstrom, 1964). The four dates, coupled with limiting minimum ages on the Elmendorf Moraine (13 and 15), place an age bracket of about 13,500 and 11,700 yr B.P. on a possibly widespread late-glacial readvance. A comparable date of 12,900 yr B.P. (2) was obtained for organic sediments deformed by an ice advance farther south on Kenai Peninsula, which Karlstrom (1964: 49-50) assigned to the Naptowne advance. Stratigraphic relations suggest that glaciers also advanced down Turnagain Arm at this time but that they terminated some 35 km east of Cook Inlet (Kachadoorian et al., 1977).

Radiocarbon dates of basal peat and lacustrine deposits indicate general deglaciation of the Cook Inlet area beginning by 13,500 yr B.P. and probably continuing for several thousand years. Peat that overlies Late Wisconsin lacustrine sediments and Knik-age drift on the Kenai Peninsula (16) has been dated at 13,500 yr B.P., and basal peat on lateral moraines farther north (17) is about 11,900 years old. Organic matter in a sediment core from Hidden Lake (18) has been dated at 10,400 yr B.P., but estimates based on sedimentation rates suggest that deglaciation of the lake basin took place some time between 13,500 and 12,000 years ago (M. J. Rymer and J. D. Sims, personal communication, 1981).

Basal peat horizons from coastal areas below about 40 m altitude range in age from about 10,000 to 9000 yr B.P. (19, 21-24), an interval similar to that for basal peat farther east along the Gulf of Alaska coast. The absence of older dates on coastal peats supports the assertion of Schmoll and others (1972) that much of this region remained depressed below sea level for some time after deglaciation.

Discussion and Conclusions

The Cordilleran ice sheet in Alaska was a complex body of ice streams, intermontane ice fields, and piedmont glaciers that coalesced with ice caps on parts of the continental shelf south of the Alaska Peninsula and the Aleutian Islands. Much of southern Alaska supports large glaciers today, and some areas may have remained sufficiently glaciated during Middle Wisconsin time that earlier and later ice advances are not separable. Many sectors of the ice sheet terminated in proglacial lakes, in marine embayments, or on presently submerged parts of the continental shelf where morphological and stratigraphic records commonly are obscure and glacier fluctuations can be inferred only from indirect evidence. In addition, southern Alaska has extremely high relief and sharply contrasting moisture regimes. Different parts of the ice sheet could have fluctuated out of phase because of such factors as precipitation shadows inland from expanding ice caps on the continental shelves. Although we have focused on a central region of the ice sheet consisting of the Copper River basin, the Cook Inlet-Susitna lowland trough, and the Alaska

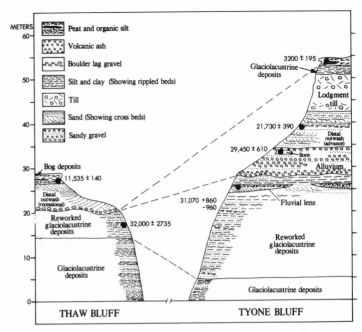

Figure 2-7. Stratigraphic sections with radiocarbon dates along the east bank of Susitna River near mouth of Tyone River.

Range, interconnected glaciers also extended throughout southeastern Alaska and through most of the Aleutian Range and the eastern Aleutian Islands. Each of these five regions differs from the others in some aspects of late-Pleistocene history and in the problems involved in its interpretation.

The stratigraphic record in the Copper River basin is difficult to assess due to the widespread presence of diamictons that may have originated from grounded glaciers, floating ice tongues or bergs, turbidity currents, subaqueous mud flows, or any combination of such agencies (Ferrians, 1963). On the basis of stratigraphic evidence, radiocarbon dates, morphological relations, and correlations with the Gulf of Alaska sediment record, we speculate that the Copper River has been able to cross the extensively glacierized Chugach Mountains only during Holocene time and during the last (Sangamon) interglaciation; it was blocked by expanded coastal glaciers during all or nearly all of Wisconsin time. We infer that the basin was filled with ice early in the Wisconsin age during maximum expansion of the late-Pleistocene Cordilleran ice sheet, that aggradation of lacustrine sediments and fine-grained alluvium prevailed through Middle Wisconsin time, and that the last major glacier advances throughout the basin took place between about 25,000 and 13,500 years ago. Glaciers that originated north of the Chugach Mountains during Late Wisconsin time were restricted by low precipitation, as were individual mountain glaciers farther north in Alaska (Porter et al., 1982).

Glacier fluctuations in the Alaska Range and the Talkeetna Mountains were controlled by regional ice levels as well as by local climatic factors. The largest north-flowing outlet glacier of Late Wisconsin age followed the Delta Valley through the Alaska Range and formed the large Donnelly moraine; a smaller outlet glacier in the Nenana Valley did not extend entirely through the range and may have been blocked by the vigorous Yanert Fork glacier during much of the Late Wisconsin (R. M. Thorson, unpublished data). This glacier read-

Table 2-3.
Late-Pleistocene and Early-Holocene Radiocarbon Dates, Cook Inlet Region

Locality	Date (yr B.P.) (and Lab Number)	Location, Material, and Stratigraphic Position	Reference
24	9300 ± 250 (W-536)	Fire Island; base of peat overlying clayey sediments and till	Karlstrom, 1964: 59
23	9350 ± 320 (I-AGS-3)	Afognak Island; base of bog organic sediments about 6 m above sea level	Karlstrom, 1964: 59
22	9440 ± 350 (I-AGS-4)	Perry Island; base of peat at 6 m above sea level	Karlstrom, 1964: 59
21	9600 ± 650 (L-137L)	Ninilchik; wood from basal peat fill in abandoned Naptowne-age channel	Karlstrom, 1964: 59
20	9870 ± 250 (W-336)	Matanuska Valley; twigs from base of bog peat above drift of late-Naptowne age	Karlstrom, 1964: 59
19	9200 ± 600 (L-163D) 9500 ± 650 (L-137C) 10,370 ± 350 (W-474)	Point Possession; W-474 is wood from organic silt above clayey sediments of Naptowne age; L-163D and L-137C are from woody peat lenses in comparable stratigraphic positions	Karlstrom, 1964: 59
18	10,380 ± 80 (USGS-317)	Hidden Lake; basal organic matter in core	M. J. Rymer and J. D. Sims, personal communication, 1981
17	11,930 ± 250 (W-360)	Willow Creek; base of peat from bog between Eklutna-age lateral moraines	
16	13,500 ± 400 (W-748)	Salamatof Creek; peat from bog filling channel cut through Naptowne lake sediments into Knik drift	Karlstrom, 1964: 59
15	11,690 ± 300 (W-2375) 13,750 ± 500 (W-2389)	Elmendorf Moraine; W-2375 is peat from colluvium associated with Elmendorf Moraine; W-2389 is marine shells from macrofossil zone of Bootlegger Cove Clay	Schmoll et al., 1972
14	14,300 ± 350 (W-2367)	Point McKenzie; marine shells from macrofossil zone of Bootlegger Cove Clay	Schmoll et al., 1972
13	11,600 ± 300 (W-540) 13,690 ± 400 (W-2151) 14,900 ± 350 (W-2369)	Point Woronzof; W-540 is peat near top of sand unit correlative to Elmendorf Moraine of Naptowne age; W-2151 and W-2369 are on marine shells from macrofossil zone of Bootlegger Cove Clay	Schmoll et al., 1972
12	30,700 ± 260 (DIC-1859)	Susitna Canyon; wood from oxidized fluvial gravels below drift of last glaciation	R. M. Thorson, unpublished field data
11	34,000 ± 2000 (W-1804)	Potter Hill (Anchorage); wood from organic zone overlain by silt, fluvial sand and gravel, and diamicton of last glaciation	R 11, p. 221
10	> 24,000 (L-137D)	Kenai; retransported log in basal Naptowne sediments	Karlstrom 1964: 59
9	8650 ± 450 (L-163B) > 25,000 (L-117M)	Boulder Point; L-163B is from bog on Naptowne Drift; L-117M is from underlying drift mapped as Eklutna in age	Karlstrom 1964: 59
8	> 30,000 (AU-36)	Kenai River; wash sample from unlogged strata at 30.8-32.0 m depth	R 18, p. 10
7	> 32,000 (W-76)	Tustumena Lake; twigs and wood from basal sands of presumed Knik age	Karlstrom 1964: 59
6	> 37,000 (W-294)	Salamatof Creek; wood from contorted organic silt deformed by either Knik or Naptowne ice advance	Karlstrom 1964: 59
5	> 38,000 (W-535)	Eagle River; peat from bog underlying advance outwash and till interpreted to be Naptowne in age and overlying till and outwash of an earlier glaciation	Karlstrom 1964: 59
4	> 38,000 (W-1806)	Anchorage; wood and peat from organic zone overlying oxidized sand and underlying gravels interpreted to be Naptowne in age	R 11, p. 221
3	> 32,000 (W-77)	Goose Bay; W-77 and W-174 are wood from base and	Karlstrom 1964: 59

Table 2-3.
continued

Locality	Date (yr B.P.) (and Lab Number)	Location, Material, and Stratigraphic Position	Reference
	> 38,000 (W-174) > 40,000 (W-644)	top, respectively, of lignitized peat layer overlying outwash of Knik (?) age and underlying gravel and till of Naptowne (?) age; W-644 is transported wood from base of advance outwash interpreted to be of Naptowne age	
2	12,900 ± 300 (W-416) 39,000 ± 2,000 (L-163A) > 44,000 (L-117-L)	East Foreland; W-416 is peat from colluvium deformed by advancing Naptowne glacier; L-117L is lignitized, transported log from base of bog overlying Knik Drift; L-163A is log in Naptowne advance outwash	Karlstrom 1964: 59
1	> 45,000 (W-2154)	Anchorage (Birchwood School); peat below gravels interpreted to be of Naptowne age and above silty clay	R 12, p. 328

Note: Localities are shown by number in Figure 2-8.

Abbreviations: R 1, R 2, etc. = volume 1, volume 2, etc., of *American Journal of Science Radiocarbon Supplement*.

vanced about 13,500 years ago (N. D. Ten Brink, personal communication, 1980), either synchronously with or slightly before the formation of the Elmendorf Moraine at Anchorage. Glaciers extended into the Mentasta Pass area but may have periodically released meltwater from the Copper River basin. We would expect that there would have been outburst floods of the type described by Post and Mayo (1971), but there seems to be little evidence for such floods in Mentasta Pass or along the Tok River (H. R. Schmoll and L. D. Carter, personal communication, 1981). Many parts of the eastern Talkeetna Mountains remained unglaciated during Late Wisconsin time, whereas valleys farther north transmitted glaciers that originated primarily in the Alaska Range and flowed southwest through the Talkeetna Mountains into the Susitna Valley. The southwestern part of the Talkeetna Mountains was closer to moisture sources in the North Pacific, and that area generated ice streams that entered the Susitna and Matanuska Valleys. Scattered radiocarbon dates from the Talkeetna region suggest a general history of Late Wisconsin glaciation similar to that of the Copper River basin. Middle Wisconsin interstadial dates are about 30,000 yr B.P. and older, Late Wisconsin outwash is bracketed by dates of about 29,500 and 11,500 yr B.P., and basal peat dates on deglaciated terrain are as old as 12,000 to 11,500 yr B.P.

The glacial history of the Cook Inlet region has been one of the more controversial problems in Alaskan geology (e.g. Miller and Dobrovolny, 1959; Karlstrom, 1964: 37-40; Schmoll et al., 1972). This glacial record is complicated by (1) extensive submergence during times of isostatic crustal depression; (2) additional sea-level changes resulting from late-Quaternary tectonism; (3) possible blockage of glaciers, meltwater, and maritime air masses by an ice cap near the mouth of the inlet; and (4) a seemingly erratic radiocarbon record that might have been affected by the redeposition of older organic detritus. Morphological evidence for interrelated ice heights and flow directions (R. M. Thorson, unpublished data), stratigraphic evidence for post-Sangamon glacier expansion across the Alaska Peninsula into the head of Bristol Bay (D. M. Hopkins, personal communication, 1981), and correlations with the Gulf of Alaska and Copper River basin records indicate to us that glacier ice filled Cook Inlet during Early Wisconsin time and that the extent of this glacier complex generally corresponds with the extent of drift assigned to the

Eklutna Glaciation by Karlstrom (1964: 54-55 and Plate 4) and other investigators. Stratigraphic relations (Miller and Dobrovolny, 1959), radiocarbon dates (Schmoll et al., 1972), and microfossil data (Péwé,

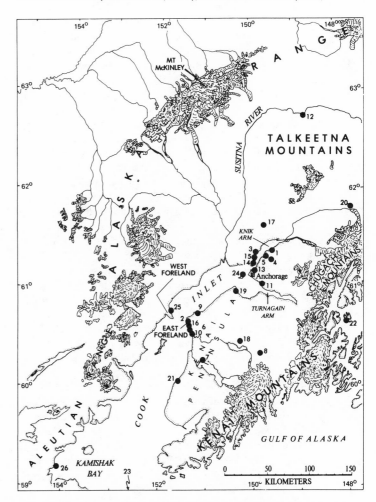

Figure 2-8. Cook Inlet-Susitna lowland region, including major geographic features and locations of radiocarbon samples.

1975: 73-74) have led us to conclude that drift assigned to the Knik and Naptowne Glaciations in the Anchorage area is largely of Late Wisconsin age and that these glacier fluctuations took place during a time of general marine submergence that we assume was mainly the result of isostatic crustal depression. Although Karlstrom (1964: 26-31) believes that a large proglacial lake filled upper Cook Inlet during Naptowne time, faunal analyses seem to show that the Bootlegger Cove Clay represents marine to estuarine conditions. Lacustrine shoreline features and deposits such as those around the Copper River basin also are rare or absent around Cook Inlet (D. M. Hopkins, personal communication, 1981).

Because radiocarbon dates (Table 2-3) provide no finite maximum age limit for the last glaciation of Cook Inlet, this ice advance could have been more extensive that what is shown in Figure 2-5, and it possibly began some time before Late Wisconsin time. We hypothesize that glaciers not only were more extensive than at present during the Middle Wisconsin interval but that they might have advanced into upper Cook Inlet and that later growth of glaciers at the mouth of the inlet possibly blocked moisture-bearing air masses and inhibited Late Wisconsin glacier advances farther to the northeast. Present evidence does not allow us to determine whether upper Cook Inlet was ever entirely deglaciated during the Wisconsin age and whether Late Wisconsin ice advances were more extensive than those of the Middle Wisconsin.

The late-glacial record from the Cook Inlet-Susitna lowland region is better dated and less ambiguous. Glaciers evidently retreated beyond the head of the inlet by about 15,000 yr B.P., and the well-dated macrofossil zone of the Bootlegger Cove Clay formed during the succeeding 1500 years (Schmoll et al., 1972). Deposition of the Elmendorf Moraine took place some time between 13,500 and 11,700 yr B.P., and dates of 13,500 and 12,900 yr B.P. on organic sediments overridden by glacier ice indicate that glaciers may have readvanced synchronously on the Kenai Peninsula and in the Nenana Valley. The destruction of the glacier at the mouth of Cook Inlet under conditions of moderating climate and rising eustatic sea level would have allowed moist air masses to penetrate inland and possibly caused these brief but widespread glacier readvances. The general absence of datable fossils in uplifted marine sediments between 13,500 and 10,000 years in age possibly reflects low salinity and turbid water conditions during the final readvance and subsequent rapid disintegration of the Late Wisconsin glaciers.

An ice cap may have been present at the mouth of Cook Inlet during much of Wisconsin time, as Karlstrom suggests (1964: 10), but little direct evidence exists for its extent, history, or flow patterns. The postulated ice cap evidently was localized by the gap in the coastal mountain chain between Kodiak Island and the Kenai Peninsula, and it must have been nourished by direct access to moisture-bearing winds from the Gulf of Alaska. We hypothesize that an ice cap like that envisioned by Karlstrom developed on the continental shelf only during Early Wisconsin time; it blocked the Cook Inlet ice stream and deflected much of it westward through the basin of Iliamna Lake into the head of Bristol Bay. The development of a similar ice cap later in Wisconsin time seems to be disproved by evidence for the eastward flow of glaciers in the Kamishak Bay area. Erratics evidently were transported to Augustine Volcano from source areas in the Aleutian Range (Johnston, 1979), and other ice-flow indicators such as grooves in bedrock indicate eastward movements into the bay during the last glaciation (R. L. Detterman, personal communication, 1981). End moraines around Iliamna Lake that are assigned to the Late Wisconsin by Detterman and Reed (1973: A54) contain only local rock types and lack erratics from source areas more distant than adjoining parts of the Aleutian Range (R. L. Detterman, personal communication, 1981). These lines of evidence suggest to us that a major ice stream, rather than an ice cap, was generated by exceptionally high amounts of precipitation opposite the gap in the outer coastal mountains. The ice stream was nourished also by tributary glaciers from Kodiak Island and the Kenai Peninsula; it flowed southeastward across the mouth of Cook Inlet and may have extended to the edge of the continental shelf. This ice stream might coincide in age with the Late Wisconsin glacial maximum, but we cannot disprove the alternative possibility that it developed in Middle Wisconsin time or that it represented a final stage in the destruction of a Late Wisconsin ice cap.

True ice caps evidently existed on the continental shelf southwest of Kodiak Island during the Late Wisconsin, although the time of their formation is uncertain. Outlet glaciers flowed northward through low passes on the Alaska Peninsula and spread as piedmont lobes along the coast of Bristol Bay (Péwé, 1975: 21-24). Other ice tongues extended into unglaciated parts of southwestern Kodiak Island (Karlstrom and Ball, 1969: 44-49). Farther to the southwest, glaciers originating south of present uplands flowed northward across Umnak Island and the southwestern tip of the Alaska Peninsula into the Bering Sea, and separate ice caps buried many other islands of the Aleutian chain (Black, 1976, 1980). Limiting dates on deglaciation include 10,625 ± 500 yr B.P. for the Cold Bay area (Funk, 1973) and 10,705 ± 295 yr B.P. for the disintegration of the Adak Island ice cap (Black, 1980). The general deglaciation of the Aleutian chain probably occurred between about 12,000 and 10,000 years ago (Black, 1980).

According to marine records, glaciers reached the edge of the continental shelf around virtually all of the Gulf of Alaska during Late Wisconsin time and glacier recession from the outer shelf may have commenced as recently as 12,000 years ago (B. F. Molnia, personal communication, 1981). Abundant evidence also seems to exist for glacier flow across islands that lie beyond the present outer coastline (e.g., Sirkin and Tuthill, 1971). These data seem to contradict earlier observations that, owing to extremely rapid Holocene uplift rates, glaciers through much of this region are nearly as extensive today as they were during the Late Wisconsin (Reid, 1970; summary in Péwé, 1975: 15-16). Most of the present coastline of the Gulf of Alaska was ice free and emergent by 10,500 to 9500 yr B.P. Older dates of 14,430 ± 160 and 10,730 ± 160 yr B.P. on marine clay have been published by Sirkin and Tuthill (1971), who infer that the Katalla Valley near Controller Bay (locality 34 in Figure 2-6) was deglaciated before 14,000 yr B.P. However, these authors also report a much younger date (6030 ± 150 yr B.P.) for sediments between the two older samples, and basal dates from geologically older surfaces in the same valley are 9180 ± 150 yr B.P. and younger. The two older age determinations must be considered questionable until confirmed by further dating.

Most of the late-Pleistocene Cordilleran ice sheet of southeastern Alaska originated as glaciers in the Alaska-Canada Boundary Ranges (Miller, 1976), which coalesced and flowed through now-submerged valleys and terminated on the continental shelf. The history of Late Wisconsin glacier expansion is not known, and stratigraphic records through most of this region are restricted to the time of the final fluctuations of retreating glaciers about 13,000 to 9500 yr B.P. Glaciomarine deposits began forming before about 13,000 yr B.P.

as glaciers retreated from the Juneau area, which was depressed isostatically at least 150 to 200 m below modern sea level at that time (Miller, 1973). Deglaciation of the Glacier Bay region began some time before 11,000 yr B.P., when peat formation began on emergent glaciomarine deposits (McKenzie and Goldthwait, 1971). The ice had retreated at least as far as its present limits through much of Glacier Bay by 10,400 ± 260 yr B.P. (Haselton, 1966: 28), and basal peats of comparable ages (10,300 ± 400 and 10,300 ± 600 yr B.P.) suggest that the lower parts of mountain valleys inland from Juneau also were free of ice by about the same time (Miller, 1973).

Note

1. Dates of >42,000 (W-3251) and >37,000 (W-3252) yr B.P., not shown in Table 2-3, were obtained by R.L. Detterman (personal communication, 1981) from West Foreland (25) and Kamishak Bay (26).

References

Benson, C. S., and Motyka, R. J. (1979). Glacier-volcano interactions on Mt. Wrangell, Alaska. *In* "Annual Report, 1977-78," pp. 1-25, Geophysical Institute, University of Alaska, Fairbanks.

Black, R. F. (1976). Geology of Umnak Island, eastern Aleutian Islands, as related to the Aleuts. *Arctic and Alpine Research* 8, 7-35.

Black, R. F. (1980). Isostatic, tectonic, and eustatic movements of sea level in the Aleutian Islands, Alaska. *In* (N. A. Morner, ed.), "Earth Rheology, Isostasy and Eustasy" pp. 231-48. John Wiley and Sons, New York.

Carlson, P. R., Burns, T. R., Molnia, B. F., and Hampton, J. C., Jr. (1977). Submarine valleys on the continental shelf, northeastern Gulf of Alaska. *Geological Society of America, Abstracts with Programs* 9, 921.

Clark, J. A., (1978). An inverse problem in glacial geology: The reconstruction of glacier thinning in Glacier Bay, Alaska, between A.D. 1910 and 1960 from relative sea-level data. *Journal of Glaciology* 18, 481-503.

Coulter, H. W., Hopkins, D. M., Karlstrom, T. N. V., Péwé, T. L., Wahrhaftig, C, and Williams, J. R. (1965). Map showing extent of glaciations in Alaska. U.S. Geological Survey Miscellaneous Geologic Investigations Map I-415.

Denton, G. H. (1974). Quaternary glaciations of the White River valley, Alaska, with a regional synthesis for the northern St. Elias Mountains, Alaska, and Yukon Territory. *Geological Society of America Bulletin* 85, 871-92.

Denton, G. H., and Armstrong, R. L. (1969). Miocene-Pliocene glaciations in southern Alaska. *American Journal of Science* 267, 1121-42.

Detterman, R. L., and Reed, B. L. (1973). "Surficial Deposits of the Iliamna Quadrangle, Alaska." U.S. Geological Survey Bulletin 1368-A, pp. A1-A64.

Ferrians, O. J. (1963). "Glaciolacustrine Diamicton Deposits in the Copper River Basin, Alaska." U.S. Geological Survey Professional Paper 475-C, pp. C121-C123.

Funk, J. M. (1973). Late Quaternary geology at Cold Bay, Alaska, and vicinity. M.S. thesis, University of Connecticut, Storrs.

Haselton, G. M. (1966). "Glacial Geology of Muir Inlet, Southeast Alaska." Ohio State University Institute of Polar Studies Report 18.

Heusser, C. J. (1960). "Late-Pleistocene Environments of North Pacific North America." American Geographical Society Special Publication 35.

Hudson, T., Plafker, G., and Rubin, M. (1976). "Uplift rates of Marine Terrace Sequences in the Gulf of Alaska." U.S. Geological Survey Circular 733, pp. 11-13.

Johnson, P. R., and Hartman, C. W. (1969). "Environmental Atlas of Alaska." University of Alaska Institute of Arctic Environmental Engineering, Fairbanks.

Johnston, D. A. (1979). "Onset of Volcanism at Augustine Volcano, Lower Cook Inlet." U.S. Geological Survey Circular 804-B, pp. B78-B80.

Kachadoorian, R., Ovenshine, A. T., and Bartsch-Winkler, S. (1977). "Late Wisconsin History of the South Shore of Turnagain Arm, Alaska." U.S. Geological Survey Circular 751-B, pp. B49-B50.

Karlstrom, T. N. V. (1964). "Quaternary Geology of the Kenai Lowland and Glacial

History of the Cook Inlet Region, Alaska." U.S. Geological Survey Professional Paper 443.

Karlstrom, T. N. V. (1976). Quaternary and upper Tertiary time-stratigraphy of the Colorado Plateaus, continental correlations, and some paleoclimatic implications. *In* "Quaternary Stratigraphy of North America" (W. C. Mahaney, ed.), pp. 275-82. Dowden, Hutchinson, and Ross, Stroudsburg, Pa.

Karlstrom, T. N. V., and Ball, G. E. (eds.) (1969). "The Kodiak Island Refugium: Its Geology, Flora, Fauna, and History." University of Alberta Boreal Institute, Edmonton.

Karlstrom, T. N. V., and others (1964). Surficial geology of Alaska. U.S. Geological Survey Miscellaneous Geologic Investigations Map I-357.

Kent, D., Opdyke, N. D., and Ewing, M. (1971). Climatic change in the North Pacific using ice-rafted detritus as a climatic indicator. *Geological Society of America Bulletin* 82, 2741-54.

McKenzie, G. D., and Goldthwait, R. P. (1971). Glacial history of the last eleven thousand years in Adams Inlet, southeastern Alaska. *Geological Society of America Bulletin* 82, 1767-82.

Miller, M. M. (1976). Quaternary erosional and stratigraphic sequences in the Alaska-Canada Boundary Range. *In* "Quaternary Stratigraphy of North America" (W. C. Mahaney, ed.), pp. 463-92. Dowden, Hutchinson, and Ross, Stroudsburg, Pa.

Miller, R. D. (1973). "Gastineau Channel Formation, a Composite Glaciomarine Deposit near Juneau, Alaska." U.S. Geological Survey Bulletin 1394-C, pp. C1-C20.

Miller, R. D. (1975). Surficial geologic map of the Juneau urban area and vicinity, Alaska. U.S. Geological Survey Miscellaneous Geologic Investigations Map I-885.

Miller, R. D., and Dobrovolny, E. (1959). "Surficial Geology of Anchorage and Vicinity, Alaska." U.S. Geological Survey Bulletin 1093.

Molnia, B. F., and Carlson, P. R. (1975). Surface sediment distribution, northern Gulf of Alaska. U.S. Geological Survey Open-File Map 75-505.

Molnia, B. F., Levy, W. P., and Carlson, P. R. (1980). Map showing Holocene sedimentation rates in the northeastern Gulf of Alaska. U.S. Geological Survey Miscellaneous Field Studies Map MF-1170.

Nelson, S. W., and Reed, B. L. (1978). Surficial deposits map of the Talkeetna quadrangle, Alaska. U.S. Geological Survey Miscellaneous Field Studies Map MF-870-J.

Nichols, D. R., and Yehle, L. A. (1969). Engineering geologic map of the southeastern Copper River basin, Alaska. U.S. Geological Survey Miscellaneous Geologic Investigations Map I-524.

Péwé, T. L. (1975). "Quaternary Geology of Alaska." U.S. Geological Survey Professional Paper 835.

Péwé, T. L. (1976). Late Cenozoic history of Alaska. *In* "Quaternary Stratigraphy of North America" (W. C. Mahaney, ed.), pp. 493-506. Dowden, Hutchinson, and Ross, Stroudsburg, Pa.

Péwé, T. L., Ferrains, O. J., Jr., Nichols, D. R., and Karlstrom, T. N. V., (1965). "Guidebook for Field Conference F, Central and South-Central Alaska." Seventh Congress of the International Association for Quaternary Research, Boulder, Colo.

Péwé, T. L., Hopkins, D. M., and Giddings, J. L. (1965). The Quaternary geology and archeology of Alaska, *In* "The Quaternary of the United States" (H. E. Wright, Jr., and D. G. Frey, eds.), pp. 355-374. Princeton University Press, Princeton, N.J.

Plafker, G. (1969). "Tectonics of the March 27, 1964, Alaska Earthquake." U.S. Geological Survey Professional Paper 543-I, pp. I1-I174.

Plafker, G. and Addicott, W. O. (1976). "Glaciomarine Deposits of Miocene through Holocene Age in the Yakataga Formation along the Gulf of Alaska Margin, Alaska." U.S. Geological Survey Open-File Report 76-84.

Plafker, G., Hudson, T., and Bruns, T. (1978). Late Quaternary offsets along the Fairweather fault and crustal plate interactions in southern Alaska. *Canadian Journal of Earth Sciences* 15, 805-16.

Porter, S. C., Pierce, K. L், and Hamilton, T. D. (1983). Late Wisconsin mountain glaciation in the western United States. *In* "Late-Quaternary Environments of the United States," vol. 1, "The Late Pleistocene" (S. C. Porter, ed.), pp. 71-111, University of Minnesota Press, Minneapolis.

Post, A., and Mayo, L. R. (1971). Glacier dammed lakes and outburst floods in Alaska. U.S. Geological Survey Hydrological Investigations Atlas HA-455 (3 sheets).

Reid, J. R. (1970). Late Wisconsin and Neoglacial history of the Martin River Glacier, Alaska. *Geological Society of America Bulletin* 81, 3593-3603.

Rubin, M., and Alexander, C. (1958). U.S. Geological Survey radiocarbon dates IV. *Science* 129, 1476-87.

Schmoll, H. R., Szabo, B. J., Rubin, M., and Dobrovolny, E. (1972). Radiometric dating of marine shells from the Bootlegger Cove Clay, Anchorage area, Alaska. *Geological Society of America Bulletin* 83, 1107-13.

Schweger, C. E. (in press). Chronology of late glacial events from the Tangle Lakes, Alaska Range, Alaska. *Arctic Anthropology*.

Sirkin, L. A. and Tuthill, S. (1971). Late Pleistocene palynology and stratigraphy of Controller Bay region, Gulf of Alaska. *In* "Etudes sur le Quaternaire dans le monde" (M. Ters, ed.), pp. 197-208. Eighth Congress of the International Association for Quaternary Research, Paris, 1969.

Thorson, R. M., Dixon, E. J., Jr., Smith, G. S., and Batten, A. R. (1981). Interstadial Proboscidean from south-central Alaska: Implications for biogeography, geology, and archeology. *Quaternary Research* 16, 404-17.

von Huene, R., Crouch, J., and Larson, E. (1976). Glacial advance in the Gulf of Alaska implied by ice-rafted material. *In* "Investigations of Late Quaternary Paleoceanography and Paleoclimatology" (R. M. Cline and J. D. Hayes, eds.), pp. 411-22. Geological Society of America Memoir.

von Huene, R., Shor, G. G., Jr., and Reimnitz, E. (1967). Geological interpretation of seismic profiles in Prince William Sound, Alaska. *Geological Society of America Bulletin* 78, 259-68.

Wahrhaftig, C. (1965). "Physiographic Divisions of Alaska." U.S. Geologic Survey Professional Paper 482.

Weber, F. R., Hamilton, T. D., Hopkins, D. M., Repenning, C. A., and Haas, H. (1981). Canyon Creek: A late Pleistocene vertebrate locality in interior Alaska. *Quaternary Research* 16, 167-80.

Williams, J. R., and Coulter, H. W. (1981). "Deglaciation and Sea-Level Fluctuations in Port Valdez, Alaska." U.S. Geological Survey Circular 823-B, pp. B78-B80.

Williams, J. R., and Johnson, K. M. (compilers) (1980). Map and description of late Tertiary and Quaternary deposits, Valdez quadrangle, Alaska. U.S. Geological Survey Open-File Report 80-892-C (2 sheets).

CHAPTER 3

The Cordilleran Ice Sheet
in Washington, Idaho, and Montana

Richard B. Waitt, Jr., and Robert M. Thorson

Introduction

During the Fraser (Late Wisconsin) Glaciation, the Cordilleran ice sheet advanced southward from source areas in British Columbia and terminated in the United States between the Pacific Ocean and the Continental Divide (Figure 3-1). The ice sheet extended farthest along major south-trending valleys and lowlands that traverse the international boundary; it formed several composite lobes segregated by highlands and mountain ranges. Each lobe dammed sizable lakes that drained generally southward or westward along ice margins and across divides.

In 1965 the entire Pleistocene stratigraphy of the Cordilleran ice sheet in and west of the Cascade Range was summarized by Crandell (1965) and that east of the Cascade Range by Richmond and others (1965). Our subject is a more detailed summary of the Fraser Glaciation, and this chapter focuses on concepts developed since 1965. To resolve some apparent discrepancies in the published literature, the senior author in 1981 investigated in reconnaissance the principal valleys between the Cascade Range and the Continental Divide. Field evidence warrants considerable revision of the ice margin in northern Idaho and northeastern Washington (R. B. Waitt, Jr., et al., unpublished report).

LANDFORMS, WEATHERING, AND SOILS

The deposits of the Fraser Glaciation are recognized by moraines and outwash bodies whose morphology is only slightly eroded. Bleached rinds on fine-grained surface stones are less than 1 mm thick; iron oxide rinds on medium-grained crystalline rocks are less than 1 cm thick. Few stones are deeply pitted or grussified, although the exteriors of some biotitic crystalline clasts have disintegrated. Soils have brown nonargillic B-horizons or nonplatey calcic horizons less than 75 cm deep (Flint, 1935, 1937; Alden, 1953; Crandell, 1963; Richmond et al., 1965; Waitt, 1972, and unpublished report; Baker, 1973, 1978b; Porter, 1976; Colman and Pierce, 1981, and personal communication, 1981; Parsons et al., 1981.)

CORRELATION, CHRONOLOGY, AND NOMENCLATURE

During the Fraser Glaciation, alpine glaciers in western Washington and southeastern British Columbia reached maximum positions between 22,000 and 18,000 yr B.P. and had greatly diminished before Cordilleran ice occupied the Puget lowland between 17,000 and 13,000 yr B.P. (Mackin, 1941; Crandell, 1963; Armstrong et al., 1965; Mullineaux et al., 1965; Halstead, 1968; Heusser, 1974; Hibbert, 1979; Clague, 1980; Clague et al., 1980; Barnosky, 1981). Just north of the international boundary east of the Cascade Range, limiting ages on ice-sheet glaciation (about 17,500 yr B.P.) and on deglaciation (about 11,000 yr B.P.) are broadly similar to limiting ages on the western lobes of Cordilleran ice near the international boundary (Fulton and Smith, 1978; Clague, 1980; Clague et al., 1980). The invasion and maximal stand of ice-sheet lobes east of the Cascade Range therefore were broadly synchronous with the ice-sheet glaciation of the Puget lowland.

We do not subdivide the Fraser Glaciation, and we suggest finer correlation only where direct evidence is available. Richmond and others (1965) correlated their subdivisions of Cordilleran ice-sheet drift east of the Cascade Range with subdivisions of Laurentide ice-sheet drift in Montana and with the type Rocky Mountain alpine-glacial sequence in Wyoming. The alpine-glacial terms *early, middle,* and *late* stages of the Pinedale Glaciation were applied to specific deposits of the Cordilleran ice sheet and the Lake Missoula floods. But successive belts of ice-sheet moraines indicate only that there were successive stillstands of the ice terminus: these moraines may not correlate with the three classic stages of alpine glaciation in the Rocky Mountains. Recent evidence casts doubt on several of the earlier correlations.

The subdivision of the Fraser Glaciation in the Puget lowland and the Pinedale Glaciation (as applied to the ice sheet east of the Cascade Range into stades and interstades (Armstrong et al., 1965; Richmond et al., 1965) implies that stratigraphic boundaries were caused by climatic fluctuations (American Commission on Stratigraphic Nomenclature, 1970, Articles 39, 40). But some stillstands or rapid

We thank individuals who by sharing some of their unpublished findings have helped this chapter to be as up-to-date as possible. In addition to the references cited, several abstracts, guidebooks, open-file reports, theses, and unpublished reports and maps also provided information. Additional references, especially to geologic maps, can be found in Thorson (1981) and Waitt (1982a). Most fieldwork in the region by both authors since 1975 has been done for the U.S. Geological Survey. The manuscript has profited from reviews by B. F. Atwater, Fred Pessl, Jr., and S. C. Porter.

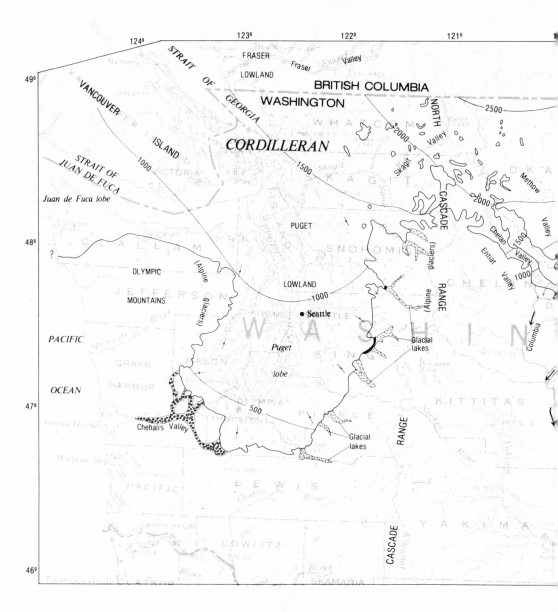

EXPLANATION

————— Inferred contour (m) on ice sheet surface

—·— Orientation of drumlin

 Large moraine

 Delta moraine

 Flowpath of Lake Missoula floods (dotted beneath ice dam)

 Outlet of lake

Figure 3-1. Generalized map of maximum extent of the Cordilleran ice sheet about 15,000 ± 1000 yr B.P. Ice-surface contours are in meters. Lakes shown at their inferred maximum limits; arms adjacent to glaciers are partly filled with outwash. Heavy arrows denote various paths of Missoula floodwater (dotted where subglacial); northwest segment of Columbia valley and Moses Coulee operated only prior to maximum icesheet stand. (Data from Alden, 1953; Crandell, 1963, 1965; Easterbrook, 1963, 1979; Richmond et al., 1965; Mullineaux, 1970; Waitt, 1972, and unpublished maps; Heusser, 1973; Alley and Chatwin, 1979; Thorson, 1980, 1981; and observations by authors).

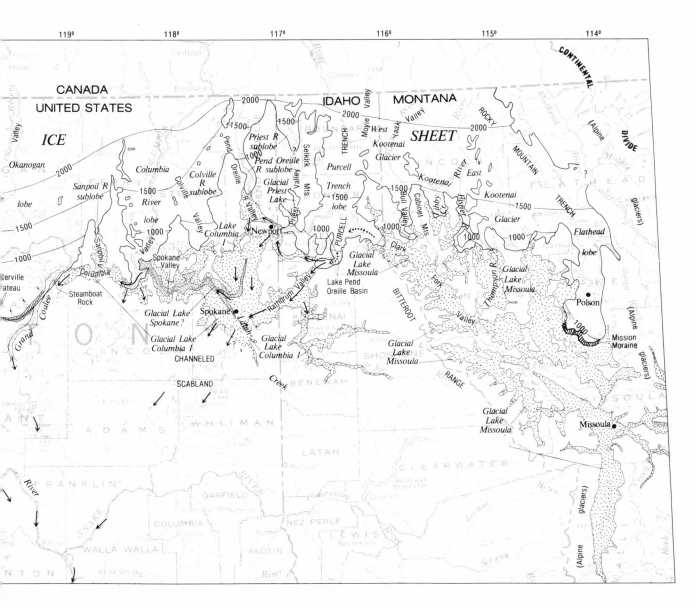

retreats of the Cordilleran ice sheet may have had nonclimatic causes. Calving into seawater influenced the Juan de Fuca and Puget lobes, and topography probably influenced advances and retreats east of the Cascade Range. It is now clear that lobes of the Cordilleran ice sheet did not all fluctuate exactly in phase with each other or with small alpine glaciers. We therefore neither apply alpine-glacial nomenclature to ice-sheet deposits nor apply stades defined in the Puget lowland to other parts of the region.

PRE-FRASER GLACIATIONS

Pre-Fraser glaciations influenced the region. Stratigraphy indicates at least three pre-Fraser ice-sheet glaciations in the Puget lowland (Crandell et al., 1958; Armstrong et al., 1965; Easterbrook, 1976) and at least two glaciations east of the Cascade Range in British Columbia (Fulton and Smith, 1978). Stratigraphy of loess and paleosols in eastern Washington suggests episodes of glaciation preceding the last

interglaciation (Richmond et al., 1965). Caliche-capped catastrophic-flood deposits indicate that at least one pre-Fraser ice sheet dammed glacial Lake Missoula (Bretz et al., 1956; Bretz, 1969; Baker, 1973, 1978b; Patton and Baker, 1978; Waitt, 1982b). The mountains of the Pacific Northwest south of the ice-sheet margin display nested moraines that reveal as many as two pre-Fraser alpine glaciations, each followed by a long nonglacial interval (Page, 1939; Alden, 1953; Richmond, 1960; Crandell, 1967; Schmidt and Mackin, 1970; Weber, 1971; Crandell and Miller, 1974; Porter, 1976; Waitt, 1979a, 1982b). South of the limit of the Fraser-age Cordilleran and Laurentide ice sheets in Montana, stratigraphic sections and eroded moraines give evidence of early ice-sheet glaciations (Alden, 1932, 1953; Richmond, 1960). Pre-Fraser ice-sheet glaciations thus clearly influenced topographic development in the region. Some previously inferred pre-Fraser drift, however, is of Fraser age, and some such ''drift'' in northern Idaho and northeastern Washington is the deposit of catastrophic floods.

Fraser Glaciation: Ice-Sheet Advance
and Maximum Stand

PUGET AND JUAN DE FUCA LOBES

Chronology

The Fraser Glaciation in western Washington consisted of an alpine phase followed in the lowlands by an advance and retreat of an ice-sheet lobe, a glaciomarine phase, and an ice-sheet readvance. Before 18,000 yr B.P. and before ice-sheet glaciation of the lowlands, mountain glaciers in the western Cascade Range, Olympic Mountains, and Vancouver Island advanced to valley mouths, built moraines, and retreated far into the mountains (Mackin, 1941; Crandell, 1963; Halstead, 1968; Carson, 1970; Williams, 1971; Heusser, 1974; Heller, 1980). Cordilleran ice advanced southward into the northern Puget lowland and across southeastern Vancouver Island about 18,000 yr B.P. (Fulton, 1971; Alley and Chatwin, 1979; Clague et al., 1980). Bifurcating around the Olympic Mountains, the ice sheet advanced to the western Strait of Juan de Fuca by about 17,000 yr B.P. and to the Seattle area after 15,000 yr B.P. (Anderson, 1968; Mullineaux et al., 1965).[1]

Glacial Dynamics and Deposits

The Puget lobe, a piedmont glacier that probably had a broadly arcuate front, advanced and blocked northward drainage, forming proglacial lakes that drained southward to the Chehalis River valley and Pacific Ocean. During its advance the Juan de Fuca lobe terminated in the Strait of Juan de Fuca and probably calved into seawater. The ablation rate probably was much greater there than along the grounded front of the Puget lobe. Consequently, much of the ice entering the northern Puget lowland from Canada may have been drawn westward toward the rapidly ablating sector.

The Juan de Fuca lobe did not advance rapidly until after about 18,000 yr B.P. sea level had dropped eustatically to low levels (MacIntyre et al., 1978), although this sequence may be coincidental. The Puget lobe did not advance rapidly southward until the Juan de Fuca lobe had extended far enough west to minimize the difference in ablation rates between the lobes. Whereas the Juan de Fuca lobe may have reached the western end of the strait as early as about 17,000 yr B.P., the Puget lobe did not reach Seattle until after 15,000 yr B.P. (Figure 3-2). The Juan de Fuca lobe had reached its outer limit and had begun to retreat before 14,500 yr B.P. (Heusser, 1973); the Puget lobe did not reach its limit until 14,500 to 14,000 yr B.P. (Porter, 1970; Hibbert, 1979) and did not linger to build a voluminous moraine (Figure 3-2).

We have reconstructed the Puget lobe from data on its maximum horizontal extent, on its basal flow lines, and on its altitude against bordering mountain ranges. The maximum horizontal extent is known from decades of field mapping; basal flow lines are inferred from the orientation of drumlinoid landforms (Thorson, 1981, and references therein); the ice-surface altitude is approximated by contouring the upper limits of Fraser-age drift at the bordering highlands (Figure 3-1). At its maximum extent the surface of the lobe descended from altitude 1900 m at the international boundary to 1100 to 1200 m at the northeastern corner of the Olympic Mountains, to about 200 to 300 m at its terminus (Easterbrook, 1963; Thorson, 1980).[2] Drumlinoid landforms indicate that flow along the southern and eastern margins was generally southward but divergent; flow along

the western margin was parallel to the steep east front of the Olympic Mountains. In valleys of the western Cascade Range and eastern Olympic Mountains, the Puget lobe dammed lakes that overflowed southward through ice-marginal channels cut across successive spurs.

Basal flow lines in the eastern part of the topographically confined Juan de Fuca lobe were parallel to the steep north front of the Olympic Mountains. The upper limit of the lobe descended from 1100 or 1200 m at its effluence from the Puget lobe to sea level near its terminus (Heusser, 1973). Alley and Chatwin (1979) indicate that during its advance and retreat the Juan de Fuca lobe was a west-flowing ice stream, but that at its maximum it was continuous with valley glaciers from southern Vancouver Island and with ice flowing southward across Vancouver Island from the Strait of Georgia. These glaciers composed an ice sheet that sloped southward or southwestward from the Coast Mountains in British Columbia to the Olympic Mountains in Washington. The Juan de Fuca lobe extended onto the continental shelf and formed part of a huge tidewater glacier complex along the west coast of Vancouver Island (Alley and Chatwin, 1979).

NORTH-CENTRAL WASHINGTON

Northeastern North Cascade Range

During its maximum stand the Cordilleran ice sheet buried much of the northeastern North Cascade Range (Barksdale, 1941; Waitt, 1972, 1977, and unpublished report). It crossed a high-relief landscape and converged into thick ice streams along the Skagit, Chelan, Methow, Okanogan, and Columbia valleys. As reconstructed from the upper limit of ice-beveled divides and erratics, the ice surface descended from above altitude 2600 m at the international boundary to 250 m in the Columbia valley (Waitt, 1972, and unpublished report). The upper limit of the ice-marginal coulees, small moraines, and terraces that descend along the lower Methow valley and Columbia valley has been traced to the head of the downvalley-sloping segment of the "great" terrace, which apparently is the outwash train built during the maximum stand of the ice sheet (Russell, 1900; Waters, 1933; Waitt, 1980b, and unpublished report and maps). The ice-sheet drift limit *descending* the Chelan valley merges downvalley with the drift limit *ascending* the lowermost Chelan valley from the

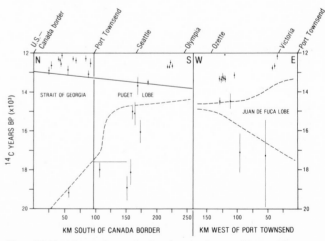

Figure 3-2. Time-distance diagram showing advance and retreat of Puget and Juan de Fuca lobes. Error bars on radiocarbon dates represent ± 1σ.

Columbia valley. High mountains on the southwest side of the Chelan drainage basin prevented ice from spilling farther southward into the Entiat drainage basin.

The ice-sheet margin as compiled by Crandell (1965: Figure 2) is now extended to include much of the Skagit-Chelan and Skagit-Methow divides and the Chelan drainage basin (Figure 3-1). The ice surface at the international boundary was approximately 700 m higher east of the Cascade crest than in the Puget lowland because the rugged, high-altitude alpine topography along the structural crest of the range restrained the westward movement of ice.

Ice Limits North of Columbia River

Ice limits in the highlands east of the Methow-Okanogan divide (Figure 3-1) have been examined only in reconnaissance (Flint, 1935). The terminus of the Sanpoil River sublobe, an eastern appendage of the Okanogan lobe, is shown (Figure 3-1) where placed by Pardee (1918) and Richmond and others (1965). Flint (1936, 1937) and Flint and Irwin (1939) had suggested that it terminated 10 to 30 km farther south, but such a limit is inconsistent with the absence of erratics and abundance of grussified granite in the lower Sanpoil valley (B. F. Atwater, unpublished data, 1980).

Okanogan Lobe on the Waterville Plateau

A thick ice stream that flowed down the Okanogan valley crossed the Columbia River valley and extended as a broad lobe across the Waterville Plateau (Figure 3-1). The drift border (Salisbury, 1901; Bretz, 1923, 1928; Flint, 1937; Hanson, 1970; Easterbrook, 1979) is marked by a large end moraine, which suggests that the lobe remained at its terminal position for a long time. Westward the moraine divides into smaller moraines, but the abrupt drift limit can readily be traced to the east side of the Columbia valley (Waters, 1933; Flint, 1935), where it descends to the "great" terrace (R. B. Waitt, Jr., unpublished maps). Eastward the moraine broadens toward the Grand Coulee, where it was truncated by catastrophic floods. Drumlinoid forms trend roughly perpendicular to the arcuate margin of the lobe (Easterbrook, 1979), a pattern showing that at its maximum stand the flow of the glacier was divergent.

As the Okanogan lobe advanced, it diverted the Columbia River into successively more southern valleys (Flint, 1935). During the maximum stand of the ice sheet, the Columbia River spilled along the course of the Grand Coulee, but the 250-m depth of the tandem gorges composing the coulee was mostly excavated by recession of enormous cataracts during catastrophic floods (Bretz, 1923, 1932, 1959, 1969; Bretz et al., 1956). Because the Fraser-age terminal moraine crosses the floor of Moses Coulee—also a great recessional gorge carved by catastrophic floods—that coulee could predate the Fraser Glaciation (Bretz, 1930, 1969). But the similarity of slide-rock accumulation and the degree of weathering in the coulee to those in the Grand Coulee indicates that Moses Coulee was last swept clean (or cut?) during the advance of the Fraser-age Okanogan lobe before it had reached its terminal position (Hanson, 1970).

Glacial Lake Columbia and the Grand Coulee

The Okanogan lobe dammed glacial Lake Columbia in the Columbia valley east of the Grand Coulee intake (Figure 3-1, Table 3-1) (Bretz, 1923, 1932; Flint, 1935, 1936; Flint and Irwin, 1939). Before the upper Grand Coulee flood-cataract alcove had retreated to the

Columbia valley and while the Okanogan lobe at its near-maximum position blocked the completed coulee, the lake spillway was as much as 250 m higher (Lake Columbia I) than the 470-m threshold altitude of the late-glacial and present coulee (Lake Columbia II). This development could have occurred either in pre-Fraser or Fraser time. Sparsely weathered drift atop Steamboat Rock (a midcoulee butte) and fresh scabland on the unglaciated eastern upland at the head of the coulee indicate that during the Fraser Glaciation the Okanogan lobe dammed a lake at least briefly at the level of Lake Columbia I. Lake Columbia I extended eastward to the Spokane area (Figure 3-1), where clayey lake sediment is intercalated with Fraser-age bed-load flood sediment.

Table 3-1.
Fraser-Age Glacial Lakes in the Columbia Drainage Basin

Glacial Lake	Approximate Maximum Altitude (m)	Maximum Depth (m)	Damming Ice Lobe (or Drift)	Spillway
Missoula	1260	670	Purcell	None; drained beneath ice dam
Columbia I	730	455	Okanogan	Upper scabland of Grand Coulee
Priest	710	100	Purcell (and drift)	Around ice dam and moraine
Clark	645	150	Pend Oreille	Scotia channel
Kootenay	645	120	Purcell	Lake Clark
Spokane (?)	640	180	Columbia	Beneath ice dam (?)
Columbia II	515	240	Okanogan	Grand Coulee
Brewster	400	220	(Drift)	Knapp Coulee

Lacustrine sediment underlying the principal till layer in the Columbia valley far north of the Fraser-age drift border is evidence of the early phase of Lake Columbia that predated the glacial maximum (Flint and Irwin, 1939; Jones et al., 1961: 27-29, Plates 4, 5). Regular thin bedding and clay beds, which do not characterize slack-water sediment of catastrophic floods (Waitt, 1980a), indicate a long-lived lake dammed by ice; overlying till indicates that the glacier later advanced across the lake basin.

NORTHEASTERN WASHINGTON

The Columbia River and its tributaries east of the Grand Coulee, an area mostly known only in reconnaissance, is stratigraphically complex because several ice lobes built outwash trains that aggraded into glacial lakes and because parts of the area were repeatedly swept by Lake Missoula floods. The maximum ice limits suggested by Richmond and others (1965) have been considerably modified (Figure 3-1) (R. B. Waitt, Jr., et al., unpublished report).

Columbia River and Colville River Lobes

The drift limit of the Columbia River lobe is sharply defined by a terminal moraine just south of the confluence of the Columbia and Spokane Rivers (R. B. Waitt, Jr., et al., unpublished report). Richmond and others (1965) placed the pre-Fraser ice limit south of this point and the Fraser ice limit north of it; but the freshness of drift and thin, brown, nonargillic soil of the moraine reveal its Fraser age. Glacial Lake Spokane, if defined as the lake dammed by the Columbia River lobe or Colville River sublobe (Richmond et al., 1965), indeed could have been ponded by the Columbia River lobe, but to no higher than about 640 m. The effects of such a lake near Spokane are indistinguishable from those of glacial Lake Columbia I (730 m), which flooded the Spokane area deeper and much wider than could Lake Spokane. Most deposits historically referred to glacial Lake Latah or Spokane (Bretz, 1923; Large, 1924; Anderson, 1927; Richmond et al., 1965) probably accumulated in the eastern part of Lake Columbia I. On the other hand, a thick deposit of lacustrine clay and silt that occurs as high as altitude 770 m north of the Spokane valley—Flint's (1936) Lake Clayton—is deeply oxidized and argillized: it relates to some pre-Fraser Pleistocene ponding, not to glacial Lake Spokane or Lake Columbia I.

The outermost drift of the Colville River lobe lies far north of the Spokane River (Weis and Richmond, 1965; Richmond et al., 1965), not south of it as inferred by Bretz (1928), Flint (1937), and Bretz and others (1956). The outer Fraser-age glacial limit in the Colville valley (Figure 3-1) is marked by the head of a thick outwash body and by the southern limit of a conspicuous kame-kettle moraine mapped by Flint (1936) but interpreted by him as marking a recessional position. A thin, discontinuous, high-level (760 m altitude) lake fill well beyond the glacial limit (Lake Wellpinit of Flint, 1936) probably accumulated in Lake Columbia I.

Gravel accumulations along the Rathdrum valley and the Little Spokane valley at altitude 610 to 630 m, which Flint (1936) inferred to be "early" outwash dammed by the Columbia lobe and Colville sublobe, are mostly gravel bars deposited by the Missoula floods that shed into glacial Lake Columbia I (or Lake Spokane?) (R. B. Waitt, Jr., et al., unpublished report).

Pend Oreille River Sublobe

We draw the maximum extent of the Pend Oreille River sublobe[3] far north of the limits drawn by Anderson (1927), Flint (1936), and Richmond and others (1965). Ice-marginal channels that descend southward along both sides of the Pend Oreille River valley are associated with till and stratified drift. This drift is capped by thin, brown, nonargillic soils that indicate its Fraser age. There is no evidence of Fraser-age till or glacial topography beyond this limit (R. B. Waitt, Jr., et al., unpublished report): apparently the Cordilleran ice sheet occupied only the northern part of the Pend Oreille River valley. The Pend Oreille River sublobe was less extensive than the Columbia River lobe and the Colville River sublobe because the entry of ice to the Pend Oreille River valley was restricted by the high-relief, high-altitude topography in the northern part of the drainage basin.

Most coarse drift south of the Pend Oreille River consists of deposits of catastrophic floods from glacial Lake Missoula (E. P. Kiver and D. F. Stradling, personal communication, 1981; R. B. Waitt, Jr., et al., unpublished report). From the discharge site at the south end of present Lake Pend Oreille, the floodwater swept not only southwestward along the Rathdrum valley, but also northwestward into and down the Pend Oreille River valley, which remained dammed to the north by the Cordilleran ice. Between Newport and the elbow of the Pend Oreille River valley, the floodwater broadly overflowed the valley divide southward, carving a sparse scabland, coulees, and cataract alcoves and depositing great bars of gravel and

sand. But a morainelike form surrounding present Calispell Lake (Salisbury, 1901) is a glacial deposit. This moraine, designated by Richmond and others (1965) as "late Pinedale" in age is considerably eroded and veneered by deposits of Fraser-age catastrophic floods. But the high degree of weathering and cementation and the thick, reddish, argillic soil on the till and associated lacustrine deposits indicate that this moraine long predates the Fraser Glaciation.

The unglaciated upper segment of the Pend Oreille River valley probably was occupied by an arm of glacial Lake Columbia I early during the glacial maximum. The Pend Oreille River sublobe dammed the valley to the north. But a terrace at about altitude 760 to 795 m, although inferred by Park and Cannon (1943) to be lacustrine in origin, instead consists of recessional outwash to the north and catastrophic-flood deposits to the south (R. B. Waitt, Jr., et al., unpublished report). Although there are exposures of lacustrine silt above altitude 645 m, they could have accumulated in Lake Columbia I. The hypothetical "Lake Clark I" of Waitt (1980a: Table 3) is thus disregarded; the demonstrable Fraser-age Lake Clark, which persisted during the glacial advance, maximum stand, and retreat, is discussed in the section on deglaciation.

IDAHO AND MONTANA

Purcell Trench Lobe

Maximal Stand

The Purcell Trench lobe dammed the Clark Fork River against the north end of the Bitterroot Range (Figure 3-1) (Pardee, 1942; Alden, 1953; Bretz et al., 1956). The extent and age of the southern distributary of the lobe along Lake Pend Oreille and along the Rathdrum valley have been disputed. Bretz (1923, 1928, 1959), Flint (1936, 1937), Alden (1953), and Bretz and others (1956) inferred that gravel south of Spokane is a glacial deposit, but Weis and Richmond (1965) inferred that this material is a Missoula-flood deposit and consequently that the ice limit was east of Spokane, approximately as drawn by Anderson (1927). Anderson (1927), Alden (1953), Richmond and others (1965), Savage (1967), and Weis (1968) assigned the outer 40 km of this "drift" to pre-Fraser glaciations. Several reconnaissance field investigations since 1968, however, each independently inferred that some of the deposits blocking reentrants along the Rathdrum valley are flood deposits (P. L. Weis; K. L. Pierce and S. M. Colman; R. B. Waitt, Jr.; R. B. Parsons and others; E. P. Kiver and D. F. Stradling; all personal communications, 1981). Shallow soils on these deposits lacking argillic B-horizons are similar to soils on Fraser-age Missoula-flood deposits at the head of the Channeled Scabland (Baker, 1973, 1978b; K. L. Pierce, personal communication, 1981) and throughout a large area to the north (Parsons et al., 1981; R. B. Waitt, Jr., et al., unpublished report. Enigmatic, apparently Fraser-age "drift" and erratics in the northern part of the Coeur d'Alene Lake basin probably are bed-load and ice-rafted flood deposits; they do not require damming by the Purcell Trench lobe as suggested by Dort (1962).

We therefore have considerably modified the limit of Fraser-age ice suggested by Richmond and others (1965) (Figure 3-1); some gravel to the south is bedload carried by great Fraser-age floods from glacial Lake Missoula. Like the Okanogan lobe, the Purcell Trench lobe may have achieved its maximum extent during the Fraser Glaciation. South of the Pend Oreille River, the upper limits of gravel and sand and of inconspicuous scabland show that the maximal flood submerged most of that area. Between Mount Spokane on the east and highlands on the west, the surface of the flood sloped steeply southward and gently westward and merged north of Spokane with the conspicuous effects of the floods that swept in more directly along the Rathdrum valley (R. B. Waitt, Jr., et al., unpublished report).

Priest River Valley and Glacial Priest Lake

The high Selkirk Mountains that surround the Priest River drainage basin greatly restricted the entry of Cordilleran ice into the valley from the north. The topography of the lower valley, moreover, is nonglacial. Evidently there was not a large "Priest River lobe" of Cordilleran ice as formerly inferred. Erosional topography and reports of high-level erratics on the northwest (Park and Cannon, 1943), however, suggest that Cordilleran ice occupied the headward parts of the Priest River drainage basin. We refer to this appendage of the Columbia River and Purcell Trench lobes as the "Priest River sublobe" (Figure 3-1).

A moraine near the mouth of the Priest River valley is not an end moraine from upvalley ice as inferred by Anderson (1927), Alden (1953), Richmond and others (1965), and Savage (1967). The alien provenance of drift stones and the upvalley progression from massive cobble till to pebble-gravel outwash to bedded silt and clay indicate that the moraine was built by the arm of the Purcell Trench lobe that flowed westward along the upper Pend Oreille River valley and thereby dammed the mouth of the Priest River valley (R. B. Waitt, Jr., et al., unpublished report). The upvalley gradation from outwash to laminated silt and clay reveals that a glacial lake, here named glacial Priest Lake, was dammed by the Purcell Trench lobe at the valley mouth—much as lakes were dammed in the western Cascade Range during the maximum stand of the Puget lobe (e.g., Mackin, 1941).

Glacial Lake Missoula and Jökulhlaups

Glacial Lake Missoula, an immense water body dammed in the upper Clark Fork valley by the Purcell Trench lobe, was the source of floodwater that catastrophically swept across the Channeled Scabland (Bretz et al., 1956; see also Baker, 1982). Shorelines near and west of Missoula (Pardee, 1942) record successive stands of the lake as much as 300 m below its highest level. The many sparsely developed shorelines argue that no one level records a lengthy stand of the lake; consistent with this relation is the absence of evidence for a lake spillway around the ice dam (Bretz et al., 1956).

Until recently there were thought to have been only a few Fraser-age fillings and emptyings of the lake (Pardee, 1942; Bretz et al., 1956; Richmond et al., 1965; Bretz, 1969; Baker, 1973, 1978b), but Chambers (1971) and Waitt (1980a) inferred as many as 40 successive fillings and drainings. Lake Missoula bottom sediment is rhythmically bedded (Chambers, 1971; Curry, 1977; Waitt, 1980a). Thin-bedded silt at the base of a typical rhythmite passes upward to progressively thinner varves, the record of a gradually deepening lake. The superposition of about 40 such unweathered rhythmites records about 40 Fraser-age fillings and drainings of the lake. Complementary rhythmic slack-water sediment reveals evidence for about 40 separate catastrophic backfloodings of Columbia River tributaries in southern Washington and northwestern Oregon (Waitt, 1980a). These stratigraphic and sedimentologic relations indicate that glacial Lake Missoula was hydrostatically controlled by the thickness of the ice dam; the lakes periodically emptied as great jökulhlaups that dis-

charged down the Columbia River valley and Channeled Scabland.[4] Just south of Spokane, 16 beds of gravel deposited by floods that ascended Latah Creek valley are each capped by a bed of varved clay at least as high as altitude 610 m, evidence that at least 16 of the jökulhlaups that swept into the Spokane area from the Rathdrum valley emptied into glacial Lake Columbia I (or Lake Spokane?). In several places far north of Spokane, the intercalation of similar flood deposits with thin clay beds at least as high as altitude 690 m indicates that some of the floods that swept through the highlands south of the Pend Oreille River valley also disgorged into glacial Lake Columbia I (Table 3-1).

Relation of Glacial Lake Missoula to Alpine Glaciers

Alpine glaciers terminated in and near the east and south margins of glacial Lake Missoula. Lake sediment interfingers with Fraser-age alpine-glacial terminal moraines in the southern Bitterroot Range (Weber, 1971), and shorelines are cut into them. These relations indicate that the moraines were contemporaneous with high stands of glacial Lake Missoula and perhaps predated them. The maximum stand of the alpine glaciers therefore was roughly contemporaneous with the maximum stand of the Purcell Trench lobe. Lake Missoula shorelines etched across the sharp Fraser-age terminal moraines of alpine glaciers that flowed from mountains on the east side of the lake (Alden, 1953: 107-13) similarly indicate that the alpine-glacial maximum there occurred before or during the higher stands of the lake.

Relation of Lake Sediment to Ice-Sheet Moraines

Alden (1953) suggested that certain ice-sheet moraines are of pre-Fraser age because they underlie lake sediment and that other moraines are of Fraser age because they are not overlain by lake sediment. Richmond and others (1965) similarly classed the outermost moraine capped by lake sediment in several valleys as "early Pinedale" and moraines not thus capped as "middle Pinedale." If, however, lakes formed and drained throughout the Fraser-age episode of damming by the Purcell Trench lobe (Waitt, 1980a; R. B. Waitt, Jr., unpublished report), a silt-veneered moraine may be of any age during the Fraser interval of ponding.

West Kootenai Glacier

From the Purcell Trench, Moyie valley, and Yaak valley, a thick ice stream ascended the western (lower) Kootenai valley and terminated in the Bull River valley.[5] This ice lobe was partly nourished by a distributary of the East Kootenai glacier that diverged westward around the north end of the Cabinet Mountains. The landforms and deposits in the Bull River valley are complicated, but Alden's (1953) identification of the terminal moraine seems correct.

East Kootenai Glacier

Much of northwestern Montana was overwhelmed by a thick ice stream guided from British Columbia along the Rocky Mountain Trench and the upper Kootenai valley. In Montana the western part of this glacier overrode a general upland area but was guided by the Kootenai valley. Stratified sand and gravel underlying drumlinoid till (Alden, 1953: 138) indicate that the western sector of the glacier advanced southward up Libby Creek valley over proglacial lacustrine and outwash sediment; the southern limit of glacial landforms and drift approximately corresponds to the ice limit mapped by Alden (1953). Farther south, moraines and drift indicate that an arm of the ice sheet that channeled along Fisher River valley east of Libby Creek valley terminated in the southern end of the Libby Creek trough about 1 km beyond Alden's limit.

The maximum extent of the eastern part of the East Kootenai glacier is delineated by the southern extent of numerous moraines, deposits of till, and deranged drainage and ice-marginal channels (Alden, 1953). Alden's "terminal moraine" in the Thompson River valley, however, is a deposit of high-gradient tributary streams; the southern limit of Fraser-age drift seems to lie along a chain of lakes 15 km to the north.

Flathead Lobe

The Flathead lobe was nourished mainly by the eastern part of the voluminous ice stream that invaded from Canada along the Rocky Mountain Trench and that may also have been fed by alpine glaciers on the northeast (Alden, 1953). Of three large moraines in the Flathead Lake valley, Alden (1953) and Richmond and others (1965) assigned the outer (Mission moraine) to a pre-Fraser glaciation and the middle (Polson moraine) and inner (Kalispell moraine) to the Fraser Glaciation. Neither work justifies the pre-Fraser assignment of the Mission moraine, and Richmond and others (1965) state that the principal break in weathering lies beyond it. The Mission moraine exhibits many closed depression and other sharp landforms (Alden, 1953; Curry, 1977; Stoffel, 1980). Everywhere the moraine is capped by a thin, brown, nonargillic soil; none of the exposed contacts between the several sedimentary facies composing the moraine seems distinctly weathered or shows a buried paleosol (Curry, 1977). The Mission moraine, a composite of two individual moraines, is therefore the outer Fraser-age moraine (Figure 3-1). Lake sediment overlying the outermost of the individual moraines (Alden, 1953; Curry, 1977) indicates that some inundations by glacial Lake Missoula postdated the maximum stand of the Flathead lobe. From its outer limit on the present south slope of the composite Mission moraine, the Flathead lobe retreated less than 2 km before pausing to build the broad ridge forming the crest of this large moraine (Curry, 1977).

DISCUSSION OF GLACIATION

Most of the ice that entered the Puget lowland originated from the western Coast Mountains and Vancouver Island in southwestern British Columbia; Cordilleran ice east of the Cascade Range flowed from a broad, high-level basin in interior British Columbia fed by mountains on the west and east. Ice from the eastern Coast Mountains bifurcated around the Cascade Range, but the west-flowing ice streams were constrained by the narrowness of the Fraser and Skagit valleys across the Cascade Range. The modified limit of the Fraser-age Cordilleran ice sheet shows clearly the influence of lowlands in southern British Columbia that channeled ice into the United States. The ice sheet extended south of latitude 48° N only where guided by great topographic troughs — the Fraser lowland, Okanogan valley, Columbia valley, Purcell Trench, and Rocky Mountain Trench.

The global climatic cooling and subsequent warming during the Late Wisconsin caused the general growth and subsequent disappearance of the Cordilleran ice sheet. But local physiography and changes in relative sea level (the combined effects of regional isostasy and global eustasy) influenced particular advances and stillstands of

glaciers west of the Cascade Range. The glacier probably advanced slowly across the easternmost part of the Strait of Juan de Fuca because the perimeter of marine calving was wide. When the ice sheet impinged against the Olympic Mountains and the tidewater perimeter was reduced, the rate of advance probably increased. The rise of eustatic sea level 17,000 to 14,000 yr B.P. (Curray, 1965; MacIntyre et al., 1978) may have precluded the Juan de Fuca lobe from lingering at its maximum position in tidewater.

Cordilleran ice in and east of the Cascade Range was not influenced by sea level. The terminal moraines of the Okanogan and Flathead lobes are voluminous and the record of Lake Missoula jökulhlaups apparently lengthy. Cordilleran ice east of the Cascade Range probably reached terminal positions early and lingered there for millennia, uninfluenced by the maritime processes that affected Cordilleran ice west of the Cascade Range. Unlike the area west of the Cascade Range, where the alpine-glacial maximum preceded the ice-sheet maximum, in Montana alpine glaciers seem to have been at near-maximum positions when nearby lobes of the Cordilleran ice sheet were at or near maximum positions. This relation further suggests that the eastern lobes of the ice sheet may have advanced more rapidly to their terminal positions, where they lingered much longer, than did the Juan de Fuca and Puget lobes.

Deglaciation

PUGET AND JUAN DE FUCA LOBES

Chronology

The Juan de Fuca lobe began retreating from its outer limit before 14,500 yr B.P. (Figure 3-2) (Heusser, 1973); the western end of the strait was ice free by 14,400 yr B.P. and the central and eastern segments by 13,600 yr B.P. (dates cited by Anderson, 1968; Heusser, 1973; Alley and Chatwin, 1979; D. P. Dethier, personal communication, 1981). The Puget lobe began to retreat about 14,500 or 14,000 yr B.P. (Thorson, 1980; Barnosky, 1981). Glaciomarine sediment in the northern Puget lowland that is radiocarbon dated as old as about 13,650 to 12,500 yr B.P. indicates that the Juan de Fuca and Puget lobes had retreated into a single lobe in the northern Puget lowland by 13,600 yr B.P.; afterward ice continued to recede rapidly northward, but it continued to calve and thus to supply abundant icebergs to the northern Puget Sound (Easterbrook, 1969; Armstrong, 1981; D. P. Dethier, personal communication, 1981). The Fraser lowland in turn became deglaciated by about 13,300 yr B.P. Near the end of the glaciomarine interval 11,500 to 11,000 yr B.P., the ice readvanced to Sumas just south of the international boundary (Easterbrook, 1969; Clague, 1980; Armstrong, 1981).

Terminal Zone in the Puget Lowland

During the initial retreat of the Puget lobe, broad areas of hummocky stagnant-ice terrain formed near the terminus (Figure 3-3) (Bretz, 1913). Small, hummocky end moraines lie at the head of extensive outwash terraces along the southwest margin of the Puget lowland (Carson, 1970); a larger moraine formed near Nisqually between retreating sublobes (Noble and Wallace, 1966); expansive kettled terraces and outwash plains lie along the southeast and east margin of the Puget lowland (Figure 3-3) (Bretz, 1913; Mackin, 1941; Crandell, 1963; Mullineaux, 1970). Within some of these kame-kettle deposits, stagnating ice apparently lingered for millennia (Porter and Carson, 1971). Ice-marginal drainage from lakes impounded in alpine valleys descended southward across successive spurs and merged with the outwash plains near the glacier terminus.

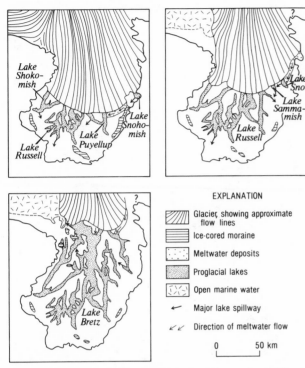

EXPLANATION

⬚	Glacier, showing approximate flow lines
⬚	Ice-cored moraine
⬚	Meltwater deposits
⬚	Proglacial lakes
⬚	Open marine water
←	Major lake spillway
↙↙	Direction of meltwater flow

0   50 km

Figure 3-3. Schematic maps of ice recession in the southern and central Puget lowland. (After Thorson, 1980.)

Southern and Central Puget Lowland and Strait of Juan de Fuca

Glacial Lake Russell, ponded beyond the north-retreating ice front, drained southward through a spillway at altitude 50 m to the Chehalis River valley (Figure 3-3) (Bretz, 1913). The lake enlarged northward as the ice receded, and it enlarged laterally as deglaciation allowed the independent marginal lakes to join it. At the glacial maximum, glacial Lake Puyallup on the southeast margin of the Puget lobe had drained south over a spillway at altitude 162 m, but it successively fell to 136, 110, and 70 m as glacial recession exposed lower outlets (Crandell, 1963); eventually the lake joined Lake Russell. Glacial Lake Skokomish on the southwest border similarly drained successively to Lake Russell through spillways at 73 m and 67 m (Bretz, 1913). Glacial Lakes Sammamish and Snohomish on the east drained across high spurs through successively lower spillways to the south and west. Meanwhile gradual incision of the spillway caused Lake Russell to fall from altitude 50 m to 41 m.

Deglaciation of the northeastern Olympic Peninsula drained Lake Russell. The spillway of the succeeding proglacial lake, here named glacial Lake Bretz (Lake Leland of Thorson, 1980), probably was about 35 m below the level of glacial Lake Russell; deltas and other evidence of the lake surface have since been isostatically upwarped (Thorson, 1981). As the Puget lobe continued to recede, all the major troughs of the Puget lowland eventually joined Lake Bretz and drained through its spillway to the Strait of Juan de Fuca (Figures 3-3 and 3-4).

On the east margin of the Puget lobe, a sequence of proglacial lakes

drained through a succession of spillways (Bretz, 1913; Mackin, 1941; Crandell, 1963; Mullineaux, 1970; Thorson, 1980; R. B. Waitt, Jr., and D. B. Booth, unpublished maps). The systematic westward decrease in altitude of these ice-marginal channels and of attendant outwash trains and deltas indicates that the eastern half of the lobe retreated roughly parallel to its margins and that during retreat it remained generally active rather than stagnating regionally.

During the succession of lakes, the land-based Puget lobe retreated only about 80 km while the Juan de Fuca lobe retreated about 200 km to the eastern end of the strait (Figure 3-2). The Juan de Fuca lobe probably retreated rapidly because it calved into seawater while sea level rose eustatically. A readvance of the Juan de Fuca lobe (Alley and Chatwin, 1979) has no recognized counterpart in the Puget lowland (Thorson, 1981). After rapidly retreating to the eastern part of the strait, the Juan de Fuca lobe may have temporarily stabilized at a broad transverse shoal (Chrzastowski, 1980), about when proglacial-lake drainage shifted from south (Lake Russell) to north (Lake Bretz). A deepening calving bay east of the transverse shoal probably then caused the northwest sector of the Puget lobe to retreat rapidly and perhaps to stagnate (Thorson, 1981).

Northern Puget Lowland

Continued eustatic rise of sea level and continued retreat of the Juan de Fuca ice front caused seawater to enter Puget Sound. High-level outwash deltas indicate that the Puget lobe had retreated to north of Seattle before the marine incursion. Farther northeast a paucity of ice-contact stratified drift and the presence of numerous ice-marginal channels suggest that the northward retreat was steady. Glaciomarine dates as old as about 13,600 yr B.P. in the northern Puget lowland and in the Fraser lowland indicate that the western part of the ice sheet retreated rapidly, probably aided by continued calving into seawater (Easterbrook, 1968, 1969; Armstrong, 1981; Fred Pessl, Jr., and D. P. Dethier, personal communication, 1981). The 2000-year glaciomarine interval indicates that, after retreating to the international boundary or beyond, a floating ice shelf continued to calve profusely and thus to supply debris-laden icebergs to the northern Puget Sound.

The rapid recession of ice during the early part of the glaciomarine interval was interrupted by stillstands. Moraines, concentrations of ice-contact stratified drift having kame-kettle topography, and submarine ridges extending from the San Juan Islands southeastward (Figure 3-4) (Bretz, 1913; Chrzastowski, 1981) suggest two brief stabilizations of the ice margin. A third and lengthy interruption in the recession of ice, the Sumas readvance and stillstand of about 11,500 to 11,000 yr B.P., built a lobate moraine and voluminous outwash train just south of the international boundary (Figure 3-4) (Armstrong et al, 1965; Easterbrook, 1969; Armstrong, 1981). Although some palynologic evidence suggests a regional climatic cooling during Sumas time (Hansen and Easterbrook, 1974; Heusser, 1977), the similarity of the youngest dates on glaciomarine sediment to those on Sumas outwash suggests that Cordilleran ice may have stabilized and readvanced because isostatic rebound grounded the glacier terminus (Armstrong et al., 1965).

Glacial Isostasy

Ice-contact deltas reveal the height of proglacial lakes in the southern Puget lowland. Reconstructed planes of glacial Lakes

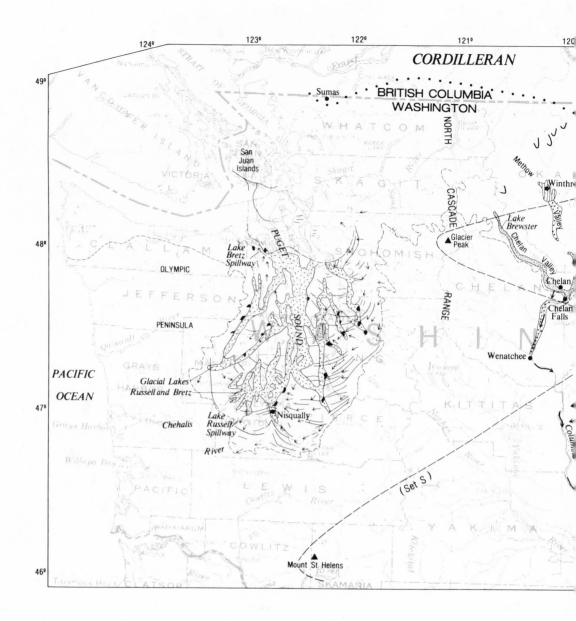

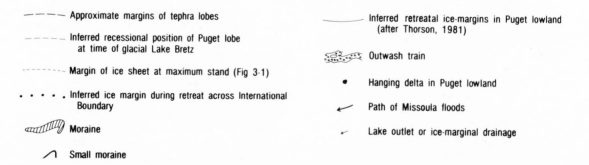

EXPLANATION

- – – – Approximate margins of tephra lobes

- – – – Inferred recessional position of Puget lobe
  at time of glacial Lake Bretz

- – – – Margin of ice sheet at maximum stand (Fig 3-1)

• • • • Inferred ice margin during retreat across International
  Boundary

Moraine

Small moraine

———— Inferred retreatal ice-margins in Puget lowland
  (after Thorson, 1981)

 Outwash train

• Hanging delta in Puget lowland

↙ Path of Missoula floods

↙ Lake outlet or ice-marginal drainage

Figure 3-4. Inferred generalized margin of Cordilleran ice sheet and glacial lakes about 12,000 ± 1000 yr B.P.

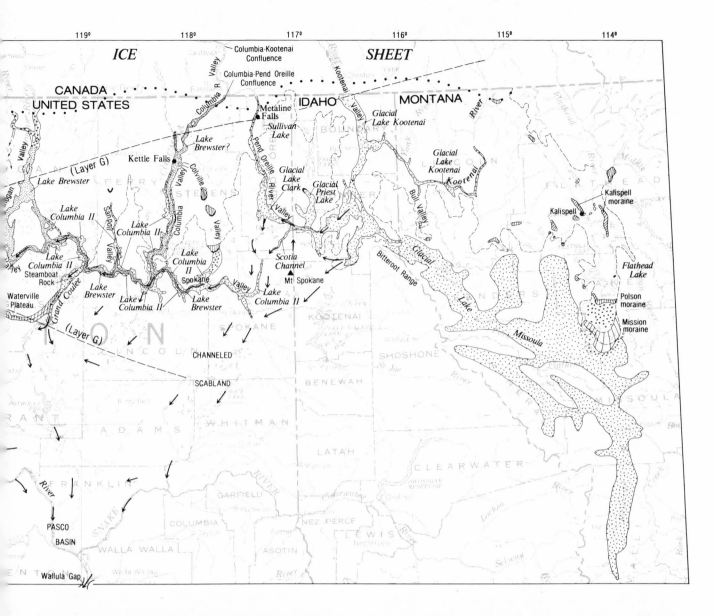

Russell and Bretz are each progressively warped up as much as 90 m northward as a consequence of differential recovery of a region isostatically depressed by the ice sheet (Thorson, 1981).

Marine sediments along the northwestern Washington coast extend only as high as altitude 10 m (Heusser, 1973); in the Puget lowland at about latitude 48° N they extend as high as 80 m and at about 48°30′ as high as 120 m and perhaps 185 m (Easterbrook, 1963); at about latitude 49°15′ in the Fraser lowland they are higher than 200 m (Mathews et al., 1970; Armstrong, 1981). The height of the marine limit reflects the combined effects of (1) rate of global eustatic sea-level rise and (2) rate of local isostatic recovery, which is influenced by local rate of deglaciation. Because eustatic sea level probably was 50 to 75 m below present sea level during deglaciation of western Washington (MacIntyre et al., 1978), the amount of isostatic recovery should be everywhere at least 50 to 75 m higher than the altitude of the marine limit. The minimum isostatic recovery thus may have been as much as 250 to 275 m near the international boundary (Armstrong, 1981); it decreased southward to nil in the southern Puget lowland (Thorson, 1981) and westward to 60 to 85 m at the northwestern Washington coast.

Apparent evidence of submergence and reemergence between 12,000 and 11,000 yr B.P. by as much as 120 m in the northern Puget lowland (Easterbrook, 1963; Mathews et al., 1970; Armstrong, 1981) was caused by some combination of tectonism, glacial isostasy, and eustasy (Mathews et al., 1970) and may partly have been caused by shifting locales of rapid glaciomarine sedimentation. Isostatic recovery was mostly complete by about 10,000 yr B.P. (Mathews et al., 1970).

NORTH-CENTRAL WASHINGTON: OKANOGAN LOBE

Waterville Plateau

Behind the terminal moraine on the Waterville Plateau, a bouldery till plain is surmounted by many small moraines of till and by kames, eskers, and deltas (Flint, 1935; Hanson, 1970; Easterbrook, 1979; R. B. Waitt, Jr., unpublished maps). Small moraines, intervening channels, and associated deltas concentric to the terminal moraine record several successive ice margins during retreat; eskers and deltas trending approximately normal to the terminal moraine record meltwater that drained outward through a marginal zone of stagnant ice.

Glacial Lake Columbia and the Upper Grand Coulee

Dammed by the Okanogan lobe, glacial Lake Columbia dropped from its high level (Columbia I) to its low level (Columbia II) (Table 3-1) when the upper Grand Coulee was suddenly deepened by recession of a huge Niagara-like cataract during catastrophic Missoula floods (Bretz, 1932) or when the ice lobe receded from the east side of the completed coulee. Deposits of the subsequent Lake Columbia II are mostly silt, evidenced in the Nespelem terrace (Pardee, 1918; Bretz, 1923, 1932; Flint, 1935). The terrace occupies the floor of the upper Grand Coulee and is graded to the controlling 470-m-altitude spillway at its southern end, but it has been isostatically raised to about 510 m in the Columbia valley (Flint, 1935).

There is no conclusive evidence of a late-glacial readvance of the ice sheet into the head of the coulee after the coulee was completed. A scabland across the summit of Steamboat Rock is discontinuously veneered by a drift of sparsely weathered stones. This drift, best explained by glaciation of the coulee after the recessional gorge was completed (Bretz, 1932; Bretz et al., 1956), probably dates to the Fraser-age ice-sheet maximum. Aerial photographs and field reconnaissance, however, indicate that the possible moraine on the coulee floor (Bretz, 1932) is a pendant flood bar; silt of the Nespelem terrace at the north end of the coulee is only locally deformed, is not overlain by till, and shows no landforms suggesting glaciation. Lake Columbia II must have occupied the upper Grand Coulee long after the ice margin last retreated from the coulee. Delta surfaces graded to the Nespelem level in the upper Grand Coulee (Bretz, 1932: Figures 52, 53) indicate, however, that glacier ice lingered on the Waterville Plateau just west of the coulee during the time of Lake Columbia II. Lake Columbia II grew northward down the Columbia River valley behind the retreating Cordilleran ice, as evidenced by lacustrine sand, silt, and clay in terraces at altitudes 500 to 540 m that overlie till as far downvalley as the Nespelem valley (Figure 3-4) (Jones et al., 1961: 27-29, Plates 4, 5).

Lower Methow, Chelan, and Adjacent Columbia valleys

Scores of small moraines distributed through 900 m of altitude in the lower Methow valley and through 350 m in the Chelan and Columbia valleys originated as ablation till bordering ice-marginal lakes and streams (Waitt, 1972, and unpublished report and maps). The moraines indicate that the thinning ice tongues continuously supplied debris to the ice margins. Flat-surfaced and hummocky stratified drift indicate that abundant meltwater streams accompanied stagnation during downwasting of the ice sheet.

Ice-marginal channels associated with the stratified drift descend along both sides of the lower Methow and Columbia valleys (Waitt, 1972, and unpublished report and maps). A suite of coulees and small recessional moraines in the Chelan valley *descend* downvalley to meet a suite of channels and moraines that *ascend* the lower end of the Chelan trough. The opposed slopes delineate the separation during downwasting of the formerly merged ice streams. Meltwater from the Okanogan, Methow, and Chelan valleys flowed along the western side of the downwasting Columbia ice stream and deepened coulees across spurs on the west side of the Columbia valley (Waters, 1933). From the Chelan valley this drainage spilled across the divide into Columbia River tributaries, thereby forming coulees.

"Great" Terrace and Lake Brewster

In the Columbia valley above Chelan Falls the "great" terrace con-

sists largely of thin-bedded fine sand and silt (Waters, 1933; Flint, 1935; Waitt, 1980b, and unpublished report and maps). The terrace is at altitude 360 to 400 m for tens of kilometers up the Columbia valley nearly to the Nespelem River (Flint, 1935; Jones et al., 1961; 68, Plate 6) and up the Okanogan valley almost to Canada (Flint, 1935). It may include a prominent terrace at about altitude 425 m (isostatically uplifted?) that extends far up the upper Columbia River valley to beyond Kettle Falls. Below Chelan Falls the terrace is a sloping outwash surface; above Chelan Falls the fine-grained sediment and flat surface of the terrace indicate that during deglaciation a series of marginal lakes merged into a single lake, here named Lake Brewster, and expanded upvalley against the retreating glacier (Figure 3-4). At the mouth of some tributaries the terrace consists of gravel derived from those tributaries (Flint, 1935; Waitt, 1972). The large volume of bed load (gravel) as well as of suspended load (silt) of the terrace deposits indicates that Lake Brewster occupied the Columbia and Okanogan valleys during a long interval while Cordilleran ice wasted back to Canada. The lake was dammed by the outwash train (and terminal moraine?) below Chelan Falls; the lake spillway apparently was Knapp Coulee on the south side of lower Chelan valley (Waitt, 1980b, and unpublished maps).

Catastrophic Floods

At least one catastrophic flood as deep as 215 m swept down the Methow-Chelan segment of the Columbia valley after retreat of the Columbia ice stream (Waitt, 1972, 1980b, 1982b, and unpublished report and maps). Flood gravel that overlies the lacustrine sediment of the "great" terrace suggests that the Missoula floods caused the breach in the valley train that drained Lake Brewster. The flood-swept "great" terrace below Chelan Falls is overlain by the 11,250-year-old Glacier Peak tephra layer B (Porter, 1978); the low-altitude bars of at least two smaller (non-Missoula?) floods are not covered by the tephra and therefore postdate it (Waitt, 1980b, and unpublished maps).

Methow Valley and Northeastern Cascade Range

Many meltwater coulees enter the Methow drainage basin over its northern and eastern divides; many marginal lakes must have developed as the divides emerged through the thinning ice. Hundreds of similar coulees record a succession of small ice-marginal streams and ice-dammed lakes through which water crossed successive spurs as it descended the Methow valley (Waitt, 1972, and unpublished report). At Winthrop a voluminous flat-topped moraine of sandy gravel marked by kettles and eskers grades downvalley to kettled terraces associated with ice-marginal channels. The moraine and terraces reveal an episode of alluviation caused by a stillstand and subsequent stagnation at Winthrop while base level remained high in the Columbia River valley, evidently during the time of Lake Brewster.

After ice-sheet tongues had retreated to the upper reaches of the drainage basin, a second period of stabilization resulted in several crested moraines. Erratics on the floors of relatively low-altitude cirques in the northeastern North Cascade Range signify that the last active ice in the valley heads was the Cordilleran ice sheet, for renewed alpine glaciation would have removed the erratics (Waitt, 1972, 1979b, and unpublished report).

Okanogan and Sanpoil Valleys

In the Okanogan valley stratified drift is distributed from high divides down to the valley bottom, more abundantly at lower altitudes (Flint, 1935). In the eastern tributaries these deposits include voluminous terraces of thin-bedded silt, sand, and clay that record a series of marginal lakes dammed to successively lower altitudes by downwasting Okanogan valley ice. Some of these terraces are pitted by kettles and are associated with coulees, small moraines, and kames, features that suggest successive stagnation of the margins of the downwasting ice. The lowest and most voluminous terrace deposit is that of the "great" terrace.

Ice was thicker to the north, and therefore the glacier terminus must have retreated northward during downwasting. In the lower Okanogan valley, Glacier Peak tephra-layer G overlies drift as far as 60 km north of the terminal moraine on the Waterville Plateau (Porter, 1978); the tephra seems to be absent farther north, although that area may be beyond its fallout area. This distribution indicates that by about 13,000 to 12,500 yr B.P. the ice-sheet margin had retreated to a position considerably north of the Columbia River (Figure 3-4).

Interrupting the silt of the Nespelem terrace at the mouth of the Sanpoil River valley, a sand-silt body is overlain by gravel that, according to Flint (1936) and Flint and Irwin (1939), coarsens up the Sanpoil valley as it merges with an outwash train. However, B. F. Atwater (personal communication, 1981, 1982) reinterprets these coarse beds as the deposits of great floods down the Columbia River valley. Outwash gravel graded to the Nespelem terrace (Lake Columbia II) extends some 20 km upvalley from the drift limit, a relation indicating that the Sanpoil River sublobe retreated at least 20 km while the Okanogan lobe remained astride the Columbia valley — until late in the history of Lake Columbia II. The Sanpoil River sublobe then disappeared by downwasting, as evidenced by stagnant-ice topography (Flint, 1936). The generally sparse drift in the upper valley indicates that the lobe wasted back steadily except when it paused to form a recessional moraine 30 km north of the glacial limit. Varved lake sediment of Lake Columbia II (?) in the lower Sanpoil valley is periodically interrupted by thick beds of sand having upvalley-directed paleocurrents (B. F. Atwater, personal communication, 1982). This alternating sequence of sand and varved mud beds is remarkably similar to deposits in the lower Priest River valley (see below). The sand beds in the Sanpoil valley evidently record multiple-catastrophic discharges from glacial Lake Missoula that swept into glacial Lake Columbia II (B. F. Atwater, personal communication, 1982).

NORTHEASTERN WASHINGTON

Upper Columbia and Colville Valleys

The silt of the Nespelem terrace, mainly sediment of glacial Lake Columbia (Bretz, 1923, 1932; Flint, 1935), is interrupted at the confluence of the Spokane and upper Columbia valleys and at the mouth of Chamokane Creek (southern Colville valley) by wedges of sand (Flint, 1936). The sand bodies may be distal outwash from the Colville River sublobe during its maximum stand of the Columbia River lobe during its initial retreat; alternatively they are deposits of great floods down the Spokane-Columbia valley. There are no known recessional moraines in the upper Columbia River valley, but the Colville River sublobe paused twice long enough to accumulate substantial debris and then to form stagnant-ice moraines (Figure 3-4) (R. B. Waitt, Jr., et al., unpublished report). As the Cordilleran ice retreated, arms of Lake Columbia II grew northward, as evidenced by

laminated mud in the upper Columbia valley and in the Colville valley (Figure 3-4) (Flint, 1936; Jones et al., 1961). Thin bedding and abundant clay beds and varves (Jones et al., 1961: 13-14, Plate 1, Figures 2 to 14 and 32 to 34) show that the lake was long lasting rather than having been briefly hydraulically dammed. In the Columbia valley about 40 km above the Spokane River confluence, sand-silt rhythmites with upvalley-directed paleocurrent indicators suggest that about 12 of the last Lake Missoula floods backflooded up the recently deglaciated valley. Terraces at the Lake Columbia II level (515 m or isostatically rebounded higher) are absent in the upper Columbia, but terraces of silt and clay at 400 to 430 m (probably isostatically rebounded) are common. This relation suggests that in the upper Columbia valley Lake Columbia II was succeeded by an arm of Lake Brewster that extended upvalley to beyond Kettle Falls.

Pend Oreille Drainage Basin

Lacustrine silt as high as altitude 1700 m in the uplands of the Pend Oreille drainage (Park and Cannon, 1943) indicates that the downwasting ice sheet ponded meltwater against emerging divides, as it did in the Methow and Okanogan drainage basins far to the west.

A low terrace of sparsely weathered sand having a nonargillic soil is nested inside the maximum Fraser-age drift in the lower Pend Oreille River valley. Sand at the south end of the terrace gradually coarsens downvalley (upglacier) and to higher altitude; below Metaline Falls the outwash is thick gravel, and 10 km south of the international boundary it merges with till. The prominent outwash terrace, which dams Sullivan Lake east of Metaline Falls, evidently records a recessional stand of the Pend Oreille River lobe.

The distal (sand) end of the recessional-outwash terrace merges with a nearly continuous silt terrace at altitude 645 m, the deposit of glacial Lake Clark (Table 3-1). The silt terrace corresponds to the bedrock floor of the Scotia channel south of Newport (Figure 3-4), the spillway of Lake Clark (Large, 1924; Flint, 1936; Park and Cannon, 1943).

IDAHO AND MONTANA

Purcell Trench Lobe

As the Purcell Trench lobe retreated to north of the Bitterroot Range, glacial Lake Missoula was succeeded by glacial Lake Clark. During further retreat of the ice margin, Lake Clark merged to the north with the expanding glacial Lake Kootenay. Alden (1953) recognized no significant recessional moraines built during retreat of the lobe.

Glacial Lake Missoula

Superposed rhythmites of individual floods in southern Washington and the absence of weathering between the rhythmites indicate that the huge jokulhlaups from glacial Lake Missoula occurred repeatedly during the entire episode of ponding (Waitt, 1980a). Because the lake was ponded near the terminus of the Purcell Trench lobe, the long interval of ponding implies that the lobe maintained its near-maximal position for millennia. As the ice dam eventually began to thin, Lake Missoula filled to successively lower levels after each discharge (Waitt, 1980a). During the glacial maximum and initial retreat, glacial Lake Missoula lapped against tongues of the West Kootenai and East Kootenai glaciers. The paucity of sand and silt in the upper Lake Missoula rhythmites suggests that these glaciers receded from the north margin of the lake while the Purcell Trench ice dam was thinning (Waitt, 1980a). As these glaciers retreated, the successively lowering Lake Missoula eventually became separated from the lake basins to the north by the emerging divide.

Varved clay and silt deposits of moraine-dammed glacial Priest Lake are regularly interrupted by 14 beds of sand having upvalley-directed paleocurrents, apparently the record of 14 large Lake Missoula jökulhlaups that backflooded into a long-lived glacial lake (R. B. Waitt, Jr., unpublished report). Varved lake deposits near Spokane are interrupted by as many as 16 beds of bed-load flood gravel (E. P. Kiver and D. F. Stradling, personal communication, 1981; R. B. Waitt, Jr., unpublished report), evidence that at least 16 of the last Lake Missoula jökulhlaups discharged into the eastern end of glacial Lake Columbia II. As many as 11 of these floods in southern Washington postdated the 13,000 yr B.P. (Mullineaux et al., 1978) Mount St. Helens set-S tephra (Figure 3-4) (Waitt, 1980a, 1980b). Lake Missoula drained for the last time when the Purcell Trench lobe at last withdrew from the north end of the Bitterroot Range. The Clark Fork drainage remained ponded, as glacial Lake Clark, by the retreating Pend Oreille River sublobe. Glacial Lake Clark drained when the combined Colville River sublobe and Columbia River lobe withdrew a few kilometers north of the international boundary and thus deglaciated the lower Pend Oreille River valley.

Glacial Lake Kootenay

An arm of glacial Lake Clark grew northward along the Purcell Trench behind the retreating Purcell Trench lobe. This arm became continuous with the developing glacial Lake Kootenay, which drained southward through glacial Lake Clark and through the Scotia channel spillway. This flow is evidenced by southward paleocurrents in silt and sand that almost filled the connecting arm to the level of the lakes. Although the surface of glacial Lake Kootenay was continuous with glacial Lake Clark, the water in the Kootenai valley below the level of the shallow connecting arm was separately dammed to the north by the retreating Purcell Trench lobe (Figure 3-4). The confluence of the Kootenai River with the Columbia River is north of the confluence of the Pend Oreille and Columbia Rivers; the lower Kootenai valley probably became deglaciated after the lower Pend Oreille valley. Glacial Lake Kootenay therefore probably outlasted glacial Lake Clark.

West Kootenai Glacier

As the West Kootenai glacier receded along the Bull Lake trough downslope, a small ice-dammed lake that discharged southward into the Bull River expanded northward behind the retreating ice front (Alden, 1953). When the ice terminus receded to the main Kootenai valley, this tributary-valley lake became an arm of glacial Lake Kootenay. The glacial deposits in the Bull River valley generally are not capped by lake sediment, evidence suggesting that glacial Lake Missoula was no longer filling above altitude 700 m by the time the valley became deglaciated.

East Kootenai Glacier

As the broad, digitate margin of the East Kootenai glacier receded along an anastomosis of valleys, several small recessional moraines formed during the first 20 km of retreat (Figure 3-4) (Alden, 1953). During retreat, glacial lakes formed in several valleys north of the divide that separates the Clark Fork and Kootenai drainage basins.

These lakes may have at first been northern ramifications of glacial Lake Missoula, but as Lake Missoula successively lowered during thinning of the Purcell Trench lobe they became separate lakes that discharged southward over the divide. When the ice margin retreated to the main Kootenai valley, the individual lakes abruptly lowered and coalesced to become southern ramifications of glacial Lake Kootenay.

Flathead Lobe

From the crest of the Mission moraine the Flathead lobe retreated partially by marginal stagnation, as evidenced by an extensive topography of shallow kettles. After retreating about 30 km, the ice terminus paused to build the huge Polson moraine, which dams Flathead Lake and obstructs valley reentrants west of the lake (Alden, 1953). The Polson moraine and outwash interfinger with and are overlain by lake sediment; shorelines are etched in the outer slope of the moraine (Alden, 1953; Bretz et al., 1956). The moraine was thus contemporaneous with some of the later stages of glacial Lake Missoula.

Lake sediment also overlies glacial sediment north of the Polson moraine (Alden, 1953; Konizeski et al., 1965), stratigraphy indicating that late stages of glacial Lake Missoula (or an early stage of Flathead Lake?) outlasted the withdrawal of Cordilleran ice from the Polson moraine. North of the Polson moraine system, the absence of Lake Missoula sediment and of shoreline terraces above altitude 1050 m (Alden, 1953) suggests that ice withdrew from the Polson moraine while Lake Missoula was filling to less than its maximum level — that is, while the Purcell Trench lobe was thinning and retreating from the north end of the Bitterroot Range.

About 60 km north of the Polson moraine, outwash graded to the recessional Kalispell moraine (Alden, 1953) is also overlain by lake silt, a relation indicating that the receding glacier continued to terminate in or near glacial Lake Missoula, or perhaps in an early stage of Flathead Lake ponded behind the Polson moraine (Alden, 1953). Glacier Peak ash (probably layer G of Porter, 1978) occurs north of the Kalispell moraine, as far as 85 km north of the Polson moraine and 110 km north of the Mission moraine. The ash occurs approximately at the contact between drift and eolian silt (Konizeski et al., 1965). At a new exposure this ash is at the contact between glaciolacustrine sediment and overlying postglacial loess; the lower part of the loess contains an ash couplet suggestive of the Mount St. Helens set-S ash (R. B. Waitt, Jr., unpublished data). The margin of the Flathead lobe therefore had retreated north of Kalispell by about 13,000 to 12,500 yr B.P., which is comparable to the minimum distance of retreat of the Okanogan lobe prior to the layer-G airfall (Figure 3-4).

North of the Kalispell moraine, several younger recessional moraines (Figure 3-4) dammed lakes along the margins of the retreating ice (Alden, 1953: 126-29, Plate 1). Outwash associated with the moraines of alpine glaciers to the east grades downvalley beneath the silt of these lakes.

SUMMARY AND DISCUSSION OF DEGLACIATION

After reaching their maximum stands, the Juan de Fuca and Puget lobes west of the Cascade Range retreated back to or beyond the international boundary by about 13,000 yr B.P. The retreat was hastened by rapid calving of the ice into seawater that was rising into the isostatically depressed Strait of Juan de Fuca and northern Puget lowland. A 2000-year-long glaciomarine interval ended with a re-advance at Sumas caused either by regrounding of the ice or by regional climatic changes.

East of the Cascade Range, Cordilleran ice formed many small moraines and glaciofluvial features as it steadily wasted down and back from terminal positions on the Waterville Plateau and in the Columbia and Chelan valleys. The glaciers remained active enough to supply debris continuously to ice margins that frequently stagnated. A recessional stillstand occurred halfway up the Methow valley, where the ice margin subsequently stagnated. A second episode of stillstand occurred after ice tongues had retreated high into the north-eastern Cascades, after which there was no late-glacial rebirth of alpine glaciers much below the level of present glaciers. The Sanpoil River, Colville River, and Pend Oreille River sublobes had one or more recessional stillstands each; the Flathead lobe had at least four recessional stillstands, and that which formed the huge Polson moraine must have been lengthy. On the other hand, a great Lake Missoula flood does not signify a recession of the Purcell Trench lobe as has been sometimes inferred, for the floods were repeated jökulhlaups determined only by the depth of the water relative to the thickness of the ice dam.

Both the Puget and Pend Oreille River lobes retreated across the international boundary about 11,000 years ago. The *general* retreat of ice on opposite sides of the Cascade Range, like the general advance, was broadly synchronous. But the eastern lobes of Cordilleran ice could have maintained near-maximum positions long after the Juan de Fuca and Puget lobes had retreated in response to marine calving. Rhythmites of the Lake Missoula floods overlying the 13,000 yr B.P. Mount St. Helens set-S tephra indicate that the Purcell Trench lobe lingered within 30 km of its maximum position for centuries after deglaciation of the southern Puget lowland. The presence of Glacier Peak layer-G (and Mount St. Helens set-S?) ash, however, indicates considerable recession of some lobes by about 13,000 to 12,500 yr B.P.

The different behavior and timing between individual lobes during their retreat resulted partly from the effects of eustasy and isostasy on the western lobes. The Puget and Juan de Fuca lobes retreated rapidly, though not exactly in phase with each other, because the glaciomarine influence did not affect the two lobes equally. Ice lobes east of the Cascades retreated more gradually, completely uninfluenced by the marine processes. The eastern lobes, however, were restricted by high-relief topography and confined valleys. Although the entire ice margin generally retreated from maximum limits to the international boundary between about 14,000 and 11,000 years ago, the temporary stillstands and rapid retreats of some lobes were influenced by local conditions that did not affect each lobe equally.

Summary

During the Fraser Glaciation the Cordilleran ice sheet occupied parts of the Fraser and Puget lowland and Strait of Juan de Fuca between about 18,000 and 13,000 yr B.P., after the maximum stand of nearby alpine glaciers. At its maximum extent about 14,500 to 14,000 yr B.P., the ice-sheet surface sloped from about altitude 1900 m at the international boundary to between 0 and 300 m at the ice terminus on the continental shelf and in the southern Puget lowland. Drainage from deglaciated alpine valleys in the Cascade Range and Olympic Mountains flowed southward along both ice margins and coalesced into meltwater streams that built broad outwash trains southward and westward to the Pacific Ocean. In the North Cascade Range, Cordilleran ice overrode high divides and inundated major drainage basins. The ice-sheet surface descended from above 2600 m near the international boundary to 270 m in the Columbia River valley. East of the Cascade Range, the Okanogan lobe extended southward as a broad lobe that dammed the Columbia River valley to form glacial Lake Columbia. The lake discharged along the course of the Grand Coulee, whose tandem gorges developed by recession of great cataracts beneath catastrophic floods from glacial Lake Missoula. The Columbia River lobe dammed the Spokane valley to form a shallow glacial Lake Spokane. The Pend Oreille River sublobe, an eastern appendage of the Columbia River lobe, was less extensive than formerly inferred. The Priest River valley remained unglaciated except for a distributary of the Purcell Trench lobe that dammed the valley mouth. The Purcell Trench lobe dammed the 2000-km³ glacial Lake Missoula, which successively discharged as huge jökulhlaups that flowed to Spokane along the Rathdrum valley and from the upper Pend Oreille River valley. From Spokane the great floods swept across the Channeled Scabland and down the Columbia River valley. The West Kootenai and East Kootenai glaciers flowed across a high-relief landscape, terminating within a general upland. The Flathead lobe was more extensive than formerly inferred. Both the Flathead lobe and nearby alpine glaciers reached near-maximum positions during high stands of Lake Missoula and thus during the maximum stand of the Purcell Trench lobe. Topographic lows trending south from southern British Columbia fed each of the major lobes of the Cordilleran ice sheet east and west of the Cascade Range, but the secondary lobation of the ice margins was determined by the configuration of local valleys.

As the Puget lobe retreated northward, ice-marginal streams and proglacial lakes progressively expanded northward. Glacial Lake Russell drained southward during initial retreat; glacial Lake Bretz later drained northward. Calving into seawater, the Juan de Fuca lobe retreated rapidly and perhaps thereby caused the northwestern part of the Puget lobe to stagnate. Continued ice retreat permitted the sea to enter Puget Sound, and a glaciomarine interval ensued from 13,500 to 11,500 yr B.P. Stillstands or readvances of the ice margin occurred during and near the end of the glaciomarine interval. In the northeastern Cascade Range and Waterville Plateau, deglaciation occurred by progressive downwasting and backwasting of ice whose margins frequently stagnated. Most lobes east of the Cascade Range built one or more small recessional moraines; the Flathead lobe built huge recessional moraines. As ice tongues retreated, glacial Lakes Columbia and Missoula fell to successively lower levels as they grew northward behind retreating ice. At length the early lakes were succeeded by glacial Lakes Brewster, Clark, and Kootenay. The apparent absence of the Glacier Peak layer-G tephra within the northern part of its projected fallout area along with the occurrence of several jökulhlaups from glacial Lake Missoula after the Mount St. Helens set-S air fall suggest that much of the glaciated terrain east of the Cascade Range remained glaciated until about 13,000 years ago. In the North Cascade Range, erratics transported by the ice sheet *up* valleys to cirque floors indicate that, as the ice sheet disappeared, alpine glaciers did not rejuvenate much below the limits of modern glaciers. Although ice lobes both east and west of the Cascade Range generally retreated from terminal positions to the international boundary during the interval 14,000 to 11,000 yr B.P., the lobes were not exactly in phase with each other. Particular stillstands and retreats were influenced by local conditions such as topography or seawater that did not affect all lobes equally.

Notes

1. Anderson (1968) contended that the Juan de Fuca lobe terminated in the eastern part of the strait. Subsequent analyses (Heusser, 1973; Alley and Chatwin, 1979; Thorson, 1981) suggest that Anderson's radiocarbon dates of about 17,200 yr B.P. represent the maximum stand of a lobe that terminated far to the west.

2. The value at the international boundary includes 250 m or more of postglacial isostatic rebound. Ice-surface gradients during the ice-sheet maximum probably were slightly gentler across the entire region than suggested by modern ice-limit data.

3. The name "Little Spokane lobe" of Richmond and others (1965) is abandoned because the revised glacier limit lies north of the Little Spokane drainage basin. We call this lobe the "Pend Oreille River sublobe" after the trough it occupied. This lobe is not to be confused with the "Lake Pend Oreille lobe" of Richmond and others (1965), which we herein name the "Purcell Trench lobe" after the trough it occupies.

4. Some investigators have held that some or all of the rhythmites in southern Washington could be a consequence of intraflood hydraulic surging (e.g., Baker, 1973; Patton et al., 1979; Bjornstad, 1980; Bunker, 1980). However, recent discoveries of varved clay beds regularly intercalated with catastrophic-flood deposits in the Spokane valley, Priest River valley, and lower Sanpoil valley are evidence that separate floods periodically emptied into long-lived contemporaneous lakes in northern Idaho and northeastern Washington.

5. Richmond and others (1965) designated the "Bull River lobe" and the "Thompson River lobe" after drainages at the maximum limit of the ice sheet. These drainages were beyond the glacier margin during advance and retreat, and the ice may not have extended into the Thompson River valley at all. For the ice mass east of the Cabinet Mountains, we use Alden's (1953) designation "East Kootenai glacier"; for the ice mass west of the Cabinet Mountains, we introduce the parallel term "West Kootenai glacier."

Dedication

We dedicate this chapter to the late J Harlen Bretz, whose durable works both west and east of the Cascade Range contribute significantly to our present understanding of the Cordilleran ice sheet. His doctoral thesis, the first comprehensive study of glaciation of the Puget Sound, has weathered seven decades of subsequent investigations with only modest modification. We name after him the last of the great glacial lakes in the southern Puget Sound, as he had honored I. C. Russell. East of the Cascade Range, Bretz's perceptive observations led to an unorthodox interpretation of the origin of the Channeled Scabland. His unique, "outrageous" hypothesis ignited and fueled one of the great debates in American geology; his own meticulously documented field evidence and cogent reasoning, augmented and acidly reargued years later, decided the issue in his favor. Recent research along and south of the drift border has been influenced far more by the publications of the independent, irascible "Doc" Bretz than by those of any other early investigator.

References

Alden, W. C. (1932). "Physiography and Glacial Geology of Eastern Montana and Adjacent Areas." U.S. Geological Survey Professional Paper 174.

Alden, W. C. (1953). "Physiography and Glacial Geology of Western Montana and Adjacent Areas." U.S. Geological Survey Professional Paper 231.

Alley, N. F., and Chatwin, S. C. (1979). Late Pleistocene history and geomorphology, southwestern Vancouver Island, British Columbia. *Canadian Journal of Earth Sciences* 16, 1645-57.

American Commission on Stratigraphic Nomenclature (1970). "Code of Stratigraphic Nomenclature." American Association of Petroleum Geologists, Tulsa, Okla.

Anderson, A. L. (1927). "Some Miocene and Pleistocene Drainage Changes in Northern Idaho." Idaho Bureau of Mines and Geology Pamphlet 18.

Anderson, F. E. (1968). Seaward terminus of the Vashon continental glacier in the Strait of Juan de Fuca. *Marine Geology* 6, 419-38.

Armstrong, J. E. (1981). "Post-Vashon Wisconsin Glaciations, Fraser Lowland, British Columbia." Geological Survey of Canada Bulletin 322.

Armstrong, J. E., Crandell, D. R., Easterbrook, D. J., and Noble, F. R. (1965). Late Pleistocene stratigraphy and chronology in southwestern British Columbia and northwestern Washington. *Geological Society of America Bulletin* 76, 321-30.

Baker, V. R. (1973). "Paleohydrology and Sedimentology of Lake Missoula Flooding in Eastern Washington." Geological Society of America Special Paper, 144.

Baker, V. R. (1978a). Paleohydraulics and hydrodynamics of Scabland floods. *In* "The Channeled Scabland" (V. R. Baker and D. Nummedal, eds.). pp. 59-79. National Aeronautics and Space Administration, Washington, D.C.

Baker, V. R. (1978b). Quaternary geology of the Channeled Scabland and adjacent areas. *In* "The Channeled Scabland" (V. R. Baker and D. Nummedal, eds.), pp. 17-35. National Aeronautics and Space Administration, Washington, D.C.

Baker, V. R. (1983). Late-Pleistocene fluvial systems. *In* "Late-Quaternary Environments of the United States," vol. 1, "The Late Pleistocene" (S. C. Porter, Ed.), pp. 115-29. University of Minnesota Press, Minneapolis.

Barksdale, J. D. (1941). Glaciation of the Methow valley, Washington. *Journal of Geology* 49, 721-37.

Barnosky, C. W. (1981). A record of late Quaternary vegetation from Davis Lake, southern Puget lowland, Washington. *Quaternary Research* 16, 221-39.

Bjornstad, B. N. (1980). Sedimentary and depositional environment of the Touchet Beds, Walla Walla River basin, Washington. Rockwell Hanford Operations Document RHO-BWI-SA-44, Richland, Wash.

Bretz, J H. (1913). "Glaciation of the Puget Sound Region." Washington Geological Survey Bulletin 8.

Bretz, J H. (1923). Glacial drainage on the Columbia Plateau. *Geological Society of America Bulletin* 34, 573-608.

Bretz, J H. (1928). The Channeled Scabland of eastern Washington. *Geographical Review* 18, 446-77.

Bretz, J H. (1930). Valley deposits immediately west of the Channeled Scabland. *Journal of Geology* 38, 385-422.

Bretz, J H. (1932). "The Grand Coulee." American Geographical Society Special Publication 15.

Bretz, J H. (1959). "Washington's Channeled Scabland." Washington Division of Mines and Geology Bulletin 45.

Bretz, J H. (1969). The Lake Missoula floods and the Channeled Scabland. *Journal of Geology* 77, 505-43.

Bretz, J H., Smith, H. T. U., and Neff, G. E. (1956). Channeled Scabland of Washington: New data and interpretations. *Geological Society of America Bulletin* 67, 957-1049.

Bunker, R. C. (1980). Catastrophic flooding in the Badger Coulee area, south-central Washington. M.A. thesis, University of Texas, Austin.

Carson, R. J., III (1970). Quaternary geology of the south-central Olympic Peninsula, Washington. Ph.D. dissertation, University of Washington, Seattle.

Chambers, R. L. (1971). Sedimentation in glacial Lake Missoula. M.S. thesis, University of Montana, Missoula.

Chrzastowski, M. J. (1980). "Submarine Features and Bottom Configuration in the Port Townsend Quadrangle, Puget Sound Region, Washington." U.S. Geological Survey Water Resources Investigations 80-14.

Clague, J. J. (1980). "Late Quaternary Geology and Geochronology of British Columbia: Part 1. Radiocarbon Dates." Geological Survey of Canada Paper 80-13.

Clague, J. J., Armstrong, J. E., and Mathews, W. H. (1980). Advance of the Late Wisconsin Cordilleran ice sheet in southern British Columbia since 22,000 yr B.P. *Quaternary Research* 13, 322-26.

Colman, S. M., and Pierce, K. L. (1981). "Weathering Rinds on Andesite and Basaltic Stones as a Quaternary Age Indicator, Western United States." U.S. Geological Survey Professional Paper 1210.

Crandell, D. R. (1963). "Surficial Geology and Geomorphology of the Lake Tapps Quadrangle, Washington." U.S. Geological Survey Professional Paper 338-A.

Crandell, D. R. (1965). The glacial history of western Washington and Oregon. *In* "The Quaternary of the United States" (H. E. Wright, Jr., and D. G. Frey, eds.), pp. 341-53. Princeton University Press, Princeton, N.J.

Crandell, D. R. (1967). "Glaciation of Wallowa Lake, Oregon." U.S. Geological Survey Professional Paper 575-C, pp. 145-53.

Crandell, D. R., and Miller, R. D. (1974). "Quaternary Stratigraphy and Extent of Glaciation in the Mount Rainier Region, Washington." U.S. Geological Survey Professional Paper 847.

Crandell, D. R., Mullineaux, D. R., and Waldron, H. H. (1958). Pleistocene sequence in the southeastern part of the Puget Sound lowland, Washington. *American Journal of Science* 256, 384-97.

Curray, J. R. (1965). Late Quaternary history, continental shelves of the United States. *In* "The Quaternary of the United States" (H. E. Wright, Jr., and D. G. Frey, eds.), pp. 723-35. Princeton University Press, Princeton, N.J.

Curry, R. R. (1977). "Glacial Geology of Flathead Valley and Catastrophic drainage of Glacial Lake Missoula, Discussion," pp. 31-38. Field Guide 4, Geological Society of America, Rocky Mountain Section Meeting. Department of Geology, University of Montana.

Dort, W., Jr. (1962). Glaciation of the Coeur d'Alene district, Idaho. *Geological Society of America Bulletin* 73, 889-906.

Easterbrook, D. J. (1963). Late Pleistocene glacial events and relative sea level changes in the northern Puget lowland, Washington. *Geological Society of America Bulletin* 74, 1465-84.

Easterbrook, D. J. (1968). "Pleistocene Stratigraphy of Island County, Washington." Washington Division of Water Resources Water-Supply Bulletin 25, (1).

Easterbrook, D. J. (1969). Pleistocene chronology of the Puget lowland and San Juan Islands, Washington. *Geological Society of America Bulletin* 80, 2273-86.

Easterbrook, D. J. (1976). Geologic map of western Whatcom County, Washington. U.S. Geological Survey Miscellaneous Investigations Map I-854-B.

Easterbrook, D. J. (1979). The last glaciation of northwest Washington. *In* "Cenozoic Paleogeography of the Western United States" (J. M. Armentrout, M. R. Cole, and H. TerBest, Jr., eds.), pp. 177-89. Society of Economic Paleontologists and Mineralogists, Pacific Section, Los Angeles.

Flint, R. F. (1935). Glacial features of the southern Okanogan region. *Geological Society of America Bulletin* 46, 169-94.

Flint, R. F. (1936). Stratified drift and deglaciation in eastern Washington. *Geological Society of America Bulletin* 47, 1849-84.

Flint, R. F. (1937). Pleistocene drift border in eastern Washington. *Geological Society of America Bulletin* 48, 203-32.

Flint, R. R., and Irwin, W. H. (1939). Glacial geology of Grand Coulee Dam, Washington. *Geological Society of America Bulletin* 50, 661-80.

Fulton, R. J. (1971). "Radiocarbon Geochronology of Southern British Columbia." Geological Survey of Canada Paper 71-37.

Fulton, R. J., and Smith, G. W. (1978). Late Pleistocene stratigraphy of south-central British Columbia. *Canadian Journal of Earth Science* 15, 971-80.

Halstead, E. C. (1969). The Cowichan ice tongue, Vancouver Island. *Canadian Journal of Earth Science* 5, 1409-15.

Hansen, B. S., and Easterbrook, D. J. (1974). Stratigraphy and palynology of the late Quaternary sediments in the Puget lowland, Washington. *Geological Society of America Bulletin* 85, 587-602.

Hanson, L. G. (1970). The origin and deformation of Moses Coulee and other scabland features on the Waterville Plateau, Washington. Ph.D. dissertation, University of Washington, Seattle.

Heller, P. L. (1980). Multiple ice flow directions during the Fraser Glaciation in the lower Skagit River drainage, northern Cascade Range, Washington. *Arctic and Alpine Research* 12, 299-308.

Heusser, C. J. (1973). Environmental sequence following the Fraser advance of the Juan de Fuca lobe, Washington. *Quaternary Research* 3, 284-300.

Heusser, C. J. (1974). Quaternary vegetation, climate, and glaciation of the Hoh River Valley, Washington. *Geological Society of America Bulletin* 85, 1547-60.

Heusser, C. J. (1977). Quaternary palynology of the Pacific slope of Washington. *Quaternary Research* 8, 282-306.

Hibbert, D. M. (1979). Pollen analysis of late-Quaternary sediments from two lakes in the southern Puget lowland, Washington. M.S. thesis, University of Washington, Seattle.

Jones, F. O., Embody, D. R., and Peterson, W. L. (1961). "Landslides along the Columbia River Valley, Northeastern Washington." U.S. Geological Survey Professional Paper 367.

Konizeski, R. L., Brietkrietz, A., and McMurtrey, R. G. (1965). "Geology and Ground Water Resources of the Kalispell Valley, Northwestern Montana." Montana Bureau of Mines and Geology Bulletin 68.

Large, T. (1924). Drainage changes in northeastern Washington and northern Idaho since extravasation of Columbia basalts. *Pan American Geologist* 41, 259-70.

MacIntyre, I. G., Pilkey, O. H., and Stuckenrath, R. (1978). Relict oysters on the United States Atlantic continental shelf: A reconsideration of their usefulness in understanding late Quaternary sea-level history. *Geological Society of America Bulletin* 89, 277-82.

Mackin, J. H. (1941). Glacial geology of the Snoqualmie-Cedar area, Washington. *Journal of Geology* 49, 449-81.

Mathews, W. H., Fyles, J. G., and Nasmith, H. W. (1970). Postglacial crustal movements in southwestern British Columbia and adjacent Washington state. *Canadian Journal of Earth Science* 7, 690-702.

Mullineaux, D. R. (1970). "Geology of the Renton, Auburn, and Black Diamond Quadrangles, King County, Washington." U.S. Geological Survey Professional Paper 672.

Mullineaux, D. R., Waldron, H. H., and Rubin, M. (1965). "Stratigraphy and Chronology of the Late Interglacial and Early Vashon Glacial time in Seattle Area, Washington." U.S. Geological Survey Bulletin 1194-O.

Mullineaux, D. R., Wilcox, R. E., Ebaugh, W. F., Fryxell, R., and Rubin, M. (1978). Age of the last major scabland flood of the Columbia Plateau in eastern Washington. *Quaternary Research* 10, 171-80.

Noble, J. R., and Wallace, F. F. (1966). "Geology and Ground-Water Resources of Thurston County, Washington." Washington Division of Water Resources Water-Supply Bulletin 10.

Page, B. M. (1939). Multiple glaciation in the Leavenworth area, Washington. *Journal of Geology* 47, 787-815.

Pardee, J. T. (1918). "Geology and Mineral Deposits of the Colville Indian Reservation, Washington." U.S. Geological Survey Bulletin 677.

Pardee, J. T. (1942). Unusual currents in glacial Lake Missoula, Montana. *Geological Society of America Bulletin* 53, 1569-99.

Park, C. F., Jr., and Cannon, R. S., Jr. (1943). "Geology and Ore Deposits of the Metaline Quadrangle, Washington." U.S. Geological Survey Professional Paper 202.

Parsons, R. B., Weisel, C. J., Logan, G. H., and Nettleton, W. D. (1981). The soil sequence of late Pleistocene glacial outwash terraces from Spokane floods in the Idaho Panhandle. *Soil Science Society of America Bulletin* 45, 925-30.

Patton, P. C., and Baker, V. R. (1978). New evidence for pre-Wisconsin flooding in the Channeled Scabland of eastern Washington. *Geology* 6, 567-71.

Patton, P. C., Baker, V. R., and Kochel, R. C. (1979). Slack-water deposits: A geomorphic technique for the interpretation of fluvial paleohydrology. *In* "Adjustments of the Fluvial System" (D. D. Rhodes and G. P. Williams, eds.), pp. 225-253. Kendall/Hunt, Dubuque, Iowa.

Porter, S. C. (1970). Glacier recession in the southern and central Puget lowland, Washington, between 14,000 and 13,000 years B.P. *In* "American Quaternary Association, Abstracts for 1970," p. 107.

Porter, S. C. (1976). Pleistocene glaciation in the southern part of the North Cascade Range, Washington. *Geological Society of America Bulletin* 87, 61-75.

Porter, S. C. (1978). Glacier Peak tephra in the North Cascade Range, Washington: Stratigraphy, distribution, and relationship to late-glacial events. *Quaternary Research* 10, 30-41.

Porter, S. C., and Carson, R. J., III (1971). Problems of interpreting radiocarbon dates from dead-ice terrain, with an example from the Puget lowland of Washington. *Quaternary Research* 1, 410-14.

Richmond, G. M. (1960). "Correlation of Alpine and Continental Glacial Deposits of Glacier National Park and Adjacent High Plains, Montana." U.S. Geological Survey Professional Paper 400-B, pp. 223-24.

Richmond, G. M., Fryxell, R., Neff, G. E., and Weis, P. (1965). The Cordilleran ice sheet of the northern Rocky Mountains, and related Quaternary history of the Columbia Plateau. *In* "The Quaternary of the United States" (H. E. Wright, Jr., and D. G. Frey, eds.), pp. 231-42. Princeton University Press, Princeton, N.J.

Russell, I. C. (1900). A preliminary report on the geology of the Cascade Mountains in northern Washington. U.S. Geological Survey Annual Report 20 (2), 83-210.

Salisbury, R. D. (1901). Glacial work in the western mountains in 1901. *Journal of Geology* 9, 718-31.

Savage, C. N. (1967). "Geology and Mineral Resources of Bonner County." Idaho Bureau of Mines and Geology, County Report 6.

Schmidt, D. L., and Mackin, J. H. (1970). "Quaternary Geology of Long and Bear Valleys, West-central Idaho." U.S. Geological Survey Bulletin 1311-A.

Stoffel, K. L. (1980). Glacial geology of the southern Flathead Valley, Lake County, Montana. M.S. thesis, University of Montana, Missoula.

Thorson, R. M. (1980). Ice-sheet glaciation of the Puget lowland, Washington, during the Vashon Stade (late Pleistocene). *Quaternary Research 13, 303-21.*

Thorson, R. M. (1981). "Isostatic Effects of the Last Glaciation in the Puget Lowland, Washington." U.S. Geological Survey Open-File Report 81-370.

Waitt, R. B., Jr. (1972). Geomorphology and glacial geology of the Methow drainage basin, eastern North Cascade Range, Washington. Ph.D. dissertation, University of Washington, Seattle.

Waitt, R. B., Jr. (1977). Evolution of glaciated topography of upper Skagit drainage basin, Washington. *Arctic and Alpine Research* 9, 183-92.

Waitt, R. B., Jr. (1979a). "Late Cenozoic Deposits, Landforms, Stratigraphy, and Tectonism of Kittitas Valley, Washington." U.S. Geological Survey Professional Paper 1127.

Waitt, R. B., Jr. (1979b). Rockslide-avalanche across distributary of Cordilleran ice in Pasayten Valley, northern Washington. *Arctic and Alpine Research* 11, 33-40.

Waitt, R. B., Jr. (1980a). About forty last-glacial Lake Missoula jökulhlaups through southern Washington. *Journal of Geology* 88, 653-79.

Waitt, R. B., Jr. (1980b). "Cordilleran Icesheet and Lake Missoula Catastrophic Floods, Columbia River Valley, Chelan to Walla Walla." Guidebook for West Coast Friends of the Pleistocene Field Conference.

Waitt, R. B., Jr. (1982a). Quaternary research in the Northwest 1805-1979 by early government surveys and the U.S. Geological Survey, and prospects for the future. *In* "Frontiers of Western Geological Exploration" (P. U. Rodda, A. O. Leviton, E. L. Yochelson, and M. Aldrich, M., eds.), pp. 167-192. American Association for the Advancement of Science, Pacific Division, San Francisco.

Waitt, R. B., Jr. (1982b). Surficial geology. *In* Geologic map of the Wenatchee 1:100,000 quadrangle, Washington. (R. W. Tabor et al.). U.S. Geological Survey Miscellaneous Investigations Map MI-1311.

Waters, A. C. (1933). Terraces and coulees along the Columbia River near Lake Chelan, Washington. *Geological Society of America Bulletin* 44, 783-820.

Weber, W. M. (1971). "Correlation of Pleistocene Glaciation in the Bitterroot Range, Montana, with Fluctuations of Glacial Lake Missoula." Montana Bureau of Mines and Geology Memoir 42.

Weis, P. L. (1968). Geologic map of the Greenacres quadrangle, Washington and Idaho. U.S. Geological Survey Geologic Quadrangle Map GQ-734.

Weis, P. L., and Richmond, G. M. (1965). "Maximum Extent of Late Pleistocene Cordilleran Glaciation in Northern Washington and Northern Idaho." U.S. Geological Survey Professional Paper 525-C, pp. 128-32.

Williams, V. S. (1971). Glacial geology of the drainage basin of the Middle Fork of the Snoqualmie River." M.S. thesis, University of Washington, Seattle.

CHAPTER *4*

Late Wisconsin Mountain Glaciation in the Western United States

Stephen C. Porter, Kenneth L. Pierce, and Thomas D. Hamilton

Introduction

Late Wisconsin mountain glaciation in the United States was confined almost entirely to high alpine ranges in the western half of the country and Alaska. Two major mountain arcs, the near-coastal Pacific Mountain system and the interior Rocky Mountain system, constitute the North American cordillera, which extends from Alaska to Mexico.

In Alaska, Pleistocene glaciers were most extensive in the south, where the Aleutian Range, the Alaska Range, and various coastal ranges form a nearly continuous succession of high mountains that stretches for more than 2000 km close to the North Pacific source of precipitation (Figure 4-1), Individual ice tongues occupied the northern valleys of the Alaska Range, but on the southern flank of the range glaciers coalesced to form the northern part of the Cordilleran ice sheet (Hamilton and Thorson, 1983). The Brooks Range, a northern extension of the Rocky Mountain system, supported a large glacier system nearly equal in size to that in the European Alps. Smaller glaciers developed in central Alaska on parts of the Ahklun Mountains, Seward Peninsula, and the Yukon-Tanana upland (e.g., Péwé et al., 1967), but two-thirds of the state (the northern and central areas) remained free of ice.

Numerous areas of mountain glaciation existed south of the Cordilleran and Laurentide ice sheets in the western United States (Figure 4-2). Although most of these areas were small and contained glaciers less than 15 km long, large glacier complexes formed in the Cascade Range of Washington and Oregon, in the Sierra Nevada, and in the Rocky Mountains (especially in the San Juan Mountains, the Front Range, the Wind River Mountains, the Uinta Mountains, the Yellowstone Plateau, and the mountains of northwestern Montana). The distribution and character of former glaciers primarily reflect topography and climate, with the areas of the most extensive glacier cover lying close to the maritime Pacific coast and in ranges that reach altitudes of 3000 m or more.

The extent of Pleistocene glaciers in the western United States and Alaska has been portrayed on the Glacial Map of North America (Flint et al., 1945) and subsequently in numerous regional reports that also discuss the stratigraphy and chronology of the glacial deposits (e.g., Coulter et al., 1965; Crandell, 1965; Richmond, 1965,

1976; Wahrhaftig and Birman, 1965; Birkeland et al., 1971; Crandell and Miller, 1974; Péwé, 1975), many of which appeared at the time of the Seventh Congress of the International Association for Quaternary Research in 1965 or shortly thereafter.

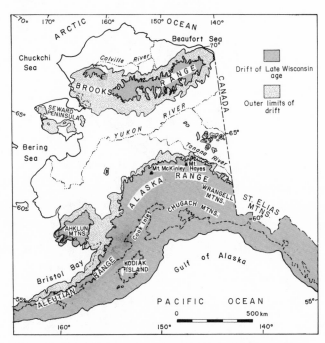

Figure 4-1. Map of Alaska showing major geographic features. Limits of Late Wisconsin drift and outermost known glacial deposits are shown in northern and western Alaska.

This chapter focuses on selected mountain regions within the western United States, including Alaska and Hawaii, where field studies during the past 15 years have produced new evidence, revised interpretations, and fresh insights into the environmental conditions

In this chapter, Porter was primarily responsible for sections on the Cascade Range, the Sierra Nevada, and Hawaii; Pierce, for the section on the Rocky Mountains; and Hamilton, for the sections on the Brooks Range and the Alaska Range.

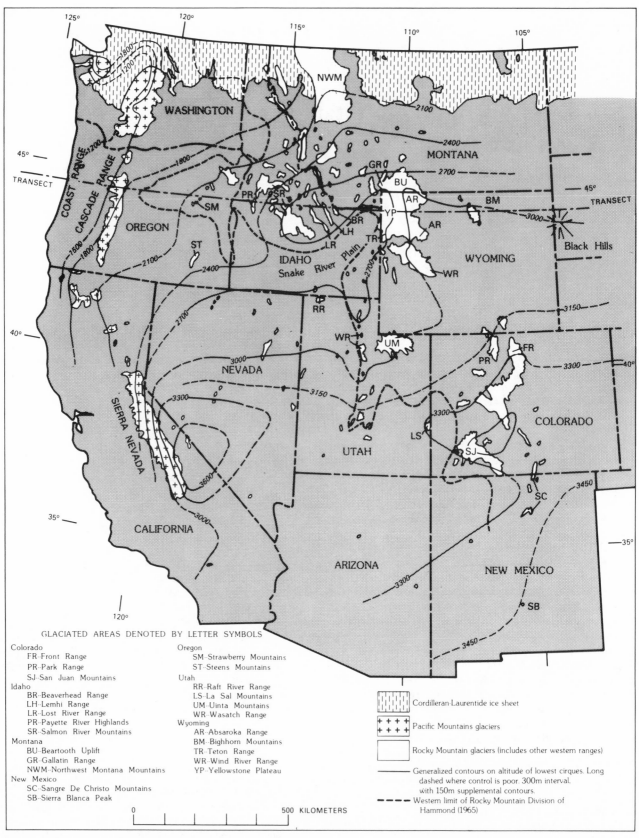

GLACIATED AREAS DENOTED BY LETTER SYMBOLS

Colorado
  FR–Front Range
  PR–Park Range
  SJ–San Juan Mountains
Idaho
  BR–Beaverhead Range
  LH–Lemhi Range
  LR–Lost River Range
  PR–Payette River Highlands
  SR–Salmon River Mountains
Montana
  BU–Beartooth Uplift
  GR–Gallatin Range
  NWM–Northwest Montana Mountains
New Mexico
  SC–Sangre De Christo Mountains
  SB–Sierra Blanca Peak

Oregon
  SM–Strawberry Mountains
  ST–Steens Mountains
Utah
  RR–Raft River Range
  LS–La Sal Mountains
  UM–Uinta Mountains
  WR–Wasatch Range
Wyoming
  AR–Absaroka Range
  BM–Bighorn Mountains
  TR–Teton Range
  WR–Wind River Range
  YP–Yellowstone Plateau

Cordilleran-Laurentide ice sheet

Pacific Mountains glaciers

Rocky Mountain glaciers (includes other western ranges)

Generalized contours on altitude of lowest cirques. Long dashed where control is poor. 300m interval. with 150m supplemental contours.

Western limit of Rocky Mountain Division of Hammond (1965)

0   500 KILOMETERS

Figure 4-2. Map showing extent of Late Wisconsin glaciers in the Rocky Mountains and other areas of the western United States. For Sierra Nevada and Cascade Range contours on lowest cirque floors are from Flint (1971: Figure 18-4), and Porter (1964a). For Rocky Mountains and adjacent areas, contours are based on about 100 altitudes of lowest cirques as defined by altitude of break in slope between cirque floor and headwall. (Map was compiled and modified from Colman and Pierce, 1979; Montagne, 1972; Flint, 1971; and Hollin and Schilling, 1981.)

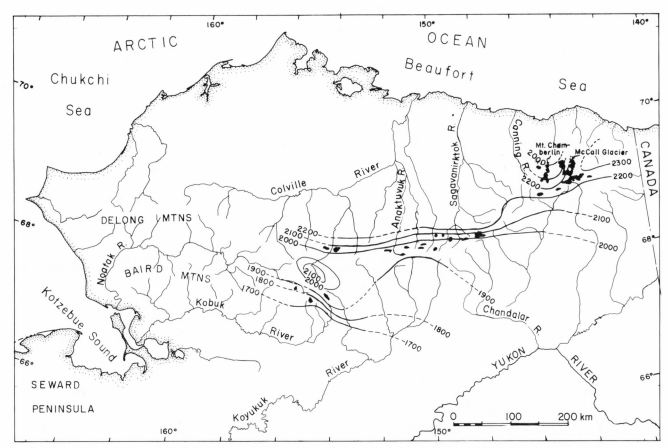

Figure 4-3. Map of Brooks Range showing major drainages, existing glaciers (size somewhat exaggerated), and modern glaciation threshold (contours at 200-m intervals).

of the last glacial age. This survey, therefore, is not exhaustive but reviews available evidence from important key areas (or type areas) and provides examples of the kinds of glacial-geologic research on late-Pleistocene environments currently in progress in the alpine regions of the western United States.

Brooks Range

The Brooks Range forms an arc 180 to 200 km wide and slightly concave to the north that extends across northern Alaska from the Canadian border to the Chukchi Sea (Figure 4-3). The western part of the range, which divides into the De Long and Baird Mountains, is generally low in altitude, with most summits reaching 900 to 1200 m and the principal peaks standing as high as 1500 m. In the central part of the range, peaks commonly reach 2000 to 2300 m, with the highest summits at 2400 to 2500 m; small glaciers are present in many of the north-facing cirques that indent the highest ridges and peaks. The range rises to its highest altitudes (about 2400 to 2700 m) close to the Beaufort Sea, where the largest glaciers attain lengths as great as 10 km.

The climate of the Brooks Range varies from arctic in the northern valleys to subarctic in the south and from continental in the eastern and central portions to maritime in the west. These gradients are reflected in the modern and fossil vegetation of the region and in the present and past snow lines (Péwé, 1975: 27-32). The forested southern valleys of the central and eastern Brooks Range are subject to the extreme continental climate characteristic of central Alaska: long and cold winters, short but warm summers, and generally light precipitation. Northern valleys and much of the western Brooks Range are subject to a moister climate, with summers that are short and cool. These valleys generally are carpeted with arctic tundra vegetation, but small stands of poplar trees occur in a few sheltered areas, and shrubs of alder and willow commonly attain heights of 2 to 3 m along stream courses. Precipitation is derived mainly from the south and west. For this reason, both modern and ancient snow lines rise northward and eastward through the range (Péwé and Reger, 1972; Péwé, 1975: 28-31).

The distribution of glaciers within the Brooks Range is controlled primarily by regional patterns of altitude and precipitation (Figure 4-3). Glaciers are absent from the low western part of the range except on prominent granitic peaks near the heads of the Noatak and Kobuk Valleys, where peaks and ridges as low as 1650 m support small cirque glaciers and peaks as high as about 2500 m support ice tongues 2 to 3 km long. Cirque glaciers and short (2 to 3 km) valley glaciers are also present in higher parts of the central Brooks Range within a zone 20 to 30 km wide that extends north from the Continental Divide. Glaciation thresholds in this belt are lowest (1850 to 1900 m) in its southern part; they generally are 100 to 150 m higher to the north, where precipitation is relatively light. ("Glaciation threshold," as used here, is equivalent to the terms "glaciation limit" and "glaciation level" used by other authors. It can be defined as the lowest altitude in a given locality at which glaciers can develop. It usually is placed below the minimum summit altitude of mountains on which glaciers occur but above the maximum summit altitude of mountains that lack glaciers but have topography favorable for them. See the discussion in Porter [1977: 102]). In the eastern Brooks

Range, glaciation thresholds are still higher (up to 2250 m) and rise 150 to 200 m northward across the range. Despite the generally high snow lines, the largest ice streams and most intensive glacierization occur on high peaks in the Mount Chamberlin area, where glaciers are especially large and abundant close to the northern flank of the range. They are nourished in part by moisture from the nearby Beaufort Sea during the summer months, when the pack ice is covered by meltwater or when it retreats from the coast. The annual precipitation of approximately 50 cm on McCall Glacier, one of the longest (8 km) in the Brooks Range, appears to be considerably greater than the annual precipitation measured in other parts of the range (Wendler et al., 1974, 1975).

LATE WISCONSIN GLACIATION

Drift of the Itkillik Glaciation, the last major Pleistocene glaciation, has been mapped extensively through the central Brooks Range (Porter, 1964b, 1966; Hamilton, 1969, 1979b, 1979c, 1980b; Hamilton and Porter, 1975). The drift was correlated by other workers with late-Pleistocene glacial advances farther to the east (Holmes and Lewis, 1965; Farrand, 1973; Sable, 1977: 24-30) and in the uplands beyond the southern flank of the Brooks Range (Yeend, 1971; Reger, 1979). In the western Brooks Range, a limited area of Itkillik-age deposits was mapped by Williams and others (1977), but glacial features dating from the late Pleistocene in general have been little studied within this region.

Field studies show that the Itkillik Glaciation can be divided into two major advances designated the Itkillik I and II phases (Hamilton

and Porter, 1975). Radiocarbon determinations (Hamilton, 1982) show that the Itkillik II advance occurred between about 25,000 and 11,500 yr B.P. and, hence, that it is of Late Wisconsin age. The Itkillik I phase lies beyond the range of conventional radiocarbon dating. Because the Itkillik I and II advances clearly represent separate glaciations, the use of these terms is being discontinued (Hamilton, 1982). The designation Walker Lake Glaciation is being extended from its type area in the Kobuk Valley to cover the last major glaciation in the southern Brooks Range. The name Itkillik Glaciation, originally employed for the next-older glaciation of the Brooks Range (Detterman et al., 1958), is being retained as it was initially defined.

The geographic pattern of Late Wisconsin (Walker Lake) glaciers within the Brooks Range (Figure 4-4) reflects regional variations in altitude and precipitation, previous history of glaciation, and, locally, late-Quaternary tectonism. Glaciers were relatively small in the low western part of the range, and ice streams did not extend to the margins of the mountain belt. Glaciers up to about 4 km long developed at the heads of some of the higher valleys in the Baird Mountains, and a small ice cap over part of the De Long Mountains may have generated outlet glaciers that extended south in the Noatak Valley. East of about longitude 158°W, a more extensive complex of valley glaciers was generated along the crest of the Baird Mountains and along the Continental Divide. Ice streams flowed northward and southward to positions along the flanks of the range and formed sharp-crested moraines that commonly nest within more extensive older end moraines (Hamilton and Porter, 1975).

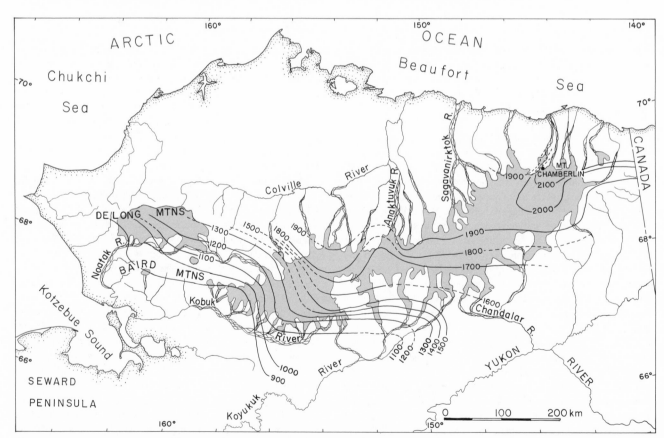

Figure 4-4. Map of Brooks Range during Late Wisconsin glaciation showing distribution of glaciers (shaded pattern), major proglacial drainage systems (braided), and inferred glaciation threshold (contours at 200-m intervals). (Data in part from Hamilton and Porter, 1975.)

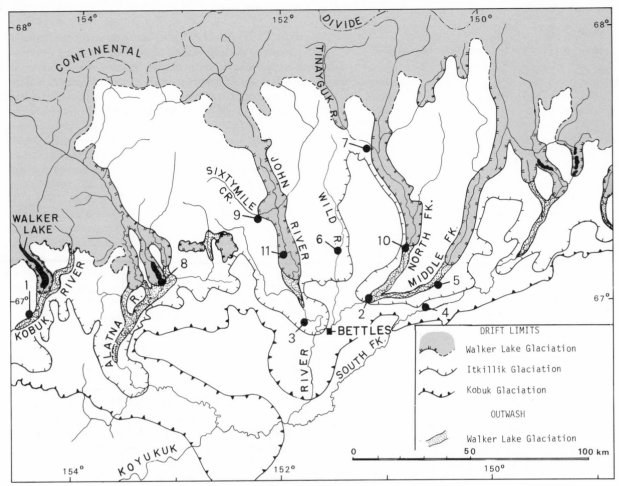

Figure 4·5. Map of younger Pleistocene drift sheets in the Koyukuk River region, south-central Brooks Range. (From Hamilton, 1982, Fig. 2.)

Within the central Brooks Range, glaciers were most extensive between longitude 158°W and 156°W, where high granitic peaks occur locally and where relatively low glaciation thresholds along the southern flank of the range allowed glaciers to develop on uplands that reach only 1100 to 1200 m. Farther east, glaciers were generated mainly along a restricted belt extending from the Continental Divide to positions near the northern flank of the range. North-flowing glaciers were relatively short, flowing through U-shaped troughs to terminate near the northern flank of the range. South-flowing glaciers had gentler gradients; they only partially occupied and weakly eroded the extensive glacial drainage systems that had been created during earlier ice advances. Ice streams reached the southern flank of the range only along the John Valley, the South Fork of the Koyukuk, and the North Fork of the Chandalar; other long ice streams in the North Fork and Middle Fork of the Koyukuk and the Middle Fork of the Chandalar terminated 10 to 20 km inside the range (Figures 4-4 and 4-5). Both north- and south-flowing glaciers formed well-integrated master ice streams spaced approximately 50 km apart, which incorporated glacier ice from neighboring valleys through a long history of drainage diversions and piracies.

In the generally higher eastern Brooks Range, glaciation was particularly extensive between about longitude 147°W and 144°W, where ice was generated in highlands both north and south of the Continental Divide. The altitudes are slightly lower farther east, and the glaciers were more restricted . The pattern of well-integrated

master glacial valleys begins to deteriorate east of about longitude 149°W in the northern Brooks Range and east of about longitude 146°W in the southern drainages. Modern glaciers and late-Pleistocene drift sheets form an almost chaotic pattern in the Mount Chamberlin area, where glaciation thresholds, earthquake epicenters, and deformation of late-Pleistocene shorelines along the adjacent Arctic coast indicate that Quaternary tectonism has remained active up to the present (A. Grantz and D. M. Hopkins, personal communication, 1980).

STRATIGRAPHY AND CHRONOLOGY

The best stratigraphic record of late-Pleistocene glaciation in the Brooks Range is in the Koyukuk region, where three major drift sheets are exposed in bluffs along the Koyukuk River and the upper Kobuk River (Hamilton, 1982). The two outermost drifts, pre-Itkillik and Itkillik in age, were deposited by generally coalescent piedmont lobes (Figure 4-5). The youngest drift, which formed during the Walker Lake Glaciation (Late Wisconsin), defined separate and much smaller ice tongues within mountain valleys or at their mouths. Radiocarbon-dated stratigraphic sections of Late Wisconsin age include distal outwash (exposures 1 through 3 in Figure 4-5 and Table 1-1), periglacial solifluction and fan deposits (exposures 4 through 7), sediments that filled glacier-dammed tributary valleys (exposure 8), and deposits that lie stratigraphically above or below till, outwash,

Table 4-1.
Radiocarbon Dates from Deposits of Walker Lake Glaciation,
Koyukuk Region, Alaska

Exposure Number	Radiocarbon Date and Laboratory Number	Material Dated	Stratigraphic Position
1	22,650 ± 220 (USGS-1044)	Peat	Channel fill in outwash
	24,300 ± 250 (USGS-1043)	Willow twigs	Near base of outwash
2	13,160 ± 170 (SI-1877)	Wood	Loess above outwash
	15,455 ± 130 (SI-1876)	Willow wood	Near base of outwash
	23,500 ± 380 (SI-1875)	Sedge and grass	Base of outwash
3	17,420 ± 180 (SI-1882)	Wood	Near top of outwash
	19,700 ± 360 (SI-1881)	Willow wood	Near base of outwash
	29,000 ± 700 (SI-1880)	Willow wood	Above lacustrine beds
4	20,600 ± 400 (I-10,573)	Peaty silt	Paleosol in colluvium
5	28,450 ± 950 (I-10,816)	Wood	Beneath fan deposit
6	22,740 ± 560 (I-11,238)	Wood	Paleosol in fan deposit
7	10,700 ± 190 (I-10,600)	Wood	Alluvium above fan deposit
8	31,000 ± 800 (W-1427)	Wood	Beneath end-moraine complex (?)
9	27,700 ± 950 (USGS-413)	Wood	Redeposited (?)
10	11,660 ± 170 (I-10,714)	Wood	Above outwash; below Holocene alluvium
11	11,560 ± 170 (I-10,471)	Wood	Basin-filling deposit behind moraine

Source: Hamilton, 1982.

and lacustrine sediments of Walker Lake age (exposure 9 through 11).

Distal outwash (sandur) deposits along the Koyukuk and John Rivers and near the head of the Kobuk River are contemporaneous with fluvial sand accumulations described by Schweger (1971, 1976) in the Kobuk Valley. Alluviation began in three of these valleys about 24,000 years ago (Figure 4-6); it was interrupted about 22,000 to 20,000 years ago, and then it continued until some time between

about 17,500 and 13,200 years ago. Concurrent solifluction on a presently stable and forested slope along the South Fork of the Koyukuk occurred in two episodes separated by a soil-forming interval about 20,600 years ago. Steep alpine fans comparable to those forming today in northern valleys of the Brooks Range also developed at that time. Along the Middle Fork, coarse gravel containing involutions, wedge casts, and stones with vertical fabric formed above fine-grained alluvium that has been dated 28,500 yr B.P. A paleosol within a comparable fan succession along Wild River formed a few thousand years later (22,700 yr B.P.), and fan formation ceased by 10,700 yr B.P. at another locality near the North Fork.

The John Valley glacier, which extended nearly 100 km south from its main source areas, dammed a series of large tributary valleys that had been deeply eroded during previous glaciations but remained unglaciated during the Late Wisconsin. Lacustrine beds in several of these valleys lack organic deposits, but a massive accumulation of sand in the valley of Sixtymile Creek contains a lens of detrital wood fragments dating to 27,700 yr B.P. This sample might provide a direct date on the maximum advance of the Walker Lake glacier that filled the John Valley, but its unique occurrence more likely is explicable in terms of the erosion and redeposition of older interstadial beds rather than the deposition of shrubs that grew during the last glacial maximum. The wood sample, therefore, may provide only a maximum limiting date on the ice dam.

Wood from the inner flank of an end-moraine complex in the Alatna Valley has been dated 31,000 yr B.P. The host sediments initially were interpreted as lacustrine beds of early-postglacial age and the date was rejected as too young (Hamilton, 1969: 217-18). A reexamination of the field notes, however, reveals that the host sediments may uncomformably underlie sediments of glacial age. The radiocarbon date, therefore, may provide a valid maximum age for the advance of Late Wisconsin glaciers in the Alatna Valley.

Limiting minimum dates on the retreat of glaciers from their terminal moraines have been obtained from organic matter from three sections in the John Valley and along the North Fork of the Koyukuk (Hamilton, 1982). A thin peat layer dated to 11,700 yr B.P. occurs above outwash gravel and below postglacial fan depostis downvalley from the North Fork moraine. A comparable limiting date (11,600 yr B.P.) was obtained from the John Valley, where large willow shrubs grew on a floodplain 30 m above modern stream level in sandy alluvium aggrading behind the Late Wisconsin end moraine. Lower (16 m above modern stream level) Holocene terraces downvalley from the John Valley moraine contain 9900-year-old wood fragments, indicating that substantial downcutting took place near the end of the Pleistocene.

Additional Pleistocene radiocarbon dates from the north-central Brooks Range (Table 4-2) document a readvance of Itkillik glaciers about 12,800 years ago in that region. Three dates from Anaktuvuk Valley that cluster around 13,100 yr B.P. were obtained from fluvial sand that formed between thicker beds of outwash gravel (Hamilton, 1980a). A date of 10,600 yr B.P. farther upvalley places a minimum limiting age on the readvance. A contemporaneous glacial readvance has been dated in the Sagavanirktok Valley, where alluviation behind a moraine dam began about 12,800 years ago and ended less than 1000 years later (Hamilton, 1979a). Three other radiocarbon dates of about the same age on wood fragments and peat beds in the Sagavanirktok Valley (Hamilton, 1979a) indicate that shrubs were growing and that peat was accumulating throughout much of the valley at that time. This evidence for increased moisture about

12,800 to 12,500 years ago indicates that the glacial readvance probably was caused by increased snowfall rather than by colder temperatures. Greater precipitation should have resulted from the widespread flooding of the Bering platform that was taking place at that time (Hopkins, 1973, 1979) and from the northward diversion of Yukon River waters into the Chukchi Sea (McManus et al., 1974; Nelson and Creager, 1977).

The history of Late Wisconsin glaciation in the central Brooks Range is summarized in Figure 4-7. Glacial advances began some time after about 30,000 yr B.P., and glaciers were advancing in most valleys by 24,500 yr B.P. The ice tongues attained at least two maxima between 24,000 and 17,000 yr B.P., punctuated by a mild interstade about 22,000 to 19,500 yr B.P. A readvance about 12,800 yr B.P. may have been most prominent in the western valleys, where climatic changes associated with the flooding of the Bering platform would have been most pronounced.

GLACIATION THRESHOLD AND ENVIRONMENTAL INFERENCES

The glaciation threshold shown in Figure 4-4 was reconstructed from the lowest altitudes of highlands associated with cirques that were active during Late Wisconsin time. (Small-scale topographic maps were used for this reconstruction, and small cirques that were weakly active may not be evident at this scale. The reconstruction [Figure 4-4]

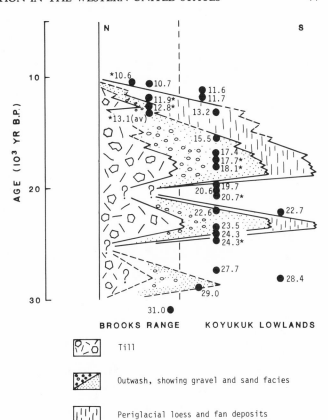

Figure 4-7. Generalized time-distance diagram showing history of Walker Lake (Late Wisconsin) glaciation in central Brooks Range. All the radiocarbon dates in thousands of years are from the Koyukuk region except where indicated by an asterisk. (From Hamilton, 1982, Fig. 13a.)

probably is a good guide to the general configuration of the glaciation threshold during Late Wisconsin time, but it may provide only minimum values for computations of depression below the modern threshold.) The threshold rises northward across the range; this reflects the general southerly precipitation sources for these glaciers. Gradients were particularly steep near the southern flank of the range, especially west of about longitude 150°W, where the broad Kobuk and Koyukuk Valleys facilitated the northward movement of moist air masses. The glaciation threshold was also deflected downward in the Anaktuvuk Pass area. Farther east, the threshold shows a more regular northward rise except in the Mount Chamberlin area, where contours are abruptly deflected as though surrounding a local high. Modern glaciation thresholds in this area are somewhat depressed (Figure 4-3), and the extensive Itkillik glaciation of the Canning Valley does not suggest any precipitation deficit in this locality during either the late Pleistocene or the Holocene. Late-Quaternary tectonism may account for the locally high glaciation threshold around Mount Chamberlin.

Depression of the glaciation threshold was fairly regular throughout the Brooks Range during Late Wisconsin time (compare Figures 4-3 and 4-4). Minimum values of threshold depression are 300 m or more in the western half of the mountain belt and generally about 200 m in the eastern half. The principal anomaly is around Mount Chamberlin, where depression might have been as little as

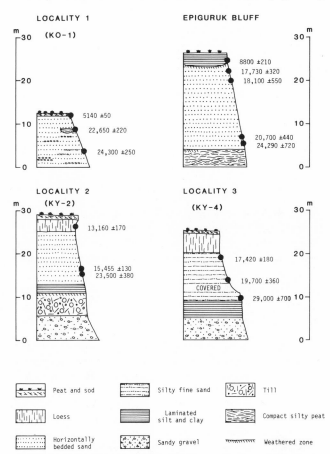

Figure 4-6. Comparison of the radiocarbon-dated distal outwash sequences in Kobuk (KO-1), John (Epiguruk Bluff), and Koyukuk Valleys (KY-2, KY-4). (Data from Hamilton, 1982, and Schweger, 1976.)

Table 4-2.
Radiocarbon Dates from Deposits of Late Wisconsin Age,
North-Central Brooks Range

Radiocarbon Date and Laboratory Number	Material Dated	Stratigraphic Position and Location	Reference(s)
13,000 ± 140 (USGS-695)	Wood	Interstadial alluvium within outwash; Anaktuvuk Valley	Hamilton, 1980a; Porter, 1964
13,170 ± 70 (USGS-694)	Peat and plant fragments		
13,270 ± 160 (Y-1084)	Wood (willow)		
10,580 ± 150 (I-11,010)	Wood and plant fragments	Postglacial alluvium; Anaktuvuk Valley	Hamilton, 1980a
11,890 ± 200 (AU-70)	Wood (willow)	Postglacial alluvium; Sagavanirktok Valley	Hamilton, 1979a
12,770 ± 180 (I-10,468)	Roots (willow)	Alluvium deposited during glacial readvance; Sagavanirktok Valley	
12,840 ± 160 (USGS-47)	Roots and wood fragments		
12,780 ± 440 (AU-72)	Wood (willow?)	Base of sand above outwash; Ribdon Valley	Hamilton, 1979a
12,170 ± 270 (AU-71)	Wood	Base of sandy lake beds; Sagavanirktok Valley	Hamilton, 1979a
12,690 ± 180 (I-10,567)	Peat	Peaty paleosol within alluvium; Sagavanirktok Valley	Hamilton, 1979a

100 m. Part of this anomaly could be the result of uplift since the time of Late Wisconsin glaciation.

If a lapse rate of about 0.6 °C per 100 m were assumed for the Brooks Range (Porter, 1966: 93) and if temperature were the only controlling variable, the general depression of the glaciation threshold would suggest a decrease in temperature of only 1 ° C or 2 °C. (Calculations based on air-temperature data for the McCall Glacier area [Wendler et al., 1974] suggest that the lapse rate in mountain valleys is more variable due to inversions and other local factors but that an average value of about 0.6 °C per 100 m is fairly representative.) This value is almost certainly too low, because much greater cooling, especially in the higher latitudes, occurred worldwide during Late Wisconsin time (Flint, 1971: 437-41; CLIMAP Project Members, 1976). In addition, periglacial features beyond the limits of Late Wisconsin glaciers in the central Brooks Range indicate strongly intensified frost climates (Hamilton, 1982). The anomalously high Late Wisconsin glaciation threshold may, in part, be an artifact of the method used in its determination, which emphasized strongly developed, large cirques and ignored smaller cirques that may have been only weakly or briefly forming at lower altitudes. However, it is likely that much of the discrepancy was caused by generally decreased precipitation during the Late Wisconsin. Northern Alaska may have experienced extreme moisture deficiencies at that time because of (1) the increase in sea-ice cover over the Arctic Ocean, (2) the exposure of the broad Bering platform by the glacial-eustatic depression of sea level, (3) the chilling of the North Pacific and Gulf of Alaska by

declining temperatures and increased discharge of glacier ice, and (4) the intensification of the Alaska Range's topographic barrier through the expansion of local glaciers.

Alaska Range

The Alaska Range forms an arc nearly 1000 km long and 50 to 100 km wide. It extends northeastward from the Alaska Peninsula to the vicinity of Mount McKinley, curves gently eastward to the Mount Hayes area, and then trends southeastward to the Canadian border (Hamilton and Thorson, 1983: Figure 2-2). The mountains form a major climatic barrier between the extreme continental conditions of central Alaska and the moister transitional-to-maritime environment farther south (Johnson and Hartman, 1969: 60-61). Mean annual precipitation and mean annual snowfall are significantly lower to the north of the range than to the south; winters are colder and summers warmer. Both flanks of the range are forested, with the treeline declining in altitude from about 1200 m in valleys on the inland side to 900 m and lower on the oceanic side.

Most of the larger streams that drain the Alaska Range are fed from glaciers and are heavily braided in at least their upper courses (Hamilton and Thorson, 1983: Figure 2-2). The largest north-flowing rivers in the region east of Mount McKinley originate south of the Alaska Range and follow broad troughs cut entirely across the mountains by Pleistocene glaciers. No major valleys cross the range west of Mount McKinley.

Near its eastern and western ends and for a length of about 80 km in the Nenana Valley area, the Alaska Range reaches only 1800 to 2100 m altitude and supports relatively small glaciers (Hamilton and Thorson, 1982: Figure 2-2). The two intervening segments of the range are higher, rising to maxima in the Mount Hayes and Mount McKinley areas. At least 14 peaks rise about 3200 m around Mount Hayes, and valley glaciers up to 30 or 35 km long radiate from summit ice fields. Near Mount McKinley, where many peaks are higher than 4000 m, an almost continuous series of summit ice caps and ice fields extends about 90 km along the crest of the range. The largest glacier in the Alaska Range, the 70-km-long Kahlitna Glacier, flows from Mount McKinley.

The modern glaciation threshold in the Alaska Range (Hamilton and Thorson, 1983: Figure 2-2) is controlled mainly by moisture transported northward from the Gulf of Alaska and northeastward up Cook Inlet from the North Pacific. Some additional precipitation is derived from the Bering Sea. The glaciation threshold typically is 300 to 400 m higher along the northern flank of the range than near the southern flank; it also declines about 300 m from the relatively continental eastern part of the range to the more maritime western sector. The threshold is lowest close to the Susitna Valley, which serves as an effective pathway for moisture moving to the western half of the range. The glaciation threshold is highest on the inland flank of the highest mountains, where precipitation shadows are most effective. It probably is 2200 m or higher north of the McKinley massif and the Mount Hayes area, and it also is unusually high north of the Wrangell Mountains/St. Elias Mountains complex.

LATE WISCONSIN GLACIATION

Data on former glaciations of the Alaska Range have been summarized by Coulter and others (1965), Péwé, Hopkins, and Giddings (1965), and Péwé (1975: 17-25, Table 2). The relative extent of glaciers and the preservation of their geologic record have been controlled by regional patterns of climate and tectonism. The glaciation of the southern flank of the range was more extensive than that of the more arid northern valleys, but most of the older glacial record in the southern drainages has been obliterated by expansions of the Cordilleran ice sheet (Hamilton and Thorson, 1983). Older portions of the glacial record also tend to be obscured in northern valleys east of Wood River (about longitude 148°W), where late-Quaternary tectonism has been especially active along the northern flank of the range (e.g., Hudson and Weber, 1977). End moraines of late-Quaternary age form the outermost drifts in most of the northeastern valleys, and older glacial deposits exist primarily as isolated remnants on plateau segments into which modern valleys have incised deeply.

During Late Wisconsin time, glaciers flowed northward and southward from the crest of the Alaska Range. Those flowing southward merged with ice of the Cordilleran ice sheet. Those flowing northward (or westward in valleys west of Mount McKinley) formed separate ice tongues that generally terminated short distances beyond the flank of the range (Hamilton and Thorson, 1983: Figure 2-4).

The limits of Late Wisconsin drift illustrated in Figure 4-1 are based on (1) published maps (Karlstrom et al., 1964; Péwé and Holmes, 1964; Coulter et al., 1965, and references cited therein; Fernald, 1965a; Holmes and Foster, 1968; Richter, 1976); (2) more recent unpublished maps, stratigraphic sections, and radiocarbon dates (Hamilton, 1973; Ten Brink, 1980; P. M. Thorson, unpublished

reports); and (3) interpretations of aerial photographs and satellite images. As defined by their drift limits, Late Wisconsin glaciers east of Mount McKinley tended to be small; they evidently were inhibited by deficient precipitation along the inland flank of the Alaska Range and the Cordilleran ice sheet. The most extensive glaciers, such as those in the valleys of the Chisana, Nabesna, Delta, and Nenana Rivers, originated south of the range as outlet glaciers from the Cordilleran ice sheet. West of Mount McKinley, moisture derived from the southern Bering Sea evidently was able to penetrate to the source areas of the west-flowing glaciers, which generally were much larger than glaciers farther east.

STRATIGRAPHY AND CHRONOLOGY

Drift assignable to the last glaciation of the Alaska Range has been given various local names (Péwé, 1975: Table 2). Recent radiocarbon dates (Tables 4-3 and 4-4) indicate that most of these deposits formed synchronously and are of Late Wisconsin age. In the upper Tanana River region, the last major Pleistocene advance of the Nabesna Glacier formed a complex series of moraines assigned to the Jatahmund Lake Glaciation (Fernald, 1965a), which was assigned a Wisconsin age on the basis of correlations with dated alluviation of the Tanana Valley (Fernald, 1965b, 1965c). Four sets of moraines were mapped within the Jatahmund Lake drift sheet (Table 4-3). The Stuver moraine, oldest in the series, is crosscut by the younger moraines and is much more subdued and dissected than any of them. On aerial photographs, it appears similar to drift of Delta age farther east, and it is shown accordingly on Table 4-3. The Takomahto Lake and Lick Creek moraines, next youngest in the series, form distinct, continuous, concentric belts. On aerial photographs, Lick Creek moraines are morphologically sharper and more distinct than moraines of Takomahto Lake age and lie closer to the modern valley floor; they must represent a glacial readvance significantly younger than the Takomahto Lake advance. The Pickerel Lakes moraines, as mapped by Fernald (1965a) near the front of the Alaska Range, appear on aerial photographs to be primarily ice-stagnation features, and they may not represent a significant glacier fluctuation. Bluff exposures along the Tanana River, as interpreted by Fernald (1965a, 1965b, 1965c) and later by L. D. Carter (personal communication, 1981), show a thick succession of alluvial and eolian sand that began forming at a level close to the modern river sometime before 42,000 yr B.P. and had accumulated to a thickness of about 20 m by 25,800 yr B.P. Massive deposits of eolian sand began to form locally about that time. The sand deposits began to stabilize about 12,000 yr B.P. and became completely inactive in one locality by about 10,200 yr B.P. As Table 4-3 shows, the stabilization of the eolian sand may correlate with retreat of the Nabesna Glacier from the Jatahmund Lake area to positions near the front of the Alaska Range.

The Donnelly Glaciation of the Delta River region was named by Péwé (1953), who initially inferred its age to be Late Wisconsin. Subsequent field studies (Péwé and Holmes, 1964; Péwé, 1965: 36-62, 88-89; Holmes and Foster, 1968) extended the mapped distribution of these glacial deposits through a broad sector of the northern Alaska Range. The Donnelly Glaciation generally has been regarded as of Wisconsin age (Péwé et al., 1965; Péwé, 1975: 21-26, Table 2), but recent radiocarbon dating indicates that it is of Late Wisconsin age (Weber et al., 1981). Limiting maximum dates of about 25,000 years have been obtained beneath till and outwash of the maximum advance in Gerstle Valley (Table 4-4) and Little Delta

Table 4-3.
Correlation of Wisconsin-Age Glacial Sequences in the Northern Alaska Range

Upper Tanana Valley[a]		Delta-Johnson Rivers Region[b]	Nenana Valley[c]		McKinley Park Region[d]	
•10,230 ± 300		•10,370 ± 150			•9580 ± 100	
JATAHMUND LAKE	Pickerel Lakes (?) •11,250 ± 250 •12,400 ± 450	III (*DONNELLY*)	Carlo readvance (?) (*CREEK*)	IV (?) III (?) (*LAKE*)	•13,270 ± 40	
	Lick Creek	II	II •13,500 ± 420	II		
		—14,800 ± 650—				
	Takomahto Lake	•18,000 ± 790 to 19,825 ± 875 I (*RILEY*)	I	I (*WONDER*)	•17,800 ± 290 •19,700 ± 200	
	—25,800 ± 800—	{24,900 ± 200 / 25,300 ± 950}				
	Stuver >42,000	DELTA	HEALY			

[a]Based on Fernald (1965a, 1965b, 1965c), with additional interpretations from aerial photographs. Moraines cannot be directly related to dated deposits in Tanana Valley, so placement of radiocarbon dates in the glacial succession is speculative.

[b]Basic Delta-Donnelly sequence is from Péwé (1953); Donnelly subdivisions and dates are from T. D. Hamilton and from N. W. Ten Brink and C. F. Waythomas (personal communication). See Weber and others (1981) for arguments supporting Early Wisconsin age for Delta Glaciation.

[c]Healy-Riley Creek-Carlo sequence is from Wahrhaftig (1958); Phase I and II subdivisions are from R. M. Thorson and from N. W. Ten Brink and C. F. Waythomas (personal communication, 1980). Phase II readvance may have lagged behind that dated near Delta River, as suggested here, but the large counting errors alternatively would permit the two radiocarbon dates to be nearly synchronous. Recent studies by Bowers (1980: 50-56) cast doubt on the validity of the Carlo readvance.

[d]Fourfold sequence was proposed by Ten Brink and associates (Ten Brink and Ritter, 1980); N. W. Ten Brink and C. F. Waythomas (personal communication, 1980), who suggest that the succession be termed "McKinley Park I, II, III, and IV."

Valley (N. W. Ten Brink and C. F. Waythomas, personal communication, 1980). During its maximum extent, the Donnelly glacier in its type locality in the Delta Junction area generated an outwash train whose upper portions are between 20,000 and 18,000 years old (Table 4-4). A younger date from interstadial deposits beneath a moraine on the Little Gerstle River indicates that the Donnelly glacier retreated from its terminal zone and then readvanced shortly after 14,800 years ago following an interval of weathering and plant growth. A third advance of a Donnelly-age glacier has been mapped in the Delta-Johnson Rivers area (T. D. Hamilton, unpublished data), and it also has been defined in valleys farther west (N. W. Ten Brink and C. F. Waythomas, personal communication, 1980). This phase is undated except for a limiting minimum age of 10,400 yr B.P. obtained by Ten Brink and Waythomas from a kettle fill on drift assigned to the Donnelly III phase along the Little Delta River.

The last glaciation of the Nenana Valley was called the Riley Creek Glaciation by Wahrhaftig (1958), who reported a date of 10,560 ± 200 yr B.P. for a readvance near the close of this glacial episode. (This solid-carbon date, determined in 1953, probably is too young [see footnotes in Karlstrom, 1964: 59, and in Thorson and Hamilton, 1977: 168]. It is not shown in Tables 4-3 and 4-4 for this reason.) Subsequent studies in the Nenana Valley have shown a complex record of glacier fluctuations during Riley Creek time (Thorson and Hamilton, 1977; R. M. Thorson, personal communication, 1979; N. W. Ten Brink and C. F. Waythomas, personal communication, 1980; C. Wahrhaftig, personal communication, 1981). During a major readvance, the Nenana Valley glacier extended northward to the

mouth of the Yanert Fork, where a lake basin was dammed and then overridden by glacier ice about 13,500 years ago (N. W. Ten Brink and C. F. Waythomas, personal communication, 1980). Ten Brink and Waythomas correlate this advance with phase II advances farther east and west, but the Nenana Valley glacier, an outlet tongue of the Cordilleran ice sheet, may not have fluctuated in phase with the independent alpine glaciers of the northern Alaska Range (Hamilton and Thorson, 1983). Fragmentary drift in Nenana Canyon lies north of the Riley Creek ice limit mapped by Wahrhaftig. This drift may represent an early Riley Creek ice limit, as proposed by Thorson and Hamilton (1977), or it may represent a slightly older deposit of Wisconsin age.

The outer limit of probable Late Wisconsin deposits in the McKinley Park area coincides with the mapped limits of drift assigned to the Wonder Lake Glaciation (Fernald, 1960; Reed, 1961; Karlstrom et al., 1964; Péwé et al., 1965; Péwé, 1975: Table 2). For the McKinley area, recent detailed mapping and stratigraphic studies by N. W. Ten Brink and associates have demonstrated an apparently consistent sequence of four moraine systems assignable to the Wonder Lake Glaciation (R. M. Thorson, personal communication, 1979; Ten Brink and Ritter, 1980; N. W. Ten Brink and C. F. Waythomas, personal communication, 1980). The two oldest moraines represent extensive ice advances under full-glacial conditions; the two younger features were formed during brief interruptions of general glacial retreat under conditions of climatic warming and widespread ice disintegration. The oldest advance culminated about 19,700 or 17,800 yr B.P., according to dates on outwash that was overridden by glacier ice during a final minor fluctuation. A minimum age limit for phase II is provided by a date of 13,300 yr B.P. on soil above till in the Toklat Valley. Independent confirmation of Ten Brink's phase I and II succession is provided by a bluff exposure along the Teklanika River, where two tills separated by slightly weathered alluvium can be related to moraines of the two oldest phases farther downvalley. Three concordant radiocarbon dates ranging from 12,200 to 12,000 yr B.P. from a kettle (Table 4-4) in the upper till place a minimum age on the Teklanika succession. Moraine successions, comparative morphology, and weathering characteristics also define two younger ice advances in the McKinley Park area (N. W. Ten Brink and C. F. Waythomas, personal communication, 1980). The earlier (phase III) was relatively extensive in many valleys but apparently absent in others. Maximum and minimum limiting dates do not closely bracket this geologic event, and its estimated age of about 12,800 to 11,800 yr B.P. is based on assumed wet and/or cool conditions associated with an apparent peat-forming episode dated in several valleys of the McKinley region. The youngest advance (phase IV), restricted to the heads of the highest mountain valleys, has been similarly dated by only a minimum age of 9600 yr B.P. from a kettle deposit and by the age range of an inferred peat-forming interval.

The Late Wisconsin glacial succession outlined in Table 4-3 is similar to the glacial successions described for the Brooks Range and for the St. Elias Mountains (Denton, 1974); it also agrees with the record of late-Pleistocene climatic fluctuations in unglaciated central Alaska. Six dates ranging from about 34,000 to 30,700 yr B.P. define a Middle Wisconsin unconformity in the ice-rich organic mucks exposed in the U.S. Army Cold Region Research and Engineering Laboratory permafrost tunnel near Fairbanks (Sellman, 1967, 1972),

and pollen data from the Fairbanks area indicate interstadial conditions during the same period (Matthews, 1974). The Boutellier nonglacial interval in the St. Elias region dates from more than 49,000 to 29,600 yr B.P. on the basis of seven radiocarbon dates (Denton, 1974). The onset of Late Wisconsin glacial conditions about 25,000 years ago is indicated by two of three dates (24,300 and 23,500 yr B.P.) near the base of distal outwash in the Brooks Range, which show close agreement with limiting dates of 25,800 and 24,900 yr B.P. on organic deposits beneath till and outwash in the Alaska Range. Full-glacial conditions of the Late Wisconsin ended about 14,000 years ago in the Tanana Valley according to pollen data discussed by Ager (1975: 33-57). Herb tundra (reflecting a climate significantly drier, cooler, and more continental than the present climate) was replaced by shrub tundra under conditions that probably were both warmer and moister. This climatic shift appears to have been synchronous with the end of full-glacial conditions in the Alaska Range described by Ten Brink and Waythomas and with the end of phase II glaciation as defined by bracketing dates in the Delta-Johnson Rivers and McKinley Park regions. It may be slightly out of phase with the expansion of the Cordilleran ice sheet in south-central Alaska as reflected in lake sediments overridden by the Nenana outlet glacier about 13,500 yr B.P. and by marine sediments of about that age that underlie the Elmendorf moraine near Anchorage (Schmoll et al., 1972; Hamilton and Thorson, 1983). The phase III readvance, which has an inferred age of 12,800 to 11,800 years in the McKinley area, may correspond in part with the glacial readvance that culminated about 12,800 to 12,500 yr B.P. in at least some valleys of the Brooks Range. General deglaciation dates of about 13,300 to 9600 yr B.P. from the Alaska Range are similar to those from other regions of Alaska, but these are mostly minimum dates on soils,

Table 4-4.
Radiocarbon Dates from Deposits of Late Wisconsin Age, North-Central Alaska Range

Radiocarbon Date (yr B.P.)	Laboratory Number	Material, Stratigraphic Position, and Location
9580 ± 100	USGS-655	Peat above youngest drift of Late Wisconsin age; McKinley Valley.
10,230 ± 300	W-980	Organic bed above dunal sand; upper Tanana Valley
11,950 ± 170	I-10,507	Wood and peat near top of kettle-filling deposit; Teklanika Valley
12,050 ± 165	I-10,254	Wood and peat near top of kettle-filling deposit; Teklanika Valley
12,300 ± 120	USGS-161	Wood fragments at base of kettle filling; Teklanika Valley
12,400 ± 450	W-1212	Thin organic bed near top of dune; upper Tanana Valley
13,270 ± 40	QL-1368	Basal peat above drift of MPII readvance; Toklat Valley
14,800 ± 650	GX-2177	Wood fragments beneath till; Little Gerstle Valley
17,800 ± 290	I-11,228	Organic silt beneath outwash; McKinley Valley
19,700 ± 200	USGS-656	Wood fragments beneath outwash; McKinley Valley
18,000 ± 790	GX-4342	Unspecified organic matter near top of outwash gravel (GX-4342 and
19,035 ± 780	GX-4344	GX-4343) and eolian sand (GX-4344); Delta Junction area
19,825 ± 875	GX-4343	
24,900 ± 200	QL-1367	Paleosol beneath outwash; Little Delta Valley
25,300 ± 950	GX-2179	Wood fragments beneath till; Gerstle Valley
25,800 ± 800	W-1174	Organic zone at base of eolian sand; upper Tanana Valley

Source: Modified from Hamilton (1982), Table 3.

peats, and kettle fillings and do not provide exact ages or rates of ice wastage.

The Late Wisconsin glaciation threshold illustrated in Figure 2-4 of Hamilton and Thorson (1983) was reconstructed through the use of the same procedures employed for its determination in the Brooks Range, but the Alaska Range reconstruction is considered less accurate because of the more active tectonism of southern Alaska, the inundation of low-lying cirques by ice of the Cordilleran ice sheet, and the widespread presence of ice caps over the higher mountain masses. However, the Late Wisconsin reconstruction shows general trends influenced by precipitation patterns and probably is an acceptable approximation. The glaciation threshold was lowest in the western part of the Alaska Range, where moisture remained readily available from maritime sources south of the emergent Bering platform, and in the Cook Inlet-Susitna lowland region, which evidently continued to serve as a pathway for moisture during the Late Wisconsin. The higher sectors of the Alaska Range and the highest ranges farther south formed effective precipitation barriers, and the glaciation threshold rose on their inland flanks. Lower segments of the Alaska Range, such as around the Delta and Nenana Valleys and the South Fork of the Kuskokwim River, served as local pathways for the movement of moisture northward across the range. Discrepancies from the general precipitation-controlled pattern are most obvious south of the Alaska Range, where late-Quaternary tectonism was most active (Hamilton and Thorson, 1983).

Depression of the Late Wisconsin glaciation threshold relative to that of today averaged 200 m or more along the northern flank of the Alaska Range east of the Nenana Valley, where precipitation was most strongly inhibited by mountains and ice fields. This figure should be considered a minimum value for reasons discussed previously and also because the eastern Alaska Range was subject to especially active late-Quaternary tectonism (Stout et al., 1973; Lanphere, 1978). The uplift of glacier source areas may have increased the present altitude of Late Wisconsin glaciation thresholds in some places. Farther west, the glaciation threshold was depressed about 300 to 350 m in valleys north and west of the range crest; this probably reflects the greater availability of moisture for glacier expansion and the greater influence of Late Wisconsin temperature depressions. Apparent depressions along the southern flank of the Alaska Range average about 200 m, but these definitely are minimum values because most lower-lying cirques were overridden by ice of the Cordilleran ice sheet during the Late Wisconsin glacial maximum.

Cascade Range

The Cascade Range owes its rugged scenic topography to a combination of recent uplift, Quaternary volcanism, and repeated glaciation. Widespread glacial sediments and landforms indicate that ice mantled the higher parts of the range and major valleys during successive Pleistocene glaciations (Crandell, 1965; Porter, 1976). The cool maritime climate of the crest and western slope of the Washington Cascades has led to the preservation of organic matter that permits the youngest deposits to be dated by radiocarbon. Widespread tephra layers, the products of explosive volcanism, constitute time-stratigraphic marker horizons for the latest Quaternary that aid in making regional correlations and in subdividing glacial and nonglacial sediments.

Extending from latitude 40°N in northern California northward into southern British Columbia beyond latitude 49°N, the Cascade Range constitutes one of the major mountain systems of the western United States, and along most of its length it causes orographic precipitation from maritime air flowing eastward off the Pacific Ocean (Figure 4-2). It divides two contrasting climatic regions, a zone on the west characterized by strongly maritime conditions and a zone east of the crest characterized by increasingly continental conditions. Although Cenozoic volcanic rocks are prominent throughout the range, the Cascade uplands also consist of older metamorphic, sedimentary, and granitic terranes, most widely exposed in Washington State. Young volcanic cones of Quaternary age rise well above the general level of the Cascade crest, some reaching altitudes of more than 4000 m. The range is between 100 and 150 km wide, with mean altitudes increasing northward.

Strong climatic gradients are found within the Cascades. In the North Cascades, for example, mean annual precipitation ranges from more than 2000 mm at sites along and west of the drainage divide to less than 500 mm at some east-slope locations (Porter, 1977: Table 1). At high altitudes, most precipitation falls as snow, nearly 85% of the annual total being recorded during the winter and spring months. Mean annual temperatures range from 1.9°C at 1850 m altitude to about 10°C at stations close to sea level; mean July temperatures are about 10°C higher.

The distribution of glaciers reflects the strong regional topographic and climatic controls. In Washington, for example, the mean altitude of small glaciers (less than 0.5 km²) rises eastward and southward across the Cascades from 1600 m in the northwest to 2400 m in the southeast (Figure 4-8) (Meier, 1961a). The glaciation threshold shows a similar trend, rising from about 1800 m to 2500 m across the North Cascades and from about 1900 m to 2200 m in the south (Porter, 1977), as do recent equilibrium-line altitudes (Figure 4-8) (Meier and Post, 1962). Most glaciers in Washington lie in the high, rugged North Cascades (some 750 glaciers) (Post et al., 1971), whereas in the generally lower mountains of southern Washington, Oregon, and northern California glaciers are restricted almost exclusively to the slopes of isolated high volcanic peaks. Most glaciers are of the cirque or valley type, and of these nearly 80% are less than 0.5 km² in area; the largest glaciers, on the slopes of Mount Rainier, are Emmons Glacier (6.7 km long, with an area of 10.7 km²) and the Carbon-Russell glacier system (9.7 km long, with an area of 13 km²).

The Cascades contain numerous Pleistocene cirques, the distribution of which reflects the climatic gradient. The altitude of cirque floors, taken as the break in slope at the base of the headwall or the level of water in tarn-filled cirques, shows considerable variation (about 500 m) owing to local orographic effects, but a regional gradient is apparent (Figure 4-8) (Porter, 1964a). The mean altitude of 118 north-facing (N ± 30°) cirques between latitude 47°15′N and 47°30′N, for example, rises progressively eastward across the range (Figure 4-9). The slope of the regression line for this population (10.3 m per kilometer) is nearly identical to that derived for two groups of north-facing glaciers within this belt but is some 500 m lower. The cirque-floor surface has no special chronologic significance, however, for cirque-floor morphology must reflect cumulative erosion during successive glacial and interglacial ages.

The Cascade Range has been among the most volcanically active regions within the United States during the late Quaternary. Until recently, most of the large volcanic cones were regarded as being extinct or dormant, but studies have shown that many have been

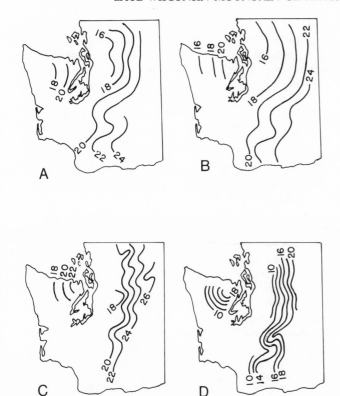

Figure 4-8. Maps of western Washington showing (A) the mean altitude of existing glaciers (Meier, 1960), (B) the 1961 equilibrium-line altitudes of glaciers (Meier and Post, 1962), (C) the calculated modern glaciation threshold (Porter, 1977), and (D) the regional pattern of cirque-floor altitudes (Porter, 1964).

repeatedly active during most or all of postglacial time. Although eruptions have often had profound effects on the local environment, several volcanoes noted for their extremely explosive eruptions have had regional or even global effects; acid fallout from the Mount Mazama eruption of about 6700 yr B.P., for example, has been tentatively identified in an ice core from central Greenland (Hammer et al., 1980). Among the most important of the explosive volcanoes are Glacier Peak, Mount St. Helens, and Mount Mazama. Airborne products of their eruptions have been distributed widely throughout the Pacific Northwest and constitute important stratigraphic marker horizons in Quaternary sediments (Figure 4-10, Table 4-5). Other tephra-producing volcanoes, such as Mount Rainier, have generated numerous pyroclastic units that aid in the local subdivision and correlation of young glacial deposits (Mullineaux, 1974).

LATE WISCONSIN GLACIATION

During the culminating ice advances of the Late Wisconsin, glaciers mantled the crest of the Cascade Range and extended down major valleys draining the eastern and western slopes. Two large ice fields in the Washington Cascades were separated by a belt of discontinuous ice cover near the head of the Naches River, where the crest is relatively low. In Oregon, a glacier system on Mount Hood volcano lay north of a nearly continuous ice field that extended south some 270 km from near latitude 45°N to Mount McLaughlin at nearly 42.5°N. Other isolated glacier systems were centered around Mount Shasta and Mount Lassen in northern California. The glacierized zone in Oregon ranged between about 10 and 50 km wide; in southern Washington, it averaged about 60 km wide and was almost 75 km wide in the North Cascades. Most large glaciers headed near the drainage divide, from which tributary ice streams flowed, eventually coalescing in major valleys. In parts of the Oregon Cascades and the southern Washington Cascades, outlet glaciers drained broad summit ice caps.

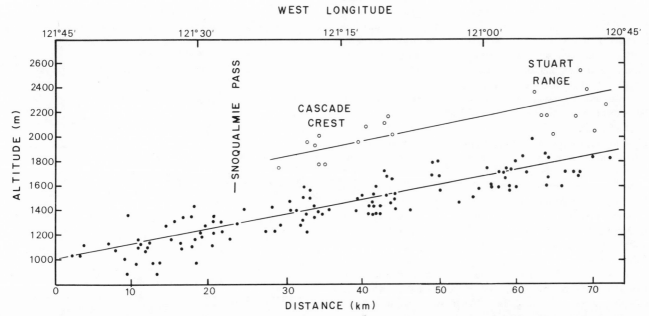

Figure 4-9. Altitudinal distribution of north-facing (N ± 30° cirques (solid circles; N = 118) along a transect across the North Cascade Range between latitude 47°20'N and 47°35'N compared with the mean altitude of north-facing glaciers (open circles; N = 20). The slopes of the regression lines are parallel, slope at 10.3 m per kilometer, and are separated by 550 m.

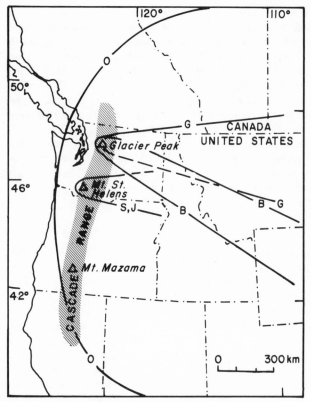

Figure 4-10. Distribution of major tephra layers produced by Glacier Peak (B, G), Mount St. Helens (S, J), and Mount Mazama (O), which are useful in dating and correlating deposits of the last glaciation. (See Table 4-5 and the text for additional data and discussion.)

The glacier cover in the southern North Cascades of Washington was asymmetrical, reflecting the present east-west asymmetry of the range. On the western slope, many valley floors lie at altitudes of less than 600 m within 10 km of the crest; on the eastern slope, such low altitudes are reached only 30 to 50 km east of the divide. Despite an eastward-rising snow line, large areas east of the crest lay within the accumulation zone, and east-slope glaciers typically were larger and longer than those in west-draining valleys.

Few detailed reconstructions of the Late Wisconsin glaciers have been attempted. In the upper Yakima River drainage basin of the North Cascades, three converging ice streams joined during the earliest (Bullfrog) advance of the last glaciation but divided into two tongues during a subsequent advance (Ronald) and into three during a still-later advance (Domerie) (Porter, 1976). During the earliest advance, the longest flow line of the glacier system extended nearly 50 km. Probably the largest single glacier in the range was in the Cowlitz Valley. This glacier, which originated on the southern flank of Mount Rainier and along the Cascade crest southeast of the volcano, flowed a maximum distance of at least 80 km (Crandell and Miller, 1974).

Late Wisconsin alpine glaciers in the northern part of the North Cascades were overwhelmed by the Cordilleran ice sheet, which expanded southward from Canada (Waitt and Thorson, 1982). Stratigraphic and morphological data indicate that valley glaciers in the North Cascades were in an expanded state prior to the arrival of the ice sheet and that at the time of the culminating advance of the

Puget lobe alpine glaciers along most or all of the western front of the range had retreated. In many valleys west of the crest, lakes were ponded between thick ice in the Puget lowland and the Cascade mountain glaciers. Although this relationship has been inferred to indicate that the advances of Cascade glaciers were out of phase with those of the ice sheet, it now seems more likely that advances were broadly synchronous but did not reflect the same relative extent of ice in the mountains and the lowland. Stratigraphic studies and regional correlations suggest, for example, that the Domerie advance in the upper Yakima River basin east of the crest was essentially contemporaneous with the culminating Vashon advance of the Puget lobe (Porter, 1976). A late-glacial readvance (Hyak) of the Puget lobe also seems to have occurred at about the same time as a readvance (Sumas) of the Puget lobe near the Canadian border (Porter, 1978). If the Bullfrog and Ronald advances are of Late Wisconsin age (this is not yet known with certainty), then they may merely reflect the shorter response time and distance of flow of the relatively short valley glaciers compared to those of the massive ice sheet, the southern terminus of which advanced many hundreds of kilometers from its primary source areas in the coastal mountains of British Columbia. Alternatively, the relationship may reflect the existence of a precipitation shadow in the Washington Cascades during Vashon time that was produced by the extensive and thick Puget lobe to the west, as suggested by Crandell (1965). Still another possible factor may have been the isostatic effect of the ice sheet on the North Cascades, for the Cordilleran ice reached a thickness of more than 1000 m in northwestern Washington and caused an isostatic depression of the Puget lowland (Thorson, 1981) and most likely of the adjacent Cascades as well. Such a depression would have led to an apparent relative rise in the equilibrium-line altitudes of mountain glaciers, thereby resulting in ice advances of smaller magnitude than would have occurred if the ice sheet had not been present.

Table 4-5.

Important Tephra Layers of Late-Pleistocene and Early-Holocene Age in the Pacific Northwest That Are Useful for Dating Glacial Deposits

Source Volcano	Layer(s)	Age (^{14}C yr B.P.)	Reference(s)
Mount Mazama	O	6700	Mullineaux, 1974
Mount Rainier	R	> 8750	Mullineaux, 1974
Mount St. Helens	J	> 8000 < 12,000	Mullineaux et al., 1975
Glacier Peak	B	> 11,300 ± 230 < 11,200 ± 100	Porter, 1978
Glacier Peak	G, M	> 11,200 ± 200 < 12,750 ± 350	Porter, 1978
Mount St. Helens	So, Sq	> 13,800 ± 210 < 15,700 ± 260	Mullineaux et al., 1975; M. Stuiver, personal communication, 1981

STRATIGRAPHY AND CHRONOLOGY

The extent of drift of the last glaciation has been mapped in many valleys of the Washington Cascades, and in some it has been subdivided into units representing successive readvances. The probable extent of ice in the Oregon Cascades was mapped by Crandell (1965), but in few areas have detailed studies been carried out. Consequently, at present the sequence and chronology of drifts in the Washington Cascades are better known.

Because of the diverse rock types and complex structure of the Cascades, drift units typically differ in lithology from one valley to the next. This makes it impractical to define regional lithostratigraphic units. Furthermore, stratigraphic exposures displaying more than one drift are rare. Nevertheless, mappable units can be defined on the basis of differences in morphology, relative position, and weathering characteristics (Porter, 1975b; Colman and Pierce, 1981). Where such studies have been made, it is apparent that the last glaciation was marked by a succession of ice advances. Moraines and associated outwash valley trains provide primary information about these events (e.g., Porter, 1976). The maximum advances have been inadequately dated, but radiocarbon analyses and tephra layers of known age provide limiting ages for the latest Pleistocene glacier fluctuations.

Although different local stratigraphic names have been employed in each valley studied, it is common practice to refer to the last glaciation in the Washington Cascades as the Fraser Glaciation, following the geologic-climate nomenclature of the adjacent Puget-Fraser lowland. However, it is not known with certainty whether the alpine and ice-sheet deposits represent exactly the same time interval. Glacier readvances and intervals of ice recession primarily reflect climatic variations and could, therefore, be regarded as a basis for defining stades and interstades, but no formal geologic-climate subdivision of this rank is in use throughout the range. For convenience, the events of the last glaciation are here divided into one or more early maxima, a later prominent advance that was broadly synchronous with the culminating Vashon advance of the Puget lobe, and one or more late-glacial readvances near the end of the Pleistocene.

MAXIMUM ADVANCE(S)

The outermost drift of the last glaciation in Cascade valleys is commonly marked by one or more end moraines and associated terraced outwash deposits (Figure 4-11). On the western side of the Washington Cascades north of the Nisqually River Valley, such deposits are not seen, because during a subsequent advance the Puget lobe presumably overrode and either destroyed or buried them. Near Mount Rainier, the maximum alpine advance is named for Evans Creek (Crandell and Miller, 1974). In the eastern North Cascades, deposits of the greatest advance are assigned to the Bullfrog Member of the Lakedale Drift (Yakima River valley) (Porter, 1976) and to the informally designated Leavenworth I drift (Wenatchee River valley) (Porter, 1969). Deposits having a comparable relative position in the northeastern Oregon Cascades were designated the Suttle Lake Member of the Cabot Creek Formation by Scott (1977); Carver (1973) named the outermost deposits of the last glaciation in the southern Oregon Cascades the Waban drift. In each case, the deposits are differentiated from still-older glacial deposits primarily on the basis of relative-age parameters and their position within the glacial succession, but they may not all be of identical age.

Generally, only a single moraine and related valley train have been mapped, but, in some cases multiple moraine crests are discernible,

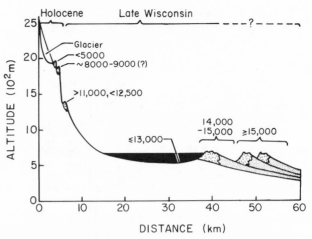

Figure 4-11. Generalized sequence of glacial deposits in east-draining valleys of the Washington Cascade Range. Inferred ages are based on available radiocarbon dates, associated tephra layers, and correlations with ice-sheet deposits on the western slope of the range. Moraines and related outwash valley trains are shown by stippled patterns; moraine-dammed lakes are shown in solid black.

implying either a fluctuating terminus or secondary halts of readvances during ice recession. In the upper Yakima River valley, a double moraine and related outwash train apparently represent a significant advance (Ronald Member of Lakedale Formation) (Porter, 1976); in the Wenatchee River-Icicle Creek drainage, a moraine occupying a comparable position is referred to as Leavenworth II (Porter, 1969). Although the Bullfrog, Ronald, and Domerie advances are inferred to be phases of the Late Wisconsin (Porter, 1976), a strong case for such an age can only be made for the Domerie. Colman and Pierce (1981), on the basis of weathering-rind data, suggest that the Bullfrog deposits may be of Early Wisconsin age and the Ronald deposits of Middle Wisconsin age. This alternative working hypothesis, although as yet unsupported by other evidence, implies that the Domerie advance may represent the Late Wisconsin maximum in the southern North Cascades.

Deposits of the maximum advance have not been dated directly in the Washington Cascades, where their age is inferred mainly from indirect evidence (Figure 4-12). In some western-slope valleys, alpine drift of the last glaciation demonstrably antedates the overlying Vashon drift of the Puget lobe. Along the South Fork of the Snoqualmie River, for example, wood in Vashon lake sediments that overlie alpine drift believed to represent the earliest Late Wisconsin alpine advance is dated 13,570±130 yr B.P. (UW-35) (Porter, 1976). Alpine drift on southeastern Vancouver Island, inferred by Halstead (1968) to be equivalent to Evans Creek drift, was deposited shortly after 19,150±250 yr B.P. (GSC-210). Near Vancouver on the adjacent mainland, a till (Coquitlam Drift) that lies beneath Vashon drift contains wood dated 21,500 ± 240 yr B.P. (GSC-2536); related outwash is dated 21,600 ± 200 yr B.P. (GSC-2203) to 21,700 ± 240 yr B.P. (GSC-2335) (Clague et al., 1980). A nearby alpine (?) till is bracketed by dates of 24,800 ± 310 yr B.P. (GSC-2273) and 18,700 ± 170 yr B.P. (GSC-2344). The area was free of ice from at least 18,700 ± 170 yr B.P. until 17,800 ± 150 yr B.P. (GSC-2297). The Coquitlam Drift and associated deposits are believed to be possible correlatives of Evans Creek drift (Clague et al.,

1980); if this is true, the maximum alpine advance of the last glaciation probably culminated sometime between about 22,000 and 18,500 yr B.P. (Figure 4-12).

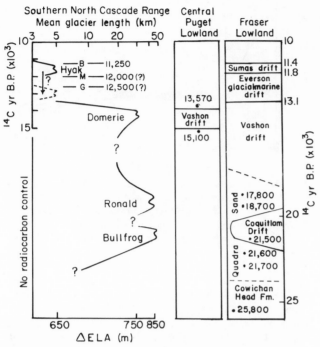

Figure 4-12. Provisional time-distance curve of glacier terminal variations and changes of equilibrium-line altitude in the southern North Cascade Range, together with tentative correlations with deposits in the Fraser and Puget lowlands west of the range. (See the text for a discussion of radiocarbon dates and tephra stratigraphy.)

VASHON-AGE ADVANCE

The retreat of glaciers in Cascade Valleys led to the trenching of end moraines and terracing of valley trains by axial streams. A subsequent advance led to the deposition of new end-moraine systems behind the outermost ridges, and younger valley trains were inset within the terraced older ones. In a number of eastern North Cascade valleys, moraines that impound deep lakes (e.g., Lake Wenatchee, Lake Cle Elum, Lake Keechelus, and Lake Kachess) apparently are composed largely of sheared and folded rhythmically laminated lake sediments containing abundant dropstones. These relationships suggest that before the advance glaciers retreated well upvalley and that some terminated in proglacial lakes ponded behind moraines of the maximum advance. In some valleys the younger moraines are closely nested within those of the older advances (e.g., the Icicle Creek-Wenatchee River and Peshastin River valleys), whereas in others they lie 5 to 30 km upvalley (e.g., the upper Yakima River and Wenatchee River valleys). The contrasting distances are due mainly to differences in the hypsometric properties of the drainage basins. Although deposits of this advance have not been well studied in the southern Cascades of Washington, multiple outwash terraces in the Cowlitz River valley, mapped before they were submerged beneath a new reservoir, point to a comparable record of glacier fluctuations. Moraines of similar character and relative position have been found in both the northern and southern Oregon Cascades (Carver, 1973); (Scott, 1977).

The younger moraines have not been directly dated, nor have they been mapped widely enough to warrant giving them a single regional name. In the upper Yakima River drainage basin, they are called Domerie moraines (Porter, 1976); in the Icicle Creek-Wenatchee River area, a possibly correlative moraine is informally designated Leavenworth III (Porter, 1969). Moraines along the South Fork of the Snoqualmie River that are regarded as equivalent to Domerie moraines on the basis of ice-limit data are believed to correlate with drift of the culminating advance of the Puget lobe (15,000 to 14,000 yr B.P.) (Porter, 1976), for they lie adjacent to laminated sediments of a lake ponded by the ice sheet. The date 13,570 ± 130 yr B.P. on wood from these sediments marks a time when the Puget lobe may still have blocked the valley and when alpine glaciers may have just begun to recede from moraines built during the readvance. That equivalent moraines east of the crest in the upper Wenatchee River drainage basin are older than about 12,500 years is shown by the presence of Glacier Peak tephra layer G (as well as younger layers) on their surface (Table 4-5) (Porter, 1978).

Widespread deglaciation followed the Domerie advance. The deep lake basins behind some moraine systems imply that the retreat was too rapid for the basins to fill with sediment as the ice receded. Furthermore, airfall tephra of the earliest major Glacier Peak eruption (layer G) has been found in valley heads as much as 40 km behind moraines. These relationships indicate that major ice streams south of the Cordilleran ice sheet largely disappeared between 14,000 and about 12,500 years ago. Much of the dissection of end moraines by axial streams and terracing of the outwash valley trains in the main valleys may have occurred during this interval.

LATE-GLACIAL READVANCE

A final phase of glacier resurgence near the end of the Pleistocene left end moraines and small outwash trains near the heads of many valleys in the Washington Cascades south of the limit of the Cordilleran ice sheet, as well as in the Oregon Cascades. Where identified and studied in the North Cascades, these deposits are referred to the Hyak advance (upper Yakima River drainage basin) (Porter, 1976) and the Rat Creek advance (Wenatchee River drainage basin) (Porter, 1969). In the northern Oregon Cascades, Scott (1977) identifies possibly comparable moraines, which he calls Canyon Creek; Carver (1973) describes late-glacial moraines in southern Oregon, which he refers to as Zephyr Lake drift. Typically, the moraines are found in or near cirque basins or in valley-head positions but generally 500 m or more lower than the termini of modern glaciers. Many cirque basins in the southern North Cascades at altitudes of 1000 to 1800 m last contained glacier ice during this late-glacial readvance.

Although often only a single moraine is present in a valley, in many places two (and sometimes more) moraines form a nested set. Among the best examples is the type Rat Creek moraine set in Icicle Creek, which consists of a pair of arcuate morainal loops (Porter, 1969). Chronologic control is not yet good enough to tell whether these moraines represent minor fluctuations of the ice margin during a single readvance or two distinct readvances.

The closest minimum date for the late-glacial drift comes from Snoqualmie Pass in the central Washington Cascades, where wood in fine gravel that overlies till inside an arcuate moraine dates to 11,050 ± 50 yr B.P. (UW-321). In this area and in many others, Mazama tephra (layer O) dating to about 6700 yr B.P. (Table 4-5) mantles the moraines. The only available maximum limiting age comes from the eastern North Cascades, where tephra layers of the late-glacial Glacier Peak eruptions mantle the land surface. Near Stevens Pass tephra layer M, which is older than 11,250 years and is inferred to be about 12,000 years old, has been found beyond, but not on, late-glacial moraines that lie within the fallout zone of this unit. This relationship implies that the moraines postdate the tephra eruption. On this basis, the late-glacial moraines were assigned a provisional age of about 12,000 to 11,000 years (Porter, 1978) and are inferred to correlate with Sumas Drift in the Fraser lowland of British Columbia, to which Armstrong (1975) assigns an age of about 11,800 to 11,400 years. However, if layer M proves to be older than estimated, then the readvance also could be older. The occurrence of Layer G tephra in cirques in the North Cascades that should have generated Rat Creek/Hyak-age glaciers (Beget, 1981; R. B. Waitt, Jr., unpublished data) suggests that the readvance may have occurred more than 12,500 years ago.

Younger moraines overlain by Mazama tephra and found only a short distance beyond the outermost Neoglacial moraines have been reported at several sites in the North Cascades (Beget, 1981; R. B. Waitt, Jr., personal communication, 1981). At Mount Rainier, such moraines are mapped as McNeeley Drift by Crandell and Miller (1974) and apparently predate a tephra (layer R) that is more than 8750 years old (Table 4-5). Although possibly of late-glacial age, available radiocarbon dates and relative-age data suggest that these moraines date to the early Holocene (Beget, 1981).

PALEOENVIRONMENTAL INFERENCES

Reconstructed equilibrium-line altitudes for Cascade glaciers are based on modern net-mass-balance curves and topographic reconstructions of former glaciers. A curve representing the mean of seven net-mass-balance curves for glaciers in northwestern North America (Figure 4-13) has been used to calculate the mass balance in both the accumulation area and the ablation area for successive altitude increments of former glaciers under assumed steady-state conditions. The equilibrium-line altitudes were adjusted until the positive net balance in the accumulation area equaled the negative net balance in the ablation area. Implicit in this method is the assumption that the mean mass-balance gradient of late-Pleistocene glaciers was similar to that of modern glaciers. Calculations for two such glaciers representing the Domerie and Hyak advances in the upper Yakima River drainage basin are shown as examples in Figure 4-14. The accumulation-area ratio (0.68) falls within the range of modern accumulation-area ratios for active glaciers in the Pacific Northwest (Meier and Post, 1962). Data for Pleistocene glaciers in a transect across the southern North Cascade Range indicate that equilibrium-line altitudes rose eastward with a gradient similar to that of the present time (Figure 4-15). If the median altitude of north-facing glaciers within this zone is assumed to approximate present-day equilibrium-line altitudes, then equilibrium-line altitudes were lowered at least 850 m below their present level during the maximum (Bullfrog) advance, about 750 m during the subsequent Domerie advance, and about 650 to 700 m during the late-glacial Hyak readvance. These values do not take into account possible glacial-isostatic, effects, which may have amounted to more than 100 m near the crest of the Cascades, where the maximum ice thickness was 600 m or more. During Vashon time (about 16,000 to 13,000 yr B.P.), the thick (more than 1000 m) and extensive Cordilleran ice sheet in and adjacent to the North Cascades would have had an additional and probably significant isostatic effect. Such isostatic corrections would increase the calculated equilibrium-line-altitude depression by a corresponding

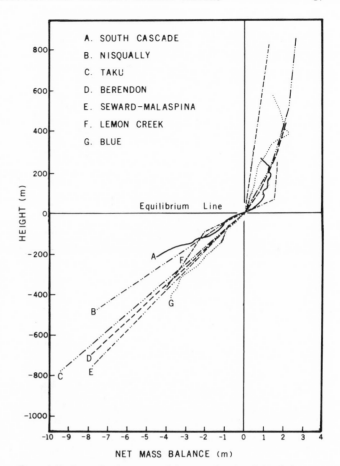

Figure 4-13. Curves showing net mass balance as a function of altitude for seven representative glaciers in the Pacific Mountain system of northwestern North America. (Data from Meier et al., 1971.)

amount.

When the change in equilibrium-line altitude is plotted as a function of the glacierized area for the upper Yakima River drainage basin (Figure 4-16), it becomes clear that once equilibrium-line altitudes are depressed by 600 m, only a modest additional decrease results in a substantial expansion of the glacier-covered area. Because such curves reflect the hypsometric characteristics of a drainage basin, different basins may have dissimilar curves (Figure 4-16). The spacing of successive moraine systems, representing different amounts of snow-line depression, reflects such hypsometric differences.

The inferred changes of equilibrium-line altitude (uncorrected for isostatic effects) through time for the southern North Cascade Range are illustrated in Figure 4-12. Only data for the culminations of advances are available; neither the extent of ice recession during intervening phases nor the configuration of glaciers at such times is known. Similar values have been obtained for the Oregon Cascades by Scott (1977), who calculates a minimum equilibrium-line-altitude depression for the last glaciation (Suttle Lake advance) of 950 m; the Canyon Creek advance represented a decrease in equilibrium-line altitudes of about 700 to 750 m.

Additional data have been obtained from the study of modern and former glaciation thresholds within the North Cascades (Porter, 1977). The present glaciation threshold has a regional west-east gradient averaging 12 m per kilometer (Figure 4-8) and can be used as an

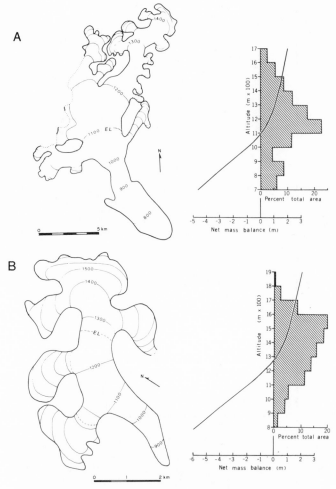

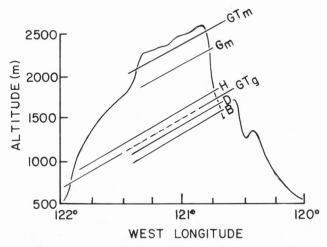

temperature was about 5.5°C ± 1.5°C lower, and the mean temperature was about 4.2°C ± 1.0°C lower (Porter, 1977). That such estimates are reasonable is suggested by independent estimates based on other types of data. For example, Heusser and others (1980) use factor and regression analysis of fossil pollen and climatic data from the Olympic Peninsula to suggest that the mean July temperature during the last glacial maximum averaged about 10.5°C, some 4°C lower than at present, and that the annual precipitation was about 1350 mm, or nearly 30% less than modern amounts. The cooler, somewhat drier climatic regime of that period resulted in the replacement of interglacial-type vegetation similar to that of the present by a flora dominated by herbs and spruce, probably indicative of a spruce parkland. Such conditions possibly were com-

Figure 4-15. Transect across the southern North Cascade Range at latitude 47°30'N showing modern (GT$_m$) and Late Wisconsin (GT$_g$) glaciation thresholds, mean altitude of modern glaciers (G$_m$), and reconstructed equilibrium-line altitudes during the Bullfrog (B), Domerie (D), and Hyak (H) advances.

Figure 4-14. Reconstructed topography and equilibrium line (EL) for glaciers in the upper Yakima River drainage basin of the North Cascade Range during the culmination of (A) the Domerie advance and (B) the Hyak advance. The mass balance/altitude curve was adjusted vertically until the positive mass balance in the accumulation area equaled the negative mass balance in the ablation area. The resulting equilibrium-line altitudes were about 1100 m and 1275 m, respectively. (See the text for a discussion.)

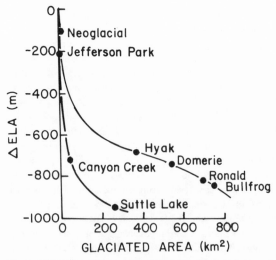

Figure 4-16. Curve of glaciated area plotted as a function of change in equilibrium-line altitude (ELA) from present level for the upper Yakima River drainage basin in the North Cascade Range of Washington (upper curve) and the upper Metolius River drainage basin of the Oregon Cascades (lower curve). (Data from Porter, 1976, and unpublished, and Scott, 1977.)

approximation for the regional snow line. A reconstructed glaciation threshold for a transect across the range representing the maximum advance of alpine glaciers during the Fraser Glaciation lies some 900 (± 100) m below that of the present (Figure 4-15), in agreement with the calculated figures for maximum equilibrium-line-altitude depression. As in the earlier example, corrections were not made for possible isostatic effects, nor was the sheltering effect of large Pleistocene cirques on the distribution of modern glaciers evaluated.

The depression of equilibrium lines and the glaciation threshold by at least 850 to 900 m at the last glacial maximum constitutes a measure of the climatic change required to generate the large glacier system that mantled the Cascades. Although unique values for temperature and precipitation regime at the glacial maximum cannot be derived from these data, estimates based on modern climatic values at the present glaciation threshold in the North Cascades suggest that the accumulation-season precipitation may not have been more than about 30% lower than present values, the ablation-season

parable to the open parkland near the timberline in the Cascades today, where the mean July temperature is close to that reconstructed for the Late Wisconsin in the adjacent Puget lowland (Heusser, 1982). The cool but above-freezing temperatures indicated by these studies are consistent with the extensive outwash deposited on both sides of the Cascades during the maximum advance, for the advance implies production of substantial meltwater from the terminal zones of warm-based glaciers.

The closest modern analogue for conditions in the Cascades at the culmination of the last glaciation probably is to be found in southern Alaska, where glaciers of comparable size now exist in a cool, moist climate. Some lie close to open spruce parkland and feed large, braided meltwater streams that flow within broad expanses of gravelly outwash.

Sierra Nevada

The Sierra Nevada extends south of the Cascade Range in California as a westward-tilted crustal block bounded along much of its complex and faulted eastern margin by a high, dissected scarp. During successive Pleistocene glaciations, ice fields formed along the crest of the range and valley glaciers descended both flanks (Wahrhaftig and Birman, 1965). The glacier system averaged about 50 km wide, extended some 450 km through nearly 4° of latitude, and formed one of the largest areas of alpine ice in the western conterminous United States during the last glaciation (Figure 4-2).

The range is composed of deformed Paleozoic and Mesozoic metasedimetary rocks intruded by a batholithic complex of quartz-diorite to granodiorite and uplifted to reach altitudes of 2500 to 4400 m along the crest. The granitic rocks underlie most of the higher, glaciated parts of the range, and the metamorphic rocks crop out mainly in a belt about 50 to 65 km wide in the northwestern part of the range. Volcanic rocks of Miocene to Holocene age are present locally within the mountains, and the younger rocks have provided important radiometric dates for associated glacial deposits.

The streams draining the relatively gentle western slope of the Sierra Nevada flow toward the Great Valley of California, the floor of which lies close to sea level. The streams draining the eastern slope are short and steep. The range forms a natural barrier to the eastward flow of maritime air masses off the Pacific Ocean; this results in a strong climatic gradient. Precipitation, which falls mainly during the winter, is highest on the western slope, where as much as 1800 mm has been recorded; east of the crest, values drop to 500 mm or less. The pronounced rain-shadow effect leads to corresponding differences in vegetation, with the deciduous and coniferous forests of the western slope contrasting the xerophytic flora found along much of the southeastern base of the range adjacent to its higher sector.

The altitudinal distribution of cirque floors reflects the climatic gradient. The altitudes of cirque floors rise from north to south along the trend of the range, and they rise steeply eastward up the slope (Figure 4-2) (Wahrhaftig and Birman, 1965). The lowest cirques are found in the northwest at altitudes close to 2400 m, and the highest lie in the southeast and reach altitudes of as much as 4000 m.

Because of the range's high snow line, present-day glaciers are confined mainly to the sheltered northern sides of high peaks in the southern half of the range and lie at altitudes between about 3400 and 4300 m. However, the Sierra Nevada has an abundance of remarkable glacial landforms at lower altitudes that indicate a former exten-

sive glacier cover. This landscape provided important examples of glacial erosional and depositional processes for geologists who were instrumental in developing glacial-geologic concepts in the western United States (e.g., Le Conte, 1873; Matthes, 1930; Blackwelder, 1931; Colby, 1950).

LATE WISCONSIN GLACIATION

The glacier complex that formed over the Sierra Nevada during the last glaciation was asymmetrically distributed, with large ice fields and long outlet glaciers flowing down the gentle western slope and short, deep valley glaciers occupying the valleys on the eastern flank. Western-slope ice streams terminated within deeply entrenched stream valleys well inside the mountains. Glaciers east of the crest typically flowed beyond bedrock canyons to downfaulted basin floors, where, confined by high lateral moraines, they formed long, narrow tongues. The largest glaciers draining the eastern slope were between 15 and 27 km long. By contrast, some on the western slope reached lengths of more than 60 km. Although the ice divide generally coincided with the present stream-drainage divide, in places ice flowed east across the crest; this implies that the ice divide was displaced locally to the west (Bateman and Wahrhaftig, 1966: 160; C. Wahrhaftig, personal communication, 1981). The smaller glaciers probably reached a maximum thickness of between 250 and 350 m, but the largest ice streams draining the western slope were more than 1000 m thick.

The extent of the ice cover has been mapped in some detail within certain drainage basins (e.g., Matthes, 1930, 1960, 1965; Birkeland, 1964; Birman, 1964). The extent of former glaciers in their terminal zones, as marked by moraines or drift limits, are shown on maps accompanying many reports (e.g., Sharp, 1969, 1972; Burke and Birkeland, 1979). Clark (1967) made a detailed reconstruction of Tioga glaciers in the West Walker River drainage basin east of the crest, where the ice was as much as 500 m thick and the calculated basal shear stress was about 1 bar.

STRATIGRAPHY AND CHRONOLOGY

Considerable effort has been directed toward differentiating and classifying the deposits of first-order glaciations (Porter, 1971) in the Sierra Nevada through the use of relative-dating methods (Blackwelder, 1931; Sharp and Birman, 1963; Birman, 1964; Sharp, 1969, 1972; Burke and Birkeland, 1979). Deposits of the youngest Pleistocene glaciation in Yosemite National Park on the western slope of the range were referred to the Wisconsin by Matthes (1930). Blackwelder (1931) recognized deposits of four glaciations on the eastern slope of the range, the youngest two of which (Tioga and Tahoe) have generally been assumed to be of Wisconsin age. Sharp and Birman (1963) later proposed the name Tenaya for moraines that lie between moraines of Tioga and Tahoe age. Although Tioga drift is widely regarded as representing Late Wisconsin glaciation, the age of the two earlier drifts is uncertain. Sharp (1972), for example, maintains that the Tenaya was a valid and distinct glaciation between the Tioga and Tahoe and that all three advances postdated the last interglaciation. Some investigators (e.g., Morrison, 1965; Smith, 1968; Birkeland and Janda, 1971) suggest that the Tenaya may be an early phase of the Tioga glaciation, but Burke and Birkeland (1979), in a regional reassessment of the stratigraphy, have been unsuccessful in differentiating the Tenaya from either Tioga or Tahoe drifts on the basis of multiple relative-dating methods. In some valleys, the weathering properties of the Tenaya seem to be close to those of Tahoe drift, but in others they more closely resemble Tioga. It may be possible that the differences could be explained by (1) subtle lithologic contrasts between basins, (2) subtle climatic differences, or (3) problems inherent in using weathering data for intervalley correlation (Birkeland et al., 1976). In the southern Sierra Nevada, the Tioga drift is much less extensive than the Tahoe; however, in the northern part of the range, the two drifts are similar in extent. If the intermediate Tenaya advance was less extensive than the Tioga in the north, its deposits may underlie Tioga drift. In the south, the Tenaya may have been a more extensive advance than the Tioga, and a more widely recognizable record may have been the result. Burke and Birkeland (1979) suggest that the Tenaya should not be regarded as a first-order glaciation unless additional field studies support such an assumption, and they propose reverting to the original twofold Tahoe-Tioga scheme. If the Tenaya does represent a major advance or readvance between the Tahoe and Tioga glaciations, its exact relationship remains obscure.

Subdividing the first-order glaciations has not been possible through the use of relative-dating methods, so little is known about the detailed history of glacier variations during the Tioga glaciation. No regional pattern of second-order fluctuations has been discerned and no nomenclature established. Furthermore, radiometric dates are not yet available to assess the chronology of such fluctuations.

Published reports suggest that there is a complex record of ice-margin fluctuations in many valleys. Birkeland (1964), working in the northeastern Sierra Nevada near Lake Tahoe, found numerous recessional moraines of Tioga age (some as much as 13 m high and clustered in groups of 3 to 7) that suggest multiple readvances during deglaciation. Matthes (1930) depicted as many as 34 nested moraine crests of late-glacial age in Little Yosemite Valley, and Birman (1964) mapped more than 12 moraine ridges of Tioga age in the Vermillion Valley, west of the crest. Putnam (1950) also found large numbers of Tioga moraines along Lee Vining Creek (29, in four main groups), at Grant Lake (30, largely in two main groups), and at June Lake (28, largely in two groups) (Figures 4-17 and 4-18). Sharp compared the spacing and estimated bulk of Tioga moraines in four canyons on the eastern slope of the range and suggested that the glaciers had responded in a similar but not identical manner during ice recession. The apparent lack of similarity in detail led him to conclude that the correlation of such moraines from canyon to canyon over any distance was likely to be unreliable.

Moraines in the headward portions of valleys in the eastern Sierra Nevada are described by Birman (1964), who refers to them as Hilgard moraines. Relative-dating studies resulted in granite-weathering ratios that implied an age older than Tioga, but the position and morphological character of the drift clearly favor a late-Tioga or post-Tioga age. Birkeland and others (1976) suggest that the anomalous relative dating resulted from comparing sites above and below the treeline. For example, weathering pits become progressively deeper with time above the treeline, but within forested areas they can be eradicated by rock spalling caused by forest fires. They suggest that the Hilgard may represent a late advance of the Tioga glaciation. The problem of the late-glacial readvance of glaciers in the Sierra Nevada clearly merits additional study.

Our present understanding of the chronology of Tioga glaciation rests on several potassium-argon dates of basalts that are interstratified with glacial drift and on a single radiocarbon date that provides an upper limiting age.

An olivine basalt flow in Sawmill Canyon on the eastern slope of the range was believed by Dalrymple (1964) to overlie pre-Tahoe

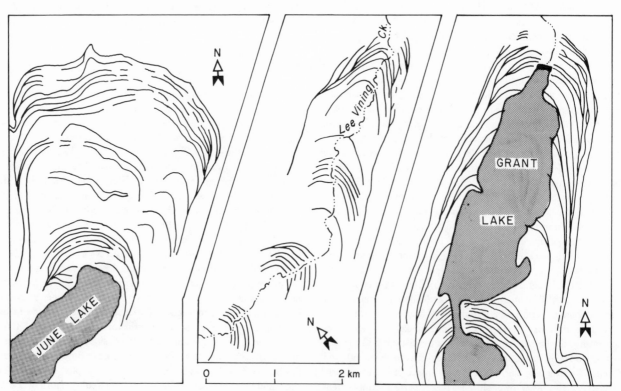

Figure 4-17. Tioga end-moraine systems along Lee Vining Creek and at Grant Lake and June Lake in the eastern Sierra Nevada. (After Putnam, 1950.)

Figure 4-18. Areal photograph of the eastern slope of the Sierra Nevada near Mono Lake showing end-moraine systems around Grant Lake (on left), Parker Creek (center), and Walker Creek (right). The type Mono Basin moraine can be seen emerging from beneath the right-lateral moraine complex of the Walker Creek Glacier. (Photograph by Mary Gillam.)

drift and to be overlain by tills of Tahoe, Tenaya, and Tioga age. However, Burke and Birkeland (1979) have reexamined the site and concluded that the underlying till is Tahoe and the overlying till is Tioga. By their interpretation, Dalrymple's dates of 90,000 ± 90,000 (KA970) and 60,000 ± 50,000 yr B.P. (KA970A), together with a new date for the basalt of 53,000 ± 44,000 yr B.P. (Dalrymple et al., 1982), provide a maximum age limit for Tioga drift and a minimum for Tahoe (Figure 4-19).

Casa Diablo till in the Mammoth Creek area, interpreted by Curry (1971) as pre-Tahoe in age, has been found to be indistinguishable from Tahoe drift of the eastern Sierra Nevada by Burke and Birkeland (1979), who reassign it to the Tahoe glaciation. Potassium-argon dates of basalt flows that bracket the till are 129,000 ± 26,000 years old (73G012) and 64,000 ± 14,000 years old (73G014) (Bailey et al., 1976; P. W. Birkeland, personal communication, 1981).

The large standard deviations for the available potassium-argon dates leave considerable room for several interpretations of the age of the Tahoe drift and its correlatives (Figure 4-19). Although the traditional view that the Tahoe may date to the Early Wisconsin and, therefore, correlate with stage 4 of the marine $\delta^{18}O$ record (Shackleton and Opdyke, 1973) has not been ruled out, the dates do not exclude the possibility that Tahoe drift antedates the last interglaciation (isotope stage 5), for at one standard deviation the two lower limiting dates for the Tahoe both encompass a substantial part of isotope stage 6. Such an interpretation would be attractive to those who favor a correlation of the Tahoe with the Bull Lake deposits of the Rocky Mountains, which have recently been interpreted as correlative with marine isotope stage 6 on the basis of combined obsidian-hydration and potassium-argon dating (Pierce et al., 1976). Additional evidence in support of this view was found on the eastern side of Cascade Lake in the Lake Tahoe area by C. Wahrhaftig (personal communication, 1981), where Tioga till rests on Tahoe till.

Grussified boulders and deep weathering on the older till suggest a long weathering interval between the two advances. Elsewhere in this area of the type Tahoe stage as defined by Blackwelder (1931), boulders on the surface of Tahoe moraines are deeply weathered and surrounded by gruss; those on Tioga moraines are virtually unweathered. These relationships imply a much longer Tahoe-Tioga nonglacial interval than is represented by post-Tioga time, possibly an order of magnitude longer. However, Colman and Pierce (1981) suggest on the basis of weathering-rind analysis that most of the drift in the Rocky Mountains that has been mapped as Bull Lake may be older than most Tahoe drift of the Sierra Nevada, which they regard as probably Early Wisconsin or younger. Until definitive radiometric dating control becomes available this problem probably will remain unsolved.

There seems to be little doubt that Tioga drift was deposited during Late Wisconsin time, and such an age is compatible with the limiting potassium-argon dates, which imply an age of less than about 50,000 years. An upper limiting age is afforded by a radiocarbon date from Osgood Swamp, a small lake dammed by a Tioga terminal moraine near the southern end of Lake Tahoe (Adam, 1967). The basal meter of a 4.4-m-long core from the basin is gray silt, inferred by Adam to have been deposited by glacial meltwater during ice recession. A sample from a depth of 3.1 m in overlying organic mud has an age of 9990 ± 800 yr B.P. (A-545) and marks a time following the deglaciation of the site when meltwater no longer contributed large quantities of silt to the basin. Lake sediments (Upper Member, Modesto Formation) on the western slope of the Sierra Nevada that are correlated with Tioga drift are bracketed by dates of about 27,000 and 9000 years and contain wood dated 14,060 ± 450 yr B.P., and age in agreement with a date of 14,100 ± 200 yr B.P. (USGS-38) for wood in outwash sand at Bakersfield (Marchand and Allwardt, 1981). These ages are also consistent with independently dated lacustrine

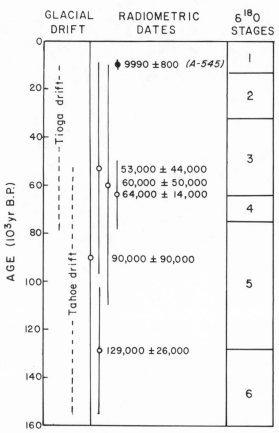

Figure 4-19. Radiometric dates related to Tahoe and Tioga drifts on the eastern flank of the Sierra Nevada. (See the text for a discussion.)

sediments east of the Sierra Nevada that indicate that the last interval of high water in Mono, China, Owens, and Searles Lakes occurred some time between about 25,000 and 10,000 years ago (Smith and Street-Perrott, 1982). Because the lakes were fed by meltwater from Sierra Nevada glaciers, this high-water interval is generally inferred to correlate with Tioga glaciation. Putnam (1950) describes field evidence that supports such an interpretation, for it indicated that the last high level of Mono Lake correlated with the maximum Tioga ice advance.

Although at present there are no dates directly associated with Hilgard or correlative moraines in the Sierra Nevada, the relative position of the moraines within the glacial sequence, the apparent snow line depression they represent, and their general relative-age characteristics all suggest that they may correlate with late-glacial moraines elsewhere in the western United States, which may be between about 13,000 and 11,000 years old (e.g., Porter, 1978: 40).

PALEOENVIRONMENTAL INFERENCES

The natural environment in the Sierra Nevada was substantially different during Late Wisconsin glaciation than it is today. Only a few very small glaciers are today scattered along the crest of the High Sierra, but large ice fields and long valley glaciers formed a vast glacier complex during the Tioga glaciation. In its overall dimensions, it was only slightly smaller than the combined modern North and South Patagonian Ice Caps of southern Chile and Argentina. The change was brought about by a substantial depression of snow line that in the southern, highest part of the Sierra Nevada was estimated by Bateman and Wahrhaftig (1966) to have been 900 m or more.

Much of the meltwater produced by glaciers draining the eastern slope was channeled into Lake Tahoe, Mono Lake, and the Owens-China-Searles-Panamint-Death Valley lakes system (Smith and Street-Perrott, 1982). Coarse, bouldery outwash fans that lie beyond end-moraine systems and that today are covered with sagebrush and trenched by small streams were actively aggrading surfaces. Meltwater on the western slope of the range was channeled down rocky canyons and spread out across the eastern margin of the Great Valley; the glacial source of alluvial units there has been inferred by Janda (1965) on the basis of lithologic and mineralogic studies.

The continued tectonic displacement of the Sierra Nevada during the late Quaternary is evidenced by offset moraines of Tioga age that were constructed across the eastern range-front fault system (e.g., Putnam, 1960). Some of this tectonic movement no doubt took place while glaciers mantled the range, and it may have accompanied deglaciation as the crust adjusted isostatically to the removal of the ice load.

Rocky Mountains

The Rocky Mountains of the United States consist of about 100 ranges distributed in a northwest-trending belt 2000 km long and 200 to 800 km wide (Figure 4-2). During Late Wisconsin time, glaciers existed in about 50 of these ranges. The contrasting characters of the Pleistocene glacial features among ranges of the Rocky Mountains are determined by local climate, altitude, and bedrock lithology.

Precipitation and humidity tend to decrease inland from the Pacific Ocean, but near the eastern front and southern end of the Rocky Mountains moisture from the Gulf of Mexico becomes important, especially during the spring and autumn. The average summer temperature decreases at a lapse rate of about 6.5 °C per 1000 m of altitude (Thompson and Dodd, 1958). At a given altitude, mean annual temperature decreases northward about 0.8 °C per degree of latitude (National Oceanic and Atmospheric Administration, 1976). The equilibrium-line altitudes of cirque glaciers reflect these climatic trends; modern equilibrium-line altitudes are about 3700 m in Colorado, about 2500 m in northwestern Montana, and about 2100 m in northern Idaho (Meier et al., 1971).

Snowpack records are available for snow courses throughout the Rocky Mountains (Figure 4-20). Differences in modern snowpack among ranges closely relate to differences in late-Pleistocene glaciation thresholds. Snowpack for a group of stations can be conveniently defined by a linear regression line of altitude against water content of the April-first snowpack (correlation coefficients average 0.90 ± 0.06). A southeast-northwest transect shows the expected increase in snowpack northward, but some ranges depart from this trend. For example, at an altitude of 2500 m, the snowpack in the western Yellowstone area has about five times as much water as that in adjacent areas (Figure 4-20A, lines 4, 5, and 6), because the Snake River Plain acts as a pathway for the inland transport of Pacific moisture to the Yellowstone area. Another departure from the trend is the San Juan Mountains, which are 3° of latitude south of the

Uinta Mountains but which at an altitude of 3000 m have a spring snowpack about 24% greater (Figure 4-20A, lines 2 and 3).

Bedrock in the glaciated cores of the mountains in Wyoming, Colorado, and New Mexico is mostly Precambrian granitic rocks. The San Juan Mountains of southern Colorado are mainly middle-Tertiary volcanic rocks. The Yellowstone Plateau consists mostly of Quaternary rhyolite flanked on the east by Eocene andesite. The mountains of northwestern Montana and the Uinta Mountains of Utah expose mostly Precambrian bedded metasedimentary rocks.

Bedrock lithology strongly affects the character of late-Pleistocene

moraines. For example, on the western side of the Wind River Mountains, the New Fork glacier flowed mostly across granitic bedrock and constructed bouldery moraines with high, steep slopes. However, in the next valley to the north, the Green River glacier headed on similar bedrock but then traversed 5 to 15 km of mudstone before depositing subdued end moraines with few boulders.

For the 1965 congress of the International Association for Quaternary Research, Richmond (1965; Richmond et al., 1965) summarized the state of knowledge about the Quaternary glaciation of the Rocky Mountains. Subsequent summary contributions include those of

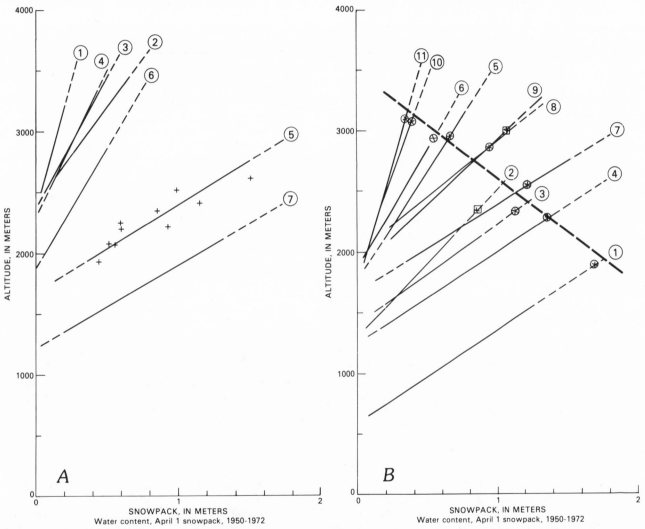

Figure 4-20. Altitudinal distribution of April-first snowpack in the Rocky Mountains from snow-course records of each state. **(A)** Transect from New Mexico (line 1) to Montana (line 7). Crosses mark values defining linear regression for line 5. Line 1, southern Sangre de Christo Range, New Mexico; line 2, southern San Juan Mountains, Colorado; line 3, eastern Uinta Range, Utah; line 4, easter Wind River Range, Wyoming; line 5, western Yellowstone Plateau-northern Teton Range, Wyoming; line 6, Lemhi Range and Beaverhead Range, central Idaho; and line 7, northwestern Montana mountains, western slope. **(B)** Transect from the Pacific Ocean to central Wyoming, numbered from west to east. Crosses mark the intercepts of the Pleistocene glaciation threshold with lines defining the altitudinal distribution of the modern snowpack. The 11 intercepts define a regression line with a correlation coefficient of 0.86. The two points enclosed by open squares were significantly wetter (line 2) or drier (line 9) in late-Pleistocene time compared to present conditions than the other sites. Heavy dashed line, linear regression for 9 sites of the Pleistocene glaciation threshold on the modern snowpack at the threshold (correlation coefficient of 0.96). Line 1, Cascade Range, Oregon; line 2, Strawberry Mountains, Idaho; line 3, westernmost Idaho; line 4, Payette River highlands, Idaho; line 5, Lost River Range, Idaho; line 6, Lemhi Range and Beaverhead Range, Idaho; line 7, western Yellowstone-northern Teton Range, Wyoming; line 8, central Yellowstone, Wyoming; line 9, Absaroka Range, Wyoming; line 10, western Bighorn Range, Wyoming; and line 11, eastern Bighorn Range, Wyoming.

Flint (1971), Porter (1971), Montagne (1972), Birkeland and others (1971), Mears (1974), Richmond (1976), Madole (1976a), Mehringer (1977), Shroba (1977), and Hollin and Schilling (1981).

LATE WISCONSIN GLACIATION

Distribution

About 50 separate ranges in the Rocky Mountains were glaciated during Late Wisconsin time (Figure 4-2). Detailed studies have been made of about 35 glaciated areas, most of which are referred to in Flint (1971: Table 18-B) and Richmond (1965: Table 1).

Contours on the altitude of the lowest cirques show an overall decrease to the northwest (Figure 4-2), a relation consistent with a poleward cooling and a moisture source from the Pacific Ocean. Departures from this overall trend generally reflect precipitation anomalies. High contours extending for hundreds of kilometers to the east of the highest parts of the Sierra Nevada and Cascade Range reflect an orographic precipitation shadow. Moisture was also extracted from air masses as they moved toward the cores of the mountains in central Colorado and Idaho. North of the Sierra Nevada, relatively low contours indicate movement of moist air eastward across southern Oregon and the Snake River Plain to the Yellowstone area. The east-west trend of contours in Montana may have resulted from the southward flow of cold air from above the Laurentide and Cordilleran ice sheets.

Types of Glaciers

Late Wisconsin glaciers ranged from small, thin, steeply sloping cirque glaciers to large, thick, gently sloping transection glaciers and ice caps. Figure 4-21 shows a subdivision based on glacier surface slope, which is inversely related to thickness by basal shear stress, according to the relation:

$$\tau_b = \varrho g h \sin \alpha \cdot F$$

as defined in Figure 4-21. Basal shear stress for modern mountain-valley glaciers is commonly about 1 bar and almost always in the range from 0.5 to 1.5 bars (Nye, 1952; Patterson, 1969: 91). Reconstructions of Rocky Mountain glaciers also have values of about 1 bar (Haefner, 1976; Gallie, 1977; Pierce, 1979). In the middle and lower reaches of a glaciated valley, the valley gradient can also be used to estimate the maximum thickness of Late Wisconsin glaciers (Figure 4-21), because the valley gradient is typically equal to or less than the surface slope.

The term "slab-type glaciers" is applied here to relatively thin glaciers with surface slopes steeper than 5° (Figures 4-21 and 4-22). This type of glacier was the most common in the Rocky Mountains and was limited to the uppermost 10 to 15 km of valleys, where valley gradients typically exceed 5°. Longitudinal cross sections of several late-Pinedale glaciers in the Gallatin Range of Yellowstone National Park (Figure 4-22) illustrate graphically that thickness tends to increase as basal slope decreases. As slab-type glaciers lengthen down steep valley floors, they tend to cease thickening when their surface slope parallels the valley gradient. Unless valley deepening has occurred between glaciations, two important consequences of this relation are these: (1) where the surface slope of a glacier parallels its basal slope, evidence of any older glaciers will not be found higher on the valley walls; and (2) slab-type glaciers record changes in equilibrium-line altitudes only by changes in their length.

Glaciers intermediate in slope and thickness between slab-type

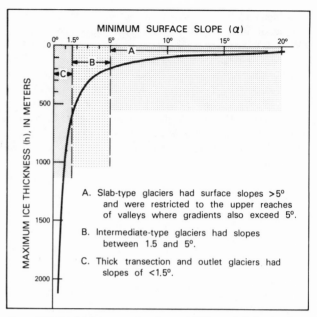

Figure 4-21. Relation between surface slope and thickness of glaciers based on basal shear stress of 1 bar. Three types of glaciers are distinguished on the basis of the slope in their midsections. Thickness (h, in centimeters) based on the relation $h = \tau_b/(\varrho g \sin \alpha F)$, where τ_b is the basal shear stress (1 bar or 10^6 dynes cm^{-2}); ϱ is the density of ice (0.9 g/cm^3); g is the acceleration of gravity (980 cm/sec/sec); α is the slope of the glacier surface; and F is the shape factor, here assumed to be 0.7 to account for drag against valley walls.

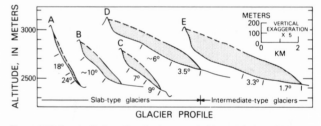

Figure 4-22. Longitudinal profiles of reconstructed slab-type glaciers of Late Pinedale age, Yellowstone National Park. Slope of valley floor is given beneath profiles. From left to right, profiles show that decreasing slope of valley-floor permits ice thickness to increase: southwest of Antler Peak (A), Indian Creek (B), the Crags (C), Panther Creek (D), and Gallatin River (E). (Based on data from Pierce, 1973.)

glaciers and transections glaciers or ice caps, here called "intermediate-type glaciers," are typified by Middle Boulder Creek glacier in Colorado (Figure 4-23). Such glaciers had surface slopes of 1.5° to 5° and maximum thicknesses of hundreds of meters, and they extended to positions where valley gradients were about 1° to 3°. These glaciers appear to have left some of the most visible erosional and depositional glacial forms in the Rocky Mountains, including steep cirque headwalls, U-shaped and staircased valleys, and high bouldery moraines. However, in studies of several intermediate-type glaciers, Haefner (1976) has found that glacial power associated with basal sliding did not correlate with quantitative measures of glacial modification.

Large ice caps or ice fields contained most of the total volume of Late Wisconsin ice in the Rocky Mountains. An ice field on the San

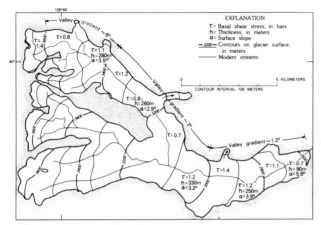

Figure 4-23. Reconstruction of intermediate-type glacier, Middle Boulder Creek, Colorado Front Range. Thin lines show modern streams. Heading in three compound cirques, this glacier extended about 15 km to where valley gradients were about 1°. Over most of its length, it was about 300 m thick, sloped about 3°, and had basal shear stress between 0.7 and 1.4 bars. (Reconstruction based on mapping of R. F. Madole, personal communication, 1981.)

Juan Mountains of southwestern Colorado was 100 km wide and reached a thickness of 1000 m; the main glaciers from this ice field flowed on valley gradients of less than 0.5° and had surface slopes as low as 1° (Figure 4-24) (Atwood and Mather, 1932). A larger ice field (30,000 km²) covered several ranges in northwestern Montana and generated outlet glaciers that extended 50 km eastward onto the plains of Montana (Figure 4-2) (Alden, 1932: Plate 2; Plate 1; Richmond et al., 1965).

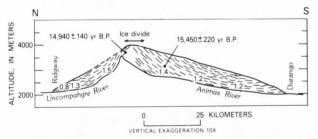

Figure 4-24. North-south cross section of the transection glacier that spanned 100 km of the crest of the San Juan Mountains, southwestern Colorado. Numbers at base of glacier are calculated basal-shear-stress values. This glacier thickened to 1000 m where the valley gradients were about 0.5° (Figure 4-21). Note that the increasing thickness of the glacier down-valley resulted in gentle slopes and thicker ice above the steep valley gradients near the topographic divide. Postglacial radiocarbon ages were projected from tributary cirques and divides on to the cross section. (Based on Atwood and Mather, 1932: Plate 3.)

Another Late Wisconsin ice field formed on the mountains in northwestern Wyoming and southwestern Montana (Figure 4-2). The northwestern part of this ice field has been reconstructed (Figure 4-25). The northern Yellowstone glacier drained most of the northern part of the ice field and had a flow line 145 km long. The lower third of this glacier flowed on valley gradients of 0.1° and reached a thickness of 1100 m.

Basal shear stress was determined for contiguous reaches of the ice field in the northern Yellowstone area (Figure 4-26). The basal shear

stress for reaches with strongly converging flow lines averages 1.2 bars, but the basal shear stress for those with strongly diverging flow lines averages 0.8 bars (Figure 4-26), thereby providing an explanation for most of the differences in calculated basal shear stress. Basal-shear-stress calculations can also be used to evaluate a contrasting reconstruction of the same ice field. This alternate reconstruction yields basal-shear-stress values of less than 0.1 bar, suggesting problems with the alternate reconstruction (squares, Figure 4-26).

Temporal Relations between Different Types of Glaciers

Studies of the northern Yellowstone area (Figure 4-25) show that the different ice sources did not build up and retreat at the same rate. Temporal relations among the different kinds of glaciers in the area (Figure 4-27) are inferred primarily from the later advance of glaciers from one source into the full-glacial domain of ice from another source. These relations show (1) that there was a strong peaking response of the ice cap on the Yellowstone Plateau, (2) that the larger mountain ice fields and the plateau ice cap grew toward maximum sizes more slowly than did the smaller valley glaciers but persisted at near-maximum size longer, and (3) that the mountain ice fields built up earlier and remained longer than the plateau ice cap, which was built on lower terrain.

Most of these differences seem to relate either to how quickly or to what extent a glacier responds to a change in equilibrium-line altitude. Small valley glaciers, which respond quickly to such changes, receded early during deglaciation. The large Yellowstone Plateau ice cap was 1000 m thick, but its crest was only 500 m above the equilibrium-line altitude (Figure 4-25). As the equilibrium-line altitude started its 1000 m rise at the end of the Pleistocene, a critical point was soon reached when the equilibrium-line altitude was above the entire plateau ice cap; thereafter, the glacier melted rapidly. A similar situation existed for the ice field on the San Juan Mountains (Figure 4-24).

Glaciers elsewhere probably responded in a similar manner to changes in equilibrium-line altitude during both glacial buildup and recession. Because of differences among source areas, end and recessional moraines that appear correlative in geographic position may not be of exactly the same age. Furthermore, the same time interval may be represented by greatly disparate amounts of relative recession.

STRATIGRAPHY AND CHRONOLOGY OF DRIFTS

The dotted line labeled 1965 model in Figure 4-28 schematically shows the chronology of the last (Pinedale) glaciation of the Rocky Mountains that was widely accepted from the mid-1950s to the mid-1970s (Richmond, 1965; Birkeland et al., 1971). New information published since 1971 and unpublished data suggest revisions of this chronology. An alternate chronology is developed here on the basis of this new information (Figure 4-28).

Stratigraphic Usage

Blackwelder (1915) introduced the names Pinedale and Bull Lake for the last two glaciations of the Wind River Mountains, Wyoming. These terms have become widely used for areas throughout the Rocky Mountains, where both glaciations have been traditionally regarded as Wisconsin in age. Recently, moraines of Bull Lake age near West Yellowstone have been dated by combined obsidian-hydration and potassium-argon techniques as about 140,000 to

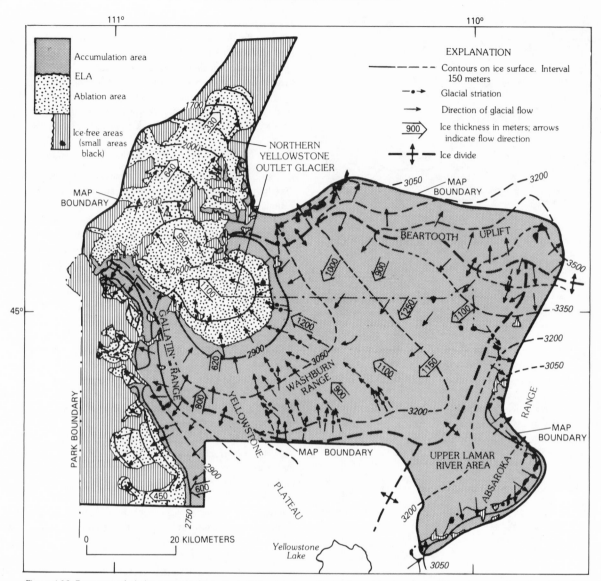

Figure 4-25. Reconstructed glacier system in the northern Yellowstone area. Glaciers from this system extended south, east, and northeast of map's boundary. The largest glacier shown within the system was the northern Yellowstone glacier; the headward limit of this glacier was an ice divide (heavy, dashed line). The altitudes of the contours on the ice surface in the accumulation area are based on nunataks and maximum altitudes of glaciated uplands; in the ablation area, they are based on morainal deposits. (After Pierce, 1979: Plate 1).

150,000 years old and, therefore, are pre-Wisconsin in age (Pierce et al., 1976; Pierce, 1979: F23-F25).

In his pioneering studies, Richmond (1960, 1962, 1965) demonstrated that soils can be used in the discrimination and correlation of Bull Lake and Pinedale deposits in the Rocky Mountains. Subsequent soil studies by Shroba (1977) and Roy and Hall (1980) show that in many areas soil development on moraines mapped by Richmond as the late stage of the Bull Lake Glaciation is similar to that on the adjacent Pinedale moraines but is significantly weaker than that on moraines assigned to the early stage of the Bull Lake. Thus, if soils are the primary basis for the Bull Lake-Pinedale distinction, many moraines assigned previously to the *late* stage of the Bull Lake Glaciation, both in the type area and elsewhere, should be grouped with the Pinedale and probably are of Wisconsin age.

Subdivisions of the Pinedale Glaciation are based on the upvalley moraine sequence (Ray, 1940; Richmond, 1960, 1965). Such subdivisions either have been given local designations or have been informally assigned to the early, middle, and late stages of the Pinedale Glaciation. Generally, the only clear evidence of age differences among the morainal subdivisions of the Pinedale is sequence and evidence of depositional breaks, such as crosscutting moraines and differing outwash levels. In the stadial classification, the outermost moraines and associated outwash are assigned to the early stage. Moraines and outwash inside or upvalley from terminal moraines of the early stage and commonly representing multiple advances are assigned to the middle stage. Moraines of the late stage lie 25% to 75% of the way upvalley from the terminal moraines to the valley heads; additional late-stage moraines commonly occur farther upvalley, in or near cirques. Among end-moraine successions, the terms "early stage," "middle stage," and "late stage" do not necessarily signify time equivalence.

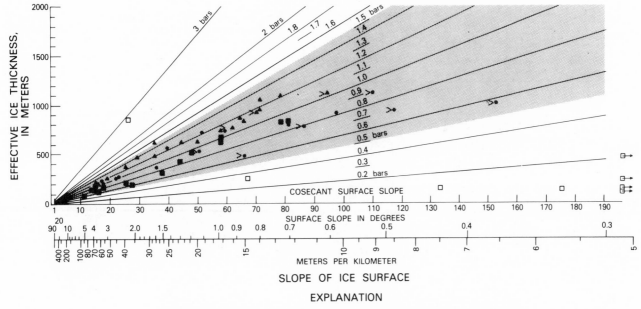

Figure 4-26. Basal shear stress of the glacier system in the northern Yellowstone area. The shaded area is the empirical range of 0.5 to 1.5 bars for modern glaciers. The solid dots are from the reconstruction of Pierce (1979). The open squares are alternate reconstructions (Richmond, 1969; U. S. Geological Survey, 1972). The arrows are points that plot to right of graph. Effective thickness is the shape factor (F, generally between 0.7 and 0.95 times the ice thickness [*h*]).

Extent of Glaciers ≥ 40,000 to 20,000 Years Ago

Recent findings cast doubt on the traditional correlation of the Pinedale Glaciation with *only* the Late Wisconsin. Some age data suggest that Pinedale glaciers were near their maximum lengths probably at least once, and perhaps twice, before 25,000 years ago; they were also at or near their maximum extent about 20,000 years ago (Table 4-6).

Studies of weathering rinds by Colman and Pierce (1979) from seven end-moraine sequences in the western United States (two in the Rocky Mountains) indicate that end moraines of the last glaciation (Pinedale) were deposited about 70,000 to 60,000, 40,000 to 30,000, and 20,000 to 15,000 years ago. Only the end moraines 20,000 to 15,000 years old are always present, and most successions do not display the entire sequence. If these conclusions are correct, then the outermost Pinedale moraines of different sequences may differ in age by as much as a factor of three. Miller's (1979: Table 5) statistical studies of relative-age criteria also suggest that end moraines of the Pinedale Glaciation span a considerable amount of time.

In the area covered by the Yellowstone Plateau ice cap, radiocarbon dates of six samples collected from beneath Pinedale drift are infinite, the oldest being older than 45,000 years (Table 4-6). In a lacustrine section 1 m above a horizon dated as older than 38,000 years, ice-rafted stones suggest that Pinedale glaciers were nearby. These infinite dates and the absence of finite dates between 45,000 and 15,000 years from deposits immediately beneath Pinedale drift may suggest that Pinedale glaciers attained and maintained at least half their full-glacial size from more than 45,000 yr B.P. to 15,000 yr B.P.

Obsidian-hydration dating of glacial abrasion cracks on pebbles from Pinedale terminal moraines near West Yellowstone indicate an age of about 40,000 to 30,000 years (Table 4-6) (Pierce et al., 1976). In addition, weathering rinds 0.4 mm thick on basalt clasts from Pinedale terminal moraines indicate that they formed during an interval more than twice as long as that represented by rinds 0.1 mm thick from Pinedale recessional moraines. The interpolation of values between those for dated Pinedale recessional deposits (14,000 years old) and Bull Lake deposits (150,000 to 140,000 years old) suggests an age of about 35,000 years for the Pinedale terminal moraines (Colman and Pierce, 1981).

In the Fraser River valley of the Colorado Front Range, an excavation just inside the Pinedale glacial limit exposed a sequence of five tills. Nelson and others (1979) conclude that the upper three tills are of Pinedale age. A weathering break is noted between the fourth and fifth tills, but no weathering break is noted between the third and fourth tills. This suggests that the latter may be of Pinedale age. Organic lake sediments between the third and fourth tills yield two finite ages of about 30,000 years (Table 4-6), but these ages are interpreted as only minimum ages.

Extensive morainal deposits along the Green River were deposited by a glacier draining the northwestern part of the Wind River ice field (Figure 4-2). Unpublished information obtained by R. C. Bright (personal communication, 1980) suggests that Pinedale ice advances culminated there both before and after about 25,000 years ago. On

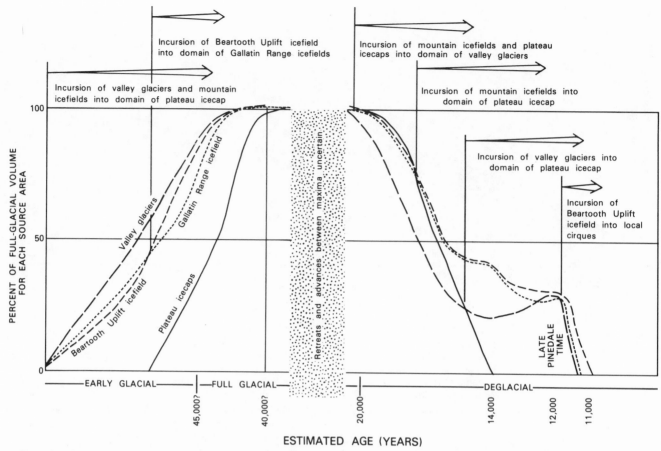

Figure 4-27. Inferred temporal relations in the northern Yellowstone area between valley glaciers, mountain ice fields, and the plateau ice cap. The points where the curves cross or diverge are based on later incursions of ice from one source into the full-glacial domain of ice from another source. The lines are drawn in an idealized form, but the line crossings and divergences are based on field data. (From Pierce, 1979: Figure 51.)

the basis of the presence of abundant lakes in undrained depressions and weak soil profiles with only cambic B-horizons, the deposits enclosing ''Dew Lake'' have been judged to be of Pinedale age. Pollen and sediment analyses of a core from ''Dew Lake'' indicate the following history: (1) glaciation (till), (2) interval of milder climate, (3) cold, glacial climate from about 21,000 to 13,000 years ago, and (4) postglacial climate after about 10,000 years ago (R. C. Bright, personal communication, 1980). The base of the core is estimated by extrapolation to be about 25,000 to 23,000 years old (Table, 4-6).

The last advance of a Pinedale glacier close to its terminal limit is well dated at Devlin Park in the Front Range of Colorado (Madole, 1980a). A glacier dammed a lake in the park starting about 23,000 years ago; this estimate is based on extrapolation to the base of the last lake sequence from radiocarbon dates as old as 22,400 ± 1070-1320 yr B.P. Pollen analysis shows tundra conditions throughout the deposition of lake sediments from about 23,000 to less than 15,000 years ago (Legg and Baker, 1980).

At Bells Canyon at the foot of the Wasatch Range, Utah, Madsen and Currey (1979) obtained a radiocarbon age on the B-horizon of a soil buried by a Pinedale terminal moraine. On the basis of radiocarbon ages of modern B-horizons of surface soil (Sharpenseel 1971), the age of the soil at the time of burial may have been as much as 5000 years. Because contamination is a possibility, the 26,080 yr B.P. age obtained is here considered a minimum age; combining the poten-

tial soil age with the size of the analytical uncertainty (5000 ± 1100 years) indicates that a Pinedale terminal moraine was deposited more than 20,000 years ago.

Chronology of Pinedale Deglaciation

At Lake Emma, a tarn in the source area of the San Juan ice field (Figure 4-24), Carrara and Mode (1979) obtained a series of radiocarbon ages of between 15,000 and 14,000 yr B.P. for layers rich in moss fragments from the lowermost 10 to 15 cm of lake sediments (Table 4.7). Carbonate bedrock is not present, so a hard-water effect is not likely to have given apparent ages that are too old. The Lake Emma dates suggest that the transection glacier had wasted away by 14,700 years ago. An age of 15,450 ± 220 yr B.P. from Molas Lake (Figure 4-25) (Maher, 1972) also suggests early deglaciation, although this age may be too old due to the hard-water effect of carbonate bedrock. The relation of the Molas Lake site to the transection glacier (Figure 4-24) shows that at least half the glacier had melted by the time this site was deglaciated.

The 1000-m-thick ice cap on the Yellowstone Plateau was deglaciated by about 14,000 years ago (Table 4-7, A-E). Lake sediments adjacent to Yellowstone Lake are as old as 14,490 ± 350 and 14,130 ± 375 yr B.P. (Richmond, 1976: 370). An age of 14,360 ± 400 yr B.P. was obtained on a composite sample from both above and below an ash provisionally identified as Glacier Peak

(Waddington and Wright, 1974); most layers of Glacier Peak ash are younger than about 12,750 years (Porter, 1978).

These ages of more than 14,000 years for the deglaciation of the Yellowstone Plateau ice cap and the San Juan Mountains ice field reflect the sensitivity of these areas to a rise in the equilibrium-line altitude. A rise of 250 to 500 m, only 25% to 50% of the Late Wisconsin-Holocene rise in equilibrium-line altitude, resulted in these areas having little or no terrain above the equilibrium-line altitude and led to their early deglaciation.

During the last decade, the deglaciation of Pinedale valley glaciers has also been dated as being significantly older than previously thought (Mehringer et al., 1971; Benedict, 1973, 1981; Andrews et al., 1975; Madole, 1976a, 1976b, 1980b; Madsen and Currey, 1979; Pierce, 1979). Deglaciation of mountain valleys started well before 14,000 years ago (Figure 4-28, Table 4-8: sites X, Y, Z). Pinedale till just beyond large terminal moraines at the base of the Wasatch Range, Utah, is overlain by lake sediments deposited about 17,000 years ago during the transgression of the last cycle of Lake Bonneville (W. E. Scott, personal communication, 1981).

By 12,000 years ago, equilibrium-line altitudes had risen more than 400 m at site U and more than 300 m at site V (Figure 4-28, and Table 4-8), and many valley glaciers had probably disappeared by this time. For high alpine sites, the oldest dates for deglaciation are about 9000 years (sites Q, R, S). Because these are minimum ages, ice may either have lasted until this time or have disappeared at an earlier date, possibly as much as 2000 years earlier. In a recent report on an analysis of young glacial deposits in the Rocky Mountains of Alberta,

Canada, Luckman and Osborn (1979) conclude that by 10,000 years ago glaciers had receded to positions close to or upvalley from the late-Neoglacial maximum (A.D. 1750-1850).

For the Front Range of Colorado, Benedict (1973) dates the Satanta Peak advance as older than 9915 ± 165 years; he concludes that this advance postdated the late Pinedale, which was 10 times larger. In the Wind River Mountains, the type Temple Lake moraines have been dated as older than 6500 ± 250 years (GX3166D), which makes them pre-Neoglacial rather than early Neoglacial (Currey, 1974; Miller and Birkeland, 1974).

Pollen records can be interpreted as suggesting that by about 11,500 years ago Pinedale glaciers had probably disappeared or were about as small as subsequent Neoglacial glaciers. The end of glacial conditions is inferred to correlate with the major zone boundaries between cold-climate (late-glacial) and Holocene-type assemblages. These boundaries are about 11,500 years old in westernmost central Montana (Mehringer et al., 1977), about 11,800 years old in northern Yellowstone Park (Gennett, 1977), about 11,700 years old east of Yellowstone Lake (Waddington and Wright, 1974), about 11,600 years old south of Yellowstone Lake (Baker, 1976), and about 10,300 years old in southeastern Idaho (Bright, 1966). After this change, no oscillations in pollen spectra indicating cooler climate exceed those correlated with Neoglacial advances; consequently, glacier advances after about 11,500 years ago were probably not much larger than those of the Neoglacial.

Between 9500 and 9000 yr B.P., trees were growing at or above the modern treeline; clearly, the Pinedale Glaciation had ended by this time. In the Colorado Front Range, wood dated 9200 ± 135 yr B.P. (I-6520) reveals that the treeline was then at or above its modern level (Benedict, 1973: 590). In the San Juan Mountains, P. E. Carrara (personal communication, 1980) found sediments above the treeline containing coniferous wood 9220 ± 120 years old (W-4524).

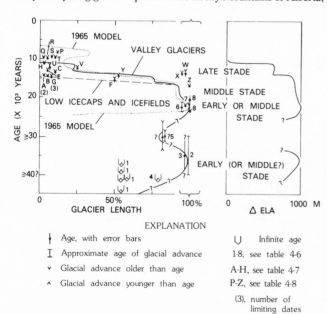

Figure 4-28. Composite diagram of relative extent versus time for Pinedale glaciers in the Rocky Mountains. (See Tables 4-6, 4-7, and 4-8 for data points). Dotted line: 1965 model; solid and dashed lines: chronology advocated here. A separate deglaciation curve (dashed line) is plotted for the San Juan Mountains and Yellowstone areas because they appear to have been deglaciated with an equilibrium-line-altitude change only 25% to 50% of the total Late Wisconsin equilibrium-line altitude (ELA) depression. On the right side of graph is a crude estimate of the change in ELA with time; in the construction of this curve, valley glaciers were assumed to have ELAs 300 m lower than adjacent ice caps, and very small cirque glaciers were assumed to have ELAs 200 to 300 m lower than sizable valley glaciers.

PALEOENVIRONMENTAL INFERENCES

Mass Balance of the Northern Yellowstone Glacier

The mass balance of the northern Yellowstone glacier can be estimated from the field reconstruction (Figures 4-25 and 4-29). At equilibrium, the net volume gained in the accumulation area should be offset exactly by the net loss in the ablation area. The resulting values for both net accumulation and ablation are about 3 km³ (Figure 4-29). Although calculated ablation exceeded accumulation by 15%, an exact balance is achieved when the estimated equilibrium-line altitude is lowered by about 50 m. If it is assumed that precipitation at that time was similar to that of the present, the annual flow across the equilibrium line was about 3 km³. This flow must have been accomplished largely by basal sliding, for flow by ice deformation could account for only about 10% of the needed discharge (Pierce, 1979: Table 7).

Climatic Differences along an East-West Transect

Along a transect from the Pacific Ocean to northern Wyoming, the altitude of the Pleistocene glaciation threshold rises stepwise inland upon crossing major orographic barriers (Figure 4-30) (K. L. Pierce, unpublished data). Along the same transect, modern snowpack shows an overall decrease inland (Figure 4-20B, lines 1-11). Reversals in both these trends occur locally. For example, the

Table 4-6.

Ages Indicating Pinedale Advance(s) ≥ 40,000 to 20,000 Years Ago

Map Location	Locality	Age (Laboratory Number)	Remarks	Reference(s)
1	Yellowstone Lake basin, Wyoming	> 45,000 (W-2411) > 42,000 (W-2264) > 42,000 (W-2197) > 40,000 (W-2955) > 38,000 (W-2012) > 29,000 (W-2582)	Ages of plant debris in lake sediments that underlie Pinedale glacial deposits; ice-rafted stones noted 1 m above horizon dated > 38,000 years	Richmond, 1976: 369; Birkeland et al., 1971; Pierce, 1979: F30
2	West Yellowstone,	40,000–30,000	Obsidian-hydration age of glacial abrasion leading to deposition of Pinedale terminal moraines of the Yellowstone Plateau ice cap. Age of ~30,000 if Pleistocene *soil* temperature was 5 °C colder than at present; age of ~40,000 if soil was 10 °C colder. Age of abrasion of samples from youngest recessional moraines is 15,000 to 10,000 years.	Pierce et al., 1976; Pierce, 1979
3	West Yellowstone, Montana	~35,000	Age of Pinedale terminal moraines of the Yellowstone Plateau ice cap estimated from weathering-rind thicknesses on basalt stones, as calibrated by weathering-rind thickness for 140,000-year-old Bull Lake moraines and for 14,000-year-old Pinedale recessional moraines	Colman and Pierce, 1981
4	Southwest of Yellowstone Park, Wyoming	> 42,000 (W-2264)	Age of compressed wood in sediment 0.5 m below Pinedale till and 0.3 m below change in pollen and sediment type indicating cooling and possible approach of glaciers; a Pinedale glacier overrode this site, probably eroding the up-permost sediment	Baker and Richmond, 1978: Figure 2
5	Fraser River Valley Front Range, Colorado	30,050 ± 1200 (SI-2912) 30,480 + 2800/− 4300 (DIC-482)	Concentrated organic material and bulk sample from lake sediments separating third and fourth tills below surface at Mary Jane site; ages are finite but interpreted as minimum ages; pollen from lake sediments indicates warm conditions followed by cooling	Nelson et al., 1979
6	Glacial Lake "Devlin" Front Range, Colorado	22,400 + 1070/− 1320 (DIC-870)	Microscopic organic matter concentrated from near base of youngest cycle of lake sediments of glacially dammed lake; projected age at base of youngest cycle is 23,000 years; glacier was near its terminus by this time	Madole, 1980a
7	Wasatch Mountains, Utah	26,080 + 1200/− 1100 (GX-4737)	Organic matter in B2ltb-horizon of buried soil at mouth of Bells Canyon; an age of > 20,000 years is estimated for the time of burial of the soil by Pinedale terminal moraines	Madsen and Currey, 1979: 257
8	Wind River Mountains, Wyoming	20,800 ± 1200 (W-998)	Age of organic matter in clay about 1 m above till at bottom of core from "Dew Lake"; projected age to base of lake sediment inferred to be > 23,000 years, and an early phase of the Pinedale Glaciation predates this age; pollen studies from the core date a later phase of the Pinedale Glaciation as between about 21,000 and 13,000 yr B.P.	R. C. Bright, personal communication, 1980

Note: All ages are in ^{14}C years unless otherwise stated.

Table 4-7.
Minimum Ages for Deglaciation of Relatively Low Ice Caps and Ice Fields

Map Location	Locality	Age (Laboratory Number)	Remarks	Reference(s)
A	Solution Creek, Yellowstone Plateau, northwestern Wyoming	14,490 ± 350 (W-3183) 14,130 ± 375 (W-2739)	Organic material in lake sediments from the southwestern part of Yellowstone Lake basin; minimum ages for deglaciation of 1000-m-thick ice cap on Yellowstone Plateau	Richmond, 1976: 370
B	Cub Creek Pond, Yellowstone Plateau, northwestern Wyoming	14,360 ± 400 (W-2780)	Age of slightly organic silt near base of pond sediment provides minimum age for wastage of Pinedale ice cap 1000 m thick from Yellowstone Plateau; ELA about 250 to 500 m higher than at Pinedale maximum	Waddington and Wright, 1974: 170; Richmond and Pierce, 1972
C	Swan Lake, Yellowstone Plateau, northwestern Wyoming	13,530 ± 130 (WIS-432)	Core ended in laminated lake sediment; age on overlying banded lake sediment from depth of 402 to 418 cm; marly sediment indicates "hard-water" effect could cause sample to date as much as 2000 years too old; Pinedale ice cap > 650 m thick on northwestern part of Yellowstone Plateau was deglaciated before deposition of laminated lake sediments	Pierce, 1979: F61
D	Buckbean fen, Yellowstone Plateau, northwestern Wyoming	>> 11,550 ± 350 (W-2285)	Age of gyttja 2.6 m above base of core from near center of Yellowstone Plateau ice cap; pollen analysis suggests correlation of base of this core with 14,360 ± 400 yr B.P. age at Cub Creek Pond; a ~14,000-year age is consistent with the rate of sedimentation for the rest of the core	Baker, 1976; Figure 16
E	Rocky Creek, Yellowstone Plateau, northwestern Wyoming	13,140 ± 700 (W-2037)	Age of peaty mud in area covered by Yellowstone Plateau ice cap; Pinedale ice cap 1000 m thick was deglaciated by this time; dated horizon overlain by gravel thought to be outwash from late-Pinedale glacier up-valley	Richmond and Pierce, 1972: strat. section 14; Pierce, 1979: F65.
F	Molas Divide, San Juan Mountains, Colorado	15,450 ± 220 (Y-1147)	Age of sediment with low organic content in lower part of core; minimum age for recession of glaciers to less than half their full-glacial length (Figure 4-24); hard-water effect could occur because of carbonate bedrock	Maher, 1972
G	Lake Emma, San Juan Mountains, Colorado	14,940 ± 140 (W-4525) 14,900 ± 250 (W-4209) 14,130 ± 150 (W-4289)	Age of moss fragments from near base of lake sediments in cirque; minimum age for deglaciation of south-facing cirque near center of San Juan ice field (Figure 4-24); north-facing cirques may have had cirque glaciers; ELA increased 300 to 500 m from Pinedale maximum	Carrara and Mode, 1979; P. C. Carrara, personal communication, 1981
H	Buffalo Pass, Park Range, Colorado Rockies	11,180 ± 150 (W-4245)	Age of clay-humus fraction extracted from lower 50 cm of core; by this time, source area of Pinedale ice cap was deglaciated and ELA was > 300 m higher	Madole 1980b

Table 4-8.
Minimum Ages for Deglaciation of Valley Glaciers

Map Location	Locality	Age (yr B.P.) (Laboratory Number)	Remarks	References
Z	Bells Canyon, Wasatch Mountains, Utah	~16,000	Moraines overlain by Bonneville shoreline deposits whose age is about 16,000 yr based on ^{14}C dates of wood from transgressive deposits	W. E. Scott, personal communication, 1982
Y	Northern Yellowstone glacier, Wyoming-Montana	14,360 ± 400 (W-2780) 13,140 ± 700 (W-2037)	Valley glaciers tributary to northern Yellowstone outlet glacier > 50% deglaciated when outlet glacier > 75% full-glacial size; dates are for deglaciation of part of outlet-glacier source area and provide minimum ages for initial recession of valley glaciers	Pierce, 1979: 61; see Figure 4-27
X	Fraser River valley, Front Range, Colorado	13,740 ± 160 (DIC-671)	Age of base of peat above lake sediments above Pinedale till; minimum age for recession from terminal moraines; pollen analysis from higher part of section indicates conditions similar to present by 12,380 ± 180 (DIC-516)	Nelson et al., 1979
W	Colorado River valley, Front Range, Colorado	13,820 ± 810 (GaK-4537)	Age of clay-humus fraction from lower 12 cm of sediment in kettle on Pinedale terminal moraines.	Madole, 1976a: 165
U	Little Cottonwood Creek, Wasatch Mountains, Utah	12,300 ± 330 (GX-3481)	Age of woody forest litter at till-bog interface; valley glacier < 33% its full-glacial length of 16 km by this time; ELA was > 300 m higher than at Pinedale maximum	Madsen and Currey, 1979: 258
V	Curelom Cirque, Raft River Mountains, Utah	12,200 ± 800 (A-1112) 12,120 ± 1250 (A-1111)	Basal clayey sediment in core from cirque; valley glacier <5% its full-glacial length by this time; ELA was > 400 m higher than at Pinedale maximum	Mehringer et al., 1971; Mehringer, 1977: Figure 7
T	Poudre Pass, Front Range, Colorado	9850 ± 300 (W-4083) 9800 ± 400 (W-4086)	Basal silty peat and peat overlying stratified drift provide minimum ages for recession of Pinedale glacier to < 10% its full-glacial length	Madole, 1980b: 121
S	Little Cottonwood Creek, Wasatch Mountains, Utah	9560 ± 240 (RL-695) 9430 ± 260 (GX-4736)	Wood and peat near base of Albion Basin bog dates recession of valley glacier to < 8% its full-glacial length	Madsen and Currey, 1979: 258
R	Caribou Lake Cirque, Front Range, Colorado	9915 ± 165 (I-6335)	Age of basal peat overlying outwash from Satanta Peak moraine in type area; the Satanta Peak advance may be latest Pinedale or post-Pinedale	Benedict, 1973: 588-90
Q	San Juan Mountains, Colorado	9620 ± 440 (St-3909)	Age of basal clay containing organic material; high, northeast-facing cirque was ice free by this time	Andrews et al., 1975
P	Fourth of July Cirque, Front Range, Colorado	9280 ± 150 (I-11,092A)	Basal muck overlying outwash of Satanta Peak advance, which is latest Pinedale or post-Pinedale in age; Glaciers < 10% full-glacial length at this time	Benedict, 1981: 59

Yellowstone area has a lower Pleistocene glaciation threshold (Figure 4-30) and a greater modern snowpack (Figure 4-20B, line 7) than areas to the east or west, because both in modern and late-Pleistocene times the Snake River Plain has acted as a route for moist air masses moving far inland (Figure 4-2).

A close inverse relation exists between the Pleistocene glaciation threshold and the modern snowpack at the altitude of glaciation threshold such that, as the altitude of the threshold increases inland by 1000 m, the water content of modern snowpack decreases about 1 m (Figure 4-20B). Because modern snowpack relates so closely to Pleistocene glaciation threshold, the Pleistocene pattern of air-mass movements apparently was similar to that at present.

The large differences in modern snowpack at the Pleistocene glaciation threshold suggest that Late Wisconsin glaciers had similar differences in their budgets. Annual accumulation at the equilibrium-line is an index to the activity of glaciers (Meier, 1961a; Andrews, 1972). Pleistocene precipitation may have been different from present-day precipitation, but the areas along the transect are likely to have had similar relative differences, as suggested by the linear relation between modern snowpack and the Pleistocene glaciation threshold (Figure 4-20B). Figure 4-31 shows three arbitrary classes of glaciers along the transect defined on the basis of the modern snowpack at the Pleistocene glaciation threshold. The activity of glaciers in the Yellowstone area was two or three times greater than the activity of glaciers of comparable size in the Bighorn Mountains and Lemhi Range. The activity of glaciers in the Cascade Range was five times greater than the activity of glaciers in the Bighorn Mountians (Figure 4-31). Low-activity glaciers occurred at higher altitudes and were surrounded by drier, colder air than were high-activity glaciers.

Differences between Late Wisconsin and Modern Temperatures

As Flint (1971: 72-73) notes, the difference between modern and Pleistocene snow lines has been widely used as a basis for calculating

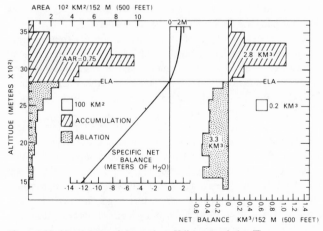

Figure 4-29. Mass balance of the northern Yellowstone glacier. The equilibrium-line altitude for this ice field is inferred to be the same as the Pleistocene glaciation threshold (Pierce, 1979: F74). The bar graph on the left shows the area-altitude distribution and indicates that the accumulation area ratio (AAR) was about 0.75. In the middle graph, the inferred specific net balance is based on modern glaciers thought to be analogous in geometry and climatic setting. On the right-hand graph is the net balance, which is the product of the area at each the altitude increment multiplied by the specific net balance of ice gained or lost for that altitude increment.

glacial-age temperatures. These estimates use the change between a present and a Pleistocene snow-line parameter, such as equilibrium-line altitude, multiplied by a lapse rate of about 6 °C per 1000 m. In the western United States, Pleistocene snow lines were about 1000 m lower than they are at present (Richmond, 1965; Flint, 1971: 468; Porter, 1977; Scott, 1977; P. E. Carrara, personal communication, 1980). For a 1000-m depression in snow line, a simple lapse-rate calculation yields an estimated Pleistocene mean annual temperature 6 °C colder than the present mean annual temperature.

An implicit assumption in this lapse-rate calculation is that the same annual accumulation at the equilibrium-line altitude of a modern glacier occurred at the equilibrium-line altitude of a late-Pleistocene glacier, which was about 1000 m lower. However, snowpack, vegetational, and hydrologic studies in the Rocky Mountains all show that modern precipitation decreases markedly with decreasing altitude. The water content of the April-first snowpack increases at gradients of 1.2 m per 1000 m for major orographic barriers and about 0.5 m per 1000 m for more isolated mountains (Figure 4-20). Mean annual precipitation also increases at about 1 m per 1000 m between altitudes of 2000 m and 3000 m just north of the Yellowstone National Park (Farnes, 1971; personal communication, 1980) and at about 0.4 m per 1000 m between 2600 m and 3700 m on the eastern slope of the Colorado Front Range (Barry, 1973: Table 4), or about 0.6 m per 1000 m overall (Landsberg, 1958: 176).

Consequently, by not considering this altitudinal change in precipitation, the traditional lapse-rate calculation actually implies an increase of about 0.5 to 1.4 m in Pleistocene precipitation (Figures 4-20 and 4-31, Table 4-9). Such a large increase in Pleistocene precipitation does not seem reasonable for the northern Rocky Mountains, because the source of moisture for the Rockies was primarily the North Pacific Ocean, where Pleistocene sea-surface temperatures were 2 °C to 4 °C colder than present temperatures (CLIMAP, Project Members, 1976).

If we assume that the altitudinal distribution of precipitation in Late Wisconsin time was the same as it is at present, then the lower amount of precipitation at lower Pleistocene equilibrium-line altitudes requires that temperatures were colder than indicated by the lapse-rate calculation alone. The slope of the dashed regression line in Figure 4-20B indicates the magnitude of this effect; for places that have 1 m less snowpack, the altitude of the Pleistocene glaciation threshold was 1000 m higher. Table 4-9 shows that the effect of the precipitation decrease may be of the same magnitude as the altitudinal lapse rate; this results in estimates of Pleistocene ablation-season temperatures 10 °C to 15 °C colder than present temperatures.

Because we do not know whether Pleistocene precipitation was similar to, less than, or greater than present-day precipitation, snow-line studies cannot determine a unique glacial-age temperature. But it appears that the lapse-rate estimates of only a 6 °C cooling have been unduly restrictive and that temperature changes twice this amount could be reasonably considered. Such values would be consistent with the results of global climate-modeling studies (Gates, 1976: Figure 7) and with the recognition of relict permafrost features indicating that temperatures in the Rocky Mountains could have been 10 °C to 15 °C colder than they are at present (Mears, 1981).

For the Little Cottonwood glacier of the Wasatch Range, Utah, snowmelt calculations have been made to determine how much colder or snowier it had to have been for the Late Wisconsin glaciers to have existed (McCoy, 1981: 138). If it is assumed that Pleistocene precipitation was the same as modern precipitation,

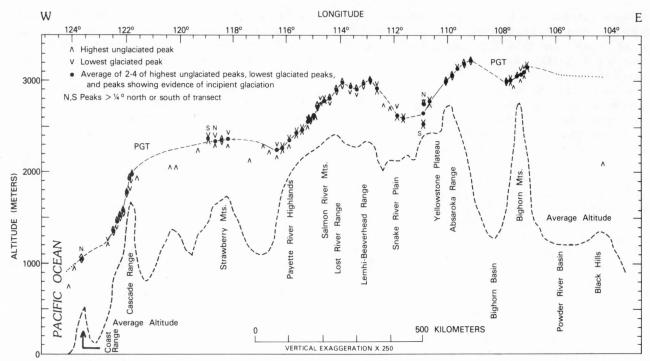

Figure 4-30. Transect from the Pacific Ocean to northern Wyoming along latitude 44°30´N showing Pleistocene glaciation threshold (PGT) and average altitude of terrain (heavy, dashed line). (Average altitude based on map provided by R. H. Godson, personal communication, 1980.)

temperatures at the Late Wisconsin glacial maximum would have been about 16 °C colder than present-day temperatures. The Little Cottonwood glacier was downwind from pluvial Lake Bonneville, so precipitation at their time may have been greater than it is now; if it is assumed that precipitation then was one and a half times what it is now, it can be concluded that temperatures would have been about 12 °C colder then. (A regional precipitation increase is not required to explain the existence of pluvial Lake Bonneville, for water-balance calculations indicate that this lake would fill to the highest shoreline under the present precipitation regime if temperatures were only 7° colder than they are at present [McCoy, 1981: 134].)

Pollen analyses of cores spanning the interval of the Late Wisconsin glacial maximum do not seem to support such a large change in temperature. For the Colorado Front Range, Legg and Baker (1980) infer a treeline lowering of only 500 m during the time interval of 21,000 to 14,000 (?) years ago, although R. G. Baker (personal communication, 1981) considers this only a minimum value. For the Chuska Mountains of New Mexico, Wright and others (1973: 1177) infer a lowering of the treeline by 900 m for cool intervals dated 18,500 to 13,500 and 34,000 to 26,000 years ago and an even greater change for cold intervals dated 26,000 to 18,500 and more than 34,000 years ago.

The effect of precipitation gradients on traditional lapse-rate calculations, therefore, suggests that the Late Wisconsin of the Rocky Mountains was either considerably colder than previously estimated (from simple lapse-rate calculations) or much wetter than the mountains are at present. The colder Late Wisconsin sea-surface temperatures in the North Pacific Ocean imply that overall precipitation in the western United States did not increase, although increased precipitation could conceivably have occurred over local, time-transgressive belts. More independent information on both

temperature and precipitation changes is needed to resolve this question.

Hawaiian Islands

Only Mauna Kea (4206 m high) among the major volcanoes of the Hawaiian Islands shows evidence of former glaciation (Figure 4-32). Its unique record of multiple dated glacial events provides important information about Pleistocene environmental conditions in the tropical mid-Pacific Ocean basin (Porter, 1979a).

The upper glaciated part of Mauna Kea consists of a complex array of cinder cones and lava flows of mainly alkalic composition, most of which were erupted during the middle and late Pleistocene (Porter, 1979b). Because of its relatively young age, the volcano is little dissected; on the glaciated upper slopes, gorges reach maximum depths of only about 70 m.

Annual precipitation locally exceeds 7500 mm on the lower windward slope of the mountain, which supports a dense tropical forest; but, on the leeward side and above the 3500-m level, precipitation falls off rapidly, and the landscape changes upslope from open grassland to alpine desert. The altitudinal decrease in precipitation is related to an atmospheric inversion that largely restricts clouds to altitudes below about 2100 m. Major cyclonic storms break the inversion and carry precipitation to the crest of the mountain, thereby accounting for much of the snowfall received in the summit region. Because of the high altitude of the summit, the mean annual temperature there is low (about 4 °C) and some snow remains in patches throughout the summer. Where cold air collects in the craters of the highest cinder cones, permafrost develops locally; patterned ground produced by frost activity is ubiquitous near the summit.

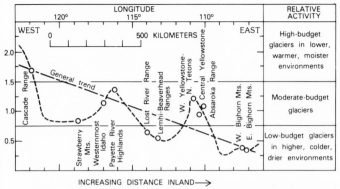

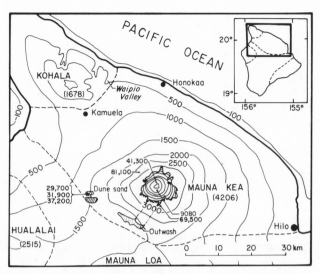

Figure 4-31. Inferred relative activity of Late Wisconsin glaciers along 44°30′N transect. The dashed line connecting the data points is based on the inversion of the Pleistocene glaciation threshold (Figure 4-30). The boundaries between high, moderate, and low activity are arbitrary.

LATE WISCONSIN GLACIATION

Morphological and stratigraphic evidence indicates that ice caps repeatedly formed and expanded at the crest of the volcano during the middle and late Pleistocene. The last ice cap, which was similar in size and configuration to earlier ones, covered about 70 km² at its greatest extent and reached a maximum thickness of about 100 m (Figure 4-32). Its margin was irregular but approximately circular, and its mean radius was close to 5 km. The glacier reached a lower limit of about 3200 m, and its median altitude was about 3700 m. The equilibrium line had an east-southeasterly gradient of about 20 m per kilometer and an estimated steady-state mean altitude of 3735 ± 25 m (3780 ± 25 m when corrected for the isostatic subsidence of the island). Although evidence of glacial erosion is widespread near the summit in the form of striated, plucked, and abraded rock surfaces and locally massive end moraines, erosion was not deep, for lava flows and cinder cones that were overwhelmed by the glacier still retain most of their preglacial topography. Ice flow was nearly radial from the summit, as indicated by the pattern of striations, but it was controlled locally by cinder cones, several of which rose above the glacier surface as small nunataks. Thin outwash fans are found along parts of the drift border, but, where tongues of ice terminated in gulches, out-

Figure 4-32. Map of Mauna Kea and adjacent parts of the island of Hawaii showing topography of late-Makanaka ice cap (contour interval 100 m) and location of radiometrically dated samples that bracket the age of the youngest drift of the Makanaka Formation.

wash is generally thin and discontinuous. Broad coalescing fans of outwash gravel are found at the southern base of the mountain in the Mauna Kea-Mauna Loa saddle, to which two major gulches directed meltwater from the southern perimeter of the glacier.

STRATIGRAPHY AND CHRONOLOGY

Deposits of the last glaciation are included in the Makanaka Formation, the youngest of three lithostratigraphic units of glacial origin that are interstratified with lavas on the upper slopes of Mauna Kea (Porter, 1979b). Two drifts, indistinguishable on the basis of lithology but differentiated by weathering criteria, morphology, distribution, and stratigraphic position, are included within the formation. The younger, representing the last ice-cap glaciation, forms distinctive bouldery end moraines around much of the summit region (Figure 4-33) and consists predominantly of medium-gray hawaiite boulders

Table 4-9.
Effect of Altitudinal Gradient of Winter Snowpack on Calculation
of Pleistocene Temperatures

(1) ΔSP/1000m (Snowpack gradient)	(2) ΔELA[a] (m)	(3) ΔT for ΔELA Alone[b] (°C)	(4) SP Change ((1) × (2)) (m)	(5) Equivalent ΔPGT[c] for (5) (m)	(6) Equivalent ΔT[b] for (5) (°C)	(7) Total Pleistocene ΔT((3) + (6)) (°C)
No gradient	− 1000	− 6.5	0	0	0	− 6.5
0.5 m per 1000 m[d]	− 1000	− 6.5	− 0.5	− 500	− 3.2	− 9.7
1.0 m per 1000 m	− 1000	− 6.5	− 1.0	− 1000	− 6.5	− 13.0
1.3 m per 1000 m[e]	− 1000	− 6.5	− 1.3	− 1300	− 8.5	− 15.0

Abbreviations: SP: water content of winter snowpack; ELA: equilibrium-line altitude; Δ: change in.

[a] Assumed to be about 1000 m; see text.

[b] Multiplied by a lapse rate of 6.5 °C per 1000 m.

[c] Column (4) divided by a precipitation gradient of − 1 m per 1000 m (Figure 4-20B).

[d] This value is similar to that for isolated mountains and may apply at altitudes of modern glaciers.

[e] This value is modern gradient across major orographic barriers at altitudes near and below the timberline.

Figure 4-33. Vertical aerial photograph showing late-Makanaka end-moraine belt in the vicinity of Pohakuloa Gulch on the southern slope of Mauna Kea.

and aa-flow rubble that weather to a light yellowish brown. Older drifts lying beyond the late-Makanaka drift limit are more weathered and topographically subdued, and the two oldest are lithologically distinct. Lavas of the Waikahalulu Formation that underlie Makanaka drift have been dated by the potassium-argon method as 81,100 ± 23,600 and 69,500 ± 2600 years old, whereas an ice-contact lava erupted before the growth of the last ice cap dates to 41,300±8300 years ago (Porter, 1979a). Tephra layers on the lower western rift zone of the volcano that underlie dune sands believed to correlate with outwash sediments in the Mauna Kea-Mauna Loa saddle of late-Makanaka age (Porter, 1975a) are associated with charcoal that is 37,200±1400 to 29,700±500 years old. The younger date gives the youngest maximum limiting age for late-Makanaka glaciation currently available. The closest minimum limiting age (9080 ± 200 yr B.P.) come from algal-rich sediments in Lake Waiau, a small lake lying close to the summit and well upslope from the innermost late-Makanaka moraines. The last ice cap, therefore, occupied the summit region sometime between about 30,000 and 10,000 years ago, or during the interval of oxygen-isotope stage 2 of the marine record (Porter, 1979a: Figure 12). Although in places late-Makanaka drift forms multiple recessional moraines, it was not possible to correlate them around the summit region, and no additional radiometric dates are currently available for assessing their possible ages.

PALEOENVIRONMENTAL INFERENCES

August sea-surface temperatures during the last glacial age (about 18,000 years ago) in the vicinity of Hawaii were estimated by the CLIMAP project members (1976) to be about 1 °C to 2 °C cooler than present temperatures. Such a change is relatively small compared to that derived for some northern oceanic regions but not much less than the average change exhibited by the world oceans. If air temperature declined a similar amount, then environmental changes in Hawaii between the glacial maximum and the present should have been minimal. However, the evidence for a large ice cap on Mauna Kea with an equilibrium-line altitude of about 3780 m implies a substantial snow-line lowering. Because no glacier exists on the mountain today, the modern snow line must lie above the sum-

mit, or at least 425 m above the glacial-age snow line. The present snow line has been estimated to lie at an altitude of 4715 ± 190 m (Porter, 1979a). If this figure is correct, then snow-line depression must have amounted to about 935 ± 190 m during the late-Makanaka advance. If only ablation-season temperature controlled the snow-line depression, then air temperature near the summit must have been lowered more than sea-surface temperatures (assuming that the CLIMAP estimate is correct). Alternatively, cooler ablation seasons may have accounted for only part of the snow-line lowering, with a change in precipitation also being involved.

At present, the summit region is very dry and receives little snow, but, if there was an increased frequency of cyclonic storms passing through this sector of the Pacific during the last glacial age, the trade-winds inversion may have been weakened and greater amounts of precipitation could have reached the summit during the accumulation season. Favorable conditions for enhanced snow accumulation may have been favored by the equatorward extension of the westerly wind belt and/or increased cloudiness in the summit region that might have decreased summer ablation.

The multiple moraines found within the limit of late-Makanaka drift imply either that the margin of the ice cap experienced a series of minor readvances after its greatest expansion or that ice recession was irregular and marked by terminal halts. However, behind the main moraine belt, which averages about 1 km wide, drift is thin and patchy and moraines are absent. This implies that a change in the pattern of deglaciation occurred after the construction of the innermost moraine and that either terminal recession was very rapid and unidirectional or the glacier had become too thin to flow and wasted away in place. At present, the timing of this change has not been dated, but the evidence from Lake Waiau, where at least 8 m of unconsolidated sediment underlies the lowest radiocarbon-dated horizon (9080 ± 200 yr B.P. [I-2636]), suggests that the deglaciation of the summit may have occurred well before the Pleistocene-Holocene transition. This supposition is supported indirectly by snow-line relationships, for a rise of the equilibrium line by 425 m, or less than half the estimated total depression, would have led to the complete deglaciation of the mountain. Late-glacial readvances in the cordillera of the western United States (e.g., Hyak, Hilgard) typically involved a snow-line depression about 50% that of the full glacial. Evidence of such readvances would, therefore, not be expected on Mauna Kea.

A sea-level lowering of about 100 m at the glacial maximum extended the coastline of some Hawaiian islands substantially seaward (Ruhe, 1964), but the island of Hawaii was little affected. The island is so young and the offshore profile generally so steep that the 100-m isobath lies within only a few kilometers of the present shore (Figure 4-32). The primary effect would have been to steepen stream gradients, thereby causing erosion in the lower reaches of major streams. The effects of the resulting incision can be seen in the deep valleys (e.g., Waipio Valley) of Kohala volcano, the lower portions of which have aggraded during the Holocene transgression.

Charcoal in sediments predating the late-Makanaka glaciation at 1645 m altitude gives a minimum upper limit of trees just prior to glaciation; charcoal in post-Makanaka sediments is found almost to the present treeline (about 2900 m). However, because no charcoal or arboreal remains have been found above the Mauna Kea-Mauna Loa saddle in sediments of last-glacial age, the treeline at that time may have been depressed below 2000 m, or at least 900 m below present treeline. If vegetational communites remained similar in composition but zonal boundaries shifted downward in response to atmospheric cooling, then the zone between the treeline and the lower limit of glacier ice may have been occupied by alpine tundra and xerophytic shrub vegetation similar to that now found in the alpine zone. However, if precipitation values were different at that time, vegetational communities may also have differed. Future palynologic studies on Mauna Kea may shed some light on this question.

Summary and Conclusions

Evidence of repreated fluctuations of alpine glaciers during the late Quaternary is found in numerous high-mountain regions of the western United States, Alaska, and Hawaii. Today, about 500 km² of glacier cover exists in the western United States and about 73,000² exists in Alaska; during the maximum of the last glaciation, these regions contained about 100,000 km² and 630,000 km² of glacier ice, representing a twentyfold and sixfold increase, respectively. Extensive mountain ice fields formed in the Brooks Range and in the Alaska Peninsula and Alaska Range, the latter being confluent with the northern part of the Cordilleran ice sheet. Mountain glacier systems developed in the Cascade Range, the Sierra Nevada, the Yellowstone Plateau, and various ranges of the Rocky Mountain system. Smaller glaciers developed in numerous other highlands that are now devoid of glacier ice. In Hawaii, a small ice cap (70 km²) formed at the summit of Mauna Kea volcano.

A chronology of glacier fluctuations is slowly developing for these different glaciated regions. In some mountains, deposits of the last glaciation are bracketed only broadly by radiometric dates (e.g., Hawaii, with radiocarbon dates indicating an age of less than 29,000 and more than 9040 yr B.P.; Sierra Nevada, with a potassium-argon date indicating an age of less than 50,000 years and a radiocarbon date indicating an age of more than 9990 yr B.P.), although in others specific ice advances and readvances are dated more closely. One or more early advances dating to between 24,000 and 17,000 years ago are recognized in the Cascade Range, the Rocky Mountains, the Alaska Range, and the Brooks Range; later readvances that culminated between about 15,000 and 12,500 years ago have been reported from the Brooks Range, the Alaska Range, and the Cascade Range. In many areas, multiple moraines of the

youngest drift sheet indicate repeated halts and/or readvances (e.g., 20 or more for some large valley glaciers in the Sierra Nevada), but dating control generally is poor or lacking. Extensive deglaciation apparently occurred well before the close of the Pleistocene; the Yellowstone Plateau and San Juan ice field were deglaciated by about 14,000 years ago, and some valleys in the Cascade Range that contained glaciers 40 to 50 km long were free of ice by about 12,000 years ago. Mountain peaks, like Mauna Kea in Hawaii, that stand well below the present snow line (500 m or more) also probably were deglaciated at an early date.

During the maximum ice advances of the last glaciation, equilibrium-line altitudes of glaciers were some 850 to 900 m lower than they are at present in the Cascade Range and about 1000 m lower in the Rocky Mountains but probably were depressed no more than about half that amount in the Brooks Range of Arctic Alaska. For the Cascades, this degree of change is thought to reflect a mean-annual temperature lowering of about 4°C to 5°C; in the Rocky Mountains, where precipitation may have been an important controlling factor, mean-annual temperature lowering may have been two to three times as great.

Twenty years ago, most investigations of alpine glaciation were primarily concerned with establishing stratigraphic sequences and developing chronologies based on relative and radiometric dating. Although such efforts continue, much current research is being directed toward glaciologic and environmental reconstructions and better chronologic control. Such studies promise to contribute important information and ideas regarding late-Quaternary history and environmental change in the western United States.

References

Adam, D. P. (1967). Late-Pleistocene and recent palynology in the central Sierra Nevada, California. In "Quaternary Paleoecology" (E. J. Cushing, and H. E. Wright, Jr., eds.), pp. 275-301. Yale University Press, New Haven, Conn.

Ager, T. A. (1975). "Late Quaternary Environmental History of the Tanana Valley, Alaska." Ohio State University Institute of Polar Studies Report 54.

Alden, W. C. (1932). "Physiography and Glacial Geology of Eastern Montana and Adjacent Areas." U.S. Geological Survey Professional Paper 174.

Alden, W. C. (1953). "Physiography and Glacial Geology of Western Montana and Adjacent Areas." U.S. Geological Survey Professional Paper 231.

Andrews, J. T. (1972). Glacier power, mass balance, velocities, and erosion potential. Zeitschrift für Geomorphologie 13, 1-17.

Andrews, J. T., Carrara, P. E., King, F. B., and Stuckenrath, R. (1975). Holocene environmental changes in the alpine zone, northern San Juan Mountains, Colorado. Quaternary Research 5, 173-97.

Armstrong, J. E. (1975). "Quaternary Geology, Stratigraphic Studies, and Revaluation of Terrain Inventory Maps, Fraser Lowland, British Columbia." Geological Survey of Canada Paper 75-1, part A, pp. 277-380.

Atwood, W. W., and Mather, K. F. (1932). "Physiography and Quaternary Geology of the San Juan Mountains, Colorado." U.S. Geological Survey Professional Paper 166.

Bailey, R. A., Dalrymple, G. B., and Lanphere, M. A. (1976) Volcanism, structure, and geochronology of Long Valley caldera, Mono County, California. Journal of Geophysical Research 81, 725-44.

Baker, R. G. (1976). "Late Quaternary Vegetation History of the Yellowstone Lake Basin, Wyoming." U.S. Geological Survey Professional Paper 729-E.

Baker, R. G., and Richmond, G. M. (1978). Geology, palynology, and climatic significance of two pre-Pinedale lake sediment sequences in and near Yellowstone National Park. Quaternary Research 10, 226-40.

Barry, R. G. (1973). A climatological transect along the east slope of the Front Range, Colorado. Arctic and Alpine Research 5, 89-110.

Bateman, P. C., and Wahrhaftig, C. (1966). Geology of the Sierra Nevada. In"Geology of Northern California," pp. 107-72. California Division of Mines and Geology Bulletin 190.

Beget, J. E. (1981). Early Holocene glacier advance in the North Cascade Range, Washington. Geology 9, 409-13.

Benedict, J. B. (1973). Chronology of cirque glaciation, Colorado Front Range. Quaternary Research 3, 584-99.

Benedict, J. B. (1981). "The Forth of July Valley." Center for Mountain Archaeology Report 2.

Birkeland, P. W. (1964). Pleistocene glaciation of the northern Sierra Nevada, north of Lake Tahoe, California. Journal of Geology 72, 810-25.

Birkeland, P. W., Burke, R. M., and Yount, J. C. (1976). Preliminary comments on late Cenozoic glaciations in the Sierra Nevada. In "Quaternary Stratigraphy of North America" (W. C. Mahaney, ed.), pp. 283-95. Dowden, Hutchinson, and Ross, Stroudsburg, Pa.

Birkeland, P. W., Crandell, D. R., and Richmond, G. M. (1971). Status of correlation of Quaternary stratigraphic units in the western conterminous United States. Quaternary Research 1, 208-27.

Birkeland, P. W., and Janda, R. J. (1971). Clay mineralogy of soils developed from Quaternary deposits of the eastern Sierra Nevada, California. Geological Society of America Bulletin 82, 2495-2514.

Birman, J. H. (1964). "Glacial Geology across the Crest of the Sierra Nevada, California." Geological Society of America Special Paper 75.

Blackwelder, E. (1915). Post-Cretaceous history of the mountains of central western Wyoming. Journal of Geology 23, 307-40.

Blackwelder, E. (1931). Pleistocene glaciation in the Sierra Nevada and Basin Ranges. Geological Society of America Bulletin 42, 865-922.

Bowers, P. M. (1980). "The Carlo Creek Site: Geology and Archeology of an Early Holocene Site in the Central Alaska Range." Fairbanks, University of Alaska Cooperative Park Studies Unit Occasional Paper 27.

Bright, R. C. (1966). Pollen and seed stratigraphy of Swan Lake, southeastern Idaho. Tebiwa 9, 1-47.

Burke, R. M., and Birkeland P. W. (1979). Reevaluation of multiparameter relative dating techniques and their application to the glacial sequence along the eastern escarpment of the Sierra Nevada, California. Quaternary Research 11, 21-51.

Carrara, P. C., and Mode, W. N. (1979). Extensive deglaciation in the San Juan Mountains, Colorado, prior to 14,000 yr BP. Geological Society of America, Abstracts with Programs 11, 399.

Carver, G. A. (1973). Glacial geology of the Mountain Lakes Wilderness and adjacent parts of the Cascade Range, Oregon. Ph.D. dissertation, University of Washington, Seattle.

Clague, J. J., Armstrong, J. E., and Mathews, W. H. (1980). Advance of the Late Wisconsin Cordilleran ice sheet in southern British Columbia since 22,000 yr B.P. Quaternary Research 13, 322-26.

Clark, M. M. (1967). Pleistocene glaciation of the drainage of the West Walker River, Sierra Nevada, California. Ph.D. dissertation, Stanford University, Stanford, California.

CLIMAP Project Members (1976). The surface of the ice-age Earth. Science 191, 1131-44.

Colby, W. E. (1950). "John Muir's Studies in the Sierra." Sierra Club, San Francisco.

Colman, S. M., and Pierce, K. L. (1979). Preliminary map showing Quaternary deposits and their dating potential in the conterminous United States. U.S. Geological Survey Field Studies Map MF-1052.

Colman, S. M., and Pierce, K. L. (1981). "Weathering Rinds on Andesitic and Basaltic Stones as a Quaternary Age Indicator, Western United States." U.S. Geological Survey Professional Paper 1210.

Coulter, H. W., Hopkins, D. M., Karlstrom, T. N. V., Péwé, T. L., Wahrhaftig, C., and Williams, J. R. (1965). Map showing extent of glaciations in Alaska. U.S. Geologic Survey Miscellaneous Geologic Investigations Map I-415.

Crandell, D. R. (1965). The glacial history of western Washington and Oregon. In "The Quaternary of the United States" (H. E. Wright, Jr., and D. G. Frey, eds.), pp. 341-53. Princeton University Press, Princeton, N.J.

Crandell, D. R., and Miller, R. D. (1974). "Quaternary Stratigraphy and Extent of Glaciation in the Mount Rainier Region, Washington." U.S. Geological Survey Professional Paper 847.

Currey, D. R. (1974). Probable pre-Neoglacial age of the type Temple Lake moraine, Wyoming. Arctic and Alpine Research 6, 293-300.

Curry, R. R. (1971). "Glacial and Pleistocene History of the Mammoth Lakes Area, California: A Geologic Guidebook." University of Montana Department of Geology, Geological Series Publication 11.

Dalrymple, G. B. (1964). Potassium-argon dates of three Pleistocene interglacial basalt flows from the Sierra Nevada, California. Geological Society of America Bulletin 75, 753-58.

Dalrymple, G. B., Burke, R. M., and Birkeland, P. W. (1982). Concerning K-Ar dating of a basalt flow from the Tahoe-Tioga interglaciation, Saw Mill Canyon, southeastern Sierra Nevada, California. Quaternary Research 17, 120-22.

Denton, G. H. (1974). Quaternary glaciations of the White River valley, Alaska, with a regional synthesis for the northern St. Elias Mountains, Alaska, and Yukon Territory. Geological Society of America Bulletin 85, 871-92.

Detterman, R. L., Bowsher, A. L., and Dutro, J. T., Jr. (1958). Glaciation on the Arctic slope of the Brooks Range, northern Alaska. Arctic 11, 43-61.

Farnes, P. E. (1971). Mountain precipitation and hydrology from snow surveys. In "Proceedings of the Western Snow Conference, 39th Annual Meeting," pp. 44-49. Colorado State University, Fort Collins.

Farrand, W. R. (1973). Observations on Pleistocene glaciation of the north slope of the Brooks Range, northern Alaska. In "Archeological Reconnaissances North of the Brooks Range in Northeastern Alaska" (R. S. Solecki, B. Salwen, and J. Jacobson, eds.), pp. 91-98. University of Calgary Department of Archaeology Occasional Papers 1.

Fernald, A. T. (1960). "Geomorphology of the Upper Kuskokwim Region, Alaska." U.S. Geological Survey Bulletin 1071-G, pp. 191-279.

Fernald, A. T. (1965a). "Glaciation in the Nabesna River Area, Upper Tanana River Valley, Alaska." U.S. Geological Survey Professional Paper 525-C, pp. C120-C123.

Fernald, A. T. (1965b). "Late Quaternary Chronology, Upper Tanana Valley, Eastern Alaska" (abstract). Geological Society of America Special Paper 82, pp. 60-61.

Fernald, A. T. (1965c). "Recent History of the Upper Tanana River Lowland, Alaska." U.S. Geological Survey Professional Paper 525-C, pp. C124-C127.

Flint, R. F. (1971). "Glacial and Quaternary Geology." John Wiley and Sons, New York.

Flint, R. F., and others (1945). "Glacial Map of North America." Geological Society of America Special Paper 60.

Gallie, T. M., III (1977). Aspects of the glacial environments of a late Pleistocene ice maximum in Little Cottonwood Canyon, Utah. M.S. thesis, University of Utah, Salt Lake City.

Gates, W. L. (1976). Modeling the ice-age climate. Science 191, 1138-44.

Gennett, J. A. (1977). Palynology and paleoecology of sediments from Blacktail Pond, northern Yellowstone Park, Wyoming. M.S. thesis, University of Iowa, Iowa City.

Haefner, B. D. (1976). Glacial valley morphometry in the Front Range of Colorado. M.A. thesis, University of Colorado, Boulder.

Halstead, E. C. (1968). The Cowichan ice tongue, Vancouver Island. Canadian Journal of Earth Sciences 5, 1409-15.

Hamilton, T. D. (1969). Glacial geology of the lower Alatna Valley, Brooks Range, Alaska. In "United States Contributions to Quaternary Research" (S. A. Schumm and W. C. Bradley, eds.), pp. 181-223. Geological Society of America Special Paper 123.

Hamilton, T. D. (1973). "Late Quaternary Glacial History, Delta-Johnson Rivers Region, Northeastern Alaska Range: Final Report." National Science Foundation Grant GS-2584, University of Alaska, Fairbanks.

Hamilton, T. D. (1979a). "Radiocarbon Dates and Quaternary Stratigraphic Sections, Philip Smith Mountains Quadrangle, Alaska." U.S. Geological Survey Open-File Report 79-866.

Hamilton, T. D. (1979b). Surficial geologic map of the Chandler Lake quadrangle, Alaska. U.S. Geological Survey Miscellaneous Field Studies Map MF-1121.

Hamilton, T. D. (1979c). Surficial geologic map of the Wiseman quadrangle, Alaska. U.S. Geological Survey Miscellaneous Field Studies Map MF-1122.

Hamilton, T. D. (1980a). "Quaternary Stratigraphic Sections with Radiocarbon Dates, Chandler Lake Quadrangle, Alaska." U.S. Geological Survey Open-File Report 80-790.

Hamilton, T. D. (1980b). Surficial geologic map of the Killik River quadrangle,

Alaska. U.S. Geological Survey Miscellaneous Field Studies Map MF-1234.

Hamilton, T. D. (1982). A late Pleistocene glacial chronology for the southern Brooks Range: Stratigraphic record and regional significance. *Geological Society of America Bulletin* 93, 700-716.

Hamilton, T. D., and Porter S. C. (1975). Itkillik Glaciation in the Brooks Range, northern Alaska. *Quaternary Research* 5, 471-97.

Hamilton, T. D., and Thorson, R. M. (1983). The Cordilleran ice sheet in Alaska. *In* "Late-Quaternary Environments of the United States," vol. 1, "The Late Pleistocene" (S. C. Porter, ed., pp. 38-52. University of Minnesota Press, Minneapolis.

Hammer, C. U., Clausen, H. B., and Dansgaard, W. (1980). Greenland ice sheet evidence of post-glacial volcanism and its climatic impact. *Nature* 288, 230-55.

Hammond, E. H. (1965). Land surface form. U.S. Geological Survey, National Atlas, sheet 61.

Heusser, C. J. (1983). Vegetational history of the northwestern United States including Alaska. *In* "Late-Quaternary Environments of the United States," vol. 1, "The late Pleistocene" (S. C. Porter, ed.), pp. 239-258. University of Minnesota Press, Minneapolis.

Heusser, C. J., Heusser, L. E., and Streeter, S. S. (1980). Quaternary temperatures and precipitation for the north-west coast of North America. *Nature* 286, 702-04.

Hollin, J. T, and Schilling, D. H. (1981). Late Wisconsin-Weichselian mountain glaciers and small ice caps. *In* "The Last Great Ice Sheets" (G. H. Denton and T. J. Hughes, eds.), pp. 179-206. John Wiley and Sons, New York.

Holmes, G. W., and Foster, H. L. (1968). "Geology of the Johnson River Area, Alaska." U.S. Geological Survey Bulletin 1249.

Holmes, G. W., and Lewis, C. R. (1965). "Quaternary Geology of the Mount Chamberlin Area, Brooks Range, Alaska." U.S. Geological Survey Bulletin 1201-B.

Hopkins, D. M. (1973). Sea level history in Beringia during the past 250,000 years. *Quaternary Research* 3, 520-40.

Hopkins, D. M. (1979). Landscape and climate of Beringia during late Pleistocene and Holocene time. *In* "The First Americans: Origins, Affinities, and Adaptations" (W. S. Laughlin and A. H. Harper, eds.), pp. 15-41. Fischer, Stuggart. (U.S. distributor, Verlag Chemie, Deerfield Beach, Fla.)

Hudson, T. and Weber, F. R. (1977). The Donnelly Dome and Granite Mountain faults, south-central Alaska. *In* "The United States Geological Survey in Alaska: Accomplishments during 1976" (K. M. Blean, ed.), pp. B64-B66. U.S. Geological Survey Circular 751-B.

Janda, R. J. (1965). Quaternary alluvium near Friant, California. *In* "Guidebook for Field Conference I, Northern Great Basin and California," pp. 128-33. Seventh Congress of the International Association for Quaternary Research, Boulder, Colo.

Johnson, P. R., and Hartman, C. W. (1969). "Environmental Atlas of Alaska." University of Alaska Institute of Arctic Environmental Engineering, College.

Karlstrom, T. N. V. (1964). "Quaternary Geology of the Kenai Lowland and Glacial History of the Cook Inlet Region, Alaska." U.S. Geological Survey Professional Paper 443.

Karlstrom, T. N. V., and others (1964). Surficial geology of Alaska. U.S. Geological Survey Miscellaneous Geologic Investigations Map I-357.

Landsberg, H. (1958). "Physical Climatology." Gray Printing Company, Dubois, Pa.

Lanphere, M. A. (1978). Displacement history of the Denali fault system, Alaska and Canada. *Canadian Journal of Earth Science* 15, 817-22.

Le Conte, J. (1873). On some of the ancient glaciers of the Sierra. *American Journal of Science* (3rd series) 5, 325-39.

Legg, T. E. and Baker, R. G. (1980). Palynology of Pinedale sediments, Devlin's Park, Boulder County, Colorado. *Arctic and Alpine Research* 12, 319-33.

Luckman, B. H., and Osborn, G. D. (1979). Holocene glacier fluctuations in the middle Canadian Rocky Mountains. *Quaternary Research* 11, 52-77.

McCoy, W. D. (1981). Quaternary aminostratigraphy of the Bonneville and Lahonton Basins, western U.S., with paleoclimatic implications. Ph.D. dissertation, University of Colorado, Boulder.

McManus, D. A., Venkatarathnam, K., Hopkins, D. M. and Nelson, C. H. (1974). Yukon River sediment on the northernmost Bering Sea shelf. *Journal of Sedimentary Petrology* 44, 1052-60.

Madole, R. F. (1976a). Bog stratigraphy, radiocarbon dates, and Pinedale to Holocene glacial history in the Front Range, Colorado. *U.S. Geological Survey Journal of Research* 4, 163-69.

Madole, R. F. (1976b). Glacial geology of the Front Range, Colorado. *In* "Quaternary Stratigraphy of North America" (W. C. Mahaney, ed.), pp. 297-318. Dowden, Hutchinson, and Ross, Stroudsburg, Pa.

Madole, R. F. (1980a). "Glacial Lake Devlin and the Chronology of Pinedale Glaciation on the East Slope of the Front Range, Colorado." U.S. Geological Survey Open File Report 80-725.

Madole, R. F. (1980b). Time of Pinedale deglaciation in north-central Colorado: Further considerations. *Geology* 8, 118-22.

Madsen, D. B., and Currey, D. R. (1979). Late Quaternary glacial and vegetation changes, Little Cottonwood Canyon area, Wasatch Mountains, Utah. *Quaternary Research* 12, 254-70.

Maher, L. J., Jr. (1972). Nomograms for computing 0.95 confidence limits of pollen data. *Review of Paleobotany and Palynology* 13, 85-93.

Marchand, D. E., and Allwardt, A. (1981). "Late Cenozoic Stratigraphic Units, Northeastern San Joaquin Valley, California." U.S. Geological Survey Bulletin 1470.

Matthes, F. E. (1930). "Geologic History of the Yosemite Valley." U.S. Geological Survey Professional Paper 160.

Matthes, F. E. (1960). "Reconnaissance of the Geomorphology and Glacial Geology the San Joaquin Basin, Sierra Nevada, California." U.S. Geological Survey Professional Paper 329.

Matthes, F. E. (1965). "Glacial Reconnaissance of Sequoia National Park, California." U.S. Geological Survey Professional Paper 504-A, pp. A1-A58.

Matthews, J. V., Jr. (1974). Wisconsin environment of interior Alaska: Pollen and macrofossil analysis of a 27 meter core from the Isabella basin (Fairbanks, Alaska). *Canadian Journal of Earth Sciences* 11, 828-41.

Mears, B., Jr. (1974). The evolution of the Rocky Mountain glacial model. *In* "Glacial Geomorphology" (D. R. Coates, ed.), pp. 11-40. State University of New York at Binghampton Publications in Geomorphology.

Mears, B., Jr. (1981). Periglacial wedges and the late Pleistocene environment of Wyoming's intermontane basins. *Quaternary Research* 15, 171-98.

Mehringer, P. J., Jr. (1977). Great Basin late Quaternary environments and chronology. *In* "Models and Great Basin Prehistory" (D. D. Fowler, ed.), pp. 113-67. Desert Research Institute Publications in the Social Sciences 12.

Mehringer, P. J., Jr., Arno, S. F., and Peterson, K. L. (1977). Postglacial history of Lost Trail Pass bog, Bitterroot Mountains, Montana. *Arctic and Alpine Research* 9, 345-68.

Mehringer, P. J., Jr., Nash, W. P., and Fuller, R. H. (1971). A Holocene volcanic ash from northwestern Utah. *Proceedings of the Utah Academy of Sciences, Arts, and Letters* 48, 46-51.

Meier, M. F. (1961a). "Distribution and Variations of Glaciers in the United States Exclusive of Alaska." International Association of Scientific Hydrology Publication 54, pp. 420-29.

Meier, M. F. (1961b). "Mass Budget of the South Cascade Glacier, 1957-60." U.S. Geological Survey Professional Paper 424-B, pp. B206-B211.

Meier, M. F., and Post, A. S. (1962). "Recent Variations in Mass Net Budgets of Glaciers in Western North America." International Association of Scientific Hydrology Publication 58, pp. 63-77.

Meier, M. F., Tangborn, W. V., Mayo, L. R., and Post, A. (1971). "Combined Ice and Water Balances of the Gulkana and Wolverine Glaciers, Alaska, and South Cascade Glacier, Washington, 1965 and 1966 Hydrologic Years." U.S. Geological Survey Professional Paper 715-A.

Miller, C. D. (1979). A statistical method for relative-age dating of moraines in the Sawatch Range, Colorado. *Geological Society of America Bulletin* 90, 1153-64.

Miller, C. D., and Birkeland, P. W. (1974). Probable pre-Neoglacial age of the type Temple Lake moraine, Wyoming: Discussion and additional relative-age data. *Arctic and Alpine Research* 6, 301-6.

Montagne, J. M. (1972). Wisconsin Glaciation. *In* "Geologic Atlas of the Rocky Mountain Region," pp. 257-60. Rocky Mountains Association of Geologists, Denver, Colo.

Morrison, R. B. (1965). Quaternary geology of the Great Basin. *In* "The Quaternary of the United States" (H. E. Wright, Jr., and D. G. Frey, eds.), pp. 265-85. Princeton University Press, Princeton, N.J.

Mullineaux, D. R. (1974). "Pumice and Other Pyroclastic Deposits in Mount Rainier National Park, Washington." U.S. Geological Survey Bulletin, 1326.

Mullineaux, D. R., Hyde, J. H., and Rubin, M. (1975). Widespread late glacial and postglacial tephra deposits from Mount St. Helens volcano, Washington. *Journal of Research, U.S. Geological Survey* 3, 329-35.

National Oceanic and Atmospheric Administration (1976). "Climatological Data: Annual Summary," Vol. 27, pp. 48-55.

Nelson A. R., Millington, A. C., Andrews, J. T., and Nichols, H. (1979). Radiocarbon-dated upper Pleistocene glacial sequence, Fraser Valley, Colorado Front Range. *Geology* 7, 410-14.

Nelson, C. H., and Creager, J. S. (1977). Displacement of Yukon-derived sediment from Bering Sea to Chukchi Sea during Holocene time. *Geology* 5, 141-46.

Nye, J. F. (1952). A method of calculating thicknesses of ice sheets. *Nature* 169, 529-30.

Patterson, W. S. B. (1969). "The Physics of Glaciers." Pergamon Press, New York.

Péwé, T. L. (1953). Big Delta area, Alaska. *In* "Multiple Glaciation in Alaska" (T. L. Péwé and others) pp. 8-12. U.S. Geological Survey Circular 289.

Péwé, T. L. (1965). Delta River area, Alaska Range. *In* "Guidebook for Field Conference F, Central and South-Central Alaska," pp. 55-98. Seventh Congress of the International Association for Quaternary Research, Boulder, Colo.

Péwé, T. L. (1975). "Quaternary Geology of Alaska." U.S. Geological Survey Professional Paper 834.

Péwé, T. L., Burbank, L., and Mayo, L. R. (1967). Multiple glaciation of the Yukon-Tanana upland, Alaska. U.S. Geological Survey Miscellaneous Geologic Investigations Map I-507.

Péwé, T. L., and Holmes, G. W. (1964). Geology of the Mt. Hayes D-4 quadrangle, Alaska. U.S. Geological Survey Miscellaneous Geologic Investigations Map I-394.

Péwé, T. L., Hopkins, D. M., and Giddings, J. L. (1965). The Quaternary geology and archaeology of Alaska, *In* "The Quaternary of the United States" (H. E. Wright, Jr., and D. G. Frey, eds.), pp. 355-74. Princeton University Press, Princeton, N.J.

Péwé, T. L., and Reger, R. D. (1972). Modern and Wisconsinan snowlines in Alaska. *In* "Proceedings of Section 12, Quaternary Geology, International Geological Congress, 24th Session, Montreal 1972, pp. 187-97.

Pierce, K. L. (1973). Surficial geologic map of the Mount Holmes quadrangle, and parts of the Tepee Creek, Crown Butte, and Miner quadrangles, Yellowstone National Park, Wyoming and Montana. U.S. Geological Survey Miscellaneous Geologic Investigations Map I-640.

Pierce, K. L. (1979). "History and Dynamics of Glaciation in the Northern Yellowstone National Park Area." U.S. Geological Survey Professional Paper 729-F.

Pierce, K. L., Obradovich, J. D., and Friedman, I. (1976). Obsidian hydration dating and correlation of Bull Lake and Pinedale Glaciations near West Yellowstone, Montana. *Geological Society of America Bulletin* 87, 703-10.

Porter, S. C. (1964a). Composite Pleistocene snow line of Olympic Mountains and Cascade Range, Washington. *Geological Society of America Bulletin* 75, 477-82.

Porter, S. C. (1964b). Late Pleistocene glacial chronology of north-central Brooks Range, Alaska. *American Journal of Science* 262, 446-60.

Porter, S. C. (1966). "Pleistocene Geology of Anaktuvuk Pass, Central Brooks Range, Alaska." Arctic Institute of North America Technical Paper 18.

Porter, S. C. (1969). "Pleistocene Geology of East-Central Cascade Range, Washington." Guidebook for Third Pacific Coast Friends of the Pleistocene Field Conference.

Porter, S. C. (1971). Fluctuations of late Pleistocene alpine glaciers in western North America. *In* "The Late Cenozoic Glacial Ages" (K. K. Turekian, ed.), pp. 307-29. Yale University Press, New Haven, Conn.

Porter, S. C. (1975a). Late Quaternary glaciation and tephrochronology of Mauna Kea, Hawaii. *In* "Quaternary Studies" (B. P. Suggate and M. M. Cresswell, eds.), pp. 247-51. Royal Society of New Zealand Bulletin 13.

Porter, S. C. (1975b). Weathering rinds as a relative-age criterion: Application to subdivision of glacial deposits in the Cascade Range. *Geology* 3, 101-4.

Porter, S. C. (1976). Pleistocene glaciation in the southern part of the North Cascade Range, Washington. *Geological Society of America Bulletin* 87, 61-75.

Porter, S. C. (1977). Present and past glaciation threshold in the Cascade Range, Washington, U.S.A.: Topographic and climatic controls, and paleoclimatic im-plications. *Journal of Glaciology* 18, 101-16.

Porter, S. C. (1978). Glacier Peak tephra in the North Cascade Range, Washington: Stratigraphy, distribution, and relationship to late-glacial events. *Quaternary Research* 10, 30-41.

Porter, S. C. (1979a). Hawaiian glacial ages. *Quaternary Research* 12, 161-87.

Porter, S. C. (1979b). Quaternary stratigraphy and chronology of Mauna Kea, Hawaii: A 380,000-year record of mid-Pacific volcanism and ice-cap glaciation. *Geological Society of America Bulletin* 90 (part 2), 980-1093.

Post, A., Richardson, D., Tangborn, W., and Rosselot, F. L. (1971). "Inventory of Glaciers in the North Cascades, Washington." U.S. Geological Survey Professional Paper 705-A.

Putnam, W. C. (1950). Moraine and shoreline relationships at Mono Lake, California. *Geological Society of America Bulletin* 61, 115-22.

Putnam, W. C. (1960). Faulting and Pleistocene glaciation in the east-central Sierra Nevada of California, U.S.A. *In* "21st International Geological Congress (Copenhagen) Report," part 21, pp. 270-74.

Ray, L. L. (1940). Glacial chronology of the southern Rocky Mountains. *Geological Society of America Bulletin* 51, 1851-1917.

Reed, J. C., Jr. (1961). "Geology of the Mount McKinley Quadrangle, Alaska." U.S. Geological Survey Bulletin 1108-A, pp. A1-A36.

Reger, R. D. (1979). Glaciation of Indian Mountain, west-central Alaska. *In* "Short Notes on Alaskan Geology: 1978," pp. 15-18. Alaska Division of Geological and Geophysical Surveys Report 61.

Richmond, G. M. (1960). Glaciation of the east slope of Rocky Mountain National Park, Colorado. *Geological Society of America Bulletin* 71, *1371-82.*

Richmond, G. M. (1962). "Quaternary Stratigraphy of the La Sal Mountains, Utah." U.S. Geological Survey Professional Paper 324.

Richmond, G. M. (1965). Glaciation of the Rocky Mountains. *In* "The Quaternary of the United States" (H. E. Wright, Jr., and D. G. Frey, eds.), pp. 217-30. Princeton University Press, Princeton, N.J.

Richmond, G. M. (1969). Development and stagnation of the Late Pleistocene icecap in the Yellowstone Lake basin, Yellowstone National Park, USA. *Eiszeitalter u. Gegenwart* 20, 196-203.

Richmond, G. M. (1976). Pleistocene stratigraphy and chronology in the mountains of western Wyoming. *In* "Quaternary Stratigraphy of North America" (W. C Mahaney, ed.), pp. 353-79. Dowden, Hutchinson, and Ross, Stroudsburg, Pa.

Richmond, G. M., Fryxell, R., Montagne, J., and Trimble, D. E. (1965). "Guidebook for Field Conference E, Northern and Middle Rocky Mountains." Seventh Congress of the International Association for Quaternary Research, Boulder, Colo.

Richmond, G. M., and Pierce, K. L. (1972). Surficial geologic map of the Eagle Peak quadrangle, Yellowstone National Park and adjoining area, Wyoming. U.S. Geological Survey Miscellaneous Geologic Investigations Map I-637.

Richter, D. H. (1976). Geologic map of the Nabesna quadrangle, Alaska. U.S. Geological Survey Miscellaneous Geologic Investigations Map I-932.

Roy, W. R., and Hall, R. D. (1980). Re-evaluation of the Bull Lake Glaciation through re-study of the type area and studies of other localities. *Geological Society of America, Abstracts with Programs* 12, 302.

Ruhe, R. V. (1964). An estimate of paleoclimate in Oahu, Hawaii. *American Journal of Science* 262, 1098-115.

Sable, E. G. (1977). "Geology of the Western Romanzof Mountains, Brooks Range, Northeastern Alaska." U.S. Geological Survey Professional Paper 897.

Schmoll, H. R., Szabo, B. J., Rubin, M., and Dobrovolny, E. (1972). Radiometric dating of marine shells from the Bootlegger Cove Clay, Anchorage area, Alaska. *Geological Society of America Bulletin* 83, 1107-13.

Schweger, C. E. (1971). Late Quaternary paleoecology of the Onion Portage region, northwestern Alaska. *Geological Society of America, Abstracts with Programs* 3, 413.

Schweger, C. E. (1976). Late Quaternary paleoecology of the Onion Portage region, northwestern Alaska. Ph.D. dissertation, University of Alberta, Edmonton.

Scott, W. E. (1977). Quaternary glaciation and volcanism, Metolius River area, Oregon. *Geological Society of America Bulletin* 88, 113-24.

Sellmann, P. (1967). "Geology of the USA CRREL Permafrost Tunnel, Fairbanks, Alaska." U.S. Army Cold Region Research Engineering Laboratory Technical Report 199, Hanover, N.H.

Sellmann, P. (1972). "Geology and Properties of Materials exposed in the USA CRREL Permafrost Tunnel." U.S. Army Cold Region Research Engineering Laboratory Special Report 177.

Shackleton, N. J., and Opdyke, N. D. (1973). Oxygen-isotope and paleomagnetic stratigraphy of equatorial Pacific core V28-238: Oxygen isotope temperatures and ice volumes on a 10^5 year and 10^6 year scale. *Quaternary Research* 3, 39-55.

Sharp, R. P. (1969). Semiquantitative differentiation of glacial moraines near Convict Lake, Sierra Nevada, California. *Journal of Geology* 77, 68-91.

Sharp, R. P. (1972). Pleistocene glaciation, Bridgeport Basin, California. *Geological Society of America Bulletin* 83, 2233-60.

Sharp, R. P., and Birman, J. H. (1963). Additions to the classical sequence of Pleistocene glaciations, Sierra Nevada, California. *Geological Society of America Bulletin* 74, 1079-86.

Sharpenseel, H. W. (1971). Radiocarbon dating of soils: Problems, troubles, hopes. *In* "Paleopedology" (D. H. Yaalon, ed.), pp. 77-87. Israel University Press, Jerusalem.

Shroba, R. R. (1977). Soil development in Quaternary tills, rock glacier deposits, and taluses, southern and central Rocky Mountains. Ph.D. dissertation, University of Colorado, Boulder.

Smith, G. I. (1968). Late Quaternary geologic and climatic history of Searles Lake, southeastern California. *In* "Means of Correlation of Quaternary Successions" (R. B. Morrison and H. E. Wright, Jr., eds.), pp. 293-310. University of Utah Press, Salt Lake City.

Smith, G. I., and Street-Perrott, A. (1983). Pluvial lakes of the western United States. *In* "Late-Quaternary Environments of the United States," vol. 1, "The Late Pleistocene" (S. C. Porter, ed.), pp. 190-211. University of Minnesota Press, Minneapolis.

Stout, J. H., Brady, J. B., Weber, F. and Page, R. A. (1973). Evidence for Quaternary movement of the McKinley strand of the Denali fault in the Delta River area, Alaska. *Geological Society of America Bulletin* 84, 939-47.

Ten Brink, N. W., and Ritter, D. F. (1980). Glacial chronology of the north-central Alaska Range and implications for discovery of early-man sites. *Geological Society of America, Abstracts with Programs* 12, 534.

Thompson, W. F., and Dodd, A. V. (1958). "Environmental Handbook for the Camp Hale and Pikes Peak Areas, Colorado." U.S. Army, Headquarters Quartersmaster Research and Engineering Command, Technical Report EP-79.

Thorson, R. M. (1981). "Isostatic Effects of the Last Glaciation in the Puget Lowland, Washington." U.S. Geological Survey Open-File Report 81-370.

Thorson, R. M., and Hamilton, T. D. (1977). Geology of the Dry Creek site: A stratified early man site in interior Alaska. *Quaternary Research* 7, 149-76.

United States Geological Survey (1972). Surficial geologic map of Yellowstone National Park. U.S. Geological Survey Miscellaneous Geologic Investigations Map I-710.

Waddington, J. C. B., and Wright, H. E., Jr. (1974). Late Quaternary vegetational changes on the east side of Yellowstone Park, Wyoming. *Quaternary Research* 4, 175-84.

Wahrhaftig, C. (1958). "Quaternary Geology of the Nenana River Valley and Adjacent Parts of the Alaska Range." U.S. Geological Survey Professional Paper 293-A.

Wahrhaftig, C., and Birman, J. H. (1965). The Quaternary of the Pacific Mountain system in California. *In* "The Quaternary of the United States" (H. E. Wright, Jr., and D. G. Frey, eds.), pp. 299-340. Princeton University Press, Princeton, N.J.

Waitt, R. B., and Thorson, R. M. (1983). The Cordilleran ice sheet in Washington, Idaho, and Montana. *In* "Late-Quaternary Environments of the United States," vol. 1, "The Late Pleistocene" (S. C. Porter, ed.), pp. 53-70. University of Minnesota Press, Minneapolis.

Watson, C. E. (1959). Climate of Alaska. *In* "Climate of the States." U.S. Weather Bureau, Climatography of the United States 60-49.

Weber, F. R., Hamilton, T. D., Hopkins, D. M., Repenning, C. A., and Haas, H. (1981). Canyon Creek: A late Pleistocene vertebrate locality in interior Alaska. *Quaternary Research* 16, 167-80.

Wendler, G., Benson, C., Fahl, C., Ishikawa, N., Trabant, D., and Weller, G. (1975). Glacio-meteorological studies of McCall Glacier. *In* "Climate of the Arctic" (G. Weller and S. A. Bowling, eds.). pp. 334-38. University of Alaska Geophysical Institute, Fairbanks.

Wendler, G., Ishikawa, N., and Streten, N. (1974). The climate of the McCall Glacier, Brooks Range, Alaska, in relation to its geographical setting. *Arctic and Alpine Research* 6, 307-18.

Williams, J. R., Yeend, W. E., Carter, L. D., and Hamilton, T. D. (1977). Preliminary surficial deposits map of National Petroleum Reserve—Alaska. U.S. Geological Survey Open-File Map 77-868.

Wright, H. E., Jr., Bent, A. M., Hansen, B. S., and Maher, L. J., Jr. (1973). Present and past vegetation of the Chuska Mountains, northwestern New Mexico. *Geological Society of America Bulletin* 84, 1155-80.

Yeend, W. (1971). "Glaciation of the Ray Mountains, Central Alaska." U.S. Geological Survey Professional Paper 750-D, pp. D122-D126.

Nonglacial Environments

Late-Pleistocene Fluvial Systems

Victor R. Baker

Introduction

Because rivers cut across diverse continental environments, they are of special interest in studies of Quaternary environmental change. The rivers of the late Pleistocene (25,000 to 10,000 yr B.P.) were affected by profound changes in climate and by other factors. Ironically, these rivers constitute "experiments" in altered regimen that are being redone today as a result of large-scale surface mining, forest clearcutting, and other artificial alterations of the environment. Moreover, the need to design nuclear-waste repositories, dam sites, and power plants for long-term stability has added new concern for the long-term activity of rivers. Our interest in late-Pleistocene rivers extends to the original events, since the juxtaposition of ecosystems along rivers allowed the first Americans to make maximum use of their resources.

River systems can be remarkably sensitive to changes in environment, especially when certain thresholds for change are exceeded (Schumm, 1979). Although relict fluvial features are a tempting source of paleohydrologic information, the current state of the art in the reconstruction of late-Pleistocene fluvial paleohydraulics and paleohydrology reveals both promise and problems. The problems derive from the chain of processes and responses that connects paleoclimatic and paleohydrologic events to discernible field evidence (Figure 5-1). As discussed more fully by Baker (1978a), the various linkages in a paleohydrologic "think back" are complicated by information losses and feedback effects. A somewhat similar model has been used to describe the problems inherent in interpreting glacial-climatic interactions (Meier, 1965).

The time period of interest in this chapter includes the key date of 18,000 yr B.P., chosen for scrutiny by the CLIMAP project members (1976) for characterizing the Earth's ice-age climate. CLIMAP produced an excellent compilation of the Earth's ocean surface temperatures. However, a corresponding analysis of continental areas found considerable variability (Peterson et al., 1979). Temperatures in the conterminous United States were cooler at 18,000 yr B.P., generally 3° to 8°C cooler, than they are at present. However, hydrologic conditions varied greatly from place to place. Sites in the Southwest were uniformly cooler and moister, while sites in the Southeast were cooler and drier. In the Midwest, by contrast, some

sites indicate drier conditions and other sites appear to have been wetter (Peterson et al., 1979). Even these conclusions must be considered tentative, for Brakenridge (1978) believes that an annual temperature depression of 7° to 8°C could explain most of the assumed indicators of moister full-glacial conditions in the southwestern United States.

Clearly, there are numerous problems both intrinsic and extrinsic to the analysis of ancient fluvial events. Late-Pleistocene terrace sequences are a case in point. The fluvial system responds to external change by shifting a number of variables (Schumm, 1977). The periods of high sediment yield that lead to valley filling can arise from such diverse causes as glacial meltwater flow, rising base level, uplift in the source area, and climatic change. It may be difficult to distinguish among these elements or to eliminate all possibilities save one. River entrenchment into the former floodplain is even more perplexing. Erosion then removes the sediments that aid in the environmental interpretation of aggradational episodes. Pitty (1971: 16) aptly summarizes the situation: "Lending confidence to the geomorphologist during his hypothetical leaps between form and process is the flimsy safety net provided by the study of sediments."

This chapter is limited in scope by necessity. Emphasis is placed on the Quaternary chronostratigraphy of the Midwest (Mickelson et al., 1983; Willman and Frye, 1970). This derives from the importance of the Laurentide ice sheet during this time interval and in the use of the Midwest chronostratigraphy in international correlations. The time of interest coincides with the terminal part of the Wisconsin stage, including the Farmdalian substage (28,000 to 22,000 yr B.P.), the Woodfordian substage (22,000 to 12,500 yr B.P.), the Twocreekan substage (approximately 12,000 yr B.P.) and the Greatlakean substage (approximately 11,000 yr B.P. as defined by Evenson et al., 1976). Other local stratigraphic terms introduced in this chapter are correlated approximately to the Midwest chronostratigraphy.

Proglacial Fluvial Systems

The last part of the Pleistocene left a legacy of relict proglacial fluvial landforms extending from all the glacial margins. Most important were the features that developed when deglaciation commenced, approximately 18,000 to 16,000 yr B.P. Church and Ryder (1972) pre-

115

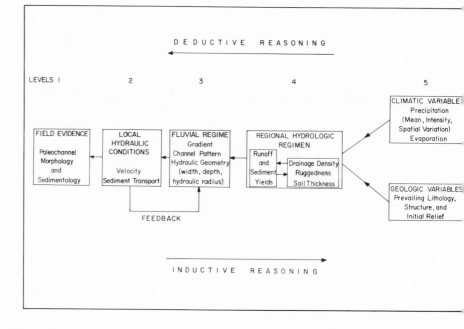

Figure 5-1. Chain of natural processes and systems interactions connecting fluvial geomorphic field evidence (level-1 information) to climatic and geologic controls (level-5 information) (Baker, 1974). The paleohydrologist must reconstruct information at different levels in this scheme by a "think-back" process involving deductive and/or inductive reasoning. The interpretation of one level of information from knowledge of another level is complicated by numerous factors, including information loss through time and feedback effects. (From Baker, 1978a.)

sent arguments that suggest that the deglacial phase of a glacial cycle is a period of profound change in fluvial sedimentation. Sediment yield rates increase to about 10 times the geologic "norm" upon the commencement of deglaciation and rapidly subside thereafter. The immense influx of sediment results in extensive outwash trains and outwash plains (sandurs). Recent advances in the study of modern proglacial sedimentation (e.g., Boothroyd and Ashley, 1975; Church, 1972; Church and Gilbert, 1975) hold promise for considerable advances in the study of late-Pleistocene outwash. Whereas earlier work concentrated on drift stratigraphy, chronology, correlation, and terminology, current research represents increasingly an attempt to analyze and even to reconstruct the dynamics of the ancient flows.

In the eastern United States, the Susquehanna and Delaware Rivers (Figure 5-2) developed prominent glaciofluvial terraces of Wisconsin age (Peltier, 1949, 1959). In New England, ice stagnation was an important process in the retreat of the Laurentide ice sheet, and outwash trains developed in complex association with ice blocks and proglacial lakes (Koteff, 1974). The Mississippi River became an immense proglacial drainage system. Mountain glaciation in the Rocky Mountains, Cascade Range, and Sierra Nevada also produced prominent outwash trains. In the Shoshone River Basin of northwestern Wyoming, for example, Moss (1974) traced the Cody terrace (Late Wisconsin) from late-Pleistocene glacial moraines of the Absaroka Range into the semiarid Big Horn Basin. Field relationships indicate contemporaneous mountain glaciation, valley aggradation in the mountain canyons, and lateral spreading of gravel in the Big Horn Basin.

The extensive sequences of late-Pleistocene outwash deposits provide a great opportunity for the reconstruction of the ancient flow events. Many of the deposits have not been altered since their deposition, and some contain evidence of paleochannel dimensions and slopes. This has led to several attempts at estimating paleoflows through the application of paleohydraulic principles. Jopling (1966) has outlined a procedure for use in sandy fluvial sediments, but he has found that the complexity of bed-form development led to several indeterminate factors. Baker (1974) and Baker and Ritter (1975) sug-

gest schemes of paleohydraulic computations for gravel-transporting systems. However, Baker (1974: 108) cautions that in applying these schemes "the magic of numbers . . . must be tempered by the reality of the assumptions that apply to such calculations." Church (1978) more fully elaborates on the assumptions in such procedures and concludes that, though the absolute magnitudes of ancient flows cannot be specified, the largest flows can be quantitatively compared within an admittedly broad error range. Nevertheless, some late-Pleistocene flows were augmented to the extent that this comparison can be useful. A case in point is that of very large proglacial flood events, such as the Lake Missoula floods (Baker, 1973; Bretz, 1969) which are discussed later in this chapter.

Numerous examples of river diversion were produced by glaciation, especially in the north-central United States (e.g., Flint, 1971: 232-36). Many of these diversions were established by pre-Wisconsin glaciations, however; the Wisconsinan events of interest here merely repeated patterns set during earlier glaciations.

For west-central Montana, Foley (1980b) describes a Late Wisconsin (Pinedale) river diversion by a piedmont lobe from the Lewis and Clark Range. The time-elevation history of the diverted channel was reconstructed, and it demonstrated as much as 60 m of bedrock incision during the past 25,000 years. Incision was most rapid during the lower and middle Pinedale stades (7 mm per year) and was least rapid during interstades and at present (0.5 mm per year). Foley (1980a) has developed a stream-incision model of bedrock removal by abrasion and compared its results with paleohydraulic calculations for the diverted river. The theoretical erosion model compares favorably with the observed incision rate.

Mountain glaciation resulted in several examples of large-scale flooding in the western United States. Birkeland (1968) describes catastrophic flooding of Tahoe (Early Wisconsin?) age along the Truckee River, which drains eastward from the Lake Tahoe region of the Sierra Nevada. Flooding up to about 25 m deep resulted from the release of about 15 km³ of water from an enlarged Lake Tahoe dammed by glacier ice. Boulders up to 12 m in diameter were transported through mountain canyons, and 3-m boulders are

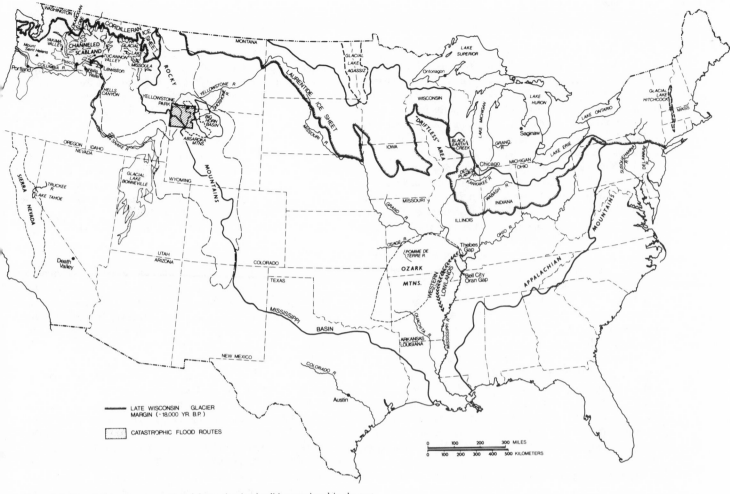

Figure 5-2. Map of the conterminous United States showing localities mentioned in the text.

relatively common in the flood debris. Pierce (1973, 1974) describes evidence for catastrophic flooding in the upper Yellowstone Valley in Wyoming. Late-Pinedale ice in the northern Yellowstone Park area dammed lakes, and their release yielded floods up to 45 m deep. At least two such floods occurred, leaving a record of scabland, longitudinal bars, and transported boulders up to 3 m in diameter. Bradley and Mears (1980) have recently refined the paleohydraulic calculation procedures for using the boulders transported by floods as indicators of the responsible flows.

Mississippi River System

The Mississippi River was the outlet of the largest proglacial drainage system in North America (Figure 5-2). During the initial stages of deglaciation, the Mississippi-Missouri-Ohio system drew its discharge from a sector of ice sheet over 2700 km long (Flint, 1971: 242).

During the Farmdalian and early-Woodfordian substages, the central part of the Mississippi River developed prominent braided stream surfaces in the Western Lowlands of Arkansas (Figure 5-2), about 100 km west of its present location (Saucier, 1974, 1978). The valley aggradation dammed the lower end of the Ouachita River of Arkansas and Louisiana (Saucier and Fleetwood, 1970). The Ouachita River subsequently trenched the lacustrine plain and developed a set of terraces, which have been correlated by Saucier and Fleetwood (1970) with the Deweyville terrace of Gagliano and Thom (1967). The Deweyville surfaces along various rivers in the south-central and

southeastern United States show enlarged meanders that imply a period of increased discharge. However, the Deweyville sequence is not well dated, perhaps having formed between 30,000 and 13,000 yr B.P. (Saucier and Fleetwood, 1970) but extending to younger ages elsewhere.

The lower Mississippi itself was gradually entrenched as sea level fell (Fisk, 1944), probably reaching its maximum low stand about 18,000 yr B.P. The deeply entrenched valley was next subjected to a massive influx of water and sediment.

The Laurentide ice sheet began to stagnate in its marginal zone about 17,000 yr B.P. (Dreimanis, 1977). After this time, a massive discharge of glacial meltwater flowed down the Mississippi River Valley into the Gulf of Mexico, as documented by Kennett and Shackleton (1975). The evidence consists of a major anomaly, dated between about 15,000 and 11,000 yr B.P., in the oxygen isotope measurements of deep-sea cores from the Gulf of Mexico. Surface salinities of the Gulf were evidently reduced by about 10% as the Laurentide ice sheet began to retreat. Emiliani (1980) suggests that the maximum discharge occurred about 11,600 yr B.P. and corresponded to a rapid global transgression between 12,000 and 11,000 yr B.P.

As the ice retreated, the newly deposited glacial drift must have been subject to extremely rapid fluvial dissection. Ruhe (1969) compares drainage development on Late Wisconsin drift surfaces in northwestern Iowa. On the "Iowan" erosion surface, which formed before 20,000 yr B.P., the drainage density varies from about 6 to 7.5

(arbitrary units). On the Tazewell drift (early Woodfordian), which was laid down about 20,000 yr B.P. (I-1864A and 0-1325), the densities average about 5. However, on the Cary drift (later Woodfordian), which formed about 14,000 yr B.P., the density is only about 2. These surfaces have all had the same opportunity for drainage development since the end of the Cary sedimentation (about 13,000 yr B.P.). Therefore, clearly, much of the drainage development dates to between 20,000 and 14,000 yr B.P. The period since 13,000 yr B.P. has been one of relatively slow drainage development.

The tremendous influx of water and sediment produced thick outwash trains in the upper Mississippi Basin. Late Wisconsin outwash in the Mississippi, Ohio, and Wabash Valleys aggraded valley floors 15 to 30 m above pre-Wisconsin levels and served as a source for the Peoria Loess. In southern Illinois, slack-water lakes formed as tributary valleys were blocked by the aggrading outwash trains.

About 17,000 yr B.P., the central Mississippi experienced a major diversion. It abandoned its valley train in the Western Lowlands of Arkansas and cut through the Bell City—Oran gap (Saucier, 1974). The zone of outwash deposition shifted eastward to the present alluvial valley (Figure 5-2). The lower Mississippi River began to aggrade rapidly the irregular valley floor that it had produced during the minimum sea level (Fisk, 1944). Aggradation was most rapid during early deglaciation under the combined influences of high sediment loads and rising base level. The early phase of this aggradation involved a great braided river that shifted across a valley of higher gradient than exists today. The coarse-grained braided stream deposits of the lower Mississippi Valley are now buried beneath the fine-grained sand, silt, and clay of the Holocene meandering Mississippi.

Aggradation in the Mississippi system began to wane as the ice

margin retreated in the upper part of the basin. The glacier terminus became fringed by proglacial lakes in the vicinity of the modern Great Lakes. This allowed only finer sediments to be transported downstream. However, the complex sequence of deglacial lakes introduced a new element to the drainage development, namely proglacial floods.

The retreat of Woodfordian-substage ice in northeastern Illinois about 15,000 to 14,000 yr B.P. produced the Kankakee flood (Willman and Frye, 1970: 34-35). Valleys and outlets through the complex of moraines in front of the retreating ice sheet were inadequate to convey the local meltwater discharge. A series of lakes formed behind the moraines and spilled over gaps through them. This flood probably facilitated the diversion, near Thebes (Figure 5-2), of the Mississippi into the ancient Ohio Valley (Willman and Frye, 1970).

The complexity of the drainage from the retreating Laurentide ice sheet is well illustrated by events in the Chicago area (Bretz, 1964; Willman, 1971). Glacial Lake Chicago formed in the southern basin of present Lake Michigan as the Lake Michigan lobe retreated northward. Bretz (1964) suggests that the highest level of the lake (the 18-m shoreline) stabilized (at approximately 14,000 yr B.P.) because a boulder pavement formed at the point of outflow, thereby protecting the underlying dam of weak drift from erosion. The pavement was destroyed when the outflow discharge (through the valley of the Des Plaines River) was abruptly augmented when Lake Erie meltwater entered Lake Chicago from the east through the Grand River valley. This sudden change occurred about 13,000 yr B.P. as the Saginaw lobe retreated northward from Michigan. Lake Chicago subsequently stabilized at the 12-m shoreline level. Further influxes of meltwater, overspillings, downcutting, and uplift of northern outlets due to isostatic rebound produced a series of shorelines and adjustments that extended into the Holocene.

By 11,200 yr B.P., meltwater from the Laurentide ice sheet was no longer a major factor in the Mississippi-Ohio system. Much of the meltwater from the eastern Laurentide ice sheet flowed through outlets farther east (Prest, 1970). The western basin continued to receive meltwater, however. The last major flow may have resulted from the cutting of an outlet for Lake Agassiz at 9200 yr B.P. (Wright, 1972). Coarse sediment continued to induce braided patterns until approximately 8800 yr B.P., when the northern Mississippi flow was completely diverted to Thebes Gap. This age for the establishment of a meandering regime in the Mississippi alluvial valley derives from studies in southeastern Missouri (King and Allen, 1977). Figure 5-3 summarizes the major late-Pleistocene events along the Mississippi River and compares them to other fluvial and glacial chronologies.

A Fluvial Chronology?

The divergence of thought concerning climatic influences on river-terrace formation (Schumm, 1965) and the problems of separating climatic controls from other influences (Flint, 1976) have led to considerable skepticism concerning the interpretation of Quaternary fluvial stratigraphy. Morrison (1968) has concluded that cycles of fluvial erosion, deposition, and stability are unreliable indicators of glacial-interglacial cycles. Unfortunately, in many interior regions of the United States, fluvial deposits provide the only easily studied record of Quaternary paleoclimate.

Frye (1973) suggests that a sequential relationship exists in the cen-

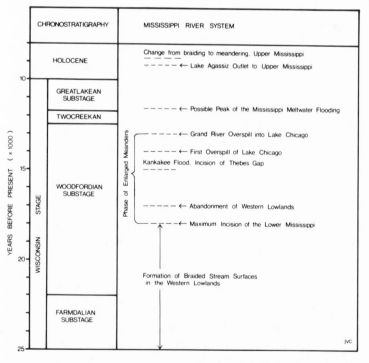

Figure 5-3. Late-Pleistocene fluvial events in the Mississippi River drainage system. (Compiled from various sources.)

tral interior of the United States among intervals of valley incision, valley alluviation, and soil formation. Erosional incision seems to have occurred in the Middle Wisconsin, that is, during the transition from the Altonian to the Farmdalian (approximately 28,000 yr B.P.). Outwash deposition and associated aggradation occurred during and after the culminating advances, that is during the Woodfordian substage (approximately 22,000 to 12,500 yr B.P.).

Perhaps the best region in which to demonstrate a precise fluvial stratigraphy is the nonglaciated central Mississippi Basin, which displays river terraces that can be related to loess, paleosols, and fossil occurrences (e.g., Brice, 1964; Schultz and Martin, 1970). Haynes (1976) and Brakenridge (1980, 1981) have developed a detailed late-Quaternary fluvial chronology for the Pomme de Terre River, which lies outside the glacial margin in southern Missouri. They have found that the alluvial valley bottom displays a complex history of cutting and filling. After a major incision phase at 32,000 yr B.P., the Pomme de Terre aggraded to nearly its present level by 27,000 yr B.P. The floodplain remained at this level until another major incision occurred at 13,000 yr B.P. The two major phases of erosion may be associated with a change of climate to warmer and wetter conditions; the episode 13,000 years ago corresponds approximately to the Twocreekan interstade in the vicinity of the ice margin. The period between 11,000 and 8000 yr B.P. was marked by very rapid aggradation. This was followed by a regime characterized by shorter and less-pronounced phases of alternating aggradation and degradation through the remainder of the Holocene.

In the arid southwestern United States, the influence of late-Pleistocene climatic change on rivers remains controversial. Lustig (1965) relates the building of alluvial fans to past wetter conditions and the current phase of fan incision to modern aridity. Hunt and Mabey (1966) attribute fan growth in Death Valley to the onset of drier conditions after a wetter phase marked by preparatory weathering. Denny (1965) claims that erosion and deposition were both integral parts of the steady-state system that generated fans; thus, there is no reason to invoke environmental change to explain alternations of process. (The reader is referred to Bull's 1977 summary of the problem.)

The Rocky Mountains provide a setting in which resistant crystalline rocks exposed in the cores of ranges contrast markedly with easily eroded siltstone and shale units of adjacent piedmont regions. This dichotomy leads to repeated stream captures in the piedmont environment. Low-gradient piedmont streams transporting fine-grained loads capture adjacent high-gradient streams, which drain the mountain cores and transport coarse sediments. The ensuing aggradation may resemble a climate-controlled depositional episode, especially when subsequent incision occurs along the sides of the valley and produces a terrace (Ritter, 1967, 1972).

Schumm's (1977) experimental studies reveal that normal processes of drainage-network development and response to rejuvenation in themselves may lead directly to episodic downcutting. Terraces form as sediment production temporarily increases beyond the transporting capability of a stream and then decreases again. Schumm's work shows that terrace formation need not always be attributed to major climatic, base-level, and tectonic changes. The removal of large quantities of stored outwash and alluvium from the upper reaches of valleys could easily produce a complex response leading to multiple terracing downstream. The thick late-glacial fills that developed in the mountains of the western United States (Alden, 1953; Howard, 1960) may have led to such events. Unusually large floods may have

served as the critical factor by exceeding the thresholds necessary to induce episodic erosion.

River Metamorphosis

Although the ultimate causes of fluvial adjustments are not always easy to discern, the adjustments themselves may leave striking landscapes. For much of the United States, the late Pleistocene was a time of immense changes in stream loads and discharges. Such changes often led to complete alterations of river morphology, a phenomenon that Schumm (1969, 1977) calls "river metamorphosis." The directions and, in some cases, the magnitudes of change can be deduced from various empirical relations and classifications recently summarized by Schumm (1977). The most spectacular examples of river metamorphosis occur on broad alluvial plains, such as the Riverine Plain of New South Wales, Australia (Schumm, 1968). Local examples of late-Pleistocene river metamorphosis occur on the Gulf Coastal Plain of the United States.

The Colorado River in central Texas flows within a late-Quaternary alluvial valley that contains nine crosscutting assemblages of fluvial channel patterns (Figure 5-4) and terraces (Figure 5-5). Detailed geomorphic mapping (Baker and Penteado-Orellana, 1977; Looney and Baker, 1977; Penteado and Baker, 1976) and sedimentologic analysis (Baker and Penteado-Orellana, 1978) show that the late Quaternary was characterized by adjustment from low sinuosity, gravel-transporting streams (Channels 6, 5 and 3 in Figure 5-4) to high-sinuosity, sand- and silt-transporting streams (Channels 6B and 4 in Figure 5-4). The low-sinuosity streams probably drained the basin during more arid phases and the high-sinuosity streams characterized humid phases, but arguments in support of this conclusion are necessarily complex (Baker and Penteado-Orellana, 1977, 1978).

The paleoclimatic implications of a fluvial sequence like the one just described proceed from the assumption that progressive changes in channel morphology and sedimentology reflect changes in the sediment load and runoff of the catchment. A large number of empirical relationships are available for calculating the flow characteristics of ancient rivers from their preserved morphology and sediments (e.g., Dury, 1965, 1976; Schumm, 1968, 1969). However, these relationships are limited by a number of severe constraints on their use (Ethridge and Schumm, 1978). Some of the problems with quantitative paleohydrology are outlined for a type of fluvial metamorphosis that developed extensively during the late Pleistocene, the phenomenon of misfitness.

MISFIT STREAMS

Misfit streams are those for which some practical measure of size, most often meander wavelength, indicates that the modern river is either too large (overfit) or too small (underfit) for the valley in which it flows. Underfit streams, the type most studied, include many variants (Dury, 1964a, 1970). Manifestly underfit streams are the most easily recognized and combine large former meanders with small present meanders. A more complex variant, the Osage-type underfitness, is named for the Osage River of the Ozark Mountains in Missouri. This variant combines the large former valley meanders with a modern stream that does not meander but possesses pools and riffles spaced appropriately for its bed width (Dury, 1966, 1970). Dury (1964a, 1964b, 1965) argues that the widespread phenomenon

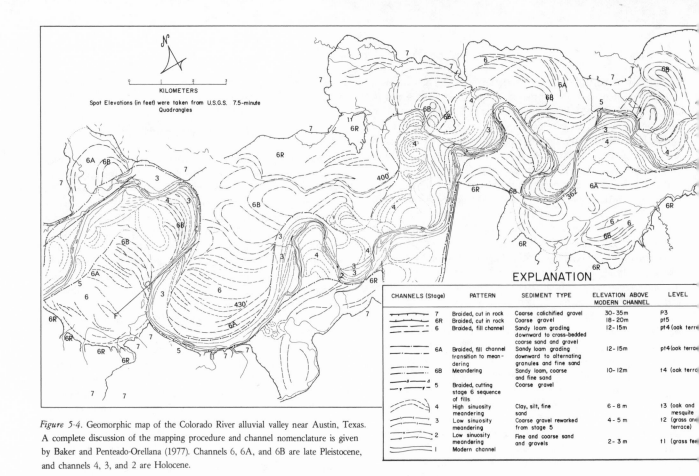

Figure 5-4. Geomorphic map of the Colorado River alluvial valley near Austin, Texas. A complete discussion of the mapping procedure and channel nomenclature is given by Baker and Penteado-Orellana (1977). Channels 6, 6A, and 6B are late Pleistocene, and channels 4, 3, and 2 are Holocene.

The explanation table within the figure reads:

EXPLANATION

CHANNELS (Stage)	PATTERN	SEDIMENT TYPE	ELEVATION ABOVE MODERN CHANNEL	LEVEL
7	Braided, cut in rock	Coarse calichified gravel	30-35m	P3
6R	Braided, cut in rock	Coarse gravel	18-20m	pt5
6	Braided, fill channel	Sandy loam grading downward to cross-bedded coarse sand and gravel	12-15m	pt4 (oak terra
6A	Braided, fill channel transition to meandering	Sandy loam grading downward to alternating granules and fine sand	12-15m	pt4 (oak terra
6B	Meandering	Sandy loam, coarse and fine sand	10-12m	t4 (oak terra
5	Braided, cutting stage 6 sequence of fills	Coarse gravel		
4	High sinuosity meandering	Clay, silt, fine sand	6-8 m	t3 (oak and mesquite
3	Low sinuosity meandering	Coarse gravel reworked from stage 5	4-5 m	t2 (grass terrace
2	Low sinuosity meandering	Fine and coarse sand and gravels	2-3 m	t1 (grass fen
1	Modern channel			

of underfitness results from profound reductions of channel-forming discharges for Quaternary rivers. Climatic change presumably has led to the discharge reductions, although the exact relationships are the subject of considerable dispute among fluvial geomorpholgists.

Distribution

Dury (1964a: A30) has mapped a broad distribution of manifestly underfit streams throughout the United States. Ratios for the former valley meander wavelength (L) to the modern river meander wavelength (l) average about 5:1. Near the former ice front in Wisconsin, the ratio was as high as 10:1; in the Ozarks, the average ratio dropped to 3:1.

Prominent underfit streams developed on the floors of late-glacial lakes, such as Lake Hitchcock in the Connecticut Valley. Others developed in upland areas that were not glaciated during the Wisconsinan, such as the Appalachians, the Ozarks, and the driftless area of

Wisconsin. Dury also describes underfit relationships in the arid southwestern United States.

Chronology

Although episodes of stream shrinkage producing underfitness undoubtedly were repeated during the Quaternary, the last phase of shrinkage occurred in the latest Pleistocene. The large valley meanders, which imply former high discharges, are more variable in age, ranging from early Pleistocene to Late Wisconsin.

Dury (1964b) argues that the latest conspicuous reduction of stream size corresponded to a transition from cool, moist conditions to warm, dry conditions. He correlates the transitions to the European Blytt-Sernander bioclimatic intervals Allerød and Boreal. For the United States, he recognizes two late-Pleistocene phases of discharge reduction. The earlier phase correlated with the beginning of the Twocreekan stage (approximately 12,500 yr B.P.); the later

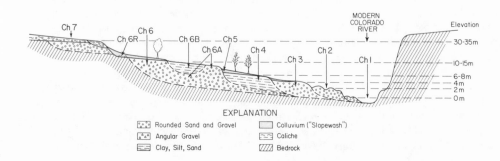

EXPLANATION

Rounded Sand and Gravel   Colluvium ("Slopewash")
Angular Gravel   Caliche
Clay, Silt, Sand   Bedrock

Figure 5-5. Schematic cross section of the Colorado River alluvial valley near Austin, Texas. (The numbers refer to the channel assemblages illustrated in Figure 5-4.)

phase correlated with the recession of Valders (Greatlakean) ice (approximately 10,000 to 9000 yr B.P.). The relationships are particularly well illustrated by Black Earth Creek, Wisconsin, where the increased valley-forming discharges occurred during the relatively short interval from 14,000 to 12,500 yr B.P. through 11,000 to 10,000 yr B.P. (Dury, 1974).

Implications

Dury (1965) has developed a quantitative theory of past stream flow for underfit streams. He reasons that, because meander wavelength is directly proportional to bankfull channel width and because bankfull width is a function of discharge to the 0.5 power (Leopold and Maddock, 1953), the wavelength of modern river meanders (l) must be directly proportional to bankfull discharge (q_b). More specifically:

$$l = 54.3 q_b^{0.5}$$

where l is measured in meters and q_b in cubic meters per second.

Because these same arguments are presumed to be applicable to the enlarged valley meanders of wavelength L formed by ancient discharges Q, then

$$\frac{Q}{q_b} = \left(\frac{L}{l}\right)^2$$

Many valley meanders in the United States were 5 times larger than the modern river meanders within those valleys, so the bankfull discharge must have been reduced by a factor of 25. For some rivers, with L/l equal to 10, the discharge reduction would be 100. A revised predictive equation (Dury, 1976, 1977) reduces these estimated discharge ratios from 25 to about 18 where the wavelength ratio is 5:1 and from 100 to about 66 where the wavelength ratio is 10:1.

Dury (1965) has calculated that the immense discharges implied by the former valley meanders were too great to have resulted solely from either reasonable increases in mean annual rainfall or reduced evaporation caused by lower temperatures. He suggests that individual rainfalls of exceptional intensity and duration were superposed on moderately increased rainfall values. Locally, the reduced infiltration capacity of frozen ground may have supported the increased runoff during the Pleistocene. Subsequent reduction of stream discharges could be explained by the decreased precipitation, increased temperature, and diminished storminess of the Holocene. Despite these arguments, however, the immense adjustment of discharge required in Dury's estimates suggests problems in the analysis.

Problems

Several aspects of Dury's theory have been seriously questioned. Schumm (1967, 1968) shows that meander wavelength does not merely depend on water discharge but also on the type of sediment transported by the stream. The meander wavelengths for coarse-sediment-transporting streams (bed-load type) will be greater for the same mean annual discharge than the meander wavelengths of fine-sediment-transporting streams (subspended-load type). Underfitness does not involve a simple reduction of discharge but usually includes a channel metamorphosis and associated change in sediment-load characteristics.

The theory of Dury assumes that the modern diminished streams and the large valleys that contain them were formed by similar processes but at different times. This assumption is derived in part from subsurface exploration (Dury, 1964b), which reveals a valley-floor relief with a pool-and-riffle pattern proportional to the valley wavelength and width by ratios similar to those observed in the modern river. Palmquist (1975) offers a theory that explains these relationships by flood scour through alluvium of the modern river in the valley floor. Because meandering streams preferentially occupy the outer half of their valley at bends but have no preferred position in the intervening reaches, they produce deeper erosion at the bends and a valley-floor relief similar to a pool-and-riffle pattern.

The differences between valley meanders and alluvial meanders are emphasized by Hack (1965), who has studied the Ontonagon area of northern Michigan, a landscape formed on the emerged bottom of a glacial lake. Meander wavelength depended solely on bed and bank materials. Where a set of valley meanders was aggraded with fine alluvium, the new meanders presumably assumed a smaller wavelength (underfitness) without changes in discharge or climate. Tinkler (1971, 1973) extends this concept to suggest that the worldwide distribution of underfit relationships resulted from completely normal fluvial processes operating in contrasting lithologic environments. With the development of a floodplain, for whatever reason, valley meanders are interpreted as becoming fossilized, and alluvial meanders develop without the necessity of a climatic change. The key difference between the valley meanders and the alluvial meanders is not merely the magnitude of the effective discharge for forming them but also the frequency with which that effective discharge recurs for different kinds of bank materials.

Implicit in Dury's theory is the hypothesis that the bankfull discharge of stream channels is most effective for geomorphic work and that this discharge has a recurrence interval of approximately 1.5 years. Although this can be demonstrated in some climatic and geologic settings (e.g., Andrews, 1980; Dury, 1973; Wolman and Miller, 1960), there is mounting evidence that the hypothesis is of limited applicability. Evidence for deviations from this hypothesis has been presented by Harvey (1969), Gupta (1975), Pickup and Warner (1976), Baker (1977), Patton and Baker (1977), and Williams (1978).

Most damaging to the theory is the requirement that bedrock valley meanders form only at times when bankfull discharges are immensely larger than at present. Bedrock meanders form actively by the action of floods that are much less common than the 1.5-year event (Tinkler, 1971). For the limestone bedrock streams of central Texas, the 1.5-year flood is ineffective. Only very large, rare floods are capable of developing significant valley meanders (Patton and Baker, 1977). The importance of large, rare floods derives from the very high threshold necessary for erosion of bedrock and coarse alluvium (Baker, 1977).

Ferguson's (1973: 43) conclusion concerning misfit streams summarizes the present state of the art: "... awareness of [various facts] ... points ... to the complexity of the whole issue and the simplifications inherent in existing quantitative treatments. We need better data, reviews of operational definitions, and studies of the hydraulic controls of meander form, but first of all a retreat from entrenched positions defending oversimple theories."

CATASTROPHIC FLOODING

The Columbia River system of the northwestern United States (Figure 5-6) was subject to perhaps the most remarkable late-Pleistocene fluvial events on the planet. The interval from approximately 18,000 to 12,000 yr B.P. was a period of phenomenal catastrophic flooding. The significance of these events was long in be-

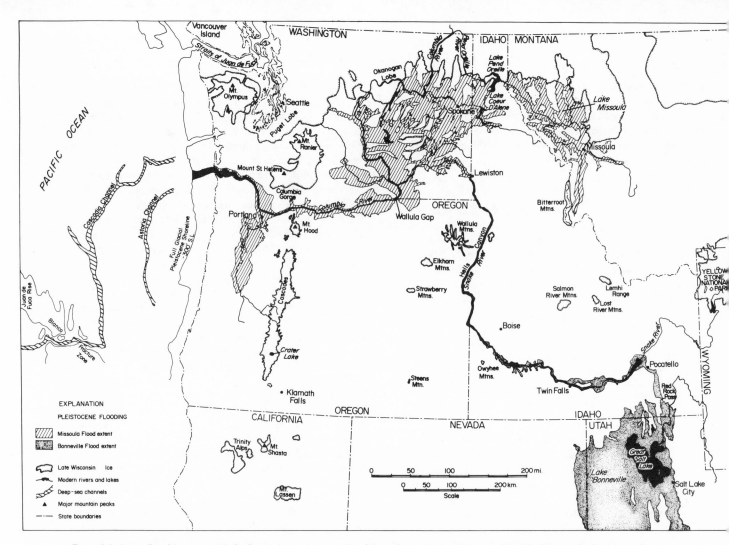

Figure 5-6. Areas affected by catastrophic flooding in the northwestern United States during the late Pleistocene. The Lake Missoula flooding occurred repeatedly during the interval from about 18,000 to 12,000 yr B.P., but only the maximum extent is shown. The Lake Bonneville flooding occurred approximately 14,000 yr B.P. (Scott and others, 1980). The area of Late Wisconsin alpine ice was more extensive in the Rocky Mountains than what is shown here, but this area has not yet received detailed study.

ing realized. Although the hypothesis was eloquently argued in a series of early papers (Bretz, 1923, 1925, 1928, 1932), the geologic community resisted the idea that an enormous plexus of proglacial channels in eastern Washington was the product of catastrophic flooding. Considering the nature and vehemence of the opposition to J Harlen Bretz's hypothesis, the eventual triumph of that idea constitutes one of the most fascinating episodes in the history of modern geology (Baker, 1978d, 1981).

Landscapes Created by Catastrophic Flooding

The great floods were released from glacial Lake Missoula (Figure 5-6), which was impounded during Late Wisconsin time by a lobe of the Cordilleran ice sheet (Pardee, 1942; Richmond et al., 1965). At its peak discharge, Lake Missoula yielded approximately 21×10^6 m^3 per second (Baker, 1973). The lake volume of 2×10^{12} m^3 probably only maintained this peak flow for several hours, but immense discharges could have been maintained for several days. The great floods completely overwhelmed the stream valleys of the Columbia Plateau region in eastern Washington. Indeed, the region of anastomosing flood channels scoured in rock and loess by the floods is a magnificent example of stream overfitness. The valleys filled to overspilling and produced divide crossings (Bretz, 1923, 1928). Bretz

named this region the Channeled Scabland (Figure 5-7). Catastrophic flooding affected the entire downstream portion of the Columbia River system (Figure 5-6), even extending its influence to the abyssal sea floor off the mouth of the Columbia River (Griggs et al., 1970).

The catastrophic-flood channels contain a variety of distinctive flood-produced landforms (Figure 5-8). Immense bars of boulders and gravel formed wherever large flow separations were generated by various flow obstructions or diversions (Baker, 1978b). Hills of loess were streamlined by the flood flows (Bretz, 1928; Patton and Baker, 1978b). The basalt bedrock was eroded to form a bizarre landscape consisting of erosional grooves, potholes, rock basins, inner channels, and cataracts. Of hydraulic significance are the giant current ripples composed predominantly of gravel and commonly over 5 m high and spaced 100 m apart. The giant current ripples are directly related to the shear stresses, mean velocities, and stream powers exhibited by the flood flows (Baker, 1973).

The easternmost system of flood-scarred channelways in the Channeled Scabland is the Cheney-Palouse scabland tract (Figure 5-9). The suggestion that the Cheney-Palouse scabland tract was the product of relatively low-discharge proglacial drainage (Flint, 1938) has failed to satisfy the field evidence, which can only be explained by the catastrophic flood hypothesis (Bretz et al., 1956; Patton and Baker, 1978b).

How Many Floods?

Bretz (1969) proposes, primarily on physiographic evidence, as many as eight separate scabland floods, of which at least four encountered an ice-blocked Columbia River and spilled over the northern margin of the Columbia Plateau. Bretz's other four floods were not diverted by ice and flowed down the Columbia River valley. Bretz (1969: 513) suggests that the earliest plateau-crossing flood occurred during the Bull Lake Glaciation, now considered correlative with the Illinoian Glaciation of the midwestern United States (Pierce et al., 1976).

Baker (1973) notes that some of the physiographic relationships described by Bretz (1969) could have been produced during the dynamic progression of a single flood. Only flood deposits recognized in a firm stratigraphic sequence can be considered unequivocal evidence for multiple flooding. The best stratigraphic information available suggests that there certainly were multiple floods in the general vicinity of the Channeled Scabland. One and possibly two floods were pre-Bull Lake in age and are overlain either by thick caliche or by Palouse Formation. The Palouse Formation is a sequence of loess units that mantles divide areas in the Channeled Scabland and was scoured by the Wisconsin-age floods of Lake Missoula (Richmond et al., 1965). Patton and Baker (1978a) describe the regional evidence for pre-Wisconsin (pre-Palouse) flooding.

Malde (1968) has studied the catastrophic flood produced by the overflow and rapid lowering of Pleistocene Lake Bonneville. He traces the course of this flood through the Snake River Plain of southern Idaho to Hells Canyon. Malde based the time of this event (about 30,000 years ago) on a radiocarbon date for molluscan fossils in flood debris and on a relict soil profile on the flood gravel (Melon Gravel). However, recent work by Scott and others (1980) demonstrates that Bonneville flooding occurred about 14,000 years ago.

Downstream from Hells Canyon at Lewiston, Idaho, probable Bonneville flood deposits are overlain by slack-water-surge deposits from the last major episode of scabland flooding. Because Bonneville flooding was confined to the Snake River Canyon, it skirted to the south of the Channeled Scabland. Nevertheless, studies in the Pasco Basin may eventually lead to the recognition of Bonneville flood deposits in association with Missoula flood deposits.

The last major episode of scabland flooding involving at least several individual floods occurred between approximately 18,000 and 12,000 yr B.P. In the eastern Channeled Scabland, the last major flood scoured channelways through the entire sequence of brown loess units that comprise the Palouse Formation. The limiting dates on the flood episode are imposed by the radiocarbon chronology of the Cordilleran ice sheet, which was responsible for the impound-

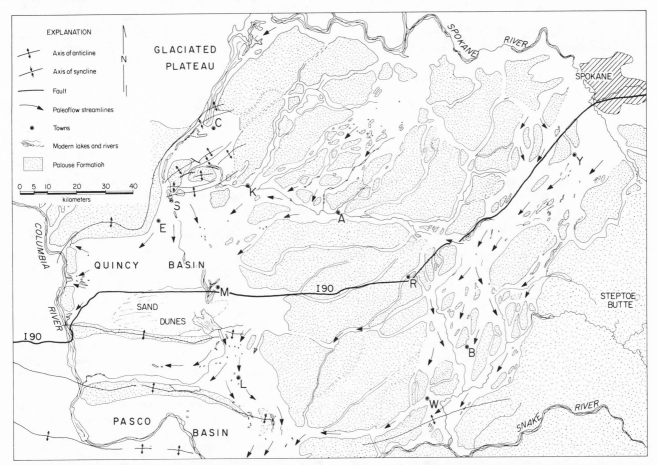

Figure 5-7. Generalized map of the Channeled Scabland showing the flood paleoflow directions. The Palouse Formation consists of loess deposited before the late-Pleistocene catastrophic floods. Towns indicated by letters are (Y) Cheney, (R) Ritzville, (B) Benge, (W) Washtucna, (A) Odessa, (K) Wilson Creek, (C) Coulee City, (S) Soap Lake, (E) Ephrata, (M) Moses Lake, and (L) Othello. (From Baker, 1978d, from *Science*, volume 202, pp. 1249-56, copyright 1978 by the American Association for the Advancement of Science.)

Figure 5-8. Typical landforms resulting from catastrophic flood flows in the Channeled Scabland: (A) pendant bar approximately 1 km long, (B) streamlined residual hills of loess, (C) scabland, and (D) giant current ripples.

ment of Lake Missoula. Ice had still not yet advanced into extensive portions of southeastern British Columbia as late as 19,000 yr B.P. (Clague et al., 1980; Fulton, 1971). The Late Wisconsin maximum for the Cordilleran ice sheet is thought to have occurred between 15,000 and 14,000 yr B.P. (Clague et al., 1980; Thorson, 1980). The recession of the Puget lobe of this ice sheet occurred between 14,000 and 13,000 yr B.P. (Thorson, 1980).

The dating of the last major scabland flooding episode is complicated by the fact that the catastrophic flood deposits occur in two facies. The main-channel facies can always be recognized as unequivocal flood deposition, for example, by its coarseness, sedimentary structures, angular boulders, and erratics. However, the slack-water facies consists of rhythmically bedded sands and silts deposited in the preflood tributaries of the scabland channels (Bretz, 1929, 1930). An unequivocal flood origin can be established by continuous tracing between main-channel areas and slack-water areas, as in the Tucannon Valley sequence (Baker, 1973), where one can follow a complete transition from chaotically deposited boulder-and-cobble gravel in the main channel (proximal facies) to rhythmites of sand and silt in slack-water areas (distal facies) 15 km up a preflood tributary of the main channel (Baker, 1973). Waitt (1980) has studied the unusually thick flood rhythmite sequences of the Walla Walla Basin and the lower Yakima Valley, where there is evidence for subaerial exposure following flood water emplacement of some of the rhythmites. At the remarkable Burlingame Canyon section, Waitt has counted at least 38 well-formed distal-facies rhythmites. One possible hypothesis to explain these new data is that each rhythmite represents a separate catastrophic flood event (Waitt, 1980). Unfor-

tunately, the proximal-facies sediments for each of these inferred floods have not yet been discovered. Thus, the possibility remains that single great floods can produce multiple distal rhythmites and that the Burlingame section represents a somewhat smaller number of Late Wisconsin flood events than is suggested by counting each individual rhythmite (Patton et al., 1979). Sedimentologic arguments for the generation of multiple rhythmites by a single flood are summarized by Bjornstad (1980).

Dating the Missoula Floods

Clear stratigraphic relationships throughout the Channeled Scabland and vicinity demonstrate that at least two major outburst floods occurred during the Late Wisconsin. The earlier flood occurred before the maximum advance of the Okanogan lobe of the Cordilleran ice sheet to its terminal position (Easterbrook, 1976; Hanson, 1970). This flood may account for Black's (1979) report of a radiocarbon date of 18,705 + 1515/− 1275 yr B.P. (Gx-5530) from fragments of wood in a clastic dike penetrating coarse sand and gravel of probable flood origin near Gable Mountain in the center of the Pasco Basin.

The late-glacial slack-water succession consists of two superposed rhythmite sequences separated by a major unconformity at sites in the Walla Walla Basin (Bjornstad, 1980) and at Badger Coulee (Bunker, 1980). At Badger Coulee, a couplet of ash layers was subaerially deposited atop the lower slack-water sequence (Bunker, 1980). The ash layers have been correlated, on the basis of trace-element chemistry, with pumice layers So and Sg of the Mount St. Helens set S tephra (Moody, 1978; Waitt, 1980). These set S tephra

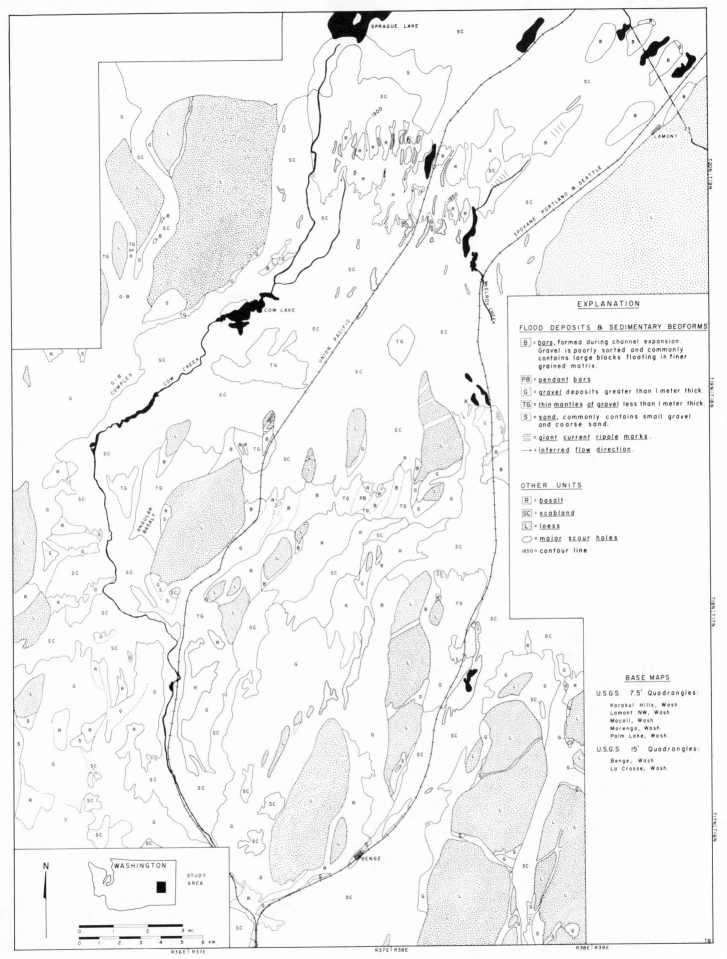

EXPLANATION

FLOOD DEPOSITS & SEDIMENTARY BEDFORMS

B = bars, formed during channel expansion. Gravel is poorly sorted and commonly contains large blocks floating in finer grained matrix.

PB = pendant bars

G = gravel deposits greater than 1 meter thick.

TG = thin mantles of gravel less than 1 meter thick.

S = sand, commonly contains small gravel and coarse sand.

≋ = giant current ripple marks.

→ = inferred flow direction.

OTHER UNITS

R = basalt

SC = scabland

L = loess

⬭ = major scour holes

1850 = contour line

BASE MAPS

U.S.G.S. 7.5' Quadrangles:

Karakul Hills, Wash.
Lamont NW, Wash.
Macall, Wash.
Marengo, Wash.
Palm Lake, Wash.

U.S.G.S. 15' Quadrangles:

Benge, Wash.
La Crosse, Wash.

WASHINGTON

STUDY AREA

N

0   1   2   3 mi
0 1 2 3 4 5 6 km

Figure 5-9. Geomorphic map of the central portion of the Cheney-Palouse scabland tract. (From Patton and Baker, 1978b.)

layers have been dated at between 12,120 ± 350 yr B.P. (W-3133) and 13,130 ± 350 yr B.P. (W-2983) (Mullineaux et al., 1975, 1978). Mullineaux and others (1978) present additional arguments that the Mount St. Helens set S tephra dates the last major scabland flood. They support the concept of a 13,000 yr B.P. flood with a radiocarbon date of 13,080 ± 350 yr B.P. (W-2983) on peat directly overlying the "Portland delta," a Lake Missoula flood deposit at Portland, Oregon.

Additional evidence for two late-Pleistocene catastrophic floods comes from the mounds of ice-rafted till in the western Pasco and Quincy Basins (Fecht and Tallman, 1978). The berg ice was probably transported by the second of the two floods. Till from the stranded ice blocks mantles slack-water sediments of the earlier flood. The contact zone between the two contains an ash that is tentatively correlated with the Mount St. Helens set S tephra (Fecht and Tallman, 1978).

Excellent stratigraphic evidence for the limiting age of the second flood (or flood episode) comes from slack-water sediments in tributaries of the lower Snake River canyons (Foley, 1976; Hammatt, 1977). The slack-water deposits overlie nonflood alluvium and colluvium that contains probable Mount St. Helens set S ash and charcoal dated at 14,300 ± 200 yr B.P. (WSU-1499) and 13,000 ± 220 yr B.P. (WSU-1615). Deposits of the earlier flood may be represented in this area as great flood bars in the Snake River Canyon itself (Hammatt, 1977).

The present stratigraphic information supports a concept of two distinct phases of late-Pleistocene catastrophic floods (Baker, 1978c). The earlier phase preceded the maximum advance of the Cordilleran ice sheet (approximately 15,000 to 14,000 yr B.P.) and may have occurred approximately 18,000 yr B.P. The later phase occurred approximately 13,000 yr B.P. and was probably associated with the stagnation and breakup of the Okanogan lobe of the Cordilleran ice sheet in the Columbia River valley northwest of the Channeled Scabland (Waitt, 1977). Although the number and timing of floods within these phases have not yet been resolved, the effects of those floods on the Channeled Scabland are manifest (Baker, 1981).

Conclusions

(1) The late Pleistocene was a period of immense adjustments in fluvial systems, many of which were associated with deglaciation. (2) Powerful techniques are now available for paleohydrologic and paleohydraulic analysis. However, the complexities of fluvial response must be appreciated in the application of these methods. (3) Classical models of river incision and aggradation are much too simple for universal application. Although detailed fluvial chronologies may be usefully developed in some local settings, regional correlation requires great caution. (4) River metamorphosis, especially underfitness, was widely achieved during the late Pleistocene. However, this phenomenon involved the adjustment of multiple variables in addition to stream flow. Sediment loads, bank materials, and magnitude-frequency relationships all were involved. (5) Large-scale flooding was locally important around the many ice margins during the late Pleistocene. In the northwestern United States, catastrophic flooding played a dominant role in the development of landscapes such as the Channeled Scabland.

References

Alden, W. C. (1953). "Physiography and Glacial Geology of Western Montana and Adjacent Areas." U.S. Geological Survey Professional Paper 321.

Andrews, E. D. (1980). Effective and bankfull discharges of streams in the Yampa River Basin, Colorado and Wyoming. *Journal of Hydrology* 46, 311-30.

Baker, V. R. (1973). "Paleohydrology and Sedimentology of Lake Missoula Flooding in Eastern Washington." Geological Society of America Special Paper 144.

Baker, V. R. (1974). Paleohydraulic interpretation of Quaternary alluvium near Golden, Colorado. *Quaternary Research* 4, 94-112.

Baker, V. R. (1977). Stream channel response to floods with examples from central Texas. *Geological Society of America Bulletin* 88, 1057-71.

Baker, V. R. (1978a). Adjustment of fluvial systems to climate and source terrain in tropical and subtropical environments. *In* "Fluvial Sedimentology" (A. D. Miall, ed.), pp. 211-30. Canadian Society of Petroleum Geologists Memoir 5.

Baker, V. R. (1978b). Large-scale erosional and depositional features of the Channeled Scabland. *In* "The Channeled Scabland" (V. R. Baker and D. Nummedal, eds.), pp. 81-115. National Aeronautics and Space Administration Planetary Geology Program, Washington, D.C.

Baker, V. R. (1978c). Quaternary geology of the Channeled Scabland and adjacent areas. *In* "The Channeled Scabland" (V. R. Baker and D. Nummedal, eds.), pp. 17-35. National Aeronautics and Space Administration Planetary Geology Program, Washington, D.C.

Baker, V. R. (1978d). The Spokane flood controversy and the Martian outflow channels. *Science* 202, 1249-56.

Baker, V. R. (1981). "Catastrophic Flooding: The Origin of the Channeled Scabland." Dowden, Hutchinson, and Ross, Stroudsburg, Pa.

Baker, V. R., and Penteado-Orellana, M. M. (1977). Adjustment to Quaternary climatic change by the Colorado River in central Texas. *Journal of Geology* 85, 395-422.

Baker, V. R., and Penteado-Orellana, M. M. (1978). Fluvial sedimentation conditioned by Quaternary climatic change in central Texas. *Journal of Sedimentary Petrology* 48, 433-51.

Baker, V. R., and Ritter, D. F. (1975). Competence of rivers to transport coarse bedload material. *Geological Society of America Bulletin* 86, 975-78.

Birkeland, P. W. (1968). Mean velocities and boulder transport during Tahoe-age floods of the Truckee River, California-Nevada. *Geological Society of America Bulletin* 79, 137-42.

Bjornstad, B. N. (1980). "Sedimentology and Depositional Environment of the Touchet Beds, Walla Walla River Basin, Washington." Rockwell International Report RHO-BWI-SA-44, Richland, Wash.

Black, R. F. (1979). "Clastic Dikes of the Pasco Basin, Southeastern Washington." Rockwell International Report RHO-BWI-C-64, Richland, Wash.

Boothroyd, J. C., and Ashley, G. M. (1975). Processes, bar morphology, and sedimentary structures on braided outwash fans, northeastern Gulf of Alaska. *In* "Glaciofluvial and Glaciolacustrine Sedimentation" (A. V. Jopling and B. C. McDonald, eds.), pp. 193-222. Society of Economic Paleontologists and Mineralogists Special Publication 23.

Bradley, W. C., and Mears, A. I. (1980). Calculations of flows needed to transport coarse fraction of Boulder Creek alluvium at Boulder, Colorado. *Geological Society of America Bulletin* 91 (part 2), 1057-90.

Brakenridge, G. R. (1978). Evidence for a cold, dry full-glacial climate in the American Southwest. *Quaternary Research* 9, 22-40.

Brakenridge, G. R. (1980). Widespread episodes of stream erosion during the Holocene and their climatic cause. *Nature* 283, 655-56.

Brakenridge, G. R. (1981). Late Quaternary floodplain sedimentation along the Pomme de Terre River, southern Missouri. *Quaternary Research* 15, 62-76.

Bretz, J H. (1923). The Channeled Scabland of the Columbia Plateau. *Journal of Geology* 31, 617-49.

Bretz, J H. (1925). The Spokane flood beyond the Channeled Scabland. *Journal of Geology* 33, 97-115, 236-59.

Bretz, J H. (1928). The Channeled Scabland of eastern Washington. *Geographical Review* 18, 446-77.

Bretz, J H. (1929). Valley deposits immediately east of the Channeled Scabland of Washington. *Journal of Geology* 37, 393-427, 505-41.

Bretz, J H. (1930). Valley deposits immediately west of the Channeled Scabland. *Journal of Geology* 38, 385-422.

Bretz, J H. (1932). "The Grand Coulee." American Geographical Society Special Publication 15.

Bretz, J H. (1964). Correlation of glacial lake stages in the Huron-Erie and Michigan basins. *Journal of Geology* 72, 618-27.

Bretz, J H. (1969). The Lake Missoula floods and the Channeled Scabland. *Journal of Geology* 77, 505-43.

Bretz, J H., Smith, H. T. U., and Neff, G. E. (1956). Channeled scabland of eastern Washington: New data and interpretations. *Geological Society of America Bulletin* 67, 957-1049.

Brice, J. C. (1964). "Channel Patterns and Terraces of the Loup Rivers in Nebraska." U.S. Geological Survey Professional Paper 422-D.

Bull, W. B. (1977). The alluvial-fan environment. *Progress in Physical Geography* 1, 222-70.

Bunker, R. C. (1980). Catastrophic flooding in the Badger Coulee area, south-central Washington. M.A. thesis, University of Texas at Austin, Austin.

Church, M. (1972). "Baffin Island Sandurs: A Study of Arctic Fluvial Processes." Canada Geological Survey Bulletin 216.

Church, M. (1978). Palaeohydrological reconstructions from a Holocene valley fill. *In* "Fluvial Sedimentology" (A. D. Miall, ed.), pp. 743-72. Canadian Society of Petroleum Geologists Memoir 5.

Church, M., and Gilbert, R. (1975). Proglacial fluvial and lacustrine environments. *In* "Glaciofluvial and Glaciolacustrine Sedimentation" (A. V. Jopling and B. C. McDonald, eds.), pp. 22-100. Society of Economic Paleontologists and Mineralogists Special Publication 23.

Church, M., and Ryder, J. M. (1972). Paraglacial sedimentation: A consideration of fluvial processes conditioned by glaciation. *Geological Society of America Bulletin* 83, 3059-72.

Clague, J. J., Armstrong, J. E., and Mathews, W. (1980). Advance of the Late Wisconsin Cordilleran ice sheet in southern British Columbia since 22,000 yr B.P. *Quaternary Research* 13, 322-26.

CLIMAP Project Members (1976). The surface of the ice-age Earth. *Science* 191, 1131-36.

Denny, C. S. (1965). "Alluvial Fans in the Death Valley Region, California and Nevada." U.S. Geological Survey Professional Paper 466.

Dreimanis, A. (1977). Late Wisconsin glacial retreat in the Great Lakes region, North America. *Annals of the New York Academy of Science* 288, 70-89.

Dury, G. H. (1964a). "Principles of Underfit Streams." U.S. Geological Survey Professional Paper 452-A.

Dury, G. H. (1964b). "Subsurface Exploration and Chronology of Underfit Streams." U.S. Geological Survey Professional Paper 452-B.

Dury, G. H. (1965). "Theoretical Implications of Underfit Streams." U.S. Geological Survey Professional Paper 452-C.

Dury, G. H. (1966). Incised valley meanders on the Colo River, New South Wales. *Australian Geographer* 10, 17-25.

Dury, G. H. (1970). General theory of meandering valleys and underfit streams. *In* "Rivers and River Terraces" (G. H. Dury, ed.), pp. 264-275. Macmillan, London.

Dury, G. H. (1973). Magnitude-frequency analysis and channel morphometry. *In* "Fluvial Geomorphology" (M. Morisawa, ed.), pp. 91-121. State University of New York at Binghamton Publications in Geomorphology.

Dury, G. H. (1974). Meandering valleys and underfit streams in southwest Wisconsin. *In* "Late Quaternary Environments of Wisconsin" (J. C. Knox and D. M. Mickelson, eds.), pp. 93-101. Wisconsin Geological and Natural History Survey, Madison.

Dury, G. H. (1976). Discharge predition, present and former, from channel dimensions. *Journal of Hydrology* 30, 219-45.

Dury, G. H. (1977). Underfit streams: Retrospect, perspect, and prospect. *In* "River Channel Changes" (K. J. Gregory, ed.), pp. 281-93. John Wiley and Sons, New York.

Easterbrook, D. J. (1976). Quaternary geology of the Pacific Northwest. *In* "Quaternary Stratigraphy of North America" (W. C. Mahaney, ed.), pp. 441-62. Dowden, Hutchinson, and Ross, Stroudsburg, Pa.

Emiliani, C. (1980). Ice sheets and ice melts. *Natural History* 89 (11), 82-91.

Ethridge, F. G., and Schumm, S. A. (1978). Reconstructing paleochannel morphologic and flow characteristics: Methodology, limitations, and assessment. *In* "Fluvial Sedimentology" (A. D. Miall, ed.), pp. 703-21. Canadian Society of Petroleum Geologists Memoir 5.

Evenson, E. B., Farrand, W. R., Eschman, D. F., Mickelson, D. M., and Maher, L. J. (1976). Greatlakean substage: A replacement for Valderan substage in the Lake Michigan Basin. *Quaternary Research* 6, 411-24.

Fecht, K. R., and Tallman, A. M. (1978). "Bergmounds along the Western Margin of the Channeled Scablands, South-Central Washington." Rockwell International Report RHO-BWI-SA-11, Richland, Wash.

Ferguson, R. I. (1973). Channel pattern and sediment type. *Area* 5, 38-41.

Fisk, H. N. (1944). "Geological Investigation of the Alluvial Valley of the Lower Mississippi River." Mississippi River Commission, Vicksburg, Miss.

Flint, R. F. (1938). Origin of the Cheney-Palouse scabland tract. *Geological Society of America Bulletin* 49, 461-524.

Flint, R. F. (1971). "Glacial and Quaternary Geology." John Wiley and Sons, New York.

Flint, R. F. (1976). Physical evidence of Quaternary climatic change. *Quaternary Research* 6, 519-28.

Foley, L. L. (1976). Slack water sediments in the Alpawa Creek drainage, Washington. M.A. thesis, Washington State University, Pullman.

Foley, M. G. (1980a). Bed-rock incision of streams. *Geological Society of America Bulletin* 91 (part 2), 2189-2213.

Foley, M. G. (1980b). Quaternary diversion and incision Dearborn River, Montana. *Geological Society of America Bulletin* 91 (part 2), 2152-88.

Frye, J. C. (1973). Pleistocene succession of the central interior United States. *Quaternary Research* 3, 275-83.

Fulton, R. J. (1971). "Radiocarbon Geochronology of Southern British Columbia." Geological Survey of Canada Paper 71-37.

Gagliano, S. M., and Thom, B. G. (1967). Deweyville terrace, Gulf and Atlantic coasts. *Louisiana State University Coastal Studies Bulletin* 1, 23-41.

Griggs, G. B., Kulm, L. D., Waters, A. C., and Fowler, G. A. (1970). Deep-sea gravel from Cascadia Channel. *Journal of Geology* 78, 611-19.

Gupta, A. (1975). Stream characteristics in eastern Jamaica, an environment of seasonal flow and large floods. *American Journal of Science* 275, 825-47.

Hack, J. T. (1965). Postglacial Drainage Evolution and Stream Geometry in the Ontonagon Area, Michigan." U.S. Geological Survey Professional Paper 504-B.

Hammatt, H. H. (1977). Late Quaternary stratigraphy and archaeological chronology in the lower Granite Reservoir area, lower Snake River, Washington. Ph.D. dissertation, Washington State University, Pullman.

Hanson, L. G. (1970). The origin and development of Moses Coulee and other scabland features on the Waterville Plateau, Washington. Ph.D. dissertation, University of Washington, Seattle.

Harvey, A. M. (1969). Channel capacity and adjustment of streams to hydrologic regime. *Journal of Hydrology* 8, 82-98.

Haynes, C. V. (1976). Late Quaternary geology of the lower Pomme de Terre Valley. *In* "Prehistoric Man and His Environments" (W. R. Wood and R. B. McMillan, eds.), pp. 47-61. Academic Press, New York.

Howard, A. D. (1960). "Cenozoic History of Northeastern Montana and Northwestern North Dakota with Emphasis on the Pleistocene." U.S. Geological Survey Professional Paper 326.

Hunt, C. B., and Mabey, D. R. (1966). "Stratigraphy and Structure, Death Valley, California." U.S. Geological Survey Professional Paper 494-A.

Jopling, A. V. (1966). Some principles and techniques used in reconstructing the hydraulic parameters of a paleoflow regime. *Journal of Sedimentary Petrology* 36, 5-49.

Kennett, J. P., and Shackleton, N. J. (1975). Laurentide ice sheet meltwater recorded in Gulf of Mexico deep-sea cores. *Science* 188, 147-50.

King, J. E., and Allen, W. H., Jr. (1977). A Holocene vegetation record from the Mississippi River valley, southeastern Missouri. *Quaternary Research* 8, 307-23.

Koteff, C. (1974). The morphologic sequence concept and deglaciation of southern New England. *In* "Glacial Geomorphology" (D. R. Coates, ed.), pp. 121-44. State University of New York at Binghamton Publications in Geomorphology.

Leopold, L. B., and Maddock, T. (1953). "The Hydraulic Geometry of Stream Channels and Some Physiographic Implications." U.S. Geological Survey Professional Paper 252.

Looney, R. M., and Baker, V. R. (1977). Late Quaternary geomorphic evolution of the Colorado River, inner Texas Coastal Plain. *Transactions of the Gulf Coast Association of Geological Societies* 27, 323-33.

Lustig, L. K. (1965). "Clastic Sedimentation in Deep Springs Valley, California." U.S. Geological Survey Professional Paper 352-F, pp. 131-92.

Malde, H. E. (1968). "The Catastrophic Late Pleistocene Bonneville Flood in the Snake River Plain, Idaho." U.S. Geological Survey Professional Paper 596.

Meier, M. F. (1965). Glaciers and climate. *In* "The Quaternary of the United States" (H. E. Wright, Jr., and D. G. Frey, eds.), pp. 795-805. Princeton University Press, Princeton, N.J.

Mickelson, D. M., Clayton, L. Fullerton, D. S., and Borns, H. W., Jr. (1983). The Late Wisconsin glacial record of the Laurentide ice sheet in the United States. *In* "Late-Quaternary Environments of the United States," vol. 1, "The Late Pleistocene" (S. C. Porter, ed.), pp. 3-37. University of Minnesota Press, Minneapolis.

Moody, U. L. (1978). Microstratigraphy, paleoecology, and tephra chronology of the Lind Coulee site, central Washington. Ph.D. dissertation, Washington State University, Pullman.

Morrison, R. B. (1968). Means of time-stratigraphic division and long-distance correlation of Quaternary successions. *In* "Means of Correlation of Quaternary Successions" (R. B. Morrison and H. E. Wright, Jr., eds.), pp. 1-113. University of Utah Press, Salt Lake City.

Moss, J. H. (1974). The relation of river terrace formation to glaciation in the Shoshone River Basin, western Wyoming. *In* "Glacial Geomorphology" (D. R. Coates, ed.), pp. 293-314. State University of New York at Binghamton Publications in Geomorphology.

Mullineaux, D. R., Hyde, J. H., and Rubin, M. (1975). Widespread late glacial and post glacial tephra deposits from Mount St. Helens volcano, Washington. *U.S. Geological Survey Journal of Research* 3, 329-35.

Mullineaux, D. R., Wilcox, R. E., Ebaugh, W. F., Fryxell, R., and Rubin, M. (1978). Age of last major scabland flood of the Columbia Plateau in eastern Washington. *Quaternary Research* 10, 171-80.

Palmquist, R. C. (1975). Preferred position model and subsurface symmetry of valleys. *Geological Society of America Bulletin* 86, 1392-98.

Pardee, J. T. (1942). Unusual currents in glacial Lake Missoula, Montana. *Geological Society of America Bulletin* 53, 1569-1600.

Patton, P. C., and Baker, V. R. (1977). Geomorphic response of central Texas stream channels to catastrophic rainfall and runoff. *In* "Geomorphology in Arid Regions" (D. O. Doehring, ed.), pp. 189-217. State University of New York at Binghamton Publications in Geomorphology.

Patton, P. C., and Baker, V. R. (1978a). New evidence for pre-Wisconsin flooding in the Channeled Scabland of eastern Washington. *Geology* 6, 567-71.

Patton, P. C., and Baker, V. R. (1978b). Origin of the Cheney-Palouse scabland tract. *In* "The Channeled Scabland" (V. R. Baker and D. Nummedal, eds.), pp. 117-30. National Aeronautics and Space Administration Planetary Geology Program, Washington.

Patton, P. C., Baker, V. R., and Kochel, R. C. (1979). Slack-water deposits: A geomorphic technique for the interpretation of fluvial paleohydrology. *In* "Adjustments of the Fluvial System" (D. D. Rhodes and G. P. Williams, eds.), pp. 225-53. Kendall/Hunt, Dubuque, Iowa.

Peltier, L. C. (1949). "Pleistocene Terraces of the Susquehanna River, Pennsylvania." Pennsylvania Geological Survey, Series 4, Bulletin G23.

Peltier, L. C. (1959). "Late Pleistocene Deposits" [of Bucks County, Pennsylvania]. Pennsylvania Geological Survey, Series 4, Bulletin C9, pp. 163-84.

Penteado, M. M., and Baker, V. R. (1976). Evolução no Quaternário Recente de Rio Colorado, no Texas Central. *Anais da Academia Brasileira de Ciências* 47 (suplemento), 71-80.

Peterson, G. M., Webb, T., III, Kutzbach, J. E., Van Der Hammen, T., Wijmstra, T. A., and Street, F. A. (1979). The continental record of environmental conditions at 18,000 yr B.P.: An initial evaluation. *Quaternary Research* 12, 47-82.

Pickup, G., and Warner, R. F. (1976). Effects of hydrologic regime on magnitude and frequency of dominant discharge. *Journal of Hydrology* 29, 51-75.

Pierce, K. L. (1973). Surficial geologic map of the Mammoth quadrangle and part of the Gardiner quadrangle, Yellowstone National Park, Wyoming and Montana. U.S. Geological Survey Miscellaneous Geological Investigations Map I-641.

Pierce, K. L. (1974). Surficial geologic map of the Tower Junction quadrangle and part of the Mount Wallace quadrangle, Yellowstone National Park, Wyoming and Montana. U.S. Geological Survey Miscellaneous Geological Investigations Map I-647.

Pierce, K. L., Obradovich, J. D., and Friedman, I. (1976). Obsidian hydration dating and correlation of Bull Lake and Pinedale Glaciations near West Yellowstone, Montana. *Geological Society of America Bulletin* 87, 703-10.

Pitty, A. F. (1971). "Introduction to Geomorphology." Methuen and Company, London.

Prest, V. K. (1970). Quaternary geology of Canada. *In* "Geologic and Economic Mineral of Canada," Economic Geology Report 1, 5th ed., pp. 676-764. Department of Energy, Mines, and Resources, Ottawa, Canada.

Richmond, G. M., Fryxell, R., Neff, G. E., and Weis, P. L. (1965). The Cordilleran ice sheet of the northern Rocky Mountains and related Quaternary history of the Columbia Plateau. *In* "The Quaternary of the United States" (H. E. Wright, Jr., and D. G. Frey, eds.), pp. 217-30. Princeton University Press, Princeton, N.J.

Ritter, D. F. (1967). Terrace development along the front of the Beartooth Mountains, southern Montana. *Geological Society of America Bulletin* 78, 467-84.

Ritter, D. F. (1972). The significance of stream capture in the evolution of a piedmont region, southern Montana. *Zeitschrift für Geomorphologie* 16, 83-92.

Ruhe, R. V. (1969). "Quaternary Landscapes in Iowa." Iowa State University Press, Ames.

Saucier, R. T. (1974). "Quaternary Geology of the Lower Mississippi Valley." Arkansas Archeological Survey Research Series 6.

Saucier, R. T. (1978). Sand dunes and related eolian features of the lower Mississippi alluvial valley. *Geoscience and Man* 19, 23-40.

Saucier, R. T., and Fleetwood, A. R. (1970). Origin and chronologic significance of late Quaternary terraces, Ouachita River, Arkansas and Louisiana. *Geological Society of America Bulletin* 81, 869-90.

Schultz, C. B., and Martin, L. D. (1970). Quaternary mammalian sequence in the central Great Plains. *In* "Pleistocene and Recent Environments of the Central Great Plains" (W. Dort, Jr., and J. Knox Jones, Jr., eds.), pp. 341-53. University Press of Kansas, Lawrence.

Schumm, S. A. (1965). Quaternary paleohydrology. *In* "The Quaternary of the United States" (H. E. Wright, Jr., and D. G. Frey, eds.), pp. 783-94. Princeton University Press, Princeton, N.J.

Schumm, S. A. (1967). Meander wavelength of alluvial rivers. *Science* 157, 1549-50.

Schumm, S. A. (1968). "River Adjustment to Altered Hydrologic Regimen, Murrumbidgee River and Paleochannels, Australia." U.S. Geological Survey Professional Paper 598.

Schumm, S. A. (1969). River metamorphosis. *Proceedings of the American Society of Civil Engineers, Journal of the Hydraulics Division* 95, 255-73.

Schumm, S. A. (1977). "The Fluvial System." John Wiley and Sons, New York.

Schumm, S. A. (1979). Geomorphic thresholds: The concept and its applications. *Transactions of the Institute of British Geographers* 4, 485-515.

Scott, W. E., McCoy, W. D., Shroba, R. R., and Miller, R. D. (1980). New interpretations of the Late Quaternary history of Lake Bonneville, western United States. *American Quaternary Association Abstracts and Programs*, Sixth Biennial Meeting, pp. 168-69.

Thorson, R. M. (1980). Ice-sheet glaciation of the Puget lowland, Washington, during the Vashon age (late Pleistocene). *Quaternary Research* 13, 303-21.

Tinkler, K. J. (1971). Active valley meanders in south-central Texas and their wider implications. *Geological Society of America Bulletin* 82, 1783-1800.

Tinkler, K. J. (1973). Active valley meanders. *Area* 5, 41-43.

Waitt, R. B., Jr. (1977). "Guidebook to Quaternary Geology of the Columbia, Wenatchee, Peshatin and Upper Yakima Valleys, West Central Washington." U.S. Geological Survey Open-File Report 77-153.

Waitt, R. B., Jr. (1980). About forty late-glacial Lake Missoula Jökulhlaups through southern Washington. *Journal of Geology* 88, 653-79.

Williams, G. P. (1978). Bank-full discharge of rivers. *Water Resources Research* 14, 1141-54.

Willman, H. B. (1971). ''Summary of the Geology of the Chicago Area.'' Illinois State Geological Survey Circular 460.

Willman, H. B., and Frye, J. C. (1970). ''Pleistocene Stratigraphy of Illinois.'' Illinois Geological Survey Bulletin 94.

Wolman, M. G., and Miller, J. P. (1960). Magnitude and frequency of forces in geomorphic processes. *Journal of Geology* 68, 54-74.

Wright, H. E., Jr. (1972). Quaternary history of Minnesota. *In* ''Geology of Minnesota: A Centennial Volume in Honor of George M. Schwartz'' (P. K. Sims and G. B. Morey, eds.), pp. 515-47. Minnesota Geological Survey, University of Minnesota, St. Paul.

CHAPTER 6

Depositional Environment of Late Wisconsin Loess in the Midcontinental United States

Robert V. Ruhe

Introduction

Late Wisconsin loess in the midcontinental United States extends from the Rocky Mountains in Colorado (Figure 6-1) eastward to the Appalachian Mountains in Pennsylvania and from Minnesota southward to Louisiana. The region includes 25° of longitude and 20° of latitude and more than 4.2 million km². Loess-derived soils cover a large part of this region and form the agricultural heartland of the United States.

Late Wisconsin loess is currently known by various names throughout the region. In Nebraska (Figure 6-1), a lower Peorian unit and an upper Bignell unit (Reed and Dreeszen, 1965) are separated by the Brady Soil (Schultz and Stout, 1945). The same units are recognized in Kansas (Frey and Leonard, 1952). In Iowa, the term "Peorian loess" (Kay and Graham, 1943) was dropped and replaced by the informal "Wisconsin Loess" (Ruhe, 1954). In Illinois, Wood-fordian loess includes Peoria loess beyond the Wisconsin drift border, and Morton and Richland are separated by glacial drift behind the border (Frye and Willman, 1960). In Indiana, Peoria loess is recognized (Wayne, 1963). Regardless of the different names used for it, the Late Wisconsin loess buries a soil whose organic horizon ranges in age from 29,000 to 30,000 years to 16,000 to 17,000 years. The Bignell loess of Nebraska and Kansas is excluded from this definition because the radiocarbon dates of the underlying Brady Soil are 9160 and 9750 yr B.P. (Dreeszen, 1970), and so the Bignell loess may be Holocene.

Physical System

Dispersion of loess by wind from source areas has been studied in the midcontinental region for more than 100 years, and the dispersion systems are well known. In general, loess is thick adjacent to major river valleys and is thin away from them. Generalized thickness contours show the patterns adjacent to the Missouri River valley between Iowa (Figure 6-1) and Nebraska and across central Missouri; adjacent to the Mississippi River valley from Minnesota and Wisconsin southward through Illinois, Kentucky, Tennessee, and Mississippi; adjacent to the Wabash River valley between Illinois and Indiana;

and adjacent to the Ohio River valley in Indiana and Kentucky. Loess also thins away from the Sand Hills in Nebraska.

During the past 50 years, the thickness patterns of the loess have been modeled mathematically as exponential functions in Illinois (Krumbein, 1937; Smith, 1942; Frazee et al., 1970), Iowa (Hutton, 1947; Simonson and Hutton, 1954), and Indiana (Caldwell and White, 1956; Fehrenbacher et al., 1965; Frazee et al., 1970; Hall, 1973). Hyperbolic models also have been used in Iowa (Ruhe, 1969; Worcester, 1973). Recently, a more complex geometric model has been formulated for variable winds (Handy, 1976). Regardless of their mathematical bases, these models reasonably predict the downwind thinning of the loess from its presumed source area.

A second major feature of the physical system is the particle-size fractionation that takes place downwind as the loess thins. The most detailed study of this feature has been done in southwestern Iowa along a traverse of 280 km southeastward from the Missouri River valley (Worcester, 1973). There, particle-size fractions systematically change with distance (D) as:

(1) Very coarse silt $(62 - 31\,\mu m)\% = 28.33 - (1.708 \times 10^{-1})D + (1.473 \times 10^{-4})D^2$, $r = 0.96$

(2) Coarse silt $(31 - 16\,\mu m)\% = 34.25 - 0.33D$, $r = 0.94$

(3) Medium silt $(16 - 8\,\mu m)\% = 11.91 + (7.186 \times 10^{-2})D + (7.245 \times 10^{-6})D^2$, $r = 0.96$

(4) Fine silt $(8 - 4\,\mu m)\% = 4.3 + 0.041D$, $r = 0.95$

(5) Very fine silt $(4 - 2\,\mu m)\% = 3.48 + 0.016D$, $r = 0.97$

(6) Coarse clay $(2 - 1\,\mu m)\% = 3.39 + 0.017D$, $r = 0.85$

(7) Medium clay $(1 - 0.5\,\mu m)\% = 10.49 - (4.127 \times 10^{-2})D + (4.850 \times 10^{-4})D^2$, $r = 0.85$

(8) Fine clay $(\times 0.5\,\mu m)\% = 3.77 + 0.039D$, $r = 0.84$

The very coarse and coarse silt fractions decrease and all other size fractions increase as the loess thins hyperbolically with distance from the source.

A third major feature of the physical system is the time transgression of the base of the loess. Along the traverse in southwestern Iowa, the radiocarbon ages of the organic carbon or wood or peat in the A- or O-horizons of soils buried by the loess progressively decrease from about 24,500 years near the Missouri River valley to about 19,000 years at the end of the traverse of 280 km (Ruhe, 1969; Worcester,

Figure 6-1. Thickness of Late Wisconsin loess in the midcontinental region. (Contours are in feet, and a comparative metric scale is given.) Sand areas (stippled) are shown by *S*. State names are abbreviated: AL—Alabama, AR—Arkansas, CO—Colorado, IL—Illinois, IN—Indiana, IA—Iowa, KS—Kansas, KY—Kentucky, LA—Louisiana, MI—Michigan, MN—Minnesota, MS—Mississippi, MO—Missouri, MT—Montana, NB—Nebraska, NM—New Mexico, ND—North Dakota, OH—Ohio, OK—Oklahoma, PA—Pennsylvania, SD—South Dakota, TN—Tennessee, TX—Texas, WV—West Virginia, WI—Wisconsin, WY—Wyoming.

1973). Toward the source, "dark bands" containing small amounts of organic carbon are stratigraphically aligned in the thick loess (Daniels et al., 1960). Radiocarbon dates of these bands range from 22,350 years for the lowest band to 15,300 years for the highest (Ruhe et al., 1971). Consequently, time lines drawn between equivalent dates of bands and basal soils (Figure 6-2) demonstrate the overlap of younger on older loess increments. The decrease in basal age of the loess from 25,000 years to 21,000 years 20 km along a

traverse is known in Illinois (Kleiss and Fehrenbacher, 1973), and the stratigraphic overlap of younger on older loess increments is shown by clay-mineral zones. (Kleiss, 1973).

The "dark bands" are recognized in Iowa only as pauses in loess deposition or, at best, incipient soil formation (Daniels et al., 1960; Ruhe et al., 1971). In Illinois, one of the bands (Frye et al., 1974) or a group of them (McKay, 1979) is the Jules Soil, which is recognized as a major separation in the Woodfordian loess (Frye et al., 1974).

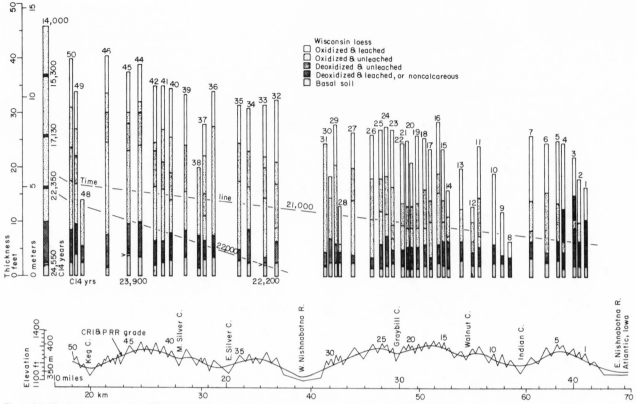

Figure 6-2. Weathering zones in Late Wisconsin loess in successive railroad cuts from west (left) to east (right) in southwestern Iowa. The lower diagram shows the topographic profile with cuts located by number. The stratigraphic columns in the upper diagram are plotted to a common datum to show the thinning of the loess. Radiocarbon dates are given for the basal soil and dark-colored bands within the loess. Time lines are constructed between correlative band and basal soil. The time line 21,000 years descends to basal soil to the east of section 1.

In Iowa, at other sites not aligned along traverses, the maximum age at the base of the loess is 29,000 years and the minimum age is 16,500 years (Ruhe, 1969). In Indiana, the range is 25,500 to 20,100 years (Ruhe and Olson, 1980).

The top of the loess in central Iowa, where it is buried beneath glacial drift, is 14,000 years old (Ruhe, 1969). In eastern Iowa, wood in terrace alluvium beneath 1.8 m of loess has been dated 11,800 ± 200 years old (I-3654) (Buckley and Willis, 1972). In Illinois the top of the loess has been dated at 12,500 years old (Willman and Frye, 1970; McKay, 1979). The top of the loess also transgresses time. Thus, the Late Wisconsin loess can be seen as wedges of sediment that thin progressively away from source areas, with a commensurate decrease in mean particle size. Deposition of the wedges was interrupted numerous times, a fact indicated by the "dark bands," but none of the bands can be considered as soil development of any great significance. The base of each wedge is progressively younger as the wedge thins, and the base crosses numerous Late Wisconsin time-stratigraphic units. In the framework of 29,000 to 17,000 years ago, the base crosses the radiocarbon zone of Farmdalian and well into the Woodfordian stage of Illinois (Willman and Frye, 1970). The top of the loess crosses the late Woodfordian into the Twocreekan substages. Consequently, where the loess is thick near source areas, all of the environments of the Late Wisconsin should have had impact on the loess. Where the loess is thin, only the later environments should have had impact.

Reconstruction of Environment

A reconstruction of environment during loess deposition requires an evaluation of the associated glacial and nonglacial phenomena, and the latter include ordinary erosion, deposition, weathering, and soil formation. In addition, nonloessial features such as the contained or associated flora and fauna may aid in reconstruction.

GLACIAL HISTORY

Loesses in the Midwest usually are correlated with the glacial succession (Frye et al., 1968), but there are some problems involved in the process. The objective here is to discuss these problems. According to the general model, glacial outwash was carried down stream valleys, where silts were deposited, and these silts were then blown onto uplands downwind. Glacier ice had to be in the heads of those watersheds that were drained by the source valleys. Glacial events and loess-deposition episodes had to be reasonably coincident.

Radiocarbon dating has established the fact that the Late Wisconsin loess in Iowa can be as old as 29,000 to 25,000 years. Westward, in Nebraska, there are comparable dates of 27,900 and 26,900 years (Dreeszen, 1970). In northwestern Iowa and across southwestern Minnesota, the maximum limit of Wisconsin glaciation, marked by the Tazewell drift (Ruhe, 1969), was reached 20,000 years ago (Figure 6-3). Tazewell is now called early Woodfordian in Illinois (Willman and Frye, 1970). A till buried beneath the drift of the Des

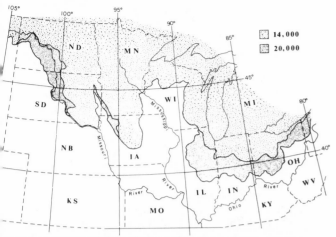

Figure 6-3. Maximum coverage of the midcontinental region by Late Wisconsin ice sheets 20,000 to 14,000 years ago. Compare Figure 6-1 for Late Wisconsin loess coverage. (The abbreviations for the state names are the same as those for Figure 6-1.)

Moines lobe in north-central Iowa is tentatively correlated with the Tazewell drift and contains wood fragments 25,390 years old in its lower 0.5 m (Kemmis et al., 1981). In comparisons of earliest loess and till dates, basal loess at places in Iowa can be 9000 to 4000 years older than the presumed correlative glacial event.

In Illinois, the earliest Peoria (Late Wisconsin) loess is considered to be 25,000 years old (McKay, 1979). Yet, the advance of the early-Woodfordian glacier has been documented at 23,000 years near Lake Michigan and 20,000 to 19,000 years at the glacial margin in west-central and south-central Illinois (Kempton and Gross, 1971). In Indiana, earliest Peoria loess is 25,000 years old, and the glacial maximum is 20,000 years old (Ruhe and Olson, 1980). Thus, in all these areas, the loess preceded the documented occurrence of presumed correlative glaciers in the watersheds.

The time precedence is also demonstrated stratigraphically. Late Wisconsin loess both underlies and overlies the 20,000-year-old drifts in northwestern Iowa, Illinois, and Indiana. In Illinois, the unit beneath the drift is Morton loess, and the unit on the drift is Richland loess (Willman and Frye, 1970). The upper loess covers the 20,000-year-old drift back to the margin of the 14,000-year-old drift, which is generally not mantled by loess (cf. Figures 6-1 and 6-3). In Iowa, the 14,000-year-old drift buries the upper loess, and a weak A-C soil profile is at the top of the loess (Ruhe et al., 1957). The 20,000-year-old drift had to be free of ice in order to permit the deposition of loess on the drift and the formation of the weak soil on the loess prior to the burial at 14,000 years. These lines of evidence indicate that two Late Wisconsin glacial advances, separated by a hiatus of less than 6000 years, occurred. During the total loess-deposition time of 29,000 to 12,000 years ago, however, the northern latitudes of the midcontinental loess region were covered by ice sheets that reached their limits 20,000 and 14,000 years ago (Figure 6-3).

A periglacial environment should have existed in the nonglacial areas. What, then, was the periglacial environment? Mean annual temperature zones that currently are latitudinally banded across the midcontinent (see Ruhe, 1982) should have been displaced southward. The present winter (January) freezing isotherm (0 °C) is along latitude 40 °N (Figure 6-1) and may be a reasonable approxima-

tion of the glacial mean annual temperature. At the Canadian border to the north, the winter isotherm is − 18 °C; to the south at the Gulf of Mexico, the isotherm is 13 °C. So, during loess deposition, mean annual temperatures in the loess region may have been 17 °C colder in the north, 14 °C colder in middle latitudes, and 8 °C in the south than they are today. These estimates are probably extreme.

WEATHERING AND SOIL FORMATION

A difficulty arises in the attempt to use features of the loess for interpreting the environment of deposition. The loess has been altered by weathering and soil formation. The increasing intensity in the development of ground soils down the sediment wedge has been known for many years (Ruhe, 1969; Fehrenbacher, 1973) and needs no further attention here. Instead, the focus is on the arrangement of weathering zones in the loess. They have been intensively studied in Iowa (Ruhe, 1969) even in routine work (Hallberg, Fenton, and Miller, 1978). Even though the zones are prevalent in the loess throughout the region, they have been essentially ignored in other states.

The zones, as specified in Figure 6-2, are well displayed along the thinning loess wedge in southwestern Iowa. In oxidized zones, 60% of the matrix has Munsell soil-color hues of 2.5Y or redder and values of 3 or higher and may have segregation of secondary iron compounds into mottles, tubules, or nodules. In deoxidized zones, 60% of the matrix has hues of 10YR, 2.5Y, or 5Y; values of 5 and 6; and chromas of 1 or 2, with segregation of iron (ferric oxides) in tubules (pipe stems) or nodules. Leached or unleached refers to the absence or presence of primary carbonates, respectively (Ruhe, 1969; Hallberg, Fenton, and Miller, 1978).

Where loess is thick in southwestern Iowa, the lowest weathering zone is called deoxidized and leached (Figure 6-2), which implies that the loess was calcareous but that the carbonates were removed by weathering. A general interpretation has been that rate of loess deposition was slow, so that carbonates were contemporaneously leached (Frye and Leonard, 1952). This analysis is implicit in the "effective age" hypothesis for the formation of loess-derived soils (Smith, 1942).

Another explanation is feasible when radiocarbon dates are examined, and the results also have a bearing on the problem of correlating units of the loess with glacial episodes. The radiocarbon dates of the soil buried by the loess and the "dark bands" in the loess permit the construction of time lines (Figure 6-2). The basal noncalcareous loess where it is overlain by calcareous loess generally is below the 21,000- and 22,000-year time lines, which predate the 20,000-year Late Wisconsin glaciation to the north in Iowa (Figure 6-3). The loess in most places buried intensively weathered pre-Wisconsin paleosols. Ordinary soil-erosion processes could have provided weathered debris from those soils, and the noncalcareous debris could have become entrained for eolian deposition during earliest loess time. Glacial supply may be marked by the overlying calcareous loess. Such a mechanism has been formulated for glacial and nonglacial sources of Wisconsin loess in Indiana (Ruhe and Olson, 1980).

The oxidized zones are in general yellowish brown to brown, and free iron oxides (chemically extractable) are diffuse throughout the matrix and range from 1.5% to 3.5% (Daniels et al., 1961). The deoxidized zones are generally gray, and free iron oxides range from only 0.3% to 0.6%. The vertically oriented pipe stems in the deoxidized loess are reddish brown and red, and free iron oxides range from 9% to 26%. The iron was mobilized and moved from the matrix of the deoxidized zone to the pipe stems, where ferric oxides were concentrated. The deoxidized zones were interpreted originally as relict zones of fluctuating water tables, where saturation would have induced the reduction and mobilization of iron and nonsaturation would have permitted the migration and precipitation of iron along aeration tubules (eventually becoming pipe stems) (Ruhe et al., 1955). Numerous recent studies have verified the relations between hydrologic regimes and deoxidized zones (Hallberg, Fenton, and Miller, 1978).

The deoxidized zone is particularly prominent where the loess buries pre-Wisconsin paleosols with clayey B-horizons on broad upland remnants of pre-Wisconsin till plains. These paleosols act as hydraulic impediments and cause the perching of water in the loess above them (Ruhe, 1969; Worcester, 1973). The deoxidized zones have been observed and studied by the author throughout Iowa and northern Missouri, southern Illinois, western Kentucky, southern Indiana, and southern Ohio. In southwestern Iowa, two of the zones are in thick loess, but, as the sediment wedge thins, one zone dominates and continues down-wedge across the broad paleosol flats. There, poor internal drainage conditions should have existed throughout the period from loess deposition to the present. Today's coincidence of perched soil water and deoxidized zones on the broad flats suggests that moisture conditions should have been similar in the past.

The pipe stems are most prominent in the deoxidized loess. They generally are considered to be evidence of the decayed vegetation that formed aeration tubules to which the iron compounds migrated and precipitated. The pipe stems are vertically oriented, and many of them are less than 1 cm in diameter. They may represent the stems and roots of grasses or other herbs. Pipe stems 3 to 8 cm in diameter (Ruhe, 1969: 48) may represent the roots of trees. None has been observed with a diameter suggesting a tree trunk. The pipe stems only imply that vegetation existed during loess deposition.

PALEOFLORA OF LOESS

The flora during loess deposition is preserved as detrital logs and trees rooted in place in poorly aerated burial sites. It is also represented by pollen in wetlands. Unfortunately, only a few localities throughout the loess region contain a preserved record. The most complete record of tree species, dated by radiocarbon methods, is in Iowa (Ruhe, 1969: 37-42). The woods are conifers, and although gaps of several thousand years exist in the record, almost all of loess-deposition time is spanned by the radiocarbon-dated samples (Table 6-1).

None of the conifers is native in Iowa today, so the wood samples indicate a different environment (Ruhe, 1982: Figures 2-1 through 2-5). The current *Picea* (spruce) and *Abies* (fir) distributions extend from northwestern Minnesota eastward across the middle of Wisconsin and Michigan. The current *Tsuga* (hemlock) distribution extends from northwestern Wisconsin eastward across Michigan and southward through the Appalachian Mountains to southeastern Tennessee. The current *Larix* (larch) distribution parallels *Picea* and *Abies*, but the southern border extends eastward along the southern boundaries of Wisconsin and Michigan.

Iowa and the coniferous forests to the north differ climatically in temperature and moisture. The mean annual temperatures are 3°C to more than 6°C cooler in the north than in Iowa, but the mean annual precipitation is similar: 60 cm to more than 80 cm in both places. The difference reflects the conservation of moisture. The mean annual pan evaporation is a third less in the north than in Iowa. During drought years, rainfall during the summer (July) is 80% to 100% of normal rainfall in the north, but in Iowa it is only 50% to 70% of the norm (Borchert, 1950). The occurrence of the fossil conifers in the loess of Iowa implies that the more northerly environment had to be present at times between 24,500 and 14,000 years ago. During those times, glacier ice was present in the state and adjoining areas, and temperatures could have been as much as 14°C colder than they are at present.

Table 6-1.
Radiocarbon Ages of Wood and Late Wisconsin Loess in Iowa

Sample	Date in yr B.P.	Location	Wood
Near Top of Loess:			
W-513	13,820 ± 400	Scranton, Greene County (central Iowa)	*Picea* (spruce)
W-517	13,910 ± 400	Scranton, Greene County	*Picea*
I-1402	14,200 ± 500	Nevada, Story County (central Iowa)	*Picea*
W-510	14,470 ± 400	Scranton 1, Greene County	*Abies* (fir), *Tsuga* (hemlock), *Larix* (larch), *Picea*
Within Loess:			
OWU-153	14,700 ± 400	Clear Creek, Story County (central Iowa)	*Tsuga*
I-1270	16,100 ± 1000	Boone, Boone County (central Iowa)	*Picea*
I-1024	16,100 ± 500	Madrid, Boone County	*Picea*
W-126	16,720 ± 500	Mitchellville, Polk County (central Iowa)	*Taxus* (yew), *Picea, Tsuga*
Just Below Base of Loess:			
W-879	19,050 ± 300	Logan, Harrison County (southwestern Iowa)	*Picea*
I-1023	23,900 ± 1100	Bentley, Pottawattamie County (southwestern Iowa)	*Picea*
W-141	24,500 ± 800	Hancock, Pottawattamie County	*Larix*

Pollen evidence for the nature of vegetation during loess deposition is as scarce as macrofossil evidence. The record is available at only a few sites in the region. In two marshes near Muscotah and Arrington in northeastern Kansas, a pollen sequence covering the last 25,000 years begins with an assemblage that indicates open vegetation, with pine, spruce, and birch as the most important tree species, along with alder and willow (Grüger, 1973). From 23,000 to 15,000 years ago, spruce forest was dominant. Unfortunately, an unconformity interrupts the sequence, and the next record at 11,500 years indicates oak-hickory forest and grassland. The spruce pollen's dominance until 15,000 years ago agrees with the macrofossil record to the northeast in Iowa.

In southeastern and east-central Iowa, two pollen sequences from 28,720 to 24,900 years ago and from 28,800 to 22,750 years ago are very similar (Hallberg et al., 1980; Van Zant et al., 1980). The spectra indicate an earlier closed coniferous forest of spruce and pine that was succeeded by spruce forest.

A discrepant history from pollen and seed studies has been reported in Illinois only 150 km to the east, strangely enough, very close to areas covered by Late Wisconsin ice sheets (Grüger, 1972a, 1972b). From 28,000 to 22,000 years ago, pine-spruce forest covered central Illinois, but a short distance away in south-central Illinois deciduous trees and pine cooccurred with areas of open vegetation. The open vegetation is presumed to have been similar to the current prairie-forest transitional zone in the northern Great Plains. While ice covered northern and central Illinois from 22,000 to 14,000 years ago, spruce and oak were the dominant trees, but pine and grassland existed in south-central Illinois and were succeeded by oak-hickory forest and prairie. Ice-free areas in northern Illinois temporarily were covered by spruce, but deciduous forest also occurred in the region. Grüger concludes that deciduous trees were close to the ice margin, that arctic climatic conditions did not exist, and that major vegetational changes did not occur during the Late Wisconsin. Meanwhile, loess was being deposited in an environment there that was strangely different from that in the adjoining states. Whitehead's (1973) regional full-glacial reconstruction of Late Wisconsin vegetational changes also sharply contrasts the history formulated in Illinois.

The problem is that during the Late Wisconsin a coniferous forest presumably existed in the West where continuous prairie exists today (Wright, 1970). Yet, in Illinois, conifers, hardwoods, and grasslands mingled in places where today grassland exists only as an extension of the Prairie Peninsula (Wright, 1968).

MOLLUSCAN FAUNA

Late Wisconsin loess includes a diverse assemblage of mollusks, mostly terrestrial gastropods. A few pulmonate gastropods with aquatic habits also occur in the faunas. In Kansas, the species are stratigraphically zoned from a lower through a transitional to an upper zone in Peoria loess. The overlying Bignell loess has its own assemblage, which is identical to the modern fauna. The latter must endure a biologically severe climate, with hot, arid summers followed by cold, dry winters (Leonard, 1952). The conclusion must be that environmental conditions during the deposition of Bignell loess were like those today on the Great Plains if radiocarbon dates of 9750 and 9160 years at the base of the Bignell loess (Dreeszen, 1970) are accepted and are not rejected as they commonly are. The Bignell loess is probably Holocene.

On the Great Plains from Texas to South Dakota, the Late Wisconsin loess is fossiliferous, as it is in Kansas, and most of the gastropods

in the faunal zone are now extinct on the Plains (Leonard and Frye, 1954). The terrestrial gastropods live elsewhere today, and their habitats are known. They require a soil cover of moist organic debris in which their eggs are deposited and their young are reared. The eggs and the young die quickly when exposed to drying. The animals feed on decaying organic matter and fungi that grow in the moist organic debris. The most favorable environment for population maintenance and growth is a moist organic cover on the soil and a climate without extreme aridity and high temperatures during the summer months. Thus, the environment during loess deposition of the Plains seemingly consisted of cooler mean temperatures and moister upland soil conditions than exist today.

In Iowa (Kay and Graham, 1943), an environmental interpretation of the gastropod fauna in the loess indicates that the plant geography during loess deposition was little different from that of the present, except for the existence of more deciduous and coniferous trees in the forests, such as in Wisconsin and Minnesota today. Denser forest conditions during the time of early loess deposition gave way to forest-border or prairie conditions during later loess deposition. The known record of macrofossil conifer flora in Iowa seemingly contradicts the environmental reconstruction from gastropod fauna. In addition, the present plant geography relates to climatic patterns of the past 8000 years following the establishment of prairie in Iowa (Ruhe, 1982). Then, too, glacier ice covered appreciable parts of northern Iowa 20,000 to 12,500 years ago, and loess was deposited in the periglacial area during the same time.

In Illinois (Leonard and Frye, 1960), the gastropods represented a woodland fauna of hardwood forests during the period 28,000 to 22,000 years ago, followed by a general vegetative cover consisting of stands of trees along streams, with large areas of open prairie, until about 15,000 years ago. The climate was cooler, as evidenced by the fact that the same mollusk species are characteristic of faunas in Canada today. Moisture was not deficient compared to today, nor were there climatic extremes during loess deposition in areas adjacent to ice fronts. At least in Illinois, there is some compatibility of reconstructions of environments based on molluscan fauna (Leonard and Frye, 1960) and on pollen and seed studies (Grüger, 1972a, 1972b). But, as noted previously, the analysis of paleoflora does not fit well within the regional reconstruction (Whitehead, 1973).

In Indiana, two faunal zones have been identified in the loess (Wayne, 1963). The upper assemblage has small species of more northerly habitat in comparison to the modern fauna. The lower assemblage has many large snails as well as smaller species that commonly exist in deciduous woodlands. Northward in the state, the two assemblages are more similar. The same problem of faunal habitat arises in Indiana. Ice covered the northern two-thirds of the state 20,000 years ago and the northern one-third of the state 14,000 years ago (Figure 6-3). Loess was deposited in the periglacial areas during the same time. Yet, the faunas are characteristic of deciduous woodlands.

To this point in reconstruction of the environment during loess deposition, conflicts exist among three lines of evidence: ice history, contained flora, and contained fauna. An additional complication becomes evident in an examination of the geomorphic history of the time.

IOWAN EROSION SURFACE

An extensive loess-mantled erosion surface covers 12,500 km² in northeastern and northwestern Iowa (Ruhe et al., 1968; Ruhe, 1969;

Ruhe, 1982: Figure 2-6). At many places, erosion remnants, locally known as "paha," stand above the erosion surface at drainage divides. The paha are elongate hills consisting of thick loess on pre-Wisconsin paleosols formed on pre-Wisconsin till. The erosion surface is cut below the level of the paleosols and into the till and other lower sediments. The loess on the lower-lying erosion surface is thinner than in the paha. The erosion surface cuts across interfluves and in many places descends along the interfluve axes in stepped levels. Since the original work (Ruhe et al., 1968), additional studies have been completed on the erosion surface (Vreeken, 1973, 1975; Miller, 1974; Hallberg, Fenton, Miller, and Lutenegger, 1978).

The erosion surface has been radiocarbon dated at many places in northeastern Iowa. Eight dates from the base of the thick loess above the paleosols in the paha range from 29,000 to 21,350 years. Eight comparative dates at the base of the thin loess and above the stone line on till that marks the adjacent lower-lying erosion surface range from 22,600 to 17,810 years (Ruhe et al., 1968; Hallberg, Fenton, Miller, and Lutenegger, 1978). These radiocarbon dates demonstrate conclusively that the Iowan erosion surface was being cut while loess was being deposited in the region. The complete loess sequence accumulated on the paha, but only a thinner and younger part of the loess was deposited on the erosion surface. At other places in Iowa, the erosion surface is dated at 12,700 years and younger (Ruhe, 1969). There is no reason to doubt the existence of the erosion surface in adjoining regions, including the driftless area of Illinois, Iowa, Minnesota, and Wisconsin.

A common feature in the loess of the paha is sedimentologic zoning. The lowest zone has a low sand content. The intermediate zone contains considerable sand, and the uppermost zone is relatively sand free. When traced from the paha to the erosion surface, the sand zone descends from its intermediate position to the base of the loess. The sand is derived from sediment produced during cutting of the erosion surface, and it was blown to its higher position on the paha.

The erosion surface of 12,500 km² and the eolian sand introduce additional complications in the environmental reconstruction during loess deposition. A common factor that controls erosion and sediment yield, whether by water or wind, is ground cover. When cover, whether grass or trees, is dense, erosion and sediment yield are minimal; but, when conditions are converse, they are maximal. Cover can readily be reduced by decreasing the amount of soil moisture to the wilting point for plants. To reach this soil-moisture requirement throughout the extensive region of the erosion surface would seem to require a relatively dry climate. The mean annual precipitation in this region in Iowa today ranges from 64 cm in the west to 86 cm in the east, with a mean annual temperature range of 7°C to 10°C. Grassland and oak-hickory forests are supported under these conditions. Under similar or lower temperatures in the past, the amount of available moisture would have had to have been less to reduce plant cover.

Yet, the conifer macroflora and pollen spectra that have the same radiocarbon ages as the erosion surface indicate the dominance of spruce forest. Time stratigraphically correlative molluscan fauna also have a northern affinity. The environmental dilemma persists.

Conclusions

Late Wisconsin loess has a well-defined sedimentologic system in

that it thins systematically with distance from a source area and has a commensurate particle-size fractionation. The base of the loess is time transgressive in that it buries a preexisting soil with a radiocarbon age of 29,000 years and a minimum age of about 17,000 years. Loess deposition terminated about 12,000 years ago, but it was not continuous throughout the entire 17,000 years. At many sites, the interruption of loess deposition is marked by "dark bands," which at best indicate only incipient soil formation. At present, no soil with significant development is recognized within the loess.

The Late Wisconsin loess has characteristic weathering zones, and the deoxidized zone with its neutral gray matrix and the iron oxide segregated in vertical tubules (pipe stems) are most important environmentally. On broad upland flats, the deoxidized zone relates to groundwater saturation zones that perch above more impermeable pre-Wisconsin paleosols. With a fluctuating water table, alternating episodes of saturation and aeration would induce the reduction, mobilization and migration of iron compounds in the loess matrix and the precipitation of iron along aeration tubules. The poor internal drainage conditions above the broad paleosol flats have existed from the beginning of loess deposition to the present. Soil-moisture conditions of the past could have been similar to those of the present. Conversely, if current hydrologic regimes do not relate to deoxidized zones, soil moisture should have been greater in the past than it is at present. The pipe stems are generally considered to represent the former presence of stems and roots of grass or other herbs or even of trees.

Although Late Wisconsin loess usually is specifically correlated with glacial events, radiocarbon dates show that the earliest loess deposition preceded the earliest glacial event by several thousand years. Then, too, a presumed Late Wisconsin loess on the Great Plains (Bignell) is really Holocene if radiocarbon dates at the base of the loess are accepted. During loess deposition, however, great ice sheets occupied the northern part of the loess region and extended as far south as latitude 40°N 20,000 years ago and almost that far south 14,000 years ago. Temperatures judged on any basis—mean annual, mean winter, or mean summer—must have been colder during those times than they are today.

Environmental reconstructions based on fossil macroflora, pollen, and molluscan faunas have many inconsistencies. The macroflora in Iowa is coniferous from the bottom to the top of the loess, and pollen spectra in Iowa and adjoining northeastern Kansas indicate spruce forest. Yet, molluscan faunas in Iowa show a plant geography little different from that of today, except for the presence of more deciduous and coniferous trees in the forests. During later loess deposition, forest border or prairie conditions are indicated.

Meanwhile, at the same latitudes in Illinois, the gastropod fauna and pollen studies first represent pine-spruce forests, followed rapidly by deciduous trees along water courses, with intervening areas of open prairie. Even while ice was present, hardwood trees were close to the ice margins, arctic conditions did not exist, and major vegetational changes did not occur.

To complicate these matters further, an extensive erosion surface was forming, and dune sand was blowing around and under stands of conifers or woodlands along water courses while glacier ice was nearby. So, the dilemma of the loess environment persists. By comparison, reconstruction of the environment of the ocean basins (CLIMAP Project Members, 1976) is a soft sea breeze!

References

Borchert, J. R. (1950). The climate of the central North American grassland. *Annals of the American Association of Geographers* 40, 1-39.

Buckley, J., and Willis, F. H. (1972). Isotopes radiocarbon measurements IX. *Radiocarbon* 14, 114-39.

Caldwell, R. E., and White, J. L. (1956). A study of the origin and distribution of loess in southern Indiana. *Soil Science Society of America Proceedings* 20, 258-63.

CLIMAP Project Members (1976). The surface of the ice-age Earth. *Science* 191, 1131-37.

Daniels, R. B., Handy, R. L., and Simonson, G. H. (1960). Dark-colored bands in the thick loess of western Iowa. *Journal of Geology* 68, 450-58.

Daniels, R. B., Simonson, G. H., and Handy, R. L. (1961). Ferrous iron content and color of sediments. *Soil Science* 91, 378-82.

Dreeszen, V. H. (1970). The stratigraphic framework of Pleistocene glacial and periglacial deposits in the central Plains. *In* "Pleistocene and Recent Environments of the Central Great Plains" (W. Dort, Jr., and J. K. Jones, Jr., eds.), pp. 9-22. University of Kansas Press, Lawrence.

Fehrenbacher, J. B. (1973). Loess stratigraphy, distribution, and time of deposition in Illinois. *Soil Science* 115, 176-82.

Fehrenbacher, J. B., White, J. L., Ulrich, H. P., and Odell, R. T. (1965). Loess distribution in southeastern Illinois and southwestern Indiana. *Soil Science Society of America Proceedings* 29, 566-72.

Frazee, C. J., Fehrenbacher, J. B., and Krumbein, W. C. (1970). Loess distribution from a source. *Soil Science Society of America Proceedings* 34, 296-301.

Frye, J. C., and Leonard, A. B. (1952). "Pleistocene Geology of Kansas." Kansas Geological Survey Bulletin 99.

Frye, J. C., Leonard, A. B., Willman, H. B., Glass, H. D., and Follmer, L. R. (1974). "The Late Woodfordian Jules Soil and Associated Molluscan Faunas." Illinois Geological Survey Circular 486.

Frye, J. C., and Willman, H. B. (1960). "Classification of the Wisconsinan Stage in the Lake Michigan Glacial Lobe." Illinois Geological Survey Circular 285.

Frye, J. C., Willman, H. B., and Glass, H. D. (1968). Correlation of midwestern loesses with the glacial succession. *In* "Loess and Related Eolian Deposits of the World" (C. B. Schultz and J. C. Frye, eds.), pp 3-21. University of Nebraska Press, Lincoln.

Grüger, E. (1972a). Late Quaternary vegetation development in south-central Illinois. *Quaternary Research* 2, 217-31.

Grüger, E. (1972b). Pollen and seed studies of Wisconsinan vegetation in Illinois, U.S.A. *Geological Society of America Bulletin* 83, 2715-34.

Grüger, J. (1973). Studies on the late Quaternary vegetational history of northeastern Kansas. *Geological Society of America Bulletin* 84, 238-50.

Hall, R. D. (1973). Sedimentation and alteration of loess in southwestern Indiana. Ph.D. dissertation, Indiana University, Bloomington.

Hallberg, G. R., Baker, R. G., and Legg, T. (1980). A mid-Wisconsinan pollen diagram from Des Moines County, Iowa. *Iowa Academy of Sciences Proceedings* 87, 41-44.

Hallberg, G. R., Fenton, T. E., and Miller, G. A. (1978). Standard weathering zone terminology for the description of Quaternary sediments in Iowa. *In* "Standard Procedures for Evaluation of Quaternary Materials in Iowa." Iowa Geological Survey Technical Information Series 8.

Hallberg, G. R., Fenton, T. E., Miller, G. A., and Lutenegger, A. J. (1978). The Iowan erosion surface: An old story, an important lesson, and some new wrinkles. *In* "Geology of East-central Iowa," pp. 1-94. Iowa Geological Survey Guidebook 2.

Handy, R. L. (1976). Loess distribution by variable winds. *Geological Society of America Bulletin* 87, 915-27.

Hutton, C. E. (1947). Studies of loess-derived soils in southwestern Iowa. *Soil Science Society of America Proceedings*, 12, 424-31.

Kay, G. F., and Graham, J. B. (1943). The Illinoian and post-Illinoian Pleistocene geology of Iowa. *Iowa Geological Survey Annual Report* 38, 1-262.

Kemmis, T. J., Hallberg, G. R., and Lutenegger, A. J. (1981). "Depositional Environments of Glacial Sediments and Landforms on the Des Moines Lobe, Iowa." Iowa Geological Survey Guidebook 6.

Kempton, J. P., and Gross, D. L. (1971). Rate of advance of the Woodfordian (Late Wisconsinan) glacial margin in Illinois: Stratigraphic and radiocarbon evidence. *Geological Society of America Bulletin* 82, 3245-50.

Kleiss, H. J. (1973). Loess distribution along the Illinois soil development sequence. *Soil Science* 115, 194-98.

Kleiss, H. J., and Fehrenbacher, J. G. (1973). Loess distribution as revealed by mineral variations. *Soil Science Society of America Proceedings* 37, 291-95.

Krumbein, W. C. (1937). Sediments and exponential curves. *Journal of Geology* 45, 577-601.

Leonard, A. B. (1952). Illinoian and Wisconsinan molluscan faunas in Kansas. *In* "Mollusca," pp. 1-38. University of Kansas Paleontology Contributions 4.

Leonard, A. B., and Frye, J. C. (1954). Ecological conditions accompanying loess deposition in the Great Plains of the United States. *Journal of Geology* 62, 399-404.

Leonard, A. B., and Frye, J. C. (1960). "Wisconsinan Molluscan Faunas in the Illinois Valley Region." Illinois Geological Survey Circular 304.

McKay, E. D. (1979). Stratigraphy of Wisconsinan and older loesses in southwestern Illinois. *In* "Geology of Western Illinois," pp. 37-67. Illinois Geological Survey Guidebook 14.

Miller, G. A. (1974). Soil parent material stratigraphy and soil development, Cedar County, Iowa. Ph.D. dissertation, Iowa State University, Ames.

Reed, E. C., and Dreeszen, V. H. (1965). "Revision of the Classification of Pleistocene Deposits of Nebraska." Nebraska Geological Survey Bulletin 23.

Ruhe, R. V. (1954). Relations of the properties of Wisconsin loess to topography in western Iowa. *American Journal of Science* 252, 663-72.

Ruhe, R. V. (1969). "Quaternary Landscapes in Iowa." Iowa State University Press, Ames.

Ruhe, R. V. (1983). Aspects of Holocene pedology in the United States. *In* "Late-Quaternary Environments of the United States," vol. 2, "The Holocene" (H. E. Wright, Jr., ed.), pp. 12-25. University of Minnesota Press, Minneapolis.

Ruhe, R. V., Dietz, W. P., Fenton, T. E., and Hall, G. F. (1968). "Iowan Drift Problem, Northeastern Iowa." Iowa Geological Survey Report of Investigations 7.

Ruhe, R. V., Miller, G. A., and Vreeken, W. J. (1971). Paleosols, loess sedimentation and soil stratigraphy. *In* "Paleopedology: Origin, nature and dating of paleosols" (D. H. Yaalon, ed.). pp. 41-60. Hebrew Universities Press, Jerusalem.

Ruhe, R. V., and Olson, C. G. (1980). Clay-mineral indicators of glacial and nonglacial sources of Wisconsinan loesses in southern Indiana, U.S.A. *Geoderma* 24, 283-97.

Ruhe, R. V., Prill, R. C., and Riecken, F. F. (1955). Profile characteristics of some loess-derived soils and soil aeration. *Soil Science Society of America Proceedings* 19, 345-47.

Ruhe, R. V., Rubin, M., and Scholtes, W. H. (1957). Late Pleistocene radiocarbon chronology in Iowa. *Iowa Journal of Science* 255, 671-89.

Schultz, C. B., and Stout, T. M. (1945). Pleistocene loess deposits of Nebraska. *American Journal of Science* 243, 231-44.

Simonson, R. W., and Hutton, C. E. (1954). Distribution curves for loess. *American Journal of Science* 252, 99-105.

Smith, G. D. (1942). "Illinois Loess: Variations in Its Properties and Distribution." Illinois Agricultural Experiment Station Bulletin 490.

Van Zant, K. L., Hallberg, G. R., and Baker, R. G. (1980). A Farmdalian pollen diagram from east-central Iowa. *Iowa Academy of Science Proceedings* 87, 52-55.

Vreeken, W. J. (1973). Soil variability in small loess watersheds: Clay and organic carbon content. *Catena* 1, 181-95.

Vreeken, W. J. (1975). Variability of depth to carbonates in fingertip loess watersheds in Iowa. *Catena* 2, 321-36.

Wayne, W. J. (1963). "Pleistocene Formations in Indiana." Indiana Geological Survey Bulletin 25.

Whitehead, D. R. (1973). Late-Wisconsin vegetational changes in unglaciated eastern North America. *Quaternary Research* 3, 621-31.

Willman, H. B., and Frye, J. C. (1970). "Pleistocene Stratigraphy of Illinois." Illinois Geological Survey Bulletin 94.

Worcester, B. K. (1973). Soil genesis on the stable primary divides of the southwestern Iowa loess province. Ph.D. dissertation, Iowa State University, Ames.

Wright, H. E., Jr. (1968). History of the Prairie Peninsula. *In* "The Quaternary of Illinois" (R. E. Bergstrom, ed.), pp. 78-88. University of Illinois College of Agriculture Special Publication 14.

Wright, H. E., Jr. (1970). Vegetational history of the central Plains. *In* "Pleistocene and Recent Environments of the Central Great Plains" (W. Dort, Jr., and J. K. Knox, Jr., eds.), pp. 157-72. University of Kansas Press, Lawrence.

CHAPTER 7

Sangamon and Wisconsinan Pedogenesis in the Midwestern United States

Leon R. Follmer

Introduction

South of the classical Wisconsinan glacier margin in the midwestern United States, a strongly developed paleosol occurs at a shallow depth beneath the upland surface, except where the surface is covered by thick loess deposits along the major rivers. This paleosol, the Sangamon Soil, is found throughout most of the region beneath deposits of the last glaciation (Wisconsinan) and in deposits of the previous glaciation (Illinoian) or in older deposits (Johnson, 1976; Leverett, 1899; Ruhe, 1974; Willman and Frye, 1970). The interglacial time known as the Sangamonian in the United States correlates in general with the Ipswichian, Eemian, and Mikulina of Europe (Flint, 1971). It is generally agreed that the Sangamonian correlates with all or part of the oxygen-isotope stage 5 of the oceanic record (cf. Bowen, 1978).

Compared to the widespread occurrence of the Sangamon Soil, the record of the interstadial soils within the Wisconsinan deposits of the Midwest is fragmentary. A significant interstadial soil (Farmdale), dated in the range of 28,000 to 22,000 yr B.P., is commonly found where it developed in early-Wisconsinan loess (Reed et al., 1965; Willman and Frye, 1970). A correlative interstadial soil (Sidney) has been found in Ohio and Indiana where it developed in till (Gooding, 1975). Soils older or younger than the Farmdale only appear locally and have been controversial (Follmer, 1978; Ruhe, 1976). The controversy stems from a combination of factors, including a weak degree of soil development or poor preservation of soil features, confusing relationships with the time-transgressive Sangamon Soil and Farmdale Soil, and differing concepts of buried soils and pedostratigraphy. Some of the problems come back from the lack of stratigraphic control on the early-Wisconsinan deposits, which are estimated to range in age from 75,000 to 28,000 yr B.P. (Frye, Follmer, et at., 1974). Tills of this age have only been described locally and are discontinuous or deeply buried. Loess of this age is widespread but is commonly thin and often included in the profile of the Sangamon Soil. In order to evaluate pedogenesis during the Wisconsinan in the Midwest and to place it in its proper perspective, the Sangamon Soil is reviewed and compared to the intra-Wisconsinan and Holocene soils.

Sangamon Soil

The Sangamon Soil was first used as a time-stratigraphic unit to separate the deposits of the Illinoian and Iowan (Wisconsinan) Stages of glaciation (Leverett, 1899). The Sangamon Soil has been found in many areas of North America where it developed on: (1) alluvial and loess deposits from the southern United States to the central Great Plains; (2) glacial deposits from the northern Great Plains through the midwestern United States to the northern (glaciated) Allegheny Plateau; (3) glacial deposits in isolated areas of southern Canada; and (4) glacial, alluvial, fluvial, and lacustrine deposits in parts of the western United States (Wright and Frey, 1965). Marine and shoreline deposits along the east coast of the United States and a variety of terrestrial deposits throughout the United States have been correlated with Sangamonian deposits. These correlations are tenuous because of the general lack of finite age determinations on deposits of this age and the inability to make detailed correlations with the Sangamon Soil in its type area of central Illinois (Figure 7-1).

The geomorphology of the type area is characterized by a nearly level till and glaciolacustrine plain that contains a distinct but subdued recessional moraine marking the limit of the youngest Illinoian till (Follmer et al., 1979). Parts of the terrain are highly dissected, giving rise to relief of 5 to 15 m over much of the area and up to about 65 m along the Illinois River valley. Former ice-walled features rise up to 60 m above the general level but are rare and generally are clustered or form linear features. Interior portions of the Illinoian landscape are poorly drained and have a poorly integrated or deranged drainage pattern. In poorly drained areas, relief is about 1 to 3 m over a gentle swell-and-swale topography with one to five swales per kilometer (Follmer, 1982).

The northeastern portion of the type area consists of late-Wisconsinan till and associated deposits that bury the Sangamon Soil to depths of 10 to 30 m. The late-Wisconsinan till here forms the terminal morainic system of the classic Wisconsinan Stage and is dated

The critical reviews and comments made on this chapter by W. H. Johnson, E. D. McKay, P. W. Birkeland, D. L. Gross, J. A. Lineback, J. P. Kempton, W. J. Morse, and R. C. Berg are gratefully acknowledged.

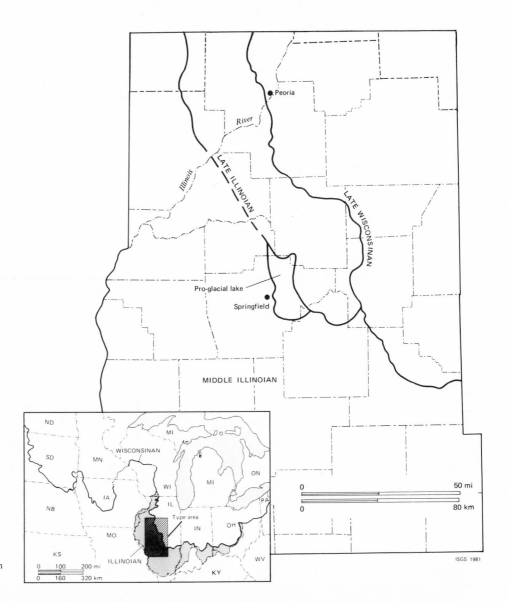

Figure 7-1. Type area of Sangamon Soil. (Modified from Follmer, 1978; inset map from Flint, 1971.)

at about 20,000 yr B.P. (Follmer et al., 1979). Beyond the morainic system, the Sangamon Soil is covered by Wisconsinan loess and accreted deposits. The youngest unit of loess covers the Wisconsinan drift, and the older loesses pass under it. In general, the Sangamon Soil is present everywhere except where it has been eroded by Wisconsinan meltwater or glaciers. Behind the terminal moraine in east-central Illinois, the Sangamon Soil is preserved below the Wisconsinan drift, but to the north the Sangamon has been eroded in most places covered by Wisconsinan drift (Horberg, 1953). Beyond the late-Wisconsinan glacial margin north of the type area, the Sangamon has been stripped from large portions of the upland by fluvial, mass-wasting, or periglacial processes (Follmer et al., 1978).

As a major stratigraphic feature, the Sangamon Soil in the area underlain by Illinoian deposits is widely recognized by interested laymen (e.g., well drillers) and earth scientists alike. But the criteria used for recognizing it have not been the same, nor have they been consistently applied. The main problem results from the time-transgressive nature of the Sangamon Soil (Follmer, 1978). The lower 1 to 2 m of the early-Wisconsinan loess, which includes the whole unit in many places, is altered and forms a pedologic continuum with the underlying Sangamon Soil developed in Sangamonian or older

deposits (Follmer, 1982; Follmer et al., 1979). Finite radiocarbon ages of the organic material interpreted by early investigations to be from the top of the Sangamon profile are in the range of about 42,000 to 25,000 yr B.P. Early studies did not recognize the early-Wisconsinan loess but included it in the Sangamon profile. Therefore, the interstadial development of organic deposits on the thin, early-Wisconsinan loess was once considered Sangamon, with respect to time stratigraphy, pedostratigraphy, and lithostratigraphy. Interpretations of the early-Wisconsinan loess in the Midwest continue to be a problem (Ruhe, 1976), because over much of the Midwest it is thin and altered. The exception is the area along the middle Mississippi River valley, where it reaches a maximum thickness of about 15 m (Willman and Frye, 1970).

Along the ancient Mississippi River valley in southwestern Illinois, the early-Wisconsinan loess contains color and mineral zones that can be traced for about 300 km (McKay, 1979; Willman and Frye, 1970). The thickest deposits are on the eastern side of the Mississippi River valley, and they rapidly thin in an easterly direction. Zonation fades away within 30 km, where the thickness decreases to less than 2 m and where pedogenic alteration is evident. At greater distances from the valley, where the loess is about a meter or less thick, it merges in-

139

to the "upper part" of the Sangamon Soil. Along other major rivers in the Midwest, the zonation has not been recognized, because the loess rarely is more than 2 m thick.

Radiocarbon ages of organic materials from within the zoned sequence of early-Wisconsinan loess range from 40,000 to 31,000 yr B.P. By extrapolation, the deposition of the main body of early-Wisconsinan loess began about 45,000 yr B.P. (McKay, 1979). Recent studies on the Sangamon Soil in its type locality described by Follmer and others (1979) support this conclusion. From an accretionary Sangamon A-horizon, the sodium hydroxide residue from carbonized wood and seeds, which combined with less than 10 μm matrix material to obtain sufficient carbon, yielded an age of 41,770 ± 1100 yr B.P. (ISGS-684). The humic acid extract yielded an age of 35,560 ± 900 yr B.P. (ISGS-688). Untreated humus from a weak soil (A/Bg) in loess 35 cm above the Sangamon surface yielded an age of 38,920 ± yr B.P. (ISGS-654).

Willman and Frye (1970) describe two minor loess deposits with soils (Chapin and Pleasant Grove, Figure 7-2) developed in each. These were interpreted to be stratigraphically above the Sangamon

Soil and below the main body of early-Wisconsinan loess. Controversial evidence for these earlier loess deposits exists in the thick loess areas, but they have not been dated and cannot be distinguished in areas in thin loess. Regardless of the age and number of increments, all of the early-Wisconsinan loess has been effectively incorporated into the upper horizons of the Sangamon Soil beyond about 30 to 50 km from the ancient Mississippi River valley. In a general sense, the Sangamon Soil continued its development into Wisconsinan time even though the pedologic conditions changed significantly (Follmer, 1978). The known changes in the soil-forming factors include partial burial and significant changes in climate and vegetation. An alternative interpretation recognizes the base of the Wisconsinan deposits as the top of the Sangamon Soil even though soil horizons developed upward and into the lower part of the Wisconsinan deposits (Follmer, 1982; Frye, Follmer, et al., 1974). The use of both concepts about the Sangamon Soil has caused correlation problems. Both concepts have utility, but the use of both causes confusion. In most stratigraphic contexts, assigning the top of the Sangamon Soil as coincident with the base of the Wisconsinan deposits is more practical, and so that approach is used in this chapter.

On the Illinoian till plain, the geomorphic surface of the Sangamon Soil is nearly parallel to the modern surface (Follmer, 1982). Because of the nature of loess deposition, the gentle swell-and-swale topography of the modern surface is a subdued reflection of the Sangamon surface. Localized redistribution of the Wisconsinan loesses by slope wash has contributed to the flatness of the Illinoian till plain. In undissected areas, most Sangamon profiles are poorly drained or gleyed. Two types of poorly drained profiles have been differentiated; one is developed in situ in nearly level glacial deposits; the other is developed in accreted or alluvial deposits during accumulation and is known as an accretion-gley (Willman and Frye, 1970).

The solum (A- and B-horizons) of the Sangamon Soil in its type area commonly ranges from 2 to 3 m thick (Follmer et al., 1979). All of the known parent materials of the Sangamon solum in the region are calcareous. Carbonate minerals are commonly leached only to the base of the solum, but leaching may extend 50 cm or more into the C-horizon. The depth of leaching in oxidized, permeable materials is often 4 to 5 m. Oxidized profiles have argillic horizons, and 50% to 75% of the weatherable minerals are depleted (Brophy, 1959; Willman et al., 1966). In contrast to in situ profiles, argillic horizons did not form in accretion-gley profiles. Mineral alteration is much lower in the poorly drained profiles, but the clay fraction is largely altered to smectite.

The pedologic classification of Sangamon Soil profiles is difficult because of the lack of information (about paleotemperature, paleochemistry, and paleohydrologic conditions) or the inability to determine the diagnostic properties required for making classifications in the schemes of modern soil taxonomy. On the basis of major horizons and degrees of mineral alteration, a general classification can be made for the common members of a catena (Table 7-1).

Post-Sangamonian Events and Soils in the Midwest

The post-Sangamonian interval as used here spans the last 75,000 years and includes the Wisconsinan and the Holocene. In terms of glacial events and soil formation in the central Midwest, it can be divided into five parts: (1) Altonian (early Wisconsinan), (2) Farm-

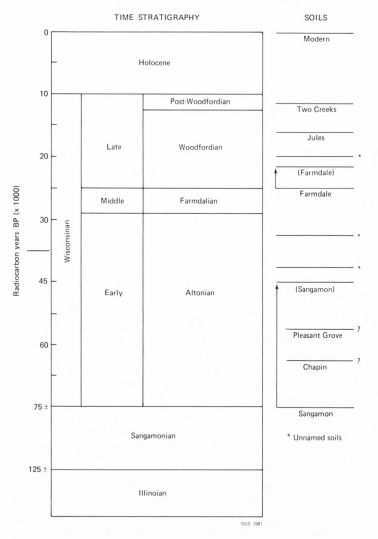

Figure 7-2. Stratigraphic terms used in this chapter. (Soil names are from Willman and Frye, 1970). Upward arrows indicate a continuation of pedogenesis that affects the age determination of buried soil surfaces.

Table 7-1.
Classification of the Sangamon Soil in Its Type Area

Drainage Class	1975 USDA Classification	1938 USDA Classification
Well to moderately well	Hapludult or Hapludalf	Red-Yellow Podzolic or Gray-Brown Podzolic
Somewhat poorly to poorly	Ochraqualf (ult?) or Albaqualf (ult?)	Gray-Brown Podzolic Planosol or Low Humic Gley
Very poorly	Cumulic Haplaquoll	Humic Gley

dalian (mid-Wisconsinan), (3) Woodfordian (classic Wisconsinan), (4) post-Woodfordian (deglaciation phase of late Wisconsinan), and (5) Holocene (Figure 7-2). A detailed time-stratigraphic classification for the Wisconsinan of the eastern Midwest is discussed by Dreimanis and Goldthwait (1973). A detailed worldwide summary of late-Wisconsinan glaciation has been published recently by Denton and Hughes (1981). The terminology used in this chapter is modified from Frye and Willman (1973).

ALTONIAN

The beginning of the Altonian is interpreted to be the time of general cooling following the Sangamonian Stage about 75,000 yr B.P. (Frye and Willman, 1973). Glacial activity is interpreted at this time from loess and till deposits found above the Sangamon Soil in Illinois (Frye and Willman, 1973) and in the eastern Midwest (Dreimanis and Goldthwait, 1973; Gooding, 1975). In thick Altonian loess deposits (Roxana Silt), the lower 1 to 2 m are pedogenically altered as shown by the leaching of carbonates, the relatively higher organic-carbon content, and the common manganese stains (Follmer et al., 1979). The ages of these depositional and pedogenic events are uncertain and controversial (Ruhe, 1976). In northern Illinois, a buried till with a well-developed soil (A/Bt/C profile) in the top has been described by Frye and others (1969) as middle Altonian, but no finite-age determinations have been made to substantiate this interpretation. Current stratigraphic studies suggest that this soil may be older than 75,000 yr B.P. (Follmer et al., 1978). A more nearly complete depositional and paleobotanical record of the early Wisconsinan has been found in the eastern Great Lakes area, where tills and lacustrine and organic deposits that indicate a cold or cool climate have been dated in the range of 67,000 to 28,000 yr B.P. (Berti, 1975; Dreimanis and Goldthwait, 1973). However, in this area few soils are described in sections with indisputable stratigraphic control. The organic deposits and the interstadial status they represent are sometimes interpreted as soils. In most studies, the pedologic characteristics of such deposits (even if they have any) have not been recognized or described. When these deposits are considered to be soils, they are much less developed, except for a few controversial profiles in Ohio, than the possible Altonian soil in northern Illinois. Few features of the early-Altonian record of the eastern Great Lakes area can be correlated to the rest of the Midwest (Johnson, 1976). During the latter part of the Altonian, from about 40,000 to 28,000 yr B.P., glacial events in Illinois are related to, but appear to be out of phase with, the events in Indiana and Ohio (Gooding, 1975) and in the eastern Great Lakes

area (Johnson, 1976). The strongest correlations between Illinois and the region to the east appear to be on the record of a glacial advance and loess deposition, which both culminated about 35,000 yr B.P. In summary, many organic soils or deposits with radiocarbon ages in the range of 67,000 to 28,000 yr B.P. have been found in the Midwest, but they do not have oxidized catena members associated with them.

FARMDALIAN

Farmdalian time in the Midwest is limited to an interval of geomorphic stability and soil formation before the advance of the Woodfordian glaciers (Willman and Frye, 1970). The climate was cool, as indicated by the pine and spruce remains found in organic deposits radiocarbon dated at about 28,000 to 25,000 yr B.P. in the type area (King, 1979). Soils formed at this time are better developed than the documented Altonian soils. The organic facies of all intra-Wisconsinan soils are similar except that those of the Farmdale Soil are often leached of carbonate minerals to greater depths. With the exception of a few controversial profiles, the Altonian soils do not have recognizable catena members, whereas the Farmdale Soil has a well-documented, oxidized member that is commonly found on early-Wisconsinan loess along the Illinois River valley and the lower Mississippi River valley (McKay, 1979). This oxidized member closely resembles Arctic Brown or Subarctic Brown Forest Soils (Cryochrepts). It has a cambic B-horizon and is leached from 1 to 3 m. The A-horizon is missing or has been highly altered by diagenesis.

In Ohio, a soil with age and characteristics similar to those of the Farmdale Soil has been described as the Sidney Soil by Forsyth (1965). At its type selection, the Sidney Soil is truncated to the B-horizon by a Woodfordian till. It appears to have a weak argillic (Bt) horizon developed in till leached to about 60 cm and oxidized to about 1.2 m. Good agreement exists on the age of the top of the Sidney Soil where it has been described in Ohio and Indiana, but the age of the underlying till and, consequently, the duration of the soil-forming interval are still in question (Gooding, 1975).

The end of the Farmdalian is marked by the beginning of Woodfordian loess deposition about 25,000 yr B.P. (McKay, 1979). Dolomitic (Peoria) loess overlies the leached Farmdale Soil in the thick-loess regions of the Midwest. In localized areas along the Mississippi River valley, organic zones that date between 23,900 and 20,900 yr B.P. occur in the lower third of the loess (McKay, 1979). These zones—which contain dolomite, wood, and muck—have been interpreted by some investigators as parts of the Farmdale Soil, but recent work by McKay (1979) shows that two soils can be distinguished. At several places described by McKay, the organic zone is separated from the Farmdale by a zone of dolomitic, nonorganic Woodfordian loess. The base of this loess defines the top of the Farmdale Soil, and the organic zone is considered to be a part of a younger soil. Where the lower loess zone is not recognized, the organic zone can be considered an upper horizon of the Farmdale Soil. This latter interpretation is supported by the pollen record in the region, which suggests that no change in the vegetational pattern occurred until the spruce-pine ratio increased significantly about 20,000 yr B.P. (King, 1979). A period of about 5000 years elapsed between the first appearance of glacially generated loess in the Mississippi River valley area and the attainment of the maximum extent of the Woodfordian glaciers in Illinois about 20,000 years ago. Approximately the basal third of the Woodfordian loess was overridden by the Woodfordian glacier along the glacial margin. In many places, the overridden loess (Morton) is relatively undisturbed. At a few locations in central Illinois, a calcareous, gleyed soil has been found in the loess, and a moss layer at the top of the loess at two sections has been dated at 19,680 ± 460 yr B.P. (ISGS-532) (Follmer et al., 1979) and 20,670 ± 280 (ISGS-828). These ages correlate with a soil under Woodfordian till in Iowa and with Connersville Interstade deposits between Woodfordian tills in Indiana and Ohio (Mayewski et al., 1981).

WOODFORDIAN

Soil formation during Woodfordian time in the Midwest was largely limited to the cold-climate type of organic-matter accumulation in poorly drained environments. Oxidized members apparently were largely inhibited or have been changed by postburial processes. Some very weakly developed, oxidized soils ("bands") have been described in the Woodfordian loess (Daniels et al., 1960; Frye and Leonard, 1965; Frye, Leonard, et al., 1974). Organic deposits, which presumably are poorly drained soils in accretionary environments, are described in Woodfordian till sequences (Gooding, 1975) but are limited in distribution except for the Connersville Interstade deposits in Indiana and Ohio (Mayewski et al., 1981). A reasonable interpretation for the lack of intra-Woodfordian soils is that most of the area was geomorphically unstable and did not allow soils to form. An alternative interpretation is that the soils were so weakly developed, except for organic-profiles, that they did not retain enough pedologic characteristics to be recognized later.

POST-WOODFORDIAN

The post-Woodfordian interval in the Midwest encompasses the deglaciation phase of the late Wisconsinan, in interval from about 12,500 to 10,000 yr B.P. Major landform construction and erosion occurred during the Woodfordian and continued while glaciers were still active in or north of the Great Lakes region. Soil formation was still largely inhibited at this time except in environments favorable for accumulation of organic matter and the development of a Bg-horizon, such as the classic Two Creeks forest bed in Wisconsin (Frye et al., 1965; Lee and Horn, 1972). No oxidized soil profiles have been found in association with the organic profiles of this age.

HOLOCENE

The Holocene record in the Midwest is well documented in terms of climatic and geologic events. The pedologic history, however, has not been resolved in much detail. Most pedologic studies in the United States focus on the classification and mapping of the Modern Soil, which includes a common assumption that the formation of the Modern Soil started about 12,000 years ago. Studies of bog stratigraphy and vegetational changes in the late Holocene show that a grassland environment gave way to an expansion of deciduous forests, particularly along stream valleys, during the last 4000 to 1000 years. The response of the soil to the expansion of the forest generally appears as a transitional zone parallel to the former forest-grassland border, where a bleached, eluvial horizon has formed in the A-horizon of the former grassland soil (relict Mollisol). The earlier vegetative influence has been masked by the continuing pedogenesis in the Holocene.

Comparison of Soil Development

In the Midwest, the Modern Soil on stable landforms with gentle slopes is well developed and is quite similar in degree of morphological development to the Sangamon Soil in similar positions. In similar parent materials on similar landforms, both soils have argillic horizons with B2t-horizons comparable in clay content, fabric, and structure. The main differences are that the Sangamon (1) is leached of carbonate minerals to greater depths and has thicker, lower horizons (B3, C1, and C2), (2) has a greater degree of mineral alteration, and (3) in nongleyed environments is always a forested type of soil profile. The Modern Soil and the Sangamon Soil in the Midwest are interglacial soils morphologically distinguishable from all other soils within the Wisconsinan deposits (Figure 7-3). The Farmdale Soil, which is best characterized in Illinois as a forested Cryochrept-Histosol catena, is probably a first-order interstadial soil in the sense described by Morrison (1978). The Farmdale and its stratigraphic equivalent, the Sidney, are clearly distinguishable from all of the other known soils in the Wisconsinan interval (about 75,000 to 10,000 yr B.P.). All of the formally named soils within the Wisconsinan are less developed than the Farmdale Soil and Sidney Soil and could be classified as second- or lower-order interstadial soils. In these soils, the horizon enriched in organic matter is the only part of the solum recognized in most studies. In fact, most studies of the interstadial organic zones describe them as deposits and do not consider whether the organic zone and the underlying deposits had pedologic characteristics. In general, solums in the oxidized facies, except for local occurrences in small areas, were not formed or preserved.

Summary of Significant Pedologic Events

The Wisconsinan pedologic record in the Midwest depends largely on the interpretation of the Sangamon Soil. The pedologic and geologic events that have contributed to the character of the Sangamon Soil are complex. Its profile appears to be polygenetic in most locations because (1) overthickened upper solum horizons occur in stable positions and (2) in sloping areas a complete profile is developed in slope sediments and the remnants of a former soil are common. In all situations, the A-horizon, where present, has been changed to some degree by diagenetic processes. Commonly, the Altonian loess thins to less than 1 m over large parts of the Midwest and eventually becomes unrecognizable as a discrete deposit. In these locations, the Farmdale Soil is superposed on the Sangamon Soil. The pedostratigraphic tops of each profile are separated by the pre-Farmdalian deposits, which are less than 1 m thick. Pedogenesis during the Farmdalian modified the Sangamon Soil, and in local situations the alterations effectively obliterated the pedostratigraphic boundary (Follmer, 1982). Except for the Farmdale Soil and Sidney Soil, very little evidence for synchronous events of pedogenesis within the Wisconsinan deposits has been found in the Midwest.

Conclusions

The main features of the Sangamonian and Wisconsinan record in the Midwest can be summarized in chronologic order as follows. (1) The Sangamon Soil formed in Sangamonian, Illinoian, or older deposits is nearly continuous across the region in front of and underlying the outer Woodfordian moraines. It is time transgressive, and detailed correlations are difficult to make. (2) The basis of the Wisconsinan record depends on the recognition and stratigraphic placement of the Sangamonon Soil. (3) General geomorphic stability prevailed during early-Altonian time, with minor glacial activity and localized loess deposition. Regional correlations of these events are poorly established. (4) A glacial advance and the deposition of loess during the late Altonian are correlated locally, but the glacial deposits are poory documented. Most tills of this age are overlain by Wood-fordian tills. (5) Formation of the Farmdale Soil is confined to an interval from around 28,000 to 25,000 yr B.P. in its type area. The Farmdale Soil and the correlative Sidney Soil are widely recognized, but their time-transgressive nature is a stratigraphic problem. (6) The advance of Woodfordian glaciers caused fluvial erosion that removed most Farmdalian and older fluvial deposits from functioning stream valleys. (7) Woodfordian glacial deposits and loess form or overlie most landforms of the present geomorphic surface. (8) The formation of the Modern Soil began when local geomorphic stability was achieved about 12,500 to 10,000 yr B.P.

In many places, Altonian and Farmdalian events are difficult to resolve from Sangamonian events because the rock-stratigraphic record is compressed and the pedostratigraphic units overlap. Only near the Illinois River valley and the lower Mississippi River valley does the Altonian loess reach an appreciable thickness (maximum about 15 m). There, the Farmdale Soil is stratigraphically separable and can be easily distinguished from the Sangamon and the lower-order soils locally found in the overlying and underlying loesses. Early Wisconsinan glacial deposits are reported in northern Illinois, the eastern Great Lake region, and other parts of the United States, but the relationships to the Sangamon Soil and Farmdale Soil are poorly known. The Late Wisconsinan events are dominated by glacier fluc-

tuations and unstable geomorphic conditions, which appear to have prohibited soil development except for organic-matter accumulation in favorable localized situations.

References

Berti, A. A. (1975). Paleobotany of Wisconsinan interstadials, eastern Great Lakes region, North America. *Quaternary Research* 5, 591-619.

Bowen, D. Q. (1978). "Quaternary Geology." Pergamon Press, Oxford, England.

Brophy, J. A. (1959). "Heavy Mineral Ratios of Sangamon Weathering Profiles in Illinois." Illinois Geological Survey Circular 273.

Daniels, R. B., Handy, R. L., and Simonson, G. H. (1960). Dark-colored bands in thick loess of western Iowa. *Journal of Geology* 68, 450-58.

Denton, G. H., and Hughes, T. J. (eds.) (1981). "The Last Great Ice Sheets." John Wiley and Sons, New York.

Dreimanis, A., and Goldthwait, R. P. (1973). "Wisconsin Glaciation in the Huron, Erie and Ontario Lobes." Geological Society of America Memoir 136, pp. 71-106.

Flint, R. F. (1971). "Glacial and Quaternary Geology." John Wiley and Sons, New York.

Follmer, L. R. (1978). The Sangamon Soil in its type area: A review. *In* "Quaternary Soils" (W. C. Mahaney, ed.), pp. 125-65. Geo Abstracts Limited, Norwich, England.

Follmer, L. R. (1982). The geomorphology of the Sangamon surface: Its spatial and temporal attributes. *In* "Spatial and temporal validity of Geomorphic Data" (C. E. Thorn, ed.), p. 9. Proceedings of the 12th Binghamton Geomorphology Symposium. Unwin and Allen, London.

Follmer, L. R., Berg, R. C., and Acker, L. L. (1978). "Soil geomorphology of northeastern Illinois." Guidebook for Joint Field Conference, Soil Science Society of America and Geological Society of America. Illinois Geological Survey, Urbana.

Follmer, L. R., McKay, E. D., Lineback, J. A., and Gross, D. L. (1979). "Wisconsinan, Sangamonian, and Illinois Stratigraphy of Central Illinois." Midwest Friends of the Pleistocene Field Conference. Illinois Geological Survey Guidebook 13.

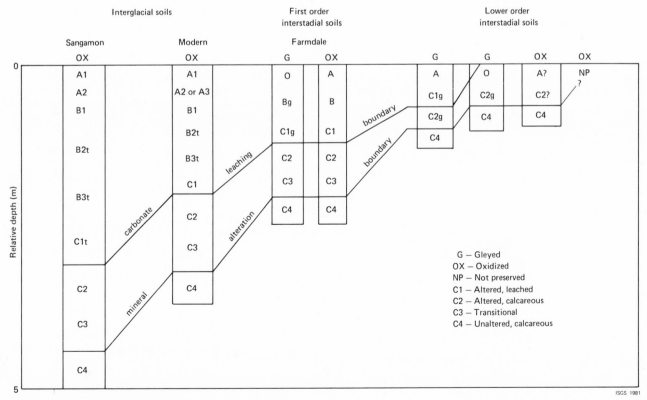

Figure 7-3. Generalized horizonation of interglacial and interstadial soils.

Forsyth, J. L. (1965). Age of the buried soil in the Sidney, Ohio, area. *American Journal of Science* 263, 571-97.

Frye, J. C., Follmer, L. R., Glass, H. D., Masters, J. M., and Willman, H. B. (1974). "Earliest Wisconsinan Sediments and Soils." Illinois Geological Survey Circular 485.

Frye, J. C., Glass, H. E., Kempton, J. P., and Willman, H. B. (1969). "Glacial Tills of northwestern Illinois." Illinois Geological Survey Circular 437.

Frye, J. C., and Leonard, A. B. (1965). Quaternary of the southern Great Plains. *In* "The Quaternary of the United States" (H. E. Wright, Jr., and D. G. Frey, eds.), pp. 203-16. Princeton University Press, Princeton, N.J.

Frye, J. C., Leonard, A. B., Willman, H. B., Glass, H. D., and Follmer, L. R. (1974). "The Late Woodfordian Jules Soil and Associated Molluscan Faunas." Illinois Geological Survey Circular 486.

Frye, J. C., and Willman, H. B. (1973). Wisconsinan climatic history interpreted from Lake Michigan lobe deposits and soils. *In* "The Wisconsinan Stage" (R. F. Black, R. P. Goldthwait, and H. B. Willman, eds.), pp. 135-52. Geological Society of America Memoir 136.

Frye, J. C., Willman, H. B., and Black, R. F. (1965). Outline of glacial geology of Illinois and Wisconsin. *In* "The Quaternary of the United States" (H. E. Wright, Jr., and D. G. Frey, eds.), pp. 43-61. Princeton University Press, Princeton, N.J.

Gooding, A. M. (1975). The Sidney Interstadial and late Wisconsin history in Indiana and Ohio. *American Journal of Science* 275, 993-1011.

Horberg, L. (1953). "Pleistocene Deposits below the Wisconsin Drift in Northeastern Illinois." Illinois Geological Survey Report of Investigations 165.

Johnson, W. H. (1976). Quaternary stratigraphy in Illinois: Status and current problems. *In* "Quaternary Stratigraphy of North America" (W. C. Mahaney, ed.), pp. 161-96. Dowden, Hutchinson, and Ross, Stroudsburg, Pa.

King, J. E. (1979). Pollen analysis of some Farmdalian and Woodfordian deposits, central Illinois. *In* "Wisconsinan, Sangamonian, and Illinoian Stratigraphy in Central Illinois," pp. 109-13. Midwest Friends of the Pleistocene Field Conference. Illinois Geological Survey Guidebook 13.

Lee, G. B., and Horn, M. E. (1972). Pedology of the Two Creeks section, Manitowoc County, Wisconsin. *Wisconsin Academy of Sciences, Arts and Letters* 60, 183-99.

Leverett, F. (1899). "The Illinois Glacial Lobe." U.S. Geological Survey Monograph 38.

McKay, E. D. (1979). Wisconsinan loess stratigraphy of Illinois. *In* "Wisconsinan, Sangamonian, and Illinoian Stratigraphy in Central Illinois," pp. 95-108. Midwest Friends of the Pleistocene Field Conference. Illinois Geological Survey Guidebook 13.

Mayewski, P. A., Denton, G. H., and Hughes, T. J. (1981). Late Wisconsin ice sheets of North America. *In* "The Last Great Ice Sheets" (G. H. Denton and T. J. Hughes, eds.), pp. 67-178. John Wiley and Sons, New York.

Morrison, R. B. (1978). Quaternary soil stratigraphy: Concepts, methods and problems. *In* "Quaternary Soils" (W. C. Mahaney, ed.), pp. 77-108. Geo Abstracts Limited, Norwich, England.

Reed, E. C., Dreeszen, V. H., Bayne, C. K., and Schultz, C. B. (1965). The Pleistocene in Nebraska and northern Kansas. *In* "The Quaternary of the United States" (H. E. Wright, Jr., and D. G. Frey, eds.), pp. 187-202. Princeton University Press, Princeton, N.J.

Ruhe, R. V. (1956). Geomorphic surfaces and the nature of soils. *Soil Science* 82, 441-55.

Ruhe, R. V. (1974). Sangamon paleosols and Quaternary environments in midwestern United States. *In* "Quaternary Environments" (W. C. Mahaney, ed.), pp. 153-67. York University Geographical Monographs 5.

Ruhe, R. V. (1976). Stratigraphy of mid-continental loess, U.S.A. *In* "Quaternary Stratigraphy of North America" (W. C. Mahaney, ed.), pp. 197-211. Dowden, Hutchinson, and Ross, Stroudsburg, Pa.

Willman, H. B., and Frye, J. C. (1970). "Pleistocene Stratigraphy of Illinois." Illinois Geological Survey Bulletin 94.

Willman, H. B., Glass, H. D., and Frye, J. C. (1966). "Mineralogy of Glacial Tills and Their Weathering Profiles in Illinois: Part II. Weathering Profiles." Illinois Geological Survey Circular 400.

Wright, H. E., Jr., and Frey, D. G. (eds.) (1965). "The Quaternary of the United States." Princeton University Press, Princeton, N.J.

CHAPTER 8

Trends in Late-Quaternary Soil Development in the Rocky Mountains and Sierra Nevada of the Western United States

Ralph R. Shroba and Peter W. Birkeland

Introduction

This chapter focuses on trends in the development of the physical and chemical properties of late-Quaternary soils in several mountain ranges of the western United States (Figure 8-1). The parent materials of these soils are coarse grained and generally are of glacial, glaciofluvial, or periglacial origin. Other soils considered here were formed in fluvial deposits along a mountain-to-plains transect in climate that range from alpine to semiarid. In studies of soil development as a function of time and in the evaluation of the possible in-

fluence of past changes in climate and vegetation, soils of a wide age range must be compared. Those soils discussed here range from about 100 years old to more than 100,000 years old; however, the ages of most of these soils are only broad estimates. Our pedologic interpretations are based on a variety of field and laboratory data; some of the morphological, textural, and mineralogical properties of soils of the Rocky Mountain region and selected chemical properties of soils of the Sierra Nevada are emphasized.

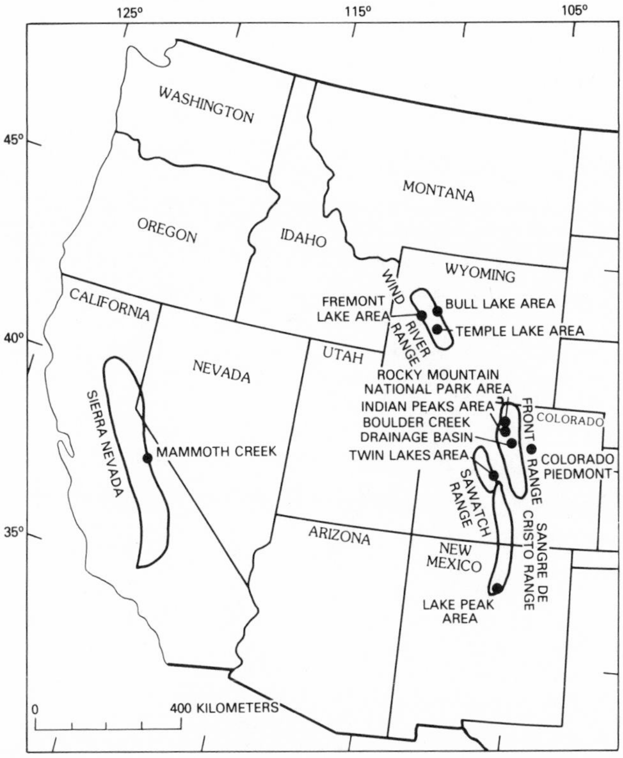

Soil-Horizon Nomenclature

The soil-horizon nomenclature used in this chapter follows that used by the United States Department of Agriculture (Soil Survey Staff, 1975), with some modifications. Soils in mountainous areas of the western United States commonly have A/B/C or A/C profiles. The most common surface horizon is the A1-horizon, a mineral horizon enriched in organic matter. Locally, it is overlain by a thin (generally less than 10 cm) layer of organic debris, the O-horizon. In some forest environments, a light-colored mineral horizon, the A2-horizon, lies beneath the A1-horizon.

The most common horizon beneath the A-horizon is the B-horizon. Two types of B-horizons are commonly recognized. Some B-horizons are characterized by moderate alteration of the parent material without a marked increase in clay content; these are informally referred to as Bs-horizons. They are equivalent to the color B-horizons of some investigations; many of them would meet all but the textural requirements of the cambic horizon (Soil Survey Staff, 1975). Other B-horizons are distinguished by a distinct increase in clay content relative to either the overlying A-horizon or the underlying C-horizon. At least part of the clay increase is considered to be the result of the downward translocation of clay from the A-horizon. These B-horizons are designated Bt-horizons; many would qualify as argillic horizons (Soil Survey Staff, 1975).

Beneath the A-horizon or the B-horizon (when it is present) is a

Figure 8-1. Map of the western United States showing soil localities mentioned in the text or cited in the tables.

We are grateful to S. M. Colman and R. F. Madole of the U. S. Geological Survey for their critical reviews of the manuscript for this chapter. The work in the Sierra Nevada was partially supported by the National Science Foundation, Grant EAR76-81241, and the U. S. Geological Survey.

slightly altered C-horizon. The term "Cox" is used to indicate slightly oxidized C-horizons that do not meet the color requirements for a B-horizon, whereas "Cn" denotes unoxidized or unaltered parent material (Birkeland, 1974). In arid and semiarid regions, secondary carbonate commonly accumulates within the C-horizon. C-horizons with secondary carbonate enrichment are referred to as Cca-horizons; the carbonate morphology of these horizons is either stage I or II of Gile and others (1966) for soils about 150,000 years old or less. Stage I morphology generally consists of thin, discontinuous coatings on pebbles in gravelly soils or faint coatings on sand grains or a few thin filaments in nongravelly soils. Stage II morphology is distinguished by continuous pebble coatings and a calcareous but loose matrix in gravelly soils or by few or numerous nodules of varying hardness and a calcareous matrix in nongravelly soils.

Soils in the Rocky Mountains and Sierra Nevada are commonly formed in nonuniform or layered parent materials, some of which are characterized by contrasting gravel content and/or texture of the fraction less than 2 mm. In some soils, especially those about 100,000 years old or older, it is often difficult to distinguish features resulting primarily from variations in the parent material from those resulting mainly from pedogenic processes. Soils with horizons developed from different materials are prefixed by Roman numerals, starting with II for the second parent material below the surface.

Holocene Soils in Alpine Areas of the Southern and Central Rocky Mountains

Soils formed in Holocene cirque deposits in areas above the present treeline from northern New Mexico to central Wyoming display systematic changes in morphology and degree of development with the increasing age of the parent material (Birkeland and Shroba, 1974; Mahaney, 1974, 1978; Mahaney and Fahey, 1976; Miller and Birkeland, 1974; Shroba, 1977). These pedologic changes are partially time dependent, but they also reflect the influence of eolian deposition and past changes in climate and vegetation. Soil parent materials include tills, rock-glacier deposits, and talus (Table 8-1) that are about 0 to 300, 1000 to 2000, 3000 to 5000, or 10,000 years old (Birkeland and Shroba, 1974). Tills and periglacial deposits of middle-Holocene age and older are overlain by a thin mantle of loess or by pebbly, silty, and clayey sand. The latter appears to be loess that is mixed with the underlying material; therefore, it is referred to as mixed loess. These fine-grained surface mantles tend to be thicker, to contain more silt and clay, and to be more widespread on deposits about 10,000 years old. Evidence from the Colorado Front Range suggests that the main episode of Holocene loess deposition in alpine areas occurred about 5000 to 6000 years ago during the latter part of the Altithermal (Benedict, 1973).

Soil properties that indicate progressive development with the increasing age of the parent material are (1) sequence and thickness of genetic horizons, (2) increase in clay content of the B-horizon, (3) depth of oxidation and color of the B- and C-horizons, and (4) degree of clay-mineral alteration. The amount of organic matter in surface horizons is less diagnostic as an index of the age of the soil.

A comparison of the sequence and thickness of soil horizons formed in deposits of different ages shows trends in pedogenesis with time. Most deposits 0 to 300 years old are unoxidized. The younger deposits within this range lack pedologic features, whereas the older deposits have minimal soils with A1/Cn profiles (Figure 8-2). Soils about 1000 to 2000 years old have A1/Cox/Cn profiles; they are easily distinguished from younger and older soils by the presence of Cox-horizons less than half a meter thick. Younger soils do not have a

Table 8-1.
Environmental Data for Holocene Soils in Alpine Areas, Southern and Central Rocky Mountains.

Locality	Parent Material		Vegetation	Climate		References
	Deposits	Dominant Lithologies		Mean Annual Temperature (°C)	Mean Annual Precipitation (cm)	
Lake Peak area, Sangre de Cristo Range, New Mexico	Talus and rock-glacier deposits mantled by 20-25 cm of loess or mixed loess	Granite	Tundra	0	70	Shroba, 1977
Twin Lakes area, Sawatch Range, Colorado	Talus, rock-glacier deposits, and tills mantled by 0-20 cm of loess or mixed loess	Biotite gneiss; some granite and granodiorite	Tundra; open spruce forest at treeline	−2 to −4	>70	Shroba, 1977
Rocky Mountain National Park and Indian Peaks areas, Front Range, Colorado	Tills and rock-glacier deposits mantled by 0-10 cm of loess or 0-25 cm of mixed loess	Granite and some biotite gneiss in park area; biotite schist and hornblende gneiss in Indian Peaks area	Tundra	−1 to −2	100	Birkeland and Shroba, 1974; Shroba, 1977
Temple Lake area, Wind River Range, Wyoming	Tills mantled by 0-10 cm of mixed loess	Biotite gneiss, granodiorite, and diorite	Tundra	−4	No data, possibly 70 to 100	Birkeland and Shroba, 1974; Miller and Birkeland, 1974

Cox-horizon, and older soils have thicker Cox-horizons (Figure 8-2). Soils about 3000 to 5000 years old have either A1/IICox/IICn or A1/IIBs/IICox/IICn profiles. Bs-horizons are slightly redder than the underlying Cox-horizons. Soils about 10,000 years old commonly have A1/IIBt/IICox profiles; the Bt-horizons are noticeably redder and more clayey but slightly thinner than the Bs-horizons of younger soils.

Soil-morphological data suggest different rates of development for the various horizons. A1-horizons can form in less than 300 years, whereas Bt-horizons generally form in about 5000 to 10,000 years. Other horizons have intermediate rates of development. Cox-horizons form within 1000 to 2000 years, and most Bs-horizons form in about 3000 to 5000 years.

The thickness of horizons in alpine soils shows several trends. A1-and Cox-horizons increase in thickness with time; Cox-horizons thicken much faster than A1-horizons (Figure 8-2). In contrast, the thickness of B-horizons does not change with the age of the parent material.

Grain-size data indicate noticeable increases in the amount of clay and silt with increasing age of the soil. This relationship, however, is not solely a result of time. Much of the variation in texture with increasing age is attributable to surface mantles that are composed partially or entirely of loess. Such surface mantles are not present in soil less than 2000 years old (Figure 8-2). The amount of clay and silt in these younger soils is low (generally less than 10% clay and less than 15% silt) and is nearly uniform with depth (Figure 8-3). Soils more than 2000 years old are typically formed in thin (5-40 cm) surface mantles of loess or mixed loess and in the underlying material (Figures 8-2 and 8-3). Horizons developed in the surface mantle (the A1-horizon and in some soils the upper part of the B-horizon) are more clayey and silty than those formed in the underlying parent material. The abundance of material less than 50 μm in diameter in surface horizons is due largely to the original grain size of the eolian

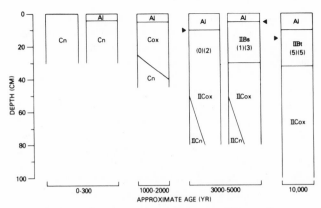

Figure 8-2. Generalized profiles of alpine soils formed in Holocene tills, periglacial deposits, and loess mantles in the southern and central Rocky Mountains. Triangular markers denote average depth to the base of the loess or mixed loess. Numbers in parentheses indicate the average increase in the percentage of clay (left) and the percentage of silt (right) relative to those in the lowest part of the profile. (Data from Birkeland and Shroba, 1974, and Shroba, 1977).

component. Soils that are about 3000 to 5000 years old and that lack B-horizons have a nearly uniform clay profile below the clay maximum in the A1-horizon, whereas soils of equivalent age with Bs-horizons have as much as a 2% increase in clay in the Bs-horizon compared to material at depth. Soils about 10,000 years old typically have Bt-horizons that contain 3% to 10% (average 5%) more clay than the underlying Cox-horizon, indicating that clay has formed in the Bt-horizon and/or has been translocated from above. Silt also accumulates in B-horizons. Typically, it increases upward from the lower part of the Cox-horizon into the B-horizon (Figure 8-3). Silt enrichment of about 1% to 10% is common in Bt-horizons compared to the underlying horizons; about half this much occurs in Bs-

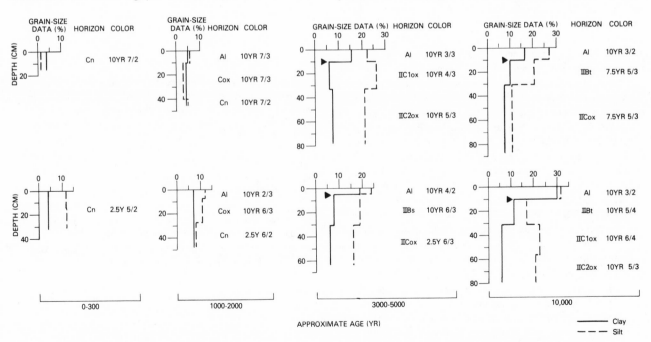

Figure 8-3. Grain-size data and dry Munsell colors of the fraction less than 2 mm of selected alpine soils formed in Holocene tills, periglacial deposits, and loess mantles in the Front Range and Sawatch Range of Colorado. Triangular markers indicate the boundary between the surface mantle and the underlying till or periglacial deposit. (Data from Shroba, 1977.)

horizons and in the upper Cox-horizons of younger soils that lack B-horizons (Figure 8-3). Although some of the textural development observed in these soils may be the result of weathering, our present interpretation is that much of the increase in clay and silt content of these B-horizons is caused by the translocation of eolian fines from the overlying surface horizon.

The depth of oxidation and the hue and/or chroma of the Cox- and B-horizons increase with the age of the parent material (Figures 8-2 and 8-3). Oxidation increases from zero in 300 years or less to greater than 100 cm in 10,000 years. Unoxidized parent materials have hues of 5Y, 2.5Y, or 10YR and chromas of 1 or 2. The colors of Cox-horizons of soils about 1000 to 5000 years old are typically 2.5Y 6/3 or 10 YR 7/3, whereas those of soils about 10,000 years old are commonly 10YR 6/3, 10YR 7/4, or 7.5YR 5/3. The Bs-horizons of soils about 3000 to 5000 years old have redder hues and/or higher chromas than do underlying Cox-horizons, and they have colors that are generally 10YR 6/3, 10YR 6/4, or 10YR 7/4. The Bt-horizons of soils about 10,000 years old are typically 10YR 5/4 or 7.5YR 5/4 to 7.5YR7/4 in color, although some are as red as 7.5YR 6/6. Color data for soils of different ages suggest that horizons with a 10YR hue and a chroma of 3 can form in 300 to 2000 years, that those with a 10YR hue and a chroma of 4 form in about 2000 to 5000 years, and that those with a 7.5YR hue and chromas of 4 to 6 form in about 5000 to 10,000 years.

X-ray-diffraction data for Holocene alpine soils in the Colorado Rockies reveal that the amount and depth of clay-mineral alteration and the formation of secondary clays tend to increase with time. The position of soils with respect to the present treeline also has a strong effect on clay mineralogy. In alpine areas 100 m or more above the present treeline, mica is more weathered and is slightly less abundant in soils more than 2000 years old than it is in younger soils (Figure 8-4). Mixed-layer clays (10-18 Å) are present in larger quantities and occur to a greater depth in the older soils (Figure 8-5). The nearly uniform content of kaolinite and mixed-layer illite-montmorillonite with depth in all soils suggests that these clay minerals are stable in high-alpine soils. Clay-mineral alteration is most extensive in soils about 10,000 years old in alpine areas at or less than 100 m above the present treeline (Figures 8-4 and 8-5). In these soils, mica is more weathered and is less abundant and secondary clays are present in greater amounts than in soils of similar age at higher elevations. A comparison of the relative amounts of clay minerals at various depths in 10,000-year-old soils near the present treeline suggests that kaolinite is stable whereas mixed-layer illite-montmorillonite and mica are unstable and are transformed to mixed-layer clays (10-18 Å), vermiculite, and hydrobiotite. Forest soils and nearby alpine soils within about 100 m above the present treeline are characterized by the presence of vermiculite and hydrobiotite, large amounts of mixed-layer clays (10-18 Å), and extensive weathering of mica and mixed-layer illite-montmorillonite, whereas alpine soils more than 100 m above the present treeline have no vermiculite or hydrobiotite, small amounts of mixed-layer clays (10-18 Å), and limited weathering of mica and mixed-layer illite-montmorillonite.

These clay-mineral trends suggest that the treeline was as much as 100 m higher at some time during the Holocene than it is at present. Although macrofossil evidence in the Colorado Front Range indicates that the treeline was at or above its present position by 9200 years ago (Benedict, 1973), the maximum expansion of the forest into alpine areas may have occurred during the middle Holocene, presumably in response to climatic conditions generally considered to

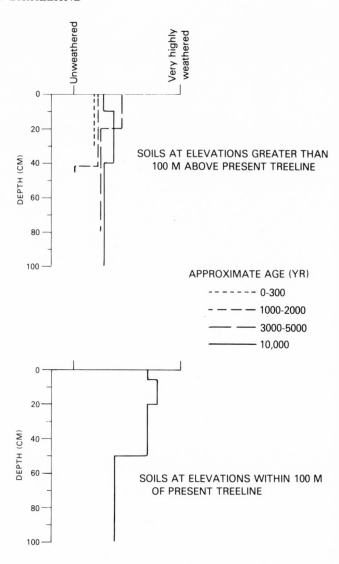

Figure 8-4. Relative degree of mica weathering in alpine soils formed in Holocene deposits in the Front Range and Sawatch Range of Colorado. Mica with a very sharp 10 Å diffraction peak after treatment with ethylene glycol is considered to be unweathered, whereas mica with a very subdued 10 Å peak or no 10 Å peak after treatment with ethylene glycol but a very sharp 10 Å peak after heating at 550°C is considered to be very highly weathered. Mica with intermediate peak characteristics are plotted between "unweathered" and "very highly weathered," depending on the relative amount of muting of the 10 Å peak. (Data from Shroba, 1977.)

be warmer than those of the present. These clay-mineral data are consistent with inferred changes in Holocene vegetation derived from pollen data for bog and lake deposits in subalpine areas of the Colorado Front Range (Maher, 1972; Pennack, 1963).

Late-Pleistocene Soils in Montane and Semiarid Areas of the Southern and Central Rocky Mountains

Soils formed in late-Pleistocene tills under montane (coniferous forest) and semiarid (sagebrush) conditions in the Rocky Mountains

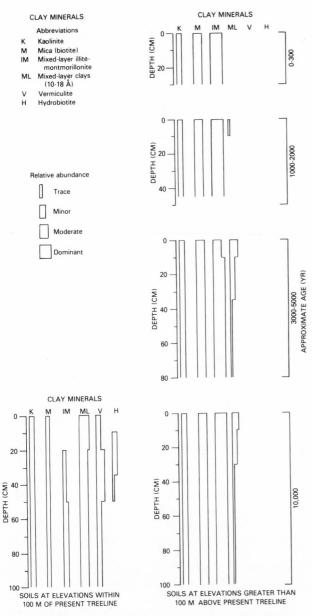

Figure 8-5. Relative abundances of clay minerals in alpine soils formed in Holocene deposits in the Front Range and Sawatch Range of Colorado. Some of the unoxidized parent materials and some of the soils at higher elevations contain trace or minor amounts of detrital chlorite. (Data from Shroba, 1977.)

age of the parent material are those related to B-horizon development. These properties include thickness, increased clay content, structural development, and thickness and abundance of clay films. Soils about 10,000 to 30,000 years old vary in degree of B-horizon development (Figure 8-6). These soils have Bs-or weak Bt-horizons or Cox-horizons in the B-horizon position. B-horizons are typically 10 to 40 cm thick and contain 1% to 7% more clay than unweathered till. The average thickness and average increase in clay are about 25 cm and 1%, respectively, for Bs-horizons, and about 25 cm and 5%, respectively, for Bt-horizons. Cox-horizons of soils that lack B-horizons have less than a 1% increase in clay. Peds in both Bs- and Bt-horizons are weakly aggregated and lack clay films.

Soils about 150,000 years old are characterized by Bt-horizons that are more strongly developed than those of the younger soils (Figure 8-6). The Bt-horizons of these soils are commonly 35 to 75 cm thick and contain 5% to 13% more clay than does the parent material. The average horizon thickness and average increase in the amount of clay are 55 cm and 9%, respectively. Peds in the Bt-horizons of these soils generally have weak or moderate subangular blocky structure and a few thin clay films (Madole and Shroba, 1979; Shroba, 1977).

Much of the variation in the amount of clay accumulation and B-horizon development in soils about 10,000 to 30,000 years old may reflect differences in the amount of infiltration of eolian material and in climate and vegetation rather than differences in ages of the tills. Bt-horizons are most common in semiarid areas, especially in areas where thin, discontinuous deposits of mixed loess are present. These horizons are less common in forested areas. In the Pine/Douglas-fir forest, Cox-horizons are typically present in the B-horizon position; at higher elevations in the spruce-subalpine fir forest, Bs-horizons are more common than Bt-horizons. In contrast to the younger soils, the similar degree of B-horizon development among soils about 150,000 years old suggests that, with increasing age of the parent material, time-dependent processes play a more important role in clay accumulation and B-horizon development and that those processes tend to mask the effects of local differences in environmental factors.

The color of the B-horizon depends partially on the age of the soil, but it also varies with the climate and vegetation. In semiarid areas, the B-horizons formed in till are similar in color regardless of age; in montane areas, the B-horizons of older soils have redder hues and/or higher chromas than those of the younger soils (Figure 8-6). Moreover, the B-horizons are redder and differences in color between the B-horizons of soils 10,000 to 30,000 years old and 150,000 years old are greater in the spruce-subalpine fir forest than in the pine/ Douglas-fir forest. Variations with altitude in the color of B-horizons in soils of similar age may be caused partially by different rates and/or intensities of chemical weathering resulting from altitudinal differences in climate and vegetation (Shroba, 1977).

In semiarid areas, the occurrence, amount, and morphology of secondary calcium carbonate are proportional to the age and carbonate content of the till. The position of the soils with respect to the lower treeline also has a significant effect. Cca-horizons are widespread in calcareous tills of the last two major glaciations where the tills are well below the lower treeline. Cca-horizons in soils about 150,000 years old contain more secondary carbonate and have a more strongly developed morphology than do soils about 10,000 to 30,000 years old (Figure 8-6). The Cca-horizons of the younger soils are about 40-80 cm thick and contain as much as 10% more carbonate than the parent material. The older soils have Cca-horizons that are about 60 to 100 cm thick and are composed of about 12% to 25%

vary considerably in morphology and degree of development (Madole and Shroba, 1979; Mahaney, 1978; Mahaney and Fahey, 1976; Pierce, 1979; Richmond, 1965; Shroba, 1977). Much of this variation can be attributed to the age of the parent material, although some is also due to environmental factors such as vegetation and climate (Table 8-2). The parent materials were deposited during the last two major glaciations (Pinedale and Bull Lake); although some controversy exists about their ages, they are considered in this chapter to be about 10,000 to 30,000 (Madole, 1976, 1980a, 1980b; Pierce, 1979) and 150,000 (Pierce, 1979) years old, respectively.

The pedologic properties that are most closely associated with the

Table 8-2.
Environmental Data for Late-Pleistocene Soils in Montane and Semiarid Areas, Southern and Central Rocky Mountains

Locality	Parent Material		Vegetation	Climate		Principal References
	Deposits	Dominant Lithologies		Mean Annual Temperature (°C)	Mean Annual Precipitation (cm)	
Lake Peak area, Sangre de Cristo Range, New Mexico	Till and rock-glacier deposits mantled by 30-40 cm of loess or mixed loess	Granite	Spruce-subalpine fir forest	0 to 1	70 to 80	Shroba, 1977
Twin Lakes area, Sawatch Range, Colorado	Till	Granodiorite, biotite gneiss, and granite	Spruce-subalpine fir forest, pine/Douglas-fir forest, and sagebrush	−3 to 2	25 to 90	Shroba, 1977
Rocky Mountain National Park area, Front Range, Colorado	Till mantled by 0-5 cm of mixed loess	Biotite gneiss and granite	Spruce-subalpine fir forest, pine/Douglas-fir forest	1 to 6	40 to 80	Shroba, 1977; Madole and Shroba, 1979
Fremont Lake and Bull Lake areas, Wind River Range, Wyoming	Till mantled by 0-20 cm of mixed loess	Biotite gneiss, granite to quartz diorite, and sedimentary rocks	Sagebrush	2 to 6	20 to 30	Shroba, 1977

more carbonate than the underlying till. Secondary carbonate in the younger soils commonly has stage I morphology that consists of thin films on the bottoms of stones as well as grain coatings and minor interstitial fillings in the matrix. In contrast, the older soils commonly have stage II morphology characterized by continuous coatings about 1 to 3 mm thick on the bottoms and sides of stones and abundant fine-grained carbonate that imparts a light gray or white color to much of the matrix (Shroba, 1977).

The clay mineralogy of soils formed in late-Pleistocene tills at lower elevations shows a close relationship to present climate and vegetation, although at higher elevations the soil-clay mineralogy also correlates with the age of the parent material. In areas with sagebrush vegetation and in the pine/Douglas-fir forest, primary clay minerals are only slightly altered and are similar to those in the underlying till. In the spruce-subalpine fir forest, however, some of the primary clay minerals are extensively altered, and the amount and depth of alteration increases with age (Figure 8-7). Under spruce and subalpine fir, mica and mixed-layer illite-montmorillonite are more highly weathered in soils that are about 150,000 years old than in those about 10,000 to 30,000 years old. Also, secondary clay minerals such as vermiculite and mixed-layer clays (10-18 Å) are slightly more abundant in the older soils.

A comparison of clay-mineral assemblages for soils from both age-groups in areas of different climate and vegetation suggest the following sequence of relative stability: kaolinite much greater than mica greater than mixed-layer illite-montmorillonite much greater than chlorite. Much of the kaolinite is inherited from the parent materials and is relatively stable in soil environments ranging from semiarid to alpine. Vermiculite, mixed-layer clays (10-18 Å), and hydrobiotite form as secondary clay minerals by the alteration of the primary clay

minerals mica, mixed-layer illite-montmorillonite, and/or chlorite. These three primary clay minerals are unstable in most soils with the exception of carbonate-rich horizons formed in calcareous tills. Vermiculite and hydrobiotite are diagnostic weathering products of primary clay minerals in forest soils; these secondary clay minerals are absent from semiarid soils (Shroba, 1977).

Soils near present vegetational boundaries in montane and semiarid areas have similar clay-mineral assemblages and show no mineralogical indication of past changes in climate and vegetation. However, other geologic and botanical data suggest major changes in environmental conditions in the Rocky Mountains during the late Quaternary (e.g., Legg and Baker, 1980; Markgraf and Scott, 1981). Clay-mineral data for soils below the present upper treeline indicate that, over the past 100,000 years, either changes in climate and vegetation were too minor and/or existed for too short a duration to produce detectable changes in soil-clay mineralogy or any diagnostic clay minerals that formed under previous environmental conditions were unstable under subsequent conditions and were not preserved.

A Transect from the Southern Rocky Mountains to the Colorado Piedmont

The major streams of the Front Range in the southern Rocky Mountains, many of which were glaciated in their headwaters, descend eastward from the Continental Divide and flow across the Colorado Piedmont to the Great Plains. Terrace remnants parallel the rivers in the mountains, and flights of terraces are discernible for short distances east of the mountain front. Farther east, all but the major terraces are concealed by eolian deposits. The stratigraphy of the sur-

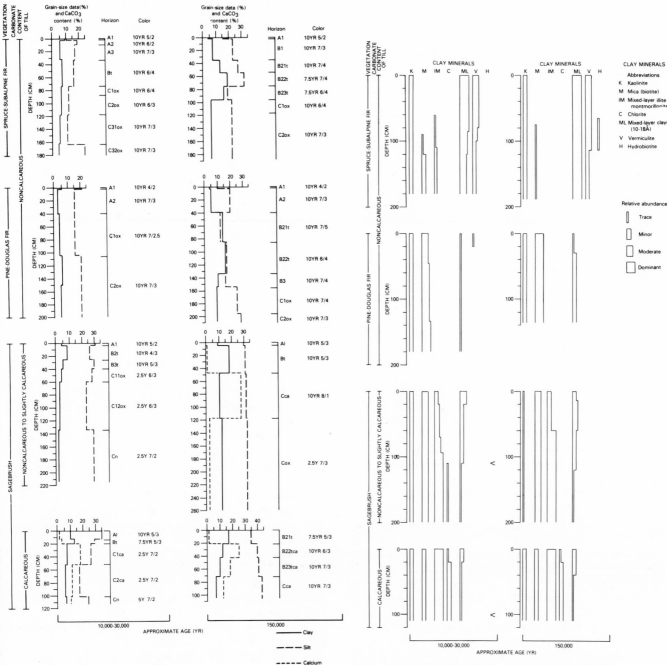

Figure 8-6. Grain-size data, calcium carbonate content, and dry Munsell colors of the fraction less than 2 mm of selected montane and semiarid soils formed in late-Pleistocene tills in the Front Range of Colorado and the Wind River Range of Wyoming. (Data from Shroba, 1977, and Madole and Shroba, 1979.)

Figure 8-7. Relative abundance of clay minerals in soils formed in late-Pleistocene tills in the Sawatch Range and Front Range of Colorado and the Wind River Range of Wyoming. Unoxidized till in the pine-Douglas-fir forest lacks mixed-layer illite-montmorillonite and chlorite. Arrowhead-shaped markers indicate depth to unoxidized till if it is present in soil pit. (Data from Shroba, 1977.)

ficial deposits and soils of this area has received much attention. Scott (1963), in particular, made extensive use of soils in formulating the current stratigraphic framework. Recent quantitative soil-geologic investigations in the region include those of Baker (1973), Machette (1975), Van Horn (1976), and Netoff (1977). Szabo (1980) discussed the ages of some of these deposits.

Netoff (1977) provides generalized environmental data for the transect. Parent materials of interest here are fluvial sands and gravels composed primarily of granitic detritus; in places, a thin surface deposit of loess is present. Vegetation varies from mixed coniferous forest in the mountains to short-grass prairie in the plains; these vegetational differences reflect the variation in climate. The mean an-

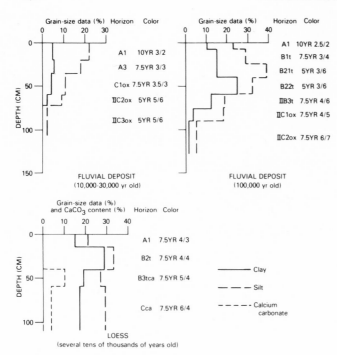

Figure 8-8. Grain-size data, calcium carbonate content, and Munsell colors for soils formed in late-Pleistocene fluvial deposits and loess in the Colorado Piedmont. (Data from Machette, 1975.)

nual temperature ranges from 100 cm in the mountains to 30 cm in the plains; the respective mean annual temperatures range from 0° to 10°C.

In the Colorado Piedmont, profile development is clearly related to age as long as the soils are developed in parent materials of similar grain size (Machette, 1975). Soils formed in coarse-grained fluvial deposits about 10,000 to 30,000 years old are characterized by A/Cox profiles that have hues as red as 7.5YR to 5YR; soils formed in alluvium in the 100,000-year-old age range have similar hues as well as Bt-horizons (Figure 8-8). In most places, however, 10YR hues are more common in the younger soils and 7.5YR hues in the older soils. Soils formed from coarse-grained parent materials usually do not contain pedogenic carbonate. In contrast, fine-grained fluvial deposits have soils with Cca-horizons. The younger fine-grained soils have pedogenic accumulations of as much as 3% carbonate with stage I morphology, whereas the older ones have as much as 8% carbonate with stage II morphology.

Loess is present locally in the Piedmont. The degree of soil development and the stratigraphic relations suggest that the loess is closer in age to the older of the two alluvial units discussed here. Soils developed in the loess have Bt-horizons with 7.5YR hues and Cca-horizons with as much as 10% pedogenic carbonate and stage II morphology (Figure 8-8) (Machette, 1975; Reheis, 1980).

Netoff (1977) has traced river-terrace remnants from moraines in the mountains eastward into the Piedmont, and he recognizes relationships between soil development and age that are similar to those just described. In addition, he demonstrates that clay mineralogy relates to the combined factors of climate and vegetation. Several clay minerals are present in the various soils along the transect; the abundance and distribution of some of them are not readily explained. The major exceptions are vermiculite, which is restricted primarily to the coniferous forests in the mountains, and smectite, which is more common in the drier soils under grassland vegetation in the plains (Figure 8-9). None of the clay-mineral trends can be ascribed to the ages of the soils.

The data for soils about 10,000 to 100,000 years old, as well as those for older soils, can be used to argue against marked long-term changes in soil-moisture conditions in the Piedmont. Most of the data on clay mineralogy, carbonate morphology, and distance from the mountain front at which pedogenic carbonate first appears in the soils seem to require soil-leaching conditions similar to those of the present. Minor climatic fluctuations could account for some of the variation in clay mineralogy at specific sites or for the overlap between the base of the Bt-horizon and the top of the Cca-horizon that is observed in some of the soils east of the mountain front.

A Transect on the Eastern Side of the Sierra Nevada, California

Soils formed in tills along a transect from cirques of the eastern Sierra Nevada (about 3300 to 3600 m in altitude) to moraines at the base of the eastern escarpment (about 2200 to 2500 m in altitude) show trends with both age and altitude. The tills are derived mainly from granitic bedrock. The mean annual temperature ranges from 3°C in the alpine areas to 8°C in the sagebrush areas, and the corresponding mean annual precipitation ranges from 60 to 40 cm (Burke and Birkeland, 1979; Curry, 1969); summers are dry, and much of the precipitation falls during the winter. The vegetation reflects the climatic differences. Sparse alpine vegetation occurs on the cirque deposits, and sagebrush with scattered pine and other conifers occurs on moraines at low elevations. A narrow band of coniferous forest separates the alpine and sagebruch communities (data are not given here for soils within this forest zone). In addition to morphological data for the soils, chemical parameters have been examined to determine whether they indicate trends with age that might not be apparent in other types of data. The chemical parameters are dithionite and oxalate extracts of iron (Fe_d and Fe_o, respectively) and aluminum (Al_d and Al_o, respectively)—the former supposedly removes total free iron and aluminum and the latter only the amorphous forms (McKeague and Day, 1966)—and the two fractions of phosphorus, acid extractable and organic bound (P_a and P_o, respectively) (Walker, 1964). Both extracts of iron and aluminum as well as organic-bound phosphorus should increase with time, and acid-extractable phosphorus should decrease.

Although the ages of the alpine Holocene soils are not well known, generalizations regarding soil development with age can be made. Bs-horizons form in 5000 to 10,000 years (Figure 8-10). Unweathered till has a 5Y hue, and soils with a 10YR hue form in less than 10,000 years. Silt and clay content increase with time but not in amounts sufficient to form Bt-horizons in Holocene time; most of the increase in fines probably results from eolian additions. Chemical trends include (1) an increase in both extractable iron and aluminum with time, (2) an increase in organic-bound phosphorus that corresponds with the amount of organic matter, and (3) little if any depletion in acid-extractable phosphorus (Figure 8-11).

Soils formed for about 10,000 to 100,000 years in tills under semiarid conditions along the eastern front of the Sierra Nevada are quite similar (Burke and Birkeland, 1979). The reddest hue of the most-weathered horizon is typically 10YR. The small variations in the distribution of clay with depth are usually attributed to subtle differences in the parent material rather than to pedogenic processes;

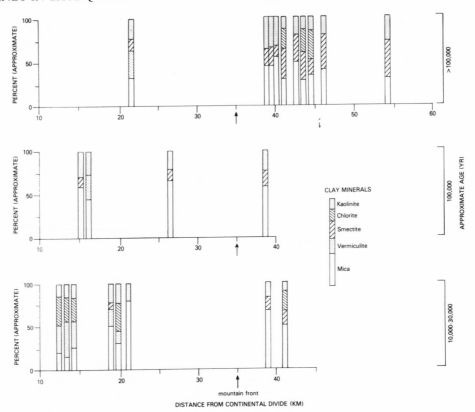

Figure 8-9. Variations in clay-mineral abundances in Bs- and Bt-horizons with distance from the Continental Divide and age of the parent material in the Boulder Creek drainage basin of Colorado. (Data from Netoff, 1977.)

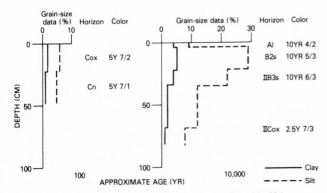

Figure 8-10. Grain-size data and Munsell colors for soils formed in Holocene tills in the alpine region of the eastern Sierra Nevada in California. (Based on unpublished data from P. W. Birkeland, R. M. Burke, and J. C. Yount.)

hence, most profiles show little evidence of clay accumulation and commonly consist of A-horizons over Bs- and Cox-horizons.

Mammoth Creek is the only locality where soil morphology correlates with the age of the till. At this locality, several soils formed in the younger (10,000 to 30,000-year-old) till lack Bt-horizons, whereas a soil formed in the older till (about 100,000 years old) has a Bt-horizon (Figure 8-12). Several factors might account for the presence of a Bt-horizon in the older till at Mammoth Creek and not elsewhere. These factors include (1) slightly higher amounts of precipitation, (2) mafic materials in the parent material, (3) surface layers of volcanic ash, (4) slightly higher clay content in the unweathered till, and (5) possibly less erosion here than at other sites. The increase in clay content in the Bt-horizon may represent the maximum amount of textural development for soils about 100,000 years old in the sagebrush region of the eastern Sierra Nevada.

Chemical and clay-mineral data for the range-front soils show little variation with age (Figure 8-13). The concentrations of iron, aluminum, and phosphorus with depth cannot be used to separate the soils into two distinct age-groups. The clay mineralogy of the soils has been examined over a large area of the Sierra Nevada (Birkeland and Janda, 1971); illite and halloysite are the most common minerals, and smectite is locally abundant. Most of the clay mineralogy does not show obvious trends that can be ascribed to the age of the soils.

The soil data do not suggest appreciable long-term changes in soil-moisture conditions over about the past 100,000 years along the eastern escarpment of the Sierra Nevada. In contrast to the lack of trends in chemical parameters in range-front soils, iron and aluminum increase with age in Holocene soils in the alpine region. The lack of chemical changes in soils at lower altitudes could be used to infer the likely range of past soil-moisture conditions. For example, if past climatic conditions were much wetter than they are at present, it seems reasonable to expect that the more effective soil moisture would be reflected in greater amounts of both forms of extractable iron and aluminum or in the depletion of acid-extractable phosphorus. In contrast, amounts of soil moisture less than those of the present, such as during a past drier climate, might not be detected through these soil properties. Alternatively, changes in soil moisture could have been either too subtle or of too short a duration to have influenced the properties of the soils.

Conclusions

The properties of soils formed in late-Quaternary deposits in the Rocky Mountains and the Sierra Nevada commonly show trends in pedologic development that correspond with the age of the parent material. These trends include (1) an increase in the number and the

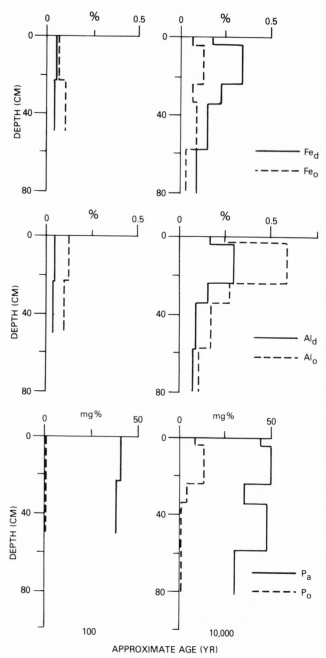

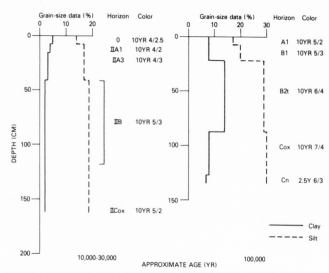

Figure 8-12. Grain-size data and Munsell colors for soils formed in late-Pleistocene tills in the eastern Sierra Nevada of California. (Data from Burke and Birkeland, 1979.)

thickness of the horizons, (2) increases in the amounts of clay and silt (partially of eolian origin), (3) redder colors, (4) carbonate accumulation in some semiarid soils, (5) clay-mineral transformations, and (6) variations in the amounts of extractable iron, aluminum, and phosphorus. In addition to varying with time, soil-clay mineralogy and soil chemistry vary with altitude. Pedogenesis is relatively rapid in alpine and adjacent high-montane environments, whereas the development of some soil properties proceeds very slowly at lower altitudes.

Except for the clay mineralogy of soils near the upper treeline, most pedologic properties do not indicate past changes in climate and vegetation. The physical and chemical properties of most soils suggest that past soil-moisture conditions were similar to those of the present and that they were not much greater than those of the present for extended periods of time. In contrast, evidence for former periods of less effective soil-moisture conditions is difficult to detect because it can be obscured by subsequent pedogenesis.

Figure 8-11. Concentration of dithionite- and oxalate-extractable iron and aluminum (Fe_d and Al_d and Fe_o and Al_o, respectively) and organic-bound (P_o) and acid-extractable (P_a) phosphorus with depth in Holocene alpine soils of the eastern Sierra Nevada of California. For phosphorus data, 1 mg% is 1 milligram (mg) per 100 g of soil. (Data from Birkeland et al., 1979.)

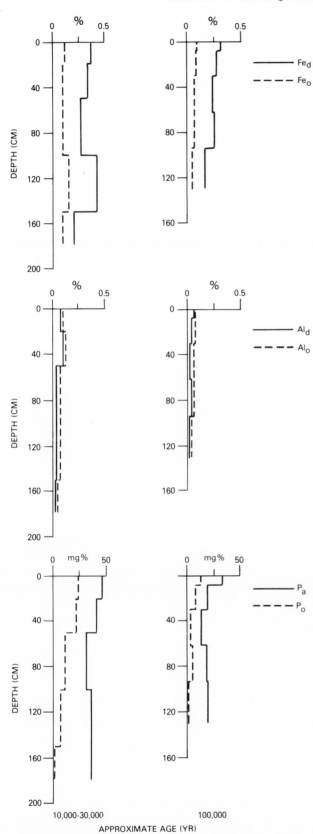

Figure 8-13. Concentration of iron, aluminum, and phosphorus in late-Pleistocene soils in the sagebrush region along the eastern escarpment of the Sierra Nevada of California. (The symbols used are the same as those used in Figure 8-11.) (Data from Birkeland et al., 1979.)

References

Baker, V. R. (1973). Paleosol development in Quaternary alluvium near Golden, Colorado. *The Mountain Geologist* 10, 127-33.

Benedict, J. B. (1973). Chronology of cirque glaciation, Colorado Front Range. *Quaternary Research* 3, 584-99.

Birkeland, P. W. (1974). "Pedology, Weathering and Geomorphological Research." Oxford University Press, New York.

Birkeland, P. W., Burke, R. M., and Walker, A. L. (1979). Variation in chemical parameters of Quaternary soils with time and altitude, Sierra Nevada, California. *Geological Society of America, Abstracts with Programs* 11, 388.

Birkeland, P. W., and Janda, R. J. (1971). Clay mineralogy of soils developed from Quaternary deposits of the eastern Sierra Nevada, California. *Geological Society of America Bulletin* 82, 2495-2514.

Birkeland, P. W., and Shroba, R. R. (1974). The status of the concept of Quaternary soil-forming intervals in the western United States. *In* "Quaternary Environments" (W. C. Mahaney, ed.), pp. 241-76. York University Geography Monographs 5.

Burke, R. M., and Birkeland, P. W. (1979). Reevaluation of multiparameter relative dating techniques and their application to the glacial sequence along the eastern escarpment of the Sierra Nevada, California. *Quaternary Research* 11, 21-51.

Curry, R. R. (1969). "Holocene Climatic and Glacial History of the Central Sierra Nevada." Geological Society of America Special Paper 123, pp. 1-47.

Gile, L. H., Peterson, F. F., and Grossman, R. B. (1966). Morphological and genetic sequences of carbonate accumulation in desert soils. *Soil Science* 101, 347-60.

Legg, T. E., and Baker, R. G. (1980). Palynology of Pinedale sediments, Devlins Park, Boulder County, Colorado. *Arctic and Alpine Research* 12, 319-33.

Machette, M. N. (1975). The Quaternary geology of the Lafayette quadrangle, Colorado. M.S. thesis, University of Colorado, Boulder.

McKeague, J. A., and Day, J. H. (1966). Dithionite- and oxalate-extractable Fe and Al as aids in differentiating various classes of soils. *Canadian Journal of Soil Science* 46, 13-22.

Madole, R. F. (1976). Bog stratigraphy, radiocarbon dates, and Pinedale to Holocene glacial history in the Front Range, Colorado, *United States Geological Survey, Journal of Research* 4, 163-69.

Madole, R. F. (1980a). Glacial Lake Devlin at the time of maximum ice expansion during the Pinedale Glaciation in the Front Range, Colorado. *Geological Society of America, Abstracts with Programs* 12, 279-80.

Madole, R. F. (1980b). Time of Pinedale deglaciation in north-central Colorado: Further considerations. *Geology* 8, 118-22.

Madole, R. F., and Shroba R. R. (1979). Till sequence and soil development in the North St. Vrain drainage basin, east slope, Front Range, Colorado. *In* "Field Guide Northern Front Range and Northwest Denver Basin, Colorado" (F. G. Ethridge, ed.), pp. 134-78. Geological Society of America Rocky Mountain Section 32nd Annual Meeting, Fort Collins, Colo.

Mahaney, W. C. (1974). Soil stratigraphy and genesis of Neoglacial deposits in the Arapahoe and Henderson Cirques, central Colorado Front Range. *In* "Quaternary Environments" (W. C. Mahaney, ed.), pp. 197-240. York University Geography Monographs 5.

Mahaney, W. C. (1978). Late-Quaternary stratigraphy and soils in the Wind River Mountains, western Wyoming. *In* "Quaternary Soils" (W. C. Mahaney, ed.), pp. 223-64. Geo Abstracts Limited, Norwich, England.

Mahaney, W. C., and Fahey, B. D. (1976). Quaternary soil stratigraphy of the Front Range, Colorado. *In* "Quaternary Stratigraphy of North America" (W. C. Mahaney, ed.), pp. 319-52. Dowden, Hutchinson, and Ross, Stroudsberg, Pa.

Maher, L. J., Jr. (1972). Absolute pollen diagram of Redrock Lake, Boulder County, Colorado. *Quaternary Research* 2, 531-53.

Markgraf, V. and Scott, L. (1981). Lower timberline in central Colorado during the past 15,000 yr. *Geology* 9, 231-34.

Miller, C. D., and Birkeland, P. W. (1974). Probable pre-Neoglacial age of the type Temple Lake moraine, Wyoming: Discussion and additional relative age data. *Arctic and Alpine Research* 6, 301-6.

Netoff, D. I. (1977). Soil clay mineralogy of Quaternary deposits in Two Front Range-Piedmont transects, Colorado. Ph.D. dissertation, University of Colorado, Boulder.

Pennack, R. W. (1963). Ecological and radiocarbon correlations in some Colorado mountain lake and bog deposits. *Ecology* 44, 1-15.

Pierce, K. L. (1979). "History and Dynamics of Glaciation in the Northern Yellowstone National Park Area." U.S. Geological Survey Professional Paper 729-F.

Reheis, M. C. (1980). Loess sources and loessial soil changes on a downwind transect, Boulder-Lafayette area, Colorado. *The Mountain Geologist* 17, 7-12.

Richmond, G. M. (1965). Glaciation of the Rocky Mountains. *In* "The Quaternary of the United States" (H. E. Wright, Jr., and D. G. Frey, eds.), pp. 217-31. Princeton University Press, Princeton, N.J.

Scott, G. R. (1963). "Quaternary Geology and Geomorphic History of the Kassler Quadrangle, Colorado." U.S. Geological Survey Professional Paper 421-A.

Shroba, R. R. (1977). Soil development in Quaternary tills, rock-glacier deposits, and taluses, southern and central Rocky Mountains. Ph.D. dissertation, University of Colorado, Boulder.

Soil Survey Staff. (1975). "Soil Taxonomy." U.S. Department of Agriculture Handbook 436.

Szabo, B. J. (1980). Results and assesments of uranium-series dating of vertebrate fossils from Quaternary alluviums in Colorado. *Arctic and Alpine Research* 12, 95-100.

Van Horn, R. (1976). "Geology of the Golden Quadrangle, Colorado." U.S. Geological Survey Professional Paper 872.

Walker, T. W. (1964). The significance of phosphorus in pedogenesis. *In* "Experimental Pedology" (E. G. Hallsworth, and D. V. Crawford, eds.), pp. 295-315. Butterworths, London.

CHAPTER 9

The Periglacial Environment in North America during Wisconsin Time

Troy L. Péwé

Introduction

A periglacial environment existed in North America during Wisconsin time throughout all of unglaciated Alaska and northwestern Canada and in the unglaciated area immediately south of the continental ice sheets in the contiguous United States, as well as near and below mountain glaciers, ice caps, and ice carapaces of the high-mountain areas still farther south. (The term ''Wisconsin'' is used in this report as a geologic-climate unit [American Commission on Stratigraphic Nomenclature, 1970]). During the last 15 to 20 years, considerable research has been done in the periglacial areas, especially in the modern periglacial environment of Alaska and Canada. Because one of the two major areas in North America with a periglacial environment in Wisconsin time still has a periglacial environment, part of this chapter is devoted to a consideration of processes and features in the modern periglacial environment. The more one knows about modern periglacial conditions, the more confidently one can interpret those of Wisconsin time.

The term ''periglacial'' (Lozinski, 1909) has not been accepted by all investigators in this field, but the concept is used widely. No single definition acceptable to everyone concerned with studying the periglacial environment has been formulated, for workers differ in their opinions of the exact climate of the periglacial zone. The term ''periglacial'' is used in various ways (Butzer, 1964; Dylik, 1964; French, 1976; Péwé, 1969, 1975b; Washburn, 1980a). For instance, Zeuner (1945) restricts the periglacial zone to an area underlain by permafrost or one where the mean annual temperature is −2°C or colder. Others state that permafrost is not necessary for a periglacial environment and that such a climate could include regions where permafrost is absent but where a considerable number of freezing and thawing cycles occur annually. Areas of glaciers are not periglacial by definition.

In this chapter the term ''periglacial'' designates nonglacial processes and features of cold climates on land characterized by intense frost action, regardless of its age or proximity to glaciers (Washburn, 1980a). Such a region is almost everywhere characteristically underlain by perennially frozen ground (Péwé, 1969: 4). The periglacial realm today must certainly extend over 20% of the Earth's land surface. In Wisconsin time, the periglacial environment

included certain areas in the temperate middle latitudes. French (1976: 3) believes that an additional 20% of the present land surface, beyond the present periglacial domain, may have had periglacial conditions at some time during the past.

A former periglacial environment is assumed when permafrost is demonstrated to have existed in the past. A history of permafrost can be identified most readily by the presence of ice-wedge casts, relict pingos, inactive cryoplanation terraces, and inactive ice-cemented (lobate) rock glaciers. Past permafrost environments also can be suggested but not proven by evidence of formerly cold-climate mass-movement phenomena such as solifluction, block fields, block streams, and a host of patterned-ground phenomena both sorted and unsorted.

Permafrost is defined as a thickness of soil or other superficial deposits, or even bedrock, that has been colder than 0°C for 2 years or longer; it is defined exclusively on the basis of temperature, irrespective of texture, degree of induration, water content, or lithologic character (Muller, 1945: 3). (The Institut Merzlotovedeniya in Yakutsk, U.S.S.R., specifies that the ground must be colder than 0°C 3 years or longer to be regarded as permafrost [Institut Merzlotovedeniya im. V. A. Obrucheva (1956)]. Brown [1967] and French [1976] believe that permafrost should be defined as ground remaining frozen throughout at least one year.) Permafrost, perennially frozen ground (Mozley, 1937; Taber, 1943), or *vechnaya merzlota* (Sumgin, 1927), is widespread in the northern part of the Northern Hemisphere (Figure 9-1), as well as in Antarctica. (*Vechnaya merzlota* means ''eternal frost'' and is being replaced in the recent Russian literature by *mnogoletnemerzlyy grunt*, which means ''perennially frozen ground.'') About 20% of the land area in the world is underlain by permafrost (Black, 1954: 842; Ferrians et al., 1969; Muller, 1945: Plate 1; Péwé, 1975b: Figure 22).

Perennially frozen ground is present throughout much of North America (Figure 9-1), but it is more widespread and extends to greater depths in the north than in the south. Permafrost in North America used to be divided arbitrarily into three general zones—the continuous zone, the discontinuous zone, and the sporadic zone (e.g., Black, 1950, 1953; Hopkins et al., 1955; Péwé, 1954)—with terms referring to the lateral continuity of the permafrost. Because the boundaries between the discontinuous zone and the sporadic per-

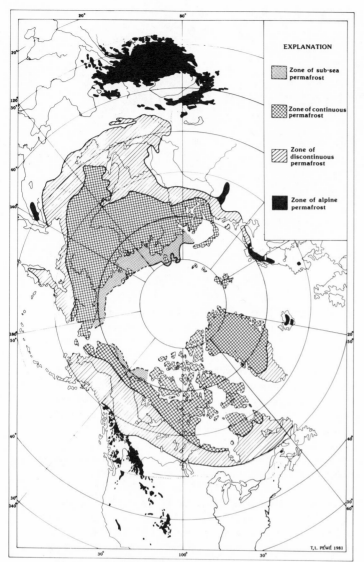

Figure 9-1. Distribution of permafrost in the Northern Hemisphere (isolated areas of alpine permafrost not shown on the map exist in high mountains and outside the map area in Mexico, Hawaii, Japan, and Europe). (The map is based on the following sources. United States land area: T. L. Péwé, published and unpublished data, 1981. United States subsea area: Hopkins et al., 1977; P. V. Sellmann, personal communications, 1981. Canadian land area; R. J. E. Brown, 1978, personal communications, 1979. Canadian subsea area: Hunter et al., 1976; A. S. Judge, personal communication, 1981. Greenland: Weidick, 1968; O. Olesen, personal communication, 1976. Iceland: T. Einarsson, personal communication, 1966; Priesnitz and Schunke, 1978. Norway: B. J. Andersen, personal communication, 1966; L. King, personal communication, 1980; H. Svensson, personal communication, 1966. Sweden: Rapp and Annersten, 1969; King, 1977. Svalbard: Liestol, 1977. Mongolia: Gravis et al., 1978. U.S.S.R. land area: Karpov and Puzanov, 1970; Gorbunov, 1978a, 1978b. U.S.S.R. subsea area: M. Vigdorchik, personal communication, 1978. China: Institute of Glaciology, Cryopedology and Desert Research, Academica Sinica 1975; S. Yafeng, personal communication, 1980. Tibet and Himalayan and adjacent mountains: Fujii and Higuchi, 1978; Y. Fujii, personal communication, 1978; Gorbunov, 1978b.

mafrost zone are difficult to place without information about temperature and because thermal data for permafrost in North America is still limited, the permafrost zones in Alaska since about the mid-1960s, and those in Canada earlier, have been listed as only continuous and discontinuous (Brown, 1960, 1966, 1969, 1970, 1978; Ferrians, 1965; Jenness, 1949; Péwé, 1966a, 1966c, 1974, 1975a, 1976, 1979a).

Although nearshore permafrost has been known under the sea (generally referred to as offshore, submarine, or subsea permafrost) for more than 60 years (Werenskiold, 1922, 1953), only recently have its distribution characteristics, and origin in North America been investigated.

The existence of perennially frozen ground in the high mountains and high plateaus south of the polar and subpolar areas has long been known, but it usually has not been shown on permafrost maps of North America because of the lack of detailed data. It is known, for example, that 20% of China is underlain by frozen ground, most of which is permafrost in the high plateau of Tibet (Péwé, 1981). This type of permafrost is referred to as "alpine" permafrost and is relatively extensive throughout the Northern Hemisphere. Fujii and Higuchi (1978) estimate that 2.3×10^6 km² exists in the Northern Hemisphere and that of that 0.45×10^6 km² lies in North America. Péwé (1983) calculates that about 100,000 km² of the contiguous United States is underlain by alpine permafrost.

Alaska and Northwestern Canada

Arctic and sub-arctic Alaska and a small part of northwestern Canada comprised an area of most rigorous periglacial environment in Wisconsin time and still today constitute an active periglacial area. Unglaciated regions in Alaska and northwestern Canada have been a periglacial area for at least a few million years, during all glacial episodes and perhaps most interglacial times. During certain parts of Wisconsin time, the Bering Shelf connecting with Asia was above sea level and was part of "Beringia" (Hopkins, 1959). This term has been expanded over the years by intensive work of Hopkins and his colleagues (Hopkins, 1967, 1972, 1982) to include all of the periglacial environment of Wisconsin time in Alaska and northwestern Canada, plus the Bering and continental shelves of northwestern North America and extending into Asia an undetermined distance.

MODERN PERIGLACIAL ENVIRONMENT

Permafrost

Permafrost is perhaps the most impressive periglacial phenomenon in Alaska and northwestern Canada. Currently, it underlies 82% of Alaska (Figure 9-2) and all of northwestern Canada (Figure 9-3). In the continuous zone of the northern part of Alaska and northwestern Canada, permafrost is almost everywhere present; 12 km south of Barrow, it extends to a calculated depth of 405 m (Brewer, 1958: 19) and to 650 m deep at Prudhoe Bay (Stoneley *In* Lachenbruch, 1970b: J3-J4). Later work shows that the base of the ice-bonded permafrost is about 630 m deep at Prudhoe Bay (Osterkamp and Payne, 1981). Lachenbruch (1970b) notes that permafrost in Prudhoe Bay is about 50% thicker than at Barrow (Figure 9-2) and 100% deeper than at Cape Simpson, although all three localities probably have similar mean surface temperatures. He explains this difference as the result

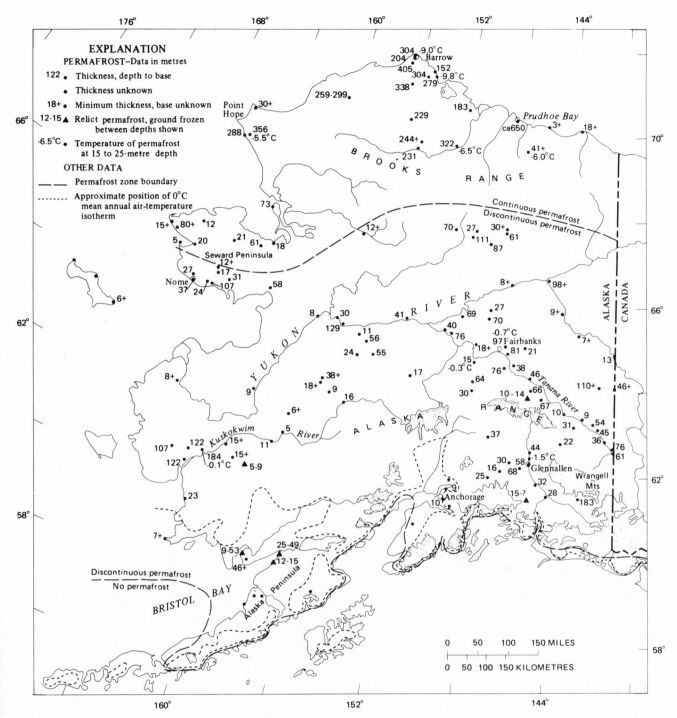

Figure 9-2. Permafrost map of Alaska. (Compiled by T. L. Péwé and L. R. Mayo; permafrost thickness from map compiled by J. R. Williams on the basis of published and unpublished sources.) (From Péwé, 1975b: Figure 23.)

of an increased proportion of siliceous sediments from Simpson to Barrow to Prudhoe, which would cause a corresponding increase in conductivity and a decrease in geothermal gradients.

Southward, in the discontinuous permafrost zone of Alaska, permafrost thickness decreases and unfrozen areas are more and more abundant, until near the southern boundary only rare patches of permafrost exist (Péwé, 1975b). The thickness of permafrost in central and southwestern Alaska ranges from a few meters to 180 m. The

southern boundary of permafrost as shown in Figure 9-2 is placed to include relict permafrost as well as small areas of permafrost existing as a result of the current favorable microclimatic conditions. This boundary lies a short distance south of the 0°C mean annual air isotherm, which is generally thought to lie today at the southern boundary of permafrost. The temperature of permafrost at the depth of 15 to 25 m ranges from −5°C in the northern part of the zone to approximately 0°C in the southern part.

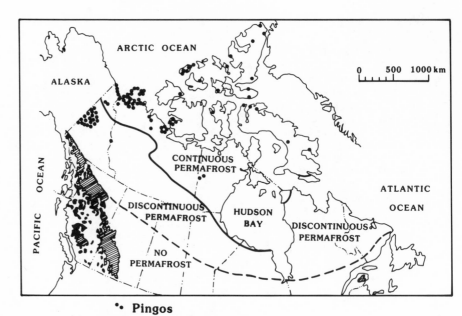

Figure 9-3. Permafrost and pingo distribution in Canada. Mainly closed-system pingos are present in the continuous permafrost zone, and mainly open-system pingos occur in the discontinuous permafrost zone. According to A. S. Judge (personal communication, 1981), the subsea ice-cored mounds may not be pingos. (Permafrost and pingo information modified from Brown and Péwé, 1973). (See Figure 9-1 for subsea permafrost and Figure 9-5 for details of pingo type and distribution in northwestern Canada.)

•• **Pingos**

▨– **Alpine permafrost zone**

Offshore permafrost probably originated on land in the periglacial environment of Wisconsin time and became submerged by the postglacial rise of sea level and coastal erosion. Whereas temperatures at depths of 15 to 25 m in continuous permafrost on land in the high Arctic range from $-10\,°C$ to $-12\,°C$, temperatures in the perennially frozen ground under the sea are relatively warm, about $0\,°C$ to $-3\,°C$.

Permafrost on the Alaskan continental shelf at the Beaufort Sea in Alaska is widespread and probably extends north to the edge of the shelf about 200 km off the coast, where the water depth is 90 m. The greatest depth of relict frozen ground recorded is 730 m in the Beaufort Sea off the western Canadian coast (Péwé, 1979c) (References relating to subsea permafrost in Alaska include Blouin et al., 1979, Chamberlain, 1979; Harrison and Osterkamp, 1978; Iskandar et al., 1978; Lewellen, 1977; Page and Iskandar, 1978; Sellmann et al., 1979; United States National Academy of Sciences, 1976; Vigdorchik; 1979. Those for Canada include Hobson et al., 1977; Hunter et al., 1976; Judge, 1977; Mackay, 1972a.)

Ice Wedges

The most conspicuous and controversial type of ground ice in permafrost is the large ice wedge or mass characterized by parallel or subparallel foliation structures. Ice wedges, and the polygonal patterns they reflect on the surface of the ground, are a characteristic feature of permafrost and the periglacial environment. The area of active ice wedges in North America, and especially Alaska, appears to coincide approximately with the continuous permafrost zone (Figure 9-4). Ice wedges occur widely to the northern reaches of the Canadian Arctic Archipelago and are reported in various areas, including Prince Patrick Island and as far north as Ellesmore Island (Brown and Péwé, 1973; Mackay, 1976d). The sub-Arctic, especially in Alaska, is a zone of inactive or weakly active ice wedges.

Apparently, cracking in ice wedges is to be expected, and wedges continue to grow where the winter temperature of the ground at the top of the permafrost is about $-15\,°C$ or colder (Péwé, 1966a); these conditions are almost limited to the tundra and the continuous permafrost zone (Black, 1969b; Jahn, 1972; Mackay, 1972c, 1972d; Pissart, 1970). From north to south across Alaska, a decreasing number of wedges crack frequently. The line dividing zones of active and inactive ice wedges is arbitrarily placed at the position where the low-center or raised-edge polygons are uncommon and where it is thought that most wedges do not frequently crack.

Few data are available from Alaska (Péwé, 1966a, 1975b: 62); this fact indicates that the minimum temperature of the ground in the winter at the permafrost table (top of the ice wedges) ranges from $-11\,°C$ to $-14\,°C$ near Kotzebue to about $-30\,°C$ near Barrow. In northwestern Alaska, the temperature at the top of the permafrost has been reported as $-18\,°C$ to $-20\,°C$ on the basis of a 2-year record (A. H. Lachenbruch, personal communication, 1962). The growth of the ice wedges does not depend on the mean annual air temperature but rather on the rate the temperature drops at the time of cracking (Lachenbruch, 1966; Péwé, 1966a).

In the discontinuous permafrost zone in northwestern North America, the wedges are essentially inactive. Inactive ice wedges have no ice seams or cracks extending upward to the surface during the spring. The top of the wedge may be flat, especially when thawing has lowered the upper surface of the wedge at some time in the past (Péwé, 1965, 1966c, 1975b; Sellmann, 1967).

Péwé has examined ice wedges in Alaska, Canada, Scandinavia, Svalbard, Siberia, and Antartica and considers all the wedges he has seen to be of Wisconsin or Holocene age. Wedges in Alaska are of several ages, but none appears older than the last glaciation. Pre-Wisconsin ice wedges have either melted or are so rare or so deeply buried that they are no longer exposed or penetrated by drilling (Péwé, 1975b: p. 53). Most wedges have been dated by geologic association; however, attempts have been made recently to date wedges by radiocarbon analysis of the organic debris in the ice. This debris is washed down into contraction cracks throughout the life of the wedge, or it is incorporated from the sides as the wedge grows. Radiocarbon dates should be no older than the enclosing sediments, and this is borne out by samples collected by Péwé. A wedge at Barrow was dated as 14,000 years old (J. Brown, 1965), wedges near

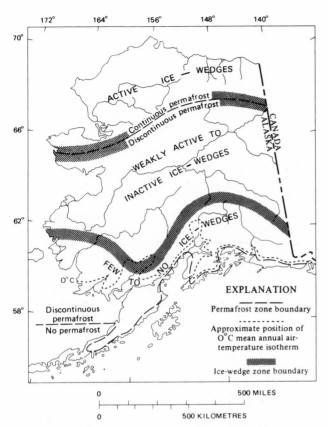

Figure 9-4. Distribution of ice wedges and permafrost in Alaska. (From Péwé, 1975b.)

Fairbanks as 31,400 and 32,300 years old (Sellmann, 1967, 1972), and wedges at Eva Creek and Ready Bullion Creek near Fairbanks as older than 25,000 years (Péwé, 1975b). By using associated organic sediments, Hopkins and Robinson (1979: B47) have obtained dates of wedges in Alaska of 3000 to 9500 years old.

Pingos

The presence of pingos is valuable not only in indicating the presence of permafrost but perhaps also in indicating the distribution of continuous or discontinuous permafrost. With one or two exceptions, the first maps of pingo distributions in Canada (Figure 9-3) and Alaska (Figure 9-5) have appeared since 1965 (Brown and Péwé, 1973; Carter and Galloway, 1979; Galloway and Carter, 1978; Holmes et al., 1968; Hughes, 1969; Mackay, 1966; Péwé, 1975b, 1982).

Considerable progress has been made in the discovery and mapping of open-system pingos in central Alaska and the Yukon Territory, as well as in the discovery of pingolike mounds in the shallow waters of the Beaufort Sea north of the mouth of the Mackenzie River (Shearer et al., 1971). The greatest advance in recent pingo research has been the development of a theory of pingo growth and measurement of growth rate by Mackay and his co-workers (Mackay, 1976b; cf. Mackay, 1972b, 1977a, 1977b, 1978a, 1978b and Mackay et al., 1979).

A restudy of photographs of the mounds in the Yukon-Kusko-kwim Delta in Alaska and an examination of palsas elsewhere in the world has led Péwé (1975b) to conclude that the 200 closed-system

pingos mentioned by Burns (1964) as occurring in the Yukon-Kuskokwim Delta are probably palsas, not pingos.

In addition to the thousands of pingos recognized in western North America, groups of pingos have been mapped on Prince Patrick Island in the Canadian Arctic Archipelago (Pissart, 1967, 1970, 1977) and on Banks Island (Pissart and French, 1976). Pingos are abundant and widespread in northwestern Alaska, where 732 pingos and pingolike features in the National Petroleum Reserve have been mapped (Carter and Galloway, 1979; Galloway and Carter, 1978).

Radiometric dating indicates that most pingos in North America are less than 4000 to 7000 years old. The famous Ibyuk pingo near Tukayaktuk is still growing, and calculations derived from the 1973 to 1975 growth rate suggest that the pingo may be about 1000 years old (Mackay, 1976a). Through careful field measurements of four pingos over several years, Mackay has shown further that many of the pingos are less than 1000 years old; some are even less than 25 years old. The highest annual growth rate that Mackay has recorded is 21.2 cm per year. Although he considers a growth rate of 1.5 m per year during the first 1 to 2 years of growth as possible, he believes that the growth rate decreases inversely as the square root of time (Mackay, 1973).

For the most part, closed-system pingos are restricted to the continuous permafrost zone, where the mean annual air temperature, at least in Alaska, is from $-6\,°C$ to $-8\,°C$ or colder. All open-system pingos in Canada appear to originate in areas where the mean annual air temperature is that cold or colder and where ice wedges are actively growing. Mackay (1978a: 145) cites a mean annual ground temperature of $-5\,°C$ or lower for closed-system pingos; the mean

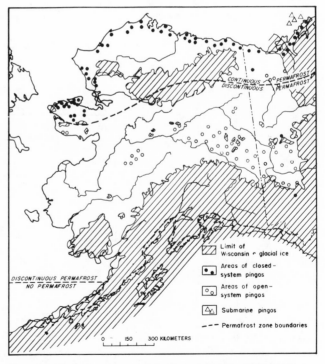

Figure 9-5. Distribution of open- and closed-system pingos in relation to permafrost zones and areas covered by Wisconsin glacier ice in northwestern North America. (From Péwé, 1975b.) The origin of the pingolike forms in the shallow sea off northwestern Canada has not yet been established (A. S. Judge, personal communication, 1981.)

annual ground temperature is generally 2°C to 6°C warmer than the mean annual air temperature (Judge, 1973).

Cryoplanation Terraces

Cryoplanation terraces are minor steplike or tablelike landforms consisting of nearly horizontal bedrock surfaces covered by a thin veneer of rock debris and bounded by ascending or descending bedrock scarps or both (Figure 9-6). They generally occur as a series of rock-cut steps on spurs of hills and mountains above the timberline, and they typically cut across lithology or structure (Reger and Péwé, 1976). The tread, or "flat," area is 10 to several hundred meters wide or long. The tread is not actually flat but slopes from 1° to 5° parallel to the ridge crest. The scarps of cryoplanation terraces range in height from 3 to 25 m and slope from about 9° to 32° when covered by rubble; nearly vertical scarps exist where bedrock is exposed or thinly buried. Cryoplanation terraces are also called altiplanation terraces (Eaken, 1916: 77-82), equiplanation terraces (Cairnes, 1912a, 1912b), and goletz terraces (Boch and Krasnov, 1951). Treads and scarps are cut into all bedrock types, but terraces are best developed on closely jointed resistant rock, such as basalt, andesite, and hornfels. The origin of cryoplanation terraces and their relationship to bedrock and climate in central and western Alaska is discussed by Reger and Péwé (Péwé and Reger, 1975; Reger and Péwé, 1976).

Scarps retreat by nivation to form a tread; surficial debris is removed across and over the tread by mass movement. The terraces form close to, but perhaps a little below, the general altitude of the snow line (Figure 9-7). On isolated ridges and peaks not large enough to support large glaciers, however, cryoplanation terraces form above the snow line. The altitudes of well-formed terraces in Alaska rise from west to east on a line that is below and parallel to past and present snow lines, as well as the present treeline.

Cryoplanation terraces are widespread in unglaciated areas of Alaska (Figure 9-6); however, it is thought that none are actively forming in Alaska today. They also have been reported from several places elsewhere in North America (Figure 9-8) (Bird, 1967; Brunn-

schweiler, 1962; Cailleux, 1967; Demek, 1969; Foster, 1967, 1969; Foster and Keith, 1969; Hughes et al., 1972; Journaux, 1969), but no distinction has been made between active and inactive forms.

Reger and Péwé (1976) have concluded that cryoplanation terraces in the temperate latitudes of North America and Europe are definitely inactive. Terraces observed in photographs and examined in the field in the sub-Arctic of Asia also appear to be inactive. Work in the sub-Arctic of Alaska indicates tht generally there are on fresh surfaces on the scarps of cryoplanation terraces; this finding suggests that nivation is no longer causing scarp retreat. Even the sharpest terraces are inactive, as indicated by photographs of the same terraces taken at intervals of 50 years. Well-developed terraces evidently take thousands of years to form. In central Alaska, according to Reger and Péwé (1976), some are cut on surfaces of Illinoian age and are younger than 50,000 years. On the Seward Peninsula, there is evidence that some terraces postdate early Wisconsin glaciation.

In general, cryoplanation terraces are distributed in areas that had or now have periglacial climates, and, therefore, they are indicators of rigorous periglacial conditions. The presence of stable cryoplanation terraces is an indication of past shallow permafrost. The sharp inactive cryoplanation terraces in Alaska today are underlain by permafrost. When they were active, shallow permafrost was undoubtedly present beneath terrace treads and promoted the over-tread movement of rock debris derived from the cutting of descending scarps.

Hundreds of sharp but inactive terraces occur in some areas of Alaska where the summer temperature is colder than 10°C. When these terraces were active, temperatures were colder. Recent work in Alaska indicates that terraces were active in some areas where the mean July temperature during Wisconsin time was about 4°C. The mean annual temperature probably was about −12°C or colder (Reger and Péwé, 1976).

Rock Glaciers

A rock glacier occurs either as debris covering a still-moving glacier that gives rise to a down-slope-moving mass of debris and ice of glacial origin or as an ice-cemented mass of talus-derived debris

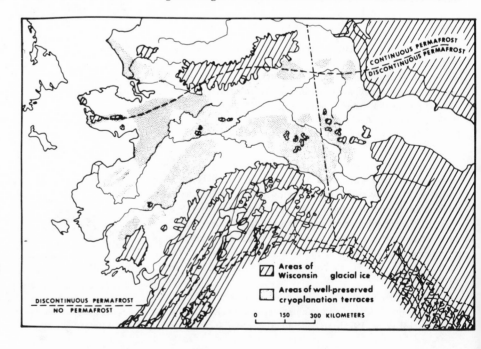

Figure 9-6. Generalized distribution of well-preserved cryoplanation terraces of Wisconsin age relative to Wisconsin glaciation in Alaska and northwestern Canada. Many small areas of terraces within tiny refugia are not shown. (Glacier extents modified from Coulter et al., 1965; Prest et al., 1968; and Yeend, 1971.) (From Reger and Péwé, 1976.)

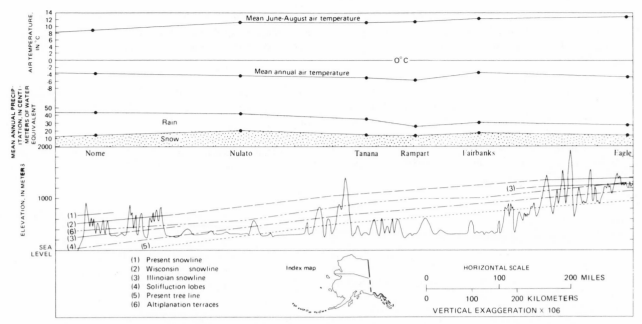

Figure 9-7. East-west section across central Alaska illustrating change of certain meteorologic parameters and elevations of present snow line, snow lines of Wisconsin and Illinoian age, well-developed cryoplanation (altiplanation) terraces, solifluction lobes, and present treeline. (Compiled by T. L. Péwé and L. R. Mayo.) (Past snow lines were determined from cirque-floor altitudes.) (From Péwé, 1975b.)

that moves slowly down slope. Rock glaciers with glacier ice are called ice cored. The term "ice-cemented rock glacier" is used for the second type (Luckman and Crockett, 1978; Potter, 1972; White, 1976: 79). The ice formed after or during the accumulation of the debris.

Tongue-shaped rock glaciers are residues of a glacier and do not require a permafrost climate in order to originate. Ice-cemented forms are lobate and need a permafrost climate in order to originate and to remain active. Harris (1981) suggests that this type exists in the colder parts of the discontinuous permafrost zone.

The typical glacier is 150 to 320 m wide and 300 to 1600 m long. In Alaska, most of rock glaciers occur in cirques. Active rock glaciers are abundant in the mountain ranges of western North America, and many inactive ones are also known.

Although rock glaciers were first named from Alaska (Capps, 1910: 360) and one of the most-detailed studies of rock glaciers was done in Alaska (Wahrhaftig and Cox, 1959), most research on active and inactive rock glaciers in North America has been done in the Rocky Mountains of southern Canada and the western contiguous United States (Benedict, 1973; Hughes, 1966; Johnson, 1967; Potter, 1972; Vick, 1981; White, 1971). (Abstracts of the complete set of papers presented at the special session on "Rock Glaciers in Canada" conducted by the Canadian Association of Geographers Conference at Laval University, May 25, 1976, appear in Proceedings of the Canadian Association of Geographers Meeting, Laval University, 1976, p. 156-67.) White (1976) presents an up-to-date, concise summary of rock-glacier research; the summary includes a discussion of the problem of classification.

Solifluction

Among the most common forms of mass wasting in the periglacial environment are frost creep and solifluction (Benedict, 1976;

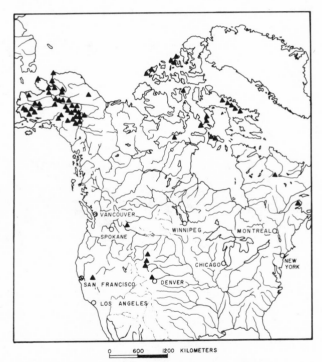

Figure 9-8. Generalized distribution of well-developed cryoplanation terraces in North America. (Modified from Demek, 1969, and Reger, 1975.)

Washburn, 1980a, 1980b). Frost creep is a ratchetlike downslope movement of particles that results from frost heaving of the ground and settling subsequent to thawing (Benedict, 1970; Washburn, 1967, 1969).

Solifluction has long been considered the slow flowing of water-saturated debris from higher to lower ground. (The original defini- tion [Andersson, 1906] was not distinctly limited to frozen ground; some authors today use the term ''gelifluction,'' which is defined by Baulig [1957: 926] as solifluction associated with frozen ground, in- cluding seasonally frozen ground as well as permafrost. Thus, gelifluction is one kind of solifluction. In this report, the term ''solifluction'' is used, and its use is restricted to the context of areas of cold climate.) The process is best developed in areas that are underlain by permafrost or long-existing seasonal frost.

The areas in which solifluction is active are almost entirely above and beyond the timberline. Areas of solifluction in North America are widespread and lie in the far-northern high mountains. In Alaska, they are well developed in the central and northern part of the state. Well-developed solifluction lobes occur at an elevation of 1200 m in central Alaska on the Canadian border and at progressively lower altitudes to the west, until on the Seward Peninsula they are only a few hundred meters above sea level (Péwé, 1975b) (Figure 9-7).

Quantitative studies of active mass wasting at various sites in North America indicate movement rates of 0.25 to 10 cm per year, the range reflecting different slope angles and climatic situations (Everett, 1966; Kerfoot and Mackay, 1972; Price, 1973; Williams, 1966).

Geobotanical Periglacial Features

Two periglacial geobotanical features that are well developed in Alaska and Canada are palsas and string bogs. These features, which

are related to permafrost and seasonal freezing of the ground, have been studied in Canada since 1965 (Zoltai, 1971; Zoltai and Tarnocai, 1971, 1974), but similar studies have yet to be done in Alaska.

WISCONSIN PERIGLACIAL ENVIRONMENT

The unglaciated part of northwestern North America (Alaska and the western part of the Yukon Territory and the Northwest Ter- ritories of Canada), the adjoining continental shelf to the north and west, and the unglaciated part of easternmost Siberia constituted a vast refugium (Beringia) that was almost completely bounded by glacier ice and marine waters during Wisconsin time.

Geography

The area of Alaska that was not covered by glacier ice in Wisconsin time is equal to about half the state, or about 700,000 km² (Figure 9-9). Contiguous with it was the unglaciated area of northwestern Canada, which is estimated to be about 150,000 km². The adjacent exposed shallow shelves of the Chukchi Sea, the Bering Sea, and the Arctic Ocean were also part of the periglacial environment. Sergin and Cheglova (1973) estimate that this exposed land increased the emergent areas within the shelves by 250% to 300%.

At the time of sea-level minimum, the shoreline of the Bering Sea lay south of the Pribilof Islands, exposing virtually all of the continen- tal shelf. In this chapter, the 90-m isobath is used to approximate the Wisconsin shoreline, although sea level did not remain at this posi- tion for very long (Hopkins, 1972).

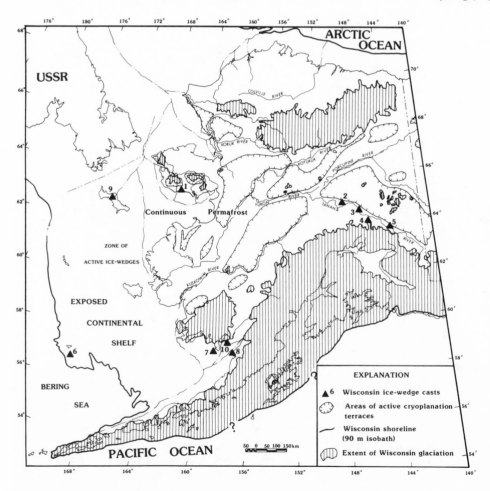

Figure 9-9. Periglacial environment in Alaska in Wisconsin time.

Periglacial Phenomena

In Wisconsin time, as today, northwestern North America displayed a vast array of periglacial phenomena, including permafrost, ice wedges, patterned ground, cryoplanation terraces, pingos, palsas, thermokarst lakes, rock glaciers, solifluction features, and many other periglacial forms. The distribution and activity of the processes, however, were different then than now. The vast unglaciated area was mostly treeless and dry (Hopkins, 1972; Matthews, 1976; Péwé, 1975b: 86); and, undoubtedly, the widespread lowlands, including Alaska (Péwé, 1975b: Figure 42), northwestern Canada, and the emerging continental shelves, were imprinted with tundra polygons. The continental climate of the interior and the arctic climate of the north undoubtedly extended much farther south (Messenger, 1977); for example, Nome, which is now on the sea coast, had a much more continental climate in Wisconsin time when the coast lay about 600 km to the south (Figure 9-9).

Although pingos (both closed and open system), palsas, and rock glaciers must have been forming actively throughout Wisconsin time, no record has yet been found of their existence. Possible exceptions are the large masses of clear, blue ice that have been exposed in placer gold-mining operations in retransported loess of Wisconsin age near Fairbanks (Péwé, 1978); these may represent pingo ice.

Solifluction

Solifluction deposits of Wisconsin age are widespread in the unglaciated areas of northwestern North America, mainly on lower hill slopes and valley bottoms. Most deposits, especially those in valley bottoms, contain some local fluvial sediments. The frozen deposits are stabilized and are 100 to 900 m lower in elevation than nearby active solifluction lobes.

Little information has been published on solifluction deposits of this age, although several maps and geologic cross sections portraying and naming solifluction deposits of middle- and late-Pleistocene age near Fairbanks have been published (Péwé, 1975c). For example, the Goldstream Formation (Wisconsin) consisting largely of retransported loess is in part a solifluction deposit (Péwé et al., 1976).

Cryoplanation Terraces

Cryoplanation terraces are the most widespread periglacial erosional form in Alaska and northwestern Canada (Figure 9-6). Striking relict forms occur throughout the central, western, northwestern, and southwestern parts of the unglaciated region of northwestern North America (Figure 9-8). The best developed and highest were certainly active during Wisconsin time. Well-developed cryoplanation terraces were being formed at elevations of about 200 m in far-western and southwestern Alaska and at progressively higher altitudes eastward to about 1000 to 1100 m at the Alaska-Canada border (Figure 9-6). They were active over an area of about 250,000 km².

Reger and Péwé (1976) suggest that a rigorous periglacial climate existed on the tops and spurs of the low mountains in glaciated Alaska, even far to the southwest. On the basis of present-day information, it is assumed that the average midday summer air temperatures were colder than 10 °C at the altitude of terrace formation and that the mean summer temperatures were probably between 2 °C and 6 °C. Locations in the lowland where the modern mean summer temperature is in the range of 2 °C to 6 °C had mean annual air temperatures of about − 12 °C (Reger and Péwé, 1976: 107-8).

Ice Wedges

Active ice wedges and their surface expression as tundra polygons were undoubtedly ubiquitous in the lowlands of Alaska and northwestern Canada and the exposed continental shelves of Beringia in Wisconsin time, except where the lowlands were covered by glaciers or lakes (Figure 9-9). Such areas, by analogy with present conditions, lay in the continuous permafrost zone (Figure 9-4).

In Wisconsin time, ice wedges were growing not only in northern and central Alaska but also far to the south, where no ice wedges exist today. The presence of former ice wedges is indicated by ice-wedge casts (Figure 9-10 and Table 9-1), which are known as far south as Bristol Bay, the Pribilof Islands, and St. Lawrence Island (Figure 9-9). The widespread occurrence of the casts indicates that a rigorous climate with a mean annual air temperature of at least − 6 °C existed throughout Beringia in Wisconsin time (Messenger and Péwé, 1977; Péwé, 1966c).

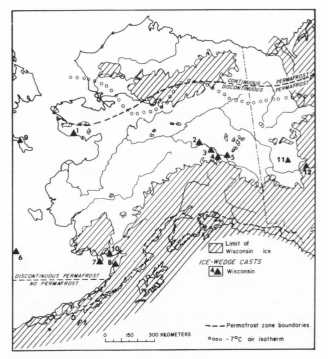

Figure 9-10. Distribution and age of ice-wedge casts in Alaska and northwestern Canada. (Numbers refer to localities listed in Table 9-1.)

Permafrost

During Wisconsin time, permafrost was more widespread and thicker than it is today in northwestern North America. Relict "patches" of permafrost in southern Alaska are part of the former zone of continuous permafrost that encompassed nearly all of the unglaciated areas of Alaska and the exposed continental shelves during the Wisconsin. The emergent shelves of the Beaufort, Chukchi, and Bering Seas were exposed to the cold glacial temperatures; this resulted in the formation of widespread continuous permafrost (Figure 9-9) that existed as far south as the Pribilof Islands, St. Lawrence Island (Figure 9-10), and Bristol Bay, where ice-wedge casts are reported. Practically all the unglaciated lowlands of Beringia were in a zone of active ice-wedge growth at that time (in contrast to present condi-

Table 9-1.
Localities of Ice-Wedge Casts of Wisconsin Age
in Alaska and Northwestern Canada

Map Location	Locality	Reference
1	Nome	Hopkins et al., 1960
2	Fairbanks	T. L. Péwé, unpublished data, 1949
3	Harding Lake Area	Blackwell, 1965
4	Big Delta	Péwé et al., 1969
5	Healy Lake	Ager, 1972
6	Pribilof Island	D. M. Hopkins, personal communication, 1972
7	Kvichak Peninsula	D. M. Hopkins, personal communication, 1972
8	Bristol Bay	Hopkins et al., 1955: 139
9	St. Lawrence Island	D. M. Hopkins, personal communication, 1981
10	Bristol Bay	D. M. Hopkins, personal communication, 1981
11	Yukon Territory	Rutter et al., 1978
12	Yukon Territory	Rutter et al., 1978

Note: Map-locality numbers correspond to numbers on Figure 9-10.

tions, in which ice wedges are actively growing only in the Arctic). The southern limit of the continuous permafrost zone is thought to have been at the Wisconsin shoreline of the Bering Shelf and to have extended eastward across Bristol Bay to the mountain ice fields of southern Alaska (Figure 9-9). Permafrost existed on isolated unglaciated hills within the mountains and probably in the southern Kenai Mountains during Wisconsin time (Bailey, 1980).

Permafrost was thicker than today in most places in Alaska and northwestern Canada (Mackay, 1976a). Many temperature profiles show that the permafrost is not now in equilibrium with the present climate at the sites of measurement (Lachenbruch, 1968: 835; Lachenbruch et al., 1962; Lachenbruch and Marshall, 1969). For example, of the 356 m of permafrost observed in the drill hole in Ogotoruk Valley, 35 km south of Point Hope in northwestern Alaska, only 260 m would exist if present surface conditions were to persist for several thousand years; that is, about 25% of the permafrost is a product of an extinct climate (Lachenbruch et al., 1966: 158).

A rigorous climate must have existed during Wisconsin time over the exposed continental shelves of the Beaufort, Chukchi, and Bering Seas, for ice-wedge casts (Figures 9-9 and 9-10) occur on islands of the Bering Sea and widespread cryoplanation terraces extend to the Bering Sea in western and southwestern Alaska. That the exposed continental shelves were perennially frozen, at least in the north, is shown by the presence of extensive relict subsea permafrost up to 730 m thick (Péwé, 1979b) (Figure 9-1). It is not known, however, whether all of the subsea permafrost originated in Wisconsin time. Subsea frozen ground off the Arctic coast of northwestern North America has warmed and thinned considerably (probably from about −33 °C or −15 °C to about 0 °C to −3 °C today) during the last 10,000 years, and permafrost formerly present on the continental shelf in the Bering Sea has apparently disappeared.

Peat deposits are widespread in central and western Alaska today in the poorly drained terrain that is underlain by permafrost. In Wiscon-

sin time, poorly drained peat deposits undoubtedly existed on the exposed continental shelves, especially in the Bering Sea and Norton Sound. These peat beds have subsequently been buried by Holocene marine sediments. Methane generated by microbial processes in this peat throughout the Bering Sea is intermittently vented to the surface (Larson et al., 1981).

Wisconsin Periglacial Environment in the Contiguous United States

In contrast to the interest and advancement of research in the field of modern periglacial features and processes in Alaska, emphasis in the temperate contiguous United States has been on the periglacial environment of Wisconsin time. Some research on the modern periglacial environment of the high mountains is under way, especially on active rock glaciers, but most work concerns an earlier, more widespread periglacial environment. Periglacial phenomena in the United States have been observed for 100 years, and at least 100 sites have been reported by more than 150 investigators. Most reported phenomena are Wisconsin or Late Wisconsin in age, although some are older.

PERMAFROST AND PERMAFROST INDICATORS

Ice-Wedge Casts

Ice-wedge casts and inactive sand wedges and the polygonal pattern they reflect are the best indicators of former permafrost and a rigorous periglacial environment. Casts form when ice wedges melt slowly as the permafrost table is lowered, generally the result of a warming climate. As the top of the ice wedge is melted down, a pronounced polygonal pattern of shallow trenches generally appears on the ground surface. Surface runoff is channeled into these furrows. The water percolates down into the melting ice wedges, carrying the sediments downward. As the ice wedge is further lowered and the surrounding sediments thaw, losing their cementing ice, they tend to collapse toward and into the area formerly occupied by the ice wedge. This is especially true in coarse, sandy gravel. In loess, the walls may stand, and the fill may originate only from above. The collapse of the bordering material, plus the downwashing of fines, results in a mixture of sediments from above and along side.

Although authentic ice-wedge casts in the United States have been described in the literature since 1949, it has only been in the last 15 years that hundreds of casts of Wisconsin age have been found near the position of the border of the former ice sheets. Most casts reported before 1965 fall into the doubtful category. Except for casts in areas of former alpine permafrost, all lie within 100 km of the Laurentide ice limit (Figure 9-11 and Table 9-2).

Ice-wedge casts have been reported from Montana (Colton, 1955; Schafer, 1949) to New Jersey (Walters, 1978). More than 100 ice-wedge casts are known to exist in southwestern North Dakota just south of the glacial limit (Clayton and Bailey, 1970). Mears (1962, 1966, 1973, 1981; personal communication, 1980) reports that more than 200 wedge casts from 26 localities are known to exist (Figure 9-12). They occur in various substrates between the high plains in eastern Wyoming and the even-higher intermountain basins in the central and western parts of the state (Figure 9-11), where a rigorous climate still exists today.

Most ice-wedge casts in the central United States have been reported near the glacial border by Black (1964a, 1964b, 1965a, 1965b) in and near the driftless area of Wisconsin (Table 9-2), where more than 200 casts are known. The casts that exist south of the glacial limit are thought to have formed between about 14,000 and 22,000 years ago, although some could have formed earlier. The large reentrant in the ice sheet was evidently a favorable locality for the formation of ice wedges during Late Wisconsin time. However, ice-wedge casts are also likely to be abundant elsewhere; the reporting of many excellent wedges from the driftless area is largely the result of the vast experience of the trained observer of ice wedges and ice-wedge casts who identified them (Black, 1965a). Periglacial phenomena have been noted in the driftless area since before the turn of the century, and yet the first ice-wedge casts were reported only within the last 20 years.

More than 250 ice-wedge casts and associated polygonal ground

Table 9-2.
Localities of Ice-Wedge Casts of Wisconsin Age
in the Contiguous United States

Map Location	Locality	Time of Ice-Wedge Formation	Distance from Late Wisconsin Ice Limit (km)	Reference
		Outside Late Wisconsin Ice Limit		
A	Near Vaughn, Montana	Wisconsin	16	Schafer, 1949
C	Southwestern North Dakota	Early Wisconsin (?)	10-100 (?)	Clayton and Bailey, 1970
D	Wyoming	Wisconsin	300-600	Mears, 1981
E	Northwestern Iowa	Wisconsin	10-100 (?)	Ruhe, 1969
E	Tama County, Iowa	Wisconsin	40 (?)	Ruhe, 1969
F	Southwestern Wisconsin	⁻20,000 yr B.P.	0.5-100	Black, 1965a
H	New Jersey	Wisconsin	10-45	Walters, 1978
S	Thurston County, Washington*	Wisconsin, ⁻14,000 yr B.P.	At border	Péwé, 1948
S	Thurston County, Washington	Wisconsin, ⁻14,000 yr B.P.	At border	Newcomb, 1952
S	Thurston County, Washington	Wisconsin, ⁻14,000 yr B.P.	At border	Ritchie, 1953
T	Central Iowa*	Wisconsin	70	Wilson, 1958
		Inside Late Wisconsin Ice Limit		
B	Wolf Point, eastern Montana	Late Wisconsin	25	Colton, 1955
F	Outagamie County and Columbia County, Wisconsin	< 12,000 yr B.P.	50	Black, 1965a
G	Central Illinois	⁻14,000 yr B.P.	50	W. H. Johnson, 1978
U	Bureau County, northern Illinois	⁻14,000 yr B.P.	30	Horberg, 1949
V	Woodford County,	Late Wisconsin	30	Frye and Willman, 1958
W	West-central Indiana*	< 14,500 yr B.P.	5	Wayne, 1965, 1967
X	Richland County, Ohio*	< 14,000 yr B.P.	5	Totten, 1973
Y	Southern Connecticut*	Late Wisconsin	⁻50	Denny, 1936
Z	Rhode Island*	Late Wisconsin	⁻70	Birman, 1952

Note: Map-locality letters correspond to letters on Figure 9-11.
*Doubtful ice-wedge casts.

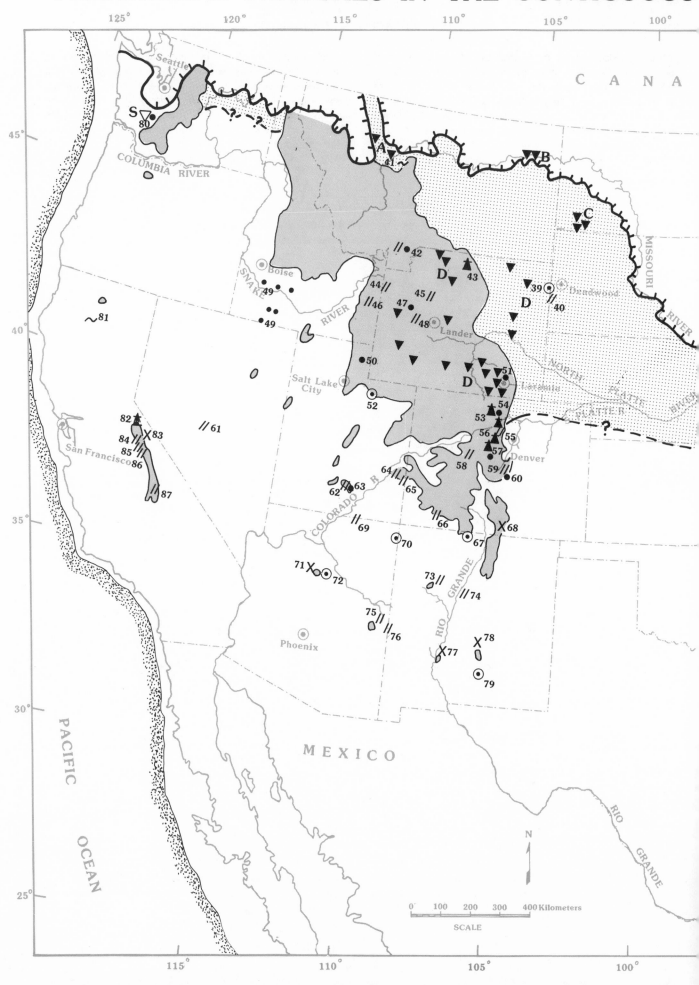

UNITED STATES IN WISCONSIN TIME

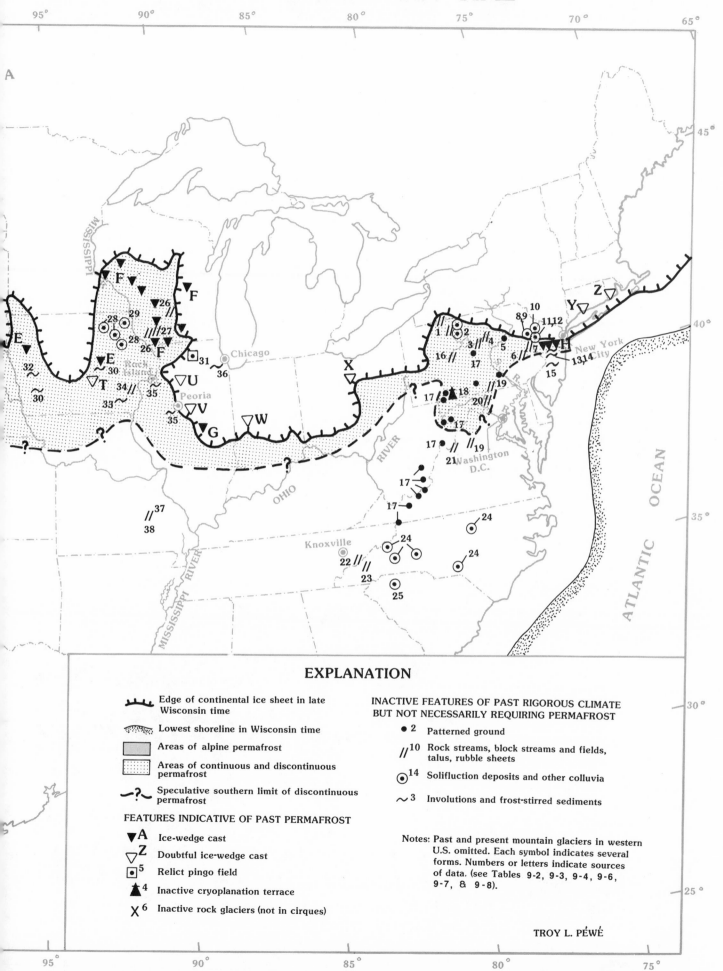

EXPLANATION

Edge of continental ice sheet in late Wisconsin time

Lowest shoreline in Wisconsin time

Areas of alpine permafrost

Areas of continuous and discontinuous permafrost

-?- Speculative southern limit of discontinuous permafrost

FEATURES INDICATIVE OF PAST PERMAFROST

▼A   Ice-wedge cast

▽Z   Doubtful ice-wedge cast

⊡5   Relict pingo field

▲4   Inactive cryoplanation terrace

X 6   Inactive rock glaciers (not in cirques)

INACTIVE FEATURES OF PAST RIGOROUS CLIMATE BUT NOT NECESSARILY REQUIRING PERMAFROST

● 2   Patterned ground

//10   Rock streams, block streams and fields, talus, rubble sheets

⊙14   Solifluction deposits and other colluvia

∼3   Involutions and frost-stirred sediments

Notes: Past and present mountain glaciers in western U.S. omitted. Each symbol indicates several forms. Numbers or letters indicate sources of data. (see Tables 9-2, 9-3, 9-4, 9-6, 9-7, & 9-8).

TROY L. PÉWÉ

Figure 9-11. Periglacial features of Wisconsin age in the contiguous United States. (Numbers refer to localities listed in Tables 9-2, 9-3, 9-4, 9-6, 9-7, and 9-8.)

patterns are reported by Walters (1978) in northern New Jersey within 10 to 45 km of the Wisconsin glacial limit. Convincing photographs and diagrams support the interpretation that they are ice-wedge casts and represent a rigorous periglacial environment at this locality in Wisconsin time, when the Atlantic shoreline probably was about 200 km east of its present position (Figure 9-11).

A number of other structures that have been reported along the ice-sheet limit and referred to as ice-wedge casts are of doubtful origin. For example, the much-discussed Mima Mounds in Thurston County, Washington, just south of the limit of the Puget lobe of the Cordilleran ice sheet (Newcomb, 1952; Péwé, 1948; Ritchie, 1953), although once thought to have resulted from the erosion of ice-wedge polygons, do not show any typical ice-wedge casts associated with the polygonal surface structures (T. L. Péwé, unpublished data, 1952, 1973).

Wedge structures in Iowa and Illinois reported by Wilson (1958), Horberg (1949), and Frye and Willman (1958), as well as those in Indiana noted by Wayne (1965, 1967) and in Ohio by Totten (1973), have been questioned by both Péwé (1973) and Black (1976).

Several decades ago, possible ice-wedge casts were reported in Connecticut (Denny, 1936) and Rhode Island (Birman, 1952). From the descriptions, however, these do not appear to have been ice-wedge casts, and Denny (personal communication, 1972) now concurs that his description of periglacial features did not include ice-wedge casts. R. F. Black (personal communication, 1972) has examined wedge forms in Connecticut and Massachusetts and concluded that they were not true ice-wedge casts. He also believes that the climate probably was too mild for ice wedges to have formed as the Late Wisconsin glacier terminus withdrew from that area (Black, 1980).

Relict Pingos

Because open- and closed-system pingos form only in perennially frozen ground, pingo scars (or relict pingos) are indisputable evidence of the presence of former permafrost and a rigorous periglacial environment. Reports of relict pingos are common in Europe, but only one locality has been relatively well documented in the temperate United States. Investigations by Flemal and others (1970, 1973; Flemal, 1972, 1976) indicate that more than 500 circular to elliptical mounds in the late-Pleistocene deposits of north-central Illinois (Figure 9-11) were formed within the lakes of pingo craters. (D. M. Hopkins [personal communication, 1981] suggests that they may be relict thaw lakes and not pingo scars. This view does not change the paleoenvironmental significance of the features.) Successive generations of pingos are thought to have developed at favorable sites and to have produced overlapping and superimposed mounds. These pingos appear to have been closed-system pingos similar to those found in the Arctic today.

The remnants of pingos in areas of former periglacial environments are probably more abundant than current awareness indicates, as has been suggested by Pissart (1968). There is no reason to believe that pingos, which are so abundant in some parts of the modern periglacial environment, were less common in similar environments in the past. Former pingos have not been identified in many areas due, in part, to the lack of attention paid to the small-scale topographic features of most of the landscape in which they are likely to occur (Flemal et al., 1973: 246-47).

Inactive Cryoplanation Terraces

Unlike ice-wedge casts, pingo scars, and inactive rock glaciers, cryoplanation terraces apparently have never been studied in the temperate United States, even though they may be widespread in the West. Before 1965, such features were mentioned at a few localities in the West (Demek, 1969; Mackin, 1947; Russell, 1933; Scott, 1965) (Tables 9-3 and 9-4), and these appear to be authentic cryoplanation terraces. Well-developed terraces occur in the Colorado Front Range (T. L. Péwé, unpublished data), and cryoplanation terraces may also exist in the Appalachian Mountains just south of the Laurentide drift border. The six sites of cryoplanation terraces recorded in the West (Figure 9-11) are in the high mountains, where other evidence also suggests a rigorous periglacial climate during Wisconsin time.

Figure 9-12. Polygonal pattern of underlying ice-wedge casts exposed along U.S. Interstate 80 about 24 km east of Rawlins and 100 km northwest of Laramie, Wyoming. (Photograph by Brainerd Mears, Jr.)

Table 9-3.
Localities of Periglacial Phenomena of Wisconsin Age in the Cordillera Area
of the Western Contiguous United States

Map Location	Feature	Location	Reference(s)
41	Involutions	Northwestern Montana	Schafer, 1949: 156
42	Patterned ground and block fields	Beartooth Mountains, Wyoming	Smith, 1950: 1503; Smith, 1949a: Plate 1, Figure 1
43	Cryoplanation terraces	Big Horn Mountains Wyoming	Mackin, 1947: 116
44	Talus	Teton Mountains, Wyoming	T. L. Péwé, unpublished data, 1974
45	Talus	Owl Creek Mountains, Wyoming	T. L. Péwé, unpublished data, 1941
46	Talus	Lemhi Range, Idaho	Ruppel and Hait, 1961: B-164
47	Patterned ground	Wind River Mountains, Wyoming	Richmond, 1949: Plate 1
48	Talus and patterned ground	Wind River Mountains, Wyoming	Holmes and Moss, 1955: 643
49	Patterned ground	Snake River Plain, Idaho	Malde, 1961, 1964
50	Patterned ground	Bear River Range, Utah	De Graff, 1976: 116
51	Patterned ground	Medicine Bow Peak, Wyoming	Mears, 1962: 48
52	Landslides	Unita Mountains, Utah	Atwood, 1909: 63
53	Cryoplanation terraces	Colorado Front Range	Scott, 1965: 15
54	Patterned ground	Rocky Mountain National Park, Colorado	Bradley, 1965: 31
55	Talus	Colorado Front Range	T. L. Péwé, unpublished data, 1945
56	Cryoplanation terraces	Colorado Front Range	Russell, 1933: 942
57	Cryoplanation terraces	Near Pikes Peak, Colorado	T. L. Péwé, unpublished data, 1969
58	Talus	Mount Elbert, Colorado	T. L. Péwé, unpublished data, 1945
59	Talus and rubble field	Colorado Front Range, Colorado Springs, Colorado	T. L. Péwé, unpublished data, 1945
60	Patterned ground	Colorado Front Range	T. L. Péwé, unpublished data, 1952
61	Talus	Western Nevada	Blackwelder, 1935: 317
62	Talus	Boulder Mountain, Utah	Flint and Denny, 1958: 133
63	Rubble sheets	Table Cliffs, Utah	Shroder, 1973: 511
64	Rock streams	La Sal Mountains, Utah	Richmond, 1952: 1292
65	Rock streams	La Sal Mountains, Utah	Richmond, 1962
66	Talus	San Juan Mountains, Colorado	Hyers, 1980
67	Landslides and solifluction	Rio Arriba County, New Mexico	Smith, 1936
68	Rock glacier	Veta Park, Colorado	Patton, 1910
69	Talus	Navajo Mountain, Utah	Blagbrough and Breed, 1967, 1969
70	Solifluction deposits	Chuska Mountains, Arizona	Blagbrough
71	Rock glacier	Kendrick Peak, Arizona	Barsch and Updike, 1971a: 102; 1971b
72	Solifluction deposits	San Francisco Peaks, Arizona	Péwé and Updike, 1970: Figure 9; 1976: 54
73	Talus	Mount Taylor, New Mexico	T. L. Péwé, unpublished data, 1970
74	Talus	Sandia Mountains, New Mexico	T. L. Péwé, unpublished data, 1972
75	Talus	White Mountains, Arizona	Merrill and Péwé, 1977: 40
76	Talus	Escudillo Mountain, Arizona	T. L. Péwé, unpublished data, 1968

Table 9-3 continued.

Map Location	Feature	Location	Reference(s)
77	Rock glacier	San Mateo Mountains, New Mexico	Blagbrough and Farkas, 1968
78	Rock glacier	Capitan Mountains, New Mexico	Blagbrough, 1976: 570
79	Solifluction deposits	Sacramento Mountains, New Mexico	Galloway, 1970: 247

Note: Map-locality numbers correspond to numbers on Figure 9-11.

Table 9-4.
Localities of Periglacial Phenomena of Wisconsin Age on the
West Coast of the Contiguous United States

Map Location	Feature	Location	Reference(s)
80	Patterned ground	Puget Sound, Washington	Eakin, 1932: 536
81	Involutions	North of Redding, California	Prokopovich, 1969: 66
82	Cryoplanation terrace	Monitor Pass, California	Demek, 1969
83	Rock streams	Head of Yosemite, Sierra Nevada	Kesseli, 1941
84	Talus and rubble sheets	Sheperd's Crest, Sierra Nevada	Fryxell, 1962: 114, 163
85	Talus and rubble sheets	Head of Yosemite, Sierra Nevada	Fryxell, 1962: 48
86	Talus and rubble sheets	Head of Sequoia, Sierra Nevada	Matthes, 1950: 74-75
87	Talus and rubble sheets	Mount Whitney, Sierra Nevada	Fryxell, 1962: 156, 157, 182

Note: Map-locality numbers correspond to numbers on Figure 9-11.

common in the higher mountains of the western United States and Canada (Table 9-5 and Figure 9-13) (Brown, 1925; P. G. Johnson, 1975, 1978; R. B. Johnson, 1967; Kesseli, 1941; Libby, 1968; Luckman and Crockett, 1978; Potter, 1972; White, 1976).

Other evidence indicating the presence of perennially frozen ground is the existence of ice wedges, active cryoplanation terraces, and pingos. None of these has been reported from the mountains of western North America, although inactive forms are known (Péwé, 1973). In addition, ground ice in the form of lenses, granules, or pore fillings at a considerable depth that has existed for 2 or more years indicates alpine permafrost. Several pieces of such evidence have been reported (Table 9-5), ranging from ice in mining and engineering excavations to ice found during trenching for alpine soil studies.

Extrapolations of the mean annual temperatures from valley weather stations to mountain tops indicate that the higher parts of the mountains have a climate with mean annual temperatures of 0 °C or colder. Since perennially frozen ground generally forms when the mean annual air temperature of an area is slightly colder than 0 °C,

Inactive Rock Glaciers

Many hundreds of inactive rock glaciers undoubtedly exist in the high mountains of the western United States. Many are in cirques and are Holocene in age; however, many others inside or outside cirques can be shown to be Wisconsin in age. Few inactive rock glaciers have been studied, except for some in the far South that denote alpine permafrost, almost certainly in Wisconsin time (Barsch and Updike, 1971a, 1971b; Blagbrough, 1976; Blagbrough and Farkas, 1968; Patton, 1910) (Table 9-3). The inactive rock glaciers about 350 m below the elevation of active rock glaciers indicate, for example, that permafrost was once present at an elevation of about 2800 m in the southern part of New Mexico (Figure 9-14).

Modern Alpine Permafrost

Prospectors, miners, and engineers in the high mountains of western North America from Colorado to Alaska have been aware for more than 100 years that mining shafts and drifts have encountered perennially frozen ground (as is indicated by the presence of ice cracks, seams, vugs, or any other opening) (Bateman and McLaughlin, 1920; Brown, 1969: 28; Brown, 1925: 466; Weiser, 1875; Wernecke, 1932), but most of this information has not been recorded in the scientific literature. Because the high-mountain regions of the world exist in cold, rigorous climates similar to the Arctic and sub-Arctic, one would expect both continuous and discontinuous permafrost to be present (Ives, 1974: 184).

Permafrost at high altitudes in polar and subpolar areas where permafrost also exists in the lowlands is generally mapped as part of the zones of continuous and discontinuous permafrost (Figure 9-1); however, in the more-southern latitudes of the Northern Hemisphere, high-altitude permafrost is referred to as alpine permafrost. According to Fujii and Higuchi (1978), 2.3% of the land area in the Northern Hemisphere and about 11% of the area of permafrost in the Northern Hemisphere are underlain by alpine permafrost.

The best evidence for the existence of perennially frozen ground is the measurement of negative temperatures over several years at a depth greater than the thickness of the active layer. Another good indicator of the presence of perennially frozen ground in alpine areas is the existence of active ice-cemented (lobate) rock glaciers, which require permafrost in order to form and to remain active. Harris (1981) places them in the colder parts of the discontinuous permafrost zone. Ice-cemented rock glaciers indicating the existence of permafrost are

Table 9-5.
Alpine Permafrost in Western North America

Map Location	Locality	Altitude (m)	N. Latitude	Reference(s)	Features
	Rocky Mountains				
1	Cassier, British Columbia	1370	59°17′	Brown, 1969: 28	Mining excavations
2	Jasper National Park, Alberta	2150	52° 50	Luckman and Crockett, 1978: 545	Ice-cemented rock glaciers
3	Banff, Alberta	2347	51°50′	Olgivie and Baptie, 1967: 744	Frozen peat
4	Banff, Alberta	2655	51°50′	Scotter, 1975: 93	Ice in the ground
5	80 km southwest of Calgary	2224	51°00′	Harris and Brown, 1978: 387	Temperature measurements
6	Beartooth Mountains, Wyoming	3230	45°00′	Johnson and Billings, 1962: 121	Bogs with ice
7	Beartooth Mountains, Wyoming	2950	44°53′	Pierce, 1961: B-155	Ice in peat
8	Yellowstone National Park, Wyoming	2600	44°50′	Pierce, 1979: F-7	
9	Yellowstone National Park, Wyoming	2400	44°50′	Pierce, 1979: F-7	
10	Absaroka Mountains, Wyoming	2700	44°38′	Potter, 1972	Ice-cemented rock glacier
11	Niwot Ridge, Colorado	3500	40°04′	Ives and Fahey, 1971; Ives, 1974: 187	Temperature measurements
12	Colorado Front Range	3500	40°05′	White, 1976: 83	Ice-cemented rock glaciers
13	McClellan Mountain, Colorado Front Range	4000	39°50′	Weiser, 1875: 77	Permanent ice in mine in bedrock
14	Sangre de Cristo Range, Colorado	2800	37°40′	R. B. Johnson, 1967: D-218	Ice-cemented rock stream
15	San Juan Mountains,	3850	37°30′	Hyers, 1980: 102	Temperature extrapolation (−2.2)
16	San Juan Mountains, Colorado	3500	37°30′	Howe, 1909: 33	Ice-cemented rock glacier
	San Juan Mountains,	3500	37°30′	S. E. White, personal communication, 1981	Ice-cemented rock glacier
	San Juan Mountains, Colorado	3500	37°30′	Spencer, 1900: 188	Ice-cemented rock glacier
	San Juan Mountains, Colorado	3500	37°30′	Brown, 1925: 465	Ice-cemented rock glacier and permanent ice in mine in bedrock
17	Tesuque Peak, New Mexico	3720	35°47′	Retzer, 1965: 38	Alpine turf (frozen)
18	White Mountains, Arizona	3475	33°50′	Merrill and Péwé, 1977: 4	Temperature extrapolation (< −1.1°C)
19	Istaccihautl, Mexico	4700	19°30′	Lorenzo, 1969: 168	Active block fields
20	Volcanic mountain, Mexico	4800	19°05′	Gorbunov, 1978b: 284	
	West Coast				
22	North Cascade Range, Washington	2200	48°13′	Libby, 1968: 318	Ice-cemented rock glacier
23	Head of Yosemite, Sierra Nevada, California	3300	38°00′	Kesseli, 1941: 205	Ice-cemented rock glacier
24	Mount Whitney, California	4300	36°35′	Retzer, 1965: 38	Temperature extrapolation

Note: Map-locality numbers correspond to numbers on Figures 9-13 and 9-15.

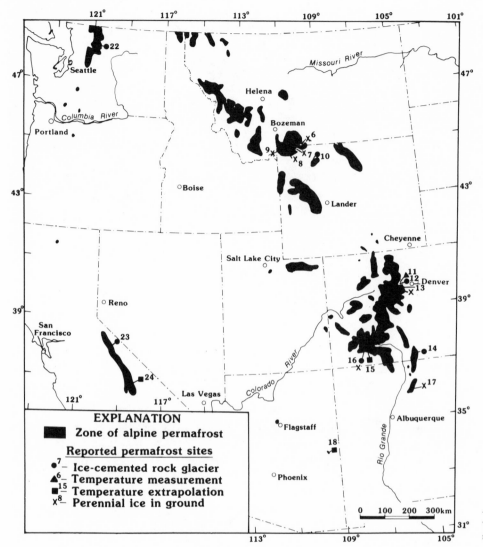

Figure 9-13. Map of alpine permafrost in the western contiguous United States. (Numbers refer to localities listed in Table 9-5.)

the existence of at least isolated bodies of permafrost can be inferred (Table 9-5).

Alpine permafrost in the contiguous United States is almost entirely limited to the high mountains of the West, although it has long been known from Mount Washington in New Hampshire (Antevs, 1932; Goldthwait, 1969; Howe, 1971; Schafer and Hartshorn, 1965: 125), where the mean annual air temperature is − 2.8 °C. Ives (1974: 185) states that 30 m of permafrost was encountered in a well on the summit of nearby Mount Katahdin in Maine.

About 100,000 km² of alpine permafrost exists in the western contiguous United States (Figure 9-13 and 9-5). Permafrost exists as low as 2500 m elevation in the northern United States to about 3500 m in Arizona (Figure 9-14). The latitudinal gradient of the elevation of the lower limit between 30° and 50° north latitude is about 80 m per degree of latitude (Figure 9-15).

Published information concerning the thickness of alpine permafrost in western North America is limited. Brown (1925: 466) states that the ice in the vugs of rock in the San Juan Mountains of Colorado extends to a depth of at least 100 m. Weiser (1875: 78) reports that a tunnel at 4000 m on the southwestern side of Mc-

Clellan Mountain in the Colorado Front Range penetrated frozen rock for more than 60 m and ended in frozen rock.

Wisconsin Alpine Permafrost

In Wisconsin time, the alpine permafrost zone of the western United States (Figure 9-11) was much larger than it is today. Permafrost extended to lower elevations and lower latitudes, and frozen ground was undoubtedly thicker. Perennially frozen ground extended in the uplands from the margin of the continental ice sheets southward to southern Arizona. Large areas existed in the North Cascades of Washington and in the Sierra Nevada. Permafrost may also have existed on many of the high, small ranges and peaks in the western states. Ice wedges were probably widespread (Mears, 1981) in the broad, intermontane basins of central and western Wyoming, which were either in the alpine- or continuous-permafrost zones. For the purposes of this chapter, the areas are placed in the alpine-permafrost zone (Figure 9-11).

The lower limit of Wisconsin alpine permafrost in northern Montana was at about 1500 m (Schafer 1949) (Table 9-6) and rose to 2000 m in Wyoming (Mears, 1981) (Figure 9-14). In southern Colorado,

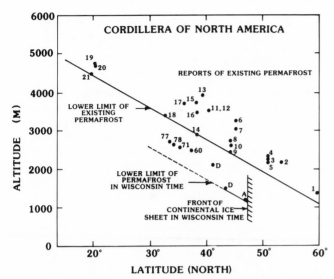

Figure 9-14. Relation between latitude and altitude of reported localities of existing alpine permafrost and alpine permafrost of Wisconsin age in the Cordillera of North America. (See Figures 9-11, 9-13, and 9-15 for maps and Tables 9-1, 9-5, and 9-7 for source data.)

inactive rock glaciers indicate a lower limit of 2500 m (Patton, 1910); in Arizona and New Mexico, their lower limit was about 2600 m (Barsch and Updike, 1971a, 1971b; Blagbrough, 1976; Blagbrough and Farkas, 1968). The mean latitudinal gradient was about 80 m per degree of latitude (Figure 9-14), which is similar to the present gradient, and the lower limit of alpine permafrost was about 1000 m lower than its modern level (Figure 9-14).

About 500,000 km² of alpine permafrost existed in the western United States during the Wisconsin. Although the Appalachian Mountains in Pennsylvania and perhaps in northern West Virginia were perennially frozen (Figure 9-11) and may be placed in the alpine-permafrost zone by some, these low mountains are placed in the continuous-discontinuous-permafrost zone in this chapter.

The evidence used so far to outline the alpine-permafrost zone of Wisconsin time (Figure 9-11 and Table 9-6) includes ice-wedge casts (Table 9-2), inactive cryoplanation terraces, and rock glaciers (Figure 9-11 and Tables 9-3 and 9-4). These data can be extrapolated to other areas by the use of contour maps, latitudinal gradient, and present climatic zonation. Reconnaissance and some detailed field examinations have also been performed in all the western states (T. L. Péwé, unpublished data).

The periglacial environment in the area of alpine permafrost was clearly colder during Wisconsin time than it is now. The 1000-m difference in the lower limit of alpine permafrost between the present and Wisconsin time (Figure 9-14 and Tables 9-2, 9-3, 9-4, and 9-6) indicates that the 0 °C mean annual isotherm was lowered approximately 1000 m, a value consistent with that of the snow-line depression in the Rocky Mountains (Porter et al., 1982). Inactive, formerly ice-cemented rock glaciers in the southwestern United States indicate that the mean annual air temperature had to be below 0 °C in order to permit permafrost to exist and rock glaciers to form. The mean annual air temperature at these localities is above freezing at present.

The mean annual air temperatures in the broad, windswept intermontane basins of southern Wyoming range from 4.8 °C at Laramie to 2.9 °C at Farson, near Lander (Mears, 1981). The widespread ice-wedge casts reported by Mears suggest that the mean annual air

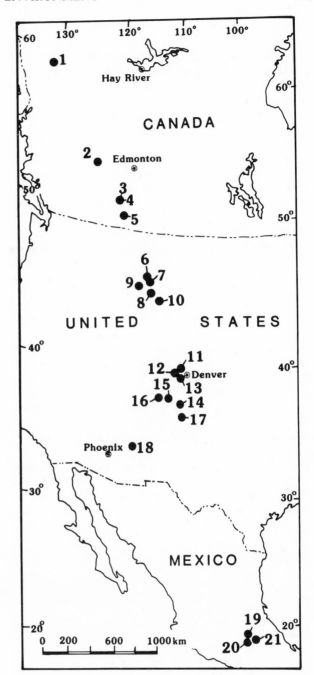

Figure 9-15. Reported localities of existing alpine permafrost in the Cordillera of North America. (Numbers refer to localities listed in Table 9-5.)

temperature during the last glaciation was about −8 °C or colder, which represents a minimum drop of 8.9 °C to 10.8 °C.

Continuous and Discontinuous Permafrost

Although it has long been assumed by many investigators that no extensive periglacial zone extended southward from the continental ice sheets in the temperate United States during Wisconsin time, recent data, together with earlier observations, permit a new interpretation of the periglacial zone (Figure 9-11). The continuous-discontinuous-permafrost zone extended from eastern Washington on the west to New Jersey on the east and from the Laurentide ice margins southward to perhaps northern Kansas, Missouri, and Kentucky (Figure 9-11).

No evidence is available to indicate permafrost in the Puget lowland of Washington State. The Mima Mounds do not prove the existence of permafrost. Perhaps the cool, wet climate supported no perennially frozen ground. Eastern Washington and western Idaho are today in a much drier continental climatic zone; during Late Wisconsin time, the area was bordered on the north by the Cordilleran ice sheet and on the east and west by glaciated mountains with alpine permafrost that extended down to an elevation of about 2000 m. In that zone, permafrost may have extended as much as 100 km southward from the ice sheet. However, no ice-wedge casts are known, and so it may have been that only discontinuous permafrost was present. Many mounds are present (Kaatz, 1959).

Widespread, nearly vertical dikes in southeastern Washington range from 1 mm to more than 1 or 2 m wide, mostly tapering downwards, and from a few centimeters to several meters long (e.g., Alwin, 1970; Jenkins, 1925; Lupher, 1941). Although most occur in the glaciofluvial deposits of the Lake Missoula flood deposits (Baker, 1982), some occur in the older Ringold Formation and in the basaltic bedrock, where the flood deposits are thin (T. L. Péwé, unpublished data). Some of the unconsolidated sediments may have been locally frozen when some of the structures were formed (R. F. Black, unpublished data; Newcomb et al., 1972). Although Alwin (1970) believes that the dikes were ice-wedge casts, after field examinations both Black (unpublished data) and Péwé reject the idea of widespread permafrost and ice wedges existing in the Pasco Basin in Late Wisconsin time.

The periglacial zone that fronted on the Laurentide ice sheet was broad in the vicinity of the Rocky Mountains and probably narrowed considerably to the Appalachian Mountains (Figure 9-11). The belt of permafrost east of the Rockies at an elevation of 1000 to 2000 m was a rigorous climatic zone and extended about 700 km south into northern Colorado and Kansas. Several hundred ice wedges are known from Wyoming and North Dakota, and continuous permafrost probably extended into Wyoming and perhaps into northern Nebraska.

In central United States, detailed investigations of ice-wedge casts by Black (1964a, 1964b, 1965a, 1965b) and of pingo scars by Flemal (1976) indicate a zone of continuous permafrost near the glacier border, especially in the reentrant between the Des Moines lobe in Minnesota and Iowa and the ice front in eastern Wisconsin and Illinois (mostly the driftless area). The discontinuous-permafrost zone extended southward, perhaps into northern Missouri and south-central Illinois (Figure 9-11).

Although physical evidence that proves the existence of permafrost adjacent to the ice sheet between Illinois and the Appalachians is lacking at the present time, such a zone (about 100 km wide) may have existed. Widespread inactive rock streams and solifluction deposits have been reported near the drift border in southern Pennsylvania (e.g., Sevon, 1975a, 1975b), and biologic studies suggest the former presence of frozen ground (Dillon, 1956: 174).

Frye and Willman (1958: 524) believe that an arctic climate could not have extended far beyond the glacier front because there is an abundant snail fauna in the Peoria Loess less than 13 km from the drift border. Some of the same species of terrestrial mollusks that occur in loess of Late Wisconsin age in Nebraska (Leonard, 1952: 8) and Illinois (Leonard and Frye, 1960), however, also are found in Holocene loess in central and far-western Alaska and in the mountains of the central Alaska Range near glacier fronts in a modern periglacial environment (Péwé, 1955: 714, 1975b: 88). Recent work by First and Fay (1980) on mollusks of the Peoria Loess indicates that a band of tundra and taiga existed south of the continental ice front.

The widespread occurrence of inactive patterned ground, rock streams, and solifluction deposits in the Appalachian Mountains at high altitudes in Pennsylvania, West Virginia, and Maryland south of the Laurentide drift border (Figure 9-11) suggests the former presence of at least discontinuous permafrost, and the presence of

Table 9-6.

Localities of Periglacial Phenomena in the Western Contiguous United States
Indicating Alpine Permafrost of Wisconsin Age

Map Location	Locality	Altitude (m)	N. Latitude	Phenomenon	Reference(s)
A	Vaughn, Montana	1300	48°00′	Ice-wedge casts	Schafer, 1949: 165
D	Big Horn Basin, Wyoming	1500	44°00′	Ice-wedge casts	Mears, 1981: 179
D	Laramie Basin, Wyoming	2200	41°00′	Ice-wedge casts	Mears, 1981: 173
68	Veta peak, Colorado	2450	37°20′	Inactive rock glacier (not in cirque)	Patton, 1910
71	Kendrick Peak, Arizona	2650	35°30′	Inactive rock glacier (not in cirque)	Barsch and Updike, 1971a, 1971b
78	Capitan Mountains, New Mexico	2560	33°40′	Inactive rock glacier (not in cirque)	Blagbrough, 1976: 570
77	San Mateo Mountains, New Mexico	2700	33°30′	Inactive rock glacier (not in cirque)	Blagbrough and Farkas, 1968: 817

Note: Map-locality numbers correspond to numbers on Figure 9-11.

Table 9-7.
Localities of Periglacial Phenomena of Wisconsin Age
in the Eastern Contiguous United States

Map Locality	Feature	Location	Reference(s)
1	"Rock cities"	Southwestern New York	Smith, 1953b: 1474
2	Colluvium and rock streams	Potter County, Pennsylvania	Denny, 1956: 33-39
3	Block fields	Bald Eagle Mountain and others, central Pennsylvania	Peltier, 1945: Kirby, 1965: 626
4	Block fields and patterned ground	Near State College, Pennsylvania	Rapp, 1967, 234
5	Patterned ground	Central Pennsylvania	Denny, 1951: 122
6	Rock streams	Blue Rocks, Pennsylvania	Potter and moss, 1966: 273-74; 1966; Wherry, 1923; Ashley, 1933: 88
7	Rock stream	Hickory Run, Pennsylvania	Smith, 1953a; Sevon, 1975b
8	Boulder field	Carbon County, Pennsylvania	Sevon, 1967
9	Boulder Field	Carbon County, Pennsylvania	Sevon, 1969: 221-27
10	Boulder colluvium	Monroe County, Pennsylvania	Sevon, 1975a: 58-59
11	Boulder colluvium	Monroe County, Pennsylvania	Sevon, 1972: 43-44
12	Boulder colluvium	Lehigh County, Pennsylvania	Sevon and Berg, 1978: 2-7
13	Involutions	Northern New Jersey	Wolfe, 1953: Plate 2A
14	Frost-stirred sediments	Princeton, New Jersey	Judson, 1965: 134
15	Involutions	Burlington, New Jersey	Richards and Rhodehamel, 1965: 13
16	"Rock cities"	Clearfield County, Pennsylvania	Ashley, 1933: 88
17	Patterned ground	The high mountains of the Appalachian Mountains, Pennsylvania, Virginia, and West Virginia	Clark, 1968: 1969
18	Cryoplanation terraces	Negro Mountain, Pennsylvania	T. L. Péwé, unpublished data, 1972
19	Rock streams	Blue Ridge Mountains, Pennsylvania and Virginia	Smith and Smith, 1945: 1198
20	Block stream	South Mountain, Maryland	Smith, 1949b: Plate 1, Figure 2
21	Block fields	West Virginia-Virginia boundary, Appalachian Mountains	Hack and Goodlett, 1960: 16
22	Block fields*	Central Great Smoky Mountains, Tennessee	King, 1964: 136
23	Block fields*	Eastern Great Smoky Mountains, North Carolina	Hadley and Goldsmith, 1963: B-108
24	Solifluction deposits*	Base of Blue Ridge Mountains and Piedmont, North Carolina	Kerr, 1881: 352
25	Colluvium*	Spartanburg County, South Carolina	Eargle, 1940: 337 Bryan, 1940: 523

Note: Map-locality numbers correspond to numbers on Figure 9-11.
*May be pre-Wisconsin in age.

Table 9-8.
Localities of Periglacial Phenomena of Wisconsin Age in
the Central Contiguous United States

Map Location	Features	Location	Reference(s)
26	Talus, block streams, frost-stirred sediments, and solifluction deposits	Driftless area, Wisconsin	Smith, 1949a, 1949b; Squire, 1897
27	Talus and block streams	Driftless area, Wisconsin	Black, 1964a, 1964b; 1969a, 1969b
28	Solifluction and colluvium	Northeastern Iowa	Hedges, 1972: 90
29	Colluvium	Northeastern Iowa	Schafer, 1962A 262
30	Involutions	Central Iowa	Ruhe, 1969: 179
31	Pingo field	Northern Illinois	Flemal, 1972, 1976; Flemal et al., 1973
32	Involutions	Crawford County, southwestern Iowa	Lees, 1927: Figure 58
33	Involutions	Southeastern Iowa	Schafer, 1953: 405
34	"Rock cities"	Wildcat Den, Iowa	T. L. Péwé, unpublished data, 1939
35	Involutions	West-central Illinois	Frye and Willman, 1958: 521
36	Involutions	Northeastern Illinois	Sharp, 1942
37	Colluvium*	St. Francois Mountains, Missouri	Peltier, 1950: 229
38	Colluvium*	St. Francois Mountains, Missouri	T. L. Péwé, unpublished data, 1939
39	Solifluction	Black Hills, near Custer, South Dakota	Norton and Redden, 1960
40	Talus	Harney Peak, Black Hills, South Dakota	T. L. Péwé, unpublished data, 1941

Note: Map-locality numbers correspond to numbers on Figure 9-11.
*May be pre-Wisconsin in age.

possible cryoplanation terraces further strengthens this possibility. However, because ice-wedge casts, pingo scars, and inactive, formerly ice-cemented rock glaciers are not known, the existence of former continuous permafrost remains unproven. Nevertheless, reported periglacial features suggest that permafrost probably was present from the ice border south to at least northern West Virginia, as indicated in Figure 9-11. On the basis of pollen studies, Maxwell and Davis (1972: 522) indicate that a broad tundra belt, similar to the modern tundra regions in the Arctic, extended 300 km south of the ice sheet along the crest of the Appalachian Mountains.

Although northwestern New Jersey is now only 10 to 50 km from the sea, it was about 200 km farther from the shoreline in Late Wisconsin time (Figure 9-11), and the region may have had a more continental climate. Walters (1978) reports more than 250 ice-wedge casts as indications of continuous permafrost conditions. Reconstructed sea-surface temperatures of 18,000 years ago indicate that polar water extended south along the east coast of the United States at the time and probably contributed to the cold, periglacial conditions south of the ice sheet's terminus (CLIMAP Project Members, 1976).

The width of the permafrost zone in the United States may have been narrower than that reported from central Europe, for in North America the permafrost zone in many areas was less than 200 km wide (Brown and Péwé, 1973; Péwé, 1973) (Figure 9-11). Perhaps the zone was restricted because the southern limit was close to latitude 40 °N, in contrast to latitude 50 °N in Europe, and because large proglacial water bodies may have influenced the extent of the permafrost zone by their ameliorating effect (Brown and Péwé, 1973; Péwé, 1973).

The zone of former active ice wedges suggests that in Late Wisconsin time tundra occupied the glacier-free parts of Montana, Wyoming, North Dakota, northern Nebraska, Iowa, western Illinois, perhaps a narrow belt adjacent to the ice front from Illinois to New Jersey, and the northern Appalachian Mountains. Studies of fossil pollen tend to support this hypothesis (Dillon, 1956; Watts, 1982). The periglacial evidence suggests that the mean annual air temperature was colder than −6 °C in this tundra zone.

Although all the information needed to reconstruct in detail the Late Wisconsin periglacial climate of the central United States is not yet available, it is reasonable to assume that the reported ice-wedge casts reflect a rigorous tundra environment in the early and middle Woodfordian of about 20,000 to 14,000 years ago. Frye and Willman (1973: 150) state that the most intense cold of Woodfordian time probably occurred about 20,000 to 19,000 years ago, and Flemal and others (1973) believe that closed-system pingos near Dekalb, Illinois, required continuous permafrost and formed 16,000 to 15,000 years ago.

OTHER PERIGLACIAL FEATURES

Patterned ground, block streams, rock streams, block fields, rubble fields, talus, "rock cities," solifluction deposits, colluvium, *grèze litèes*, frost-stirred sediments, and similar features with various other names are active and widespread in polar and high-altitude periglacial environments. They all suggest but do not prove the presence of permafrost (Black, 1969b; Goldthwait, 1976; Jahn, 1978; Péwé, 1973: 16). (Goldthwait [1976: 34] believes that patterned soil features with a diameter of more than 2 m require the presence of permafrost.) They are generally associated with numerous daily air and ground freeze-thaw cycles, poor drainage, and long-lasting seasonally frozen ground or other impermeable layers.

Relict periglacial features of these types are widespread in Europe but less so in North America. The known localities in the United States are plotted in Figure 9-11. They occur both within and outside the projected zone of past permafrost.

Eastern United States

Most reports (Table 9-7) of periglacial features of Wisconsin age south of the Laurentide ice limit are from the eastern United States (Figure 9-11). This is because more scientists interested in such phenomena have worked there for a longer time than elsewhere in the United States and because the uplands of the northern Appalachian Mountains provide a more rigorous climate and uneven topography than is present in the lowlands of the central United States.

A great concentration of periglacial features has been reported in the Appalachian Mountains in Pennsylvania and New York (Figure 9-11). Almost all types are present except inactive, formerly ice-cemented rock glaciers. Patterned ground occurs on the tops of most flat summits from Pennsylvania to southern Virginia (Figure 9-11) (Clark, 1968). Individual studies south of the glacier border include those of Hodgson (1967), Rapp (1967), Clark, (1968), Ciolcosz and others (1971), and Troutt (1971).

Striking unvegetated rock streams composed of sandstone and quartzite blocks stretch downward in linear bands from ridge tops in the Appalachian Mountains (Table 9-7) (Ashley, 1933; Kirby, 1965; Sevon, 1967, 1969; Wherry, 1923). Hickory Run, a boulder stream, is preserved as a state park in Pennsylvania (Smith, 1953a). Potter and Moss (1966) have demonstrated that the rock accumulations overlie a well-developed soil thought to be pre-Wisconsin in age. The Wisconsin, if not Late Wisconsin, age of most of these features is further suggested by Sevon (1975a), who reports periglacial features developed on Illinoian till near the Late Wisconsin drift border in Pennsylvania. (See also Berg et al., 1977; Crowl and Sevon, 1980; Sevon, 1975a, 1975b, 1975c; and Sevon and Berg, 1978.)

Inactive mass-wastage deposits from Wisconsin time (or perhaps earlier in the South) are widespread and known from New York to South Carolina (Table 9-8 and Figure 9-11) (Bryan, 1940; Denny, 1951; Eargle, 1940; Kerr, 1881; King, 1964; Sevon, 1972).

"Rock cities," large, stabilized blocks of rock that have moved laterally along joints to form narrow passageways, are thought by Ashley (1933) and Smith (1953b) to be the work of frost action adjacent to the ice sheet in Pennsylvania and New York (Figure 9-1 and Table 9-7) in Late Wisconsin time. Involutions (cryoturbations) and frost-stirred sediments are also known from the eastern United States (e.g., Denny, 1951; Judson, 1965; Wolfe, 1953) (Table 9-7).

Although the type and distribution of periglacial phenomena in the eastern United States suggest that permafrost existed in the northern Appalachian Mountains during Late Wisconsin time (Figure 9-11), in the central Appalachian Mountains patterned ground, mass-wastage deposits, and other features may document a nonpermafrost environment.

Central United States

In addition to permafrost, a variety of periglacial features were active in front of the Late Wisconsin Laurentide ice sheet in the 1500-km-wide plains and lowlands between the Rocky Mountains and the Appalachian Mountains: involutions, frost-stirred sediments, colluvium or solifluction deposits, and taluses (Table 9-8). Patterned ground and block streams were not common, and rock glaciers apparently did not form.

Except for the long-known inactive features of the driftless area of Wisconsin, such as talus, solifluction deposits, block streams, and involutions (e.g., Black, 1964a, 1964b; Chamberlain, 1897; Hedges, 1972; Schafer, 1962; Smith, 1949a, 1949b; Squire, 1897), widespread periglacial phenomena are not reported in the nonmountainous areas of the central United States. Knowledge of the distribution of these features indicates only a bunching of data for areas near the former ice front in Wisconsin, Iowa, and Illinois (Figure 9-11). Isolated persuasive examples also occur in Illinois (Frye and Willman, 1958), Iowa (Lees, 1927), and South Dakota (Norton and Redden, 1960) (Table 9-8).

Although permafrost may not have been present in the St. Francis Mountains of southern Missouri, slope processes apparently were more active there in Wisconsin time than now. Inactive colluvium there may represent the southernmost evidence of periglacial activity in the lowlands of the central United States (Figure 9-11).

Western United States

The mountains and high plateaus of the western United States provide widespread examples of active and inactive (Wisconsin) periglacial features, such as various types of patterned ground, frost-stirred sediments, involutions, rock streams, rubble sheets, talus slopes, and other mass-wasting phenomena (Figure 9-11 and Table 9-3).

Inactive patterned ground has been reported mostly from the high mountains of Wyoming (Holmes and Moss, 1955; Mears, 1962; Richmond, 1952, 1962; Smith, 1949b, 1950), Idaho (Malde, 1961, 1964), Colorado (Bradley, 1965) (Figure 9-16), and Utah (DeGraff, 1976), but it is widespread from northern New Mexico to the Canadian border.

Mass-wastage deposits such as talus, colluvium, solifluction deposits, and landslides exist on nearly every steep mountain slope. A few reports describe inactive deposits that probably were formed in Wisconsin time when a colder climate with more freeze-thaw cycles and ice growth were common at lower altitudes. Blackwelder (1935) and Smith (1936) may have been the first to report widespread, stabilized slope deposits in the West that were probably formed at a time when a more rigorous climate prevailed. Stabilized talus (Figure 9-17) has been reported from New Mexico and Arizona and in most mountains north to the Canadian border (Table 9-3). Stabilized

Figure 9-16. Inactive stone rings of Wisconsin age on top of Colorado Front Range near Rocky Mountain National Park, Colorado. (Photograph No. PK 2511 by T. L. Péwé, September 1952.)

solifluction deposits have been reported from Arizona (Péwé and Updike, 1976); the southernmost record of such deposits is in the Sacramento Mountains of New Mexico at latitude 33 °N, where they date to 23,000 to 17,000 years ago and represent a depression of the zone of active solifluction by some 1200 to 1400 m (Galloway, 1970: 247).

DeGraff (1976) has studied 22 areas of patterned ground in the Bear River Range of north-central Utah, and he concludes that the relict patterned ground represents periglacial phenomena of late-Pleistocene age.

Most investigators have attempted to relate stabilized mass-wasting debris to cold periods of the late Pleistocene (e.g., Blagbrough and Breed, 1967, 1969; Merrill and Péwé, 1977). Richmond (1962), in a detailed stratigraphic study of the La Sal Mountains in Utah, describes frost rubble of various ages, especially late Pleistocene, and mentions that such rubble is today actively forming locally at higher altitudes in this area.

West Coast of the Contiguous United States

Reports of periglacial phenomena on the west coast are fewer than for other parts of the contiguous United States, except the south-central region. This is both because of the mild climate of the area and the lack of detailed investigations in the mountainous areas.

Isolated examples of rock streams and taluses near the crest of the Sierra Nevada have been reported (Fryxell, 1962; Kesseli, 1941) (Table 9-6 and Figure 9-11), but no systematic work has been done. However, the Sierras, Cascades, and several high glaciated sectors of the coastal mountain system should provide clear evidence of both present and past periglacial activity.

Earth mounds of various sizes and shapes in the northern and central lowlands of California have been reported for many years, but they do not appear to be periglacial phenomena.

Ice-Wedge Casts in Southern Canada

As the Laurentide ice sheet in North America withdrew progressively farther northward in latest-Wisconsin time, permafrost formed in southern Canada in the deglaciated zone, even though the general trend was toward an ameliorating climate on a worldwide scale. Ice-wedge casts (Figure 9-18 and Table 9-9) indicate continuous permafrost there, at least for a brief interval. Dionne (1975) reports more than 300 ice-wedge casts in southern Quebec; these indicate that permafrost existed there after the retreat of the ice sheet between 13,000 and 11,000 years ago. The implication is that mean annual air temperature was colder than -6 °C. He gives no positive evidence of tundra conditions in southern Quebec after about 11,000 years ago.

Conclusion

Much new research has focused on the interpretation of periglacial features of Wisconsin age in North America. The current knowledge of the type and distribution of periglacial phenomena is presented in this chapter on maps of the contiguous United States and northwestern North America, and it permits refined interpretations of paleoclimatic conditions. These summaries are more detailed than previous attempts, but no doubt they will change as additional field studies generate new information.

Figure 9-17. Stabilized talus of Wisconsin age. Sandstone blocks of the Dakota Formation on Table Mountain, 30 km south of Colorado Springs, Colorado. (Photograph No. PK 83 by T. L. Péwé, June 1945.)

Table 9-9. Localities of Ice-Wedge Casts of Wisconsin Age in Southern Canada

Map Location	Locality	Time of Ice-Wedge Formation	Distance from Maximum Late Wisconsin Ice Border (km)	Reference
1	Edmonton, Alberta	>31,000 yr B.P.	600	Westgate and Bayrock, 1964
2	Edmonton, Alberta*	Early Wisconsin	600	Berg, 1969
3	Calgary, Alberta	Pre-Late Wisconsin	⁻300	Morgan, 1969
4	Toronto, Ontario	⁻13,000 yr B.P.	⁻300	Morgan, 1972
5	15 km northwest of Quebec City, Quebec	12,000 to 11,500 yr B.P.	⁻600	Dionne, 1971
6	St. Lawrence River Valley	⁻12,000 yr B.P.	⁻600	Dionne, 1969, 1975
7	Southeastern Quebec	⁻12,000 yr B.P.	⁻600	Lagarec, 1973
8	Northern Nova Scotia	⁻12,000 yr B.P.	⁻200	Borns, 1965
9	Newfoundland	⁻11,200 yr B.P.	⁻100	Brookes, 1971

Note: Map-locality numbers correspond to numbers on Figure 9-18.
*Extinct sand wedge, not ice-wedge cast.

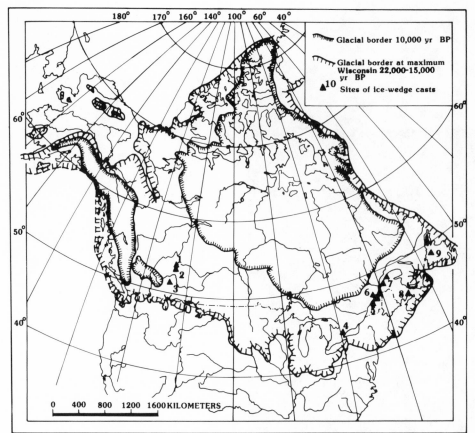

Figure 9-18. Distribution of ice-wedge casts in southern Canada in relation to position of Wisconsin glacial ice fronts. (See Table 9-9 for sources of data and age of wedges. Glacier ice borders generalized from Prest et al., 1968.) (Numbers refer to locations listed in Table 9-9.)

References

Ager, T. A. (1972). Surficial geology and Quaternary history of the Healy Lake area, Alaska. M.S. thesis, University of Alaska, Fairbanks.

Alwin, J. A. (1970). Clastic dikes of the Touchet Beds, southeastern Washington. M.S. thesis, Washington State University, Pullman.

American Commission on Stratigraphic Nomenclature (1970). ''Code of Stratigraphic Nomenclature.'' American Association of Petroleum Geologists, Tulsa, Okla.

Andersson, J. G. (1906). Solifluction, a component of subaerial denudation. *Journal of Geology* 14, 91-112.

Antevs, J. G. (1932). ''Alpine zone of Mt. Washington Range.'' Merrill and Webber, Auburn, Maine.

Ashley, C. H. (1933). ''The Scenery of Pennsylvania.'' Pennsylvania Topographic and Geologic Survey, 4th series, Bulletin G 6.

Atwood, W. W. (1909). ''Glaciation of the Uinta and Wasatch Mountains.'' U.S. Geological Survey Professional Paper 61.

Bailey, P. K. (1980). Periglacial landforms and processes in the southern Kenai Mountains, Alaska. M.S. thesis, University of North Dakota, Grand Forks.

Baker, V. R. (1983). Late-Pleistocene fluvial systems. *In* ''Late-Quaternary Environments of the United States,'' vol. 1, ''The Late Pleistocene'' (S. C. Porter, ed.), pp. 115-29. University of Minnesota Press, Minneapolis.

Barsch, D., and Updike, R. G. (1971a). Periglaziale Formung am Kendrick Peak in Nord-Arizona wärend der letzten Kaltzeit. *Geographica Helvetica* 3, 99-114.

Barsch, D., and Updike, R. G. (1971b). The Pleistocene periglacial geomorphology (rock glaciers and block fields) at Kendrick Peak, Northern Arizona. *Arizona Geological Society Digest* 9, 225-43.

Bateman, A. M., and MacLaughlin, D. H. (1920). Geology of the ore deposits of Keanecutt, Alaska. *Economic Geology* 15, 1-80.

Baulig, H. (1957). Peneplains and pediplains. *Geological Society of America Bulletin* 68, 913-30.

Benedict, J. B. (1970). Downslope soil movement in a Colorado alpine region: Rates, processes, and climatic significance. *Arctic and Alpine Research* 2, 165-226.

Benedict, J. B. (1973). Origin of rock glaciers. *Journal of Glaciology* 12, 520-22.

Benedict, J. B. (1976). Frost creep and gelifluction: A review. *Quaternary Research* 6, 55-76.

Berg, T. E. (1969). Fossil sand wedges at Edmonton, Alberta, Canada. *Biuletyn Peryglacjalny* 19, 325-33.

Berg, T. M., Sevon, W. D., and Bucek, M. S. (1977). ''Geology and Mineral Resources of the Pocono Pines and Mountain Pocono Quadrangles, Monroe County, Pennsylvania.'' Pennsylvania Topographic and Geologic Survey Atlas 204, c and d.

Bird, J. B. (1967). ''The physiography of Arctic Canada, with Special reference to the Area South of Parry Channel.'' Johns Hopkins University Press, Baltimore.

Birman, J. H. (1952). Pleistocene clastic dikes in weathered granite gneiss, Rhode Island. *American Journal of Science* 250, 721-34.

Black, R. F. (1950). Permafrost. *In* ''Applied Sedimentation'' (P. D. Trask, ed.), pp. 247-73. John Wiley and Sons, New York.

Black, R. F. (1953). Permafrost—A review. *New York Academy of Science Transactions, Series 2* 15, 126-31.

Black, R. F. (1954). Permafrost—A review. *Geological Society of America Bulletin* 65, 839-55.

Black, R. F. (1964a). Periglacial phenomena of Wisconsin, north central United States. *In* ''Proceedings of the Sixth Conference of the International Union for Quaternary Research,'' Warsaw, 1961, vol. 4, pp. 21-28.

Black, R. F. (1964b). Periglacial studies in the United States. *Biuletyn Peryglacjalny* 14, 5-29.

Black, R. F. (1965a). Ice-wedge casts of Wisconsin. *Transactions of the Wisconsin Academy of Science, Arts and Letters* 54, 187-222.

Black, R. F. (1965b). Paleoclimatologic implications of ice-wedge casts of Wisconsin. *In* ''Abstracts for the Seventh Congress of the International Association for Quaternary Research,'' Boulder, Colo., 1965.

Black, R. F. (1969a). Slopes in southwestern Wisconsin, USA, periglacial or temperate? *Biuletyn Peryglacjalny* 18, 69-82.

Black, R. F. (1969b). Climatically significant fossil periglacial phenomena in northcentral United States. *Biuletyn Peryglacjalny* 20, 225-38.

Black, R. F. (1976). Periglacial features indicative of permafrost: Ice and soil wedges. *Quaternary Research* 6, 3-26.

Black, R. F. (1980). Evidence of permafrost during the latest Wisconsinian ice advance and retreat in Connecticut. *In* "American Quaternary Association, Sixth Biennial Meeting, Abstracts and Program, 18-20 August 1980," p. 25, Institute for Quaternary Studies, University of Maine, Orono.

Blackwelder, E. (1935). Talus slopes in the Basin Range province. *In* "Proceedings, Geological Society of America for 1934," p. 317.

Blackwell, M. F. (1965). Surficial geology and geomorphology of the Harding Lake area, Big Delta quadrangle, Alaska. M.S. thesis, University of Alaska, Fairbanks.

Blagbrough, J. W. (1967). Cenozoic Geology of the Chuska Mountains. *In* "New Mexico Geological Society, Guidebook for the 18th Field Conference," pp. 70-77.

Blagbrough, J. W. (1976). Rock glaciers in the Capitan Mountains, south central New Mexico. *Geological Society of America, Abstracts with Programs* 8, 570-71.

Blagbrough, J. W., and Breed, W. J. (1967). Protalus ramparts on Navajo Mountain, southern Utah. *American Journal of Science* 265, 759-72.

Blagbrough, J. W., and Breed, W. J. (1969). A periglacial amphitheater on the northeast side of Navajo Mountain, southern Utah. *Plateau* 42, 20-26.

Blagbrough, J. W., and Farkas, S. E. (1968). Rock glaciers in the San Mateo Mountains, south-central New Mexico. *American Journal of Science* 266, 812-23.

Blouin, S. E., Chamberlain, E. J., Sellmann, P. V., and Garfield, D. E. (1979). "Penetration Tests of Subsea Permafrost, Prudhoe Bay, Alaska." U.S. Army Cold Region Research and Engineering Laboratory Report 79-7.

Boch, S. G., and Krasnov, I. I. (1951). The process of goltez-planation and the formation of altiplanation terraces. *Privoda* 5, 20-35.

Borns, H. W., Jr. (1965). Late glacial ice-wedge casts in northern Nova Scotia, Canada. *Science* 148, 1223-25.

Bradley, W. C. (1965). Glacial geology, periglacial features, and erosion surfaces in Rocky Mountain National Park. *In* "Guidebook for One-Day Conferences, Boulder Area, Colorado" (C. B. Schultz and H. T. U. Smith, eds.), pp. 27-33. Seventh Congress of the International Association for Quaternary Research, Boulder, Colo.

Brewer, M. C. (1958). Some results of geothermal investigations of permafrost in northern Alaska. *American Geophysical Union Transactions* 39, 19-26.

Brookes, I. A. (1971). Fossil ice wedge casts in western Newfoundland. *Maritime Sediments* 7, 118-22.

Brown, J. (1965). Radiocarbon dating, Barrow, Alaska. *Arctic* 18, 36-48.

Brown, J. (1966). Ice-wedge chemistry and related frozen ground processes, Barrow, Alaska. *In* "Permafrost International Conference (Lafayette, Indiana, 11-15 November 1963) Proceedings," pp. 94-98. National Academy of Sciences/National Research Council Publication 1287.

Brown, R. J. E. (1960). The distribution of permafrost and its relation to air temperature in Canada and the USSR. *Arctic* 13, 163-77.

Brown, R. J. E. (1966). Influence of vegetation on permafrost. *In* "Permafrost International 1966). Ice-wedge chemistry and related frozen ground processes, Barrow, Alaska. *In* "Permafrost International Conference (Lafayette, Indiana, 11-15 November 1963) Proceedings," pp. 94-98. National Academy of Sciences/National Research Council Publication 1287.

Brown, R. J. E. (1960). The distribution of permafrost and its relation to air temperature in Canada and the USSR. *Arctic* 13, 163-77.

Brown, R. J. E. (1966). Influence of vegetation on permafrost. *In* "Permafrost International Conference (Lafayette, Indiana, 11-15 November 1963) Proceedings," pp. 20-25. National Academy of Sciences/National Research Council Publication 1287.

Brown, R. J. E. (1967). Permafrost in Canada. Canada Geological Survey Map 1246-A, National Research Council Publication NRC-9769.

Brown, R. J. E. (1969). Factors influencing discontinous permafrost in Canada. *In* "The Periglacial Environment" (T. L. Péwé, ed.), pp. 11-53. McGill-Queen's University Press, Montreal.

Brown, R. J. E. (1970). "Permafrost in Canada." University of Toronto Press, Toronto.

Brown, R. J. E. (1978). Permafrost map. *In* "Hydrological Atlas of Canada." Department of Fisheries and Environment, Ottawa, Canada.

Brown, R. J. E., and Péwé, T. L. (1973). Distribution of permafrost in North America and its relationship to the environment: A review, 1963-1973. *In* "Permafrost: North American Contribution to the Second International Conference," pp. 71-100. National Academy of Sciences, Washington, D.C.

Brown, W. H. (1925). A possible fossil glacier. *Journal of Geology* 33, 464-66.

Brunnschweiler, D. (1962). The periglacial realm in North America during the Wisconsin glaciation. *Biuletyn Peryglacjalny* 11, 15-27.

Bryan, K. (1940). Soils and periglacial phenomena in the Carolinas. *Science* 91, 523-24.

Burns, J. J. (1964). Pingos in the Yukon-Kuskokwim Delta, Alaska; Their plant succession and use by mink. *Arctic* 17, 203-10.

Butzer, K. W. (1964). "Environment and Archeology." Aldine, Chicago.

Cailleux, A. (1967). Actions du vent et du froid entre le Yukon et Anchorage, Alaska. *Geografiska Annaler* 49A (Z-4), 145-54.

Cairnes, D. D. (1912a). Differential erosion and equiplanation of Yukon and Alaska. *Geographical Society of America Bulletin* 23, 333-48.

Cairnes, D. D. (1912b). Some suggested new physiographic terms (equiplanation, deplanation, and aplanation). *American Journal of Science*, 34, 75-87.

Capps, S. R., Jr. (1910). Rock glaciers in Alaska. *Journal of Geology* 18, 359-75.

Carter, L. D., and Galloway, J. P. (1979). Arctic Coastal Plain pingos in National Petroleum Reserve in Alaska. *In* "U.S. Geological Survey in Alaska: Accomplishments in 1978" (K. M. Johnson and J. R. Williams, eds.) pp. B33-B35. U.S. Geological Survey Circular 804B.

Chamberlain, E. J. (1979). Overconsolidated sediments in the Beaufort Sea. *The Northern Engineer* 10, 24-29.

Chamberlain, T.C .(1897). Editorial footnote. *Journal of Geology* 5, 825-27.

Ciolcosz, E. J., Clark, G. M., and Hock, J. T. (1971). "Slope Stability and Denudation Processes: Central Appalachians." Geological Society of Washington, Washington, D.C.

Clark, G. M. (1968). Sorted patterned ground: New Appalachian localities south of the glacial border. *Science* 161, 355-56.

Clark, G. M. (1969). Preliminary distribution, characteristics and origin of some Appalachian sorted patterned ground localities south of the glacial border. *Geological Society of America, Abstracts with Programs* 7, 34.

Clayton, L., and Bailey, P. K. (1970). Tundra polygons in northern Great Plains. *Geological Society of America, Abstracts with Programs* 2, 382.

CLIMAP Project Members (1976). The surface of the ice-age Earth. *Science* 191, 1131-44.

Colton, R. B. (1955). Geology of the Wolf Point quadrangle, Montana. U.S. Geological Survey Map GQ67.

Coulter, H. W., Hopkins, D. M., Karlstrom, T. N. V., Péwé, T. L., Wahrhaftig, C., and Williams, J. R. (1965). Extent of glaciations in Alaska. U.S. Geological Survey Miscellaneous Geological Investigations Map I-415.

Crowl, G. H., and Sevon, W. D. (1980). "Glacial Border Deposits of Late Wisconsin in Age in Northeastern Pennsylvania." Pennsylvania Topographic and Geologic Survey General Geology Report 71.

DeGraff, J. V. (1976). Relict patterned ground, Bear River Range, north-central Utah. *Utah Geology* 3, 111-16.

Demek, J. (1969). Cryoplanation terraces: Their geographical distribution, genesis and development. *Academi Nakladatelstvi Ceskoslovenské Akademi Ved* 79 (4), 1-80.

Denny, C. S. (1936). Periglacial phenomena in southern Connecticut. *American Journal of Science* 32, 322-42.

Denny, C. S. (1951). Pleistocene frost action near the border of the Wisconsin drift in Pennsylvania. *Ohio Journal of Science* 51, 116-25.

Denny, C. S. (1956). "Surficial Geology and Geomorphology of Potter County, Pennsylvania." U.S. Geological Survey Professional Paper 288.

Dillon, L. S. (1956). Wisconsin climate and life zones in North America. *Science* 123, 167-76.

Dionne, J. C. (1969). Nouvelles Observations de fentes de gel fossiles sur la côté sud du Saint-Laurent. *Revue Géographie Montreal* 23, 307-16.

Dionne, J. C. (1971). Fentes de cryoturbation tardiglaciaires dans la région de Québec. *Revue Géographie Montréal* 25, 245-64.

Dionne, J. C. (1975). Paleoclimatic significance of late Pleistocene ice-wedge casts in southern Quebec, Canada. *Palaeogeography, Palaeoclimatology, Palaeoecology* 17, 65-76.

Dylik, J. (1964). Eléments essentiels de la notion de "périglaciaire." *Biuletyn Peryglacjalny* 14, 111-32.

Eakin, H. M. (1916). "The Yukon-Koyukuk Region, Alaska." U.S. Geological Survey Bulletin 631.

Eakin, H. M. (1932). Periglacial phenomena in the Puget Sound region. *Science* 75, 536.

Eargle, D. H. (1940). The relations of soils and surface in the South Carolina Piedmont. *Science* 91, 337-38.

Everett, K. R. (1966). Slope movement and related phenomena. *In* "Environment of the Cape Thompson Region, Alaska" (N. J. Wilimovsky, ed.), pp. 175-220. U.S. Atomic Energy Commission, Division of Technical Information, Washington, D.C.

Ferrians, O. J., Jr. (1965). Permafrost map of Alaska. U.S. Geological Survey Miscellaneous Geological Inventory Map I-445.

Ferrians, O. J., Jr., Kachadoorian, R., and Greene, G. W. (1969). "Permafrost and Related Engineering Problems in Alaska." U.S. Geological Survey Professional Paper 676.

First, T. S., and Fay, L. P (1980). Peoria Loess mollusc faunas and Woodfordian biomes of the upper Midwest. *In* "American Quaternary Association Sixth Biennial Meeting, Abstracts and Program, 18-20 August 1980," p. 82. Institute for Quaternary Studies, University of Maine, Orono.

Flemal, R. C. (1972). Ice injection origin of the De Kalb mounds, north-central Illinois, U.S.A. *In* "Proceedings, 24th International Geological Congress, Montreal," section 12, pp. 130-35.

Flemal, R. C. (1976). Pingos and pingo scars: Their characteristics, distribution, and utility in reconstructing former permafrost environments. *Quaternary Research* 6, 37-53.

Flemal, R. C., Hesler, J. L., and Hinkley, K. C. (1970). De Kalb mounds: Possible remnants of pingos. *In* "Thirty-fourth Annual Tri-State Field Conference Guidebook" (I. E. Odom and M. P. Weiss, eds.), pp. 65-72. Northern Illinois University, De Kalb.

Flemal, R. C., Hinkley, K. C., and Hesler, J. L. (1973). De Kalb mounds: A possible Pleistocene (Woodfordian) pingo field in north-central Illinois. *In* "The Wisconsinan Stage" (R. F. Black, R. P. Goldthwait, and H. B. Willman, eds.), pp. 226-50. Geological Society of America Memoir 136.

Flint, R. F., and Denny, C. S. (1958). "Quaternary Geology of Boulder Mountain, Aquarius Plateau, Utah." U.S. Geological Survey Bulletin 1061-D.

Foster, H. L. (1967). "Geology of the Mount Fairplay Area, Alaska." U.S. Geological Survey Bulletin 1241-B.

Foster, H. L. (1969). "Reconnaissance Geology of the Eagle A-1 and A-2 Quadrangles, Alaska." U.S. Geological Survey Bulletin 1271-G.

Foster, H. L., and Keith, T. E. C. (1969). "Geology along the Taylor Highway, Alaska." U.S. Geological Survey Bulletin 1281.

French, H. M. (1976). "The Periglacial Environment." Longman Group Ltd, London.

Frye, J. C., and Willman, H. B. (1958). Permafrost features near the Wisconsin glacial margin in Illinois. *American Journal of Science* 256, 518-24.

Frye, J. C., and Willman, H. B. (1973). Wisconsin climatic history interpreted from Lake Michigan lobe deposits and soils. *In* "The Wisconsinan Stage" (R. F. Black, R. P. Goldthwait, and H. B. Willman, eds.), pp. 135-52. Geological Society of America Memoir 136.

Fryxell, F. (ed.) (1962). "François Matthes and the Marks of Time." Sierra Club, San Francisco.

Fujii, Y., and Higuchi, K. (1978). Distribution of alpine permafrost in the Northern Hemisphere and its relation to air temperature. *In* "Proceedings of Third International Conference on Permafrost," vol. 1, pp. 366-71. Canada National Research Council, Ottawa.

Galloway, J. P., and Carter, L. D. (1978). "Preliminary Map of Pingos in National Petroleum Reserve in Alaska." U.S. Geological Survey Open Field Report 78-795.

Galloway, R. W. (1970). The full-glacial climate in the southwestern United States. *Annals, Association of American Geographers* 60, 245-56.

Goldthwait, R. P. (1969). Patterned ground and permafrost on the Presidential Range.

In "Résumé des Communications," p. 150. Eighth Congress of the International Associates for Quaternary Research, Paris.

Goldthwait, R. P. (1976). Frost sorted patterned ground: A review. *Quaternary Research* 6, 27-35.

Gorbunov, A. P. (1978a). Permafrost in the mountains of central Asia. *In* "Proceedings of the Third International Conference on Permafrost," vol. 1, pp. 372-77. Canada National Research Council, Ottawa.

Gorbunov, A. P. (1978b). Permafrost investigations in high-mountain regions. *Arctic and Alpine Research* 10, 283-94.

Gravis, G. F., Zabolotnik, S. I., Lisun, A. M., and Sukhodrovskii, V. L. (1978). The geocryological characteristics of the Mongolian People's Republic and some characteristics of permafrost development in the past. *In* "Permafrost: Translation of the USSR Contribution to the Second International Conference," pp. 81-85. National Academy of Sciences, Washington, D.C.

Hack, S. T., and Goodlett, J. C. (1960). "Geomorphology and Forest Ecology of a Mountain Region in the Central Appalachians." U.S. Geological Survey Professional Paper 347.

Hadley, J. B., and Goldsmith, R. (1963). "Geology of the Eastern Great Smokey Mountains, North Carolina and Tennessee." U.S. Geological Survey Professional Paper 349B.

Harris, S. A. (1981). Distribution of active glaciers and rock glaciers compared to the distribution of permafrost landforms based on freezing and thawing indices. *Canadian Journal of Earth Science* 18, 376-85.

Harris, S. A., and Brown, R. J. E. (1978). Plateau Mountain: A case study of alpine permafrost in the Canadian Rocky Mountains. *In* "Proceedings of Third International Conference on Permafrost," vol. 1, pp. 385-91. Canada National Research Council, Ottawa.

Harrison, W. D., and Osterkamp, T. E. (1978). Heat and mass transport processes in subsea permafrost: I. An analysis of molecular diffusion and its consequences. *Journal of Geophysical Research* 83, 4704-11.

Hedges, J. (1972). Expanded joints and other periglacial phenomena along the Niagara Escarpment. *Biuletyn Peryglacjalny* 21, 87-126.

Hiene, K. (1977). Beobachtangen und Uberlegungen zun Eiszeitlichen Depression von Schneegrenze und Strukturbodengrenze inden Tropen und Subtropen. *Erkunde* 31, 161-77.

Hobson, G. D., and others (1977). Permafrost distribution in the southern Beaufort Sea as determined from seismic measurements. *In* "Proceedings of a Symposium on Permafrost geophysics, 12 October 1976," pp. 91-98. Canada National Research Council, Associations Committee on Geotechnical Research Technical Memorandum 119.

Hodgson, D. A. (1967). The morphology of an area of patterned ground in central Pennsylvania. M.S. thesis, Pennsylvania State University, University Park.

Holmes, G. E., Hopkins, D. M., and Foster, H. L. (1968). "Pingos in Central Alaska." U.S. Geological Survey Bulletin 1241-H.

Holmes, G. E., and Moss, J. H. (1955). Pleistocene geology of the southwestern Wind River Mountains, Wyoming. *Geological Society of America Bulletin* 66, 629-54.

Hopkins, D. M. (1959). Cenozoic history of the Bering Land Bridge. *Science* 129, 1519-28.

Hopkins, D. M. (ed.) (1967). "The Bering Land Bridge" Stanford University Press, Stanford, Calif.

Hopkins, D. M. (1972). The paleogeography and climatic history of Beringia during late Cenozoic time. *Internord* 12, 121-50.

Hopkins, D. M. (1982). Aspects of the Paleogeography of Beringia during the late Pleistocene. *In* "Paleoecology of Beringia" (D. M. Hopkins, J. V. Matthews, Jr., C. E. Schweger, and S. B. Young, eds.), pp. 3-28. Academic Press, New York.

Hopkins, D. M., Karlstrom, T. N. V., and others (1955). "Permafrost and Ground Water in Alaska." U.S. Geological Survey Professional Paper 264-F.

Hopkins, D. M., MacNeil, F. S., and Leopold, E. B. (1960). The Coastal Plain at Nome, Alaska: A late-Cenozoic type section for the Bering Strait region. *In* "Proceedings, 21st International Geological Congress, Copenhagen, part 4, pp. 46-57.

Hopkins, D. M., and Robinson, S. W. (1979). Radiocarbon dates from the Beaufort and Chukchi sea coasts. *In* "U.S. Geological Survey in Alaska: Accomplishments during 1978" (K. M. Johnson and J. R. Williams, eds.), pp. B44-B47. U.S. Geological Survey Circular 804B.

Horberg, L. (1949). A possible fossil ice wedge in Bureau County, Illinois. *Journal of Geology* 57, 132-36.

Howe, E. (1909). "Landslides in the San Juan Mountains, Colorado: Including a Consideration of Their Causes and Their Classification." U.S. Geological Survey Professional Paper 67.

Howe, J. (1971). Temperature test readings in test boreholes. *Mt. Washington Observatory News Bulletin* 12, 37-40.

Hughes, O. L. (1966). Logan Mountains, Yukon Territory: Measurements on a rock glacier. *Ice* 20, 5.

Hughes, O. L. (1969). "Distribution of Open-System Pingos in Central Yukon Territory with Respect to Glacial Limits." Canada Geological Survey Paper 69-34.

Hughes, O. L., Rampton, Y. N., and Rutter, N. W. (1972). Quaternary geology and geomorphology, southern and central Yukon (northern Canada). *In* "Guidebook, 24th International Geological Congress, Montreal," field excursion A 11.

Hunter, J. A. M., Judge, A. S., MacAuley, H. A., Good, R. L., Gagne, R. M., and Burns, R. A. (1976). "Permafrost and Frozen Subsea Bottom Materials in the Southern Beaufort Sea." Beaufort Sea Project Report 22, Canada Department of the Environment.

Hyers, A. D. (1980). Mesoscale relationships of talus and insolation, San Juan Mountains, Colorado. Ph.D. dissertation, Arizona State University, Tempe.

Institut Merzlotovedeniya im. V. A. Obrucheva (1956). "Osnovnyye ponyatiya i terminy geokriologii (merzlotovedeniya)." Akademia Nauk SSSR, Moskva.

Institute of Glaciology, Cryopedology and Desert Research (1975). "Permafrost." Academia Sinica, Lanzhou.

Iskandar, I. K., Osterkamp, T. E., and Harrison, W. D. (1978). Chemistry of interstitial water from subsea permafrost, Prudhoe Bay, Alaska. *In* "Proceedings of the Third International Conference on Permafrost," vol. 1, pp. 93-98. Canada National Research Council, Ottawa.

Ives, J. D. (1974). Permafrost. *In* "Arctic and Alpine Environments" (J. D. Ives and R. G. Barry, eds.), pp. 159-94. Methven and Company, London.

Ives, J. D., and Fahey, B. D. (1971). Permafrost occurrence in the Front Range, Colorado Rocky Mountains, U.S.A. *Journal of Geology* 10, 105-11.

Jahn, A. (1972). Tundra polygons in the Mackenzie Delta area. *Göttinger Geographisher Abhandlung* 60 (Hans Poser Festschrift), 285-92.

Jahn, A. (1978). Mass wasting in permafrost and non-permafrost environments. *In* "Proceedings of the Third International Conference on Permafrost," vol. 1, pp. 296-300. Canada National Research Council, Ottawa.

Jenkins, O. P. (1925). Clastic dikes of eastern Washington and their geologic significance. *American Journal of Science* 10, 234-46.

Jenness, J. L. (1949). Permafrost in Canada. *Arctic* 2, 13-27.

Johnson, P. G. (1975). Mass movement processes in Metalline Creek, southwest Yukon Territory. *Arctic* 28, 130-39.

Johnson, P. G. (1978). Rock glacier types and their drainage systems, Grizzly Creek, Yukon Territory, *Canadian Journal of Earth Sciences* 15. 1496-1507.

Johnson, P. L., and Billings, W. D. (1962). The alpine vegetation of the Beartooth Plateau in relation to cryopedogenic processes and patterns. *Ecological Monographs* 32, 105-35.

Johnson, R. B. (1967). "Rock Streams on Mount Mestas, Sangre de Cristo Mountains, Southern Colorado." U.S. Geological Survey Professional Paper 575-D, pp. D217-D220.

Johnson, W. H. (1978). Patterned ground, wedge-shaped bodies, and circular land forms in central and eastern Illinois. *Geological Society of America, Abstracts with Programs* 10, 257.

Journaux, M. A. (1969). Phenomènes périglaciares dans le Nord de L'Alaska et du Yukon. *Geographie Association du France Bulletin* 368-369, 337-50.

Judge, A. S. (1973). Deep temperature observations in the Canadian North. *In* "Permafrost: North American Contribution to the Second International Conference," pp. 35-40. National Academy of Sciences, Washington, D.C.

Judge, A. S. (1977). Permafrost, hydrates and the offshore thermal regime. *In* "Proceedings of a Symposium on Permafrost Geophysics, 12 October 1976," pp. 99-113. Canada National Research Council, Associations Committee on Geotechnical Research Technical Memorandum 119.

Judson, S. (1965). Quaternary processes in the Atlantic Coastal Plain and Appalachian Highlands. *In* "The Quaternary of the United States" (H. E. Wright, Jr., and D. G. Frey, eds.), pp. 133-36. Princeton University Press, Princeton, N.J.

Kaatz, M. R. (1959). Patterned ground in central Washington: A preliminary report. *Northwest Science* 33, 145-56.

Karpov, V. M., and Puzanov, I. I. (1970). Construction and permafrost. *In* "Popular Science Series." Publisher for Literature on Construction, Leningrad.

Kerfoot, D. E., and Mackay, J. R. (1972). Geomorphology process studies, Garry Island, N.W.T. *In* "Mackenzie Delta Area Monograph" (D. E. Kerfoot, ed.), pp. 115-30. 22nd International Geographical Congress, Brock University, St. Catharines, Ontario.

Kerr, W. C. (1881). On the action of frost in the arrangement of superficial earthy material. *American Journal of Science* 21, 345-58.

Kesseli, J. E. (1941). Rock streams in the Sierra Nevada, California. *Geographical Review* 31, 203-27.

King, P. B. (1964). Geology of the central Great Smokey Mountains, Tennessee. U.S. Geological Survey Professional Paper 349C.

King, L. V. (1977). Detection of permafrost occurrence by means of a hammer seismograph at Tarsala, Swedish Lapland. *Zeitschrift fur Gletscherkunde und Glazialgeologie*, 12, 187-204.

Kirby, A. V. T. (1965). Boulder fields on Bald Eagle Mountain, Pennsylvania (abstract). *Annals of the Association of American Geographers*, 55, 626.

Lachenbruch, A. H. (1966). Contraction theory of ice wedge polygons: A qualitative discussion. *In* "Permafrost International Conference (Lafayette, Indiana, 11-15 November 1963) Proceedings," pp. 63-71. National Academy of Sciences/National Research Council Publication 1287.

Lachenbruch, A. H. (1968). Permafrost. *In* "The Encyclopedia of Geomorphology" (R. W. Fairbridge, ed.), pp. 833-39. Reinhold, New York.

Lachenbruch, A. H. (1970a). "Some Estimates of the Thermal Effects of a Heated Pipeline in Permafrost." U.S. Geological Survey Circular 632.

Lachenbruch, A. H. (1970b). Thermal considerations in permafrost. *In* "Geological Seminar on the North Slope of Alaska Proceedings" (W. L. Adkison and M. M. Borsge, eds.), pp. J1-2-J2-5. American Association of Petroleum Geologists, Pacific Section, Al-Rio, Los Angeles.

Lachenbruch, A. H., Brewer, M. C., Green, G. W., and Marshall, B. V. (1962). Temperatures in permafrost. *In* "Temperature: Its Measurement and Control in Science and Industry 3, Part I," pp. 791-803. Reinhold, New York.

Lachenbruch, A. H., Greene, G. W., and Marshall, B. V. (1966). Permafrost and geothermal regimes. *In* "Environment of the Cape Thompson Region, Alaska" (N. J. Wilimovsky, ed.), pp. 149-63. U.S. Atomic Energy Commission, Oak Ridge, Tenn.

Lachenbruch, A. H., and Marshall, B. V. (1969). Heat flow in the Arctic. *Arctic* 22, 300-311.

Lagarec, D. (1973). Postglacial permafrost features in eastern Canada. *In* "Permafrost: North American Contribution to the Second International Conference," pp. 126-31. National Academy of Sciences, Washington, D.C.

Larson, M. C., Nelson, C. H., and Thor, D. R. (1981). Sedimentary processes and potential geologic hazards on the sea floor of northern Bering Sea. *In* "Oceanography of the Eastern Bering Sea Shelf" (D. W. Hood and J. A. Calder, eds.), vol. 1, pp. 247-62. U.S. Department of Commerce, NOAA, Washington, D. C.

Lees, J. H. (1927). Geology of Crawford County. *Iowa Geological Survey Annual Report 1925 and 1926* 32, 243-362.

Leonard, A. B. (1952). "Illinoian and Wisconsin Molluscan Faunas in Kansas." Kansas University Paleontological Contribution 9.

Leonard, A. B., and Frye, D. G. (1960). "Wisconsin Molluscan Faunas of the Illinois Valley Region." Illinois Geological Survey Circular 304.

Lewellen, R. (1977). "A Study of Beaufort Sea Coastal Erosion, Northern Alaska." Environmental Assessment of the Alaskan Continental Shelf, Annual Reports of Principal Investigators, University of Alaska, Fairbanks.

Libby, W. (1968). "Rock Glaciers in the North Cascade Range, Washington" (abstract). Geological Society of America Special Paper 101, pp. 318-19.

Leistøl, O. (1977). Pingos, springs, and permafrost in Spitsbergen. *In* "Norsk Polarinstitutt Yearbook, 1975," pp. 7-29. Norsk Polarinstitutt, Oslo.

Lorenzo, J. L. (1969). Minor periglacial phenomena among the high volcanoes of Mexico. *In* "The Periglacial Environment" (T. L. Péwé, ed.), pp. 161-75. McGill-Queen's University Press, Montreal.

Lozinski, W. (1909). Uber die mechanische Verwitterung der Sandsteine im gemassigten Klima. *Bulletin International de l'Académie Polonaise des Sciences et des Lettres, Classe des Sciences Mathématiques et Naturelles* 1, 1-25.

Luckman, B. H., and Crockett, J. K. (1978). Distribution and characteristics of rock glaciers in the southern part of Jasper National Park, Alberta. *Canadian Journal of Earth Science 15, 540-50.*

Lupher, R. L. (1941). Clastic dikes of the Columbia Basin region, Washington and Idaho. *Bulletin of the Geological Society of America* 55, 1431-62.

Mackay, J. R. (1966). Pingos in Canada. *In* "Permafrost International Conference, (Lafayette, Indiana, 11-15 November 1963) Proceedings," pp. 71-76. National Academy of Science/National Research Council Publication 1287.

Mackay, J. R. (1972a). Offshore permafrost and ground ice, southern Beaufort Sea, Canada. *Canadian Journal of Earth Sciences* 9, 1550-61.

Mackay, J. R. (1972b). Some observations on growth of pingos. *In* "Mackenzie Delta Area Monograph" (D. E. Kerfoot, ed.), pp. 141-47. 22nd International Geographic Congress, Brock University, St. Catherines, Ontario.

Mackay, J. R. (1972c). Some observations on ice-wedges, Garry Island, Northwest Territories. *In* "Mackenzie Delta Area Monograph" (D. E. Kerfoot, ed.), pp. 131-39. 22nd International Geographical Congress, Brock University, St. Catherines, Ontario.

Mackay, J. R. (1972d). The world of underground ice. *Annals of the Association of American Geographers* 62, 1-22.

Mackay, J. R. (1973). The growth of pingos, western Arctic Coast, Canada. *Canadian Journal of Earth Sciences* 10, 979-1004.

Mackay, J. R. (1976a). "The Age of Ibuyuk Pingo, Taktoyaktuk Peninsula, District of Mackenzie." Canada Geological Survey Paper 76-IB, pp. 59-60.

Mackay, J. R. (1976b). The growth of ice wedges (1966-1975), Garry Island, Northwest Territories, Canada. *In* "International Geography: I. Geomorphology and Paleogeography," pp. 180-82. 23rd International Geographical Congress, Moscow.

Mackay, J. R. (1976c). "Ice Segregation at Depth in Permafrost." Canada Geological Survey Paper 76-IA, pp. 287-88.

Mackay, J. R. (1976d). "Ice Wedges as Indicators of Recent Climatic Change, Western Arctic Coast." Canada Geological Survey Paper 76-IA, pp. 233-34.

Mackay, J. R. (1976e). On the origin of pingos: A comment. *Journal of Hydrology 30,* 295-98.

Mackay, J. R. (1976f). "Pleistocene Permafrost, Hooper Island, Northwest Territories." Canada Geological Survey Paper 76-IA, pp. 17-18.

Mackay, J. R. (1977a). "Permafrost Growth and Subpermafrost Pore Water Expulsion, Tuktoyaktuk Peninsula, District of Mackenzie." Canada Geological Survey Paper 77-IA, pp. 323-26.

Mackay, J. R. (1977b). Pulsating pingos, Tuktoyaktuk Peninsula, Northwest Territories. *Canadian Journal of Earth Sciences* 77-IB, 273-75.

Mackay, J. R. (1978a). Contemporary pingos: A discussion. *Biuletyn Peryglacjalny* 27, 133-54.

Mackay, J. R. (1978b). Sub-pingo water lenses, Tuktoyaktuk Peninsula, Northwest Territories. *Canadian Journal of Earth Science* 15, 1219-27.

Mackay, J. R., Konishchev, V. N., and Popov, A. I. (1979). Geological control of the origin, characteristics, and distribution of ground ice. *In* "Third International Conference on Permafrost," vol. 2, pp. 1-18. Canada National Research Council, Ottawa.

Mackin, J. H. (1947). Altitude and local relief of Bighorn area during the Cenozoic. *In* "Wyoming Geological Association Field Conference Big Horn Basin, Guidebook," pp. 103-20. Wyoming Geological Association, Laramie.

Malde, H. E. (1961). "Patterned Ground of Possible Solifluction Origin at Low Altitude in the Western Snake River Plain, Idaho." U.S. Geological Survey Professional Paper 424B, B170-B173.

Malde, H. E. (1964). Patterned ground in the western Snake River Plain, Idaho, and its

possible cold-climate origin. *Geological Society of America Bulletin* 75, 191-208.

Matthews, J. V., Jr. (1976). Arctic-steppe: An extinct biome. *In* "American Quaternary Association, Fourth Biennial Meeting, Abstracts," pp. 75-77. Arizona State University, Tempe.

Matthes, F. E. (1950). "Sequoia National Park: A Geological Album." University of California Press, Berkeley.

Maxwell, J. A., and Davis, M. B. (1972). Pollen evidence of Pleistocene and Holocene vegetation on the Allegheny Plateau, Maryland. *Quaternary Research* 2, 506-30.

Mears, B., Jr. (1962). Stone nets on Medicine Bow Peak, Wyoming. *University of Wyoming Contributions to Geology* 1, 48.

Mears, B., Jr. (1966). "Ice-Wedge Pseudomorphs in the Laramie Basin, Wyoming" (abstract). Geological Society of America Special Paper 87, p. 295.

Mears, B., Jr. (1973). Were Rocky Mountain intermontane basins Pleistocene tundras? (abstract). *In* "International Association for Quaternary Research Ninth Congress Abstracts," p. 233. Christchurch, New Zealand.

Mears, B., Jr. (1981). Periglacial wedges in the late Pleistocene environment of Wyoming's intermontane basins. *Quaternary Research* 15, 171-98.

Merrill, R. K., and Péwé, T. L. (1977). "Late Cenozoic Geology of the White Mountains, Arizona." Arizona Bureau of Geology and Mineral Technology Special Paper 1.

Messenger, J. A. (1977). Wisconsinan paleogeography of Alaska. M.S. thesis, Arizona State University, Tempe.

Messenger, J. A., and Péwé, T. L. (1977). Paleogeographic maps of Alaska for the Wisconsinan. *Geological Society of America, Abstracts with Programs* 9, 748-49.

Morgan, A. V. (1969). Intraformational periglacial structures in the Nose Hill gravels and sands, Calgary, Alberta, Canada. *Journal of Geology* 77, 358-64.

Morgan, A. V. (1972). Late Wisconsin ice-wedge polygons near Kitchener, Ontario, Canada. *Canadian Journal of Earth Sciences* 9, 607-17.

Mozley, A. (1937). Frozen ground in the subarctic region and its biological significance. *Scottish Geological Magazine* 53, 266-70.

Muller, S. W. (1945). "Permafrost or Permanently Frozen Ground and Related Engineering Problems." U.S. Engineers Office, Strategic Engineering Study Special Report 62. (Republished 1947, Ann Arbor, Michigan, R. W. Edwards Inc.)

Newcomb, R. C. (1952). Origin of the Mima Mounds, Thurston County region, Washington. *Journal of Geology* 60, 461-72.

Newcomb, R. C., Strand, J. R., and Frank, F. J. (1972). "Geology and Ground Water Characteristics of the Hanford Reservation of the U.S. Atomic Energy Commission, Washington." U.S. Geological Survey Professional Paper 717.

Norton, J. J., and Redden, J. A. (1960). Structure associated with rock creep in the Black Hills, South Dakota. *Geological Society of America Bulletin* 71, 1109-12.

Olgivie, R. T., and Baptie, B. (1967). A permafrost profile in the Rocky Mountains of Alberta. *Canadian Journal of Earth Sciences* 4, 744-45.

Osterkamp, T. E., and Payne, M. W. (1981). Estimates of permafrost thickness from well logs in northern Alaska. *Cold Regions Science and Technology* 5, 13-27.

Page, F. W., and Iskander, I. K. (1978). "Geochemistry of Subsea Permafrost at Prudhoe Bay, Alaska." U.S. Army Cold Region Research and Engineering Laboratory Report 78-14.

Patton, H. B. (1910). Rock streams of Veta Peak, Colorado. *Geological Society of America Bulletin* 21, 663-76.

Peltier, L. C. (1945). Block fields in Pennsylvania. *Geological Society of America Bulletin* 56, 1190.

Peltier, L. C. (1950). The geographic cycle in periglacial regions as it is related to climatic geomorphology. *Annals of the Association of American Geographers* 40, 214-36.

Péwé, T. L. (1948). Origin of Mima Mounds. *Scientific Monthly* 66, 293-96.

Péwé, T. L. (1954). "Effects of Permafrost on Cultivated Fields, Fairbanks Area, Alaska." U.S. Geological Survey Bulletin 989-F, pp. 315-54.

Péwé, T. L. (1955). Origin of the upland silt near Fairbanks, Alaska. *Geological Society of America Bulletin* 66, 699-724.

Péwé, T. L. (1965). Fairbanks area. *In* "Guidebook for Field Conference F, Central and South-Central Alaska," pp. 6-36. Seventh Congress of the International Association for Quaternary Research, Boulder, Colo.

Péwé, T. L. (1966a). Ice-wedges in Alaska: Classification, distribution, and climatic significance. *In* "Permafrost International Conference (Lafayette, Indiana, 11-15

November 1963). Proceedings,'' pp. 76-81. National Academy of Sciences/National Research Council Publication 1287.

Péwé, T. L. (1966b). Paleoclimatic significance of fossil ice wedges. *Biuletyn Peryglacjalny* 15, 65-73.

Péwé, T. L. (1966c). ''Permafrost and Its Effect on Life in the North.'' Oregon State University Press, Corvallis.

Péwé, T. L. (1969). The periglacial environment. *In* ''The Periglacial Environment: Past and Present'' (T. L. Péwé, ed.), pp. 1-9. McGill-Queen's University Press, Montreal.

Péwé, T. L. (1973). Ice wedge casts and past permafrost distribution in North America. *Geoforum* 15, 15-26.

Péwé, T. L. (1974). Permafrost. *In* ''Encyclopaedia Britannica,'' 15th ed., pp. 89-95. Encyclopaedia Britannica, Chicago.

Péwé, T. L. (1975a). Permafrost: Challenge of the Arctic. *In* ''1976 Yearbook of Science and the Future,'' pp. 90-105. Encyclopaedia Britannica, Chicago.

Péwé, T. L. (1975b). ''Quaternary Geology of Alaska.'' U.S. Geological Survey Professional Paper 835.

Péwé, T. L. (1975c). ''Quaternary Stratigraphic Nomenclature in Unglaciated Central Alaska.'' U.S. Geological Survey Professional Paper 862.

Péwé, T. L. (1976). Permafrost. *In* ''Yearbook of Science and Technology,'' pp. 30-47. McGraw-Hill, New York.

Péwé, T. L. (1978). Tyndall figures in ice crystals of ground-ice in permafrost near Fairbanks, Alaska. *In* ''Proceedings of Third International Conference on Permafrost,'' vol. 1, pp. 312-17. Canada National Research Council, Ottawa.

Péwé, T. L. (1979a). Permafrost: And its effects on human activities in Arctic and subArctic regions. *Geojournal* 3, 333-44.

Péwé, T. L. (1979b). Permafrost in western Canada: A report of the Third International Permafrost Conference. *Zeitschrift für Gletscherkund und Glazialgeologie* 15, 119-25.

Péwé, T. L. (1981). Tibetan science updated. *Geotimes* 26, 16-20.

Péwé, T. L. (1982). ''Geologic Hazards and Their Effect on Man: Fairbanks Area, Alaska.'' Alaska Division of Geology and Geophysics Special Report 15.

Péwé, T. L. (1983). Alpine permafrost in the contiguous United States. *Arctic and Alpine Research* (in press).

Péwé, T. L., Bell, J. W., Forbes, R. B., and Weber, F. R. (1976). Geologic Map of the Fairbanks D2 NE quadrangle, Alaska. U.S. Geological Survey Map I-950.

Péwé, T. L., Church, R. E., and Andresen, M. J. (1969). ''Origin and Paleoclimatic Significance of Large-Scale Patterned Ground in the Donnelly Dome Area, Alaska.'' Geological Society of America Special Paper 103.

Péwé, T. L., and Reger, R. D. (1981). Cryoplanation terraces. *In* ''Proceedings of the Symposium on the Qinghi-Xizang (Tibet) Plateau,'' pp. 1789-94. Peking.

Péwé, T. L., and Updike, R. G. (1970). Guidebook to the geology of the San Francisco Peaks, Arizona. *Plateau*, 43, 45-102.

Péwé, T. L., and Updike, R. G. (1976). ''San Francisco Peaks: A Guidebook to the Geology.'' 2nd ed. Museum of Northern Arizona, Flagstaff.

Pierce, K. L. (1979). ''History and Dynamics of Glaciation in the Northern Yellowstone National Park Area.'' U.S. Geological Survey Professional Paper 729-F.

Pierce, W. G. (1961). ''Permafrost and Thaw Depressions in a Peat Deposit in the Beartooth Mountains, Northwestern Wyoming.'' U.S. Geological Survey Professional Paper 424B.

Pissart, A. (1967). Les Pingos de l'Ile Prince Patrick (76°N-120°W). *Geographical Bulletin* 9, 189-217.

Pissart, A. (1968). Pingos, Pleistocene. *In* ''The Encyclopedia of Geomorphology'' (R. W. Fairbridge, ed.), pp. 847-48. Reinhold, New York.

Pissart, A. (1970). Les Phénomenès physiques essentielles liés au gel, les structures périglaciaires qui en résultent et leur signification climatique. *Annales de la Société géologique de Belgique* 93, 7-49.

Pissart, A. (1977). The origin of pingos in regions of thick permafrost (Canadian Arctic). *In* ''International Association for Quaternary Research 10th Congress Abstracts,'' p. 361. Birmingham, England.

Pissart, A., and French, H. M. (1976). Pingo investigations, north-central Banks Island, Canadian Arctic. *Canadian Journal of Earth Sciences* 13, 937-46.

Porter, S. C., Pierce, K. L., and Hamilton, T. D. (1983). Late Wisconsin mountain glaciation in the western United States. *In* ''Late-Quaternary Environments of the United States,'' vol. 1, ''The Late Pleistocene'' (S. C. Porter, ed.), pp. 71-111. University of Minnesota Press, Minneapolis.

Potter, D. B., and Moss, J. H. (1966). New evidence supporting the solifluction origin of the Blue Rocks Block Field in Berks County, Pennsylvania. *Geological Society of America, Abstracts with Programs*, pp. 273-74.

Potter, N., Jr. (1972). Ice-cored rock glacier, Galena Creek, northern Absaroka Mountains, Wyoming. *Geological Society of America Bulletin* 83, 3025-58.

Potter, N., Jr., and Moss, J. H. (1968). Origin of the Blue Rocks Block Field and adjacent deposits, Berks County, Pennsylvania. *Geological Society of America Bulletin* 79, 255-62.

Prest, V. K., Grant, D. R., and Rampton, V. N. (1968). Glacial map of Canada. Geological Survey Canada Map 1253A.

Price, L. W. (1973). Rates of mass wasting in the Ruby Range, Yukon Territory. *In* ''Permafrost: North American Contribution to the Second International Conference,'' pp. 235-45. National Academy of Sciences, Washington, D. C.

Priesnitz, K., and Schunke, E. (1978). An approach to the ecology of permafrost in central Iceland. *In* ''Proceedings of the Third International Conference on Permafrost,'' vol. 1, pp. 475-79. Canada National Research Council, Ottawa.

Prokopovich, N. P. (1969). Pleistocene permafrost in California's Central Valley? *Geological Society of America, Abstracts with Programs* 5, 66.

Rapp, A. (1967). Pleistocene activity and Holocene stability of hillslopes, with examples from Scandinavia and Pennsylvania. *In* ''L'Evolution des Versants,'' 40, pp. 65-91. Les Congrèes et colloques de l'Université de Lièege, Université de Lièege.''

Rapp, A., and Annersten, L. (1969). Permafrost and tundra polygons in northern Sweden. *In* ''The Periglacial Environment: Past and Present'' (T. L. Péwé, ed.), pp. 65-91. McGill University Press, Montreal.

Reger, R. D. (1975). Cryoplanation terraces of interior and western Alaska. Ph.D. dissertation, Arizona State University, Tempe.

Reger, R. D., and Péwé, T. L. (1976). Cryoplanation terraces: Indicators of a permafrost environment. *Quaternary Research* 6, 99-109.

Retzer, J. L. (1965). Present soil-forming factors and processes in arctic and alpine areas. *Soil Science* 99, 38-44.

Richards, H. G., and Rhodehamel, E. C. (1965). New Jersey Coastal Plain field trip (August 17). *In* ''Guidebook for Field Conference B-1, Central Atlantic Coastal Plain'' (C. B. Schultz and H. T. U. Smith, eds.), pp. 10-13. Seventh Congress of the International Association for Quaternary Research, Boulder, Colo.

Richmond, G. M. (1949). Stone nets, stone stripes, and soil stripes in the Wind River Mountains, Wyoming. *Journal of Geology* 57, 143-53.

Richmond, G. M. (1952). Comparison of rock glaciers and block streams in the La Sal Mountains, Utah. *Geological Society of America Bulletin* 63, 1292-93.

Richmond, G. M. (1962). ''Quaternary Stratigraphy of the La Sal Mountains, Utah,'' U.S. Geological Survey Professional Paper 324.

Ritchie, A. M. (1953). The erosional origin of the Mima Mounds of southwest Washington. *Journal of Geology* 61, 41-50.

Ruhe, R. V. (1969). ''Quaternary Landscapes in Iowa.'' Iowa State University Press, Ames.

Ruppel, E. T., and Hait, M. H., Jr. (1961). ''Pleistocene Geology of the Central Part of the Lemhi Range, Idaho.'' U.S. Geological Survey Professional Paper 424-B.

Russell, R. J. (1933). Alpine land forms in western United States. *Geological Society of America Bulletin* 44, 927-50.

Rutter, N. W., Foscolos, A. E., and Hughes, O. L. (1978). Quaternary soils. *In* ''Third York Quaternary Symposium,'' pp. 309-59. Geo Abstracts Limited, Norwich, England.

Schafer, G. M. (1953). A disturbed buried gumbotil soil profile in Jefferson County, Iowa. *Iowa Academy of Science* 60, 403-7.

Schafer, J. P. (1949). Some periglacial features in central Montana. *Journal of Geology* 57, 154-74.

Schafer, J. P. (1962). ''Pleistocene Frost Action in and near Northeastern Iowa.'' Geological Society of America Special Paper 68, p. 262, abstract.

Schafer, J. P., and Hartshorn, S. H. (1965). The Quaternary of New England. *In* ''The Quaternary of the United States'' (H. E. Wright, Jr., and D. G. Frey, eds.), pp. 113-28. Princeton University Press, Princeton, N.J.

Scott, B. W. (1965). The ecology of the alpine tundra on Trail Ridge. *In* ''Guidebook for One-Day Conferences, Boulder Colorado Area,'' (C. B. Schultz and H. T. U.

Smith, eds.), pp. 13-16. Seventh Congress of the International Association for Quaternary Research, Boulder, Colo.

Scotter, G. W. (1975). Permafrost profiles in the Continental Divide region of Alberta and British Columbia. *Arctic and Alpine Research* 7, 93-95.

Sellmann, P. V. (1967). "Geology of the USA CRREL Permafrost Tunnel Fairbanks, Alaska." U.S. Army Cold Regions Research and Engineering Laboratory Technical Report 199.

Sellmann, P. V. (1972). "Geology and Properties of Materials Exposed in the USA CRREL permafrost Tunnel." U.S. Army Cold Regions Research and Engineering Laboratory Special Report 177.

Sellmann, P. V., Chamberlain, E. J., Blouin, S. E., Iskandar, I. K., and Lewellen, R. I. (1979). Field methods and preliminary results from subsea permafrost investigations in the Beaufort Sea, Alaska. *In* "Proceedings of the Symposium on Permafrost Field Methods and Permafrost Geophysics," pp. 207-13. National Research Council of Canada Technical Memorandum 124.

Sergin, S. Y., and Cheglova, M. S. (1973). Theoretical reconstruction of the climate of the Beringian land during glacial epochs. *In* "Proceedings of All-Union Symposium on the Bering Land Bridge and Its Role for the History of Holarctic Floras and Faunas in the Late Cenezoic," pp. 63-65. U.S.S.R. Academy of Science, Moscow.

Sevon, W. D. (1967). The Bowmanstown Boulder Field, Carbon County, Pennsylvania. *Proceedings, The Pennsylvania Academy of Science* 40, 90-94.

Sevon, W. D. (1969). Sedimentology of some Mississippian and Pleistocene deposits of northeastern Pennsylvania. *In* "Geology of Selected Areas in New Jersey and Eastern Pennsylvania, and Guidebook of Excursions" (S. Subitzky, ed.), pp. 214-34. Rutgers University Press, New Brunswick, N.J.

Sevon, W. D. (1972). Late Wisconsinan periglacial boulder deposits in northeastern Pennsylvania. *Geological Society of America, Abstracts with Programs* 4, 43-44.

Sevon, W. D. (1975a). "Geology of Mineral Resources of the Christians and Pohopoco Mountain Quadrangles, Carbon and Monroe Counties, Pennsylvania." Pennsylvania Topographic and Geologic Survey Atlas 195, a and b.

Sevon, W. D. (1975b). "Geology and Mineral Resources of the Christians and Pohopoco Mountains Quadrangles, Carbon and Monroe Counties, Pennsylvania." Pennsylvania Topographic and Geologic Survey Atlas 194, c and d.

Sevon, W. D. (1975c). "Geology and Mineral Resources of the Tobyhanna and Buck Hill Falls Quadrangles, Monroe County, Pennsylvania." Pennsylvania Topographic and Geologic Survey Atlas 204, a and b.

Sevon, W. D., and Berg, T. M. (1978). "Geology and Mineral Resources of the Skytop Quadrangle, Monroe and Pike Counties, Pennsylvania." Pennsylvania Topographic and Geologic Survey Atlas 214, a.

Sharp, R. P. (1942). Periglacial involutions in northeastern Illinois. *Journal of Geology* 50, 113-33.

Shearer, J. M., MacNab, R. F., Pelletier, B. R., and Smith, T. B. (1971). Submarine pingos in the Beaufort Sea. *Science* 174, 816-18.

Shroder, J. F., Jr. (1973). Movement of boulder deposits, Table Cliffs Plateau, Utah. *Geological Society of America, Abstracts with Programs* 5, 511.

Smith, H. T. U. (1936). Periglacial landslide topography of Canjilon Divide, Rio Arriba County, New Mexico. *Journal of Geology* 44, 836-60.

Smith, H. T. U. (1949a). Periglacial features in the driftless area of southern Wisconsin. *Journal of Geology* 57, 196-215.

Smith, H. T. U. (1949b). Physical effects of Pleistocene climatic changes in nonglaciated areas: Eolian phenomena, frost action, and stream terracing. *Geological Society of America Bulletin* 60, 1485-1516.

Smith, H. T. U. (1950). Cryopedologic phenomena in the Beartooth Mountains, Wyoming-Montana. *Geological Society of America Bulletin* 16, 1503.

Smith, H. T. U. (1953a). The Hickory Run Boulder Field, Carbon County, Pennsylvania. *American Journal of Science* 251, 625-42.

Smith, H. T. U. (1953b). Periglacial frost wedging in the "rock cities" of southwestern New York. *Geological Society of America Bulletin* 64, 1474.

Smith, H. T. U., and Smith, A. P. (1945). Periglacial rock streams in the Blue Ridge area. *Geological Society of America Bulletin* 56, 1198.

Spencer, A. C. (1900). A peculiar form of talus. *Science* 11, 188.

Squire, G. H. (1897). Studies in the driftless area of Wisconsin. *Journal of Geology* 5, 825-36.

Sumgin, M. I. (1927). "Perennially Frozen Soil in the Limits in U.S.S.R." Far Eastern Geophysical Observatory, Vladivostok, U.S.S.R. (In Russian.)

Taber, S. (1943). Perennially frozen ground in Alaska: Its origin and history. *Geological Society of America Bulletin* 54, 1433-1548.

Totten, S. M. (1973). "Glacial Geology of Richland County, Ohio." Ohio Geological Survey Report of Investigations 88.

Troutt, W. R. (1971). An occurrence of large-scale, inactive, sorted patterned ground south of the glacial border in central Pennsylvania. M.S. thesis, Pennsylvania State University, University Park.

United States National Academy of Sciences (1976). "Problems and Priorities in Offshore Permafrost Research." National Academy of Sciences, Washington, D.C.

Vick, S. G. (1981). Morphology and the role of landsliding in formation of some rock glaciers in the Mosquito Range, Colorado. *Geological Society of America Bulletin* 92 (part 1), 75-84.

Vigdorchik, M. E. (1979). "Submarine Permafrost on the Alaskan Continental Shelf." Westview Press, Boulder, Colo.

Wahrhaftig, C., and Cox, A. (1959). Rock glaciers in the Alaska Range. *Geological Society of America Bulletin* 70, 383-436.

Walters, J. C. (1978). Polygonal patterned ground in central New Jersey. *Quaternary Research* 10, 42-54.

Washburn, A. L. (1967). Instrumental observations of mass-wasting in the Mesters Vig District, northeast Greenland. *Meddelelser om Gronland* 166 (4), 1-297.

Washburn, A. L. (1969). Weathering, frost action, and patterned ground in the Mesters Vig District, northeast Greenland. *Meddelelser om Gronland* 176 (4), 1-303.

Washburn, A. L. (1980a). "Geocryology." Halstead Press, New York.

Washburn, A. L. (1980b). Permafrost features as evidence of climatic change. *Earth Science Reviews* 15, 327-402.

Watts, W. A. (1983). Vegetational history of the eastern United States 25,000 to 10,000 years ago. *In* "Late-Quaternary Environments of the United States," vol. 1, "The Late Pleistocene" (S. C. Porter, ed.), pp. 294-310. University of Minnesota Press, Minneapolis.

Wayne, W. J. (1965). Western and central Indiana—Day 4, September 9. *In* "Guidebook for Field Conference, G, Great Lakes-Ohio River Valley," pp. 29-36. Seventh Congress of the International Association for Quaternary Research, Boulder, Colo.

Wayne, W. J. (1967). Periglacial features and climatic gradient in Illinois, Indiana, and western Ohio, east-central United States. *In* "Quaternary Paleoecology" (E. J. Cushing and H. E. Wright, Jr., eds.), pp. 393-414. Yale University Press, New Haven, Conn.

Weidick, A. (1968). Observations on some Holocene glacier fluctuations in west Greenland. *Meddelelser om Gronland* 165 (6), 1-202.

Weiser, S. (1875). Permanent ice in a mine in the Rocky Mountains. *Philosophical Magazine* 49, 77-78. (Also in *Silliman's American Journal*, December 1874.)

Werenskiold, W. (1922). Frozen earth in Spitsbergen. *Geofysiske Publikationer* 2 (10), 1-10.

Werenskiold, W. (1953). The extent of frozen ground under the sea bottom and glacier beds. *Journal of Glaciology* 2, 197-200.

Wernecke, L. (1932). Glaciation, depth of frost, and ice veins of Keno Hill and vicinity, Yukon Territory. *Engineering and Mining Journal* 133, 38-43.

Westgate, J. A., and Bayrock, L. A. (1964). Periglacial structures in the Saskatchewan gravels and sands of central Alberta, Canada. *Journal of Geology* 72, 641-48.

Wherry, E. P. (1923). The Blue Rocks of Greenwich Township. *Transactions of the Historical Society of Berks County, (Pennsylvania)* 3, 204-8.

White, S. E. (1971). Rock glacier studies in the Colorado Front Range, 1961 to 1968. *Arctic and Alpine Research* 3, 43-64.

White, S. E. (1976). Rock glaciers and block fields: Review and new data. *Quaternary Research* 6, 77-97.

Williams, P. J. (1966). Downslope soil movement at a sub-Arctic location with regard to variations with depth. *Canadian Geotech Journal* 3, 191-203.

Wilson, L. R. (1958). Polygonal structures in the soil of central Iowa. *Oklahoma Geological Notes, Oklahoma Geological Survey* 18, 4-6.

Wolfe, P. E. (1953). Periglacial freeze-thaw basins in New Jersey. *Journal of Geology* 61, 131-41.

Yeend, W. (1971). Glaciation of the Ray Mountains, central Alaska. *In* "Geological

Survey Research 1971,'' pp. D122-D126. U.S. Geological Survey Professional Paper 750-D.

Zeuner, F. E. (1945). ''The Pleistocene Period: Its Climate, Chronology, and Faunal Successions.'' Ray Society. London.

Zoltai, S. C. (1971). ''Southern Limit of Permafrost Features in Peat Landforms, Manitoba and Saskatchewan.'' Geological Association of Canada Special Paper 9,

pp. 305-10.

Zoltai, S. C., and Tarnocai, L. (1971). Properties of a wooded palsa in northern Manitoba. *Journal of Arctic and Alpine Research* 3, 115-19.

Zoltai, S. C., and Tarnocai, C. (1974). ''Soils and Vegetation of Hummocky Terrain.'' Environmental-Social Commission, Northern Pipelines (Canada), Task Force on Northern Development Report 74-5.

CHAPTER 10

Pluvial Lakes of the Western United States

George I. Smith and F. Alayne Street-Perrott

Introduction

More than a hundred closed basins in the western United States contained lakes during the Late Wisconsin, 25,000 to 10,000 yr B.P., but only about 10% of the lakes are perennial and of substantial size today (Figure 10-1). Changes in one or more elements of the climate—precipitation, temperature, evaporation, wind, cloud cover, and humidity—were responsible for most of these dramatic oscillations. The climatically enlarged lakes are referred to by most investigators in the United States as "pluvial lakes," although the relative importance of increased rainfall (the literal meaning of the word *pluvial*) versus changes in the other climatic controls is not specifically implied and is still being debated. The large lakes left geomorphic, stratigraphic, paleontologic, and archaeological evidence of their former existence; their careful study, along with evidence from dry phases intervening between times of lake expansion, provides a paleoclimatic record of alternating pluvial and interpluvial periods over an area that represents about a third of the conterminous United States.

Most pluvial lakes formed in basins that are still topographically closed and have internal drainage. The largest group was within the Great Basin area of Nevada, Utah, and California, but clusters of pluvial lakes also existed in the peripheral parts of the Basin and Range province in Oregon, California, Arizona, New Mexico, and northern Mexico. Numerous small lakes also occurred in the southern High Plains of Texas and New Mexico and in some externally drained valleys in California. Most of these basins in the western United States are products of late-Cenozoic tectonic activity, although some are the results of deflation or of damming by lava or landslides. The intensity of these geologic processes in the arid western states explains the concentration of closed basins in this area, especially in the Basin and Range physiographic province. Where climates are not now arid, most closed depressions contain lakes that are constantly filled and overflowing. Former more-intense pluvial conditions produced no increase in these lake levels, and more-arid periods left records that are now concealed by water. Lakes in less-arid areas also

tend to have shorter lives; they fill with sediment more rapidly or erode their outlets to the valley-floor level and drain.

The value of fluctuations in lake levels as an indicator of climate has been recognized for more than 250 years (Halley, 1715). In 1776, Velez de Escalante (translation, 1943) speculated that shells near Utah Lake recorded a former large lake, the one now known as Lake Bonneville. Pluvial lakes in the area were first well documented about a century later in a series of classic studies by Whitney (1865), Simpson (1876), King (1878), Russell (1883, 1884, 1885, 1889), and Gilbert (1885, 1890) on the two largest pluvial lakes—Lake Bonneville in Utah, Nevada, and Idaho, and Lake Lahontan in Nevada, California, and Oregon—as well as on smaller lakes in California and Oregon. The pluvial chain of lakes that existed in Owens, China, Searles, Panamint, and Death Valleys in California was later documented by Gale (1914); other former lakes in Oregon, Nevada, California, Arizona, and New Mexico were identified by Waring (1908, 1909), Meinzer (1911), Meinzer and Kelton (1913), Meinzer and Ellis (1915), and Clark and Riddell (1920). The regional paleoclimatic implications of these extinct lakes were summarized by Meinzer (1922), who first compiled a map of pluvial lakes in the Basin and Range province, and the inferred relations between these lakes and glacial events in other parts of the world were discussed by Antevs (1925). Since then, several maps showing the locations of pluvial lakes have been published, mostly with accompanying texts and lists of references; examples are those by Hubbs and Miller (1948), Feth (1961), Snyder and others (1964), Morrison (1965c), and Mifflin and Wheat (1979).

Besides being records of paleoclimate, the datable stratigraphic and geomorphic records found in many pluvial-lake basins provide the means of determining the recency of faulting, the rates of tectonic tilting, and the kinetics of landform development, and of obtaining detailed records of past variations in the positions and polarities of the Earth's magnetic field. Pluvial-lake shorelines are also favorable sites for finding traces of early human cultures. The derived paleoclimatic records also provide inferential methods of dating or correlating se-

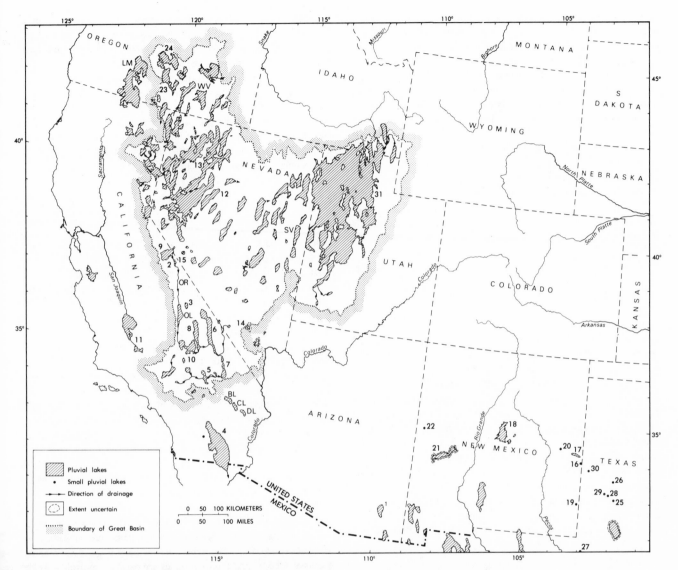

Figure 10-1. Distribution of pluvial lakes within the Great Basin (outlined) and elsewhere in the western United States known or inferred to have expanded during the period 10,000 to 25,000 yr B.P. The numbered lakes are listed in Table 10-3. The unnumbered lakes and valleys discussed in text are identified by letters—in California: OL, Owens Lake; OR, Owens River; BL, Bristol Lake; CL, Cadiz Lake; and DL, Danby Lake; in Nevada: SV, Spring Valley; in Oregon: LM, Lake Modoc; WV, Warner Valley.

quences of other climate-related geologic phenomena, such as glaciation or soil and terrace development.

Factors Affecting Pluvial-Lake Levels

CLIMATIC FACTORS

The amount of precipitation falling on the drainage area of closed basins is the most significant factor affecting the supply of water to many lakes, although the amount falling directly on a lake's surface can be significant. Substantial differences in the percentage of the tributary's precipitation that reaches the lake, however, are caused by variation in any of several factors. The most important of these are the seasonal distribution and intensity of the precipitation, the pro-

portions of snow and rain, the nature of the vegetative and soil cover, the mean annual and seasonal temperatures, the topographic relief and slope angle, and the amount of annual variability. Langbein and others (1949: Figure 2), Snyder and Langbein (1962: Figure 5), Dury (1965: C15-C40), Schumm (1965: Table 1), Galloway (1970: 250-52), Brakenridge (1978: 30-34), Mifflin and Wheat (1979: 37-49), and Benson (1981: 394-400) propose ways to estimate the net effect of these variables, but some factors are treated differently or are omitted and the results differ markedly. A major source of uncertainty in these attempts is the fact that in presently arid regions, drainage from areas lying at higher elevations and receiving annually more than about 40 cm of precipitation is required to maintain perennial lakes, and there is no present analogue of the pluvial runoff from the more-arid lower sectors. Estimates of past increases in runoff that were necessary to support maximum-sized pluvial lakes range from about double (Galloway, 1970: Figure 5, Table 1) to an order of magnitude (Dury, 1965: C22; Smith, 1976a: Table 1) (Table 10-2).

One of the most important influences on evaporation from lakes is water temperature, which determines vapor pressure. In the 0°-to-30°C range, water-temperature reductions of 5°C reduce the vapor pressure to about 72% of the original value; reductions of 10°C, about the maximum lowering of air temperatures indicated by most data on Pleistocene paleoclimates in temperate latitudes, produce vapor pressure reductions to about 50% of the original value. However, water temperature is determined by a complex interaction among about a dozen energy sources and sinks, and these vary in magnitude according to the temperature and absolute humidity of the air, the duration and intensity of solar radiation, and the velocity of the wind. Most of these factors become less effective as air temperatures become lower (Kohler, 1954: 132-36; Kohler et al., 1955: 9-19; Harbeck et al., 1958; Kohler et al., 1959: 43-50; Meyers, 1962; Helly et al., 1966: 9-19) and as elevations increase (Blaney, 1956: Figures 1 and 2; Harding, 1965; Mifflin and Wheat, 1979: Figure 24). During cooler pluvial periods, however, the changes in the factors other than the temperature that influence evaporation tend to cancel each other: evaporation (1) is increased as a result of lower salinities and probably higher wind velocities, (2) is possibly decreased as a result of higher humidity and cloudiness, and (3) is partially offset by precipitation on the enlarged lake surface.

In saline lakes, the evaporation rate and the degree of fractionation of stable isotopes during evaporation are reduced in proportion to the percentage of dissolved ions as a result of the hydration energies of those ions (Langbein, 1961; Stewart and Friedman, 1974; Friedman et al., 1976). In saturated brines dominated by sodium and potassium, evaporation rates are reduced to 60% to 80% of those of fresh water (Harbeck, 1955); saturated brines dominated by divalent magnesium and calcium cations, however, can have rates reduced to 10% of fresh water (Turk, 1970). Water color and depth, as well as current velocities and patterns, also affect surface-water temperatures, and evaporation rates are also affected by these factors.

OTHER FACTORS

Nonclimatic phenomena can also change the levels of lakes in closed basins. Tectonic events such as the raising or lowering of catchment areas and outlet sills by faulting or crustal warping can gradually lead to new lake-basin dimensions and hydrologic regimes. Erosion of the outlet barrier can quickly lower or drain a lake in a closed basin, local subaerial erosion of surrounding exposures can fill a lake basin to the level of its sill, and headward erosion of nearby streams can lead to stream capture and to a rapid increase or decrease in the inflow to a lake. Volcanic eruptions and landslides can abruptly dam water channels and temporarily or permanently block or divert flow to downstream lakes.

Most of these processes, however, result in changes in lake levels that are long lasting compared to fluctuations caused by climatic change. The Bonneville Flood, for example, apparently occurred after the capture of the externally flowing Bear River by a tributary of Lake Bonneville as a result of lava damming (Bright, 1967: 4-5); the increased tributary flow caused the lake to rise to a previously unoccupied spillway level, which was eroded so rapidly through unconsolidated sediments down to bedrock that it produced downstream flooding with an estimated discharge of more than 400,000 m³ per

second (Malde, 1968). As another example, the initial formation of pluvial Lake Las Vegas 30,000 years ago may have been tectonic (Haynes, 1967: 82), and its draining 14,000 years ago could have been a result of erosion at its outlet, although Mifflin and Wheat (1979: 27) conclude that a spring-fed marsh, rather than a lake, occupied that site. Overall, however, few changes in the levels and histories of lakes in the western United States appear to have been caused by these nonclimatic processes during the Late Wisconsin. Nevertheless, the two examples illustrate why caution must be used in translating pluvial-lake changes, especially the longer and more "permanent" changes, into climatic histories.

Processes of Pluvial-Lake Sedimentation

CLASTIC SEDIMENTATION

The sizes and shapes of pluvial lakes in the western United States were first reconstructed on the basis of the record left by nearshore clastic sedimentation. Very few studies have been made, however, of the shoreline depositional environments of fluctuating pluvial lakes, although a large literature exists on erosion and sedimentation processes along the shorelines of coasts and large stable lakes (e.g., Sly, 1978: 72-76). Gilbert (1885: 98-100, Plate XII) inferred that the V-shaped terraces and bars in the Lake Bonneville area of Utah were a result of the lake's fluctuating character; Born and Ritter (1970) attribute river terraces leading to Pyramid Lake, Nevada, to fluctuating lake levels; and Born (1972: 66-90) describes the late-Quaternary and modern depositional environments in the delta portion of the constantly changing Pyramid Lake.

Clastic sediments in deeper-water and other lower-energy environments of lakes generally consist of well-sorted fragments derived from weathered bedrock in the drainage area. Compared to freshwater-lake sediments, pluvial-lake sediments may contain larger percentages of chemical precipitates (discussed later), and diagenesis commonly is more extensive. However, many of the processes by which sediment is introduced from outside the lake and by which various organic and inorganic components are absorbed or precipitated from the lake water are essentially the same as those in freshwater lakes, and these processes are partially controlled by the specific clastic mineral phases present, especially the clay species (Jones and Bowser, 1978). Sly (1978: 77-83) discusses the interaction between energy source and sediment supply in freshwater lakes. He also notes that lake sediments are finer grained than marine sediments deposited in water of comparable depth because of the lower-energy environments of lakes; lake-sedimentation rates, however, are usually higher. The most diagnostic and frequently used physical, mineralogical, petrographic, and textural characteristics of closed-basin and freshwater lacustrine sediments are summarized by Picard and High (1972). The limnology and the bottom sediments of several present-day Lahontan Basin lakes are described by Hutchinson (1937), and the organic and inorganic composition of Pyramid Lake bottom sediments, by Swain and Meader (1958).

CHEMICAL SEDIMENTATION

Even though pluvial lakes were (by definition) larger than the lakes presently occupying the same closed basins, many had no outlet, and so they accumulated salts and deposited chemical sediments. Carbonate minerals are the most abundant precipitates in sediments deposited in relatively fresh water, although differentiating chemically precipitated carbonate from detrital, skeletal, and diagenetic carbonates can be difficult (Kelts and Hsu, 1978). Deposition of carbonate minerals from fresh water generally requires decreasing the partial pressure of carbon dioxide as a result of exsolution or removal by microorganisms, increasing the chemical activity of calcium, warming the waters, or mixing chemically dissimilar waters (Jones and Bowser, 1978: 211-15; Smith, 1979: 79-82).

Pluvial lakes that were constantly resupplied with the required components precipitated substantial percentages of aragonite, calcite, protodolomite, or gypsum along with the clastic fraction—in effect, undergoing the first stages of brine evolution (Hardie and Eugster, 1970; Eugster and Hardie, 1978: 243-47). For example, cores of pluvial Lake Bonneville sediments, interpreted as having been deposited during the latest pluvial period in a relatively freshwater environment, contain 18% to 64% total carbonate (Eardley and Gvosdetsky, 1960: Figure 1). In moderately saline, unstratified pluvial lakes, a continuous rain of chemical sediments probably occurred at rates proportional to seasonal changes in the evaporation/inflow balance (Hardie et al., 1978: 23-27). In stratified lakes, denser saline bodies of brine were overlain by lighter layers of less saline water that flowed into the basin and brought new components; these combined, on mixing, with components from the lower layer, and chemical sedimentation rates probably became more proportional to freshwater inflow volumes. Aragonitic laminae in pluvial-lake deposits 24,000 to 10,500 years old from Searles Lake cores are inferred to be products of seasonally replenished stratified perennial lakes; they now contain, on the average, about 55% carbonate minerals as a mixture of primary and diagenetic species that includes some carbonate introduced later by downward-moving brines (Smith, 1979: 111, Table 13).

Pore waters of sediments deposited in closed basins are commonly saline, either because of a brine layer along the floor of the lake at the time of initial deposition or because of a subsequent downward migration of denser brines formed during later desiccations. Reaction of these brines with both chemical sediments and clastic silicate sediments commonly leads to the development of one or more diagenetic minerals and to the partial or total destruction of some original phases. The chemical principles involved in these reactions are summarized by Berner (1971: Chapters 6-10); examples of Quaternary lake deposits that document these processes are cited by Sheppard and Gude (1968: 32-36), Eugster and Hardie (1978: 252-55), and Smith (1979: 100-106).

Late-Quaternary Geology of Selected Pluvial-Lake Areas

The geologic record left by late-Pleistocene pluvial lakes in the western United States can be divided into two broad categories, geomorphic and stratigraphic. The geomorphic evidence, such as gravel bars and beach deposits, and the erosional shorelines required as the sources of such sediments, are evident in most of the basins that contained lakes. The distinctive shapes, lithologies, and horizontal alignments of these features provided early investigators with nearly indisputable evidence of the former presence of standing water at higher levels. The classic studies of Russell (1885, 1889), Gilbert (1890), and Gale (1914) were based largely on features of these types, although the studies were supplemented by stratigraphic considerations. The more recent shoreline mapping of Nevada's pluvial lakes

by Mifflin and Wheat (1979: 10-37) and the study of Lake Bonneville shoreline zones by Currey (1980) illustrate the extent to which features of these types, used with little stratigraphic corroboration but in conjunction with radiocarbon dating, still provide a reliable basis on which to reconstruct the size, distribution, and contemporaneity of pluvial lakes and from which to infer the climatic regimes that produced them. In many basins, in fact, nearshore features provide the only exposed evidence of former lakes.

Extensive bars, deeply cut shorelines, and thick beach deposits in pluvial-lake basins provide the best-documented portions of the geomorphic record, and they are usually inferred to represent periods of lake stability caused by an escape of excess water via a spillway (which prevented the lake level from fluctuating as short-term variations in the pluvial intensity occurred), by an unusual basin configuration, or by a stable climate. Between these periods of stability, however, periods of instability probably characterized most pluvial lakes, and the less-pronounced nearshore features that developed during those times now provide a correspondingly more obscure geomorphic record of that part of a lake's history.

The stratigraphic record left by pluvial lakes can be more nearly complete and less ambiguous. However, there are geologic problems. The sedimentary record exposed around the edges of lake basins is commonly fragmentary and many have hiatuses caused by erosion or nondeposition, and the record provided by cores from areas of nearly continuous sedimentation rarely allows the reconstruction of the water depths over the site, unless salts were being deposited, in which case the deposition rate and age estimates based on it become uncertain. To solve these problems, most investigators collect stratigraphic data from numerous settings in order to gain a complete record. But the most difficult aspect of stratigraphic studies based on outcropping lacustrine sediments—and one of the most serious obstacles to the correct reconstruction of pluvial-lake histories from this type of evidence—is the reliable correlation among widely separated outcrops. Outcrops commonly consist of nearshore facies of lacustrine deposits, and the character of the clastic fraction of these sediments varies greatly, even over short distances, making long-distance lithologic correlations very difficult. The most reliable means of making correlations are distinctive volcanic ash layers, but many basins have few or none. Basins with outcrops containing some horizons characterized by abnormally high (or low) percentages of chemical sediments—usually calcite, aragonite, or gypsum—generally can be correlated over long distances, because the abnormal horizons represent a basin-wide chemical anomaly in the lake waters. The soils developed on the exposed lacustrine sediments or overlying alluvium during low stands have been used extensively by some investigators for correlating sections of lake sediments.

The clastic fraction in deep-water facies found in cores commonly shows little stratigraphic change, and correlations over long distances are also difficult. However, as in outcrops, the lithologic variations of perennial-lake deposits caused by changes in water chemistry or lake structure are generally widespread enough to provide a reliable basis for intrabasin correlation and possibly for interbasin correlation if they are caused by regional climatic events. When pluvial lakes shrink, saline layers are deposited in the basin centers, but lateral changes in the mineralogy and texture of saline beds make distant correlations difficult; correlation of the perennial-lake sediments above and below such beds usually provides a better basis.

Reconstructing lake histories from any of these types of data requires reconstructing the lake level through time. Dated geomorphic records provide reliable histories of the stable periods and elevations of high lake levels. Dated stratigraphic records of outcropping sediments permit the reconstruction of both stable and fluctuating lake periods, and the elevation range of each lacustrine and nonlacustrine stratigraphic unit allows at least minimum estimates of lake-surface elevation change. Studies of cores permit the unequivocal separation of perennial-lake periods from saline- or dry-lake periods. Stratigraphic studies of both cores and outcrops also allow supplemental data to be derived from their chemical and mineralogical nature, textural character, and fossil content.

LAKE BONNEVILLE SYSTEM

Lake Bonneville, which covered about 51,640 km² at its highest stage, was the largest of the late-Pleistocene pluvial lakes in western North America. (Areas and other data on individual lakes are mostly from Mifflin and Wheat [1979, appendix] or Snyder and others [1964].) At its highest stage, it formed an irregular oval body of open water that contained about 20 relatively small islands; its longest dimension, nearly 500 km, was oriented north-south. Water covered the lake floor to a maximum depth of about 335 m, and its total volume was nearly 7500 km³. The weight of this water caused isostatic depression of the crust; and, since the lake's desiccation about 10,000 yr B.P., the once-level Bonneville shoreline has rebounded isostatically so that the shorelines on islands near the middle of the lake are now about 65 m above the elevation of the shoreline near the edge of the lake (Crittenden, 1963a: 5520, Figures 1 and 2; 1963b).

Lake Bonneville is probably the most studied of the pluvial lakes in western North America, and so its history from 25,000 to 10,000 yr B.P. should be the best known, but substantial differences in the details of interpretation still exist. Gilbert's classic study (1890) expressed the lake's history in terms of shoreline levels, and he named the three most prominent benches as the Bonneville (1565 m elevation), Provo (1470 m), and Stansbury (1350 m) levels. (The elevations are nominal; they vary from place to place because of the differential isostatic rebound that followed evaporation of the lake water.) Ives (1951) named two lower shorelines as the Timpie (1295 m) and the Dunway (1320 m); Eardley and others (1957) renamed the Timpie as the Gilbert level. All these shorelines are considered by Currey (1980: Figure 8) to have been occupied within the time of interest here. Through the use of stratigraphic and radiometric dating methods, other reconstructions of Lake Bonneville's history have been made by Antevs (1945, 1948, 1952, 1955), Ives (1951), Williams (1962), Hunt and others (1953), Eardley and others (1957, 1973), Broecker and Orr (1958), Bissell (1963), Morrison (1961b, 1965a, 1965b), Richmond (1961), Broeker and Kaufman (1965), and Scott and others (1980). Morrison (1965c: Figure 2) summarizes his own (1965a) and seven preceding interpretations of Lake Bonneville's history. Figure 10-2 shows Morrison and Frye's (1965) interpretation, plus four others that are not included in Morrison's summary or have been published since. Two of the lake histories illustrated in Figure 10-2 have age controls based on radiocarbon dates, one on both radiocarbon and amino acid dating methods, and one on geomorphic methods. Much of the age discrepancy among the histories is attributable to the geochemical problems associated with radiocarbon dating. (See the section on Lake Lahontan.)

As might be expected with this many investigators studying Lake Bonneville over a period encompassing a century, the history of the nomenclature for the shorelines and the sediments deposited in Lake

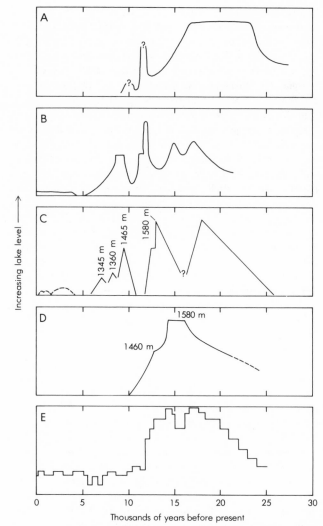

Figure 10-2. Interpretations of the history of Lake Bonneville during the period 30,000 yr B.P. to present. (A modified from Broecker and Orr, 1958: Figure 9; B modified from Broecker and Kaufman, 1965: Figure 4; C modified from Morrison and Frye, 1965: Figure 2; D modified from Scott et al., 1980; and E modified from Currey, 1980: Figure 8.)

Bonneville has become complex. Early investigators based their conclusions on combinations of geomorphic and stratigraphic criteria, and usages developed in which shore features and the sediments inferred to be their offshore equivalents were assigned ages and stratigraphic positions more on the basis of their elevation within the basin than on conventional stratigraphic criteria, such as superposition or lateral correlation of exposed sections. Morrison (1965c: 273-76) summarizes his own and previous usages; Scott (1980) presents an updated review and a significantly different version of the stratigraphic record. For brevity, no attempt is made here to explain the interrelation among the stratigraphic units defined or used by the various investigators; details about the past and present usage of the terms "Alpine Formation," "Bonneville Formation," "Draper Formation" and "Provo Formation," which appear to include sediments deposited partially or entirely within the time period here under discussion, can be found elsewhere (International Union for Quaternary Research, 1977: 89-94).

Recent studies by Currey (1980) and Scott and others (1980) probably come the closest to reconstructing a correct history for Lake Bonneville over the period 25,000 to 10,000 yr B.P., because they use the largest number of stratigraphically controlled radiocarbon dates and other methods that serve as tests of those ages. Scott's (1980) review of the stratigraphic complexities in the basin, however, illustrates the difficulties involved in compiling dates from widely spaced localities to create a lake chronology; isolated sections of lake deposits must be correlated by criteria that require assumptions, depositional environments are sometimes difficult to reconstruct with certainty, and rapid facies changes in nearshore environments make even short-range stratigraphic correlations in these areas uncertain.

When Lake Bonneville fell from the Bonneville level at 1565 m to the Provo level at 1470 m, the Escalante Desert, Rush Valley, and White Valley arms were detached, and the lakes in those arid valleys probably shrank or desiccated. Continuing recession past levels below the Stansbury level at 1350 m detached the arms of the lakes in Sevier Valley, Provo Valley, and Cache Valley. These three basins, however, continued to receive water from the perennially large Sevier, Provo, and Bear Rivers; and, during all except interpluvial periods, one or all of the basins were presumably full, overflowing, and contributing their excess water to the lower lake. Evaporation from their surfaces prior to overflow, however, effectively acted on the overall hydrologic budget as if they were all part of the same lake. The fall of Lake Bonneville below 1286 m disconnected the Great Salt Lake Desert arm; this represented 48% of the evaporation area of the full lake (Eardley et al., 1957; Figure 1), and without major inflow the arm desiccated and formed the Bonneville Salt Flats.

The area of present-day Great Salt Lake at 1280 m elevation is about 8290 km², and Utah Lake covers about 345 km²; their combined area is about a 17th of the maximum size of Lake Bonneville. It appears from this that the net effects of changes in the inflow volumes, precipitation on the surface, and evaporation rates during pluvial periods promoted lakes that were at least 17 times larger than lakes permitted by the present climate. This figure may be too low in that it does not take into account the water volume lost by the enlarged lake from its overflow or the reduced evaporation rate caused by the present lake's salinity. As a measure of relative inflow volumes, it is clearly too high because it does not take into account the precipitation directly on the increased surface area of the lake, which could offset 25% to 35% of the evaporation from it. Even if these considerations perfectly offset each other, this figure would not imply a seventeen-fold increase in the discharges of the rivers that drain from the Wasatch Range into present-day Great Salt Lake. Proposed evaporation rates during pluvial periods range from 10% to 45% below present rates (Table 10-1). With the largest estimate of evaporation reduction and with precipitation on the lake offsetting 35% of the remaining evaporation, inflow requirements would be reduced to about 36% of apparent volumes, and this would reduce the input requirements to about 6 times present river volumes. Also, during the Pleistocene, other rivers in the drainage area probably flowed into Lake Bonneville. There is no quantitative way to reconstruct the pluvial flow of the now-dry tributaries, but present runoff from arid areas north, west, and south of the former Lake Bonneville are about 5% or less of the runoff from high areas of the Wasatch Range (Langbein et al., 1949: Plate 1). The areas drained along those rela-

Table 10-1.
Published Estimates of Full-Glacial Precipitation
and Evaporation over Closed-Basin Lakes

Paleolake or Area	Reference	ΔTa (°C)	ΔTs (°C)	ΔP (cm)	ΔE (cm)	ΔE (%)
Estancia, New Mexico	Leopold, 1951	− 6.5	− 9	+18	− 38	34
	Brakenridge, 1978[a]	− 8	− 8	− 4	− 25	42
	Galloway, 1970	−10.5	−10	− 4.6	− 51	45
	Leopold, 1951	− 8	− 8	+ 2	− 42.5	38
Warner, Oregon	Weide, 1976[b]	− 5	−12	+10	− 22	20
Lahontan, Nevada	Antevs, 1952	—	—	+84	—	34
	Broecker and Orr, 1958	− 5	− 5	+27	−41	30
	Mifflin and Wheat, 1979[c]	− 3	—	+23	−18	16
Spring Valley, Nevada	Snyder and Langbein, 1962	—	− 7	+21	−33	30
	Brakenridge, 1978[a]	− 8	− 8	0	−48	43(?)
Nevada (entire state)	Mifflin and Wheat, 1979[d]	− 3	—	—	—	10
Southern High Plains, Texas	Reeves, 1965	− 5	−10	0	−41	27
	Reeves, 1966a	− 5	− 8	+39	− 4	27
	Reeves, 1973	−10	−10	+26	—	40
	Reeves, 1976	−10	−10	+37	—	—

Source: Modified from Brakenridge, 1978.

Note: ΔTa and ΔTs are the changes in annual and summer temperatures, respectively; ΔP is the change in annual precipitation; ΔE the change in annual evaporation.

[a] Recalculated.

[b] Average of two values from Weide's Table 1.

[c] Calculated from Figure 24, Table 8, and p. 44.

[d] Calculated from p. 43.

tively arid shores were not large, however, and it seems unlikely that Pleistocene runoff from them would have constituted more than 10% of the total lake inflow. Thus, these considerations imply that the pluvial-age rivers draining the Wasatch Range had flow volumes at least 5 times their present volumes, and possibly much greater.

LAKE LAHONTAN SYSTEM

Lake Lahontan was the second-largest pluvial lake in the western United States. In contrast to Lake Bonneville, it was more an interconnected series of long, narrow lakes than an open body of water. Its 22,900-km² area and 280-m maximum depth made it slightly less than half as large as Lake Bonneville, and the geometry of its basins and its multiple sources of water made it different in several hydrologic aspects. Its highest shoreline averaged 1330 m elevation. On shrinking, Lake Lahontan was fragmented into nine smaller basins (Benson, 1978: Figure 1); five of these sustain lakes or marshy areas today, and they probably existed independently as lakes of significant size during pluvial transitions and semipluvial periods.

Evidence of pluvial Lake Lahontan was first noted by a geologist with the Simpson expedition in 1858-1859, and the lake was later mapped by King (1878) and named for Baron LaHontan, an explorer. The classic study by Russell (1885) set new standards for geologic investigations of pluvial lakes by integrating data from stratigraphy, geomorphology, and water chemistry, and it used tufa morphology as a supplementary criterion. Russell identified a "lower lacustral clay" and an "upper lacustral clay" separated by a "medial gravel." Morrison (1965c: Figure 4) summarizes the lake histories inferred by Russell (1885), Antevs (1945, 1948, 1952), and himself (1961a, 1964) and presents a revised history of the lake without absolute age assignments. He names the lacustrine deposits the Eetza Formation (oldest), Sehoo Formation, and Fallon Formation; these were separated by subaerial deposits of the Wyemaha Formation and Turupah Formation. Morrison and Frye's (1965) history, which has a time scale based on radiocarbon ages, assigns the Sehoo Formation to the period 25,000 to 10,000 yr B.P. Figure 10-3 shows Morrison and Frye's chronology, along with those from four other studies based on ¹⁴C dates on tufa, shells, marl, organic material, dung, and wood; on ²³⁰Th/²³⁴U dates on shells; and on tephrochronology. As with the history of Lake Bonneville, there remain substantial differences among interpretations of Lake Lahontan's history for the period between 25,000 and 10,000 yr B.P.

The first concerted attempt to assign an absolute age scale to Lake Lahontan's (and Lake Bonneville's) history is Broecker and Orr's (1958). They consider the problems caused by the probable initial disequilibrium between the ¹⁴C in the lake and the atmosphere, and they conclude (p. 1012) that a maximum of 500 years should be added to the quoted ages. They consider the postdepositional contamination of calcium carbonate by diffusion with carbon dioxide in the air (pp. 1012-15), but the possibility of contamination by solution and redeposition is only noted. A continuation of that study done by Broecker and Kaufman (1965) includes about 80 new ¹⁴C ages, explores further the problems of contamination, assesses the magnitude of those errors by several methods, and includes 23 new dates based on ²³⁰Th/²³⁴U ratios in shells and other materials. Their published ¹⁴C ages and lake-level chronology (Figure 10-3A) include a 500-year correction of the laboratory age. As shown by Morrison and Frye (1965: Figure 4), however, a number of ¹⁴C dates from the Lake Lahontan area appear to be in conflict with one other, and a detailed

history cannot be easily reconstructed from the radiocarbon data alone.

Broecker and Walton (1959) find that the carbon in most modern lakes in the Great Basin is "old" relative to that of the carbon dioxide in the air and that the size of the discrepancy is related to the lake's area (which represents the potential exchange area) and its accumulated carbonate content (which represents the amount that needs to be exchanged). They believe (p. 33) that only freshwater lakes in drainage areas composed chiefly of siliceous rocks are likely to have dissolved carbon dioxide with radiocarbon values close to those of the atmosphere and that the water in lakes with high levels of dissolved carbonate—caused by carbonate rocks in their drainage area or by evaporative concentration— could have ¹⁴C concentrations as low as 50% of those of the atmosphere and thus have apparent ages up to almost 6000 years. They construct a method of estimating the size of this error in a given lake (pp. 30-32) and conclude that the ¹⁴C ages of samples of lacustrine materials from Lake Bonneville and Lake Lahontan should be reduced by 500 years. Their method for this calculation, however, shows that the magnitude of disequilibrium between the ¹⁴C content of the lake water and the air is inversely pro-

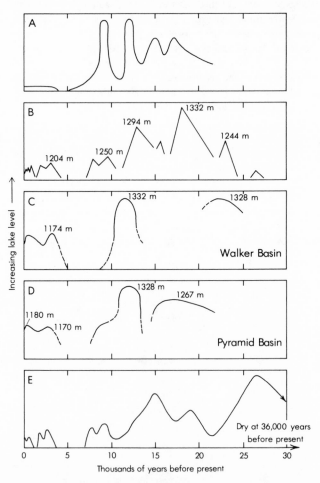

Figure 10-3. Interpretations of the history of Lake Lahontan during the period 30,000 yr B.P. to present. (A modified from Broecker and Kaufman, 1965: Figure 4; B modified from Morrison and Frye, 1965: Figure 2; C modified from Benson, 1978: Figure 6; D modified from Benson, 1978: Figure 5, and Born, 1972: Figure 28; and E modified from Davis, 1978: Figure 3).

portional to the volume/area ratio of the lake, among other factors. With this ratio constantly changing as a lake level fluctuates and with no allowance for the possibility of the existence of stratified lakes (making the effective volume to be exchanged smaller), application of a single correction factor to all dates from one lake, regardless of the lake's level and size, leads to systematically distorted versions of the lake's history.

Benson (1978) shows that with careful attention to petrographic details a case can be made for the use of certain types of dense tufa for reliable radiocarbon dating. He discusses the problem of the "age" of the sample when deposited, caused by the initial non-equilibration of ^{14}C between the atmosphere and the lake water, and problems of later contamination. He also notes that ^{13}C is preferentially incorporated in tufa and that ^{14}C is also enriched, making the sample "too young," and he notes that several factors tend to offset each other. Benson then reconstructs the history of Lake Lahontan in the Walker Basin (Figure 10-3C) and the Pyramid Basin (Figure 10-3D) on the basis of radiocarbon dates on dense lithoid tufa, and he concludes (1) that one or more high stands uniting all nine of the Lake Lahontan basins occurred between the 25,000 and 21,500 yr B.P. and about 13,600 and 11,000 yr B.P. and (2) that an intermediate-sized lake, which connected Pyramid Lake with other basins but not with the lake in Walker Basin, existed between 21,500 to 15,000 yr B.P. Tufa apparently was not forming between 15,000 and 13,600 yr B.P., but Benson offers no explanantion for this finding. Possibly, the lake level fell during this period, as inferred in Searles Valley, California. Indirect evidence suggests that Walker Lake desiccated from about 9000 to 5400 yr B.P. (Benson, 1978: 314) and that Pyramid Lake did not. Born's (1972) study of delta growth in the Pyramid Basin (Figure 10-3D) includes data derived from radiocarbon dates on wood that show that the Pyramid Lake arm of Lake Lahontan has not risen above the 1180 m level since 9700 yr B.P.

Using tephrochronology for correlations and radiocarbon dates on wood and muck, Davis (1978) proposes a different time scale for Morrison's (1964) stratigraphic study, upon which Morrison and Frye's (1965) history is based (Figure 10-3E). Davis places the base of the Sehoo Formation, which represents the younger of the two most widely preserved series of lake deposits, at an age greater than 36,000 yr B.P. In this respect, his history does not appear compatible with the others. He expresses some reservations concerning the tephro-stratigraphic isolation of the units (p. 60) that move the age of the earliest deposits of the Sehoo Formation that far back in time, but he

has good grounds for his tephrostratigraphic age assignments. His age assignments rest largely on the correct identification of the Sehoo Formation and Wyemaha Formation in two sections measured by Morrison (1964: 122) and on Davis's (1978: 80) correlation of them with the Pyramid Island locality, where wood for the two radiocarbon dates was collected. At all three localities, the Churchill Soil, which is the most frequently used basis for identifying the interpluvial Wyemaha Formation, is missing, and so it appears that the stratigraphic correlations need to be reexamined.

Considerable differences remain among the reconstructed Lake Lahontan histories. The problems facing investigators in this area are similar to those of investigators studying Lake Bonneville. Using the same reasoning, therefore, we consider the more recent age-controlled studies in the Lake Lahontan area by Born (1972) and Benson (1978) as well as most of the study by Davis (1978) to be more probably correct because they are based on more stratigraphic and chronometric data than earlier studies.

During the late 19th century, the perennial lakes in the region once covered by Lake Lahontan had a total area of about 1300 km²; the areas of individual lakes were as follows (Harris, 1970: Table 1; Rush, 1970: Figure 4; Russell, 1885: 55-81): Pyramid Lake, 495 km²; Walker Lake, 283 km²; Winnemuca Lake, 236 km²; Honey Lake, 233 km²; Humboldt Lake, 52 km²; and Soda Lake, 1 km². This is about an 18th of the surface area of Lake Lahontan at its highest stage, and it shows that climates during pluvial maxima promoted lakes that were about 18 times larger than the present-day lakes. For the reasons explained in the discussion of Lake Bonneville, however, lower evaporation rates and an increased number of tributaries reduced the volumes of water required from the main tributaries during pluvial periods. The volume of precipitation falling on the enlarged lake, which did not overflow, would have substantially reduced it further. From the same assumptions used in the Lake Bonneville estimate, it appears that maximum volumes of flow in the main rivers leading in pluvial Lake Lahontan were at least 6 times those of the present.

OWENS RIVER SYSTEM

The Owens River system consisted of a chain of pluvial lakes occupying a succession of closed basins in southeastern California. The Owens River, which drains the eastern side of the central Sierra Nevada, supplied most of the water to those lakes, although at times Mono Lake became the northernmost lake in the chain when it overflowed and added its surplus water to the headwaters of the Owens

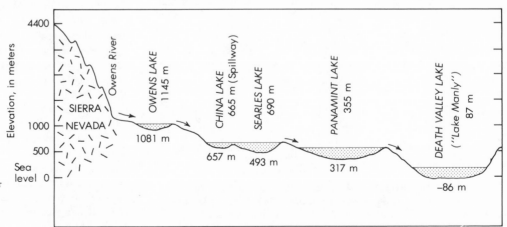

Figure 10-4. Diagrammatic cross section of pluvial lakes downstream from the Owens River showing elevations of floors of basin and lake surfaces during high stands. (Modified from Gale, 1914: Plate VII.)

River. Figure 10-4 shows a diagrammatic cross section of the five lakes downstream from the Owens River in that system.

The geology of pluvial Mono Lake was first studied in detail by Russell, who described it in an 1889 report devoted to the present and past lakes in the basin and to the glacial history of the adjacent Sierra Nevada. He noted that at least two episodes of glaciation occurred in the mountains (p. 363) and that shorelines from the latest high stand of the lake were carved on the lowest of the terminal moraines extending into the pluvial lake area (pp. 369-71). From this relation, he deduced that there were two glacial episodes, which he correlated with the two lacustral units in Lake Lahontan, separated by an interglacial and interlacustral period (pp. 370-71).

Putnam (1950) proposed that the pluvial lake in Mono Basin be named Lake Russell. He correlated the spacing and sizes of the youngest recessional moraines with the spacing and widths of the shorelines cut as the lake last receded, and he concluded that the glaciers and lake receded according to the same timetable. More recently, however, Lajoie (1968) has shown that variations in shoreline widths are more related to variations in the slopes of the surfaces on which they were cut. He (1968; personal communication, 1981) also has made a stratigraphic study of the lake sediments controlled by a tephrochronologic framework and radiocarbon dates. Eighteen radiocarbon dates on tufa and ostracodes from the lacustrine sediments of the last major pluvial cycle exposed in Mono Basin range in age from 34,900 to 12,800 yr B.P.; his 1981 reconstruction of the history of Lake Russell, revised from his 1968 version, is plotted in Figure 10-5A. He also notes that the lack of saline layers in drill holes that extend deeper than the Bishop Tuff show that this lake has not dried during the last 700,000 years.

Lake Russell overflowed at its highest level into Adobe Valley. There, a small lake 25 m deep formed and overflowed eastward into a valley leading southward to the Owens River. Lake Adobe appears to have been shallow at 11,000 yr B.P., but it increased in depth gradually until about 8000 yr B.P., when it shrank to a shallow

marsh; about 4400 yr B.P., it again expanded briefly and formed a small lake (Batchelder, 1970).

The Owens River drains an area of about 8500 km², but virtually all of its present flow comes from the 16% of its drainage area that lies on the eastern slope of the Sierra Nevada (Lee, 1912: 9). Until 1912, the Owens River terminated in a saline lake, Owens Lake, which was about 10 m deep and 290 km² in area before agricultural irrigation in the area became extensive. All of the river's water was diverted to Los Angeles in 1912, and the lake desiccated. The lack of buried salines in the upper 280 m of lake fill (Smith and Pratt, 1957: 5-14) suggests that the lake had not desiccated naturally for the last several hundred thousand years. During pluvial periods, however, Lake Owens overflowed and carried its dissolved components downstream. The present quantity of salines in Owens Lake appears to represent only about 2000 years of accumulation, suggesting that it last overflowed about 2000 years ago (Smith, 1976a: p. 99).

The succession of pluvial lakes in Owens Valley and downstream to Panamint Valley was first adequately documented by Gale (1914; see Smith, 1979, pp. 6-7, for a more detailed account of this and earlier work). Gale's study was primarily geomorphic, although he also described the mineralogy, chemistry, and subsurface stratigraphy of Searles Lake, which at that time was being developed into a valuable source of industrial and agricultural chemicals. Except for descriptions of cores (Smith and Pratt, 1957: 5-25), stratigraphic studies of the lacustrine deposits in Owens Lake and China Lake have not been published.

Stratigraphic studies of the subsurface lake deposits in Searles Valley have been published by Smith and Pratt (1957: 25-51), Flint and Gale (1958), Haines (1959), and Smith (1962, 1979: 8-68). Dates on these units, based on the radiocarbon ratios in organic carbon, carbonate minerals, and wood, are listed by Flint and Gale (1958: 704-5), Stuiver (1964: Tables 1-3), and Stuiver and Smith (1979: Table 19, Figure 30). Stratigraphic studies of the extensively exposed lacustrine sediments in Searles Valley have been made, but only preliminary accounts of that work have been published (Smith, 1966: 170-77; 1968: 297-306: 1976b: 591-94; and two informal guidebooks). The pluvial history of Searles Lake, however, has been reconstructed in some detail for the period between 25,000 and 10,000 yr B.P. The subsurface record of interbedded salt layers and mud layers provides clear and well-dated evidence of interpluvial saline or dry lakes that were shallow and pluvial freshwater lakes that were deeper. The correlative outcropping lacustrine sediments document the elevations reached by the enlarged pluvial-lake surfaces; alluvium, fossil soils, and erosional unconformities correlate with the shallow-lake or dry periods. The inferred history of Searles Lake, based on published and unpublished data, is plotted in Figure 10-5B.

Shorelines and scattered outcrops of pluvial-lake sediments in Panamint Valley were noted by Gale (1914: 312-17) but were not studied in detail until the more recent work of R. Smith (1972, 1975, 1978a, 1978b). Smith considers most of the deposits and shoreline to be older than the period of interest in this chapter. The lowest shorelines, considered by both Blackwelder (1954: 37) and Smith (1978a: 19) to represent the most recent lakes, are only 44 m above the lowest point in the valley and 248 m below its spillway. Their well-preserved character and indirect evidence implies a very late Pleistocene age; local runoff is not adequate to sustain a lake of that size without the inclusion of overflow from Searles Lake (Smith, 1978a; 1978b: 19), and that last occurred during very late Pleistocene time (Figure 10-5B). The northern part of Panamint Valley also contained a small

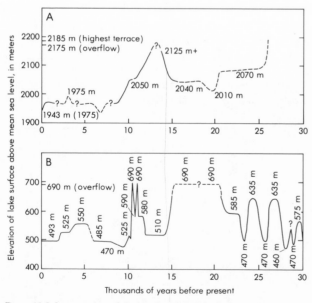

Figure 10-5. Interpretations of the histories of (A) Lake Russell in Mono Basin (from K. Lajoie, personal communication, 1981), and (B) Searles Lake (from G. I. Smith, unpublished data) during the period 30,000 yr B.P. to present.

(18-40 km²) lake that radiocarbon dates show existed at least up to about 10,000 yr B.P. (Peterson, 1980: 176-78). Drill-hole data (Smith and Pratt, 1957: 51-62) do not resolve any details of the late-Pleistocene pluvial period in Panamint Valley, but they confirm the existence of a succession of perennial lakes during that and earlier parts of the valley's history.

The existence of a pluvial lake in Death Valley was presumed long before shoreline evidence was discovered. However, this is not surprising since the sediment and erosion record is meager, and, as noted by Hunt and Mabey (1966: A71), "Whatever the cause, this California lake left one of the least distinct and most incomplete records of any Pleistocene lake in the Great Basin—another California superlative!" The evidence that does exist consists of erosional shorelines, scattered lakeshore gravels, tufa, and core samples of lake deposits recovered from the deeper valley fill (Blackwelder, 1933; Hunt and Mabey, 1966: A69-A76; Hooke, 1971; Hooke, 1972: 2086, 2092). Hooke (1972: 2086-87) proposes a lake 90 m deep in Death Valley between 11,000 and 10,000 yr B.P. on the basis of radiocarbon dates and a hypothesis involving tectonic tilting and alluvial-fan growth. The radiocarbon dates on lacustrine silt and clay, which range from about 11,900 to 21,500 yr B.P., confirm the existence of a lake in the valley during that period, but the paucity of outcropping lacustrine sediments and fresh shoreline features, which are almost uniformly found in other nearby basins with lakes of the same general age, suggests that the lakes of these ages in Death Valley were small and substantially less than 90 m deep.

The present and past balances between evaporation and inflow require each basin in the Owens River chain to be considered separately. The calculations presented in Table 10-2 assume that reductions in air temperatures of 5° and 10°C led to reductions in lake-water temperatures by the same amount and that evaporation rates were reduced to 72% and 50% of their present values (the approximate change in vapor pressures over these temperature ranges). They also assume that precipitation falling on the lakes was not significantly greater than it is at present, because more than 60% of the cumulative lake area lies in substantially more arid basins than represented by the main tributary region. This is a highly simplified approach to a complex problem, but the lake areas are probably within 10% and the inferred evaporation rates within 40% of being correct, meaning that the calculated inflow/evaporation volumes should not differ by more than 50% from the correct value, whereas the calculated changes in required inflow volumes are several hundred percent of present volumes. Inflow from areas downstream from the Owens River is not considered, and it may have compensated for part of the recharge requirement attributed in Table 10-2 to the Owens River, but present runoff from most of the downstream drainage is about 1% of that coming from the eastern Sierra Nevada to the Owens River itself (Langbein et al., 1949: Plate 1). This fact suggests that pluvial runoff volumes to downstream drainages were also relatively small. It appears likely, therefore, that during the period 25,000 to 10,000 yr B.P., when Owens, China, and Searles Lakes were full and Panamint Lake was small (Table 10-1), the flow in the Owens River and its tributaries was at least 3.5 times its present flow. This method of estimating water volumes (Table 10-2) suggests that when all of the valleys in the system were full and a maximum-sized lake existed in Death Valley—the "maximum pluvial" episode in this area—the Owens River's flow was nearly 6 times that of the present river if pluvial lake temperatures were about 10°C cooler, was more than 8 times that of the present if they were 5°C cooler,

and was 11 times greater if little or no cooling accompanied those episodes.

OTHER PLUVIAL LAKES IN THE WESTERN UNITED STATES

As is evident from Figure 10-1, many other closed basins in the western United States contained pluvial lakes. Their geologic record is of uneven quality, and about two-thirds of them have not received careful study. Table 10-3 lists 31 dated lacustrine deposits showing pluvial expansions during the period 25,000 to 10,000 yr B.P. The locations of the deposits are indicated on Figure 10-1; lakes not numbered on the figure can be identified on the previously cited larger-scale maps of pluvial lakes. Discussions of selected lake histories in these basins follow. Those selected have the best records, the most dates, and/or the greatest significance.

Arizona

Only one major pluvial lake existed within Arizona, Lake Cochise in the valley now containing Willcox Playa. Beach ridges lie 26 m above the present playa, and lacustrine sediments below the playa's surface are dark colored and have negative Eh values, documenting the former existence of pluvial lakes (Martin, 1963: 439, Figure 3; Schreiber et al., 1972: 133, 176). Radiocarbon dates on exposed and subsurface carbon and carbonate materials suggest that high-level pluvial lakes occupied the basin during much of the period from before 30,000 until 13,000 yr B.P. and that lower-level lakes existed between 11,500 and 10,500 yr B.P. (Schreiber et al., 1972: 134, 176).

The almost complete lack of pollen in the top 2.1 m of lake sediment, which have a radiocarbon age of about 23,200 yr B.P. at their base, precludes the reconstruction of the surrounding vegetation during most of the pluvial episodes of the last 25,000 years (Martin, 1963: 440-44). Pine pollen in the underlying 1.6 m of sediments below the dated material accounts for 95% to 99% of the total count; this fact reflects a pluvial interval whose intensity is not exceeded in any lower zone in the 43-m core and indicates that only one maximum-intensity pluvial episode occurred in this area during the period represented by the core. Martin and Mosimann (1965: 352-56) note, however, that there appears to have been marked climatic metastability in this area, as is indicated by detailed study of the pollen from the Wisconsin-age sediments in this core.

California

Several closed basins in California contained pluvial lakes besides those that were part of the Lake Lahontan and Owens River systems. However, dates are available for only five of them, and stratigraphic control on those dates is available for only two. The existence of pluvial Lake Tulare in the southern San Joaquin Valley is based on a complex subsurface stratigraphy of lacustrine and nonlacustrine deposits that have yielded five radiocarbon dates ranging from about 26,800 to 9000 yr B.P. (Croft, 1968). The deposits underlie an area now characterized by closed drainage caused by tectonic warping, and the past history of lacustrine deposition may have been as dependent on tectonic processes as on pluvial climates. Nevertheless, perennial lakes appear to have occupied the area intermittently during a period from about 24,000 to 8000 yr B.P. Croft (1968: Figure 4) infers a period of nonlacustrine deposition that ended about 12,500 yr B.P., although he notes that the data allow this nonlacustrine interval to have begun as early as 17,000 yr B.P. or as late as 14,000 yr B.P. and

Table 10-2.
Relative Inflow Volumes Required to Balance Evaporation at Varying Temperatures
and for Various Lengths of the Owens River System of Pluvial Lakes

Size of Last Lake in Chain	Elevation of Water Surface[a] (m)	Added (and Cumulative) Area (km²)	Evaporation[b] (m/year)			Cumulative Volume[c] (10⁹m³)			Relative Discharge of Owens River and Its Tributaries[d]		
			T = 0°	T = 5°	T = 10°	T = 0°	T = 5°	T = 10°	T = 0°	T = 5°	T = 10°
Owens Lake, historic	1095	290	1.27[e]	—	—	0.41[f]	—	—	1.0	—	—
Owens Lake, abnormal[g]	1085	—	—	—	—	0.74[h]	—	—	1.8	—	—
Lake Owens, full	1145	694 (694)	1.23	0.89	0.62	0.85	0.62	0.43	2.1	1.5	1.1
China Lake, full	665	155 (849)	1.41	1.02	0.70	1.07	0.78	0.54	2.6	1.9	1.3
Searles Lake, full	690	839 (1688)	1.65	1.19	0.82	2.45	1.78	1.23	6.0	4.3	3.0
Lake Panamint, small[i]	355[i]	175 (1863)	1.80	1.30	0.90	2.76	2.01	1.39	6.7	4.9	3.4
Lake Panamint, full	602	707 (2570)	1.65	1.19	0.82	3.62	2.62	1.81	8.8	6.4	4.4
Lake Manly, full[j]	87	533 (3103)	1.97	1.42	0.98	4.67	3.38	2.33	11.4	8.2	5.7

[a] Surface of last lake in chain.

[b] Net evaporation rate for last lake in chain; except for Owens Lake, rates for present evaporation (T = 0°) adapted from Meyers' data on present gross annual rates (1962: Plate 3), which indicate approximately 1.52 m for China, 1.78 m for Searles, 1.93 m for Panamint, and 2.13 m for Death Valley, reduced to net rates by assuming 10 cm annual precipitation on the lakes in China, Searles, and Panamint Valleys, 5 cm on lake in Death Valley; effects of reducing lakes' water temperatures by 5° and 10° (T = 5° and T = 10°) calculated by use of factors 0.72 and 0.50, the approximate reduction in vapor pressure of water with 5° and 10° reductions in the range of 0° to 30°C; corrected for changes in pluvial lake surface elevations above present valley floors using lapse rate of 6.5°C/1000 m (0.64 m/year/1000 m).

[c] Except as noted, figures are sums of losses from each lake in chain, the product of (evaporation rate for that lake) × (area of that lake), and values represent volumes of both total evaporation loss and offsetting total Owens River inflow.

[d] Relative to calculated volume of present Owens River, 0.41 × 10⁹ m³/year.

[e] Average of net rates observed in May 1939 through April 1940 (Dub, 1947: Table 3) and September 1969 through August 1970 (Friedman et al., 1976: Figure 1), both corrected for salinity effect by assuming coefficient = 0.9.

[f] Calculated conditions in 1872, prior to irrigation, from data of Gale (1914: 254, 255, 261), calculated by using ratio of lake size to river flow in 1909-1912 and determining flow necessary to balance lake size in 1872.

[g] 1968-1969, record wet season (Friedman et al., 1976: Figure 1).

[h] Flow in Owens River observed in 1969 (Friedman et al., 1976: 503) plus volume diverted into Owens Valley Aqueduct during same period.

[i] Latest-Pleistocene lake, 44 m deep (Smith 1978a).

[j] This "maximum pluvial" condition apparently did not occur during the period 25,000 to 10,000 yr B.P. Calculation uses Death Valley evaporation rate for highest shoreline level but assumes that the volume of water required from the Owens River system was only one-third of total because the Mojave River and Amargosa River systems also contributed their flow to the lake in the valley.

to have been caused by either climatic change or alterations in the drainage regime.

A stratigraphically controlled history of lake-level fluctuations in the Silver Lake area, a shallow basin near the northern end of the larger pluvial Lake Mojave, includes radiocarbon dates on shells and tufa recording three separate pluvial stages in the late Pleistocene. A major pluvial episode ended about 14,500 yr B.P., and short pluvial episodes caused overflow into Death Valley between 13,750 and 12,000 and between 11,000 and about 9000 yr B.P. (Ore and Warren, 1971: Figure 7). An episode of partial filling can be inferred from a radiocarbon date on tufa between 8500 and 7500 yr B.P., and the consistency between the other dates on tufa and the dates on shells supports the acceptance of this date and the existence of the episode. However, the basin has been partially filled to depths of 2 to 3 m

three times during the 20th century, so the climatic significance of the last stratigraphically dated flooding may not be major.

The five smaller pluvial lakes shown in Figure 10-1 lying southeast of the boundary of the Great Basin (which is drawn on the boundary between the Lahontan Hydrologic Basin and the Colorado Desert Hydrologic Basin as defined by the California Department of Water Resources) are conjectural, as are the three lakes near the northern end of the state. The three largest of the southern lakes—Bristol, Cadiz, and Danby—have extensive subsurface deposits of salts, which must have required large volumes of water, but only Danby Lake shows any reported geomorphic or stratigraphic evidence of the existence of an enlarged lake; a small terrace composed of lacustrine sediments was tentatively reported by Thompson (1929: 708), but it has not been subsequently verified. These basins may eventually be

Table 10-3.
Pluvial Lakes in the Western United States with Radiocarbon-Dated Chronologies in the Range of 25,000 to 10,000 yr B.P.

State/ Lake Number	Pluvial Lake (and Name of Modern Lake, Playa, or Valley, if Different)	Latitude (°N)	Longitude (°W)	Finite ^{14}C dates	Oldest [14]C Date (yr B.P.)	Maximum Area (km²)	Maximum Increase in Depth Relative to Present (m)	References[a] (Most Useful References in Italic)
Arizona								
1	Cochise (Willcox)	32°08′	109°51′	33	27,600 ± 900 (> 30,000)	310	26	*Schreiber et al.*, 1972; Long, 1966
California								
2	Adobe (Black Lake)	37°55′	118°36′	7	11,350 ± 350	52	24	*Batchelder*, 1970, personal communication, 1978; Snyder et al., 1964; *Hubbs and Miller*, 1948
3	Deep Spring	37°17′	118°02′	47[b]	10,000 ± 1000	ca. 44	unknown[c]	Snyder et al., 1964; *Jones*, 1965; Peterson et al., 1963 (R)[d]; Hubbs et al., 1963, 1965 (R); Hubbs and Bien, 1967 (R)
4	Le Conte (Salton Sea)	33°20′	116°00′	48	> 50,000	ca. 4600	> 85[d]	*Hubbs and Miller*, 1948; Hubbs et al., 1960 (R); Hubbs et al., 1963 (R); Hubbs and Bien, 1967 (R); Crane and Griffin, 1958 (R); Fergusson and Libby, 1962, 1963 (R); Bien and Pandorfi, 1972 (R); van de Kamp, 1973; Spiker et al., 1977 (R); Stanley, 1966; Wilson and Wood, 1980
5	Manix (Troy Lake, Coyote Lake)	35°03′	116°42′	5	30,950 ± 100	407	116	Hubbs and Miller, 1948 Snyder et al., 1964; Hubbs et al., 1962 (R); Fergusson and Libby, 1962 (R); Hubbs et al., 1965 (R); Berger and Libby, 1967 (R); Blackwelder, 1933
6	Manly (Death Valley Salt Pan, Badwater)	36°00′	116°48′	4	21,500 ± 700	1600	183	*Hooke*, 1972; Snyder et al., 1964
7	Mojave (Soda Lake, Silver Lake)	35°22′	116°08′	24	15,350 ± 240	ca. 200	> 12	*Ore and Warren*, 1971; Snyder et al., 1964; Stuiver, 1969 (R)
8	Panamint	36°18′	117°18′	10	32,900 ± 1700	722	283[e]	Snyder et al., 1964; Hubbs et al., 1965 (R); Berger and Libby, 1966 (R); Smith, 1978b; Peterson, 1980
9	Russell (Mono Lake)	38°03′	118°46′	19[f]	36,280 ± 600[g]	692	238	*Lajoie* 1968, personal communication, 1981; Hubbs et al., 1965 (R); Ferguson and Libby, 1962 (R); Snyder et al., 1964
10	Searles (China Lake, Searles Lake)	35°36′	117°42′	110	46,350 ± 1500[h]	994	196	*Smith*, 1968, 1979; Stuiver, 1964; Flint and Gale, 1958; Rubin and Berthold, 1961 (R); Ives et al., 1964 (R); Levin et al., 1965 (R); Ives et al., 1967 (R); Robinson, 1977 (R); Marsters et al., 1969 (R); Peng et al., 1978; Damon et al., 1964 (R)
11	Tulare (Kern, Buena Vista, and Tulare Lakes)	36°00′	119°40′	8	26,780 ± 600[i]	ca. 4100	?[j]	*Croft*, 1968; Davis et al., 1959; Janda and Croft, 1967; Ives et al., 1967 (R); Hubbs and Bien, 1967 (R); Buckley et al., 1968 (R)
Nevada								
12	Dixie (Humboldt Salt Marsh)	39°55′	117°60′	2	11,700 ± 180	1088	72	Buckley and Willis, 1970 (R) Snyder et al., 1964; *Hubbs and Miller*, 1948
13	Lahontan (Pyramid, Walker and Honey Lakes; Carson Sink;	40°00′	119°30′	169	40,000[k]	22,440	160[l]	*Benson*, 1978, 1981; *Mifflin and Wheat*, 1979; Broecker and Orr, 1958; Olson and Broecker, 1961 (R); Broecker

Table 10-3 continued.

State/ Lake Number	Pluvial Lake (and Name of Modern Lake, Playa or Valley, if Different)	Latitude (°N)	Longitude (°W)	Finite ^{14}C dates	Oldest ^{14}C Date (yr B.P.)	Maximum Area (km²)	Maximum Increase in Depth Relative to Present (m)	References[a] (Most Useful Reference in Italic)
	Winnemucca, Smoke Creek, Black Rock, Desert Valley, and Buena Vista Basins)							and Kaufman, 1965; Born, 1972; Fergusson and Libby, 1964 (R); Rubin and Alexander, 1960 (S); Levin et al., al., 1965 (R); Morrison and Frye, 1965; Davis, 1978; Verosub et al., 1980
14	Las Vegas	36°19'	115°11'	3^m	31,300 ± 2500	unknown	unknown	*Haynes*, 1967; *Mifflin and Wheat*, 1979
15	Teel (Teels Marsh)	38°12'	118°21'	1	10,760 ± 400	unknown	unknown	Hay, 1966; Crane and Griffith, 1965 (R)
New Mexico								
16	Arch (Big Salt Lake, Laguna Salada)	34°04'	103°07'	5	22,300 ± 700	ca. 130	>12	Reeves, 1966b; Glass et al., 1973; Leonard and Frye, 1975; Hester, 1975; Harbour, 1975; Olson and Broecker, 1961 (R)
17	Blackwater Draw	34°15'	103°20'	17	15,770 ± 440	unknown	unknown	*Haynes*, 1975
18	Estancia (Laguna del Perro, Salina Lake, etc.)	34°45'	106°00'	4	> 33,000	2860^n	90	*Bachhuber*, 1971; Bachhuber and McLellan, 1977
19	Lea County	33°27'	103°09'	2	16,010 ± 180	uncertain (small)	uncertain (small)	*Leonard and Frye*, 1975; Coleman, 1974 (R); Glass et al., 1973
20	Portales Valley	34°26'	103°49'	1	15,280 ± 210	unknown (small)	unknown (small)	Coleman, 1974 (R); Glass et al., 1973; *Leonard and Frye*, 1975
21	San Augustin (San Augustin Playa)	33°50'	108°10'	16	27,000 + 5000 − 3200	660	50°	*Powers*, 1939; Stearns, 1962; Stuiver and Deevey, 1962; Clisby and Sears, 1956; Damon et al., 1964; Long and Mielke, 1966; Schultz and Smith, 1965; Foreman et al., 1959
22	Zuni	34°27'	108°46'	1	23,000 ± 1500	5	15	*Schultz and Smith*, 1965; Haynes et al., 1967 (R); Cummings, 1968
Oregon								
23	Chewaucan (Abert Lake, Summer Lake)	42°40'	120°30'	6	30,700 + 2500 − 1900	1240	115	Allison, 1966; Buckley et al., 1968 (R); Allison, 1954; *Phillips and Van Denburgh*, 1971; *Van Denburgh*, 1975; Levin et al., 1965 (R); Ives et al., 1967 (R); Sullivan et al., 1970 (R)
24	Fort Rock (Silver Lake, Christmas Lake, Fossil Lake)	43°10'	120°45'	4	29,000 + 2000 − 1600	3885	49	*Bedwell*, 1973; Allison, 1966, 1979
Texas								
25	Guthrie	33°06'	101°48'	4	34,400 ± 3450	unknown (small)	unknown (small)	*Reeves and Parry*, 1965; Reeves, 1966b
26	Lubbock	33°38'	101°54'	3	12,650 ± 250	unknown (small)	unknown (small)	Green, 1961; Wendorf, 1970; *Black*, 1974; Hester, 1975; Broecker and Kulp, 1957 (S)
27	Monahans Dunes	31°36'	102°53'	2	19,200 ± 500	unknown	unknown	*Haynes*, 1975, citing Green, 1961; Broecker and Kulp, 1957 (S); Olson and Broecker, 1961 (R)
28	Mound	33°05'	102°05'	8	>37,000	< 20	< 15	Reeves and Parry, 1965; Reeves, 1966b; *Bates et al.*, 1970; Hester, 1975; Harbour, 1975
29	Rich	33°17'	102°12'	4	32,525 ± 2400	unknown (small)	<15	Reeves and Parry, 1965; Reeves, 1966b; Hester, 1975; *Haynes*, 1975; Olson and Broecker, 1961 (R)
30	White	ca. 33°58'	102°44'	1	19,275 ± 560	unknown (small)	unknown (small)	*Reeves and Parry*, 1965; Hester, 1975; Harbour, 1975

Table 10-3. Pluvial Lakes in the Western United States with Radiocarbon-Dated Chronologies in the Range of 25,000 to 10,000 yr B.P.

State/Lake Number	Pluvial Lake (and Name of Modern Lake, Playa, or Valley, if Different)	Latitude (°N)	Longitude (°W)	Finite ^{14}C dates	Oldest [14]C Date (yr B.P.)	Maximum Area (km²)	Maximum Increase in Depth Relative to Present (m)	References (Most Useful References in Italic)
Utah								
31	Bonneville (Great Salt, Utah, and Sevier Lakes; Great Salt Lake and Escalante Deserts; Cache, Sevier, White, and Rush Valleys)	40°30'	113°00'	114	>37,000[P]	51,640	ca. 335	Broecker and Orr, 1958; Broecker and Kaufman, 1965; Rubin and Alexander, 1958, 1960 (S); 1960 (S); Rubin and Berthold, 1961 (R); Ives et al., 1964, 1967 (R); Levin et al., 1965 (R); Jennings, 1957; Eardley, 1962; *Morrison and Frye*, 1965; Marsters et al., 1969 (R); Stuiver, 1969 (R); Morrison, 1966; Kaufman and Broecker, 1965; Eardley et al., 1973; Crittenden, 1963a, 1963b; Scott, 1980; Scott et al., 1980

[a] References to radiocarbon date lists published in *Radiocarbon* are indicated by (R), those in lists in *Science* are indicated by (S), and unless cited in text they are not included in the references at the end of this chapter.
[b] With one exception, these ages were measured on diagenetic dolomite in lacustrine muds and, therefore, should not be regarded as accurate measures of the time of sedimentation. Peterson et al., (1963) suggests that the dolomite began to nucleate after the close of the last pluvial period, about 10,000 yr B.P.
[c] The former lake depth impossible to reconstruct because of poor lacustrine record.
[d] The lake basin has been strongly affected by faulting and tilting, so the late-Pleistocene configuration is difficult to reconstruct accurately.
[e] Highest shoreline remnants dated suggest a maximum depth of 298 m.
[f] Dates are unpublished.
[g] Lacustrine deposits recovered by a core in middle of basin indicate that it has been continuously occupied by a lake since a time greater than 730,000 yr B.P., the age of the interbedded Bishop Tuff (Lajoie, 1968).
[h] Major fluctuations in lake level have occurred throughout the last 3.2 million years (Liddicoat et al., 1980).
[i] Lacustrine clay units within the Quaternary sediments of the San Joaquin Valley indicate at least nine major lake expansions beginning well before 600,000 yr B.P.
[j] Depths of large late-Pleistocene and early-Holocene lakes are uncertain because of structural downwarping and changes in the height of the alluvial-fan barriers separating the present lake basins, but they probably were shallow.
[k] Oldest [230]Th age is 250,000 yr B.P.
[l] Measured from natural (preirrigation) level of Pyramid Lake (1180 m).
[m] Three dates from lacustrine beds, numerous others from nonlacustrine beds.
[n] Maximum area includes satellite Pinos Wells and Encino Basins.
[o] Highest shoreline remnants dated suggest a maximum depth of 69 m.
[p] Oldest [230]Th age is 105,000 yr B.P. Major fluctuations of the lakes in the Bonneville Basin began before deposition of Bishop Ash, potassium-argon dated at 730,000 yr B.P. (Eardley et al., 1973).

found to have been sumps for interbasin groundwater movements during the most intense or persistent pluvial periods, with the salts introduced via that subsurface drainage. Any lakes that existed in them as perennial lakes were probably largely nourished by that water and thus can be considered exposures of the groundwater table.

The large lake near the southern end of California, Pleistocene Lake LeConte (or Holocene Lake Cahuilla) in the Imperial Valley, shows abundant evidence of Holocene high stands (e.g., a large lake existed for about 1600 years and ended 300 yr B.P. or possibly 700 to 500 yr B.P.), but its hydrology depends more on tectonic deformation (Stanley, 1966; Wilson and Wood, 1980) and on the geometry of the distributary channels of the lower Colorado River than on climate (van de Kamp, 1973: 830). Data on the Pleistocene lake's history are meager, but the history's climatic relevance would be small.

Nevada

Most of the closed basins in the northern three-fourths of Nevada appear to have contained pluvial lakes. Those outside the Lake Lahontan and Lake Bonneville basins have not received as much study as

the larger lakes have, and, in many instances, the only record of them consists of shorelines, relict fish or snail populations, or cores that document the presence of perennial lake deposits. Descriptions and discussions of the evidence for (or against) the existence of pluvial lakes in each of these basins is given by Hubbs and Miller (1948: 42-60, 90-103, 157-65, and map) and by Mifflin and Wheat (1979: 11-37, 53-57). The latter study argues for the absence of pluvial lakes in Nevada south of about latitude 37°; it interprets the fine-grained, flat-lying sediments found in some basins south of that latitude as paludal or playa deposits rather than as pluvial-lake deposits. More dated sedimentologic and stratigraphic studies are needed to resolve this question.

The climatic implications of the lake in Spring Valley, the largest of the four pluvial lakes shown in Figure 10-1 that are aligned with the central segment of the Nevada-Utah boundary, have been investigated by Snyder and Langbein (1962). Combining modern precipitation, evaporation, and runoff data with mathematical expressions of the hydrologic budget, Snyder and Langbein conclude that the expanded lake could have been sustained by a 70% increase in precipitation (from the present-day 30 to 51 cm) and a 29% decrease in evaporation (from 112 to 79 cm). These conditions could have resulted from a 550-m depression of the present elevation-precipitation relation and a decrease in summer temperatures that lowered snow lines by 1200 m.

The lacustrine sediments in Las Vegas Valley, inferred by Haynes (1967: 78) to be remnants of pluvial Lake Las Vegas, are considered by Mifflin and Wheat (1979: 27) to be paludal or playa deposits. The lack of other large pluvial lakes in this area, the absence of a geologic mechanism for restricting a smaller water body to the northern part of the valley, the relatively young geologic ages of the fine-grained deposits, and the almost complete absence of other exposures of lake sediments in the entire valley combine to support the more recent view. However, the thickness of mudstones and carbonates (4.5 m) and the details of their lithologies support the earlier conclusion. If the "lake deposits" in Las Vegas Valley are eventually shown to be paludal deposits, the correspondence between their age and the ages of pluvial-lake deposits elsewhere could simply reflect a rise in water tables and an increase in spring activity that accompanied those climatic shifts.

New Mexico

Pluvial Lake Estancia in the central part of the state consisted of two freshwater lakes as much as 90 m deep during the interval that began about 18,000 and ended 10,500 yr B.P. The lakes were separated by one low stand near the middle of this period (Bachhuber and McClellan, 1977: 254, Figure 2). Another smaller lake, Lake Willard, formed during the period between about 8500 and 6000 yr B.P. Saline stages at the beginning of the first freshwater stage and at the beginning of Lake Willard supported marine-type foraminiferal assemblages. Their faunal composition suggests that summer temperatures during the initial stages of lake development were nearly 10°C lower than present-day summer temperatures. Leopold (1951), using meteorologic data and methods, suggests a 9°C reduction in summer temperature for this area and estimates a decrease in mean annual temperature during pluvial periods of about 8°C, a decrease in evaporation to about 70% of the present rate, and an increase in precipitation of about 50%.

The San Augustin Plains, a tectonic basin in west-central New Mexico, also contained a pluvial lake, which covered 660 km² to a maximum depth of 50 m (Foreman et al., 1959: 117; Stearns, 1962: 29). Shorelines, beaches, and bars record its high levels; two cores that sample the upper 610 m of valley fill confirm lacustrine deposition for most of the last half of the time represented. The pollen in the upper 197 m of the core suggests that subalpine climates characterized the area only during the latter part of the interval sampled and that warmer and drier climates characterized the earlier times (Clisby and Sears, 1956). A series of radiocarbon dates on carbonate and organic carbon in a nearby 6-m core shows that climatic conditions during almost all of the period from 25,000 to 10,000 yr B.P. produced lakes in this depression (Stuiver and Deevey, 1962). Dates on tufa from the high terraces, though less reliable, suggest high stands in the valley during the last 5000 years of this period (Long and Mielke, 1966).

Other pluvial lakes in New Mexico also reached high levels during this interval. Very small tributary areas characterize some basins, and they, like the lakes in southern California, may reflect regional increases in the pluvial period's water-table levels rather than increased surface flow. The pluvial-lake water surface in Salt Lake, which lies in a subsidence depression in the east-central High Plains area, stood at a level about 12 m above the present lake's surface during a period that ended about 14,000 yr B.P. (Glass et al., 1973: 9). Pluvial Lake Zuni, which lies in the center of a small, steep-sided volcanic crater, left marl and *Chara* deposits, radiocarbon dated at about 22,000 yr B.P., as much as 15 m above the present salt-lake floor (Damon et al., 1964).

Oregon

Pluvial lakes occupied all of the closed basins in the Oregon segment of the Great Basin, and many of these contain lakes or marshy areas today. The largest, Lake Modoc, lay in the Basin and Range province, west of the northern Great Basin and astride the California-Oregon border. It covered 2800 km² and had a maximum depth of 64 m (Dicken, 1980: 183). Age control is unavailable, but the hydrology of the basin would make translation of a lake history into climatic terms difficult because of the high porosity of the enclosing volcanic rock and because the present outlet, the Klamath River, must have been blocked by ice, landslides, or some other mechanism.

Pluvial Lake Chewaucan, the lake that occupied the basins now containing Abert Lake and Summer Lake in south-central Oregon, covered an area of about 1240 km² to a maximum depth of about 115 m (Allison, 1954; Phillips and Van Denburgh, 1971: B12). As noted by Van Denburgh (1975: C29), the elevations of the shorelines around pluvial Fort Rock Lake to the north, which covered the present Fort Rock Lake and Christmas Lake valleys, are neary the same; this fact suggests a subsurface hydrologic connection, and any climatic reconstruction should probably combine data from both basins. The history of Fort Rock Lake is partially known on the basis of discontinuous outcroppings of lacustrine sediments that indicate two lake stands separated by a low stand that produced an unconformity (Bedwell, 1973; Allison, 1979: 44). The lower section, 60 to 200 m thick, is overlain unconformably by a few meters of lacustrine deposits that have radiocarbon dates on two samples near their base and about a meter apart averaging about 30,000 yr B.P. The upper unit locally has disconformities within it, suggesting that minor lake fluctuations occurred throughout the period. All lake deposits are overlain by the Mazama Ash (6600 yr B.P.).

The combined area of Lake Chewaucan and Fort Rock Lake is 5125 km² (Table 10-3); the present area of the only perennial lakes in the two basins, Abert and Summer, which probably receive substantial amounts of water from the Fort Rock Basin, is 350 km² (Phillips and Van Denburgh, 1971: B5, B24). The pluvial-lake area, therefore, is about 15 times that of the present area, and calculations like those applied to Lake Bonneville, Lake Lahontan, and the Owens River system suggest pluvial inflow volumes at least 5 times those of the present.

A pluvial lake in the Warner Valley covered 1310 km² and had a maximum depth of 98 m. Weide (1976) has applied the climatic reconstruction methods used by Leopold (1951) and Snyder and Langbein (1962) to this area, although it is climatically and geographically quite different, and has concluded that the Snyder and Langbein model most satisfactorily accounts for both the present hydrologic regime (which supports several small lakes) and the pluvial lake's regime. The present lakes have a 22nd the area of the pluvial lake in that valley, and Weide's calculated pluvial runoff ranges from 2.4 to 3.0 times the calculated present runoff. Pluvial runoff was the result of an estimated 17% to 37% increase in precipitation and a

16% to 23% decrease in lake evaporation caused by a 4° to 6°C decrease in mean annual temperature (Weide, 1976: Table 1).

Texas

Many small pluvial lakes existed in the southern High Plains of northwestern Texas. Most left geomorphic, stratigraphic, and paleontologic records of their existence, and there have been numerous studies of them, especially during the last two decades. The smallest lakes formed mostly in basins created at least in part by deflation; larger lakes occurred in basins attributed to the blocking of older drainage channels followed by erosional enlargement (Reeves, 1966b: 271-81). Since the last pluvial period, many of the lacustrine deposits have themselves been dissected by wind and water. The Late Wisconsin lacustrine deposits are composed of gravel, sand, black to blue-gray clay (mostly sepiolite), carbonates (mostly dolomite), and gypsum (Reeves, 1976: 220-23).

Most of the lake deposits are assigned by Reeves (1968, 1972, 1976: 220) to the Tahokan pluvial, inferred to have lasted from about 22,000 to 12,500 yr B.P. Radiocarbon dates of these sediments are not numerous, and many are on carbonates and, therefore, are questionable. Carbonates near the top and base of the lower of the two units of lacustrine deposits in Rich Lake have radiocarbon dates of about 26,500 to 17,400 yr B.P., and the upper unit is estimated to represent the period between 16,000 and 12,000 yr B.P. Shells in correlative lacustrine deposits in the Monahans dune area have dates of about 19,200 and 13,400 yr B.P. (Haynes, 1975: 80-82). Between 13,000 and 11,000 yr B.P., however, desiccation, lower water tables, and deflation characterized the region (Haynes, 1975: 83).

Pluvial lakes in northwestern Texas are considered by Reeves (1976: 214) to have been the product of a 100% increase in precipitation. Some of the basins, however, contained lakes that occupied as much as 85% of their drainage basins (Reeves, 1966a: 643), and these lake-to-drainage-area ratios are much greater than those found in other pluvial-lake basins. This ratio, designated by Snyder and Langebein (1962: 2389, Table 3) as Z, was calculated by them for eight typical pluvial lakes, and they found the values of Z to lie in the range of 0.22 to 1.12, with the highest values characteristic of lakes adjacent to high mountain ranges, as theory would suggest. The Z value for some of the Texas pluvial lakes would exceed 4.0, and the adjacent regions are definitely not characterized by high mountain ranges. It seems likely, therefore, that groundwater contributions were required. Because of the impervious nature of the regional soils, Reeves (1973: 701-2) considers subsurface flow to the pluvial lakes in the southern High Plains to be a minor part of their hydrologic budget. However, we feel that the anomalous value of Z and the disproportionate size of even a few lakes in this area relative to their basins argues more strongly for their nourishment by an elevated pluvial-period groundwater table and for a major contribution to all the lakes from groundwater. Many or all of these lakes, therefore, in combination with some of the pluvial lakes in southern Calfornia and New Mexico, may document increases in the level of the pluvial period's groundwater table over a large segment of the southwestern United States south of about latitude 35° and from Texas westward. This regional transition in hydrologic regimes may be a result of the southward increase in summer precipitation in this area along a steep gradient at this latitude, as is indicated by paleobotanical evidence (Wells, 1979: 322-24). Lake Cochise in Arizona may also have had this origin, but the evidence is inconclusive.

Areal and Temporal Variations in Pluvial-Lake Levels

Evidence that can be used to reconstruct partial histories of former lake levels is available in 31 closed basins of the West. In Figures 10-2, 10-3, and 10-5, curves representing the histories of four basins are plotted: Lake Bonneville, Lake Lahontan, and two lakes in the Owens River system. Although disagreements remain among interpretations of the details of these histories, there appear to be broad areas of agreement. There is agreement, for example, on the fact that rapid and large-amplitude fluctuations occurred in these late-Quaternary pluvial lakes. It is also generally accepted that there were one or more high lake stands during the first two-thirds of the period between 25,000 and 10,000 yr B.P. and that there were one or more very brief periods of expansion about 12,000 yr B.P. In addition, most of the evidence shows that there were significant fluctuations in the sizes of these lakes during the last 10,000 years.

Because it is possible to make reliable reconstructions of past water levels during at least part of the history of many other lakes, it is feasible to extend the spatial coverage of the dated pluvial-lake histories beyond the areas represented in Figures 10-2, 10-3, and 10-5 by pooling all the available dated lake-level evidence, assessing its validity, and assuming that regional climates affected large areas in a similar manner. To do this, we made a careful search of the literature relevant to the pluvial lakes of the western United States, following the procedures established by Street and Grove (1976, 1979) and concentrating on closed basins that have yielded radiocarbon dates. The present survey revealed 31 dated basins in the United States (including the Salton Sea), and the entire compilation has been checked and revised. The results are summarized in Figure 10-6, and the basins for which we found data are listed in Table 10-3.

A standard procedure is used for expressing the evidence in a semi-quantitative form (Street and Grove, 1979: 84-87). Each basin is considered in relation to its own internal range of variation. This procedure allows for minor downcutting of outlets or uncertainties in the elevations of the shorelines or lacustrine sediments. Each radiocarbon date and the relative elevation of the indicated lake level is assigned to 1000-year time intervals covering the period between 30,000 years ago and the present. Where dates within a 1000-year interval indicate different levels, the level chosen for mapping is defined as follows: *High*: lake level was high (70% to 100% of the maximum lake level) during all or part of this time period; *intermediate*: lake level was intermediate (15% to 70%) or low, but not high, during all or part of this time period; and *low*: lake level was low (0% to 15%) during all of this time period. This method underemphasizes arid periods shorter than 1000 years, and it may overemphasize single radiocarbon dates from a given basin, which mostly came from lacustrine rather than nonlacustrine deposits.

Where radiocarbon dates conflict, the interpretation used is based on the following inferred order of the dates' decreasing reliability: (1) charcoal and wood in basin sediments or cave sequences; (2) peat or organic carbon in lake sediments; (3) disseminated inorganic carbonate, calcareous algae, ostracods, and mollusks in lake sediments; (4) tufa; and (5) carbonate sediments. Dates on bone and on soil humus or carbonate have been treated with caution. No adjustment has been made for the initial $^{14}C/^{12}C$ ratio of the lake waters.

The information of lake levels is summarized in two ways. In Figure 10-6, the history of lake levels over the entire West is shown in the form of a histogram of lake levels against time. Figure 10-7

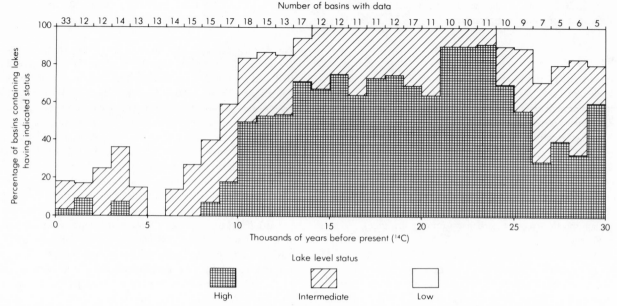

Figure 10-6. Relative percentages of basins containing dated materials at elevations indicative of low, intermediate, or high stages versus their radiocarbon ages.

shows the same information on six maps for 1000-year time periods starting at 24,000, 18,000, 16,000, 13,000, 12,000, and 10,000 yr B.P. The maps indicate the areal distribution of each lake level during that block of time and presumably reveal regional variations in the intensity of pluvial conditions. It should be noted that the symbols representing Lake Bonneville and Lake Lahontan indicate conditions over areas that cover a quarter to a third of their respective states. All of the 1000-year intervals were plotted; the time periods illustrated represent the clearest or most significant patterns.

During the period 24,000 to 14,000 yr B.P. (Figure 10-6), all of the lakes for which information is available were either high or intermediate. High levels were most widespread from 24,000 to 21,000 yr B.P., when only Lake Bonneville appears to have remained intermediate. Between 21,000 and 14,000 yr B.P., only about 70% of the data points record high levels. The intermediate-depth lakes of these ages clustered in the northwestern part of the pluvial-lake region (Figures 10-7B and 10-7C), particularly in areas adjacent to the Sierra Nevada, although there was considerable fluctuation. No information is available from Oregon to determine whether the region of intermediate levels extended that far northwest. A brief episode of partial desiccation in the southern High Plains appears to have been recorded by the widespread but thin Vigo Park dolomite identified by Reeves and Parry (1965; Reeves, 1976), although the radiocarbon dates from this lithologic unit are scattered widely and are probably not very reliable.

The period between 14,000 and 10,000 yr B.P. was characterized by rapid, large-amplitude fluctuations that may or may not have been synchronous across the region. The apparent lack of synchrony in the raw data could be the result of errors inherent in the radiocarbon dating method when applied to the record of events that were apparently short. The data may, however, reflect genuine spatial variations in lake behavior. Many lakes experienced a drop in levels centered on the period 14,000 to 13,000 yr B.P. In Arizona, New Mexico, and Texas, 13,000 yr B.P. has been considered the end of the

main lacustrine phase (Haynes, 1967), although apparent high stands of a few lake levels in these areas (Figures 10-7D and 10-7E) partially conflict with this view. The pattern of fluctuations during the period 12,000 to 10,000 yr B.P. is particularly complex. A major expansion of Lakes Mojave, Searles, Mono, and Tulare, located in southern California near the Sierra Nevada, culminated between about 13,500 and 11,000 yr B.P. Broecker and Orr (1958: Figure 9) and Broecker and Kaufman (1965: Figure 3) also infer that the first of two major expansions of Lake Bonneville and Lake Lahontan occurred within this period, but Bright's (1966) radiocarbon data from peaty lake sediments appear to preclude any overflow of Lake Bonneville after about 13,000 yr B.P. Morrison and Frye (1965: Figure 2), however, indicate several expansions of Lake Bonneville and Lake Lahontan starting about 10,000 yr B.P. on the basis of stratigraphic evidence, and we suspect that one or more modest reexpansions did occur. Distinct expansions occurred in Searles Lake and Lake Mojave between about 11,000 and 10,000 yr B.P., and a brief lacustrine recovery occurred in the pluvial lakes of Arizona, New Mexico, and Texas during this time (Wendorf, 1961).

Between 10,000 and 5000 yr B.P. (Figure 10-6), many areas were characterized by drought, which culminated between 6000 and 5000 yr B.P., when not a single lake is known to have been high. In some basins, the fall in water levels during the first part of this period was apparently rapid and unreversed, but in others there seems to have been some lake level fluctuation.

Between 5000 yr B.P. and the present, there have been significant reexpansions of some lakes, notably those on the western margins of the Great Basin and in California. The vertical amplitude of these fluctuations is surprisingly large: up to 90 m in the Walker Lake area, 85 m around the Salton Sea, and 65 m in Searles Lake. During historic time, lake levels were higher before the start of irrigation agriculture in the late 19th century, and present lake levels in many areas are unrepresentative of the average for the last few centuries or millennia (Harding, 1965).

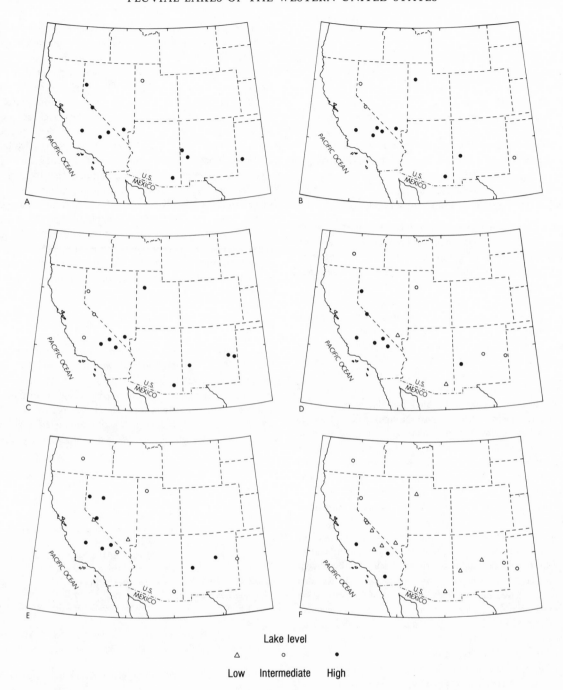

Figure 10-7. Locations of lakes having low, intermediate, or high stands during 1000-year periods that began (A) 24,000, (B) 18,000, (C) 16,000, (D) 13,000, (E) 12,000, and (F) 10,000 yr B.P.

Paleoclimatic Estimates Derived from Lake-Level Data

Lake-level fluctuations are potentially the best source of quantitative estimates of paleoprecipitation and one of the best sources for generalized paleoclimatic data in desert areas where pollen data or studies of pack-rat middens are sparse. However, attempts to derive paleoprecipitation estimates from lake-level information have limita-

tions, largely because of the difficulties involved in estimating past rates of evaporation and percentages of precipitation that became runoff. The problems encountered, described in numerous papers, have been recently summarized by Brakenridge (1978), Mifflin and Wheat (1979), and Benson (1981).

The surface area of a topographically closed lake fluctuates in order to balance the lake's water budget, as expressed in the following equation:

$$A_L P_L + A_T P_T k + G_I = A_L E + G_O,$$

where:

A_L = lake area,
P_L = precipitation falling on the lake,
A_T = tributary catchment area,
P_T = mean annual precipitation falling in the tributary area,
k = runoff coefficient (the proportion of the rain or snow that eventually becomes runoff and reaches the lake),
G_I = ground water inflow to the basin,
E = mean annual evaporation from the lake, and
G_O = groundwater outflow from the basin.

Theoretically, this equation can be used to calculate past values of P_L and P_T *provided that*: (1) the basin has been topographically and hydrographically stable, (2) A_L and A_T can be determined from geologic information, (3) G_I and G_O are negligible or can be determined from hydrologic information, and (4) k and E can be reliably estimated.

In most late-Pleistocenene pluvial basins where a good record exists, pluvial values for A_L and A_T can be measured quite accurately, but the remaining values are less readily determined. Present values for P_L can be measured, but, as emphasized by Weide (1976), even present values for P_T are not easily determined in areas where topographic relief is great. Present values of k are imperfectly known, and estimates of pluvial values of P_T and k are, therefore, even less reliable. Similarly, present values (or even the sign) of G_I and G_O are commonly not well known, and their values and signs may not have been the same during pluvial periods in the past. The fact that we know far less than we could of the present relationships among these climatic and hydrologic factors in desert areas suggests this as a topic needing more study. Especially needed is a better basis for developing estimates of pluvial runoff from areas now contributing no runoff, which constitute the largest areal percentage of many basins.

The value for E depends only in part on the temperature of the lake water. However, because the other relevant factors are difficult to reconstruct from geologic data, most studies—including the present one—assume that pluvial values for E can be satisfactorily estimated from empirical relationships between modern data on evaporation and air temperature. Significant differences of opinion remain, however, about the change in air-temperature values during Pleistocene pluvial periods (Table 10-2). Temperature estimates used in reconstructing lake histories are derived from a host of geologic and hydrologic criteria, all of which are subject to uncertainties, and some are derived from evidence in areas far removed from the site of the paleolake. Possible errors in paleotemperatures are then compounded by becoming the basis of only partially constrained assumptions concerning past atmospheric lapse rates and relative humidity, which are combined with nearly unconstrained estimates of percentages of cloud cover, velocities of winds, and the seasonal distributions of all these variables. The conclusions reached by various authors concerning pluvial values for E, P_L, and P_T have, therefore, differed greatly, both because of difference in the paleotemperature values adopted and because of their assumptions concerning the other variables.

The problems stemming from uncertainties about pluvial paleotemperatures are not avoided by the method used in this chapter, which compares present and pluvial lake surface areas and assumes that pluvial-period air temperatures bore about the same (although unspecified) relation to pluvial-lake surface-water temperatures as they do today (obviously, this method can only be used in basins that still contain lakes). It also assumes that other factors tended to cancel each other. During pluvial periods, increased evaporation was probably promoted by higher wind velocities (indicated by the clast sizes found in pluvial-lake gravel bars constructed by wind-driven waves and currents) and by decreased salinities. Decreased evaporation was promoted by higher relative humidities and lower solar radiation values (which are suggested by several meteorologic considerations). It also assumes that the decreased net evaporation caused by greater amounts of precipitation on the lake surface can be arbitrarily placed at 35%. Many important factors are not reflected in the results of this simplified approach, but for some purposes estimates of the volume of pluvial runoff—the values derived here—are of more interest than estimates of the precipitation and/or temperature that caused them, for they lead more directly to estimates of pluvial-period stream dimensions, flood magnitudes, and sedimentation and erosion rates.

Conclusions

During pluvial maxima, some (probably most) pluvial-period groundwater tables in the American Southwest rose more than 10 m. Most streams in the remaining parts of the western United States carried substantially more water during pluvial periods than they do now, and precipitation amounts probably also increased markedly. The volume of stream flow during "maximum pluvial" periods in this region possibly approached an order of magnitude greater than today's.

References

Allison, I. S. (1954). Pluvial lake levels of south-central Oregon (abstract). *Geological Society of America Bulletin* 65, 1331.

Allison, I. S. (1966). "Fossil Lake, Oregon, Its Geology and Fossil Faunas." Oregon State University Studies in Geology 9.

Allison, I. S. (1979). "Pluvial Fort Rock Lake, Lake County, Oregon." State of Oregon Department of Geology and Mineral Industries Special Paper 7.

Antevs, E. A. (1925). "On the Pleistocene History of the Great Basin." Carnegie Insititution of Washington Publication 352, pp. 53-104.

Antevs, E. A. (1945). Correlation of Wisconsin glacial maxima. *American Journal of Science* 243-A, 1-39.

Antevs, E. A. (1948). Climatic changes and pre-white man. *In* "The Great Basin, with Emphasis on Glacial and Postglacial Times," pp. 168-91. University of Utah Bulletin 38.

Antevs, E. A. (1952). Cenozoic climates of the Great Basin. *Geologische Rundschau* 40, 94-108.

Antevs, E. A. (1955). Geologic-climatic dating in the West. *American Antiquity* 20, 317-35.

Bachhuber, F. W. (1971). Paleolimnology of Lake Estancia and Quaternary history of the Estancia Valley, New Mexico. Ph.D. dissertation, University of New Mexico, Albuquerque.

Bachhuber, F. W., and McClellan, W. A. (1977). Paleoecology of marine foraminifera in the pluvial Estancia Valley, central New Mexico. *Quaternary Research* 7, 254-67.

Batchelder, G. L. (1970). Post-glacial fluctuations of lake level in Adobe Valley, Mono County, California. *In* "American Quaternary Association, 1st Biennial Meeting, Abstracts," p. 7. Bozeman, Mont.

Bates, T. R., Reeves, C. C., Jr., and Parry, W. T. (1970). Late Pleistocene history of pluvial Lake Mound, Lynn and Terry Counties, Texas. *Texas Journal of Science* 21, 245-59.

Bedwell, S. F. (1973). "Fort Rock Basin: Prehistory and Environment." University of Oregon Press, Eugene.

Benson, L. V. (1978). Fluctuation in the level of pluvial Lake Lahontan during the last 40,000 years. *Quaternary Research* 9, 300-318.

Benson, L. V. (1981). Paleoclimatic significance of lake level fluctuations in the Lahontan Basin. *Quaternary Research* 16, 390-403.

Berner, R. A. (1971). "Principles of Chemical Sedimentology." McGraw-Hill, New York.

Bissell, H. J. (1963). "Lake Bonneville: Geology of Southern Utah Valley, Utah." U.S. Geological Survey Professional Paper 257-B, pp. 101-30.

Black, C. C. (ed.) (1974). "History and Prehistory of the Lubbock Lake Site." The Museum Journal, Texas Tech University, 15.

Blackwelder, E. (1933). Lake Manly: An extinct lake of Death Valley. *Geographical Review* 23, 464-71.

Blackwelder, E. (1954). Pleistocene lakes and drainage in the Mojave region, southern California. *In* "Geology of Southern California" (R. H. Jahns, ed.), pp. 35-40. California Division of Mines Bulletin 170.

Blaney, H. F. (1956). "Evaporation from Free Water Surfaces at High Altitudes." Journal of the Irrigation and Drainage Division, Proceedings of the American Society of Civil Engineers 82 (IR3), Proceedings Paper 1104.

Born, S. M. (1972). "Late Quaternary History, Deltaic Sedimentation, and Mudlump Formation at Pyramid Lake, Nevada." Center for Water Resources Research, Desert Research Institute, Reno, Nev.

Born, S. M., and Ritter, D. F. (1970). Modern terrace development near Pyramid Lake, Nevada, and its geologic implications. *Geological Society of America Bulletin* 81, 1233-41.

Brakenridge, G. R. (1978). Evidence for a cold, dry full-glacial climate in the American Southwest. *Quaternary Research* 9, 22-40.

Bright, R. C. (1966). Pollen and seed stratigraphy of Swan Lake, southeastern Idaho: Its relation to regional vegetational history and to Lake Bonneville history. *Tebiwa* 9, 1-47.

Bright, R. C. (1967). Late-Pleistocene stratigraphy in Thatcher Basin, southeastern Idaho. *Tebiwa* 10, 1-7.

Broecker, W. S., and Kaufman, A. (1965). Radiocarbon chronology of Lake Lahontan and Lake Bonneville II, Great Basin. *Geological Society of America Bulletin* 76, 537-66.

Broecker, W. S., and Orr, P. C. (1958). Radiocarbon chronology of Lake Lahontan and Lake Bonneville. *Geological Society of America Bulletin* 70, 1009-32.

Broecker, W. S., and Walton, A. (1959). Geochemistry of C¹⁴ in freshwater systems. *Geochimica et Cosmochimica Acta* 15, 15-38.

Clark, W. O., and Riddell, C. W. (1920). "Exploratory Drilling for Water and Use of Groundwater for Irrigation in Steptoe Valley, Nevada." U.S. Geological Survey Water-Supply Paper 467.

Clisby, K. H., and Sears, P. B. (1956). San Augustin Plains: Pleistocene climatic changes, *Science* 124, 537-38.

Crane, H. R., and Griffin, J. B. (1965). University of Michigan radiocarbon dates X. *Radiocarbon* 7, 123-52.

Crittenden, M. D., Jr. (1963a). Effective viscosity of the Earth derived from isostatic loading of Pleistocene Lake Bonneville. *Journal of Geophysical Research* 68, 5517-30.

Crittenden, M. D., Jr. (1963b). "New Data on the Isostatic Deformation of Lake Bonneville." U.S. Geological Survey Professional Paper 454-E.

Croft, M. G. (1968). Geology and radiocarbon ages of late Pleistocene lacustrine clay deposits, southern part of San Joaquin Valley, California. *In* "Geological Survey Research 1968," pp. 151-55. U.S. Geological Survey Professional Paper 600-B.

Cummings, D. (1968). Geologic map of the Zuni Salt Lake volcanic crater, Catron County, New Mexico. U.S. Geological Survey Miscellaneous Investigations Map I-544.

Currey, D. R. (1980). Coastal geomorphology of Great Salt Lake and vicinity. *In* "Great Salt Lake: A Scientific, Historic, and Economic Overview" (J. W. Gwynn, ed.), pp. 69-82. Utah Geological and Mineral Survey Bulletin 116.

Damon, P. E., Haynes, C. V., and Long, A. (1964). Arizona radiocarbon dates V. *Radiocarbon* 6, 91-107.

Davis, G. H., Green, J. H., Olmsted, F. H., and Brown, D. W. (1959). "Groundwater Conditions and Storage Capacity in the San Joaquin Valley, California." U.S. Geological Survey Water-Supply Paper 1469.

Davis, J. O. (1978). "Quaternary Tephrochronology of the Lake Lahontan Area, Nevada and California." Nevada Archeological Survey Research Paper 7.

Dicken, S. N. (1980). Pluvial Lake Modoc, Klamath County, Oregon, and Modoc and Siskiyou Counties, California. *Oregon Geology* 42, 179-87.

Dub, G. D. (1947). Owens Lake: Source of sodium minerals. *Mining Technology* 11 (5), 1-13.

Dury, G. H. (1965). "Theoretical Implications of Underfit Streams." U.S. Geological Survey Professional Paper 452-C.

Eardley, A. J. (1962). "Gypsum Dunes and Evaporite History of the Great Salt Lake Desert." Utah Geological and Mineralogical Survey Special Studies 2.

Eardley, A. J., and Gvosdetsky, V. (1960). Analysis of Pleistocene core from Great Salt Lake, Utah. *Bulletin of the Geological Society of America* 71, 1232-1344.

Eardley, A. J., Gvosdetsky, V., and Marsell, R. E. (1957). Hydrology of Lake Bonneville and sediments and soils of its basin. *Bulletin of the Geological Society of America* 68, 1141-1201.

Eardley, A. J., Shuey, T. T., Gvosdetsky, V., Nash, W. P., Picard, M. D., Grey, D. C., and Kukla, G. J. (1973). Lake cycles in the Bonneville Basin, Utah. *Bulletin of the Geological Society of America* 84, 211-16.

Eugster, H. P., and Hardie, L. A. (1978). Saline lakes. *In* "Lakes: Chemistry, Geology, and Physics" (A. Lerman, ed.), pp. 237-93. Springer-Verlag, New York.

Feth, J. H. (1961). A new map of western conterminous United States showing the maximum known or inferred extent of Pleistocene lakes. *In* "Geological Survey Research 1961," pp. 110-12. U.S. Geological Survey Professional Paper 424-B.

Flint, R. F., and Gale, W. A. (1958). Stratigraphy and radiocarbon dates at Searles Lake, California. *American Journal of Science* 256, 689-714.

Foreman, F., Clisby, K. H., and Sears, P. B. (1959). Plio-Pleistocene sediments and climates of the San Augustin Plains, New Mexico. *In* "Guidebook of West-Central New Mexico 10th Field Conference" (J. E. Weir, Jr., and E. H. Baltz, eds.), pp. 117-20. New Mexico Geological Society, Socorro.

Friedman, I., Smith, G. I., and Hardcastle, K. G. (1976). Studies of Quaternary saline Lakes: II. Isotopic and compositional changes during desiccation of the brines in Owens Lake, California, 1969-1971. *Geochimica et Cosmochimica Acta* 40, 501-11.

Gale, H. S. (1914). "Salines in the Owens, Searles, and Panamint Basins, Southeastern California." U.S. Geological Survey Bulletin 580-L, pp. 251-323.

Galloway, R. W. (1970). The full-glacial climate in the southwestern United States. *Annals of the Association of American Geographers* 60, 245-56.

Gilbert, G. K. (1885). "The Topographic Features of Lake Shores." Extract from the Fifth Annual Report of the Director of the U.S. Geological Survey, 1883-1884.

Gilbert, G. K. (1890). "Lake Bonneville." U.S. Geological Survey Monograph 1.

Glass, H. D., Frye, J. C., and Leonard, A. B. (1973). "Clay minerals in east-central New Mexico." New Mexico Bureau of Mines and Mineral Resources Circular 139.

Green, F. E. (1961). Discussion of the pollen and stratigraphic data. *In* "Paleoecology of the Llano Estacado" (F. Wendorf, ed.), pp. 48-58. Santa Fe Museum, University of New Mexico Press, Albuquerque.

Haines, D. V. (1959). "Core Logs from Searles Lake, San Bernardino County, California." U.S. Geological Survey Bulletin 1045-E, pp. 138-317.

Halley, E. (1715). On the cause of the saltness of the ocean, and of the several lakes that emit no rivers. *Philosophical Transactions of the Royal Society of London* 6, 169.

Harbeck, G. E., Jr. (1955). "The Effect of Salinity on Evaporation." U.S. Geological Survey Professional Paper 272-A.

Harbeck, G. E., Jr., Kohler, M. A., Koberg, G. E., and others. (1958). "Water-Loss Investigations: Lake Mead Studies." U.S. Geological Survey Professional Paper 298.

Harbour, J. (1975). General Stratigraphy. *In* "Late Pleistocene Environments of the Southern High Plains" (F. Wendorf and J. J. Hester, eds.), pp. 33-35. Fort Burgwin Research Center Publications 9.

Hardie, L. A., and Eugster, H. P. (1970). The evolution of closed-basin brines. *In* "Fiftieth Anniversary Symposia: Mineralogy and Petrology of the Upper Mantle; Sulfides; Mineralogy and Geochemistry of Non-Marine Evaporites" (B. A. Morgan, ed.), pp. 273-90. Mineralogical Society of America Special Paper 3.

Hardie, L. A., Smoot, J. P., and Eugster, H. P. (1978). Saline lakes and their deposits: A sedimentological approach. *In* "Modern and Ancient Lake Sediments" (A. Matter and M. E. Tucker, eds.), pp. 7-41. Blackwell Scientific Publications, Oxford, England.

Harding, S. T. (1965). "Recent Variations in the Water Supply of the Western Great Basin." Water Resources Center Archives, University of California, Report 16.

Harris, E. E. (1970). "Reconnaissance Bathymetry of Pyramid Lake, Wahoe County, Nevada." U.S. Geological Survey Hydrologic Investigations Atlas HA-379.

Hay, R. L. (1966). "Zeolites and Zeolitic Reactions in Sedimentary Rocks." Geological Society of America Special Paper 85.

Haynes, C. V. (1967). Quaternary geology of the Tule Springs area, Clark County, Nevada. *In* "Pleistocene Studies in Southern Nevada" (H. M. Wormington and D. Ellis, eds.), pp. 15-104. Nevada State Museum Anthropological Papers 13.

Haynes, C. V. (1975). Pleistocene and recent stratigraphy. *In* "Late Pleistocene Environments of the Southern High Plains" (F. Wendorf and J. J. Hester, eds.), pp. 57-290. Fort Burgwin Research Center Publications 9.

Helly, A. G., Hughes, G. H., and Irelan, B. (1966). "Hydrologic Regimen of Salton Sea, California." U.S. Geological Survey Professional Paper 486-C.

Hester, J. J. (1975). The sites. *In* "Late Pleistocene Environments of the Southern High Plains" (F. Wendorf and J. J. Hester, eds.), pp. 13-32. Fort Burgwin Research Center Publications 9.

Hooke, R. LeB. (1971). "Logs of Boreholes in the Death Valley, California, Salt Pan." U.S. Department of Commerce National Technical Information Service, Washington, D.C.

Hooke, R. LeB. (1972). Geomorphic evidence for Late Wisconsin and Holocene tectonic deformation, Death Valley, California. *Geological Society of America Bulletin* 83, 2073-98.

Hubbs, C. L., and Miller, R. R. (1948). The zoological evidence: Correlation between fish distribution and hydrographic history in the desert basins of western United States. *In* "The Great Basin, with Emphasis on Glacial and Postglacial Times," pp. 17-166. University of Utah Bulletin 38.

Hunt, C. B., and Mabey, D. R. (1966). "Stratigraphy and Structure of Death Valley, California." U.S. Geological Survey Professional Paper 494-A.

Hunt, C. B., Varnes, H. D., and Thomas, H. E. (1953). "Lake Bonneville: Geology of Northern Utah Valley, Utah." U.S. Geological Survey Professional Paper 257-A.

Hutchinson, G. E. (1937). A contribution to the limnology of arid regions. *Transactions of the Connecticut Academy of Arts and Sciences* 33, 47-132.

International Union for Quaternary Research (1977). "Quaternary stratotypes of North America." Subcommission on North American Quaternary Stratigraphy, International Union for Quaternary Research.

Ives, R. L. (1951). Pleistocene valley sediments of the Dugway area, Utah. *Geological Society of America Bulletin* 62, 781-97.

Janda, R. J., and Croft, M. G. (1967). The stratigraphic significance of a sequence of noncalcic brown soils formed on the Quaternary alluvium of the northeastern San Joaquin Valley, California. *In* "Quaternary Soils" (R. B. Morrison and H. E. Wright, Jr., eds.), vol. 9, pp. 157-90. Proceedings of the Seventh Congress of the International Association for Quaternary Research. Center for Water Resources Research, Desert Research Institute, University of Nevada, Reno.

Jennings, J. D. (1957). "Danger Cave." University of Utah Department of Anthropology Anthropological Papers 27.

Jones, B. F. (1965). "The Hydrology and Mineralogy of Deep Springs Lake, Inyo County, California." U.S. Geological Survey Professional Paper 502-A.

Jones, B. F., and Bowser, C. J. (1978). The mineralogy and related chemistry of lake sediments. *In* "Lakes: Chemistry, Geology, and Physics" (A. Lerman, ed.), pp. 179-235. Springer-Verlag, New York.

Kaufman, A., and Broecker, W. S. (1965). Comparison of Th^{230} and C^{14} ages for carbonate materials from Lakes Lahontan and Bonneville. *Journal of Geophysical Research* 70, 4039-54.

Kelts, K., and Hsu, K. J. (1978). Freshwater carbonate sedimentation. *In* "Lakes: Chemistry, Geology, and Physics." (A. Lerman, ed.), pp. 295-323. Springer-Verlag, New York.

King, C. (1878). Systematic geology. *In* "U.S. Geological Exploration of the Fortieth Parallel." Vol. 1: Washington, D.C.

Kohler, M. A. (1954). Lake and pan evaporation. *In* "U.S. Geological Survey, Water Loss Investigations: Lake Hefner Studies Technical Report," pp. 127-48. U.S.

Geological Survey Professional Paper 269.

Kohler, M. A., Nordenson, T. J., and Baker, D. R. (1959). "Evaporation Maps for the United States." U.S. Department of Commerce Technical Paper 37.

Kohler, M. A., Nordenson, T. J., and Fox, W. E.(1955). "Evaporation from Pans and Lakes." U.S. Department of Commerce Research Paper 38.

Lajoie, K. R. (1968). Quaternary stratigraphy and geologic history of Mono Basin, eastern California. Ph.D. dissertation, University of California, Berkeley.

Langbein, W. B. (1961). "Salinity and Hydrology of Closed Lakes." U.S. Geological Survey Professional Paper 412.

Langbein, W. B., and others (1949). "Annual Runoff in the United States." U.S. Geological Survey Circular 52.

Lee, C. H. (1912). "An Intensive Study of the Water Resources of a Part of Owens Valley, California.' U.S. Geological Survey Water-Supply Paper 294.

Leonard, A. B., and Frye, J. C. (1975). "Pliocene and Pleistocene Deposits and Molluscan Faunas, East-Central New Mexico." New Mexico Bureau of Mines and Mineral Resources Memoir 30.

Leopold, L. B. (1951). Pleistocene climate in New Mexico. *American Journal of Science* 249, 152-68.

Libby, W. F. (1955. "Radiocarbon Dating." 2nd ed. University of Chicago Press, Chicago.

Liddicoat, J. C., Opdyke, N. D., and Smith, G. I. (1980). Palaeomagnetic polarity in a 930-m core from Searles Valley, California. *Nature* 286, 22-25.

Long, A. (1966). Late Pleistocene and recent chronologies of playa lakes in Arizona and New Mexico. Ph.D. dissertation, University of Arizona, Tucson.

Long, A., and Mielke, J. E. (1966). Smithsonian Institution radiocarbon measurements III. *Radiocarbon* 7, 245-56.

Malde, H. E. (1968). "The Catastrophic Late Pleistocene Bonneville Flood in the Snake River Plain, Idaho." U.S. Geological Survey Professional Paper 596.

Martin, P. S. (1963). Geochronology of pluvial Lake Cochise, southern Arizona: II. Pollen analysis of a 42-meter core. *Ecology* 44, 437-44.

Martin, P. S., and Mosimann, J. E. (1965). Geochronology of pluvial Lake Cochise, southern Arizona: III. Pollen statistics and Pleistocene metastability. *American Journal of Science* 263, 313-58.

Meinzer, O. E. (1911). "Report on the Geology and Waters of Estancia Valley, New Mexico." U.S. Geological Survey Water-Supply Paper 275.

Meinzer, O. E. (1922). Map of the Pleistocene lakes of the Basin-and-Range province and its significance. *Geological Society of America Bulletin* 33, 541-52.

Meinzer, O. E., and Ellis, A. J. (1915). "Groundwater in Paradise Valley, Arizona." U.S. Geological Survey Water-Supply Paper 375-B, 51-75.

Meinzer, O. E., and Kelton, F. C. (1913). "Geology and Water Resources of Sulphur Spring Valley, Arizona." U.S. Geological Survey Water-Supply Paper 320.

Meyers, J. S. (1962). "Evaporation from the 17 Western States." U.S. Geological Survey Professional Paper 272-D.

Mifflin, M. D., and Wheat, M. M. (1979). "Pluvial Lakes and Estimated Pluvial Climates of Nevada." Nevada Bureau of Mines and Geology Bulletin 94.

Morrison, R. B. (1961a). Lake Lahontan stratigraphy and history in the Carson Desert (Fallon) area, Nevada. *In* "U.S. Geological Survey Research 1961," pp. 111-114, U.S. Geological Survey Professional Paper 424-D.

Morrison, R. B. (1961b). New evidence on the history of Lake Bonneville from an area south of Salt Lake City, Utah. *In* "U.S. Geological Survey Research 1961," pp. 125-27. U.S. Geological Survey Professional Paper 424-D.

Morrison, R. B. (1964). "Lake Lahontan: Geology of Southern Carson Desert, Nevada." U.S. Geological Survey Professional Paper 401.

Morrison, R. B. (1965a). "Lake Bonneville: Quaternary Stratigraphy of Eastern Jordan Valley, South of Salt Lake City, Utah." U.S. Geological Survey Professional Paper 477.

Morrison, R. B. (1965b). New evidence on Lake Bonneville stratigraphy and history from southern Promontory Point, Utah. *In* "Geological Survery Research 1965," pp. 111-19. U.S. Geological Survey Professional Paper 525-C

Morrison, R. B. (1965c). Quaternary geology of the Great Basin. *In* "The Quaternary of the United States" (H. E. Wright, Jr., and D. G. Frey, eds.), pp. 265-86. Princeton University Press, Princeton, N.J.

Morrison, R. B. (1966). Predecessors of Great Salt Lake. *In* "The Great Salt Lake" (W. L. Stokes, ed.), pp. 75-104. Utah Geology Society, Salt Lake City.

Morrison, R. B., and Frye, J. C. (1965). "Correlation of the Middle and Late Quaternary Successions of the Lake Lahontan, Lake Bonneville, Rocky Mountain (Wasatch Range), Southern Great Plains, and Eastern Midwest Areas." Nevada Bureau of Mines Report 9.

Ore, H. T., and Warren, C. N. (1971). Late Pleistocene-early Holocene geomorphic history of Lake Mojave, California. *Geological Society of America Bulletin* 82, 2553-62.

Peng, T-H., Goddard, J. G., and Broecker, W. S. (1978). A direct comparison of ^{14}C and ^{230}Th ages at Searles Lake, California. *Quaternary Research* 9, 319-29.

Peterson, F. F. (1980). Holocene desert soil formation under sodium salt influence in a playa-margin environment. *Quaternary Research* 13, 172-86.

Peterson, M. N. A., Bien, G. S., and Berner, R. A. (1963). Radiocarbon studies of recent dolomite from Deep Spring Lake, California. *Journal of Geophysical Research* 68, 6493-6505.

Phillips, K. N., and Van Denburgh, A. S. (1971). "Hydrology and Geochemistry of Abert, Summer, and Goose Lakes, and Other Closed-Basin Lakes in South-Central Oregon." U.S. Geological Survey Professional Paper 502-B.

Picard, M. D., and High, L. R., Jr. (1972). Criteria for recognizing lacustrine rocks. *In* "Recognition of Ancient Sedimentary Environments" (J. K. Rigby and W. K. Hamblin, eds.), pp. 108-45. Society of Economic Paleontologists and Mineralogists, Special Publication 16.

Powers, W. E. (1939). Basin and shore features of extinct Lake San Augustin, New Mexico. *Journal of Geology* 11, 345-56.

Putnam, W. C. (1950). Moraine and shoreline relationships at Mono Lake, California. *Geological Society of America Bulletin* 61, 115-22.

Reeves, C. C., Jr. (1963). The full-glacial climate of the southern High Plains, west Texas. *Journal of Geology* 81, 693-704.

Reeves, C. C., Jr. (1965). Pleistocene climate of the Llano Estacado. *Journal of Geology* 73, 181-88.

Reeves, C. C., Jr. (1966a). Pleistocene climate of the Llana Estacado II. *Journal of Geology* 74, 642-47.

Reeves, C. C., Jr. (1966b). Pluvial lake basins of west Texas. *Journal of Geology* 74, 269-91.

Reeves, C. C., Jr. (1968). "Introduction to Paleolimnology." Developments in Sedimentology 2, Elsevier, New York.

Reeves, C. C., Jr. (1972). Tertiary-Quaternary stratigraphy and geomorphology of west Texas and southeastern New Mexico. *In* "Guidebook, 23rd Field Conference," pp. 108-17. New Mexico Geological Society, Albuquerque.

Reeves, C. C., Jr. (1973). The full-glacial climate of the southern High Plains, west Texas. *Journal of Geology* 81, 693-704.

Reeves, C. C., Jr. (1976). Quaternary stratigraphy and geologic history of southern High Plains, Texas and New Mexico. *In* "Quaternary Stratigraphy of North America" (W. C. Mahaney, ed.), pp. 213-34. Dowden, Hutchinson, and Ross, Stroudsburg, Pa.

Reeves, C. C., Jr., and Parry, W. T. (1965). Geology of west Texas pluvial lake carbonates. *American Journal of Science* 263, 606-15.

Richmond, G. M. (1961). New evidence of the age of Lake Bonneville from the moraines in Little Cottonwood Canyon, Utah. *In* "U.S. Geological Survey Research 1961," pp. 127-28. U.S. Geological Survey Professional Paper 424-D.

Rush, F. E. (1970). "Hydrologic Regimen of Walker Lake, Mineral County, Nevada." U.S. Geological Survey Hydrologic Investigations Atlas HA-415.

Russell, I. C. (1883). "Sketch of the Geological History of Lake Lahontan." Third Annual Report of the U.S. Geological Survey pp. 189-235.

Russell, I. C. (1884). "A Geological Reconnaissance in Southern Oregon." Fourth Annual Report of the U.S. Geological Survey pp. 431-64.

Russell, I. C. (1885). "Geological History of Lake Lahontan, a Quaternary Lake of Northwestern Nevada." U.S. Geological Survey Monograph 11.

Russell, I. C. (1889). "Quaternary History of Mono Valley, California." Eighth Annual Report of the U.S. Geological Survey pp. 261-394.

Schreiber, J. F., Jr., Pine, G. L., Pipkin, B. W., Robinson, R. C., and Wilt, J. C. (1972). Sedimentologic studies in the Willcox Playa area, Cochise County, Arizona. *In* "Playa Lake Symposium" (C. C. Reeves, Jr., ed.), pp. 133-84. International Center for Arid and Semi-Arid Land Studies Publication 4.

Schultz, C. B., and Smith, H. T. U. (eds.) (1965). "Guidebook for Field Conference H, Southwestern Arid Lands." Seventh Congress of the International Association for Quaternary Research. The Nebraska Academy of Sciences, Lincoln.

Schumm, S. A. (1965). Quaternary paleohydrology. *In* "The Quaternary of the United States" (H. E. Wright, Jr., and D. G. Frey, eds.), pp. 783-94. Princeton University Press, Princeton, N.J.

Scott, W. E. (1980). New interpretations of Lake Bonneville stratigraphy and their significance for studies of earthquake-hazard assessment along the Wasatch Front. *In* "Proceedings of Conference X, Earthquake Hazards along the Wasatch and Sierra-Nevada Frontal Fault Zones," pp. 548-76. U.S. Geological Survey Open-File Report 80-801.

Scott, W. E., McCoy, W. D., Shroba, R. R., and Miller, R. D. (1980). New interpretations of late Quaternary history of Lake Bonneville, western United States. *In* "American Quaternary Association, Sixth Biennial Meeting, Abstracts and Program, 18-20 August 1980," pp. 168-69. Institute for Quaternary Studies, University of Maine, Orono.

Sheppard, R. A., and Gude, A. J., III (1968). "Distribution and Genesis of Authigenic Silicate Minerals in Tuffs of Pleistocene Lake Tecopa, Inyo County, California." U.S. Geological Survey Professional Paper 597.

Simpson, J. H. (1876). "Explorations across the Great Basin of the Territory of Utah." U.S. Army Engineering Department, Washington, D.C.

Sly, P. G. (1978). Sedimentary processes in lakes. *In* "Lakes: Chemistry, Geology, and Physics" (A. Lerman, ed.), pp. 65-89. Springer-Verlag, New York.

Smith, G. I. (1962). "Subsurface Stratigraphy of the Late Quaternary Deposits, Searles Lake, California: A Summary." U.S. Geological Survey Professional Paper 450-C, pp. 65-69.

Smith, G. I. (1966). Geology of Searles Lake: A guide to prospecting for buried continental salines. *In* "Second Symposium on Salt" (J. L. Rau, ed.), vol. 1, pp. 167-80. Northern Ohio Geological Society, Cleveland.

Smith, G. I. (1968). Late-Quaternary geologic and climatic history of Searles Lake, southeastern California. *In* "Means of Correlation of Quaternary Successions" (R. B. Morrison and H. E. Wright, Jr., eds.), vol. 8, pp. 293-310. Proceedings of the Seventh Congress of the International Association for Quaternary Research. University of Utah Press, Salt Lake City.

Smith, G. I. (1976a). Origin of lithium and other components in the Searles Lake evaporites, California. *In* "Lithium Resources and Requirements by the Year 2000" (J. D. Vine, ed.), pp. 92-103. U.S. Geological Survey Professional Paper 1005.

Smith, G. I. (1976b). Paleoclimatic record in the upper Quaternary sediments of Searles Lake, California, U.S.A. *In* "Paleolimnology of Lake Biwa and the Japanese Pleistocene" (S. Horie, ed.), vol. 4, pp. 577-604. Private publication, Kyoto, Japan.

Smith, G. I. (1979). "Subsurface Stratigraphy and Geochemistry of Late Quaternary Evaporites, Searles Lake, California. U.S. Geological Survey Professional Paper 1043.

Smith, G. I., and Pratt, W. P. (1957). "Core Logs from Owens, China, Searles, and Panamint Basins, California." U.S. Geological Survey Bulletin 1045-A, pp. 1-62.

Smith, R. S. U. (1972). Tentative correlation of pluvial events in Panamint Valley, California, with Sierra Nevada Pleistocene glaciations. *Geological Society of America, Abstracts with Programs* 4, 672.

Smith, R. S. U. (1975). Late Quaternary pluvial and tectonic history of Panamint Valley, Inyo and San Bernardino counties, California. Ph.D. dissertation, California Institute of Technology, Pasadena.

Smith, R. S. U. (1978a). Late Pleistocene paleohydrology of pluvial Lake Panamint, California. *Geological Society of America, Abstracts with Programs* 10, 148.

Smith, R. S. U. (1978b). "Pluvial History of Panamint Valley, California: A Guidebook for the Friends of the Pleistocene, Pacific Cell." University of Houston, Houston.

Snyder, C. T., Hardman, G., and Zdenek, F. F. (1964). Pleistocene lakes in the Great Basin. U.S. Geological Survey Miscellaneous Geologic Investigations Map I-416.

Snyder, C. T., and Langbein, W. B. (1962). The Pleistocene lake in Spring Valley, Nevada, and its climatic implications. *Journal of Geophysical Research* 67, 2385-94.

Stanley, G. M. (1966). "Deformation of Pleistocene Lake Cahuilla Shoreline, Salton Sea Basin," (abstract). Geological Society of America Special Paper 87, 165.

Stearns, C. E. (1962). "Geology of the North Half of the Pelona Quadrangle, Catron County, New Mexico." New Mexico Institute of Mining and Technology, State Bureau of Mines and Mineral Resources, Bulletin 78.

Stewart, M., and Friedman, I. (1974). Deuterium fractionation between aqueous salt solutions and water vapor. *Journal of Geophysical Research* 80, 3812-18.

Street, F. A., and Grove, A. T. (1976). Environmental and climatic implications of late Quaternary lake-level fluctuations in Africa. *Nature* 261, 385-90.

Street, F. A., and Grove, A. T. (1979). Global maps of lake-level fluctuations since 30,000 B.P. *Quaternary Research* 12, 83-118.

Stuiver, M. (1964). Carbon isotopic distribution and correlated chronology of Searles Lake sediments. *American Journal of Science* 262, 377-92.

Stuiver, M., and Deevey, E. S., Jr. (1962). Yale natural radiocarbon measurements VII. *Radiocarbon* 4, 250-62.

Stuiver, M., and Smith, G. I. (1979). Radiocarbon ages of stratigraphic units. *In* "Subsurface Stratigraphy and Geochemistry of Late Quaternary Evaporites, Searles Lake, California," pp. 68-75. U.S. Geological Survey Professional Paper 1043.

Swain, F. M., and Meader, R. W. (1958). Bottom sediments of southern part of Pyramid Lake, Nevada. *Journal of Sedimentary Petrology* 28, 286-97.

Thompson, D. G. (1929). "The Mohave Desert Region, California." U.S. Geological Survey Water-Supply Paper 578.

Turk, L. J. (1970). Evaporation of brine: A field study on the Bonneville Salt Flats, Utah. *Water Resources Research* 6. 1209-15.

van de Kamp, P. C. (1973). Holocene continental sedimentation in the Salton Basin, California: A reconnaissance. *Geological Society of America Bulletin* 84, 827-48.

Van Denburgh, A. S. (1975). "Solute Balance at Abert and Summer Lakes, South-Central Oregon." U.S. Geological Survey Professional Paper 502-C.

Velez de Escalante, S. (1776; 1943 English translation). Journal and itinerary of the route from Presidio de Santa Fe del Nuevo-Mexico to Monterey, in Northern California. *Utah Historical Quarterly* 11, 27-113.

Verosub, K. L., Davis, J. O., and Valastro, S. (1980). A paleomagnetic record from Pyramid Lake, Nevada, and its implications for proposed geomagnetic excursions. *Earth and Planetary Science Letters* 49, 141-48.

Waring, G. A. (1908). "Geology and Water Resources of a Portion of South-Central Oregon." U.S. Geological Survey Water-Supply Paper 220.

Waring, G. A. (1909). "Geology and Water Resources of the Harney Basin Region, Oregon." U.S. Geological Survey Water-Supply Paper 231.

Weide, D. L. (1976). The Warner Valley, Oregon: A test of pluvial climatic conditions. *In* "American Quaternary Association, Fourth Biennial Conference, Abstracts," p. 23. Arizona State University, Tempe.

Wells, P. V. (1979). An equable glaciopluvial in the west: Pleniglacial evidence of increased precipitation on a gradient from the Great Basin to the Sonoran and Chihuahuan Deserts. *Quaternary Research* 12, 311-25.

Wendorf, F. (1961). "Paleoecology of the Llano Estacado." Fort Burgwin Research Center Publications 1.

Wendorf, F. (1970). The Lubbock subpluvial. *In* "Pleistocene and Recent Environments of the Central Great Plains" (W. Dort, Jr., and J. K. Jones, Jr., eds.), pp. 23-57. Department of Geology, University of Kansas, Special Publication 3.

Whitney, J. D. (1865). "Report of Progress and Synopsis of the Fieldwork from 1860-1864." Vol. 1. Geological Survey of California.

Williams, J. S. (1962). "Lake Bonneville: Geology of Southern Cache Valley, Utah." U.S. Geological Survey Professional Paper 257-C.

Wilson, M. E., and Wood, S. H. (1980). Tectonic tilt derived from lake-level measurements, Salton Sea, California. *Science* 207, 183-86.

Coastal and Marine Environments

Sea Level and Coastal Morphology of the United States through the Late Wisconsin Glacial Maximum

Arthur L. Bloom

Introduction

The nature of sea-level research requires looking at areas beyond the coasts and continental shelves of a single nation. The surface of the interconnected world ocean is continuous today and was so during glacially controlled sea-level minima, with the exception of minor mediterranean and shelf seas. Therefore, information about sea-level changes anywhere in the world is, in principle, applicable to the study of coastal changes in the United States. Complications arise because of the middle-latitude Northern Hemisphere location of the United States, which was partially covered by and otherwise adjacent to the largest continental ice sheets of the Wisconsin Glaciation. Both the Atlantic and Pacific coasts show strong latitudinal contrasts in their record of Pleistocene sea-level changes, with postglacial isostatic emergence dominant in the north and postglacial submergence of varying amounts elsewhere. Furthermore, the Pacific coast has been strongly affected by tectonic movements, even within the last 25,000 years. The very large Alaskan continental shelf, which merged with the Asian continent during times of lower sea level to form part of the region known as Beringia, may also have been affected by tectonic movements along its southern margin. The only tropical regions of the United States are islands along tectonic island arcs or mid-oceanic volcanic ridges. The islands offer special opportunities for research by virtue of their coral reefs and tectonic activity.

This review of sea-level changes and coastal morphology through the Late Wisconsin glacial maximum is divided into two parts. The first is a review of sea-level changes at the end of Middle Wisconsin time, from about 28,000 to 23,000 yr B.P. That was a time of generally lower global sea level falling from a relatively high interstadial position. Questions about that Middle Wisconsin high sea level continue to bother researchers.

The second part of the chapter reviews the paleogeography of the continental margins of the United States during the sea-level minimum associated with the Late Wisconsin glacial maximum (23,000 to 15,000 yr B.P.) and the subsequent rapid rise of sea level until about 10,000 yr B.P. Almost all the evidence for full-glacial and late-glacial paleogeographic description is now submerged. Nevertheless, the extent of the sea-level minimum and the conditions during the early stages of the late-glacial transgression are of special significance to the discovery and exploitation of resources along the continental shelf and to studies of faunal (including human) and floral migrations.

The Question of a Middle Wisconsin High Sea Level

Imbedded in the literature of sea-level research are several hundred radiocarbon dates, all near the upper age limit of the method, that have been cited as proof that sea level stood above its present position at least once during Middle Wisconsin time. The event is claimed to have been brief, even though the ages assigned to it cover a long time span. The dates are usually in the range of 40,000 to 25,000 yr B.P. Samples that give radiocarbon ages in this time range have been reported from almost every coastal region, including Eurasia, Africa, and Australia, but most of the data are from the middle and southern Atlantic Coast and the Gulf Coast of the United States. The abundance of such dates on the southeastern coasts of the United States may only be an artifact caused by the ready availability of radiocarbon dating laboratories. Nevertheless, the persistent and recurring reports of radiocarbon-dated evidence for a Middle Wisconsin high sea level, especially on the coasts of the southeastern United States, are so well known that the subject must be reviewed in some detail for the international audience of this volume.

Critics have been harsh in their evaluation of the evidence for the age of the alleged high sea level. About 20 years ago, Broecker (1965: 746) asserted that there were then no reliable pre-Holocene radiocarbon dates that placed sea level within 15 m of its present level, back to the limits of the dating method. In 1973, Thom reviewed "the dilemma of high interstadial sea levels during the last glaciation" in detail. He compiled and evaluated 188 radiocarbon dates that had been quoted in support of a high interstadial sea level. He proposed five criteria that, if satisfied, would qualify a sample for a rating of "superior." He separately evaluated the dates on marine carbonate (mollusks, corals, algae, oolites, etc.) and organic carbon samples.

This chapter is a contribution to the U.S. National IGCP Project 61, the Sea-Level Project. Cornell University Department of Geological Sciences contribution No. 733.

None of the carbonate samples that he evaluated satisfied more than two of his five criteria, and therefore, all were judged "inferior" or "not sufficiently documented." Some of the organic carbon samples satisfied three of the five criteria and were judged "questionably satisfactory." None of the 188 samples received the "superior" rating because none had been adequately (by Thom's criteria) pretreated, fractionated, cross-verified by other dates, and stratigraphically documented (Thom, 1973: 183).

In spite of Broecker's and Thom's critical and pessimistic reviews, others have continued the effort to document a Middle Wisconsin high sea level. Caution is usually expressed, although a misleading impression of support for the idea might be gained by a reader who is not familiar with the problem. For example, Field and others (1979: 625) have published 8 new radiocarbon dates and 2 more selected from published reports that "provide supportive evidence for a late Pleistocene high stand of sea level." They assert that the ages they report "appear to be fairly reliable. Samples from the same general area have similar ages, and only two of the ten could not be dated precisely." However, their tabulations list only two finite dates on marine shells, one of 23,550 ± 850 yr B.P. (GX-2152) from a depth of 25 m below sea level and the other of 25,200 + 3000/− 2000 yr B.P. (GX-2573) from a depth of 8.3 m below sea level. The first of the two samples of intertidal pelecypod shells (*Donax variabilis*) had been redated according to an earlier report (Field, 1974: 59) and "did in fact produce an anomalous age, thus serving as a reminder that all dates on carbonates in this age range are suspect." Five of the remaining 10 supportive dates are "infinite" (beyond the range of reliable counting statistics), 1 has no counting error reported, and the remaining 2 finite dates (of 32,730 ± 1650 yr B.P. (I-7438) and 33,000 ± 600 yr B.P. [Wor-Dg 13]) are from nonmarine peats 9.7 m below sea level. Elsewhere in the article (Field et al., 1979: 625), the authors state that, "although [the samples] do not preclude the possibility that sea level rose during that period to near its present level, they do indicate that it did not attain its present level, at least not prior to 27,000 yr B.P." It is to be regretted that the major illustration in their article shows all of the "dead" samples plotted as finite points at the minimum age assigned by the dating laboratories in the same manner as the finite ages. Thus, an unwary reader could easily assume that the results support a high sea level when only 1 of the 10 samples cited in support of the idea has a reproducible finite date and even it had been labeled "suspect" in an earlier article.

Belknap and Kraft (1977) have compiled a list of 24 radiocarbon samples older than 12,000 yr B.P. that seem to correlate with an earlier sea-level curve published by Milliman and Emery (1968). As shown by Figure 11-1, 10 of the 24 samples had finite radiocarbon ages in the range of 34,000 to 16,000 yr B.P., including 2 that indicate that the sea was 4 to 6 m above its present level in an unacceptable age range of 20,000 to 16,000 yr B.P. The other 14 samples were too old for radiocarbon dating and were plotted as a group at the upper end of the time scale. One locality, at 5.5 m above sea level, provided 4 inconsistent radiocarbon ages of 16,970 ± 290 (I-6052), 34,000 ± 2000 (I-749), 31,900 ± 1400 (I-7524), and >37,000 yr B.P. (L-819). Belknap and Kraft note that the interpretation of a Middle Wisconsin high sea level contradicts the deep-sea oxygen-isotope record, and they declare the conflict to be beyond the scope of their article (Belknap and Kraft, 1977: 619).

Regrettably, the few finite radiocarbon dates from lists such as those of Belknap and Kraft (1977) and Field and others (1979) are too frequently selectively cited in other publications; this gives a false impression about the weight of the evidence that supports the alleged Middle Wisconsin high sea level. The origins of the hypothesis can be traced to publications by Milliman and Emery (1968) and Curray (1960). Milliman and Emery (1968) list only three radiocarbon ages that ranged from 36,000 to 30,000 yr B.P., all from previously published sources, in support of the idea. One was from a peat sample and two were determined on mollusk shells. The peat, which they call "salt-marsh peat," is actually described in their cited source (Trautman and Willis, 1966: 174, sample I-1745) as a freshwater peat. The error is perpetuated in a more recent article (Blackwelder et al., 1979: 619). One of the two shell dates cited by Milliman and Emery is the 34,000 ± 2000 yr B.P. noted in the preceding paragraph, one of a group of four inconsistent age analyses from the same locality. The other shell date has not been further verified.

Two decades ago, Curray (1960, 1965) cautiously proposed a high Middle Wisconsin sea level in the northwestern Gulf of Mexico on the basis of two radiocarbon dates (26,900 ± 1800 [J-383] and 32,500 ± 3500 yr B.P. [J-526]) on mollusk shells dredged from a depth of about 15 m from the submerged Freeport Rocks off the Texas coast. The younger sample was slightly recrystallized, and its actual age was judged to be somewhat older (Curray, 1960: 255). A project colleague (Parker, 1960: 328) suggested that both dates may be too young because of carbonate replacement and contamination. Both samples were of cemented beachrock and coquina, and the ages are suspect by current standards of sample selection. They have not been verified by any later analyses. Curray commented (1960: 255-56; 1965: 724) that, "on the basis of these two dates, the possibility is suggested that sea level may have been somewhere between the present level and minus 8 fathoms (ca. 15 m) at approximately 30,000 years B.P." and that "this portion of the curve . . . is largely speculation. It is based on a few dates suggesting an interstadial high stand of sea level sometime between about 22,000 and 35,000 B.P. and on correlation with continental Pleistocene events. This interstadial stand of sea level probably lay slightly below present sea level." The caution with which he hypothesized the event and the weak evidence with which he supported it have been lost in two decades of repeated and frequently careless citation.

What are the observations that so compel researchers to cling to a few dubious radiocarbon dates and to conclude that sea level on the southeastern Atlantic and Gulf coasts was above its present level during an interstadial interval within the last 40,000 to 20,000 years of the Wisconsin Glaciation? Most of the field evidence is geomorphic. On the southeastern Coastal Plain of the United States are a series of emerged fossil barriers, lagoonal sediments, and related depositional and erosional landforms. The Coastal Plain demonstrates a long Neogene history of emergence during fluctuating sea levels (Cronin, 1981; Oaks and DuBar, 1974). Hoyt and others (1968) first proposed that the Princess Anne shoreline at +4.5 m was approximately 48,000 to 40,000 years old and that the Silver Bluff shoreline at +2 m was approximately 30,000 to 25,000 years old. Ten radiocarbon dates from Georgia were discussed in support of the proposal, of which 5 were finite. Two of the finite dates were replicate samples from the same locality, and a third was from a shallow boring and was bracketed by "dead" samples, 0.2 m above and below it. All dates were on mollusk shells from bore holes. Hoyt and others were firm but cautious in supporting their hypothesis, but much of the supporting evidence they cited from other regions has been subsequently revised or refuted. In a later article (Hoyt and Hails, 1974), the beginning of Silver Bluff deposition was extended back to 37,000 yr B.P. or

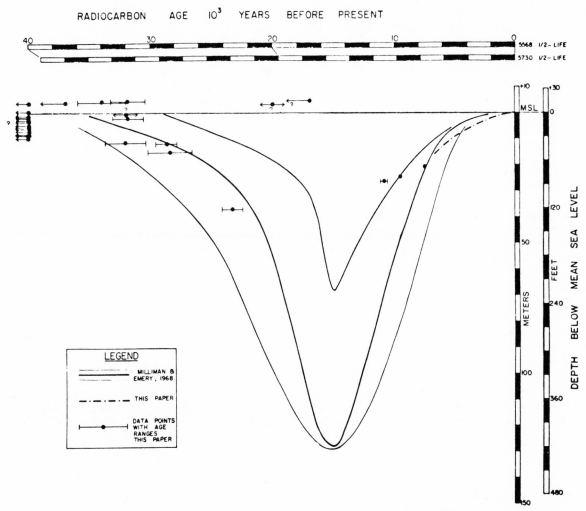

Figure 11-1. Radiocarbon-dated samples from Delaware that relate to sea-level history. Samples older than 9000 yr B.P. are compared to the sea-level curve of Milliman and Emery (1968). Upper and lower lines are the envelope of values compiled by Milliman and Emery. (From Belknap and Kraft, 1977: Figure 5.)

earlier on the basis of additional radiocarbon ages, including some infinite ones. Many comments similar to those in the preceding paragraphs could be made about the interpretation of old radiocarbon dates published by Hoyt and others (1968). However, criticism is better reserved for those who have subsequently cited the work without noting the honest caution it expressed.

Most of the emerged coastal landforms and sediments of the southeastern United States are too old or are unsuitable for any available dating method. A few uranium-thorium ages on ahermatypic corals suggest that some of the emerged coastal sediments associated with the landforms were deposited within the Wisconsin Glaciation, perhaps from 94,000 to 62,000 yr B.P. (Cronin et al., 1981; Flint, 1971: 589; Oaks et al., 1974: 75, 80). Ages calculated from amino acid racemization rates are less generally accepted, but at least a few of them also imply Middle Wisconsin ages (50,000 to 30,000 yr B.P.) for the youngest Pleistocene Coastal Plain sediments and landforms (Belknap and Wehmiller, 1980).

Sea level during the last interglacial was a few meters above its present level, +6 m being a commonly accepted estimate (Bloom et al., 1974; Matthews, 1973). Deposits of this age are widely known on

stable and rising coasts all over the world. The coastal plains of the southeastern United States Atlantic and Gulf coasts have had net emergence and seaward tilting since late-Mesozoic time. Thus, it is not surprising to find there emerged littoral marine deposits and landforms of last-interglacial age, approximately 125,000 years old. But, if sea level was not above its present level at any time from about 120,000 years ago until the Holocene (Figure 11-2), any valid intra-Wisconsin dates on emerged littoral samples must imply tectonic uplift. Samples of Early Wisconsin age, such as the coral samples with uranium-thorium dates of 94,000 to 62,000 yr B.P., that could correlate with oxygen-isotope stages 3 or 5a in deep-sea cores or perhaps with the Barbados I terrace (Matthews, 1973; Mesolella et al., 1969), could have been brought above present sea level by a slow tectonic uplift rate of 0.1 to 0.3 m per 1000 years even if they formed during an Early Wisconsin interstadial sea-level maximum that terminated at the hypothesized level of −15 m (Bloom et al., 1974; Cronin, 1981; Cronin et al., 1981; Matthews, 1973). However, to bring Middle Wisconsin interstadial littoral deposits above present sea level from their inferred levels of formation at −38 to −42 m (Figure 11-2), tectonic uplift of 1 to 2 m per 1000 years would have been required.

Such rates are known only from active orogenic zones and are totally inconsistent with the tectonic setting of the southeastern United States (Cronin, 1981).

A slight and unevaluated possibility remains that the southeastern coast of the United States was close enough to the Laurentide ice margin of 28,000 to 23,000 yr B.P. to have been uniquely affected by isostatic and geoidal deformation. It is argued in subsequent paragraphs that the Middle and Late Wisconsin ice masses were highly dynamic and unstable, so that isostatic compensation was never achieved during that interval. It is barely possible that some combination of delayed isostatic depression along the southeastern coasts of the United States combined with an abrupt interstadial rise of sea level could have produced a littoral record that is now above sea level. Unless such a unique set of circumstances can be demonstrated, however, the validity of the Middle Wisconsin high sea level must be doubted. With the St. Lawrence lowland filled with glacier ice throughout Middle Wisconsin time (Dreimanis and Raukas, 1975), sea level probably was below its present level on all coasts of the United States. Isostatic effects determined the level from place to place, but, until evidence can be obtained to substantiate specific sea levels at specific times and places, the estimates of − 38 m for the interstadial interval of about 40,000 yr B.P. and − 40 to − 42 m for the interstadial interval of 30,000 to 28,000 yr B.P. can be accepted as a working hypothesis (Bloom et al., 1974; Chappell and Veeh, 1978). The Middle Wisconsin high sea level alleged for the southeastern coast of the United States and elsewhere is invalid; the hypothesis is based on contaminated or misinterpreted radiocarbon dates and is not supported by other dating methods, by documented ice-marginal positions, by the deep-sea oxygen-isotope record, or by the radiometric dating of emerged coral reefs.

The Late Wisconsin Sea-Level Minimum

One would expect that after 150 years of discussion and research the extent of sea-level lowering at the time of the latest and best-known advance of the youngest ice age would be well known, but it is not. The thickness of the layer of water that was removed from the world ocean to form the ice sheets of the maximum Wisconsin Glaciation is currently estimated to be in the range of 120 ± 60 m. The error range of 50% represents realistic variations in the range of reputable opinions, whether based on reconstructed ice sheets, oxygen-isotope fractionation ratios, submarine geologic research, or theoretical modeling of the isostatic responses of a viscoelastic Earth.

We have not progressed far from the convenient and often-quoted estimate of 100 m (300 ft) used by R. A. Daly and his contemporaries of 50 years ago. Indeed, the numbers we debate today are within the range suggested by Maclaren (1842) when he proposed the testing of Agassiz's then new and radical theory of Pleistocene ice sheets by measuring the depth of the shorelines that would have resulted from the removal of so much water from the sea. Maclaren guessed that ice sheets as large as those proposed by Agassiz would have required a sea level at least 213 m (700 ft) lower than the modern level. He noted that, even if his estimate were too large by a factor of two, a global (eustatic) lowering of sea level by 107 m (350 ft) would have been an excellent confirmation of the glacial theory. Although we no longer seek evidence in support of continental glaciation, we have yet to document the magnitude of the obvious and important perturbation of the global hydrologic cycle that it must have represented.

The wide range of contemporary estimates of sea-level lowering during the Late Wisconsin glacial maximum results from valid reasoning based on a variety of acceptable premises. For instance,

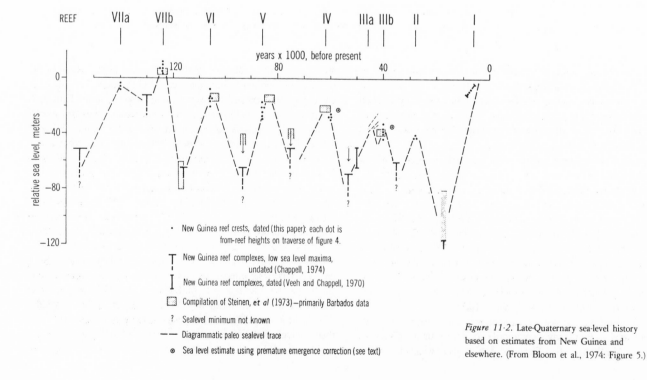

Figure 11-2. Late-Quaternary sea-level history based on estimates from New Guinea and elsewhere. (From Bloom et al., 1974: Figure 5.)

Hughes and others (1981: 274, 308) compiled maps of the areal extent of ice cover at 18,000 yr B.P. and calculated ice thicknesses from equations of plastic flow and basal sliding at appropriate temperatures. Their maximum and minimum estimates, equivalent to 163 and 127 m of sea-level lowering, are at the upper limit of volume estimates, because they assume that enough time had elapsed at the glacial maximum for all ice sheets to equilibrate to ideal flow-law configurations. Isostatic depression of the lithosphere under all ice sheets was also assumed to be complete at 18,000 yr B.P.

Peltier and Andrews (1976) also estimate the areal distribution and thickness of Northern Hemisphere ice sheets at 18,000 yr B.P. Instead of assuming theoretical ice-surface gradients and subice isostatic compensation, they attempt to approximate realistic ice thicknesses from evidence such as the distance from rapidly calving ice margins. Their model estimates an ice-mass equivalent to 77 m of sea-level lowering at 18,000 yr B.P., excluding changes in the volume of the Antarctic ice sheet and a few minor mountain ice caps. A refinement of their method worked out by Clark and others (1978: 276) has resulted in an estimate of a 75.6-m rise of sea level between 16,000 and 5000 yr B.P. because of the ablation of Northern Hemisphere ice sheets. To these estimates, Clark and Lingle (1979) add an estimated 25 m of sea-level rise from the late-glacial volume reduction of the Antarctic ice sheet for a total of about 100 m of sea-level change. The 25 m of sea-level change from Antarctica is derived from Hughes and others (1981) and is included in their range of sea-level-change estimates of 163 to 127 m. Thus, two current estimates of the volume of vanished Northern Hemisphere ice sheets range from sea-level equivalents of 77 m up to 100 or even 150 m.

Estimates of the maximum lowering of sea level during full-glacial time have also been made from the deviation of the $^{18}O/^{16}O$ ratios in benthonic Foraminifera samples from deep-sea cores. Deep-ocean waters were enriched about 1.65‰ in ^{18}O during the last glacial maximum, which has been calculated as the equivalent of a sea-level lowering of 165 m (Shackleton, 1977: 174). This estimate is close to the upper range of glaciologic estimates. An earlier and more frequently cited reference to this method (Shackleton and Opdyke, 1973: 45) suggests a maximum sea-level lowering of 120 m on the basis of a maximum ^{18}O enrichment of a 1.2‰ in benthonic Foraminifera, but that work is based on a deep-sea core with a rather slow sedimentation rate, and bioturbation had probably mixed the sediment layers sufficiently to reduce the amplitude of the ^{18}O enrichment (Shackleton, 1977: 177).

It is possible that the worldwide average depth of the break in slope between continental shelves and continental slopes, carefully compiled by Shepard (1973: 277) as being at −130 m, is actually a good median value for the estimation of sea-level lowering by growth of continental ice sheets. The value is based on a compilation from hydrographic charts and probably is as free from bias as any such compilation can be. Of course, there is no proof of whether the shelf break is a consequence of the Late Wisconsin sea-level lowering or of an earlier low sea level. Nevertheless, reputable submarine geologists place heavy emphasis on the relict character of the modern continental shelves and their genesis as a consequence of Pleistocene lower sea levels (Curray, 1965; Emery, 1968; Southard and Stanley, 1976: 356). Presumably, Shepard's average shelf-break depth of −130 m has a very large standard deviation, which reflects erosional and depositional changes as well as isostatic, tectonic, geoidal, and other perturbations that can equal one-third of the change in height of the water column over a point on a continental shelf.

An instructive illustration of the isostatic and other deforming influences on water depth over a continental shelf is provided by Clark (1981). Figure 11-3 shows his predicted late-Pleistocene and Holocene sea-level curves for four parts of the Atlantic coast of the United States. He assumes uniform mantle viscosity and a total eustatic sea-level rise of 75 m (near the lower end of the wide range of opinions about eustatic sea-level change). From the figure, the magnitude of shelf emergence 16,000 years ago could have ranged from less than 50 m in New Jersey, relatively close to the ice margin, to −90 m farther south in Georgia and Florida. The two northern sites, New Jersey and Virginia, would have experienced net emergence from 16,000 to about 10,000 yr B.P., during the same time interval when the southern sites were beginning to submerge. Another expression of the same data (Figure 11-4) predicts that the 16,000 yr B.P. shoreline on the Atlantic continental shelf of the United States tilts downward to the south from less than −50 m in New Jersey to −90 m in Florida whereas various Holocene shorelines deepen toward the north. Blackwelder (1981) emphasizes that the radiocarbon-dated record of sea-level change on the middle-Atlantic Coast shows much less total late-glacial submergence and much less variation from place to place than Clark predicts. The absolute range of the predicted sea-level changes is less important than the form of the curves in Figures 11-3 and 11-4, however. If the prediction is reasonably correct in direction, even if it is not accurate in magnitude, it explains the difficulty and dangers of trying to estimate the age of a particular submerged fossil shoreline from its depth or of trying to predict the depth of a shoreline of any late-glacial or postglacial age. Clark's analysis and Blackwelder's reply are an excellent case history of the problems that have been encountered in trying to establish the extent of "eustatic" sea-level change. We are now reconciled to the probability that every coast has its own unique sea-level history that results from the interaction of variables that are as yet only hypothesized and not measured. Perhaps the current range of estimates of 120 ± 60 m for the Late Wisconsin sea-level change is an accurate expression of the extreme complexity of a variable that earth scientists previously and naively expected to be unique and easily measured.

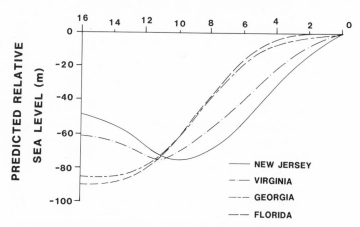

TIME (1000 YRS BP)

NEW JERSEY
VIRGINIA
GEORGIA
FLORIDA

Figure 11-3. Predicted relative sea-level changes of the United States mid-Atlantic coastal region. The model assumes uniform mantle viscosity and a total eustatic sea-level rise of 75 m. (From Clark, 1981: Figure 1; used with the permission of the Geological Society of America.)

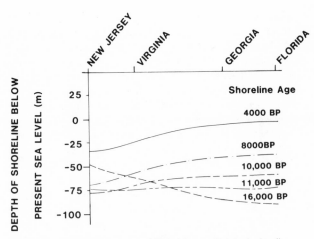

Figure 11-4. Predicted deformation of United States mid-Atlantic shorelines caused by glacial isostasy. Shorelines older than 11,000 yr B.P. become deeper southward, whereas younger shorelines become deeper northward. (From Clark, 1981: Figure 2; used with the permission of the Geological Society of America.)

Paleogeography of the Full-Glacial and Late-Glacial Coasts

On the basis of the theoretical and observational considerations discussed in the preceding paragraphs, it is possible to comment briefly on the paleogeography of the exposed continental margins of the United States during the time of maximum glacial advance and the beginning of the rapid retreat of the ice margins and the corresponding rise of sea level. The review proceeds geographically from the northeastern glaciated states southward and westward along the Atlantic and Gulf coasts and then northward along the Pacific coast to Alaska. A few comments are also included concerning the tropical island shelves of Hawaii, Puerto Rico, and the Pacific territories of the United States.

ATLANTIC COAST

The Atlantic Coastal Plain was glaciated north of latitude 40°30′N from the lower New York Harbor eastward and northward. In this region, subdued cuestas of poorly indurated Cretaceous and Tertiary sedimentary rocks form the cores of Long Island, Cape Cod, and the submerged Georges Bank. The drowned intercuesta lowlands form Long Island Sound, Block Island Sound, Buzzards Bay, Nantucket Sound, and Cape Cod Bay. The larger Gulf of Maine is nearly enclosed by the Georges Bank cuesta, but it is a complex structural lowland eroded in late Paleozoic metamorphic rocks and Mesozoic sandstones and basalts.

At the maximum of the Late Wisconsin glaciation, the ice margin lay on the rims of the Coastal Plain cuestas. Moraines were built that now accentuate the relief of the largely submerged ridges. Almost all of the subaerial relief of Long Island, Cape Cod, and associated islands is morainal. Extensive outwash plains of sand and gravel extended south and east from the terminal moraines across the emerged Coastal Plain at least as far east as Georges Bank, beyond which the ice margin may have calved directly into the sea (Pratt and Schlee, 1969).

Numerous mastodon and mammoth teeth have been collected from the continental shelf of the northeastern United States by fishermen; this confirms that the ice-marginal outwash plains were not sterile (Figure 11-5). It is reasonable to suppose that paleo-Indians hunted on the plains, although no direct evidence has been found.

The sandy, braided outwash plains probably were subject to strong wind action, with dune fields and possibly loess deposits forming the surface relief. Only small remnants of such terranes are now found above sea level on Long Island and the islands south of Cape Cod. The remainder of the vast plain is now submerged and was so reworked during the marine transgression that it surface relief, vegetation, and soils can only be postulated. Minor evidence of periglacial structures in the glacial drift of Rhode Island and southern Massachusetts (Kaye, 1960: 383; 1964) suggests that part of the now-submerged Coastal Plain could have been underlain by permafrost.

East of Cape Cod toward Nova Scotia, Canada, the Georges Bank cuesta and associated moraine are entirely submerged, although some of the gravel shoals are only a few meters below sea level. Strong tidal currents have scoured and reworked the morainal sediments.

The Northeast Channel is a submerged, U-shaped trough with a sill depth of −220 m that cuts the submerged Coastal Plain cuesta at the northeastern end of Georges Bank (Figure 11-5). It is nearly 40 km wide and 70 km long, and its floor is a series of closed basins with depths to 370 m. It may have been shaped by an outlet glacier flowing out of the Gulf of Maine and calving at the continental margin directly across the shelf break. It is similar in form and perhaps in origin to the much larger St. Lawrence Channel under the Gulf of St. Lawrence and the Grand Banks of Canada. It seems likely that the Gulf of Maine was full of grounded glacier ice during full-glacial time, for the maximum depths of the structurally controlled basins in the gulf are generally less than 300 m (Uchupi, 1968).

The deglaciation of the northeastern continental shelf created some interesting and unique geomorphic conditions. As the ice margin retreated from the terminal moraines that had been built on the cuesta rims, the intercuesta lowlands were flooded, first by ice-marginal lakes and then by the sea. Especially in the Gulf of Maine, a calving basin must have developed behind the Georges Bank cuesta while it was still an exposed gravel-capped ridge. The Gulf of Maine is almost 150 km wide, and its only access to the sea until Holocene time was the deep outlet-glacier trough of the Northeast Channel. It may have deglaciated rapidly, causing the glacier ice on the adjacent New England uplands to flow with stong convergence toward it. Late-glacial moraines and flow-direction indicators in the vicinity of Boston, Massachusetts, show that there was a strong eastward flow of ice into the basin.

A short distance north of Boston, the first late-glacial marine sediments appear above the present sea level. The inland limit of the emerged marine deposits rises steadily to a maximum of 100 to 150 m in Maine, reflecting the strong postglacial isostatic uplift of the region. In Massachusetts, mollusk shells from the emerged marine sediments give radiocarbon ages of about 14,000 yr B.P. (Kaye and Barghoorn, 1964). Farther north, the ages are younger and cluster closely at about 12,500 ± 700 yr B.P. (Stuiver and Borns, 1975). The late-glacial marine submergence was brief and nearly synchronous, following the retreating ice margin inland across Maine and New Hampshire and producing a series of moraine banks, delta kames, and other features diagnostic of an ice margin calving into the sea. Until isostatic uplift exceeded the rate of late-glacial sea-level rise at about 12,000 yr B.P., there was no land between the retreating ice margin and the transgressing sea across the extreme northeastern United States. Then, within a thousand years, the ice margin

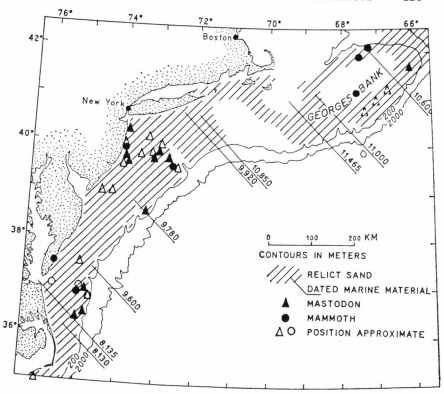

Figure 11·5. Localities on the Atlantic continental shelf where mastodon or mammoth teeth have been accidentally collected by fishermen. Note their distribution relative to the area of relict sand. Radiocarbon ages of shallow-water oyster shells and intertidal peat beds are also shown. (From Whitmore et al., 1967: Figure 2; used from *Science* volume 156, pp. 1477-81; copyright 1967 by the American Association for the Advancement of Science.)

retreated to the north, and the remnant ice sheet disintegrated over the highlands as the sea withdrew southward because of rapid isostatic recovery. A large area of newly deglaciated or emerged terrain was open for postglacial occupation. Paleo-Indians were in the area at least as early as 9000 yr B.P. (Byers, 1959).

The lower New York Harbor is formed in part by a moraine loop from Long Island westward across Staten Island into New Jersey. The submerged Hudson River Channel crosses the continental shelf from New York Harbor southeastward about 170 km to a depth of about −80 m, where it terminates in a delta-shaped submarine plain (the Hudson Apron) at the head of the Hudson submarine canyon (Veatch and Smith, 1939: 40). At least two other former shorelines at −40 and −110 m have been identified in the vicinity of the lower termination of the Hudson River Channel (Dillon and Oldale, 1978). Clearly, glacial meltwater from the Hudson River eroded a broad valley across the continental shelf until about 12,000 yr B.P., when the ice margin retreated north to the St. Lawrence lowland and the glacier-marginal drainage was diverted to the northeast. Final verification of the age of the Hudson River Channel and its associated delta would add to our knowledge of the deglaciation of eastern North America. The channel probably carried one or more catastrophic floods from central New York State as the ice margin retreated from the northern Appalachian Plateau and very large ice-marginal lakes suddenly drained eastward down the Mohawk and Hudson Rivers (Hand, 1975).

An intriguing concentration of mastodon and mammoth fossils has been found in extreme southern New York State and adjacent New Jersey and on the continental shelf just south of the submerged Hudson River Channel (Figure 11-5). We can speculate that the large channel and floodplain of the extended late-glacial Hudson River was a barrier to the northward migration of large mammals and that the relict population died as a result of rising sea level, changing vegeta-

tional patterns, and possibly human predation.

The submerged Hudson River Channel forms a convenient northern boundary for the middle-Atlantic continental shelf, which is remarkably uniform in structure, morphology, and sedimentology for more than 1200 km southward to the carbonate depositional province of Florida. Figures 11-3 and 11-4 predict the late-glacial sea-level history for various transects across the shelf, although most published dates indicate that the net amount of Holocene submergence has been less than that predicted, usually not more than about 10 m in the past 7000 years. The predicted 4000 yr B.P. shoreline of Figure 11-4 probably should be drawn nearly horizontal within a few meters below present sea level, for instance. Nevertheless, Figures 11-3 and 11-4 are important in that they illustrate the dynamic potential of late-glacial and postglacial sea-level change on the Atlantic continental margin of the United States and the danger of attempting to correlate submerged shorelines by depth alone.

Authorities agree that most of the Atlantic continental shelf was exposed during Late Wisconsin time. The Holocene sediments on it are thin and discontinuous and include a variety of reworked deltaic, estuarine, and lagoonal units that demonstrate the former emergence of shallow coastal environments. Emery (1968) estimates that 70% of all shelf sediments are relicts from times of lower sea level. Questions remain about the extent of sea-level lowering, as noted earlier. Milliman and Emery (1968) estimate a sea-level minimum of −130 m on the Atlantic continental shelf, close to Shepard's −130 m estimate of the global average depth of the shelf break. Others have proposed shallower depths. For instance, Dillon and Oldale (1978) conclude that all samples from the Atlantic continental shelf from depths greater than −85 m that have been used to construct sea-level curves are probably invalid, and they revise the full-glacial sea-level minimum of Milliman and Emery (1968: Figure 5) (Figure 11-1) upward to −85 m. A series of recent publications (Emery and Merrill,

1979; Macintyre et al., 1978, 1979) express agreement that, in spite of differences in opinion about the timing and amount of the full-glacial lowering of sea level, the valid facts are so few for depths greater than −60 m that only future work can resolve the issue.

Today, the rivers of the Atlantic Coastal Plain terminate in estuaries such as New York Harbor, Delaware Bay, Chesapeake Bay, Albemarle Sound, and Pamlico Sound. During full-glacial times, the estuaries were the valleys of the extended rivers. Some of the rivers have prominent shelf valleys, such as the Hudson River Channel discussed previously. Others have indistinct or nonexistent shelf valleys or shelf valleys that diverge from seismically determined buried valleys. These complexities have been attributed to the marine reworking of surficial shelf deposits during the late-glacial marine transgression (Swift, 1973; Swift et al., 1980). The modern shore-face scarp extends from sea level to depths of −15 to −20 m and is believed to be an erosional feature in equilibrium, as outlined by Zenkovich (1967: 252-56). The landward migration of the shore-face scarp has reworked the top 10 to 15 m of shelf sediment and obliterated most of the former subaerial relief. One component of the marine transgression was the landward retreat of the estuaries at river mouths, and Swift and others (1980) have proposed that the submarine valleys on the New Jersey continental shelf are the transgressive retreat paths of estuarine-mouth scour channels rather than drowned river valleys. By this reasoning, few subaerial or fluvial landforms survived marine alteration during late-glacial time in spite of the high percentage of relict sediments on the shelf.

South on the Atlantic continental shelf of the United States, the biogenic carbonate content of the sediments increases and detrital minerals decrease in grain size and total abundance. Some of the problems in interpreting the late-glacial paleogeography of the shelf are related to the southerly transition from detrital quartz sand to limestone. For instance, low ridges on the shelf or at the shelf break have been interpreted as dune or beach ridges and have been used to define late-Pleistocene sea levels. However, Macintyre and Milliman (1970) and Macintyre and others (1975) have shown that a variety of erosional and depositional processes, some submarine and some littoral, can produce carbonate-cemented shelf ridges. North of Cape Hatteras (latitude 35°N), the carbonate cement of the shelf-break sandstone ridges is aragonitic with a peculiar carbon-isotope composition that implies a methane source associated with buried tidal-marsh organic muds. A relict late-Pleistocene nearshore environment is correctly inferred for these sediments. From Cape Hatteras south to Cape Canaveral (latitude 28°N), the shelf-edge sandstone ridges are cemented with magnesium-rich calcite, probably through submarine diagenetic lithification. South of Cape Canaveral, the shelf-edge ridges are mostly oolitic or coralline debris of former shallow-water algal or coral reefs. It is obvious that not all of the morphological ridges indicate former shorelines, even though they are typically at depths of −70 to −90 m, where late-Pleistocene shorelines are to be expected. The confusion of deeper-water lithified deposits with shore-zone landforms, the great range of radiocarbon ages that has resulted from diagenesis, and the complexities of isostatic deleveling (Figure 11-4) helps explain the current disagreement about the extent and age of sea-level changes on the Atlantic continental shelf of the United States.

South from Miami Beach, Florida, the marine sedimentary environment is totally carbonate. Corals, algae, Foraminifera, and other calcitic and aragonitic organisms dominate sedimentary patterns. Coral reefs are not well developed, but corals are common sediment suppliers. The shelf is a relict carbonate bank, with karst depressions that developed subaerially during glacial times. During the Wisconsin Glaciation, Florida was a low emerged karst plateau, somewhat like the modern Yucatan Peninsula. Dolines similar to the cenotes of the Yucatan are abundant in Florida, and they extend well below present sea level as "blue holes" in the reef. Evidence of early human activity has been found in one Florida doline that extends to at least 55 m below present sea level (Clausen et al., 1979). The shell of a tortoise that had been killed and cooked by paleo-Indians, dated by radiocarbon at about 12,000 yr B.P., was found almost 21 m below sea level on a ledge within the doline. Pollen of sclerophylous plants and an extensive relict sand sheet on the Florida peninsula suggest that the area was arid during full- to late-glacial time, probably an edaphic condition caused by the radical lowering of the water table within the permeable karstic limestone that underlies the entire peninsula. The relative effects of water-table lowering and climate-induced aridity in creating the full-glacial dry conditions in Florida have not been fully determined (Watts, 1969; Watts and Stuiver, 1980).

A very shallow carbonate bank sweeps westward and northward from the Florida Keys into the Gulf of Mexico. The carbonate bank ranges from 125 to 225 km in width seaward to the −120 m isobath, which has been taken as the edge of the inner shelf (Bergantino, 1971). The shallower parts of the shelf have drowned karst topography and some living patch reefs. A full-glacial sea-level lowering of 120 m, hypothesized only from poorly documented shoreline features at the shelf break, would have doubled both the area and the maximum elevation of the Florida peninsula, but the monotonous limestone lithology and low relief of the region would not have been much different from the present landscape.

GULF COAST

The Gulf of Mexico continental shelf of the United States is relatively narrow in the east, seaward of Mississippi and Alabama, but much wider west of the Mississippi Delta plain seaward of Louisiana and Texas. The dominant feature of the Gulf Coast is the Mississippi Delta plain, which has prograded entirely across the shelf in the late Holocene and now discharges directly down the continental slope (Figure 11-6). During full-glacial time, the Mississippi River carried voluminous discharges of glacial meltwater and sediment from the southern quadrant of the Laurentide ice sheet. Although outwash terraces are obvious features of the upper Mississippi Valley, these pass below the present alluvial plain in the lower valley, where the coarse late-glacial sediments are found only beneath many meters of Holocene mud and peat.

The late-glacial and postglacial evolution of the lower Mississippi River valley and the Mississippi Delta plain has been studied in great detail and is well summarized by Fisk (1947), Fisk and McFarlan (1955), and Morgan (1970). Figure 11-7, from an early but still excellent account of the development of the lower Mississippi Valley, shows that during full-glacial time the river flowed in a very deep valley, often referred to as the Mississippi Trench. During the retreat of the southern margin of the ice sheet and while sea level rose, the trench was initially aggraded by outwash gravel and sand of a braided river system. Later, when the extensive proglacial Great Lakes developed along the southern ice margin, the coarse alluvium was trapped and only silt and clay were carried by the meandering, aggrading river. The sequence of deep erosion, coarse-sediment aggradation, and fine-sediment aggradation is the complex result of sea-level rise, ice-margin retreat, climatic change, isostatic uplift,

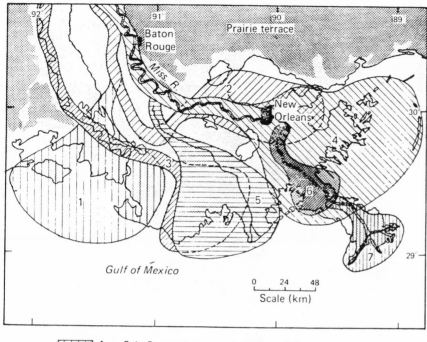

Figure 11-6. Chronology of the deltas that comprise the Mississippi River Delta plain. (From Morgan, 1970: Figure 12.)

1	Sale Cypremort		>4600 yr. B.P.
2	Cocodrie	ca.	4600-3500 yr. B.P.
3	Teche	ca.	3500-2800 yr. B.P.
4	St. Bernard	ca.	2800-1000 yr B.P.
5	Lafourche	ca.	1000-300 yr. B.P.
6	Plaquemine	ca.	750-500 yr B.P.
7	Balize		<550 yr

drainage diversion, and proglacial-lake formation in the headwaters of the river.

An axiom of United States Gulf Coast Quaternary history is that the glacial-age lowering of sea level caused other rivers to entrench their valleys across the Coastal Plain and their extended consequent valleys across the exposed continental shelf in the same manner as the Mississippi River. However, rivers that originate on the Gulf Coastal Plain or the adjacent hinterlands were much less affected by glacial-age sea-level lowering than was the Mississippi River. The obvious vertical lowering of base level was largely balanced by the horizontal extension of river courses across the flat gradient of the emerged continental shelf. The floodplains of most Gulf Coast rivers have gradients of 2 or 3×10^{-4}. The gradient of the Gulf of Mexico continental shelf ranges from 5×10^{-4} seaward from the Louisiana-Texas border to 1 or 1.5×10^{-3} seaward from the Texas coast. By comparing the very gentle and nearly equivalent gradients of the modern floodplains and the gradients of the continental shelf, it can be seen that the gradients of the glacial-age extended consequent rivers were only moderately increased, and valley entrenchment was only one-quarter to one-half the possible emergence. Some rivers even may have been forced to aggrade their extended segments across the exposed shelf, especially when climatic and vegetational changes in their catchments resulted in an increased proportion of coarse alluvium.

The extent of the Late Wisconsin maximum sea-level lowering on the Gulf of Mexico continental shelf has not been established. Fisk and McFarlan (1955) estimate a lowering of 140 m on the basis of their adjusted projection of the buried and downwarped Mississippi Trench. Curray (1960: 258) proposes that the maximum lowering

was only about 120 m on the basis of the depth of the shelf break in the northwestern Gulf of Mexico, far from the subsiding area of the Mississippi Delta plain. Depositional and erosional features at about −120 m support his inference, although the deepest samples submitted for dating by the radiocarbon method were nearshore or shelf mollusks that lived in depths as great as 18 m. Their radiocarbon age is about 17,000 yr B.P. and they were collected at a depth of −88 m., and so a sea-level lowering of only 70 m is the minimum that can be hypothesized. Buried river valleys cross the present Texas coast at known depths of −39 m (Sabine River) (Nelson and Bray, 1970: 55) and more than −30 m (Colorado River) (Wilkinson and Basse, 1978: 1594). The Appalachicola Delta region shows a record of Holocene valley filling of only about 23 m (Schnable and Goodell, 1968). As noted in the preceding paragraph, these valleys were so extended across the emerged Gulf of Mexico continental shelf that their deepest entrenchment was much less than the actual lowering of sea level. Nevertheless, their profiles suggest that sea-level lowering was not as great as had been proposed in earlier publications, and perhaps −80 m is the lowest estimate that can be confidently supported at present.

The exposed Gulf of Mexico shelf sediments were subaerially weathered for most of the Wisconsin glacial stage, so a distinctive yellow, mottled clay horizon or soil marks the base of the Holocene transgression. This is the Prairie or Beaumont surface or soil, which rises inland above present sea level and the Holocene deposits of the Gulf Coastal Plain. Progressive subsidence of the Gulf Coast geosyncline is hypothesized to have been accompanied by an upward flexure inland of a "hinge line" near the present Texas-Louisiana

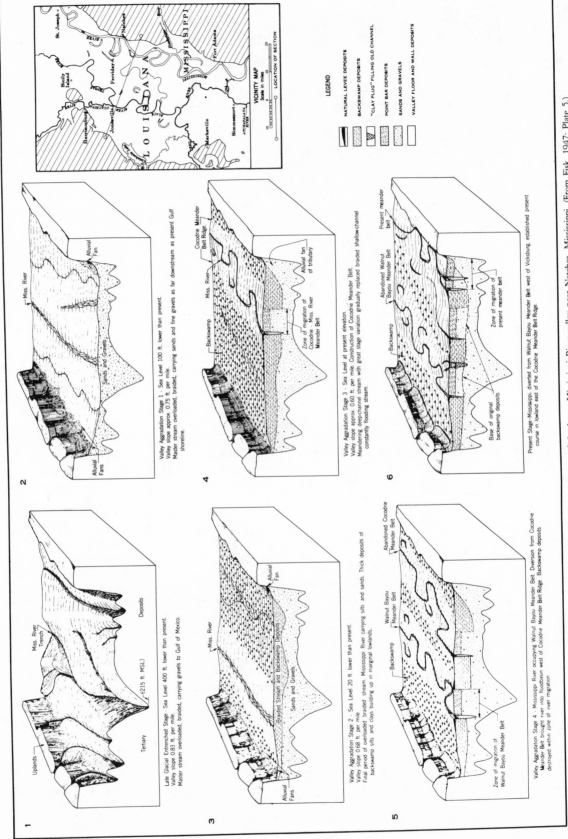

Figure 11-7. Stages in the late-glacial and postglacial development of the lower Mississippi River valley near Natchez, Mississippi. (From Fisk, 1947: Plate 5.)

coastline. The warped surface rises inland with a gradient of about 4×10^{-4} until it intersects older and more strongly tilted terraces of greater Pleistocene age (Bernard et al., 1962).

The Prairie-Beaumont surface shows evidence of warm subtropical weathering in a climate similar to that found in the region today. In Texas, the weathered horizon includes caliche nodules and layers; again, this implies a climate not much different from today's. If sea level was as low as the shelf break at -120 m, most rivers would have delivered their sediments directly to the head of the continental slope through river valleys that were only moderately incised across the otherwise low-relief extended Coastal Plain. As sea level rose during late-glacial time, or if it did not sink as low as formerly believed, active coastal processes would have moved sediments along the shelf shorelines. Winds, currents, and littoral drift were the same during full-glacial time as now (Curray, 1960: 259). Mississippi River sediment was transported far to the west, and Rio Grande sediment was transported northward. During the marine transgression, river valleys were probably modified by the retreat of estuaries along their axes, as suggested for the morphological evolution of the Atlantic continental shelf. Changes in sediment-transport directions along the shoreline during the transgression have been documented (Curray, 1960: 259-63). Barrier-island systems must have formed, moved landward, and either been overtopped or incorporated into the modern barriers during the transgression. Knolls or ''reefs'' of mollusks, corals, and algae grew on the outer edge of the Gulf of Mexico continental shelf during the period of low sea level. Their loci may have been rising salt diapirs.

There is no know evidence that paleo-Indians occupied the exposed Gulf of Mexico continental shelf. We can assume that the climate of the area was cooler than it is now and perhaps somewhat drier. The geomorphic apprearance of the region would have been much the same as it is along the Gulf Coast today, with perhaps somewhat greater erosional relief, better drainage, and fewer coastal swamps. Low, flat interfluves would have separated broad river lowlands, with regional relief of about 30 m or less.

PACIFIC COAST

The active continental margin along the Pacific coast of the United States is a total contrast to the trailing continental margin with coastal plains of the Atlantic and Gulf coasts. Active tectonics associated with regional strike-slip faulting characterize the California coast. Farther north, postglacial isostatic movements related to the Puget lobe of the Cordilleran ice sheet add additional complexity. In general, a continental shelf is absent or trivial along the Pacific coast, so a full-glacial lowering of sea level of any reasonable amount would have caused a seaward regression of the shoreline of only a few kilometers. The few islands off southern California were more exposed during times of low sea level, but none is positively known to have been connected to the mainland. No major changes in the coastal outline of California occurred as a result of the glacial lowering of sea level except for the drainage of San Francisco Bay and a few smaller bays. Longshore transport may have intensified, however.

During late-Cenozoic time, tectonic uplift of 0.1 to 1.0 mm per year has been in progress along much of the California coast (Lajoie et al., 1979). Where right-lateral strike slip dominates, as along the great San Andreas Fault system, the vertical uplift of the coast is not great. However, where east-west transverse structures cross the general north-northwest trend of the San Andreas system, rates of vertical uplift increase. In the region of Ventura, California, average rates of late-Cenozoic uplift may approach 10 mm per year (Lajoie et al., 1979).

The mountainous coast and offshore islands of California have conspicuous sets of emerged marine terraces. Most are erosional rather than constructional. The thin layer of marine strata on some terraces is invariably covered by alluvium and colluvium, so that good exposures are rare and marine fossils are even less common. Some progress in interpreting the age of the California terraces has been made with radiometric and amino acid dating methods, but no regional correlation has been achieved. The lowest terraces, where best dated, give ages in the range of 125,000 to 80,000 yr B.P. (Kern, 1977; Ku and Kern, 1974; Lajoie et al., 1979; Wehmiller and Belknap, 1978). Holocene terraces have emerged as much as 40 m in areas of strongest uplift.

In spite of general late-Cenozoic tectonic uplift, the last 15,000 years of California coastal evolution have been dominated by submergence coincident with deglaciation. High-gradient, mostly intermittent rivers that reach the arid southern California coast now end in saline lagoons that break through barrier beaches only during floods. During most of the year, the strong south-trending currents and high levels of wave energy keep the inlets closed. However, borings through the lagoonal and floodplain sediments show at least one episode of erosion to a sea level probably 90 m lower than the present level and subsequent aggradation (Upson, 1949). This cycle clearly postdates the emerged terraces of last-interglacial age. Where the valley fills have been dated, as in San Francisco Bay, the deposits to depths of 40 m below present sea level have radiocarbon ages of about 9000 yr B.P. or younger (Atwater et al., 1977).

The Puget-Fraser lowland between Seattle, Washington, and Vancouver, British Columbia, has had a complex late-Quaternary history involving glacier advance and retreat, sea-level change, crustal isostatic response, and active tectonism. The lowland was deglaciated but colder than present from about 30,000 to 20,000 yr B.P. (the Olympia interval) (Easterbrook, 1969). During the last glaciation, a large ice lobe entered the lowland from the mountain ice sheet north of the Canadian border. The lobe of Canadian ice, known as the Puget lobe, was about 1800 m thick at the international boundary and 1000 m thick at Seattle. It probably caused rapid but local isostatic adjustments. It reached its maximum expansion at about 15,000 to 14,000 yr B.P. During deglaciation, seawater entered the depressed lowland, and so the late-glacial deposits are a complex of glacial and marine sediments. By about 13,000 yr B.P., the lowland was largely ice free but extensively submerged. By 10,000 yr B.P., postglacial uplift had decanted marine waters into the present Puget Sound waterways and the Holocene phase of weathering, soil formation, and stream incision had begun. The late-glacial marine deposits are now found to an altitude of at least 200 m (Easterbrook, 1969). In the Puget lowland and along the coast to the north in British Columbia and Alaska, the emerged late-glacial marine clay may be thixotropic and subject to catastrophic landslides.

BERINGIAN COAST

Beringia, including the presently submerged continental shelf between Alaska and Siberia, is of great interest to American and Soviet Quaternary scientists because it is the most likely route for the immigration of the aboriginal population to the Americas. Since the international symposium on the subject at the Seventh Congress of the

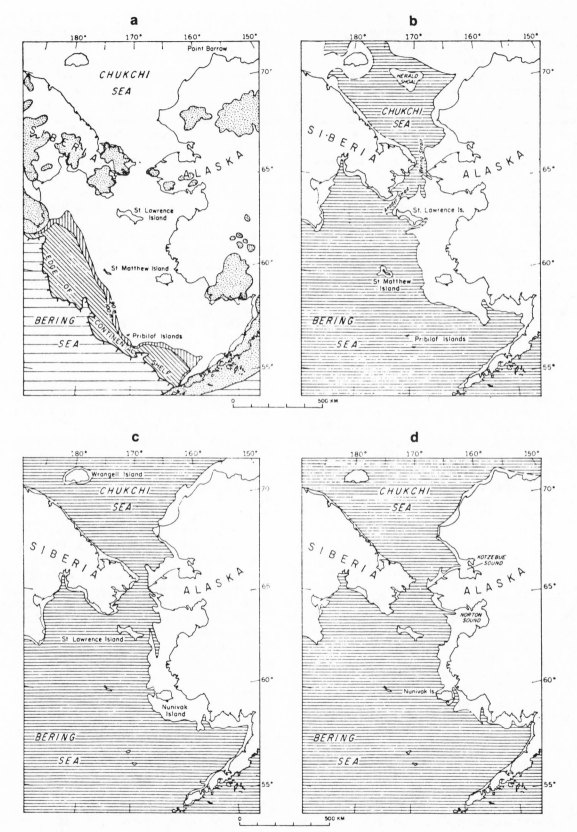

Figure 11-8. Paleogeography of the Bering Sea region during the maximum Wisconsin Glaciation and through late-glacial time: (A) full-glacial time, sea level between −90 to −100 m; (B) sea level near −38 m; (C) sea level near −30 m; (D) sea level near −20 m. (From Hopkins, 1973: Figures 2-5.)

International Association for Quaternary Research in 1965 and the subsequent publication of the symposium papers (Hopkins, 1967), the steady flow of new reports has continued. It seems to be the consensus that full-glacial sea level did not fall below − 120 m in the region and that a depth range of − 90 to − 100 m is more likely to mark the maximum Wisconsin sea-level lowering (Figure 11-8). During the full-glacial interval, most of Beringia was a monotonous plain emerged no more than about 50 m, above which were scattered a few hills such as the present Pribilof Islands (Colinvaux, 1981).

Earlier literature supported the idea of a Middle Wisconsin sea level on the Alaskan coast that rose to within a few meters of the present level. Named the Woronzofian transgression (Hopkins, 1967: 79), it was believed to date within the range of 35,000 to 25,000 yr B.P. Subsequently, the type section of glaciomarine Bootlegger Cove Clay that documents the Woronzofian transgression has been shown to be only about 14,000 years old and to have been deposited during the late-glacial marine transgression (Hopkins, 1973: 523). Nevertheless, emerged marine sediments on the Arctic Coastal Plain near Point Barrow are still referred to the hypothesized Middle Wisconsin transgression, although their present altitude of up to + 7 m is attributed to tectonism. A series of radiocarbon dates from a sedimentary formation informally called the Barrow unit range from more than 44,000 to 25,300 yr B.P., including four finite ages and four beyond the range of the method. On the basis of three finite ages ranging between 37,000 and 31,400 yr B.P., the marine unit is assigned a Middle Wisconsin age (Sellman and Brown, 1973: 178). The dates were obtained from detrital organic bands in sandy marine beds that have been extensively reworked in their upper portions. The wide range of dates, the unusually large counting errors associated with them, and the lack of interlaboratory agreement on one pair of samples all suggest that the Middle Wisconsin age should be questioned until better documentation is provided. When the dates are questioned, the inferred late-Pleistocene tectonic uplift of the Point Barrow region is also open to debate.

Even if the Middle Wisconsin sea level rose to only − 38 or − 41 m, as inferred on Figure 11-2, the Bering Land Bridge would have been broken, for the limiting depth of the Anadyr Strait is − 48 m. The uncertainties continue to be of great interest because of the questions they raise about the timing of the immigration of humans into the New World from Siberia.

The climate of Beringia during full-glacial time was of a cold and dry continental type, even on the southern coast of the land bridge. Colinvaux (1981: 33) describes the landscape as "herb tundra, . . . bare ground, perhaps loess fields or frost boils, . . . a remarkable homogeneity over the spread from north coast to south, a distance of some 1500 km." Only a few dwarf willows and possibly some birches could have been found on the essentially treeless tundra. The environment was similar to that of the Arctic Coast around Point Barrow today, without evidence of the maritime climate and forests that might have been expected along the southern coast of the region. Furthermore, St. Paul Island had been isolated from the mainland by the late-glacial rise of sea level by 13,000 yr B.P., but maritime forest vegetation did not reach the island until 11,000 yr B.P.; this demonstrates that the succession from tundra to forest was not the result of the maritime influence of rising sea level but of a more fundamental climatic change (Colinvaux, 1981: 35).

The late-glacial flooding of Beringia was rapid because of the low relief of the region and the generally low altitude. From a full-glacial

low sea level of − 90 to − 100 m sometime before 17,000 yr B.P., sea level had risen to about − 70 m by 17,000 to 16,000 yr B.P., to − 38 m by shortly after 13,000 yr B.P., and to − 30 m by about 11,800 yr B.P. Freshwater peat was still accumulating at a depth of − 20 m as recently as 10,000 yr B.P., but present sea level had been reached by 5000 yr B.P. (Figure 11-8). Several pauses or regressions in the rise have been inferred from drowned beach ridges and other shoreline forms. Submerged shorelines at − 38 m, − 30 m, − 24 m to − 20 m, and − 12 to − 10 m are notable (Hopkins, 1973). The land connection between Siberia and Alaska was first broken at the Anadyr Strait and through the Bering Strait at present depths of − 46 to − 48 m (Creager and McManus, 1967: 13).

TROPICAL ISLANDS

Hawaii, Puerto Rico, and several territories of the United States (the Virgin Islands, Guam, and American Samoa) are all tropical, and so coral-reef growth is an integral feature of their Pleistocene sea-level history. Some representative recent descriptions include those of Adey (1975), Emery (1962), and Tracey and others (1964). Over most of the coral-reef regions of the tropical oceans, the vigorous growth of Holocene coral reefs has mantled and obscured the older erosional and depositional surfaces on which they grew. Therefore, the lowering of sea level during the Late Wisconsin glacial maximum is not well known. Shepard (1970) describes closed basins on the floor of Truk lagoon (Federated States of Micronesia) that he infers to be dolines formed by karst solution during times of low sea level. Their depth exceeds 73 m, a suggested minimum for the corresponding sea-level drop. No evidence is available that can relate that sea level to the last glacial maximum, although both the sharpness of the drowned topography and the thin sedimentary cover suggest that the latest sea-level minimum was at least partially responsible for the karst episode. By 7000 yr B.P., sea level had risen to at least − 10 m on Oahu, Hawaii (Easton and Olson, 1976). Similar results have been established from bore holes in some Micronesian atolls. Corals to a depth of − 11 m at Eniwetak atoll are less than 6000 years old, but below a depth of − 14 m they are 120,000 years old (Thurber et al., 1965). In general, the late-glacial rise of sea level flooded a karst landscape, and coral growth was reestablished on a coral limestone substrate early in Holocene time. Since then, fringing and barrier reefs have been building upward and outward over antecedent karst platforms.

References

Adey, W. H. (1975). "Algal Ridges and Coral Reefs of St. Croix." Atoll Research Bulletin 187.

Atwater, B. F., Hedel, C. W., and Helley, E. J. (1977). "Late Quaternary Depositional History, Holocene Sea-Level Changes and Vertical Crustal Movement, Southern San Francisco Bay, California." U.S. Geological Survey Professional Paper 1014.

Belknap, D. F., and Kraft, J. C. (1977). Holocene relative sea-level changes and coastal stratigraphic units on the northwest flank of the Baltimore Canyon trough geosyncline. *Journal of Sedimentary Petrology* 47, 610-29.

Belknap, D. F., and Wehmiller, J. F. (1980). Amino acid racemization in Quaternary mollusks: Examples Delaware, Maryland and Virginia. *In* "Biogeochemistry of Amino Acids" (P. E. Hare, T. C. Hoering, and K. King, Jr., eds.), pp. 401-14. John Wiley and Sons, New York.

Bergantino, R. N. (1971). Submarine regional geomorphology of the Gulf of Mexico. *Geological Society of America Bulletin* 82, 741-52.

Bernard, H. A., LeBlanc, R. J., and Major, C. F. (1962). Recent and Pleistocene geology of southeast Texas. *In* "Geology of the Gulf Coast and Central Texas and Guidebook of Excursions" (E. H. Rainwater and R. P. Zingula, eds.), pp. 175-224. Houston Geological Society, Houston.

Blackwelder, B. W. (1981). Reply (to Clark on "Late Wisconsin and Holocene tectonic stability of the United States mid-Atlantic coastal region"). *Geology* 9, 439.

Blackwelder, B. W., Pilkey, O. H., and Howard, J. D. (1979). Late Wisconsinan sea levels on the southeast U.S. Atlantic shelf based on in-place shoreline indicators. *Science* 204, 618-20.

Bloom, A. L., Broecker, W. S., Chappell, J. M. A., Matthews, R. K., and Mesolella, K. J. (1974). Quaternary sea level fluctuations on a tectonic coast: New ^{230}Th/^{234}U dates from the Huon Peninsula, New Guinea. *Quaternary Research* 4, 185-205.

Broecker, W. S. (1965). Isotope geochemistry and the Pleistocene climatic record. *In* "The Quaternary of the United States" (H. E. Wright, Jr., and D. G. Frey, eds.), pp. 737-53. Princeton University Press, Princeton, N.J.

Byers, D. S. (1959), Radiocarbon dates for the Bull Brook site, Massachusetts. *American Antiquity* 24, 427-29.

Chappell, J., and Veeh, H. H. (1978). ^{230}Th/^{234}U age support of an interstadial sea level of -40 m at 30,000 yr B.P. *Nature* 276, 602-3.

Clark, J. A. (1981). Comment (on "Late Wisconsin and Holocene tectonic stability of the United States mid-Atlantic coastal region"). *Geology* 9, 438.

Clark, J. A., Farrell, W. E., and Peltier, W. R. (1978). Global changes in postglacial sea level: A numerical calculation. *Quaternary Research* 9, 265-87.

Clark, J. A., and Lingle, C. S. (1979). Predicted relative sea-level changes (18,000 years B.P. to present) caused by late-glacial retreat of the Antarctic ice sheet. *Quaternary Research* 11, 279-98.

Clausen, C. J., Cohen, A. D., Emiliani, C., Holman, J. A., and Stipp, J. J. (1979). Little Salt Spring, Florida: A unique underwater site. *Science* 203, 609-14.

Colinvaux, P. (1981). Historical ecology in Beringia: The south land bridge coast at St. Paul Island. *Quaternary Research* 16, 18-36.

Creager, J. S., and McManus, D. A. (1967). Geology of the floor of Bering and Chukchi Seas: American studies. *In* "The Bering Land Bridge" (D. M. Hopkins, ed.), pp. 7-31. Stanford University Press, Stanford, Calif.

Cronin, T. M. (1981). Rates and possible causes of neotectonic vertical crustal movements of the emerged southeastern United States Atlantic Coastal Plain. *Geological Society of America Bulletin* 92, 812-33.

Cronin, T. M., Szabo, B. J., Ager, T. A., Hazel, J. E., and Owens, J. P. (1981). Quaternary climates and sea levels of the U.S. Atlantic Coastal Plain. *Science* 211, 233-40.

Curray, J. R. (1960). Sediments and history of Holocene transgression, continental shelf, northwest Gulf of Mexico. *In* "Recent Sediments, Northwest Gulf of Mexico" (F. P. Shepard, F. B. Phleger, and Tj. H. van Andel, eds.), pp. 221-66. American Association of Petroleum Geologists, Tulsa, Okla.

Curray, J. R. (1965). Late Quaternary history, continental shelves of the United States. *In* "The Quaternary of the United States" (H. E. Wright, Jr., and D. G. Frey, eds.), pp. 723-35. Princeton University Press, Princeton, N.J.

Dillon, W. P., and Oldale, R. N. (1978). Late Quaternary sea-level curve: Reinterpretation based on glaciotectonic influence. *Geology* 6, 56-60.

Dreimanis, A., and Raukas, A. (1975), Did Middle Wisconsin, Middle Weichselian, and their equivalents represent an interglacial or an interstadial complex in the Northern Hemisphere? *In* "Quaternary Studies: Selected Papers from IX INQUA Congress" (R. P. Suggate and M. M. Cresswell, eds.), pp. 109-20. Royal Society of New Zealand Bulletin 13.

Easterbrook, D. J. (1969). Pleistocene chronology of the Puget lowland and San Juan Islands, Washington. *Geological Society of America Bulletin* 80, 2273-86.

Easton, W. H., and Olson, E. A. (1976). Radiocarbon profile of Hanauma reef, Oahu, Hawaii. *Geological Society of America Bulletin* 87, 711-19.

Emery, K. O. (1962). "Marine Geology of Guam." U.S. Geological Survey Professional Paper 403B.

Emery, K. O. (1968). Relict sediments on continental shelves of the world. *American Association of Petroleum Geologists Bulletin* 52, 445-64.

Emery, K. O., and Merrill, A. S. (1979). Relict oysters on the United States Atlantic continental shelf: A reconsideration of their usefulness in understanding late Quaternary sea-level history: Discussion. *Geological Society of America Bulletin* 90, 689-92.

Field, M. E. (1974). Buried strandline deposits on the central Florida inner continental shelf. *Geological Society of America Bulletin* 85, 57-60.

Field, M. E., Meisburger, E. P., Stanley, E. A., and Williams, S. J. (1979). Upper Quaternary peat deposits on the Atlantic inner shelf of the United States. *Geological Society of America Bulletin* 90, 618-28.

Fisk, H. N. (1947). "Fine-Grained Alluvial Deposits and Their Effects on Mississippi River Activity." vols. 1 and 2. U.S. Army Corps of Engineers, Mississippi River Commission, Vicksburg, Miss.

Fisk, H. N., and McFarlan, E., Jr. (1955). Late Quaternary deltaic deposits of the Mississippi River. *In* "Crust of the Earth" (A. Poldervaart, ed.), pp. 279-302. Geological Society of America Special Paper 62.

Flint, R. F. (1971). "Glacial and Quaternary Geology." John Wiley and Sons, New York.

Hand, B. M. (1975). Paleohydraulics of catastrophic discharge from proglacial lake near Syracuse, New York. *Geological Society of America Abstracts with Programs* 7, 69-70.

Hopkins, D. M. (ed.) (1967). "The Bering Land Bridge." Stanford University Press, Stanford, Calif.

Hopkins, D. M. (1973). Sea-level history in Beringia during the past 250,000 years. *Quaternary Research* 3, 520-40.

Hoyt, J. H., and Hails, J. R. (1974). Pleistocene stratigraphy of southeastern Georgia. *In* "Post-Miocene Stratigraphy, Central and Southern Atlantic Coastal Plain" (R. Q. Oaks and J. R. DuBar, eds.), pp. 191-205. Utah State University Press, Logan.

Hoyt, J. H., Henry, V. J., and Weimer, R. J. (1968). Age of late-Pleistocene shoreline deposits, coastal Georgia. *In* "Means of Correlation of Quaternary Successions: VII INQUA Congress proceedings" (R. B. Morrison and H. E. Wright, Jr., eds.), vol. 8, pp. 381-93. University of Utah Press, Salt Lake City.

Hughes, T. J., Denton, G. H., Andersen, B. G., Schilling, D. H., Fastook, J. L., and Lingle, C. S. (1981). The last great ice sheets: A global view. *In* "Last Great Ice Sheets" (G. H. Denton and T. J. Hughes, eds.), pp. 263-317. John Wiley and Sons, New York.

Kaye, C. A. (1960). "Surficial Geology of the Kingston Quadrangle, Rhode Island." U.S. Geological Survey Bulletin 1071-I, pp. 341-96.

Kaye, C. A. (1964). "Outline of Pleistocene Geology of Martha's Vineyard, Massachusetts." U.S. Geological Survey Professional Paper 501-C, pp. 134-39.

Kaye, C. A., and Barghoorn, E. S. (1964). Late Quaternary sea-level change and crustal rise at Boston, Massachusetts, with notes on the autocompaction of peat. *Geological Society of America Bulletin* 75, 63-80.

Kern, J. P. (1977). Origin and history of upper Pleistocene marine terraces, San Diego, California. *Geological Society of America Bulletin* 88, 1553-66.

Ku, T.-L., and Kern, J. P. (1974). Uranium-series age of the upper Pleistocene Nestor Terrace, San Diego, California. *Geological Society of America Bulletin* 85, 1713-16.

Lajoie, K. R., Kern, J. P., Wehmiller, J. F., Kennedy, G. L., Mathieson, S. A., Sarna-Wojcicki, A. M., Yerkes, R. F., and McCrory, P. F. (1979). Quaternary marine shorelines and crustal deformation, San Diego to Santa Barbara, California. *In* "Geological Excursions in the Southern California Area" (P.L. Abbott, ed.), pp. 1-15. Department of Geological Sciences, San Diego State University, San Diego, Calif.

Macintyre, I. G., Blackwelder, B. W., Land, L. S., and Stuckenrath, R. (1975). North Carolina shelf-edge sandstone: Age, environment or origin, and relationship to pre-existing sea levels. *Geological Society of America Bulletin* 86, 1073-78.

Macintyre, I. G., and Milliman, J. D. (1970). Physiographic features on the outer shelf and upper continental slope, Atlantic continental margin, southeastern United States. *Geological Society of America Bulletin* 81, 2577-97.

Macintyre, I. G., Pilkey, O. H., and Stuckenrath, R., (1978). Relict oysters on the United States Atlantic continental shelf: A reconsideration of their usefulness in understanding late Quaternary sea-level history. *Geological Society of America Bulletin* 89, 277-82.

Macintyre, I. G., Pilkey, O. H., and Stuckenrath, R. (1979). Relict oysters on the United States Atlantic continental shelf: A reconsideration of their usefulness in understanding late Quaternary sea-level history: Reply. *Geological Society of America Bulletin* 90, 692-94.

Maclaren, C. (1842). The glacial theory of Professor Agassiz. *American Journal of Science*, series 1, 42, 346-65.

Matthews, R. K. (1973). Relative elevation of late Pleistocene high sea level stands: Barbados uplift rates and their implications. *Quaternary Research* 3, 147-53.

Mesolella, K. J., Matthews, R. K., Broecker, W. S., and Thurber, D. L. (1969). Astronomical theory of climatic change: Barbados data. *Journal of Geology* 77, 250-74.

Milliman, J. D., and Emery, K. O. (1968). Sea levels during the past 35,000 years. *Science* 162, 1121-23.

Morgan, J. P. (1970). Deltas: A résumé. *Journal of Geological Education* 18, 107-17.

Nelson, H. F., and Bray, E. E. (1970). Stratigraphy and history of the Holocene sediments in the Sabine-High Island area, Gulf of Mexico. *In* "Deltaic Sedimentation, Modern and Ancient" (J. P. Morgan and R. J. Shaver, eds.), pp. 48-77. Society of Economic Paleontologists and Mineralogists Special Publication 15.

Oaks, R. Q., Coch, N. K., Sanders, J. E., and Flint, R. F. (1974). Post-Miocene shorelines and sea levels, southeastern Virginia. *In* "Post-Miocene Stratigraphy, Central and Southern Atlantic Coastal Plain" (R. Q. Oaks and J. R. DuBar, eds.), pp. 53-87. Utah State University Press, Logan.

Oaks, R. Q., and DuBar, J. R. (1974). "Post-Miocene Stratigraphy, Central and Southern Atlantic Coastal Plain." Utah State University Press, Logan.

Parker, R. H. (1960). Ecology and distributional patterns of marine macro-invertebrates, northern Gulf of Mexico. *In* "Recent Sediments, Northwest Gulf of Mexico" (F. P. Shepard, F. B. Phleger, and Tj. H. van Andel, eds.), pp. 302-37. American Association of Petroleum Geologists, Tulsa, Okla.

Peltier, W. R., and Andrews, J. T. (1976). Glacial-isostatic adjustment: I. The forward problem. *Geophysics Journal of the Royal Astronomical Society* 46, 605-46.

Pratt, R. M., and Schlee, J. (1969). Glaciation on the continental margin off New England. *Geological Society of America Bulletin* 80, 2335-41.

Schnable, J. E., and Goodell, H. G. (1968). "Pleistocene-Recent Stratigraphy, Evolution, and Development of the Appalachicola Coast, Florida." Geological Society of America Special Paper 112.

Sellmann, P. V. and Brown, J. (1973). Stratigraphy and diagenesis of perennially frozen sediments in the Barrow, Alaska, region. *In* "North American Contribution, Permafrost, Second International Conference," pp. 171-81. National Academy of Sciences, Washington, D.C.

Shackleton, N. J. (1977). Oxygen-isotope stratigraphic record of the late Pleistocene. *Philosophical Transactions of the Royal Society of London*, series B, 280, 169-82.

Shackleton, N. J., and Opdyke, N. D. (1973). Oxygen isotope and paleomagnetic stratigraphy of equatorial Pacific core V28-238: Oxygen isotope temperatures and ice volumes on a 10^5 year and 10^6 year scale. *Quaternary Research* 3, 39-55.

Shepard, F. P. (1970). Lagoonal topography of Caroline and Marshall Islands. *Geological Society of America Bulletin* 81, 1905-14.

Shepard, F. P. (1973). "Submarine Geology." 3rd ed. Harper and Row, New York.

Southard, J. B., and Stanley, D. J. (1976). Shelf-break processes and sedimentation. *In* "Marine Sediment Transport and Environmental Management" (D. J. Stanley and D. J. P. Swift, eds.), pp. 351-77. John Wiley and Sons, New York.

Stuiver, M., and Borns, H. W., Jr. (1975). Late Quaternary marine invasion in Maine: Its chronology and associated crustal movement. *Geological Society of America Bulletin* 86, 99-103.

Swift, D. J. P. (1973). Delaware shelf valley: Estuary retreat path, not drowned river valley. *Geological Society of America Bulletin* 84, 2743-48.

Swift, D. J. P., Moir, R., and Freeland, G. L. (1980). Quaternary rivers on the New Jersey shelf: Relation of seafloor to buried valleys. *Geology* 8, 276-80.

Thom, B. G. (1973). Dilemma of high interstadial sea levels during the last glaciation. *Progress in Geography* 5, 167-246.

Thurber, D. L., Broecker, W. S., Blanchard, R. L., and Potratz, H. A. (1965). Uranium-series ages of Pacific atoll coral. *Science* 149, 55-58.

Tracey, J. I., Jr., Schlanger, S. O., Stark, J. T., Doan, D. B., and May, H. G. (1964). "General Geology of Guam." U.S. Geological Survey Professional Paper 403A.

Trautman, M. A., and Willis, E. H. (1966). Isotopes, Inc. radiocarbon measurements V. *Radiocarbon* 8, 161-203.

Uchupi, E. (1968). "Atlantic Continental Shelf and Slope of the United States: Physiography." U.S. Geological Survey Professional Paper 529-C.

Upson, J. E. (1949). Late Pleistocene and recent changes of sea level along the coast of Santa Barbara County, California. *American Journal of Science* 247, 94-115.

Veatch, A. C., and Smith, P. A. (1939). "Atlantic Submarine Valleys of the United States and the Congo Submarine Valley." Geological Society of America Special Paper 7.

Watts, W. A. (1969). A pollen diagram from Mud Lake, Marion County, north-central Florida. *Geological Society of America Bulletin* 80, 631-42.

Watts, W. A., and Stuiver, M. (1980). Late Wisconsin climate of northern Florida and the origin of species-rich deciduous forest. *Science* 210, 325-27.

Wehmiller, J. F., and Belknap, D. F. (1978). Alternative kinetic models for the interpretation of amino acid enantiomeric ratios in Pleistocene mollusks: Examples from California, Washington, and Florida. *Quaternary Research* 9, 330-48.

Whitmore, F. C., Jr., Emery, K. O., Cooke, H. B. S., and Swift, D. J. P. (1967). Elephant teeth from the Atlantic continental shelf. *Science* 156, 1477-81.

Wilkinson, B. H., and Basse, R. A. (1978). Late Holocene history of the central Texas coast from Galveston Island to Pass Cavallo. *Geological Society of America Bulletin* 89, 1592-1600.

Zenkovich, V. P. (1967). "Processes of Coastal Development" (D. G. Fry, trans.). John Wiley and Sons, New York.

CHAPTER 12

The Ocean around North America at the Last Glacial Maximum

John Imbrie, Andrew McIntyre, and T. C. Moore, Jr.

Introduction

Modern terrestrial climates are significantly influenced by the distribution of continental and maritime air masses. Over much of the Soviet Union, for example, the climate at all seasons is dominated by continental air masses originating in polar or tropical regions. In contrast, the climate of North America, particularly the climate of the United States, is significantly influenced by air masses of maritime origin.

Because the ocean exerts such an important control on modern climate, it is reasonable to assume that a knowledge of the ocean during the last ice age would improve our understanding of ice-age climates. Toward this end, members of the CLIMAP project have used micropaleontologic evidence to reconstruct the surface characteristics of the global ocean during the last glacial maximum, about 18,000 years ago. The results of that global reconstruction and the methods on which it is based are fully described elsewhere (Brunner and Cooley, 1976; CLIMAP Project Members, 1976, 1981; Imbrie and Kipp, 1971; McIntyre et al., 1976; Moore et al., 1980, 1981; Prell et al., 1976). The objectives of this chapter are (1) to focus attention on one part of this reconstruction, namely, the ocean around North America and (2) to draw inferences from the oceanic pattern about the ice-age climate of North America.

In order to achieve these objectives, we examine maps of fossil plankton that reflect the distribution of water masses in the modern ocean and the ocean at the last glacial maximum. These maps summarize the microfossil data that were used by CLIMAP investigators to infer the surface temperatures of the ice-age ocean. After discussing the biotic maps, we present maps showing CLIMAP's reconstruction of sea-surface temperatures for the ice-age ocean. Next, we examine anomaly maps that highlight temperature differences between the modern and ice-age oceans and present temperature transects that show the nature of nearshore oceanic changes. Finally, on the basis of these oceanic changes we make inferences about the terrestrial climate of North America during the last glacial maximum.

Biotic Patterns

CLIMAP's study of microfossils from the modern seabed shows that the modern ocean can be divided into five different regions, each dominated by a particular assemblage of planktonic organisms (Figure 12-1A). These are called the polar, subpolar, transitional, western tropical, and eastern tropical assemblages (CLIMAP Project Members, 1981; Moore et al., 1981). Each is defined by an objective technique, the varimax rotation of Q-mode eigenvectors (Imbrie and Kipp, 1971), and each appears to be related to a particular combination of near-surface physical and chemical properties that defines a particular water mass (Moore et al., 1981). Strong gradients in these properties occur near the boundaries between adjacent assemblage regions. In the modern ocean, there is a conspicuous asymmetry between the east and west coasts of North America, with polar faunas extending much farther south along the eastern continental margin. In addition, the biotic gradients are much stronger along the east coast than they are along the west coast. These observations of planktonic distributions are clearly a reflection of the oceanographic differences between the strong western boundary currents in the Atlantic (northward-flowing Gulf Stream and southward-flowing Labrador Current) and the weaker flows of the eastern boundary currents of the Pacific.

Figure 12-1B shows the distribution of the same biotic assemblages in the ice-age ocean. A comparison of the two faunal dominance maps can be directly interpreted in terms of changes in the areal distribution of water masses between modern and ice-age times. One conspicuous feature of the ice-age map is the considerable increase in the area dominated by the polar assemblage in both oceans, but particularly in the eastern side of the Atlantic Ocean. Another feature is the increased biotic gradients in the middle-latitude region of both oceans. Again, the steepening of the gradients is particularly striking in the Atlantic. The areal coverage of the eastern tropical assemblage expanded its area of dominance during ice-age times, primarily at the expense of the western tropical assemblage in the Pacific Ocean. However, the total area covered by the two tropical assemblages exhibited relatively little change.

Having made broad comparisons between Figures 12-1A and 12-1B, we can now turn our attention to specific features of the two maps. On the basis of the faunal distributions, the surface waters in

All the authors were members of the CLIMAP project.

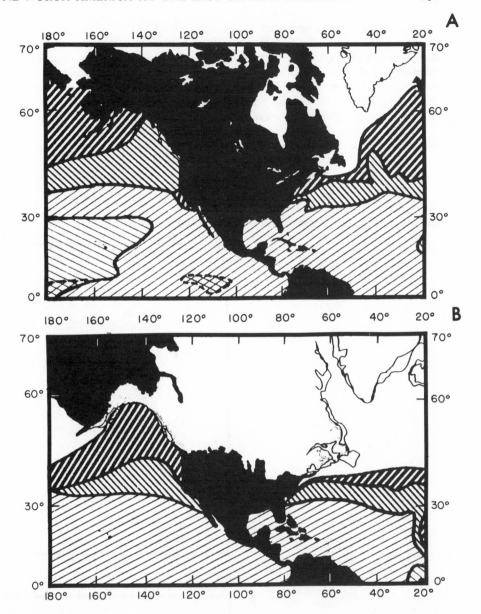

Figure 12-1. (A) Areas of dominance for planktonic faunas in the modern oceans as indicated by the distribution of microfossil remains in surface sediments. The distribution for the Atlantic Ocean is based on Foraminifera; the distribution for the Pacific Ocean is based on Radiolaria. (B) Areas of dominance for planktonic faunas during the ice age. Faunas are defined in surface sediment samples, but their distribution in ice-age (18,000-year-old) samples is used to construct this map. The distribution of the sample control is shown on Figure 11-3. Continental outlines reflect an assumed 150 m lowering of sea level. The area of dominance is defined by regions containing samples in which the assemblage (factor) has a value of 0.6 or higher. Regions dominated by polar assemblages are shown in white. The areas dominated by subpolar assemblages have heavy diagonal hatching; the transitional-subtropical regions have intermediate hatching; the Atlantic and eastern Pacific tropical areas of dominance have light right-diagonal hatching; the western Pacific tropical areas have light, left-diagonal hatching. Parts of the tropical Pacific contain samples in which both the western and eastern tropical assemblages have values greater than 0.6. These areas are shown by cross-hatching. Continental areas that contain large glaciers are shown in white. (From, Moore et al., 1981, used with the permission of the Geological Society of America.)

contact with North America at the last glacial maximum were sharply partitioned along the 40th parallel. North of this dividing line, colder subpolar conditions existed in the Pacific, and polar waters were present in the Atlantic. Today in the Pacific, subpolar assemblages lie offshore of North America from latitude 68°N to 59°N, transitional assemblages from latitude 58°N to 32°N, and subtropical-tropical assemblages south of latitude 32°N. At the last glacial maximum, the polar assemblage expanded southeastward into the Gulf of Alaska at latitude 57°N. The subpolar and transitional assemblages were compressed, lying offshore from latitude 57°N to 38°N and 38°N to 28°N, respectively. These changes in the biotic assemblages indicate the presence of colder surfaces waters in the eastern boundary current during the last glacial maximum.

In the open Atlantic Ocean, the biotic assemblages that define surface-water masses show even greater changes that in the Pacific between glacial and interglacial climates. But, along the eastern shore of North America, the glacial-interglacial difference is actually less pronounced than that found along the Pacific coast. The southern

boundaries of the polar, subpolar, and transitional assemblages are shifted southward from latitude 44° to 36°, 36° to 35°, and 35° to 30°, respectively. This relatively minor change is readily explained by reference to modern geography and oceanography. Today in the Atlantic, the cold subpolar water of the Labrador Current flows southward along the coast between the land and the Gulf Stream. The Gulf Stream follows the coastline until it reaches the broad headland defined by Cape Hatteras and then angles across the North Atlantic. Although the lowering of sea level during the last glacial maximum altered the position of the shorelines, it did not change the geometry of the continental margin, which is responsible for this boundary configuration and the control of the boundary current flow. During the last glacial maximum, cold surface waters still moved south along the east coast, the Gulf Stream veered northeastward north of Cape Hatteras, and the resulting shear zone was maintained between the two water bodies moving in opposite directions. Thus, the water-mass patterns as delineated by the subpolar and transitional biotic assemblages show less variation along the

Atlantic continental margin between climatic extremes than is found along the Pacific margin. The distribution of the polar assemblage, however, shows how strong the overall oceanographic change was, for at the last glacial maximum it dominated the inshore surface waters from the northernmost tip of the continent to Cape Hatteras. In short, from the cape northward, the shores of eastern North America experienced polar conditions at the last glacial maximum. Polar conditions were present about 21° of latitude farther south along the east coast that along the west coast of North America.

Sea-Surface Temperature Patterns

CLIMAP's reconstruction of sea-surface temperatures at the last glacial maximum shows a number of striking changes from today's

ocean (Figures 12-2 and 12-3). These changes include (1) a large increase in the area of winter sea ice, (2) an expansion of polar water, which presumably accompanied equatorward shifts of polar fronts to just above latitude 40°N in both the Atlantic and the Pacific, and (3) an increased equatorward extent of cool water in the North Pacific eastern boundary current. This reconstruction reveals that the ice-age ocean is strikingly similar to the present ocean in at least one respect: large areas of the tropics and subtropics were as warm as today or only slightly cooler.

Temperature Anomaly Patterns

In order to facilitate the study of the ice-age ocean, temperature-difference maps were prepared; modern sea-surface temperatures

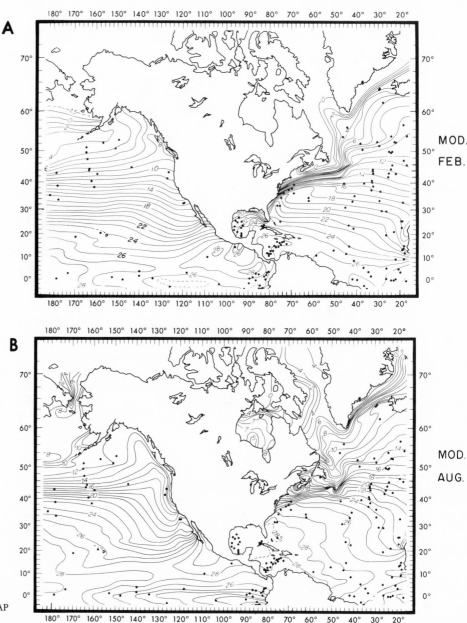

Figure 12-2. Modern sea-surface temperatures (°C). (A) February. (B) August. (From CLIMAP project members, 1981.)

were subtracted from values given on the ice-age map. The resulting anomaly maps (Figure 12-4) show areas of maximum and minimum change between the two climatic extremes.

The following significant features can be identified. (1) Both the Atlantic and the Pacific Oceans exhibit substantial cooling north of about latitude 40°N, with much greater cooling in the Atlantic. As could be expected from the preceding discussion, regions of maximum change occur in areas of the North Atlantic where polar fronts moved equatorward at the last glacial maximum. Here, the largest anomalies (more than 10°C in February and more than 14°C in August) occur around latitude 45°N. The largest anomalies in the western Pacific (about 4°C) occur at about the same latitude. (2) The centers of the subtropical gyres in both oceans show little or no cooling during the ice age. For the summer in the Atlantic and for both seasons in the Pacific, the ice-age ocean was actually a degree or so warmer. In the winter, cooling in the subtropical Atlantic did not exceed about 2°C. (3) Significant cooling, reaching about 4°C, occurred in tropical latitudes along the eastern sides of the oceans. This feature may be the result of increased upwelling and equatorward transport

from the subpolar regions. (4) Modest cooling, reaching about 2°C in August, occurred along the equator. This feature may be the result of increased flow and divergence.

Coastal Transects

To investigate the role of the coastal ocean during the last ice age, we have prepared a set of temperature profiles along two north-south transects (Figure 12-5). Both profiles are drawn along lines parallel to and 2° longitude offshore from the present coastline. Although the ice-age ocean was cooler than the ocean today along both coasts, at all latitudes and at both seasons, there were nevertheless significant differences between the westcoast's and the eastcoast's responses to climatic change. The main difference is that the largest divergence in the temperature patterns along the modern and ice-age transects occurs in winter along the west coast and in summer along the east coast. In both transects, the difference between modern and last-glacial-maximum conditions takes the form of a sharp increase in gradient near latitude 40°N.

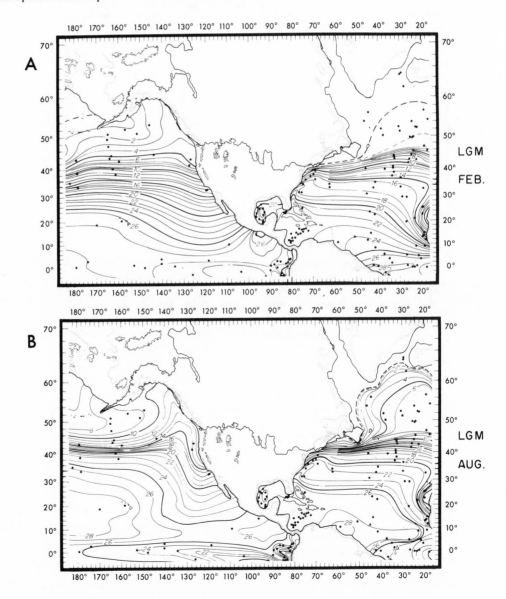

Figure 12-3. Sea-surface temperatures (°C). (A) February and (B) August of the last glacial maximum (LGM), 18,000 yr B.P. Temperature estimates were made from the distribution of modern and ice-age planktonic microfossils. Sea ice occupies areas north of 0°C isotherms. The sample control is shown by black dots. (From CLIMAP Project Members, 1981.)

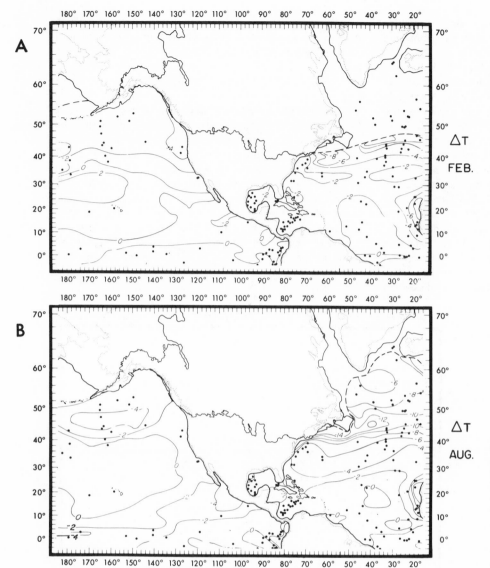

Figure 12-4. Sea-surface tempearture (°C) anomaly map for (A) February and (B) August of the last glacial maximum, 18,000 yr B.P. Temperature differences were obtained by subtracting modern temperatures from ice-age values. Sea ice occupies areas north of the dashed lines. Note that low-latitude areas within zero isopleths were somewhat warmer during ice-age times than they are today. The sample control is shown by black dots.

Along the eastern transect, the inflection point of the curves is essentially fixed by the configuration of the continental margin, and it represents the boundary between subtropical (Gulf Stream) and transitional-polar (Labrador Current) water. Along the western transect, the controlling influences are apparently dynamic rather than topographic. Today, the waters transported across the Pacific in the northern limb of the subtropical gyre (North Pacific Current) are deflected northward and southward to become the Alaska and California Currents. In winter, sea-surface-temperature gradients are low and uniform. Sharp water-mass boundaries are absent, and there are no well-defined oceanic fronts. In summer, the export of cool surface waters northward and southward results in a very low sea-surface-temperature gradient in the north and a steeper one in the south. The inflection point is at approximately latitude 35°N. As already discussed, the striking difference for the Pacific during the last ice age occurred during winter north of about latitude 42°N. This suggests an increased flow in the Alaska Current during the winter season. Similarly, in the Atlantic the greatest cooling occurs in summer, when presumably the Labrador Current carried colder waters to the south.

Discussion

It is well known that precipitation patterns over North America today are significantly influenced by the temperature of the surrounding oceans. Throughout the south-central and southeastern United States, for example, the climate is dominated by the flow of moist, maritime air originating at tropical latitudes in the Atlantic, Caribbean, and Gulf of Mexico (Bryson et al., 1970). This effect is seen during all seasons but is particularly evident during the summer rainy season, which occurs when the westerlies are relatively weak and the area occupied by continental air masses of polar origin is relatively small (Barry and Chorley, 1970). In addition, air masses of Pacific origin exert a significant influence on the precipitation patterns of the western half of the United States, but there the effect is strongly seasonal. Summer rains, typical of the western interior, reflect the penetration inland of maritime polar air; winter rains, which are typical of the Pacific coast and the American Southwest, reflect the expansion of maritime, tropical air masses (Barry and Chorley, 1970).

Therefore, it seems reasonable to assume that the precipitation patterns over North America during the last ice age would have been influenced by the temperatures of the surrounding oceans. Specifically, given the anomaly patterns of the ice age oceans shown in Figure 12-4, we should be able to deduce certain changes in low-latitude rainfall patterns. As already discussed, one major feature of the ice-age ocean was a warming of the Pacific subtropical gyre, particularly in winter. Another feature was the cooling of the Atlantic at subtropical latitudes, particularly in summer. Therefore, because the moisture content of the air overlying the ocean is a strong function of sea-surface temperature, we would expect (1) a wetter ice-age climate in the American Southwest, particularly in winter, and (2) a drier

climate in the Southeast, particularly in summer. These interpretations seem to be supported by independent geologic observations reported by Peterson and others (1979).

Summary and Conclusions

Many previous studies of ice-age climates, based largely on terrestrial data from the middle latitudes and emphasizing the conspicuous advance of the ice sheets, describe the global pattern of climatic change in terms of an equatorward compression of all climatic zones. When

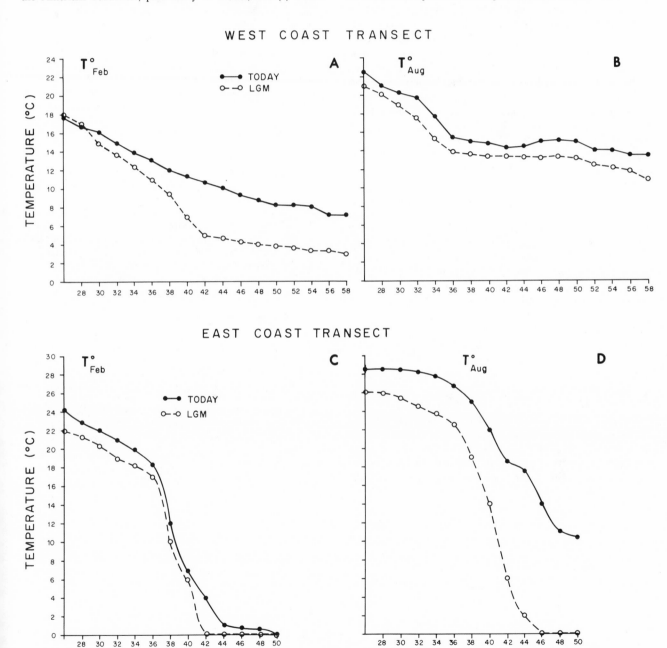

Figure 12-5. Near-shore transects of sea-surface temperature (°C) for waters off North America today and at the last glacial maximum. Transects were constructed along a line drawn parallel to 2° longitude from the coast. (A and B) Temperatures along the West Coast transect for February and August, respectively. (C and D) Temperatures along the East Coast transect for February and August, respectively. (Data from CLIMAP Project Members, 1981.)

the oceanic data are fully taken into account, however, it is clear that the ice-age map is characterized by a compression of climatic belts toward the centers of the subtropical oceanic gyres. It is these gyres that form the thermally and geographically stable part of the climate system. Surrounding them were areas of modest cooling in the equatorial and boundary currents and areas of greater cooling in the middle-latitude zonal flows. These boundary currents, in turn, are surrounded by areas in which ice-age cooling was even more extreme: continents, ice sheets, and ice-covered oceans.

The Northern Hemisphere, particularly the North Atlantic, underwent the greatest change between Pleistocene climatic extremes. In general, the Northern Hemisphere's glacial ocean was characterized by an expansion of the region dominated by the polar faunas (water masses) and a compression of the subpolar and transitional zones. These changes in the distribution of the planktonic biota can be used to infer the following major changes in the oceanographic and thermal patterns of the ice-age ocean: (1) a marked steepening of thermal gradients along frontal systems, particularly in the North Atlantic Ocean but also in the North Pacific Ocean; (2) an equatorward displacement and areal contraction of the subpolar regions of high thermal gradient; in the central North Atlantic, this displacement was about 15° to 20°; (3) a large shift in the path of the Gulf Stream, which flowed with increased intensity due east at latitude 40°N across the North Atlantic at the last glacial maximum but did not greatly shift the position at which it diverged from the North American continental margin; (4) a small shift and reorientation in the frontal zone associated with the North Pacific Drift; (5) an apparently intensified flow of the Alaska Current during the last-glacial-maximum winter in the Pacific and a stronger influence of the Labrador Current along the east coast of North America during the last-glacial-maximum summer; (6) an increased upwelling along equatorial divergences and along eastern boundary currents; (7) nearly stable positions of the central gyres in the subtropical Atlantic and Pacific; in the Pacific, the temperatures in the central gyre were higher during ice-age times than they are today, especially in winter; in the Atlantic, the temperatures in subtropical latitudes were somewhat cooler than they are today, especially in summer.

Modern precipitation patterns over North America are strongly influenced by air masses originating in high-pressure regions over the subtropical ocean. In the southeastern United States, the climate is dominated by the flow of moist, maritime air originating at tropical latitudes in the Atlantic, Caribbean, and Gulf of Mexico. The result is a summer rainy season. In the Southwest, winter rains result from

the expansion of maritime Pacific tropical air masses. Given the features of the ice-age ocean, we would expect a wetter ice-age climate in the Southwest, particularly in winter, and a drier climate in the Southeast, particularly in summer.

References

Barry, R. G., and Chorley, R. J. (1970). "Atmosphere, Weather, and Climate." Holt, Rinehart and Winston, New York.

Brunner, C. A., and Cooley, J. F. (1976). Circulation in the Gulf of Mexico during the last glacial maximum 18,000 years ago. *Geological Society of America Bulletin* 87, 681-86.

Bryson, R. A., Baerreis, D. A., and Wendland, W. M. (1970). The character of late-glacial and post-glacial climatic changes. *In* "Pleistocene and Recent Environments of the Central Great Plains," pp. 53-74. University of Kansas Special Publication 3, University of Kansas Press, Lawrence.

CLIMAP Project Members (1976). The surface of the ice-age Earth. *Science* 191, 1131-37.

CLIMAP Project Members (1981). Seasonal reconstructions of the Earth's surface at the last glacial maximum. Geological Society of America Map and Chart Series 36.

Imbrie, J., and Kipp, N. G. (1971). A new micropaleontological method for quantitative paleoclimatology: Application to a late Pleistocene Caribbean core. *In* "Late Cenozoic Glacial Ages" (K. K. Turekian, ed.), pp. 71-181. Yale University Press, New Haven, Conn.

McIntyre, A., and Kipp, N. G. (with Bé, A. W. H., Crowley, T. J., Kellogg, T. B., Gardner, J. V., Prell, W. L., and Ruddiman, W. F.) (1976). Glacial North Atlantic 18,000 years ago: A CLIMAP reconstruction. *In* "Investigation of Late Quaternary Paleoceanography and Paleoclimatology" (R. M. Cline and J. D. Hays, eds.), pp. 43-76. Geological Society of America Memoir 145.

Moore, T. C., Jr., Burckle, L. H., Geitzenauer, K., Luz, B., Molina-Cruz, A., Robertson, J. H., Sachs, H., Sancetta, C., Thiede, J., Thompson, P., and Wenkam, C. (1980). the reconstruction of sea-surface temperatures in the Pacific Ocean of 18,000 B.P. *Marine Micropaleontology* 5, 215-47.

Moore, T. C., Jr., Hutson, W. H., Kipp, N. G., Hays, J. D., Prell, W. L., Thompson, P., and Boden, G. (1981). The biological record of the ice-age ocean. *Palaeogeography, Palaeoclimatology, Palaeoecology* 35, 357-70.

Peterson, G. M., Webb, T., III, Kutzbach, J. E., van der Hammen, T., Wijmstra, T. A., and Street, F. A. (1979). The continental record of environmental conditions at 18,000 yr B.P.: An initial evaluation. *Quaternary Research* 12, 47-82.

Prell, W. L., Gardner, J. V., Bé, A. W. H., and Hays, J. D. (1976). Equatorial Atlantic and Caribbean foraminiferal assemblages, temperatures, and circulation: Interglacial and glacial comparisons. *In* "Investigations of Late Quaternary Paleoceanography and Paleoclimatology" (R. M. Cline and J. D. Hays, eds.), pp. 247-66. Geological Society of America Memoir 145.

Pleistocene Biota

CHAPTER 13

Vegetational History of the Northwestern United States Including Alaska

Calvin J. Heusser

Introduction

The northwestern United States is a region of contrasting climate and vegetation caused by the mountain provinces' influence on eastward-moving maritime air from the Pacific Ocean. Forests today clothe the humid western mountain slopes, while interior intermontane plateaus and basins in the rain shadows of the mountains are covered by semiarid steppe vegetation. During the Late Wisconsin, glaciers spread into the northern states along a continuous front and developed as independent ice caps and glacier complexes at higher altitudes in the mountains to the south. Preexisting vegetation was overrun along the advancing edge of the ice front, and the colder climate of the ice age effected a redistribution of the vegetational units. The dense coniferous forests containing trees of great size and age in present-day coastal Washington are totally unlike the treeless tracts that occupied the same ground during the Late Wisconsin glaciation.

The Quaternary vegetational history of this vast region $(1.26 \times 10^6 \text{ km}^2)$ has been studied in only a few palynologic investigations compared to the extensive treatment of the subject for the Midwest and the Northeast. Data from the Northwest are far fewer in number and tend to be concentrated in sectors, and so large areas have been left unstudied. Moreover, the data apply to the entire Late Wisconsin only in western Washington; elsewhere, the last several millennia are covered. These same limitations apply to Alaska, an even larger area $(1.46 \times 10^6 \text{ km}^2)$.

This chapter's treatment of the vegetational history of the region updates previous surveys from different parts of the Northwest that were based on the pioneer studies published by Hansen (1947) earlier in this century. Surveys by Colinvaux (1967b), Wright (1971), McAndrews and King (1976), and Heusser (1957, 1960, 1965, 1969a, 1977b) apply generally or partially to the Northwest and Alaska.

The Northwestern Conterminous United States

LANDSCAPE

The Pacific Mountain System, Intermontane Plateaus, Rocky Mountain System, and Interior Plains are the physiographic units distinguishable from west to east in the region (Fenneman, 1931). The two cordilleras stand out in the landscape, alternating broadly with the more subdued bounding upland. The mountains trend predominantly north-south, with the Rocky Mountains dominating the eastern sector (Figure 13-1).

The Pacific Mountain System bordering the Pacific Ocean in western Washington and western Oregon consists of the Coast Range, the Olympic Mountains, and the Klamath Mountains on the west; the Cascade Range on the east; and, wedged between the Coast Range and the Cascade Range, the Puget-Willamette Lowland. The Oregon-Washington Coast Range, with moderate relief over its extent of almost 500 km, rises to only 1249 m in altitude. This contrasts the Olympic Mountains, which reach 2425 m in the far-northern coastal border, and the Klamath Mountains, which reach 2259 m in southwestern Oregon. The Puget-Willamette Lowland, a partially drowned trough, runs with irregular width for over 500 km to west-central Oregon.

Most formidable is the Cascade Range, with altitudes averaging between 1500 m and somewhat over 2100 m. The crest of the range features a number of volcanoes, including Mount Rainier (4393 m), Mount Adams (3752 m), Mount Hood (3422 m), Mount Baker (3277 m), and Glacier Peak (3181 m). Eruptions of these and other volcanoes in the range have taken place at various times during the Quaternary; the latest is the current series from Mount St. Helens (2948 m high before the 1980 eruptions) in southwestern Washington. Most celebrated of all was the eruption of Mt. Mazama, the site of present-day Crater Lake (1882 m), which occupies the caldera. The eruption spewed pumiceous ash over a vast part of the northwestern United States and southwestern Canada about 6600 years ago (Powers and Wilcox, 1964). Glacier Peak is known for its widespread Late Wisconsin tephra, which was deposited during eruptions between about 12,750 and 11,250 yr B.P. (Porter, 1978).

The Columbia Plateau province, extending over much of eastern Washington, eastern Oregon, and southern Idaho, makes up most of the Intermontane Plateaus physiographic unit that drains to the

This chapter is dedicated to Henry P. Hansen, professor emeritus, Oregon State University, who was the first to undertake a reconstruction of the Quaternary vegetational history of the Pacific Northwest. The support of the National Science Foundation is gratefully acknowledged.

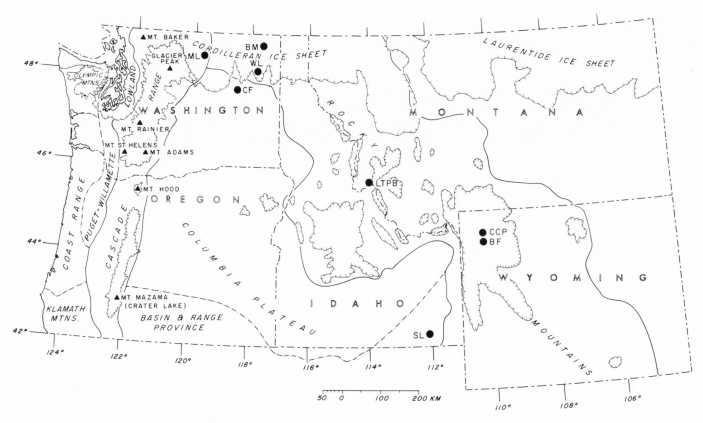

Figure 13-1. Map of the northwestern United States showing the limits of the physiographic provinces, the extent of the ice sheets and glacier complexes (hachured), and the locations of Late Wisconsin fossil pollen sites east of the Cascade Range. (ML, Mud Lake; CF, Creston Fen; WL, Waits Lake; BM, Big Meadow; LTPB, Lost Trail Pass Bog; CCP, Cub Creek Pond; BF, Buckbean Fen; and SL, Swan Lake.)

Pacific in the north; the Basin and Range province, located south of the Columbia Plateau, is characterized by interior drainage. In eastern Washington, landforms average between 300 and 600 m in altitude; east of the Cascade Range, altitudes reach 1000 m in Oregon and about 1800 m in southern Idaho; in mountainous northeastern Oregon, the highest summit is 2997 m. Much of the Columbia Plateau was formed by successive lava flows. In eastern Washington, the lava has been fluvially eroded by the discharge of Pleistocene ice-dammed lakes, which gave rise to the Channeled Scabland (Baker, 1982). Some closed basins in Oregon still contain lakes that formed under the pluvial climate of the Pleistocene.

The Rocky Mountain System is a northwest-trending complex of mountain ranges, basins, and plateaus in the region between Wyoming and northeastern Washington, northern Idaho, and western Montana. Much of southern Wyoming is a broad basin lying at altitudes of 2000 to 2300 m; a lava plateau in the northwest reaches 2600 m. Neighboring ranges contain numerous peaks over 3000 m and several above 4000 m; the highest peak is at 4202 m. In the north, summit altitudes in the Rockies are lower and only occasionally exceed 3000 m. Intermontane valleys are distinctive features of the northern sector; some, called "trenches," exhibit broad floors hundreds of kilometers long at altitudes 600 to 1500 m below the mountain crests.

The Great Plains province along the eastern side of the Rockies covers almost two-thirds of Montana and a third of Wyoming. The High Plains extend southward from southeastern Wyoming. Altitudes along the boundary with the Rocky Mountains average

close to 1700 m. The High Plains are striking in their flatness, whereas the rest of the Great Plains contain several isolated mountains and occasional badlands characterized by sharp erosional sculpture.

GLACIATION

Late Wisconsin glaciers originating in Canada as two independent ice sheets flowed southward in the form of huge, coalescent lobes to as far as latitude 47°N in northern Washington, Idaho, and Montana (Flint, 1971). The lobate character of the drift border (Figure 13-1) is the result of ice moving along unobstructed intermontane valleys. The Cordilleran ice sheet in the Rocky Mountains provinces and the provinces to the west originated in the mountains of western Canada (Hamilton and Thorson, 1982; Waitt and Thorson, 1982); the Laurentide ice sheet, which covered the northern Great Plains province, was centered in east-central Canada (Mickelson et al., 1982). Beyond the limits of the ice sheets, glaciers also formed extensive complexes in the Cascade Range of Washington and Oregon and in the Rocky Mountains of central Idaho, northwestern Wyoming, and adjacent Montana (Porter et al., 1982).

Glaciation of Late Wisconsin age in the Rocky Mountains, designated Pinedale Glaciation (Richmond, 1965; Richmond et al., 1965), occurred between about 30,000 yr B.P. (most ages by obsidian hydration date it between 35,000 and 20,000 yr B.P.) and the end of a late-glacial readvance estimated to have occurred between 13,000 and 11,000 yr B.P. (Pierce, 1979; Pierce et al., 1976). The Late Wisconsin in the northern part of the Puget-Willamette Lowland and adjoining southwestern Canada, represented by Fraser Glaciation (Armstrong et al., 1965; Crandell, 1965), began about 25,000 yr B.P. and lasted through the time of readvancing ice in southwestern Canada between 11,800 and 11,400 yr B.P. (Armstrong, 1975) and in the northern Cascades some time between about 12,000 and 11,000 yr B.P. (Porter, 1976, 1978).

CLIMATIC REGIMEN

The northwestern United States, situated in a zone of westerly winds, is strongly influenced by air masses that move inland from the Pacific Ocean. Continental air flowing southward from the high latitudes of Canada and from the interior east of the region only occasionally and temporarily alters this dominant regimen. The maritime air is cooled as it crosses the ocean, with the result that cool temperate and humid conditions are widespread in coastal areas. The tendency is for the ocean to modify temperatures both in summer and winter.

Storms move across Washington and Oregon in winter when the Aleutian low-pressure system controls circulation over the northern ocean. Storms are less frequent in summer, when the North Pacific high-pressure cell replaces the low. This seasonal change underlies the winter-wet/summer-dry feature of the climate that is so distinctive on the western side of the Cascade Range. Less than 10% of the region's annual precipitation falls during June, July, and August (Phillips, 1974; Sternes, 1974). The effect of the north-south-trending ranges is to lift the eastward-moving maritime air to condensation levels. Precipitation in the western Olympic Mountains, Washington Cascades, and Oregon Coast Range reaches over 3000 mm annually, with amounts close to 4700 mm recorded from the western Olympics. The proportion falling as snow at or near sea level is negligible, but in the mountains it is considerable. Amounts of snowfall in the mountains, over 25 m measured on Mount Rainier (record snowfall about 33 m) and 22 m at Crater Lake, for example, are some of the highest on record in the United States.

Precipitation decreases sharply to minimal amounts in the lee of the Cascades and in the intermontane valleys, but it rises on the western slopes of the Rocky Mountains. In places, yearly average precipitation is as low as 200 mm. Values in the Rockies increase to about 1500 mm at higher elevations, and in the Great Plains of eastern Wyoming and Montana the average is 300 to 400 mm (Cordell, 1974; Lowers, 1974; Rice, 1974). Annual snowfall in the Rockies averages 4.6 to 7.6 m, but some years it reaches 12.3 m or more.

Temperatures along the coast average about 15 °C to 16 °C in July and 4 °C to 5 °C in January. These values contrast with averages in the Columbia Plateau province of 19 °C to 23 °C for July and about −1.7 °C for January and make apparent the moderating influence of the ocean. The effect of continentality is more evident farther east in the Rockies, where temperatures at lower elevations average over 18 °C in July and between −5 °C and −6 °C in January.

MODERN VEGETATION

The climatic regimen of winter storms (with protracted periods of moisture and cloudiness, summers with relatively short intervals of soil-water loss, and moderate temperatures over much of the year) is very favorable for tree growth and accounts for the dense coniferous forests of the region. The forests with massive trees of great longevity are unrivaled among temperate forest regions of the world (Waring and Franklin, 1979).

The vegetation of the northwestern conterminous United States is predominantly forest west of the Cascades and in the interior mountains east of the range. The remainder of the region is covered mainly by steppe vegetation and to a small extent by alpine tundra and barren ground. The distribution of the vegetational units (Figure 13-2) is patterned in a general way on the physiographic character of the region. Units of potential natural vegetation are shown in Figure 13-2 in modified collective form from the classifications by Küchler (1964) and Franklin and Dyrness (1973); range maps of selected coniferous tree species (Figure 13-3) are from Fowells (1965). Plant nomenclature follows Hitchcock and Cronquist (1973).

Pacific slope coniferous forest is spread west of the Cascade crest and occurs as an outlier in the northern Rocky Mountains. Daubenmire (1943) points out that the outlier centered in northern Idaho is maintained by a climate similar to that of the western slope of the Cascades and is a consequence of its location along storm tracks from the Pacific. Western hemlock (*Tsuga heterophylla*) and western red cedar (*Thuja plicata*) are the most widely distributed dominants (Figure 13-3). Sitka spruce (*Picea sitchensis*), also a dominant, is found in the more immediate coastal belt. Douglas-fir (*Pseudotsuga menziesii*) occurs throughout the region covered by the forest and is mostly seral. Where hemlock does not compete, as in large sections of the Rocky Mountains, Douglas-fir is not seral and its position in the forest is maintained. Grand fir (*Abies grandis*) is seral for the most part, but it is less extensive than Douglas-fir.

Transects between sea level and the alpine tundra illustrate altitudinal changes in community composition and structure in the Pacific slope coniferous forest (Figure 13-4). In the western Olympic Mountains, lowland forest consists of western hemlock and western red cedar (Figure 13-5) with only occasional Douglas-fir, grand fir, western white pine (*Pinus monticola*), and lodgepole pine (*P. contorta*) up to altitudes of about 600 m (Fonda and Bliss, 1969). Sitka spruce (Figure 13-6) in the lowland forest is found mostly below 150 m on floodplains in the river valleys and in a fringe along the ocean where fog and drizzle prevail. Red alder (*Alnus rubra*), bigleaf maple (*Acer macrophyllum*), vine maple (*A. circinatum*), western yew (*Taxus brevifolia*), cottonwood (*Populus trichocarpa*), and willow (*Salix hookeriana, S. lasiandra, S. sitchensis,* and *S. scouleriana*) are also typical of stands in the river valleys.

The undergrowth in the mature forest, exemplified by *Polystichum munitum, Gaultheria shallon,* and *Oxalis oregana,* is variable and increases where shading from the canopy lessens. The bracken (*Pteridium aquilinum* var. *pubescens*), a species dependent on adequate light, is a frequent invader of open space. Shrubs, particularly light-dependent plants, include *Rubus spectabilis, R. parviflorus, Cornus stolonifera,* and *Sambucus racemosa.* The vascular cryptogam *Selaginella oregana* is a common epiphyte hanging from the branches of vine maple and other small trees. Bryophytes are luxurious in the forest.

Montane forest of predominantly Pacific silver fir (*Abies amabilis*) and western hemlock occurs above 600 m to about 1100 m on the Pacific side of the Olympic Mountains. The stature and girth of the trees decrease with increasing altitude and also vary as a function of exposure. The undergrowth consists of species distinctive of this forest community (*Vaccinium ovalifolium, Rubus pedatus, Coptis laciniata,* and *Streptopus roseus* var. *curvipes*) and includes plants of the lowland forest; mountain alder (*Alnus sinuata*) and vine maple follow the stream courses. Above 1100 m to an altitude of 1400 m, closed subalpine forest of Pacific silver fir and mountain hemlock

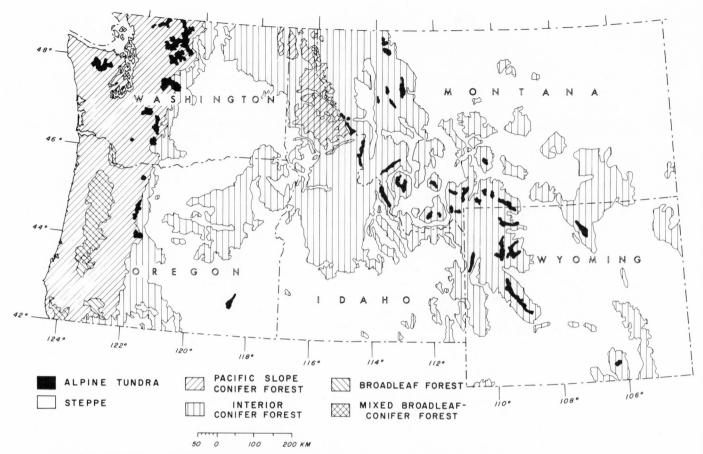

Figure 13-2. Distribution of vegetational units in the northwestern United States. (Generalized from a map by Kü [1964].)

(*Tsuga mertensiana*) covers the slopes; subalpine fir (*Abies lasiocarpa*) forest, structurally closed to about 1525 m, comprises the highest arboreal community to an average altitude of 1600 m. At the upper limit of tree growth, islands and upward-narrowing vertical files of mostly subalpine fir are ecotonal with subalpine meadow and tundra communities. Shrubs in the Pacific silver fir/mountain hemlock forest are *Vaccinium deliciosum* and *Rhododendron albiflorum*; *Vaccinium membranaceum* is dominant beneath subalpine fir. Alaska yellow cedar (*Chamaecyparis nootkatensis*) is an arboreal associate restricted to the uppermost forest.

The transect of the Pacific slope coniferous forest between sea level and the alpine tundra in the Cascades (Figure 13-4) first encounters the forest of the Puget lowland. Western hemlock and western red cedar continue to be consistently representative at low altitudes, whereas Sitka spruce is infrequent and only occurs locally. Less humid and warmer conditions (compared with the western side of the Olympics) favor Douglas-fir, and so its numbers in the forest are considerably greater. In the drier northwestern part of the lowland, located in the rain shadow of the Olympic Mountains, Douglas-fir maintains itself as the dominant in mature communities (Fonda and Bernardi, 1976). Climatic conditions favor grand fir to a certain extent and are also favorable to a broad sclerophyllous element, namely, Oregon oak (*Quercus garryana*) and madrone (*Arbutus menziesii*), which ranges into the region from California. Red alder and other broad-leaved trees noted along the Olympic coastal rivers also grow on the floodplains in the Puget lowland.

Lowland forest rises to an altitude of 900 m in the western Cascades, above which montane forest of Pacific silver fir and western hemlock continues to 1500 m. The dominants are the same species as in the montane forest of the Olympics; Douglas-fir, however, becomes common, and noble fir (*Abies procera*), not encountered previously, is frequently the dominant in successional stands. Subalpine forest at higher altitudes reaches about 2000 m, dissolving into parkland at the edge of the tundra (Figure 13-7). Subalpine fir, mountain hemlock, Pacific silver fir, and Alaska yellow cedar, noted previously in the subalpine forest of the Olympic Mountains, are leading arboreal species; whitebark pine (*Pinus albicaulis*), dwarfed and contorted, is the common timberline tree. Most of the shrubs and herbs characteristic of the montane and subalpine forests of the Olympics are also present in the Cascades.

Profiles of the mean July temperature for the western slope of the mountains (Figure 13-4) show a gradient of 0.4 °C per 100 m, which approximates the wet lapse rate of humid air rising up the windward side by expansion. (See Heusser, 1977a, for data.) The Olympic Mountains profile, over 4 °C cooler near sea level than the Cascade Range-Puget lowland profile, reveals the influence of the cool ocean in affecting the distribution of Sitka spruce. Temperatures at corresponding forest boundaries show good agreement; the forest-tundra ecotone occurs close to 10 °C. An increase in the altitudes of the vegetational boundaries inland is readily apparent from the two profiles.

Parts of the region west of the Cascades are mixed broadleaf/conifer

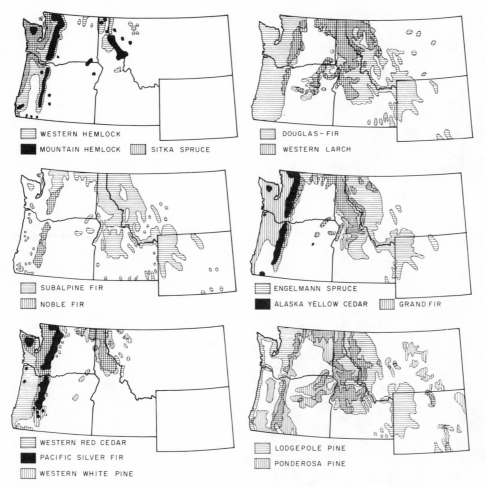

WESTERN HEMLOCK

MOUNTAIN HEMLOCK   SITKA SPRUCE

DOUGLAS-FIR

WESTERN LARCH

SUBALPINE FIR

NOBLE FIR

ENGELMANN SPRUCE

ALASKA YELLOW CEDAR   GRAND FIR

WESTERN RED CEDAR

PACIFIC SILVER FIR

WESTERN WHITE PINE

LODGEPOLE PINE

PONDEROSA PINE

Figure 13-3. Range maps of representative coniferous trees in the northwestern United States. (Data from Fowells, 1965.)

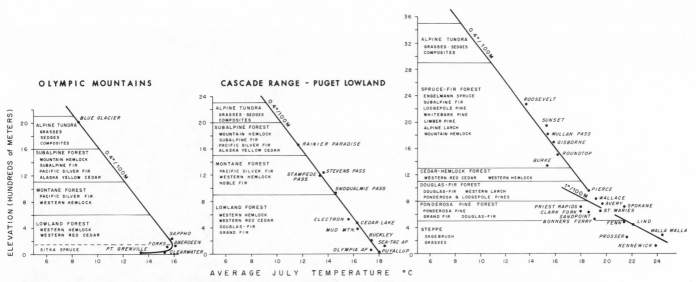

Figure 13-4. Vegetational zonation and mean July temperature profiles for the Olympic Mountains, the Cascade Range-Puget Lowland, and the northern Rocky Mountains. Meteorologic stations are indicated by dots. (Data from numerous sources.)

Figure 13-5. Giant western red cedar of the Pacific slope coniferous forest in coastal western Washington.

forest (Figure 13-2). The Oregon sector of the Willamette-Puget Lowland exhibits the greatest area covered by this unit. Oregon oak is the dominant, mixed largely with Douglas-fir; stands are often open, with a ground cover of grasses and some shrubs. Forest in parts of the Klamath Mountains of southwestern Oregon is a mixture of Douglas-fir, madrone, chinquapin (*Castanopsis chrysophylla*), tanoak (*Lithocarpus densiflorus*), California laurel (*Umbellularia californica*), and oak. Broadleaf forest (Figure 13-2), commonly consisting of Oregon oak, occurs in small areas in the southwest; in northwestern Oregon, lowland is locally forested by red alder and Oregon ash (*Fraxinus latifolia*).

East of the Cascades, semiarid vegetation broadly classified as steppe is extensive across the Columbia Plateau, much of the southwestern Rocky Mountains of Wyoming, and in the Great Plains of Montana (Figure 13-2). Most of the region is a shrub steppe (Figure 13-8) of sagebrush (*Artemisia tridentata*) and bunchgrass (*Agropyron spicatum*). The species composition in the shurb steppe is variable because of rainfall and soil patterns; some associates include *Artemisia tripartita*, *Chrysothamnus nauseosus*, *Purshia tridentata*, *Balsamorhiza sagittata*, *Eurotia lanata*, and *Stipa comata*. Communities of shadscale (*Atriplex confertifolia*) and greasewood (*Sarcobatus vermiculatus*) are found in the drier alkaline areas of southeastern Oregon. Steppe bunchgrasses, typically *Agropyron spicatum* and *Festuca idahoensis*, blanket the loess-covered plateau of southeastern Washington. Northwest of the mountains in northeastern Oregon, another grass, *Poa secunda*, mixes with these species. The grassland of the Great Plains consists of relatively short grasses (*Agropyron smithii*, *Bouteloua gracilis*, and *Stipa comata*, for the most part) and composites (e.g., *Artemisia frigida*, *gutierrezia sarothrae*, and *Liatris punctata*).

The steppe is at the lower altitudinal limit of the interior coniferous forest communities of ponderosa pine (*Pinus ponderosa*). In open stands, ponderosa pine forest runs along the eastern base of the Cascades, locally interrupted by groves of aspen (*Populus tremuloides*), and becomes widely extended in south-central Oregon. Transitional areas of western juniper (*Juniperus occidentalis*) border

Figure 13-6. Sitka spruce characteristic of the lower Hoh River valley of western Washington.

Figure 13-7. View north down debris-covered Carbon Glacier on Mount Rainier in the Washington Cascades. Parkland of subalpine fir and mountain hemlock in the foreground is bounded downglacier by closed subalpine forest.

the forest to the east. In the Rockies, ponderosa pine is more discontinuous; the steppe at these higher altitudes is usually bounded by Douglas-fir and grand fir forests and, in the highest parts, by forests of spruce and subalpine fir.

A transect of the western slope of the northern Rockies at about latitude 48°N (Figure 13-4) intercepts the units of the interior coniferous forest, which begins at an altitude of 500 m (Daubenmire, 1952, 1968, 1970; Habeck, 1968). Ponderosa pine forest abruptly borders the steppe in open stands that exhibit a cover of shrubs (*Symphoricarpos albus, Physocarpus malvaceus,* and *Purshia tridentata*) or of grasses (*Agropyron spicatum, Festuca idahoensis,* and *Stipa comata*). Relatively mesic localities usually include numbers of Douglas-fir and grand fir. Douglas-fir dominates the forest at higher altitudes between 800 and 1100 m, where, under less xeric conditions, ponderosa pine, lodgepole pine, and western larch (*Larix occidentalis*) occur as seral constituents; the grass *Calamagrostis rubescens* or the low shrub *Arctostaphylos uva-ursi* prevails in pure stands of Douglas-fir, whereas *Symphoricarpos* and *Physocarpus* are frequent in mixed stands.

Western red cedar/western hemlock forest is found at altitudes between 1100 and 1300 m. Although the average level of humidity is high, past fires have favored the presence of western larch and western white pine. Species in the undergrowth are *Pachistima myrsinites, Oplopanax horridum, Athyrium filix-femina, Ribes lacustre,* and *Linnaea borealis.* Subalpine fir/Engelmann spruce (*Picea engelmannii*) forest occupies a broad altitudinal zone, much of

Figure 13-8. Sagebrush steppe on the Columbia Plateau in east-central Oregon.

Figure 13-9. Biogenic sediments of Pleistocene age exposed in the sea cliff near Kalaloch, Washington.

which is parkland, over a range of 1600 m to some 2900 m in altitude. Communities contain mountain hemlock and alpine larch (*Larix lyallii*) and lodgepole, white bark, and limber (*Pinus flexilis*) pines, with a ground cover contributed by *Vaccinium membranaceum*, *V. scoparium*, *Menziesia ferruginea*, and *Xerophyllum tenax*.

The temperature profile for the northern Rocky Mountains transect (data mostly from Daubenmire, 1956) reveals an average July depression of around 6 °C between the lower steppe and the Douglas-fir forest (Figure 13-4). This amount roughly conforms with a dry lapse rate of 1 °C per 100 m. Values for altitudinally higher zones, following a wet lapse rate (0.4 °C per 100 m), indicate an additional temperature depression of about 8 °C. The 10 °C-to-11 °C average for the forest-tundra transition agrees with values obtained from the ecotone in the Olympics and Cascades.

Alpine vegetation is distributed throughout the Rockies and in the Cascade Range and Olympic Mountains (Figure 13-2). Communities consist of grasses, sedges, composites, heaths, and a variety of shrubs and herbs of diverse taxonomic affinities. Exposure, snow cover, soil depth and texture, slope angle, and available moisture are some of the factors relating to the makeup, extent, and distribution of plant cover in this zone, where frost processes are active. Upper altitudes, which appear to coincide with a mean July temperature of 8 °C to 9 °C (Figure 13-4), range between about 2100 m in the Olympic Mountains and 3500 m in the Rocky Mountains.

On the high plateau above 3300 m in south-central Montana and adjacent Wyoming, Johnson and Billings (1962) recognized communities of *Geum rossii* and *Carex rupestris* on exposed sites, *Deschampsia caespitosa* and *Carex scopulorum* in well-drained protected localities, *C. scopulorum* on wet soils, and *Salix planifolia* as the dominant along drainage courses; also frequently encountered were *Phlox caespitosa*, *Artemisia scopulorum*, *Polygonum bistortoides*, *Festuca ovina*, and *Erigeron simplex*. Along environmental gradients in the northern Cascades in Washington, Douglas and Bliss (1977) classified snow-bed communities (*Saxifraga tolmiei-Luzula piperi*, *Erigeron pyrolaefolium-Luzula piperi*, *Carex nigricans*, *C. breweri*, *C. capitata*, and *Antennaria lanata*), mesic herb communities (*Lupinus latifolius* and *Festuca viridula*), dwarf-shrub communities (*Cassiope mertensiana*, *Phyllodoce empetriformis*, *P. glanduliflora*, *Arctostaphylos uva-ursi*, *Empetrum nigrum*, *Salix nivalis*, *S. cascadensis*, and *Dryas octopetala*), dry grass and sedge communities (*Danthonia intermedia*, *Calamagrostis purpurascens*, *Carex spectabilis*, *C. phaeocephala*, *C. scirpoidea* var. *pseudoscirpoides*, *C. nardina*, and *Kobresia myosuroides*), and herb-field, fellfield, boulder-field, and vegetation-stripe communities with a diversity of plants (of which *Phlox diffusa*, *Potentilla diversifolia*, *Oxytropis campestris* var. *gracilis*, and *Solidago multiradiata* are representative).

VEGETATIONAL HISTORY

Pollen data for a history of Late Wisconsin vegetation west of the Cascades are from lake and bog deposits and exposed sections of sediments (Figure 13-9) located in Washington. No radiocarbon-dated pollen records of Late Wisconsin age have been demonstrated in Oregon because of an apparent paucity of sedimentary basins in this time range. In Washington, the sites are located in the Olympic Mountains west of Seattle and in the Puget-Willamette Lowland between Portland and the Canadian border (Figure 13-10). All sites ex-

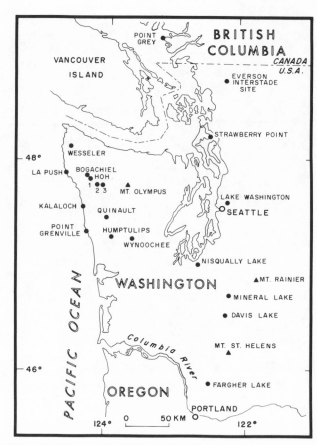

Figure 13-10. Locations of Late Wisconsin fossil pollen sites west of the Cascade Range in Washington.

cept those at La Push, Kalaloch, and Point Grenville on the Pacific coast have been glaciated at some time.

Modern pollen-rain data from over 150 stations in western Washington and Oregon serve as a means for interpreting the vegetational history from the fossil records (Heusser, 1969b, 1973c, 1978a, 1978b, 1978c). Observations at these stations indicate overall relationships between the surface pollen and the vegetational units; the assemblages constitute potential analogues for fossil assemblages. Pollen of alder, a major seral species in the disturbed existing vegetation, dominates stations from lowland to upper-montane altitudes in much of the Coast Range, Olympics, and Cascades. In Douglas-fir stands in the lowland, pollen of Douglas-fir is underrepresented; stands in drier areas where alder does not flourish show a more representative proportion of Douglas-fir. Amounts of western hemlock and Sitka spruce pollen are highest in the humid valleys of the Olympics, and pine pollen is overrepresented at stations on the dry eastern side of the Cascades. Pollen of mountain hemlock is associated with the subalpine forest; herbaceous pollen types—grasses, sedges, and composites (including *Artemisia*)—increase with mountain hemlock at stations near the timberline and in the alpine tundra. Grasses, along with oak, are also indicative of the existing vegetation of the Puget-Willamette Lowland of western Oregon.

Two bog sites are singled out from a number on the west side of the Olympic Mountains that contain Late Wisconsin pollen records. They are located 1 km apart and lie close to an end moraine formed

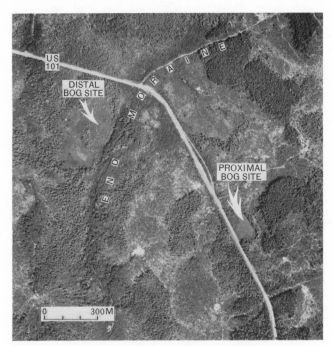

Figure 13-11. Vertical aerial view of end moraine on Bogachiel-Hoh drainage divide in relation to distal and proximal box sites. (Aerial photograph taken in 1972 by Washington Department of Natural Resources [OLY-71 362-24B-26].)

by an alpine glacier more than 30,000 years ago (Figure 13-11). One site on the distal side of the moraine is in the drainage of the Bogachiel River; the other, situated proximally, is in the Hoh River drainage (Bogachiel and Hoh 1 in Figure 13-10).

In a pollen diagram of a section of the distal bog (Figure 13-12), pollen assemblage zones DB-2A through DB-2C are between approximately 28,000 and 10,500 years old. This time span is based on the regional radiocarbon-age control of correlative pollen stratigraphy in addition to the ages in the section (8830 ± 260 [RL-1252], 16,850 ± 630 [RL-1253], 20,100 ± 750 [RL-578], and 30,000 ± 800 yr B.P. [Y-2453]). Zone DB-2C, the oldest, is an assemblage of herbs, *Tsuga mertensiana*, and *Pinus*; zone DB-2B, dated at 20,100 to 16,850 yr B.P., is predominantly herbs; and zone DB-2A, the youngest, consists of herbs, *Alnus*, and *Pteridium*. Herbs in this sequence are the Gramineae (grass), Cyperaceae (sedge), and Compositae (composite tribes Liguliflorae and Tubuliflorae including *Artemisia*) and also the subalpine/high-latitude indicators *Polygonum bistortoides* type and *Selaginella selaginoides*. Timberline vegetation is inferred from the modern pollen-rain data; it became tundralike for a time (zone DB-2B) and was later replaced by successional communities of alder and bracken. Holocene forest trees, *Picea sitchensis, Tsuga heterophylla*, and *Thuja plicata* (zones DB-1A through DB-1D), were climatically and edaphically restricted during the Late Wisconsin. The numbers of tree taxa recorded are low (minimum 14%) and probably reflect long-distance wind transport. Pollen influx (grains per square centimeter per year) was only 350 in zone DB-2B compared with a Holocene maximum of approximately 12,000 (Heusser, 1978d); nearby (site Hoh 2 in Figure 13-10), values during the Late Wisconsin were less than 100 and during the Holocene were as much as 5000 (Heusser, 1974). The low rates for the Hoh 2 site compare with rates of 5 to 169 grains per square centimeter per year determined in Holocene sediments in the tundra of

the Alaskan Aleutian Islands (Heusser, 1973d).

The proximal bog (Figure 13-11) is younger than the distal one and is believed to be less than 20,000 years old because the pollen diagram (Figure 13-13) does not contain in its lower part a proportion of *Tsuga heterophylla* similar to that in zone DB-3A (Figure 13-12). To the contrary, the amounts of nonarboreal types correspond with values in DB-2B, which dates between 20,100 and 16,850 yr B.P. Furthermore, the radiocarbon ages of 15,400 ± 550 (RL-1251) and 10,910 ± 280 yr B.P. (RL-577) for the proximal bog section cover the Late Wisconsin interval postdating the nonarboreal pollen maximum.

In the diagram for the proximal bog, zone PB-2, which is considered to be approximately equivalent to zones DB-2A and DB-2B, contains a detailed sequence of assemblages. Zones PB-2E and PB-2G are assemblages consisting mostly of herbs, with PB-2F dominated by *Pinus* and herbs; zones PB-2B and PB-2D are of *Tsuga mertensiana, Pinus*, and herbs, and PB-2C is of *Tsuga heterophylla, T. mertensiana, Pinus*, and Tubuliflorae; PB-2A is a *Pinus*-herb-*Pteridium* assemblage. This record generally reveals a greater quantity of pine, the result of local overrepresentation, than was found in the distal bog sequence; also, the increase of *Tsuga mertensiana* and *T. heterophylla* apparently was local. Thus, pollen data from the two sites reflect mostly treeless vegetation consonant with tundra from about 20,000 to 16,850 years ago (see also Heusser, 1973a), with pine increased over part of the interval; later, from 15,400 to 10,910 years ago, mountain hemlock and, for a time, western hemlock developed a community of limited local extent in what apparently was parkland; and, finally, pine and alder succeeded with open space occupied by bracken.

A pollen diagram (Figure 13-14) of a peat exposure in the Quinault River drainage, about 40 km southeast of the Bogachiel-Hoh sites (Figure 13-10), shows a Late Wisconsin *Pinus*-herb assemblage in the beginning (zone Q-2) succeeded by an early-Holocene assemblage made up largely by alder (zone Q-1). The boundary between the assemblage zones is dated at close to 10,500 yr B.P. from radiocarbon ages of 10,475 ± 240 (GX-3544) and 10,150 ± 325 yr B.P. (GX-3543). An age of 12,100 ± 350 yr B.P. (GX-3545) dates the bottom of the section, which rests on outwash gravel of a glacier in the Quinault River valley. Pine, although overrepresented, formed a maximum of 55% of the pollen sum and apparently was the characteristic arboreal component of the vegetation that succeeded in this part of the western Olympics during the last two millennia of the Wisconsin.

Pollen data from the Bogachiel, Hoh, and Quinault sites are diagrammed (Figure 13-15) in relation to data from Point Grenville (L. E. Heusser, personal communication, 1972), Humptulips (C. J. Heusser, unpublished data), Kalaloch (Heusser, 1972), La Push (Florer, 1972), and Wesseler (Heusser, 1973b) to provide an overview of western Olympic vegetation. (The locations of all sites are shown in Figure 13-10.) Assemblages are primarily herbs and secondarily pine and mountain hemlock during the interval from approximately 25,000 to 20,000 years ago. From 20,000 to about 15,000 years ago, these assemblages continued to be evident except at southerly coastal sites where pine, mountain hemlock, spruce, and western hemlock were of primary importance. Between 15,000 and 10,500 years ago, conifers, notably pine, along with alder became dominant throughout.

The landscape during the early part of the Late Wisconsin is conceived as having been treeless except for scattered islands of trees and as having resembled present-day subalpine parkland. Tundra vegeta-

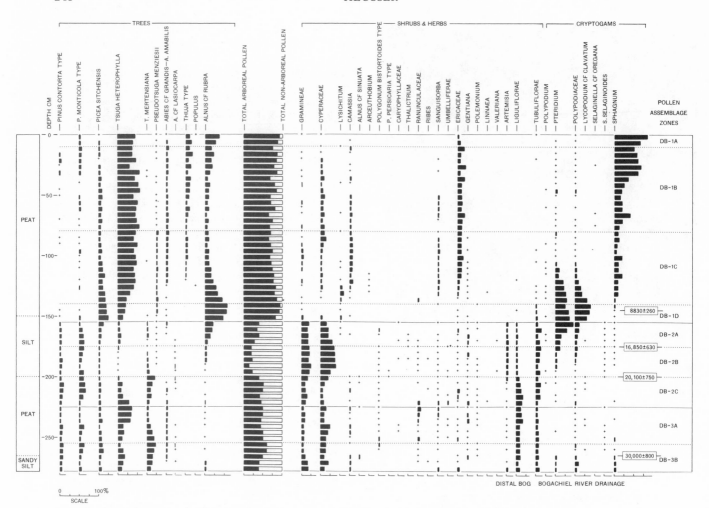

Figure 13·12. Pollen diagram of Bogachiel distal bog section showing pollen assemblage zones with radiocarbon ages to the right and sediment types to the left; plus symbols denote percentages of less than 2%.

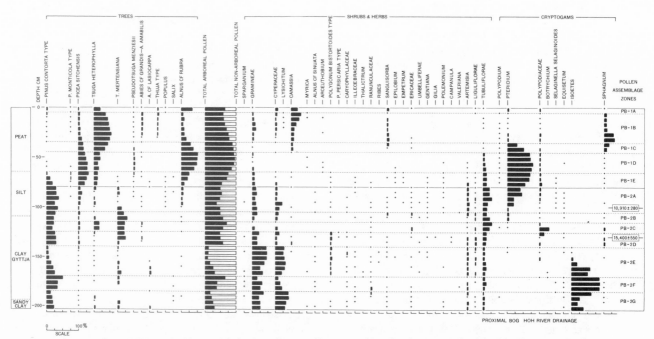

Figure 13·13. Pollen diagram of Hoh proximal bog section showing pollen assemblage zones with radiocarbon ages to the right and sediment types to the left; plus symbols denote percentages of less than 2%.

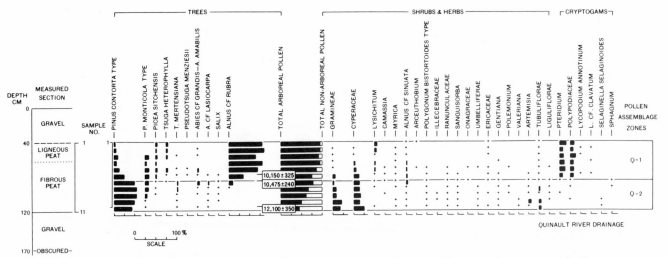

Figure 13-14. Pollen diagram of peat exposure in Quinault River drainage showing pollen assemblage zones to the right, radiocarbon ages, and sediment types to the left; plus symbols denote percentages of less than 2%.

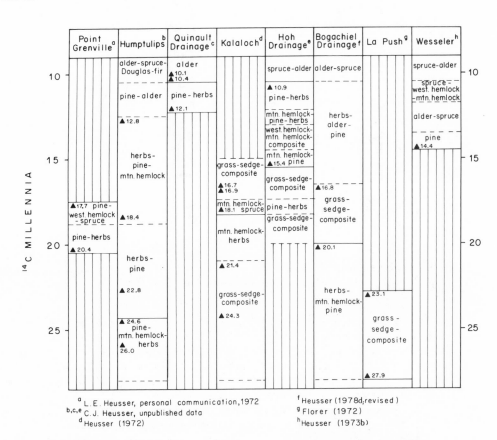

Figure 13-15. Stratigraphic relations of pollen assemblage zones for sites west of the Olympic Mountains. (See Figure 13-10 for locations.) Triangles indicate levels of radiocarbon ages (10³ yr B.P.).

[a] L. E. Heusser, personal communication, 1972
[b,c,e] C. J. Heusser, unpublished data
[d] Heusser (1972)
[f] Heusser (1978d, revised)
[g] Florer (1972)
[h] Heusser (1973b)

tion during the mid-Late Wisconsin apparently became more extensive in the north. Point Grenville, on the southern coast, at that time featured pine, but later western hemlock and spruce appear, and they may show a forest boundary during an episode centered around 18,000 years ago that favored the increase of conifers as far north as the Hoh River drainage. Pine was also important at about this time in the Wynoochee River valley on the southern side of the Olympic Mountains (L. E. Heusser, personal communication, 1972). During the last several millennia of the Wisconsin, the conifers (especially pine) spread in the beginning, and, subsequently, alder spread and

continued to develop in seral communities during the early part of the Holocene.

The pollen diagram from Fargher Lake (Figure 13-16), which is located just north of the Columbia River (Figure 13-10), is used to interpret the vegetation of the central part of the Puget-Willamette Lowland. Fargher Lake, just south of latitude 46°N, is the southernmost of the western sites in the region. Of the five pollen-assemblage zones recognized in a stratigraphic section, the Late Wisconsin covers the two upper zones FL-1 and FL-2. Applicable radiocarbon ages for this part of the section are 17,100 ± 650 (RL-1243), 20,500

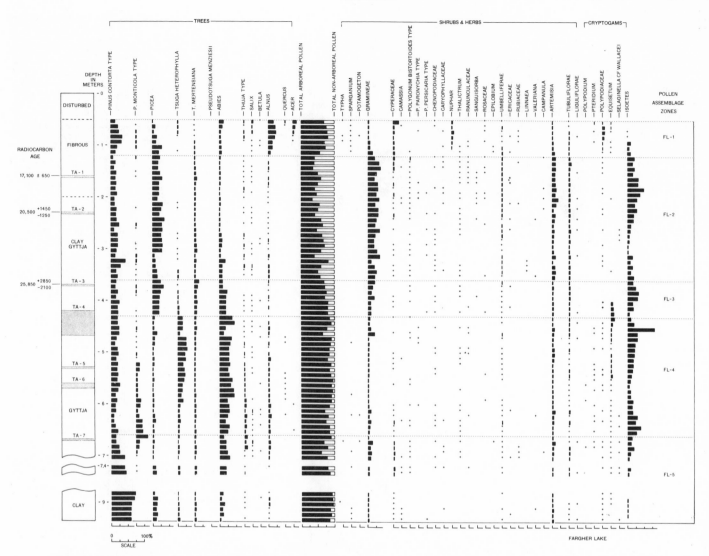

Figure 13-16. Pollen diagram of a section from Fargher Lake showing pollen assemblage zones to the right and sediment types with radiocarbon ages to the left; tephra layers are designated TA; plus symbols denote percentages of less than 2%.

+ 1450/ − 1250 (RL-1244), and 25,850 + 2850/ − 2100 yr B.P. (RL-1245). The pollen assemblage of zone FL-2, between about 25,000 and 12,500 years old, consists of *Picea, Pinus, Tsuga mertensiana,* and a significant amount of herbs contributed by the Gramineae and Tubuliflorae, including *Artemisia;* zone FL-1 from 12,500 to 10,000 years old is a *Pinus-Abies-Alnus* assemblage.

The vegetation around Fargher Lake appears to have been spruce parkland during the Late Wisconsin after having been high-montane forest, in which true fir and western hemlock and some mountain hemlock were important constituents. Pine may have grown in the vicinity, but the moderate amount of pine pollen implies transport from communities some distance away. No analogue for FL-2 spectra is evident in the surface-pollen data set. The best approximation is made from data for subalpine localities in the nearby Cascades of Washington and Oregon, where the components are best portrayed (Heusser, 1978b, 1978c). Engelmann spruce, which ranges today into the high Cascades, probably was the dominant species in the parkland. Pine and alder invaded the parkland toward the close of the Wisconsin, thus following the pattern of vegetational change that is

observed in the western Olympics.

At Fargher Lake, the percentages of nonarboreal taxa are not as high as in the Bogachiel and Hoh diagrams (Figures 13-12 and 13-13). Also, whereas spruce is a leading arboreal type, reaching 36% of the pollen sum, values of spruce in the Bogachiel and Hoh data are less than 10%. Reconstructed mean July temperatures during the Late Wisconsin fall to around 11 °C at Fargher Lake and to 10 °C in the western Olympic records; amounts of total annual precipitation are, respectively, 1600 and 1300 mm (Heusser and Heusser, 1980; Heusser et al., 1980). The reconstructed temperatures agree with the temperatures at the timberline in the present-day Olympics and Cascades (Figure 13-4). Respective temperatures today for the Fargher Lake and Olympic sites are about 17 °C and 15 °C; precipitation amounts to 1400 and 3200 mm.

Fossil-pollen data from the Puget-Willamette Lowland can be compared with the assemblages from Fargher Lake (Figure 13-17). Sites from south to north (Figure 13-10) are Davis Lake (Barnosky, 1979), Mineral Lake (Hibbert, 1979), Nisqually Lake (M. Tsukada, personal communication, 1974), and Strawberry Point (Hansen and Easterbrook, 1974). Assemblages covering the early part of the record (25,000 to 17,000 yr B.P.) consist of herbs with spruce and pine. Percentages of arboreal types decrease in assemblages north of Fargher Lake. At Point Grey in British Columbia (Figure 13-10), almost within the geographic limits of data coverage, arboreal types amount to only 3% to 13% in deposits about 24,500 years old (Mathewes, 1979). Later assemblages (17,000 to 10,000 yr B.P.) are spruce, pine, occasional mountain hemlock, and herbs until about 13,000 yr B.P.; afterward, pine is dominant until dominance shifts to alder during the early Holocene. The pollen stratigraphy of the Everson interstade type locality (Hansen and Easterbrook, 1974) and a section from Lake Washington east of Seattle (Hansen, 1974) also show the prominence of pine during the last three millennia of the Wisconsin. These data imply a vegetational gradient resembling the gradient to the west, that is, treeless conditions prevailing in the north changing southward to parkland, which becomes forest near Fargher Lake. The replacement of tundra and parkland by pine and other conifers and by alder is a late event common to the region west of the Cascade Range.

Vegetational history interpreted from pollen assemblages east of the Cascades provides detail for the Holocene, whereas only partial reconstructions apply to the Late Wisconsin, for few data are as old as 14,000 yr B.P. Localities studied (Figure 13-1) are Cub Creek Pond (Waddington and Wright, 1974) and Buckbean Fen (Baker, 1976) in northwestern Wyoming; Lost Trail Pass Bog (Mehringer et al., 1977) in western Montana; Swan Lake (Bright, 1966) in southeastern Idaho; and Waits Lake, Creston Fen, Big Meadow, and Mud Lake in northeastern Washington (Mack et al., 1976; 1979; Mack, Rutter, Bryant, and Valastro, 1978; Mack, Rutter, Valastro, and Bryant, 1978).

Pollen in surface samples relates to regional vegetation in the eastern part of the Rocky Mountains and the northern Great Plains (Baker, 1976; McAndrews and Wright, 1969). Pollen of spruce, fir, and Douglas-fir is found concentrated in the forested areas of the mountains and travels only a few tens of kilometers beyond its place of origin. Pine pollen also is concentrated both within and outside pine forest and was observed in quantities reaching 20% at stations over 300 km from the nearest stands of pine. Pollen of *Artemisia* occurs in large amounts on the plateaus and in the intermontane valleys of the Rockies where sagebrush dominates and also in the Great Plains province in relation to species of *Artemisia* other than sagebrush. Pollen of grasses and goosefoot (Chenopodiaceae) characterizes the plains, with goosefoot percentages higher in the basins than in their surroundings.

Surface pollen data collected in eastern Washington and northern Idaho (Mack and Bryant, 1974; Mack, Bryant, and Pell, 1978) tend to confirm these observations. In the Columbia Plateau province, pine pollen at stations about 10 km from stands of pine was found to be over 50% of the total; at stations more than 100 km distant, pine was usually at least 10%. In forests along the western border of the Rockies, increases in pollen from western red cedar and western hemlock, as well as from spruce, fir, and Douglas-fir, was directly related to the proportion of these trees in the forest communities.

Fossil-pollen assemblages at plateau sites in northwestern Wyoming between about 2300 and 2500 m altitude (Baker, 1976; Waddington and Wright, 1974) are largely of sagebrush, grass, and pine but may include birch and juniper as well as other taxa from before 14,000 to 12,000 yr B.P.; later assemblages consist of pine and sagebrush until about 10,000 yr B.P. (Figure 13-18). The common

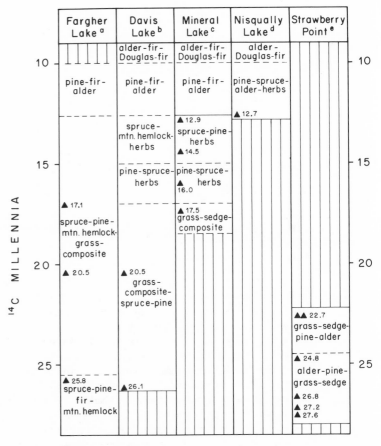

[a] Heusser & Heusser (1980; revised)
[b] Barnosky (1979)
[c] Hibbert (1979)
[d] Tsukada, personal communication, 1974
[e] Hansen & Easterbrook (1974)

Figure 13-17. Stratigraphic relations of pollen assemblage zones for sites in the Puget-Willamette Lowland, Washington. (See Figure 13-10 for locations.) Triangles indicate levels of radiocarbon ages (10^3 yr B.P.).

name of sagebrush used here may include one or more species of *Artemisia*, and sagebrush (*A. tridentata* and other species) may or may not be included. The source of *Artemisia* contributing to the pollen in the assemblages is problematical. Percentages of *Artemisia* in present-day subalpine and tundra communities in the area are higher than in the sagebrush steppe at lower altitudes. Local species of *Artemisia* appear to be low pollen producers, and so much of the pollen may originate in the steppe.

Pollen data indicate that tundra lay in proximity to the treeline of whitebark pine, spruce, subalpine fir, juniper, willow, and bog birch (*Betula glandulosa*). The tundra was replaced before about 11,500 yr B.P. by a spruce/subalpine fir/whitebark pine parkland that subsequently developed into a lodgepole and whitebark pine forest with openings of *Artemisia*. At Lost Trail Pass Bog (Figure 13-18), located 300 km northwest of the Wyoming plateau at an altitude of 2152 m (Mehringer et al., 1977), a somewhat similar set of assemblages is found between 12,000 and 10,000 yr B.P., for a sagebrush steppe after about 11,500 yr B.P. gave way to forest dominated by whitebark pine.

The group of sites in northeastern Washington portrays a parallel

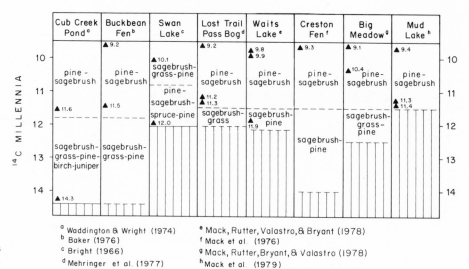

Figure 13-18. Stratigraphic relations of pollen assemblage zones for sites east of the Cascade Range. (See Figure 13-1 for locations.) Triangles indicate levels of radiocarbon ages (10³ yr B.P.).

[a] Waddington & Wright (1974)   [e] Mack, Rutter, Valastro, & Bryant (1978)
[b] Baker (1976)   [f] Mack et al. (1976)
[c] Bright (1966)   [g] Mack, Rutter, Bryant, & Valastro (1978)
[d] Mehringer et al. (1977)   [h] Mack et al. (1979)

vegetational history (Figure 13-18). At Waits Lake (Mack, Rutter, Valastro, and Bryant, 1978), mostly treeless communities of *Artemisia*, grass, and buffalo berry (*Shepherdia*) with some pine were established initially about 12,000 yr B.P. A decrease of *Artemisia* and an increase of pine followed, accompanied by an increase in alder and birch and by slight increases in spruce and fir until 10,000 yr B.P. A shift in the Haploxylon/Diploxylon pine ratio takes place over this interval, indicating that whitebark pine might have been prevalent at the beginning and lodgepole pine later. The similarity of Creston Fen assemblages (Mack et al., 1976) to Waits Lake assemblages implies the same vegetational changes to the southwest. The proportions of taxa vary, however, and buffalo berry is not recorded. Low values of pollen influx at Big Meadow (Mack, Rutter, Bryant, and Valastro, 1978) establish that the *Artemisia*-dominated assemblage represents tundralike vegetation, which lasted until almost 10,000 yr B.P., at which time the influx increased. Finally, Mud Lake (Mack et al., 1979), dating from 11,400 yr B.P., reveals the importance of pine (reaching 80% of the pollen sum) just east of the Cascades.

These records of tundralike vegetation near latitude 48°N to 49°N stand in contrast to the record at Swan Lake (Bright, 1966) in the sagebrush steppe of southeastern Idaho near latitude 42°N. A spruce-pine assemblage at Swan Lake dates from 12,090 yr B.P. (Figure 13-18), when tundra spread across northeastern Washington after the retreat of the Late Wisconsin glacier. Spruce (at 15%), some fir, and both Haploxylon- and Diploxylon-type pines account for about 85% of the pollen sum. These data point to more mesic conditions 12,000 years ago than exist today at Swan Lake and indicate that a refugium for spruce-fir forest must have existed there during the Late Wisconsin. This evidence is in keeping with the hypothesis (Flint, 1971; Wells, 1979) that precipitation/evaporation ratios in the Basin and Range province, in which Swan Lake is located, were greater during pluvial times than they are now.

An overview of the taxa and the concentration of pollen produced by Late Wisconsin vegetation in the northwestern conterminous United States is provided by records of pollen in marine cores with high sedimentation rates from the northeastern Pacific Ocean (Florer-Heusser, 1975; Heusser and Florer, 1973; Heusser and Shackleton, 1979). Cores taken at depths between 1500 and nearly 3000 m are from the continental slope and rise off the coast of Washington and Oregon. Several radiocarbon-dated horizons, the Mount Mazama volcanic ash layer, and, in one case, accompanying $^{18}O/^{16}O$ stratigraphy serve to establish the Late Wisconsin time range of the sediments. Pollen in regional marine core tops reflects the distribution of taxa in the modern coastal vegetation and shows the greatest concentration near the mouth of the Columbia River (Heusser and Balsam, 1977). These data imply that fluvial transport is the major means by which pollen is carried to the marine depositional environment. The predominance of westerly winds apparently minimized the aerial transport of pollen from the continent.

Late Wisconsin intervals in the cores reveal the primary importance of pine and the secondary importance of herbaceous pollen types in contrast to the greater amounts of arboreal taxa, spruce, western hemlock, and alder characteristic of the Holocene. Pine averages about 50%, with maxima over 75%, while herbs are close to 25% but reach over 40%. Additional arboreal taxa include spruce, western hemlock, fir, Douglas-fir, alder, and oak. These, present in significant numbers along with pine, indicate that coastal vegetation of some diversity was dominated by open forest communities that graded into parkland and ultimately into a narrow band of tundra adjacent to the glacier front in northwestern Washington. This interpretation corresponds to the reconstruction made from pollen data from continental western Washington. The striking increase in the prevalence of herbs in the southernmost core off Oregon (latitude 43°15′N) from the Late Wisconsin (Heusser and Shackleton, 1979) suggests increased parkland in the nearby Klamath Mountains.

Alaska

LANDSCAPE

Four major physiographic provinces recognized in Alaska are, from south to north, the Pacific Mountain System, Intermontane Plateaus, Rocky Mountain System, and Arctic Coastal Plain (Wahrhaftig, 1965). The Pacific Mountain System covers the Alaska Range and the regions along the Alaska Peninsula and Aleutian Islands and along the Gulf of Alaska and the Alaska panhandle (Figure 13-19). The highest peaks are found in the Pacific Mountain System and include Mount McKinley (6158 m), the highest in North America, and Mount St. Elias (5457 m). The Intermontane Plateaus coincide in a general way with the drainage of the Yukon River and its tributaries

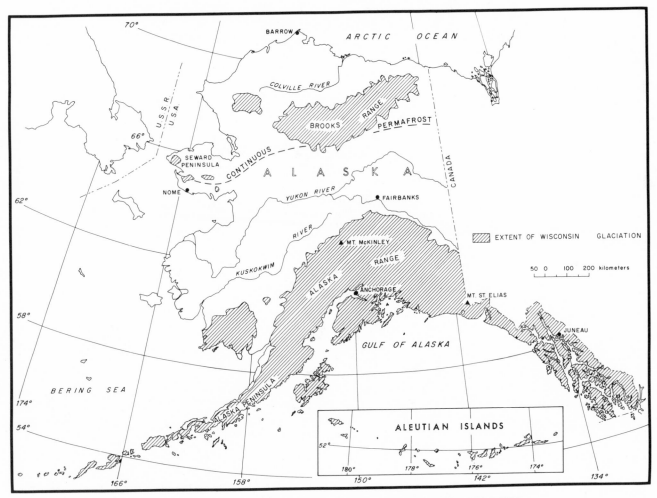

Figure 13-19. Extent of Wisconsin Glaciation and continuous permafrost in Alaska. (Data from Coulter et al., 1965, and Péwé, 1975.)

and the drainage of the Kuskokwim River, which flow into the Bering Sea. Permafrost, although discontinuous in the province, contributes to the widespread occurrence of sluggishly drained ground; thaw lakes occur where the permafrost has locally disappeared and in some sectors are numerous. The Rocky Mountain System to the north generally follows a latitudinal belt across Alaska that is identified by the Brooks Range, where summits in the northeastern part reach a maximum altitude of 2800 m. The Colville River drains the northern portion of the province and crosses the Arctic Coastal Plain before reaching the Arctic Ocean. The Arctic Coastal Plain, imperceptibly sloping and nearly featureless, is underlain by continuous permafrost and marked by polygons, pingos, and other features common to frozen ground.

GLACIATION

Wisconsin glaciation was extensive in the Brooks Range, the Alaska Range, and adjoining regions bordering the Pacific Ocean (Figure 13-19); in addition, the mountains southeast of the Kuskokwim River delta and in parts of the Seward Peninsula were areas of limited local glaciation (Coulter et al., 1965; Hamilton and Thorson, 1982; Péwé, 1975; Porter et al., 1982). Central and western Alaska as well as northern Alaska north of the Brooks Range remained virtually unglaciated. These regions contain thick sequences of nonglacial

sediments of Wisconsin age, including loess. In the Brooks Range, Late Wisconsin glacier advances culminated between approximately 24,000 and 17,000 yr B.P. and at about 12,800 to 12,500 yr B.P. (Hamilton and Thorson, 1982; Porter et al., 1982); in southern Alaska, the glaciation dated from after about 25,000 to about 14,000 yr B.P. with a later culmination about 12,800 to 11,800 yr B.P. (Black, 1976; Denton, 1974; Porter et al., 1982).

CLIMATIC REGIMEN

Storms moving eastward along the Aleutian low-pressure trough in winter affect the Pacific coastal district, causing heavy precipitation, strong wind, and cloudiness (Watson, 1974). Central Alaska at this time of year is dominated by high pressure, which occasionally extends southward to the southern coast. Also, winter cyclonic storms from the Aleutians occasionally move northeastward over the Bering Sea toward the Seward Peninsula or up the Yukon River drainage. In summer, under the influence of high pressure over the North Pacific, the frequency of storms diminishes along the southern coast; pressure over the interior in summer is low, by comparison, when heating of the land occurs.

The average precipitation decreases from the Pacific coastal district northward to the Arctic Ocean. Measurements for the years from 1931 to 1960 (Watson, 1974) are highest in the panhandle (Juneau,

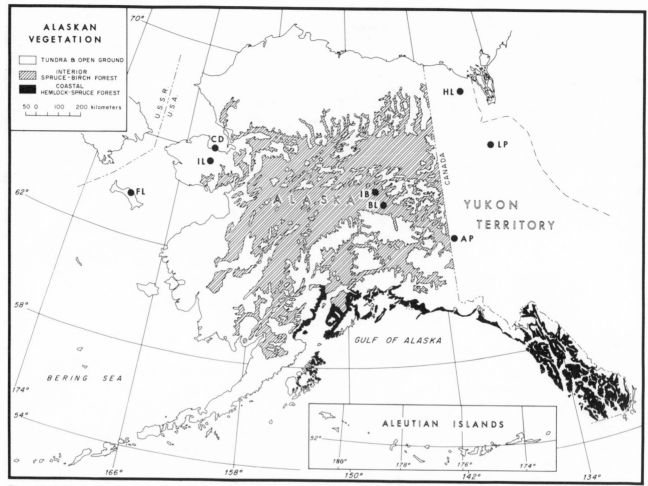

Figure 13-20. Distribution of vegetational units in Alaska from a map by Sigafoos (1958) showing locations of Late Wisconsin fossil pollen sites in Alaska (FL, Flora Lake; IL, Imuruk Lake; CD, Cape Deceit; IB, Isabella Basin; and BL, Birch Lake) and Yukon Territory (HL, Hanging Lake; LP, Lateral Pond; and AP, Antifreeze Pond).

2772 mm), lower for south-central, central, and western Alaska (Anchorage, 374 mm; Fairbanks, 352 mm; Nome, 454 mm), and lowest in the far north (Barrow, 108 mm). Precipitation increases at Pacific stations during the cooler part of the year and at stations to the north during the summer months. Snow accumulation is heavy in mountainous regions of southern Alaska; measurements of the 1964-1965 snowpack in the vicinity of Mount St. Elias (Figure 13-19) indicate as much as 5.8 m at the 1765-m level (Marcus and Ragle, 1970). Temperature data indicate relatively warm conditions near the southern coast in winter and in the northern interior in summer. This is shown by average January and July temperatures for the years 1931 to 1960 (Watson, 1974), respectively, at Juneau (−5.2°C, 12.7°C), Anchorage (−11.1°C, 14.6°C), Fairbanks (−23.9°C, 15.4°C), Nome (−15.3°C, 9.7°C), and Barrow (−26.8°C, 3.9°C).

MODERN VEGETATION

The modern vegetation of Alaska is represented by coastal hemlock-spruce forest, interior spruce-birch forest, and tundra (Sigafoos, 1958) (Figure 13-20). The coastal forest is a continuation of the Pacific slope coniferous forest of Washington and Oregon, and it includes tree species that range northward along the coast from that region. It covers much of southeastern Alaska to altitudes of 450 m. The unit

narrows in the west, and the treeline at the heads of the fjords lies at 100 m or less. Sitka spruce, the widest-ranging tree, forms the edge of the forest in the southwest; western hemlock, mountain hemlock, and Alaska yellow cedar range to the northwestern border of the Gulf of Alaska; and western red cedar, Pacific silver fir, western yew, subalpine fir, and lodgepole pine occur only in the southeastern district.

Interior spruce-birch forest extends irregularly southwestward from between latitude 62°N and 69°N along the Canadian border to south-central Alaska. Only along the southern edge is contact made with the coastal hemlock-spruce forest (Figure 13-21); along the northern border, the forest runs for varying distances up the valleys of southward-flowing rivers. The treeline on mountain slopes in the interior lies at altitudes of between 180 and 300 m. White spruce (*Picea glauca*) is the principal upland conifer and occurs in pure stands or with white birch (*Betula papyrifera*). Black spruce (*Picea mariana*) communities are coincident with poorly drained ground and muskegs; less commonly, they are found on the upland and may contain white birch and aspen (*Populus tremuloides*). Balsam poplar (*Populus balsamifera*) typically occupies the floodplains of rivers and grows beyond the limits of spruce in the north and west. Recurrent fires account for the preponderance of white birch and aspen over large sections of the interior.

Tundra and barren ground, at first transitional with the forest communities, lie beyond the limits of trees. Tundra is broadly classified as herb tundra—typified by cottongrass (*Eriophorum*), other sedges, and scattered dwarf birch (*Betula nana*) and crowberry (*Empetrum nigrum*) in places of impeded drainage—and as shrub tundra—formed of willow, alder (subspecies of *Alnus crispa* and *A. incana*, according to Hultén, 1968), dwarf birch, and heaths on better-drained sites. Tundra becomes expansive in northern and western Alaska; in central and southern Alaska, it forms part of a complex mosaic with the forest and muskeg communities.

VEGETATIONAL HISTORY

Quaternary deposits of glacial and nonglacial provenance are a familiar feature of Alaska. Although pollen in Holocene sections has been widely studied, few Late Wisconsin sections have been investigated for pollen. Pollen-stratigraphic data of Late Wisconsin age ascertained by radiocarbon dating are from sections located between latitude 65°N and 66°N (Figure 13-20) at Cape Deceit (Matthews, 1974a) and Imuruk Lake (Colinvaux, 1964) in the western tundra and in the Isabella Basin (Matthews, 1974b) and at Birch Lake (Ager, 1975) in the east-central part of the interior spruce-birch forest; an undated sequence at Flora Lake (Colinvaux, 1967a, 1967b) on St. Lawrence Island at the northern edge of the Bering Sea is also believed to be of Late Wisconsin age.

Wormwood, grass, and sedge are consistent components of the early pollen assemblages (Figure 13-22). In addition, three of the assemblages spanning the interval ending about 14,730 yr B.P. contain willow. Later assemblages, until about 10,000 yr B.P., consist mostly of birch, willow, and herbs; in the Isabella Basin, wormwood persisted in place of willow. The designation wormwood includes the many regional arctic and subarctic species of *Artemisia*. No finite radiocarbon ages mark the lower stratigraphic boundary of the wormwood assemblages that are in evidence for an estimated 10 millennia or more. Percentages of wormwood pollen in Late Wisconsin assemblages are much higher than in present-day surface samples. Colinvaux (1964) believes that loess deposited at the time may have provided a substrate more suitable for wormwood than any that exists today. Also, wormwood may be better adapted than the other plants in the tundra to enduring times of excessive desiccation and wind.

Figure 13-21. Parkland in south-central Alaska where interior spruce-birch forest and coastal hemlock-spruce forest are in contact.

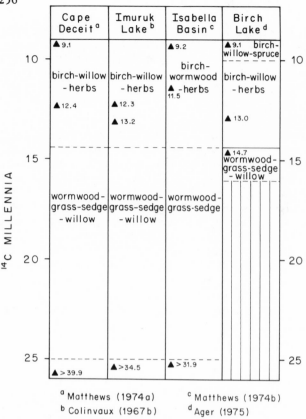

Figure 13-22. Stratigraphic relations of pollen assemblage zones for sites in Alaska. (See Figure 13-20 for locations.) Triangles indicate levels of radiocarbon ages (10^3 yr B.P.).

Matthews (1974a, 1974b) and Ager (1975) refer to the vegetation of the Late Wisconsin as steppe-tundra, which is comparable to the Siberian steppe-tundra of grass and *Artemisia*. Their data imply a shift in community makeup as well as structure after about 14,730 yr B.P., when the herbaceous character of the tundra took on a shrubby aspect. In the southwestern Yukon Territory at Antifreeze Pond (Figure 13-20), some 350 km southeast of Birch Lake, the shift did not take place until about 10,000 yr B.P. (Rampton, 1971) and apparently was delayed because of the influence of high altitudes in this mountainous part of the Yukon.

The concept of steppe-tundra has been rejected by Cwynar and Ritchie (1980) from pollen-influx studies at two northern Yukon sites (Figure 13-20). Influx values of less than 5 grains per square centimeter per year for herbs during the early part of the Late Wisconsin imply sparse herbaceous tundra resembling present-day fell-field vegetation. The rapid rise of birch between about 14,000 and 12,000 yr B.P. that has been interpreted as marking the change from herb tundra to shrub tundra is believed to be exaggerated because birch is overrepresented. The preferred interpretation is that, despite this increase, herb tundra remained in areas too xeric to be covered by shrub tundra of birch and other frutescent species.

The evidence for the presence of spruce in unglaciated interior Alaska and the adjacent Yukon Territory during the Late Wisconsin remains inconclusive. Spruce pollen in sediments of this time range occurs in amounts of only a few percent, and so transport from afar or redeposition from older deposits cannot be ruled out. Radiocarbon-dated spruce-wood samples are no older than 13,600 yr B.P. (Péwé, 1975). Spruce during the Middle Wisconsin formed a comparatively

distinctive member of an assemblage that included alder, thus providing a firmer base for interpreting local spruce communities from that time (Matthews, 1974b; Rampton, 1971). Nevertheless, several investigators contend that limited areas of spruce woodland existed in east-central Alaska during the Late Wisconsin, when, after the Middle Wisconsin interval of spruce expansion, the altitude of the treeline was lowered an estimated 400 to 600 m (Colinvaux, 1967b; Hopkins, 1967; Matthews, 1970, 1979; Péwé, 1975). It is perhaps from these refugia or from the conterminous United States or western Canada via an ice-free corridor between the Laurentide and Cordilleran ice sheets that spruce migrated and expanded in Alaska during the early Holocene (Hopkins et al., 1981).

Conclusions

Vegetation of the northwestern conterminous United States today consists of coniferous forests on the Pacific slope and in the mountains of the interior, steppe of sagebrush and grasses on the semiarid intermontane plateaus, and alpine tundra with barren ground above the upper altitudinal limit of trees. Distribution of the units results primarily from the influence of physiography on the humid westerly winds, which lose their moisture on windward mountainsides and become relatively dry crossing the interior plateaus.

Late Wisconsin vegetation, greatly modified by a contrasting set of climatic conditions, occupied lower altitudes and lower latitudes beyond the glacier margins. The Cordilleran and Laurentide ice sheets encroached in the north and spread south as far as latitude 47°N; glacier complexes also developed independently in the cordilleras. Evidence from fossil pollen preserved in radiocarbon-dated sediments, when interpreted in the context of modern pollen-vegetation relationships, indicates that Late Wisconsin tundralike vegetation stood close to the glacier margins. Latitudinal vegetational gradients along which tundra changed to parkland and eventually to coniferous forest are evident west of the Cascade Range and in the interior. Toward the end of the Wisconsin, as glaciers wasted away, invasion largely by conifers brought about replacement of the tundra and parkland by forest in which relatively mesic conditions prevailed.

Pollen records from interior Alaska indicate that herb tundra was widespread during most of the Late Wisconsin and only during the last several millennia did shrub tundra expand. Today, spruce-birch forest occupies the central lowland, forming a vegetation complex with muskeg and tundra. Fire and permafrost are the important factors influencing the distribution of the vegetational units.

References

Ager, T. A. (1975). "Late Quaternary Environmental History of the Tanana Valley, Alaska." Ohio State University Institute of Polar Studies Report 54.

Armstrong, J. E. (1975). "Quaternary Geology, Stratigraphic Studies and Revaluation of Terrain Inventory Maps, Fraser Lowland, British Columbia." Geological Survey of Canada Paper 75-1, part A, pp. 377-80.

Armstrong, J. E., Crandell, D. R., Easterbrook, D. J., and Noble, J. B. (1965). Late-Pleistocene stratigraphy and chronology in southwestern British Columbia and northwestern Washington. *Geological Society of America Bulletin* 76, 321-30.

Baker, R. G. (1976). "Late Quaternary Vegetation History of the Yellowstone Lake Basin, Wyoming." U.S. Geological Survey Professional Paper 729-E.

Baker, V. R. (1983). Late-Pleistocene pluvial systems. *In* "Late-Quaternary Environments of the United States," vol. 1, "The Late Pleistocene" (S. C. Porter, ed.), pp. 115-29. University of Minnesota Press, Minneapolis.

Barnosky, C. L. (1979). Late Quaternary vegetation history of the southern Puget lowland: A long record from Davis Lake, Washington. M.S. thesis, University of Washington, Seattle.

Black, R. F. (1976). Geology of Umnak Island, eastern Aleutian Islands, as related to the Aleuts. *Arctic and Alpine Research* 8, 7-35.

Bright, R. C. (1966). Pollen and seed stratigraphy of Swan Lake, southeastern Idaho: Its relation to regional vegetation history and to Lake Bonneville history. *Tebiwa* 9, 1-47.

Colinvaux, P. A. (1964). The environment of the Bering Land Bridge. *Ecological Monographs* 34, 297-329.

Colinvaux, P. A. (1967a). A long pollen record from St. Lawrence Island, Bering Sea (Alaska). *Palaeogeography, Palaeoclimatology, Palaeoecology* 3, 29-48.

Colinvaux, P. A. (1967b). Quaternary vegetational history of Arctic Alaska. *In* "The Bering Land Bridge" (D. M. Hopkins, ed.), pp. 207-31. Stanford University Press, Stanford, Calif.

Cordell, G. V., Jr., (1974). The climate of Montana. *In* "Climates of the States," vol. 2, "Western States" (F. van der Leeden and T. L. Troise, eds.), p. 743-62. Water Information Center, Port Washington, N.Y.

Coulter, H. W., Hopkins, D. M., Karlstrom, T. N. V., Péwé, T. L., Wahrhaftig, C., and Williams, J. R. (1965). Extent of glaciation in Alaska. U.S. Geological Survey Miscellaneous Geological Investigations Map I-415.

Crandell, D. R. (1965). The glacial history of western Washington and Oregon. *In* "The Quaternary of the United States" (H. E. Wright, Jr., and D. G. Frey, eds.), pp. 341-53. Princeton University Press, Princeton, N.J.

Cwynar, L. C., and Ritchie, J. C. (1980). Arctic steppe-tundra: A Yukon perspective. *Science* 208, 1375-77.

Daubenmire, R. F. (1943). Vegetation zonation in the Rocky Mountains. *The Botanical Review* 9, 325-93.

Daubenmire, R. F. (1952). Forest vegetation of northern Idaho and adjacent Washington, and its bearing on concepts of vegetation classification. *Ecological Monographs* 22, 301-30.

Daubenmire, R. F. (1956). Climate as a determinant of vegetation distribution in eastern Washington and northern Idaho. *Ecological Monographs* 26, 131-54.

Daubenmire, R. F. (1968). "Forest Vegetation of Eastern Washington and Northern Idaho." Washington Agricultural Experiment Station Technical Bulletin 60.

Daubenmire, R. F. (1970). "Steppe Vegetation of Washington." Washington Agricultural Experiment Station Technical Bulletin 70.

Denton, G. H. (1974). Quaternary glaciations of the White River valley, Alaska, with a regional synthesis for the northern St. Elias Mountains, Alaska and Yukon Territory. *Geological Society of America Bulletin* 85, 871-92.

Douglas, G. W., and Bliss, L. C. (1977). Alpine and high subalpine plant communities of the North Cascade Range, Washington and British Columbia. *Ecological Monographs* 47, 113-50.

Fenneman, N. M. (1931). "Physiography of the Western United States." McGraw-Hill, New York.

Flint, R. F. (1971). "Glacial and Quaternary Geology." John Wiley and Sons, New York.

Florer, L. E. (1972). Quaternary paleoecology and stratigraphy of the sea cliffs, western Olympic Peninsula, Washington. *Quaternary Research* 2, 202-16.

Florer-Heusser, L. E. (1975). Late Cenozoic marine palynology of northeastern Pacific Ocean cores. *In* "Quaternary Studies" (R. P. Suggate and M. M. Cresswell, eds.), pp. 133-38. Royal Society of New Zealand Bulletin 13.

Fonda, R. W., and Bernardi, J. A. (1976). Vegetation of Sucia Island in Puget Sound, Washington. *Torrey Botanical Club Bulletin* 103, 99-109.

Fonda, R. W., and Bliss, L. C. (1969). Forest vegetation of the montane and subalpine zones, Olympic Mountains, Washington. *Ecological Monographs* 39, 271-301.

Fowells, H. A. (1965). "Silvics of Forest Trees of the United States." U.S. Department of Agriculture Forest Service Agriculture Handbook 271.

Franklin, J. F., and Dyrness, C. T. (1973). "Natural Vegetation of Oregon and Washington." U.S. Department of Agriculture Forest Service General Technical Report PNW-8.

Habeck, J. R. (1968). Forest succession in the Glacier Park cedar-hemlock forest. *Ecology* 49, 872-80.

Hamilton, T. D., and Thorson, R. M. (1983). The Cordilleran ice sheet in Alaska. *In*

"Late-Quaternary Environments of the United States," vol. 1, "The Late Pleistocene" (S. C. Porter, ed.), pp. 38-52. University of Minnesota Press, Minneapolis.

Hansen, B. S., and Easterbrook, D. J. (1974). Stratigraphy and palynology of late Quaternary sediments in the Puget lowland, Washington. *Geological Society of America Bulletin* 85, 587-602.

Hansen, H. P. (1947). Postglacial forest succession, climate, and chronology in the Pacific Northwest. *American Philosophical Society Transactions* 37, 1-130.

Hansen, H. P. (1974). "Pollen Analysis of a Deep Section from Lake Washington, Seattle." Birbal Sahni Institute of Paleobotany Special Publication 5, pp. 1-4.

Heusser, C. J. (1957). Pleistocene and postglacial vegetation of Alaska and the Yukon Territory. *In* "Arctic Biology" (H. P. Hansen, ed.), pp. 62-72. Eighteenth Annual Biology Colloquium, Oregon State College, Corvallis.

Heusser, C. J. (1960). "Late Pleistocene Environments of North Pacific North America." American Geographical Society Special Publication 35.

Heusser, C. J. (1965). A Pleistocene phytogeographical sketch of the Pacific Northwest and Alaska. *In* "The Quaternary of the United States" (H. E. Wright, Jr., and D. G. Frey, eds.), pp. 469-83. Princeton University Press, Princeton, N.J.

Heusser, C. J. (1969a). Late-Pleistocene coniferous forests of the northern Rocky Mountains. *In* "Coniferous Forests of the Northern Rocky Mountains" (R. D. Taber, ed.), pp. 1-23. University of Montana Foundation, Missoula.

Heusser, C. J. (1969b). Modern pollen spectra from the Olympic Peninsula, Washington. *Torrey Botanical Club Bulletin* 96, 407-17.

Heusser, C. J. (1972). Palynology and phytogeographical significance of a late-Pleistocene refugium near Kalaloch, Washington. *Quaternary Research* 2, 189-201.

Heusser, C. J. (1973a). Age and environment of allochthonous peat clasts from the Bogachiel River valley, Washington. *Geological Society of America Bulletin* 64, 797-804.

Heusser, C. J. (1973b). Environmental sequence following the Fraser advance of the Juan de Fuca lobe, Washington. *Quaternary Research* 3, 284-306.

Heusser, C. J. (1973c). Modern pollen spectra from Mount Rainier, Washington. *Northwest Science* 47, 1-8.

Heusser, C. J. (1973d). Postglacial vegetation on Umnak Island, Aleutian Islands, Alaska. *Palaeobotany and Palynology Review* 15, 277-85.

Heusser, C. J. (1974). Quaternary vegetation, climate, and glaciation of the Hoh River valley, Washington. *Geological Society of America Bulletin* 85, 1547-60.

Heusser, C. J. (1977a). Quaternary palynology of the Pacific slope of Washington. *Quaternary Research* 8, 282-306.

Heusser, C. J. (1977b). A survey of Pleistocene pollen types of North America. *In* "Contributions of Stratigraphic Palynology," vol. 1, "Cenozoic Palynology" (W. C. Elsik, ed.), pp. 111-29. American Association of Stratigraphic Palynologists Contribution Series 5A.

Heusser, C. J. (1978a). Modern pollen rain in the Puget lowland of Washington. *Torrey Botanical Club Bulletin* 105, 296-305.

Heusser, C. J. (1978b). Modern pollen rain of Washington. *Canadian Journal of Botany* 56, 1510-17.

Heusser, C. J. (1978c). Modern pollen spectra from western Oregon. *Torrey Botanical Club Bulletin* 105, 14-17.

Heusser, C. J. (1978d). Palynology of Quaternary deposits of the lower Bogachiel River area, Olympic Peninsula, Washington. *Canadian Journal of Earth Science* 15, 1568-78.

Heusser, C. J., and Florer, L. E. (1973). Correlation of marine and continental Quaternary pollen records from the Northeast Pacific and western Washington. *Quaternary Research* 3, 661-70.

Heusser, C. J., and Heusser, L. E. (1980). Sequence of pumiceous tephra layers and the late Quaternary environmental record near Mount St. Helens. *Science* 210, 1007-9.

Heusser, C. J., Heusser, L. E., and Streeter, S. S. (1980). Quaternary temperatures and precipitation for the north-west coast of North America. *Nature* 286, 702-4.

Heusser, L. E., and Balsam, W. L. (1977). Pollen distribution in the Northeast Pacific Ocean. *Quaternary Research* 7, 45-62.

Heusser, L. E., and Shackleton, N. J. (1979). Direct marine-continental correlation:

150,000-year oxygen isotope-pollen record from the North Pacific. *Science* 204, 837-39.

Hibbert, D. M. (1979). Pollen analysis of late-Quaternary sediments from two lakes in the southern Puget lowland, Washington. M.S. thesis, University of Washington, Seattle.

Hitchcock, C. L., and Cronquist, A. (1973). "Flora of the Pacific Northwest." University of Washington Press, Seattle.

Hopkins, D. M. (1967). The Cenozoic history of Beringia: A synthesis. *In* "The Bering Land Bridge" (D. M. Hopkins, ed.), pp. 451-4. Stanford University Press, Stanford, Calif.

Hopkins, D. M., Smith, P. A., and Matthews, J. V., Jr. (1981). Dated wood from Alaska and the Yukon: Implications for forest refugia in Beringia. *Quaternary Research* 15, 217-49.

Hultén, E. (1968). "Flora of Alaska and Neighboring Territories." Stanford University Press, Stanford, Calif.

Johnson, P. L., and Billings, W. D. (1962). The alpine vegetation of the Beartooth Plateau in relation to cryopedogenic processes and patterns. *Ecological Monographs* 32, 105-35.

Küchler, A. W. (1964). "Potential Natural Vegetation of the Conterminous United States." American Geographical Society Special Publication 36.

Lowers, A. R. (1974). The climate of Wyoming. *In* "Climates of the States," vol. 2, "Western States" (F. van der Leeden and F. L. Troise, eds.), pp. 961-75. Water Information Center, Port Washington, N.Y.

McAndrews, J. H., and King, J. E. (1976). Pollen of the North American Quaternary: The top twenty. *Geoscience and Man* 15, 41-49.

McAndrews, J. H., and Wright, H. E., Jr. (1969). Modern pollen rain across the Wyoming basins and the northern Great Plains (U.S.A.). *Palaeobotany and Palynology Review* 9, 17-43.

Mack, R. N., and Bryant, V. M., Jr. (1974). Modern pollen spectra from the Columbia Basin, Washington. *Northwest Science* 48, 183-94.

Mack, R. N., and Bryant, V. M., Jr., and Fryxell, R. (1976). Pollen sequence from the Columbia Basin, Washington: Reappraisal of postglacial vegetation. *The American Midland Naturalist* 95, 390-97.

Mack, R. N., Bryant, V. M., Jr., and Pell, W. (1978). Modern forest pollen spectra from eastern Washington and northern Idaho. *Botanical Gazette* 139, 249-55.

Mack, R. N., Rutter, N. W., Bryant, V. M., Jr., and Valastro, S. (1978). Late Quaternary pollen record from Big Meadow, Pend Oreille County, Washington. *Ecology* 59, 956-65.

Mack, R. N., Rutter, N. W., and Valastro, S. (1979). Holocene vegetation history of the Okanogan Valley, Washington. *Quaternary Research* 12, 212-25.

Mack, R. N., Rutter, N. W., Valastro, S., and Bryant, V. M., Jr. (1978). Late Quaternary vegetation history at Waits Lake, Colville River valley, Washington. *Botanical Gazette* 139, 499-506.

Marcus, M. G., and Ragle, R. H. (1970). Snow accumulation in the Icefield Ranges, St. Elias Mountains, Yukon. *Arctic and Alpine Research* 2, 277-92.

Mathewes, R. W. (1979). A paleoecological analysis of Quadra Sand at Point Grey, British Columbia, based on indicator pollen. *Canadian Journal of Earth Sciences* 16, 847-58.

Matthews, J. V., Jr. (1970). Quaternary environmental history of interior Alaska: Pollen samples from organic colluvium and peats. *Arctic and Alpine Research* 2, 241-51.

Matthews, J. V., Jr. (1974a). Quaternary environments at Cape Deceit (Seward Peninsula, Alaska): Evolution of an ecosystem. *Geological Society of America Bulletin* 85, 1353-84.

Matthews, J. V., Jr. (1974b). Wisconsin environment of interior Alaska: Pollen and macrofossil analysis of a 27 meter core from the Isabella Basin (Fairbanks, Alaska). *Canadian Journal of Earth Science* 11, 828-41.

Matthews, J. V., Jr. (1979). Tertiary and Quaternary environments: Historical background for an analysis of the Canadian insect fauna. *In* "Canada and Its Insect Fauna" (H. V. Danks, ed.), pp. 31-86. Entomological Society of Canada Memoir 108.

Mehringer, P. J., Jr., Arno, S. S., and Petersen, K. L. (1977). Postglacial history of Lost Trail Pass Bog, Bitterroot Mountains, Wyoming. *Arctic and Alpine Research* 10, 171-80.

Mickelson, D. M., Clayton, L., Fullerton, D. S., and Borns, H. W., Jr. (1983). The Late Wisconsin glacial record of the Laurentide ice sheet in the United States. *In* "Late-Quaternary Environments of the United States," vol. 1, "The Late Pleistocene" (S. C. Porter, ed.), pp. 3-37. University of Minnesota Press, Minneapolis.

Péwé, T. L. (1975). "Quaternary Geology of Alaska." U.S. Geological Survey Professional Paper 835.

Phillips, E. L. (1974). The climate of Washington, *In* "Climates of the States," vol. 2, "Western States" (F. van der Leeden and F. L. Troise, eds.), pp. 935-60. Water Information Center, Port Washington, N.Y.

Pierce, K. L. (1979). "History and Dynamics of Glaciation in the Northern Yellowstone National Park Area." U.S. Geological Survey Professional Paper 729-F.

Pierce, K. L., Obradovich, J. D., and Friedman, I. (1976). Obsidian hydration dating and correlation of Bull Lake and Pinedale Glaciations near West Yellowstone, Montana. *Geological Society of America Bulletin* 87, 703-10.

Porter, S. C. (1976). Pleistocene glaciation in the southern part of the North Cascade Range, Washington. *Geological Society of America Bulletin* 87, 61-75.

Porter, S. C. (1978). Glacier Peak tephra in the North Cascade Range, Washington: Stratigraphy, distribution, and relationship to late-glacial events. *Quaternary Research* 10, 30-41.

Porter, S. C., Pierce, K. L., and Hamilton, T. D. (1983). Late Wisconsin mountain glaciation in the western United States. *In* "Late-Quaternary Environments of the United States," vol. 1, "The Late Pleistocene" (S. C. Porter, ed.), pp. 71-114. University of Minnesota Press, Minneapolis.

Powers, H. A., and Wilcox, R. E. (1964). Volcanic ash from Mount Mazama (Crater Lake) and from Glacier Peak. *Science* 144, 1334-36.

Rampton, V. (1971). Late Quaternary vegetational and climatic history of the Snag-Klutlan area, southwestern Yukon Territory, Canada. *Geological Society of America Bulletin* 82, 959-78.

Rice, K. A. (1974). The climate of Idaho. *In* "Climates of the States," vol. 2, "Western States" (F. van der Leeden and F. L. Troise, eds.), pp. 640-56. Water Information Center, Port Washington, N.Y.

Richmond, G. M. (1965). Glaciation of the Rocky Mountains. *In* "The Quaternary of the United States" (H. E. Wright, Jr., and D. G. Frey, eds.), pp. 217-30. Princeton University Press, Princeton, N.J.

Richmond, G. M., Fryxell, R., Neff, G. E., and Weis, P. L. (1965). The Cordilleran ice sheet of the northern Rocky Mountains, and related Quaternary history of the Columbia Plateau. *In* "The Quaternary of the United States" (H. E. Wright, Jr., and D. G. Frey, eds.), pp. 231-42. Princeton University Press, Princeton, N.J.

Sigafoos, R. S. (1958). "Vegetation of Northwestern North America, as an Aid in Interpretation of Geologic Data." U.S. Geological Survey Bulletin 1061-E, pp. 165-83.

Sternes, G. L. (1974). The climate of Oregon. *In* "Climate of the States," vol. 2, "Western States" (F. van der Leeden and F. L. Troise, eds.), pp. 841-60. Water Information Center, Port Washington, N.Y.

Waddington, J. C. B., and Wright, H. E., Jr. (1974). Late Quaternary vegetational changes on the east side of Yellowstone Park, Wyoming. *Quaternary Research* 4, 175-84.

Wahrhaftig, C. (1965). "Physiographic Divisions of Alaska." U.S. Geological Survey Professional Paper 482.

Waitt, R. B., and Thorson, R. M. (1983). The Cordilleran ice sheet in Washington, Idaho, and Montana. *In* "Late-Quaternary Environments of the United States," vol. 1, "The Late Pleistocene" (S. C. Porter, ed.), pp. 53-70. University of Minnesota Press, Minneapolis.

Waring, R. H., and Franklin, J. F. (1979). Evergreen coniferous forests of the Pacific Northwest. *Science* 204, 1380-86.

Watson, C. E. (1974). The climate of Alaska. *In* "Climate of the States," vol. 2, "Western States" (F. van der Leeden and F. L. Troise, eds.), pp. 481-502. Water Information Center, Port Washington, N.Y.

Wells, P. V. (1979). An equable glaciopluvial in the West: Pleniglacial evidence of increased precipitation on a gradient from the Great Basin to the Sonoran and Chihuahuan Deserts. *Quaternary Research* 12, 311-25.

Wright, H. E., Jr. (1971). Late Quaternary vegetational history of North America. *In* "The Late Cenozoic Ice Ages" (K. K. Turekian, ed.), pp. 421-64. Yale University Press, New Haven, Conn.

CHAPTER 14

Late Wisconsin Paleoecology
of the American Southwest

W. Geoffrey Spaulding, Estella B. Leopold,
and Thomas R. Van Devender

Introduction

The southwestern United States, with its diverse climate and biota, has a complex Pleistocene history. In this mostly arid region the impact of continental glaciation was indirect but nevertheless pronounced. The question of the effect of glacial climates on desert ecosystems has been raised since at least the early 20th century, when Spalding (1909) suggested that long-term movements of desert plants were related to geologic processes. An elaborate fossil record now provides abundant evidence for these movements, and it shows an environment and ecology remarkably different from today's. The Southwest is of particular interest to paleoecologists because of the new literature on plant macrofossils from ancient middens of the packrat *Neotoma*. The combination of pollen-stratigraphic and plant-macrofossil data provides a unique view of a landscape now lost.

In this chapter, we describe advances in paleoecology and offer a state-of-the-art view of the Late Wisconsin biota of the American Southwest. Our study area encompasses the interior West south of latitude 42°N, west of longitude 100°W, and east of the Sierra Nevada (Figure 14-1). We revise and expand Martin and Mehringer's (1965) work on the late-Pleistocene biogeography of this region, which provides a baseline for our review. It and other paleoecologic studies that rely primarily on pollen analyses also provide hypotheses that are testable with additional techniques and data.

Southwestern Phytogeography

Shmida and Whittaker (1979) note that the deserts of North America are smaller than those of the Old World and, moreover, are not dry enough to be considered deserts in a strict sense. Only the valleys of the Mojave and Chihuahuan Deserts and land along the lower Colorado River and upper Gulf of California are "true" deserts, with less than 150 mm average annual precipitation and less than 10% perennial plant cover. Floristic similarities among the steppe deserts of Eurasia and North America are marked; they share a relatively large number of identical or vicarious species (Yurtsev, 1963, 1972). Other plant taxa important to both New World and Old World steppes appear to have common ancestry (e.g., *Artemisia* spp.) (MacArthur and Plummer, 1978). On the other hand, few forms are common to

the warm deserts of both North America and the Old World. This reflects the general pattern of biotic exchanges that occurred across the Bering Land Bridge; cold-tolerant forms were involved, but thermophilous taxa remained relatively isolated in the southern latitudes.

Whether true desert or semidesert in a global perspective (and we attempt no distinction, following local tradition instead), the prime limiting ecologic factors in interior western North America are available moisture and minimum winter temperatures. Consequently, minor variations in geography, topography, and substrate often result in major differences in the composition of plant communities (Beatley, 1975; Shelford, 1978; Whittaker and Niering, 1965). Succession plays a minor role in most Southwest plant communities, and most associations can be considered climax vegetation (Shreve, 1942a).

COOL DESERTS

Steppe vegetation occurs north of latitude 37°N in the Great Basin Desert (Shreve, 1942a; the Intermountain Region of Cronquist et al., 1972), where basin floors usually exceed 1400 m elevation and where annual rainfall is less than about 460 mm (Figure 14-1). Steppe covers vast areas in Washington, Oregon, Idaho, Nevada, and Utah. Great Basin desert scrub (Brown et al., 1979) occurs outside the physiographic Great Basin in the Columbia River basin (Daubenmire, 1970), on the eastern flank of the Rocky Mountains, and on the Colorado Plateau of northern Arizona, Utah, and western Colorado.

Shrub communities dominated by various species of tridentate-leaved sagebrushes (*Artemisia* sec. Tridentatae) (Beetle, 1960; MacArthur and Plummer, 1978) characterize the valley flanks over much of the Intermountain Region. Species of saltbush (e.g., *Atriplex confertifolia* S. Wats., *A. canescens* Nutt., *A. polycarpa* S. Wats., *A. torreyi* S. Wats.) are often the dominant perennials near the basin floors. Other important shrubs include the composites rabbitbrush (*Chrysothamnus* spp.) and horsebrush (*Tetradymia* spp.) and the

We wish to thank K. L. Cole, R. S. Thompson, J. L. Betancourt, and P. J. Mehringer, Jr., for sharing their unpublished data. Critical reviews and helpful discussions were also provided by R. G. Baker, P. S. Martin, D. K. Grayson, V. M. Bryant, G. R. Brakenridge, and V. Markgraf. Funding for much of this research was provided by the National Science Foundation.

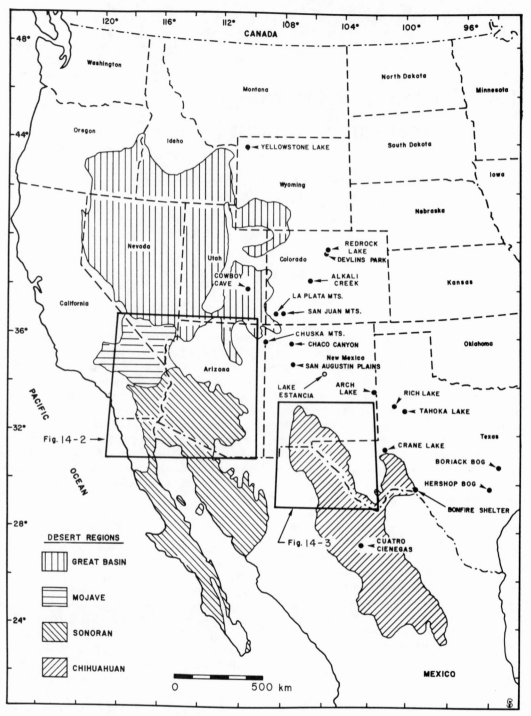

Figure 14-1. Map of the deserts of North America showing localities mentioned in the text. A solid circle represents selected pollen sites; and an open circle, other localities, excluding packrat-midden sites. (After Shreve, 1942a.)

chenopods winterfat (*Ceratoides lanata* J. T. H.), greasewood (*Sarcobatus vermiculatus* Torr.), and hopsage (*Grayia spinosa* Moq.). Above the shrub communities of the central and southern Great Basin Desert is a zone of pinyon-juniper woodland (*Pinus monophylla* Torr. and Frem. and *Juniperus osteosperma* Little) (Tueller et al., 1979; West et al., 1978). Woodland is absent on mountain flanks north of about latitude 40°N and the vegetation above desert scrub is grassy steppe or montane coniferous forest.

WARM DESERTS

The vegetation of warm deserts is more diverse than that of the steppe in terms of perennial species, growth forms, climatic regimes, and geographic entities (Shmida and Whittaker, 1979; Shreve, 1942a, 1964). The one plant that is an important element of all North American warm deserts is the creosote bush (*Larrea divaricata* Cav. subsp. *tridentata* Felger and Lowe) (Mabry et al., 1977; Wells and Hunziker, 1976).

The Sonoran Desert of Arizona, California, and adjacent Mexico (Figure 14-2) is a subtropical warm desert that rarely experiences sustained freezing temperatures. The Mojave Desert of southern Nevada, adjacent Arizona, and eastern California (Figure 14-2) and the Chihuahuan Desert of southern New Mexico, western Texas, and the Mexican Plateau (Figure 14-3) are warm temperate deserts with lower winter temperatures. Basin floors in the Sonoran and Mojave Deserts extend to, or nearly to, sea level; in the Chihuahuan Desert, they lie above 800 m elevation.

Creosote-bush desert scrub is typical of the low desert plains in the Chihuahuan Desert. Mixed desert scrub, dominated variously by acacia (*Acacia neovernicosa* Isely), mesquite (*Prosopis glandulosa* Torr.), tarbush (*Flourensia cernua* DC.), mariola (*Parthenium incanum* H. B. K.), and resinbush (*Viguiera stenoloba* S. F. Blake), oc-

curs at higher altitudes (Johnston, 1977). Arborescent plants are generally absent from the northern Chihuahuan Desert. Small succulents are important in many desert scrub communities, particularly agave (*Agave lechuguilla* Torr.). Grasslands play an important role at intermediate elevations, as well as in edaphic communities on poorly drained soils and on igneous outcrops (Shreve, 1942b; Johnston, 1977). Intense grazing by domestic stock in the northern Chihuahuan Desert appears to have caused the development of "disclimax" desert scrub in what used to be grassland (York and Dick-Peddie, 1969).

Shreve (1964) recognizes two subdivisions of the Sonoran Desert in the United States (Figure 14-2). The section along the lower Colorado River and the head of the Gulf of California is the largest contiguous area of "true" desert in North America. This "Colorado Desert" of

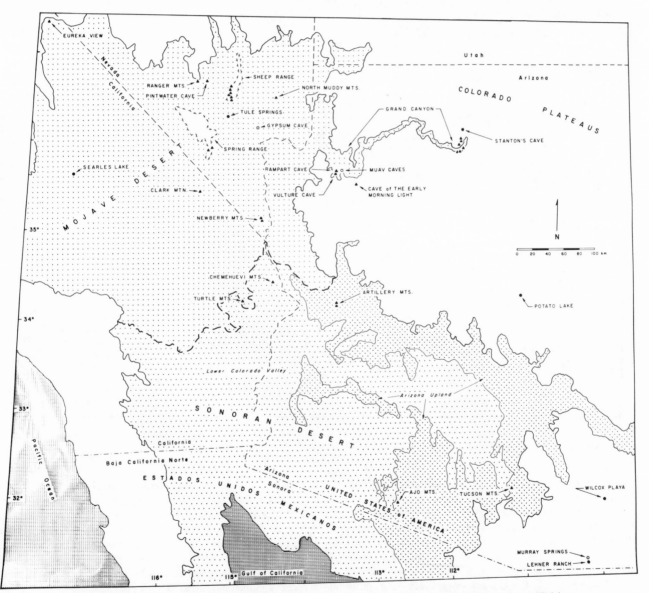

Figure 14-2. Map of the northern Sonoran and Mojave Deserts showing localities mentioned in the text. The vegetational distribution is simplified from Brown and Lowe (1980). A solid triangle represents packrat-midden sites; other symbols follow Figure 14-1. The bold dashed line separates the Sonoran and Mojave Deserts, and the Arizona Upland and the lower Colorado Valley sections of the Sonoran Desert are separated by the lighter line. (See Figure 14-1 for the relative position of this map.)

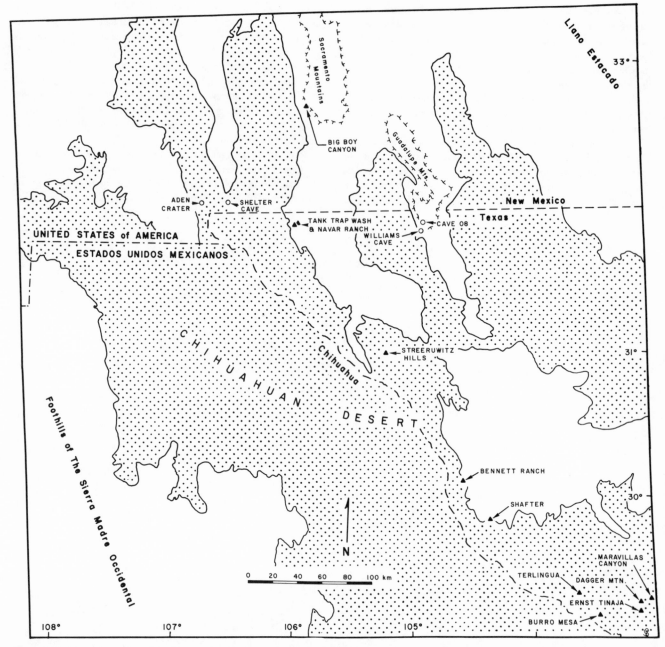

Figure 14-3. Map of the northern Chihuahuan Desert showing localities mentioned in the text. Vegetational distribution is simplified from Brown and Lowe (1980). Symbols for fossil sites follow Figures 14-1 and 14-2. (See Figure 14-1 for the relative position of this map.)

some authors is biogeographically part of the Sonoran Desert (Shreve, 1942a). Mountains occupy a smaller proportion of this lower Colorado River valley section than any other part of the Sonoran Desert, and creosote-bush desert scrub is the dominant vegetational type. Vegetation is frequently restricted to small drainages, and the interfluves may be nearly devoid of perennial plant cover. Common plants on rocky slopes include white bursage (*Ambrosia dumosa* Payne), blue paloverde (*Cercidium floridum* Benth.), and brittlebush (*Encelia farinosa* Gray).

Creosote-bush desert scrub is not as widespread a community type in the Arizona Upland section of the Sonoran Desert. The region lies at higher elevations characterized by coarser soils and greater precipitation than in the lower Colorado River valley (Figure 14-2). The paloverde-saguaro (*Cerdicium microphyllum* Rose & Johnst.—*Carnegia gigantea* Brit. & Rose) association typifies Arizona Upland vegetation and appears lush when compared to vegetation in the lower Colorado River valley. Common shrubs or small trees include creosote bush, ocotillo (*Fouquieria splendens* Engelm., also common in the Chihuahuan Desert), whitethorn acacia (*Acacia constricta* Benth.), mesquite (*Prosopis velutina* Woot.), jojoba (*Simmondsia chinensis* Schneid.), triangle-leaf bursage (*Ambrosia deltoidea* Payne), and brittle bush (*Encelia farinosa* Gray).

A greater array of high-elevation vegetational types surrounds the Sonoran and Chihuahuan Deserts than the Intermountain Region to the north. Grassland, chaparral, pine-oak or oak woodland, and pinyon-juniper woodland are found above warm-desert scrub communites (Brown and Lowe, 1980; Martin, 1963b). The great variety of upland communities in the highlands of the southern deserts may be due to mild winter temperatures and an increased proportion of summer precipitation.

The Mojave Desert is the smallest of North America's deserts, and floristically it is the least well defined. Staggered distributional limits of a suite of Mojave and Sonoran Desert plants places the southern Mojave Desert boundary at about latitude 34°N (Figure 14-2) (Brown and Lowe, 1980; Hastings et al., 1972). Decreasing amounts of summer precipitation to the northwest and lower winter temperatures are important determinants of this boundary. The northern limit of creosote bush (betweeen latitudes 37°N and 37°30′N) provides a marker for the Mojave Desert/Great Basin Desert boundary (Cronquist et al., 1972; Shreve, 1942a). Basin floors increase in elevation markedly to the north in this area, augmenting the latitudinal decline in thermal regimes (Beatley, 1974, 1975).

The low areas of the Mojave Desert support saltbush or creosote-bush vegetation, and mixed desert scrub vegetation occurs on the slopes at intermediate elevations. Shrubs such as creosote bush, hopsage (*Grayia spinosa*), wolfberry (*Lycium pallidum* Miers., *L. andersonii* Gray), pepperweed (*Lepidium fremontii* Wats.), Mormontea (*Ephedra californica* Wats., *E. nevadensis* Wats.) occur both in the lowlands and on the valley flanks. Blackbrush (*Coleogyne ramosissima* Torr.) desert scrub covers extensive areas on alluvial fans between about 1300 m and about 1850 m elevation. Joshua-tree and Mojave yucca (*Yucca brevifolia* Engelm. and *Y. schidigera* Roezl.) are often prominent members of the blackbrush community, the former being the only arborescent species common in the Mojave. Blackbrush vegetation is rich in shrub and semishrub associates such as pepperweed, ratany (*Krameria parvifolia* Benth.), ground thorn (*Menodora spinescens* Gray), and desert almond (*Prunus fasciculata* Gray). The upland vegetation of the Mojave Desert is similar to that of the Great Basin Desert. Interior chaparral frequently dominated by mountain mahogany (*Cercocarpus ledifolius* Wats. or *C. intricatus* Nutt.) occurs on xeric aspects, and *Pinus monophylla* and *Juniperus osteosperma* woodlands dominate the mesic slopes.

Chronology and Evidence of Wisconsin Climates

Extensive lake deposits in the American deserts correlate in time with continental glacial deposits and establish the fact that mesic conditions prevailed during the Wisconsin Glaciation in North America (Figure 14-4) (Morrison, 1965; Morrison and Frye, 1965). The correspondence between times of deep lakes in the southwestern basins and times of continental glaciation in the North is the chief basis for correlating "pluvial" with glacial events. The radiocarbon dating of ancient shoreline features provides evidence for a complex chronology spanning the last 45,000 years (Benson, 1978; Mehringer, 1977; Smith, 1977).

The major Wisconsin lake systems of the Great Basin (Figure 14-4), outlined by Snyder and others (1964), are the Lahontan and Bonneville lakes in the Great Basin and smaller systems in the Mojave Desert of California, including Searles Lake and others formerly connected by the Owens River (Smith, 1977; Smith and Street-Perrott, 1983). Although lake levels were generally high during the Late Wisconsin, records of major lake-level fluctuations of different basins are not in full agreement (Benson, 1978; Broecker and Orr, 1958; Smith, 1968; Smith and Street-Perrott, 1983).

The chronologic divisions we utilize are based on the radiocarbon ages for major climatic changes in the Great Basin and Mojave Deserts. The transition from Middle to Late Wisconsin is set at 24,000 yr B.P., the date for the basal Parting Mud from Searles Lake (Smith, 1977) (Figure 14-17) and the approximate time of a marked rise in *Pinus* and a decline in Chenopodiineae pollen in pluvial Lake Bonneville (Mehringer, 1977) (Figure 14-13), as well as a significant rise in Lake Lahontan (Benson, 1978). We follow the CLIMAP project members (1976) in using 18,000 yr B.P. as the date for the Late Wisconsin maximum and arbitrarily choose 18,000 ± 3000 yr B.P. as the full glacial in the American West. The latest Wisconsin extends from 15,000 to 10,000 yr B.P.; the date of 11,000 yr B.P. marks the approximate time of an important fall in levels of pluvial lakes and, biotically, the end of Late Wisconsin conditions in the Southwest (Van Devender and Spaulding, 1979). Plant communities of a transitional or somewhat mesic character persisted in many desert areas until the close of the early Holocene about 7800 yr B.P. (Van Devender, 1977b).

Paleoecologic Research

Studies of late-Quaternary vegetational changes in the arid West began in earnest during the late 1950s, when palynologic techniques were applied to alluvial and lacustrine sediments. Paul B. Sears was among the first to recognize the potential of such research, and the efforts of P. S. Martin and P. J. Mehringer brought pollen analysis to its zenith in the American West during the mid-1960s (Martin, 1963a, 1963b; Martin and Mehringer, 1965; Mehringer, 1967). Palynology in the Southwest has lagged since then, although there are important exceptions. The bulk of the new evidence comes from plant macrofossils, and we devote most of this chapter to these more recent findings.

ANALYSIS OF PACKRAT MIDDENS

Packrats, or woodrats, are cricetid rodents of the genus *Neotoma* Say and Ord. More than 15 species range throughout North America and Central America. All share the trait of gathering prodigious amounts of plant debris for den construction and food (Finley, 1958; Wells, 1976). A packrat forages mostly within 30 m of its den; distances exceeding 100 m are virtually unknown (Bleich and Schwartz, 1975; Cranford, 1977; Raun, 1966; Stones and Hayward, 1968). Packrat houses in open areas lie protected in heavy brush or cactus patches; in rugged topography, dens are found in rock shelters and caves. Functionally specific areas within a packrat den include trash middens, which also serve as urination and defecation points (Finley, 1958; Van Devender, 1973).

The analysis of the plant fragments found in packrat middens has led to a major advance in the understanding of paleoecology in the arid West. Ancient middens sheltered in caves or overhangs may remain intact for tens of thousands of years. Midden deposits are cohesive, covered with a dull-gray to brown convoluted weathering rind (Figure 14-5) and saturated with crystallized urine, called amberat. Desiccation preserves middens, and the physical and

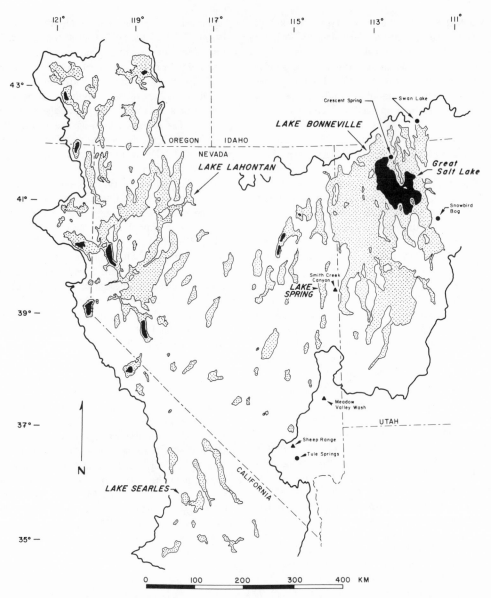

Figure 14-4. Map of the glacial-age lakes of the Great Basin and Mojave Deserts. Existing bodies of water are solid, and stippling indicates the maximum extent of Wisconsin lakes (after Mehringer, 1977). The heavy line indicates the boundary of the physiographic Great Basin (Morrison, 1965). (Data from Snyder et al., 1964; modified after Mifflin and Wheat, 1979, and Spaulding's field observations.) Symbols for fossil sites follow Figures 14-1 and 14-2. Smith Creek Canyon in the Snake Range of Nevada drains into the Bonneville Basin (Thompson, 1979).

chemical properties of amberat augment the middens' preservation. Amberat impregnation renders middens cohesive and nearly rock hard, enabling them to adhere to a ceiling or wall after a supporting shelf collapses. It may provide an unfavorable environment for insects (Wells, 1976, 1977), and it contributes to the formation of a weathering rind. Amberat is hygroscopic and will rehydrate during infrequent periods of high atmospheric humidity. Within months, a fresh break on an indurated midden is sealed by a glossy layer of crystallized urine drawn by capillary action from within the midden. Such a surface is sticky whenever it is humid, and over the centuries it incorporates dust and debris and becomes a lusterless, convoluted, encasing rind (Figure 14-5) that is no longer subject to rehydration.

Ancient packrat middens contain thousands of perfectly preserved plant macrofossils, all gathered from the local plant community and most identifiable to the species level (Phillips, 1977; Van Devender, 1973). For analysis, discrete layers are collected, cleaned of weathering rind, disaggregated in water, flushed through soil sieves, and dried. The plant fragments and animal remains are then sorted for identification. When samples are carefully selected in good stratigraphic context, it is reasonable to associate a radiocarbon date with other species in the assemblage. An extralocal plant type (a species not occurring in the immediate vicinity today) is preferred for radiocarbon dating (Van Devender, 1977a). Identifiable plant fragments from each macrofossil assemblage are counted, and the most abundant species are assumed to have been the most important in the paleovegetation.

Figure 14-5. Fossil packrat middens from the Sheep Range of southern Nevada. (A) A midden radiocarbon dated at more than 48,000 yr B.P. (the vertical bar equals 1 m). (B) the interior of a midden showing a mat of *Pinus flexilis* and *P. longaeva* needles radiocarbon dated at 11,860 ± 160 yr B.P. (coin is 22 mm in diameter).

The increment of time encompassed by a single midden sample is unknown, and deposition of a single unit may take a season or a century. For this reason, Wells (1977) uses a technique in which leafy horizons of limited extent are collected from a midden and the fragments imbedded on its surface are examined. A portion is then submitted intact for radiocarbon dating. Although tighter chronologic control is achieved by this technique, the observed diversity of the fossil flora is restricted, the macrofossils in an amberat matrix cannot be quantified, and the sample cannot be used to obtain multiple dates.

Ancient middens are plentiful in many areas of the Southwest and provide an elaborate, high-resolution view of local vegetation at particular times in the past. But assumptions regarding the sample area and the completeness of the assemblage are rarely tested. Although most workers have examined modern packrat deposits and are confident that plant macrofossils in ancient middens provide a fairly representative picture of the local paleovegetation, there are few studies of the biases inherent in midden analysis. The technique offers unparalleled potential for paleoclimatic and paleoecologic reconstructions, but, even if the basic assumptions of midden analysis are verified by future testing, it will not supplant palynology as the only technique for determining regional paleovegetation in the arid West. We feel that interdisciplinary studies combining both packrat-midden

and pollen analysis offer the greatest promise.

POLLEN ANALYSIS

In the Southwest, palynologists have recovered pollen in sediments from extant lakes, dry lakes (playas), alluvium, spring deposits, and cave fill. Pollen also has been obtained from fossil coprolites (Martin et al., 1961), although it is not clear to what extent the assemblages represent the regional vegetation or seasonal phenology (King, 1977). Pollen tends to be best preserved in lake sediments, but the relative pollen frequencies from spring-mound sediments may be better for reconstructing local vegetation. Most studies utilize percentages of total pollen, but pollen influx (amount of pollen per square centimeter per year) has been calculated in only a few instances. Macrofossils tend to be rare or absent in sediments in desert lakes. Few lake-basin sites are near packrat-midden localities, and only by radiocarbon dating can the pollen results be correlated with macrofossil data. In a few cases, pollen and macrofossils from the same suite of middens were studied (e.g., King and Van Devender, 1977).

The advantage of having both fossil-pollen evidence from lakes or playas and contemporaneous midden records upslope in the same region is clear; a lacustrine pollen record represents regional vegetation and provides a perspective for results from macrofossils representing strictly local vegetation. Well-dated lacustrine sequences can provide continuous records of climatic change that are normally not available from isolated midden samples. Conversely, macrofossil assemblages from middens yield information on precisely which plant species were growing upslope, thereby allowing the calibration of the pollen data.

Dominant pollen types appearing in the Pleistocene record of the Southwest include the following groups: (1) Chenopodiineae (*Amaranthus*, *Sarcobatus*, and other Chenopodiaceae); (2) Asteraceae (separable into *Artemisia*, Liguliflorae, Tubuliflorae, and other short-spine and long-spine pollen types); (3) Poaceae (grasses); and (4) wind-pollinated trees, the most common being *Pinus* (pine), *Abies* (fir), *Picea* (spruce), *Juniperus* (juniper), *Pseudotsuga* (Douglas-fir), and *Quercus* (oak). In the Southwest, pollen of groups 1 and 2 generally represent shrubs.

Late Wisconsin Phytogeography

The macrofossil record from the warm deserts shows that during the Middle and Late Wisconsin pygmy-conifer woodland ranged across regions now occupied by desert scrub (Table 14-1) (Van Devender and Spaulding, 1979). At higher latitudes in the Great Basin Desert, subalpine conifers dominated to near the level of the pluvial lakes (Thompson and Mead, 1982). Although desert scrub or steppe vegetation may have existed on the alluvial fans unsampled by ancient middens (Thompson and Mead, 1982), neither is known in the Southwest before 10,500 yr B.P. (Van Devender, 1977b). (Recently dated packrat middens from the northern Mojave Desert [latitude 36°34′N, longitude 116°05′N, 900 m elevation] provide records of glacial-age desert scrub on a xeric slope at 14,810 ± 400 yr B.P. and, in a mesic wash, at 11,680 ± 650 yr B.P.)

WOODLANDS

We use the term "woodland" to refer to communities with open canopies dominated by small trees (Brown et al., 1979; Whittaker, 1975). Paleowoodland is commonly inferred from assemblages containing the abundant macrofossils of pygmy conifers (*Juniperus* spp., *Cupressus* spp., pinyon pines) and oaks. Discriminating the species of juniper that dominated the Wisconsin woodlands to the south is often difficult, but much is known of the taxonomy of the pinyon pines (Lanner and Van Devender, 1974, 1981). Single-needle pinyon (*Pinus monophylla*) ranged to areas south and east of its present limits through the southern Mojave and Sonoran Deserts (Van Devender and Spaulding, 1979). Following Little (1968), Wells (1979: 319) claims that the pinyon of the glacial-age Sonoran woodlands is actually *Pinus edulis* Engelm. var. *fallax* Little. Since the taxonomic validity of that variety is open to question (Lanner, 1974), we refer the material to *P. monophylla*. Single-needle pinyon pine was not common in the Mojave Desert north of about latitude 36°N, and records of pinyon north of latitude 36°40'N are unknown. Juniper (in this case probably *Juniperus osteosperma* Little) apparently extended farther north than pinyon. Thompson (1979) reports small amounts in a Middle Wisconsin midden from the Snake Range at latitude 39°20'N.

In contrast to single-needle pinyon, neither Colorado pinyon pine (*Pinus edulis*) nor Mexican pinyon (*P. cembroides* Zucc.) expanded into the warm deserts during the Wisconsin (Lanner and Van Devender, 1981; Van Devender and Spaulding, 1979). Although Colorado pinyon occurred at lower elevations than it does today on the northern periphery of the Chihuahuan Desert, Texas pinyon (*P.*

remota Bailey and Hawksworth) appears to have been the common pine in the Wisconsin woodlands of trans-Pecos Texas and adjacent Mexico (Table 14-1, Figure 14-3). Glacial-age fossils of Mexican pinyon are unknown; it appears to have dispersed over much of its present range during the Holocene, displacing Texas pinyon (Lanner and Van Devender, 1981).

Throughout the Southwest, woodland and forest vegetation expanded at the expense of desert scrub and steppe, the reverse of the expansion proposed in the CLIMAP (1976) reconstruction of full-glacial vegetation in the Southwest. South of about latitude 37°N, woodland occurred at elevations both above and below those postulated by Martin and Mehringer (1965) and Wells and Berger (1967) (Table 14-1). On rocky slopes, it ranged from close to the basin floors upward into a transition with communities dominated by montane trees between 1500 and 2000 m elevation.

FORESTS

Forest or parkland vegetation is inferred from packrat middens containing the abundant remains of subalpine or montane trees. Some of these trees, such as the Great Basin bristlecone pine (*Pinus longaeva* D. K. Bailey) and limber pine (*P. flexilis* James), are normally of short stature and often occur in open stands that could be construed as woodland. Others, such as Douglas-fir (*Pseudotsuga menziesii* Franco) and white fir (*Abies concolor* Hoopes), are taller trees that comprise forests or, when in open stands, parkland.

Closed canopies are not implied here by the occurrence of forest trees; the structure could be open or closed. In any case, these assemblages (Table 14-2) do contrast those dominated by pygmy conifers (juniper and pinyon pine) (Table 14-1). Full-glacial montane and subalpine trees are known only from highlands in the Mojave and Great Basin Deserts. Subalpine species such as limber pine, bristlecone pine, and spruce (*Picea* sp. or *P. engelmannii* Parry) are the most important trees in the fossil record above about 1800 m elevation (Table 14-2). Most mesophytic montane species, such as Rocky Mountain juniper (*Juniperus scopulorum* Sarg.) and ponderosa pine (*Pinus ponderosa* Laws) are rare or absent. White fir (*Abies concolor*) and Douglas-fir appear to have been important components of full-glacial vegetation at some localities to the south (e.g., the Grand Canyon in Arizona and the Meadow Valley Wash in southeastern Nevada) (Cole, 1982; Madsen, 1973) (Figures 14-2 and 14-4 and Table 14-2).

Midden records of abundant forest trees in the Sonoran and Chihuahuan Deserts are restricted to a single locality at 2000 m elevation in the Guadalupe Mountains of western Texas, where a mixed coniferous "forest" with spruce, dwarf juniper (*Juniperus communis* L.), Rocky Mountain juniper, Douglas-fir, and southwestern white pine (*Pinus strobiformis* Engelm.) is dated between 13,000 and 12,000 yr B.P. (Van Devender et al., 1979).

MOUNTAINS AS BIOTIC ISLANDS

During interglacials, isolated desert mountains serve as refugia for less drought-tolerant species, and their islandlike nature has long intrigued biogeographers (Merriam, 1890; Wallace, 1880). By the mid-1960s, the biotic consequences of glacial climates, in terms of lowered vegetation zones and reduced barriers to migration, had received some attention (Clokey, 1951; Martin and Mehringer, 1965; Mehringer, 1965, 1967). MacArthur and Wilson's (1967) theories of island biogeography have stimulated new research into the nature of

Table 14-1.
Packrat-Midden Records of Full-Glacial Woodland in Present Desert vegetation in the Southwestern United States

Area	Site	Latitude N	Longitude W	Elevation (m)	Radiocarbon Date (yr. B.P.)	Laboratory Number	Abundant Fossil Plants	Common or Occasional Fossil Plants	References
Mojave Desert									
Eureka Valley, California	Eureka View No. 5A	37°20'	117°47'	1430	14,720 ± 530	GX-6230	Atriplex confertifolia, Juniperus osteosperma	Artemisia sec. Tridentatae, Ephedra cf. viridis, Hecastocleis shockleyi	Spaulding, 1980
Ranger Mountains, Nevada	Ranger Mountains No. 1	36°45'	116°52'	1130	16,800 ± 300	Not given	Juniperus osteosperma	—	Wells and Berger, 1967
Pintwater Range, Nevada	Pintwater Cave	36°45'	116°35'	1280	16,400 ± 250	UCLA-1099	Atriplex confertifolia, Juniperus osteosperma*	Chrysothamnus sp.*	Wells and Berger, 1967
Sheep Range, Nevada	Basin Canyon No. 1	36°42'	115°16'	1635	15,610 ± 260	WSU-1856	Juniperus osteosperma	Artemisia sec. Tridentatae, Pinus flexilis, Tetradymia sp.	Spaulding, 1981
	Basin Canyon No. 2B	36°42'	115°16'	1630	19,200 ± 580	A-1741	Juniperus osteosperma	Chrysothamnus sp., Atriplex confertifolia, Jamesia americana	Spaulding, 1981
	Eyrie No. 3 (1)	36°38'	115°17'	1855	16,490 ± 220	WSU-1853	Juniperus osteosperma	Ephedra viridis, Opuntia polyacantha, Pinus flexilis	Spaulding, 1981
	Eyrie No. 3 (2)	36°38'	115°17'	1855	18,890 ±340	WSU-2042	Juniperus osteosperma	Forsellesia nevadensis, Opuntia polyacantha, Prunus fasciculata	Spaulding, 1981
	Deadman No. 1 (2)	36°37'	115°17'	1970	16,800 ± 245	WSU-1860	Juniperus osteosperma	Jamesia americana, Salvia dorrii, Opuntia polyacantha, Tetradymia sp.	Spaulding, 1981
	Deadman No. 1 (4)	36°37'	115°17'	1970	18,680 ± 280	WSU-1857	Juniperus osteosperma	Pinus flexilis, P. longaeva, Tetradymia sp., Prunus fasciculata	Spaulding, 1981
	Flaherty Mesa No. 1	36°30'	115°14'	1770	20,380 ± 340	WSU-1862	Juniperus osteosperma		Spaulding, 1981
	Flaherty Mesa No. 2	36°30'	115°14'	1720	18,790 ± 340	WSU-1855	Juniperus osteosperma, Ephedra sp.	Forsellesia nevadensis, Symphoricarpos sp.	Spaulding, 1981
	Willow Wash No. 4A	36°28'	115°15'	1585	21,350 ± 440	WSU-1858	Juniperus osteosperma	Pinus monophylla, Ribes cf. velutinum	Spaulding, 1981
	Willow Wash No. 4C (2)	36°28'	115°15'	1585	19,020 ± 750	UCR-729	Juniperus osteosperma	Forsellesia nevadensis, Ribes cf. velutinum, Symphoricarpos sp.	Spaulding, 1981
	Willow Wash No. 4D	36°28'	115°15'	1585	17,700 ±740	UCR-730	Juniperus osteosperma	Chrysothamnus sp., Tetradymia sp., Symphoricarpos sp.	Spaulding, 1981

Table 14-1 continued.

Area	Site	Latitude N	Longitude W	Elevation (m)	Radiocarbon Date (yr. B.P.)	Laboratory Number	Abundant Fossil Plants	Common or Occasional Fossil Plants	Reference(s)
					Mojave Desert (continued)				
	Penthouse No. 1 (3)	36°28'	115°15'	1600	19,400 ± 300	A-1772	*Juniperus osteosperma*	*Cercocarpus intricatus, Forsellesia nevadensis, Pinus monophylla*	Spaulding, 1981
	Penthouse No. 3 (2)	36°28'	115°15'	1600	21,210 ± 440	A-1775	*Juniperus osteosperma*	*Symphoricarpos* sp., *Tetradymia* sp.	Spaulding, 1981
North Muddy Mountains, California	Muddy Mountains No. 1	36°40'	114°34'	530	17,750 ± 200	UCLA-1218	*Juniperus osteosperma**	*Purshua glandulosa, Ephedra viridis*	Wells and Berger, 1967
Lower Grand Canyon, Arizona	Rampart Cave Unit B	36°06'	113°56'	535	18,890 ± 500	A-1356	*Juniperus* sp., *Lycium andersonii*	*Atriplex confertifolia, Fraxinus anomala, Salvia dorrii*	Phillips, 1977
	Rampart Cave Pit B (f)	36°06'	113°56'	535	16,330 ± 270	A-1569	*Atriplex confertifolia, Fraxinus anomala, Juniperus* sp.	*Chrysothamnus* sp., *Salvia dorrii*	Phillips, 1977
	Vulture Cave No. 6	36°06'	113°56'	645	17,610 ± 290	A-1603	*Atriplex confertifolia, Juniperus* sp.	*Coleogyne ramosissima, Symphoricarpos* sp.	Phillips, 1977
	Vulture Cave No. 10	36°06'	113°56'	645	19,050 ± 390	A-1606	*Atriplex confertifolia, Juniperus* sp., *Sphaeralcea* sp.	*Coleogyne ramosissima, Symphoricarpos* sp.	Phillips, 1977
Grand Canyon, Arizona	Horseshoe Mesa No. 6	36°02'	111°59'	1450	18,630 ± 310	A-1798	*Juniperus* cf. *osteosperma, Rosa* cf. *stellata*	*Ribes* sp., *Opuntia* sp.	Cole, 1981
	Horseshoe Mesa No. 11	36°02'	111°59'	1450	20,630 ± 470	A-2337	*Juniperus* cf. *osteosperma*	*Atriplex confertifolia, Ribes* sp., *Rosa* cf. *stellata, Agave utahensis*	Cole, 1981
	Hance Canyon No. 4	36°02'	111°58'	1100	17,400 ± 450	WK-179	*Juniperus* cf. *osteosperma*	*Atriplex confertifolia, Opuntia erinacea*	Cole, 1981
	Grapevine Canyon No. 1	36°03'	112°02'	1100	16,400 ± 190	WK-165	*Juniperus* cf. *osteosperma*	*Atriplex confertifolia, Equisetum* cf. *hymale, Opuntia* sp., *Quercus* sp.	Cole, 1981
Spring Range, Nevada	Blue Diamond Road No. 3	36°02'	115°23'	1050	15,040 ± 650	UCR-725	*Juniperus osteosperma*	*Atriplex canescens, Cercocarpus intricatus, Opuntia* sp., *Pinus monophylla*	Spaulding, 1981
	Blue Diamond Road No. 5	36°02'	115°23'	1100	15,800 ± 680	UCR-726	*Juniperus osteosperma*	*Forsellesia nevadensis, Opuntia* sp., *Pinus monophylla*	This report
Peach Springs Wash, Arizona	Cave of the Early Morning Light**	35°43'	113°23'	1300	16,580 ± 460	A-1718	*Pinus edulis, Juniperus* sp.	*Agave utahensis, Artemisia* sec. Tridentatae, *Cercocarpus intricatus, Opuntia whipplei*	This report

Table 14-1 continued.

Area	Site	Latitude N	Longitude W	Elevation (m)	Radiocarbon Date (yr. B.P.)	Laboratory Number	Abundant Fossil Plants	Common or Occasional Fossil Plants	Reference(s)
Mojave Desert (continued)									
Newberry Mountains, Nevada	Sacaton Wash No. 1	35°15'	114°37'	730	19,690 ± 600	I-3659	Juniperus sp., Pinus monophylla*	Not given	Leskinen, 1975
	Newberry Mountains No. 4	35°16'	117°37'	850	15,000 ± 1600	A-1136	Juniperus sp.*	Not given	Mead et al., 1978
Turtle Mountains, California	Turtle Mountains No. 1	32°20'	114°50'	850	19,500 ± 380	Not given	Juniperus osteosperma	Pinus monophylla, Opuntia erinacea	Wells and Berger, 1967
Sonoran Desert									
Chemehuevi Mountains, California	Chemehuevi Mountains No. 1	34°40'	114°40'	260	16,900 ± 190	Not given	Juniperus sp.	Not given	Wells and Hunziker, 1976
Bill Williams River, Arizona	Artillery Mountains No. 2	34°22'	113°37'	725	18,320 ± 400	A-1101	Juniperus sp.	Pinus monophylla	Van Devender, 1973
	Artillery Mountains No. 3	34°22'	113°37'	725	21,000 ± 400	USGS-196	Juniperus sp., Pinus monophylla	Quercus dunnii, Acacia greggii, Ephedra nevadensis	Van Devender, 1973
Tucson Mountains, Arizona	Tucson Mountains No. 3	32°19'	111°12'	740	21,000 ± 700	A-994	Ericameria cuneata, Juniperus sp.	Agave, sp., Opuntia chlorotica, Pinus monophylla	Van Devender, 1973
Ajo Mountains, Arizona	Montezuma Head No. 1A	32°07'	112°42'	975	20,490 ± 510	A-1695	Juniperus sp., Pinus monophylla	Artemisia sec. Tridentatae, Berberis sp., Quercus turbinella, Yucca brevifolia	Van Devender and Spaulding, 1979
	Montezuma Head No. 1C	32°07'	112°42'	975	17,830 ± 870	A-1697	Juniperus sp., Pinus monophylla	Artemisia sec. Tridentatae, Berberis sp., Quercus turbinella, Yucca brevifolia	Van Devender and Spaulding, 1979
Chihuahuan Desert									
Sacramento Mountains, New Mexico	Big Boy Canyon No. 2	32°53'	105°54'	1495	18,300 ± 420	A-2064	Juniperus scopulorum, Pinus edulis, Rhus aromatica-type	Gutierrezia sp., Juniperus cf. monosperma, Philadelphus microphyllus, Quercus cf. undulata	This report
	Big Boy Canyon No. 3	32°53'	105°54'	1495	16,260 ± 350	A-2065	Same as above	Echinocereus sp., Juniperus cf. monosperma, Quercus cf. undulata	This report

Table 14-1 continued.

Area	Site	Latitude N	Longitude W	Elevation (m)	Radiocarbon Date (yr. B.P.)	Laboratory Number	Abundant Fossil Plants	Common or Occasional Fossil Plants	Reference(s)
					Chihuahuan Desert (continued)				
Hueco Mountains, Texas	Tank Trap Wash No. 1 (5)	31°54'	106°09'	1340	19,610 ± 1150	A-1710	*Pinus* cf. *remota*, *Juniperus* sp.	*Cercocarpus montanus*, *Mortonia scabrella*, *Quercus pungens*	Lanner and Van Devender, 1981
	Tank Trap Wash No. 2	31°54'	106°09'	1340	21,200 ± 990	A-1722	*Pinus* cf. *remota*, *Juniperus* sp.	*Cercocarpus montanus*, *Forsellesia spinescens*, *Quercus pungens*	Lanner and Van Devender, 1981
	Navar Ranch No. 3B	31°54'	106°09'	1370	16,240 ± 430	A-1645	*Pinus remota*, *Juniperus* sp.	*Berberis trifoliolata*	Lanner and Van Devender, 1981
Sierra Diablo, Texas	Streeruwitz Hills No. 1 (P4)	31°07'	105°09'	1430	18,060 ± 1320	A-1623	*Juniperus* sp., *Pinus remota*	—	This report
	Streeruwitz Hills No. 1	31°07'	105°09'	1430	15,050 ± 800	A-1846	*Juniperus* sp., *Pinus* cf. *remota*	—	This report
Chinati Mountains, Texas	Bennett Ranch No. 1	30°37'	104°59'	1035	18,190 ± 380	A-1831	*Berberis haematocarpa*,	*Pinus remota*, *Prosopis glandulosa*, *Echinocereus* sp.	This report
	Shafter No. 1A	29°47'	104°22'	1310	15,950 ± 900	A-1845	*Juniperus* sp., *Pinus remota*	*Lithospermum* sp., *Quercus binckleyi*, *Sophora* nr. *gypsophila*	Van Devender et al., 1978
	Shafter No. 1B	29°47'	104°22'	1310	15,700 ± 230	A-1581	*Juniperus* sp., *Pinus remota*	*Castela stewarti*, *Quercus pungens*, *Sophora* nr. *gypsophila*	This report
Big Bend, Texas	Maravillas Canyon TRV No. 1	29°33'	102°49'	600	20,600 ± 1530	A-1860	*Juniperus* sp., *Pinus remota*	*Agave* sp., *Nolina* sp., *Quercus* sp., *Berberis* cf. *trifoliolata*, *Cirsium* sp.	This report
	Maravillas Canyon TRV No. 3	29°33'	102°49'	600	16,160 ± 130	A-1842	*Juniperus* sp., *Pinus remota*	*Opuntia* cf. *phaeacantha*, *Quercus* sp.	This report
	Dagger Mountain No. 1	29°32'	103°06'	880	20,000 ± 390	Not given	*Juniperus* sp., *Opuntia* cf. *macrocentra*, *Pinus remota*	—	Wells, 1966; Lanner and Van Devender, 1981
	Dagger Mountain No. 3	29°32'	103°06'	850	16,250 ± 240	Not given	*Juniperus* sp., *Pinus remota*	—	Wells, 1966; Lanner and Van Devender, 1981
	Terlingua No. 1	29°18'	103°41'	915	15,000 ± 440	WK-175	*Juniperus* sp.	—	This report
	Burro Mesa No. 1	29°16'	103°23'	1200	18,750 ± 360	Not given	*Juniperus* sp., *Junglans microcarpa*, *Pinus remota*, *Opuntia* cf. *macrocentra*		Wells, 1966; Lanner and Van Devender, 1981
	Ernst Tinaja No. 1	29°16'	103°01'	760	15,300 ± 670	WK-174	*Juniperus* sp., *Pinus remota*	*Koeberlinia spinosa*, *Quercus binckleyi*	This report

*Relative abundance values are not available.
**Site located in woodland vegetation.

Table 14-2. Packrat-Midden Records of Full-Glacial Forest in Highlands of the Great Basin Desert and the Mojave Desert

Area	Site	Current Vegetation	Latitude N	Longitude W	Elevation (m)	Radiocarbon Date (yr B.P.)	Laboratory Number	Abundant Fossil Plants	Common or Occasional Fossil Plants	Reference(s)
Snake Range, Nevada	Streamview No. 2	wd	39°20'	114°10'	1860	17,350 ± 435	GX-5866	*Pinus longaeva*	*Atriplex confertifolia, Cercocarpus intricatus, Juniperus communis, Picea engelmanii*	Thompson and Mead, 1982
	Ladder Cave No. 2B	wd	39°20'	114°10'	2060	17,960 ± 1100	A-2092	*Pinus longaeva*	*Ribes montigenum*	Thompson and Mead, 1982
Meadow Valley Wash, Nevada	Etna No. 1	wd/gds	37°33'	114°37'	1350	20,000 ± 400	RL-221	*Pinus flexilis, Pseudotsuga menziesii*	*Juniperus osteosperma, Opuntia* sp.	Madsen, 1973; this report
	Etna No. 3	wd/gds	37°33'	114°37'	1350	15,190 ± 260	RL-292	*Pinus flexilis*	*Juniperus scopulorum, Pseudotsuga menziesii, Quercus gambeli*	Madsen, 1973; this report
Sheep Range, Nevada	Eyrie No. 5 (3)	mds	36°38'	115°17'	1860	19,750 ± 450	WK-167	*Pinus flexilis*	*Agropyron* sp., *Artemisia* sec. Tridentatae, *Epbedra* sp. *Ribes* cf. *velutinum*	Spaulding, 1981
	Deadman No. 1 (1)	wd/mds	36°37'	115°17'	1970	17,420 ± 250	LJ-3707	*Pinus flexilis*	*Cercocarpus intricatus, Juniperus osteosperma, Pinus longaeva, Tetradymia* sp.	Spaulding, 1981
	Spires No. 2 (1)	wd	36°35'	115°18'	2040	18,800 ± 130	USGS-198	*Pinus flexilis, P. longaeva*	*Cercocarpus intricatus, Juniperus osteosperma, Pinus longaeva, Tetradymia* sp.	Spaulding, 1981
Grand Canyon, Arizona	Nankoweap No. 9B	wd	36°15'	111°57'	2020	17,950 ± 600	RL-1180	*Picea* sp., *Pinus flexilis, Pseudotsuga menziesii*	*Abies concolor, Juniperus communis*	Cole, 1981
	Nankoweap No. 9C	wd	36°15'	111°57'	2020	18,130 ± 350	A-1964	*Pinus flexilis*	*Abies concolor, Juniperus communis, Picea* sp., *Pseudotsuga menziesii*	Cole, 1981
	Chuar No. 2	wd	36°11'	111°55'	1450	16,165 ± 615	GX-6302	*Abies concolor, Pseudotsuga menziesii*	*Artemisia* sec. Tridentatae, *Juniperus osteosperma, Symphoricarpos* sp.	Cole, 1981
	Chuar No. 8b	wd	36°11'	111°55'	1770	18,800 ± 800	RL-1178	*Pinus flexilis, Pseudotsuga menziesii*	*Abies concolor, Artemisia* sec. Tridentatae	Cole, 1981
	Chuar No. 8c	wd	36°11'	111°55'	1770	18,490 ± 660	A-2023	*Abies concolor, Pinus flexilis, Pseudotsuga menziesii*	*Holodiscus dumosus, Rubus* sp.	Cole, 1981
	Clear Creek No. 2	wd/mds	36°08'	112°00'	1600	15,840 ± 310	WK-176	*Abies concolor, Juniperus osteosperma, Pseudotsuga menziesii*	*Pinus flexilis, Ribes* sp., *Rosa stellata*	Cole, 1981
Spring Range, Nevada	Potosi Mountain No. 2A (2)	wd	36°00'	115°23'	1880	14,900 ± 180	LJ-4004	*Pinus flexilis, P. longaeva*	*Abies concolor, Juniperus osteosperma, Fraxinus anomala*	Thompson and Mead, 1982
Clark Mountain, California	Clark Mountain No. 2	wd	35°33'	115°37'	2140	19,900 ± 1500	Gak-1987	*Pinus longaeva*	*Artemisia* sec. Tridentatae, *Pinus flexilis*	Spaulding, 1981

Abbreviations: gds, Great Basin desert scrub; mds, Mojave desert scrub; wd, woodland.

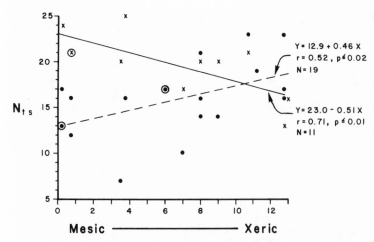

Figure 14-6. The apparent relationship between plant species abundance and site exposure during the Late Wisconsin and at present in the Sheep Range of southern Nevada. N_{ts} is the number of taxa of trees, shrubs, and succulents noted within 30 m of a midden site (each individual sample is represented by a letter x) or that occur in a given midden assemblage (represented by a solid dot). Only middens dated between 24,000 and 15,500 yr B.P. are considered. Circled symbols represent two samples. The abscissa was constructed from an arbitrary scale of effective moisture at a site, ranging from mesic northeast-facing cliff bases (0 to 1) to xeric southwest-facing ridges (12 to 13). The best fit for the modern plant data (x's) is shown by a solid line; that for the fossil data (dots) is dashed. Note the intersection of the two lines far to the right on the y-axis, which implies that most sites supported depauperate vegetation during the Late Wisconsin relative to the modern plant communities (Spaulding, 1981).

montane biota. Central to their thesis is a model relating species diversity on islands to an equilibrium between extinction and immigration rates, which are themselves a general function of island size and distance to source areas ("mainlands").

The birds that are restricted to desert mountaintops in the Great Basin conform to an equilibrium postulated by the MacArthur and Wilson model (Behle, 1978; Johnson, 1975, 1978), but the small boreal mammals do not (Brown, 1971, 1978). This is attributed to a lack of immigration by boreal mammals during the present interglacial. Noting the Wisconsin packrat-midden evidence for mesophytic vegetation between mountains, Brown attributes disequilibrium in boreal mammal populations to the stocking of these mountain ranges during the Wisconsin and a subsequent absence of immigration between massifs over the last 10,000 years. Small-mammal diversity during the interglacial would, therefore, be controlled only by extinction rates, in large part a function of island (mountain-mass) size (Thompson and Mead, 1982). The paleontologic record developed by Grayson (1977a, 1979, 1981b) confirms Brown's theory. Mesophytic animals such as pika (*Ochotona* cf. *priceps* Richardson), marmot (*Marmota flaviventris* Aud. & Bachman), and heather vole (*Phenacomys intermedius* Merriam) ocurred on many more mountain ranges during the late glacial *and* early Holocene than they do today (Grayson, 1981).

Harper and others (1978) report that the proportion of plants adapted for long-distance dispersal in Great Basin mountains does not decrease with distance from the Sierra Nevada or Wasatch Range "mainlands." But, as expected, the total percentage of Sierran and Wasatch species on isolated mountains does decrease with increasing

distance from those respective areas. Only one mountain range considered by Harper and others (1978), the Spring Range in southern Nevada, is known to have been connected to others by mesophytic vegetation during the Wisconsin (Mehringer, 1967; Spaulding, 1981; Wells and Berger, 1967). There were probably others, but packrat-midden data from interconnecting highlands are lacking.

Mehringer (1966) suggests that the increase in the size of montane islands during the Wisconsin led to greater plant-species diversity, but the packrat-midden evidence offers no confirmation of this hypothesis. The late-Pleistocene floras from three large, isolated mountain ranges (the Spring and Sheep Ranges of the Mojave Desert and the Snake Range of the Great Basin Desert) contain no plant species that do not occur somewhere in these mountains today (Spaulding, 1981; Thompson, 1979; Thompson and Mead, 1982) (Figures 14-2 and 14-4). The Late Wisconsin floras also lack important elements present today; macrofossil assemblages from these mountains suggest an increase in the number of woody species over the last 10,000 years (Figure 14-6). Wisconsin records of extralimital plants (not present in the range today) are restricted to smaller mountains lower than about 2300 m altitude. This fact reflects Holocene extinction.

CONTRASTS WITH EARLIER MODELS

Ponderosa pine (or "yellow pine") forest and parkland have figured strongly in reconstruction of the glacial-age landscape based on pollen analysis (e.g., Martin, 1963a; Martin and Mehringer, 1965; Mehringer, 1967). Although many areas remain unsampled, macrofossil evidence for that vegetational type is lacking. Ponderosa pine is indeed an important dominant of upland vegetation today, and a simple zonal model of vegetation displacement resulting from a cooler and wetter climate implies its widespread presence during the Wisconsin (Martin and Mehringer, 1965: Figure 4). But, in the Southwest, the oldest macroscopic remains of this species date to only 10,000 yr B.P. (Spaulding, 1981; Van Devender and Spaulding, 1979). According to macrofossil evidence, xerophytic woodland and subalpine trees grew where earlier hypothesis suggested relatively mesic ponderosa pine.

For reasons that will emerge in the following sections, it is apparent that several of the full-glacial vegetation zones (not simply ponderosa pine) mapped by Martin and Mehringer (1965: Figure 4) probably did not exist as major community types in this region. The assumption that major vegetation zones at 18,000 yr B.P. have direct equivalents in the modern vegetation of the Southwest (evident in the 1965 map) is crumbling (Cole, 1982).

Plant Dynamics

The shifts in plant ranges caused by climatic change can occur along any physical gradient modulating temperature and effective moisture. Vertical relief is pronounced in the West, and vegetation ranging from desert scrub to forest to alpine tundra is arrayed along mountainside gradients. Topographic and edaphic variables influence plant distribution, and certain species and vegetational types have markedly skewed distributions on different slopes (Whittaker and Niering, 1965; Whittaker et al., 1968). A combination of such sharp environmental gradients and elaborate macrofossil records provides an ideal opportunity for investigating the dynamics of vegetational change.

ELEVATION SHIFTS AND ANOMALOUS
PLANT ASSOCIATIONS

Figure 14-7 presents the current elevational range and Wisconsin fossil records of selected trees and shrubs in southern Nevada. Some species occurred more than 1000 m below their current lower limits, others less than 400 m, and some are missing from the records. Certain shrubs showed no relative decline in their upper elevational limits during the Wisconsin. In fact, shadscale (*Atriplex confertifolia*) actually occurred much higher than it does today in this region (Figure 14-7) (Spaulding, 1981; Van Devender and Spaulding, 1979). The presence of shadscale at higher elevations is explicable in light of its tolerance of low temperatures (Beatley, 1974, 1975) and the absence of potential competitors, which may restrict its upper range today. Late-Pleistocene packrat middens from above 1500 m elevation in some areas of the northern Mojave Desert suggest decreased species diversity (Figure 14-6) (Spaulding, 1981), and many of the plant species that dominate today's upper-elevation desert and woodland are rare or missing. A relatively dry and cold full-glacial climate coupled with (and perhaps bringing about) lower species diversity may account for the observed habitat expansion of *Atriplex confertifolia*.

Differences in the range changes of various plant species fit an individualistic model of community change. Differential displacement (relative to current distributions) led to ''anomalous'' species associations (Table 14-3) (Spaulding, 1981). The upward expansion of shadscale is an extreme example. Other desert shrubs occurred at relatively high elevations during the Late Wisconsin, while, at the same time, there was a relative downward displacement of a number of montane and subalpine plants. Some of the resultant associations are not known by the authors to occur today (Table 14-3), such as Apache plume (*Fallugia paradoxa* Endl.) with *Pinus flexilis* and desert almond (*Prunus fasciculata* Gray) with *P. flexilis* and *P. longaeva*. The identification of these desert shrubs is unmistakable and, in the absence of serious contamination, unexpected.

HABITAT SHIFTS

Figure 14-8 illustrates the distribution of two Mojave Desert endemics in the Sheep Range of southern Nevada and their Wisconsin fossil record. There is no observed change in the late-Pleistocene elevational range of either, but a shift from north- to south-facing

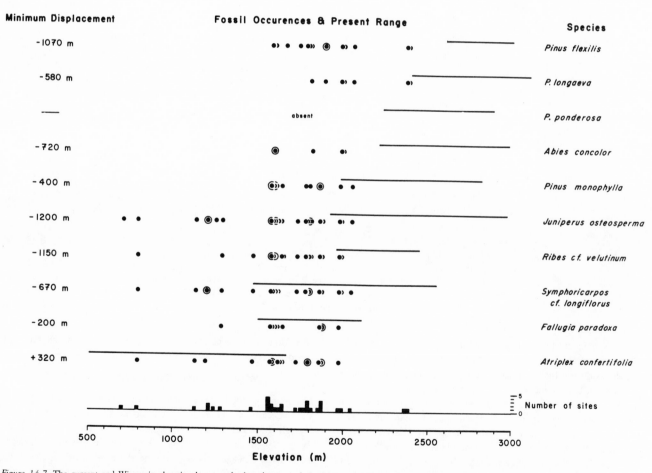

Figure 14-7. The current and Wisconsin elevational range of selected trees and shrubs on calcareous substrate in southern Nevada. Figures to the left indicate the vertical distance between the present range of a species (solid line) and its lowest-elevation fossil record. All records considered are older than 11,000 yr B.P. and are from sites that lie between latitudes 36°15′N and 37°N and between longitudes 115°W and 117°W. Circled dots indicate the occurrence of a species at more than one site at that elevation. The lowest record for *Juniperus osteosperma* is based on wood from the Tule Springs site (Mehringer, 1967).

Table 14-3. Some Anomalous Associates of Three Desert Shrubs in Late-Pleistocene Packrat
Middens from the Sheep Range of Southern Nevada

Species	Anomalous Associates	Assemblage	Elevation (m)	Radiocarbon Age (approximate yr B.P.)
Atriplex confertifolia S. Wats	Pifl, Pimo	Ph 2(1)	1580	29,000
Atriplex confertifolia S. Wats	Abco, Homi	WW 4A	1585	21,350
Atriplex confertifolia S. Wats	Pifl, Abco	WW 4C(1)	1585	24,400
Atriplex confertifolia S. Wats	Pifl, Jaam, Phmi	WW 4C(2)	1585	19,020
Atriplex confertifolia S. Wats	Jaam, Chmi	WW 4D	1585	17,700
Atriplex confertifolia S. Wats	Jaam	WW 4E	1585	> 44,600
Atriplex confertifolia S. Wats	Jaam, Homi	BC 2B	1630	19,200
Atriplex confertifolia S. Wats	Pifl, Homi, Chmi	BC 1	1635	15,610
Atriplex confertifolia S. Wats	Pifl, Homi	FM 2	1720	18,790
Atriplex confertifolia S. Wats	Chmi	WC 1	1790	> 50,400
Atriplex confertifolia S. Wats	Pifl, Chmi	Ey 3(1)	1855	16,490
Atriplex confertifolia S. Wats	Pifl, Pimo	Ey 5(2)	1860	30,470
Atriplex confertifolia S. Wats	Pifl, Pilo, Jaam	Dm 1(4)	1970	18,680
Atriplex confertifolia S. Wats	Pilo, Jaam, Phmi, Pifl	Sp 2(2)	2040	16,200*
Fallugia paradoxa Endl.	Homi, Pifl, Chmi	BC 1	1635	15,610
Fallugia paradoxa Endl.	Pifl, Chmi	Ey 3(1)	1855	16,490
Fallugia paradoxa Endl.	Pifl	Ey 5(2)	1860	30,470
Fallugia paradoxa Endl.	Pifl, Chmi	Ey 5(3)	1860	19,750
Fallugia paradoxa Endl.	Pilo, Pifl, Jaam	Dm 1(1)	1970	17,420
Prunus fasciculata Gray	Pifl	Ph 2(1)	1580	29,000
Prunus fasciculata Gray	Pifl, Abco	WW 4C(1)	1585	24,400
Prunus fasciculata Gray	Pifl, Chmi	Ey 3(1)	1855	16,490
Prunus fasciculata Gray	Pifl	Ey 5(2)	1860	30,470
Prunus fasciculata Gray	Pifl, Pilo, Jaam	Dm 1(1)	1970	17,420
Prunus fasciculata Gray	Pilo, Pifl, Jaam	Dm 1(4)	1970	18,680
Prunus fasciculata Gray	Pifl, Jaam	Dm 1(2)	1970	16,800

Source: Data from Spaulding, 1981.

Abbreviations: For plants — Abco, *Abies concolor* Hoopes; Chmi, *Chamaebatiaria millefolium* Maxim.; Homi, *Holodiscus microphyllus* Rydb.; Jaam, *Jamesia americana* Torrey & Gray; Phmi, *Philadelphus microphyllus* C.L. Hitchc.; Pifl, *Pinus flexilis* James; Pilo, *Pinus longaeva* D.K. Bailey; Pimo, *Pinus monophylla* Torr. & Frem. For the fossil assemblages — BC, Basin Canyon; Dm, Deadman; Ey, Eyrie; FM, Flaherty Mesa; Ph, Penthouse; Sp, Spires; WC, Wagon Canyon; WW, Willow Wash.

*Radiocarbon date must be considered a minimum age.

slopes is evident. Furthermore, the range of Mojave prickleleaf (*Hecastocleis shockleyi* Gray) was apparently attenuated on northern slopes during the full glacial, while *Forsellesia pungens* Heller occurred on both mesic and xeric exposures.

Similar phenomena have been documented in the Grand Canyon (Cole, 1981) as well as in the Sonoran and Chihuahuan Deserts. Plants presently restricted to washes in the low deserts, such as catclaw acacia (*Acacia greggii* Gray), occurred on rocky slopes during the Wisconsin (Van Devender, 1973; Van Devender and Spaulding, 1979). Van Devender and Riskind (1979) have recorded the presence of woodland on limestone during the late glacial in the Hueco Mountains of western Texas. These mesophytic plants persist in the area to this day but are restricted to outcrops of syenite. The

transition to Holocene climate in this area led to the development of woodland refugia on isolated hills of favorable substrate.

LATITUDINAL SHIFTS

On desert mountain ranges, the elevation of lower woodland boundaries declines with decreasing latitude from the Great Basin Desert though the Mojave Desert to the Sonoran Desert (Martin, 1963b; Wells, 1979). For example, at about 1430 m elevation, creosote bush/white bursage occurs in the northwestern Mojave Desert (at about latitude 37°N), and oak woodland occurs at similar altitudes above the northeastern Sonoran Desert (at about latitude 32°N) (Figure 14-9). An increasing proportion and amount of summer precipitation to the south and east (Bryson and Lowry, 1955; Hales,

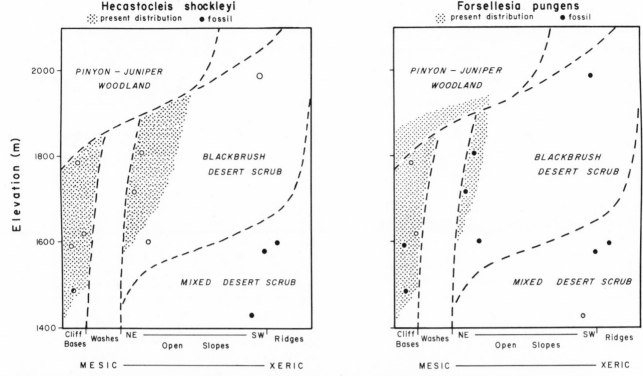

Figure 14-8. The current distribution of two Mojave Desert endemics in the Sheep Range of southern Nevada and their Wisconsin fossil records. Today, both species occur only in protected localities, although both ranged to drier slopes during the full glacial. Mojave prickle-leaf (*Hecastocleis shockleyi* Gray) may have been restricted to those xeric slopes, but the distribution of *Forsellesia pungens* Heller was wider. The single late-Pleistocene record of *Hecastocleis* from a mesic site (half-open circle) is Middle Wisconsin in age. Open circles indicate no fossil occurrence at that site.

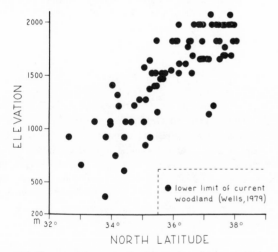

Figure 14-9. The current lower elevational limit of woodland on isolated desert mountains in the Sonoran, Mojave, and southern Great Basin Deserts. (After Wells, 1979.)

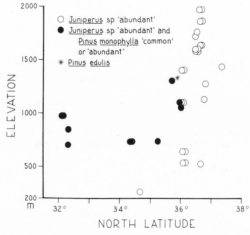

Figure 14-10. Full-glacial (21,000 to 15,000 yr B.P.) packrat-midden records of woodland in current desert scrub vegetation in the Sonoran, Mojave, and southern Great Basin Deserts between longitudes 110°W and 118°W.

1974; Hastings and Turner, 1965) may account for the descent of woodland with decreasing latitude (Wells, 1979).

During the Wisconsin, juniper woodlands ranged to near the level of the basin floors in the warm deserts, but pinyon pine occurred at much lower elevations in both the Sonoran and Chihuahuan Deserts than in the Mojave (Figures 14-10 and 14-11). A dichotomy between

the woodland records to the north and south of latitude 35°30′N is evident. Assemblages from the Sonoran Desert contain pinyon, juniper, and shrub live oak (*Quercus turbinella* Green), but those from the Mojave are usually dominated by juniper alone (Figure 14-10). Wells (1979: 320) suggests that this represents a "pluvial enhancement" of the existing monsoonal precipitation regime. We

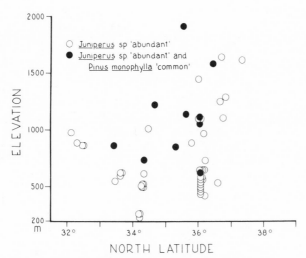

Figure 14-11. Latest Wisconsin (15,000 to 11,000 yr B.P.) packrat-midden records of woodland in desert scrub vegetation of the Sonoran, Mojave, and southern Great Basin Deserts between longitudes 110°W and 118°W. (Compare with Figure 14-9.)

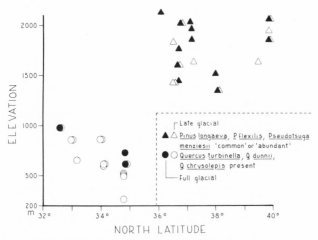

Figure 14-12. Full-glacial and latest-Wisconsin macrofossil records of evergreen oaks and montane conifers from the Sonoran, Mojave, and southern Great Basin Deserts between longitudes 111°55′W and 115°40′W. The present vegetation at most of these sites is desert scrub. (See Tables 14-1 and 14-2.)

feel that this represents the near absence of summer rainfall in the Mojave Desert and relatively warmer summer and winter temperatures to the south.

A similar phenomenon is seen in the changing distribution of evergreen oaks (e.g., *Quercus turbinella, Q. chrysolepis* Liebm., *Q. dunnii* Kell., *Q. pungens* Liebm.) on desert mountains. Oaks were scarce or absent from the full-glacial Mojave, but evidence of oak species is common in middens from the Sonoran and Chihuahuan Deserts. Although evergreen oaks range into the Mojave as far north as latitude 37°15′N today (Little, 1976), oak fossils older than 14,000 years are unknown north of latitude 34°20′N (Figure 14-12). Reports of Wisconsin records of *Q. dunnii* and *Q. chrysolepis* (Van Devender and Spaulding, 1979) from the Newberry Mountains (latitude 35°15′N) are in error. The oldest confirmed oak date is 9500 yr

B.P. (Leskinen, 1975), and it may reflect the northward migration of these species in response to increasing summer rainfall.

Figure 14-12 presents Wisconsin records of montane conifers in areas now largely occupied by Mojave and Great Basin desert scrub. The absence of high-elevation midden sites in the Sonoran Desert is striking, but it may only be a function of lack of suitable substrate and the paucity of high-elevation field searches. A salient feature of the high-altitude (more than 1500 m) record north of about latitude 35°N is the importance of subalpine conifers and the absence of temperate montane species (e.g., *Pinus ponderosa, Quercus gambelii* Nutt.).

GLACIAL AND INTERGLACIAL IMPORTANCE

An intriguing aspect of the packrat-midden record is the importance of certain species in the Wisconsin vegetation that presently have very restricted distributions; bristlecone pine (*Pinus longaeva*) and Texas pinyon pine (*P. remota*) are examples. They were widespread in the Great Basin Desert and the northern Chihuahuan Desert, respectively, during the Late Wisconsin. A transition to the present interglacial drastically reduced their ranges. On the basis of their distributional behavior, at least, these species appear better adapted to glacial climatic conditions. The fossil record and modern distribution of creosote bush and ponderosa pine may provide examples of the same phenomenon, in reverse. Both are widely distributed in the Southwest today, but the oldest reliably associated dates on each are between 10,500 and 10,000 yr B.P. They apparently were restricted during the last ice age. Hence, some plants may be better adapted to glacial climatic regimes, and some to interglacial. The cyclic nature of Quaternary climate (Imbrie and Imbrie, 1980) implies that once every 70,000 to 100,000 years each group of plant species has a chance to disperse to diverse localities, only to experience drastic range reduction at the end of a climatic phase. Plants better adapted to interglacial climates would, of course, have much shorter "residence times" at their optimum distributions.

Pollen Records

Fossil-pollen records from the Southwest support the macrofossil evidence that forest and woodland plants extended downslope during the Wisconsin, largely displacing extensive areas of desert scrub. Pollen sequences from large pluvial lakes provide general records of this downward shift of trees but do not clarify the altitudinal scale or species involved. A list of the pollen sites and the basic characteristics of Wisconsin vegetation is provided in Table 14-4. The discussion in this section emphasizes new literature concerning the area and presents key sites according to geographic region.

GREAT BASIN DESERT AND PERIPHERY

A pollen record from pluvial Lake Bonneville sediments of Middle to Late Wisconsin age at Crescent Spring (Figure 14-13) contrasts with that from Holocene sediments of the nearby Great Salt Lake (Figure 14-13). Postglacial spectra are dominated by *Artemisia*, Chenopodiineae, and *Juniperus* pollen (similar to current steppe vegetation), whereas, the Late Wisconsin lake sediments (younger than a radiocarbon date of about 26,700 yr B.P.) contain higher percentages of *Pinus, Artemisia*, and some *Picea* and *Abies*. The record suggests a significant expansion of woodland near the sites and of forests at higher elevations (Mehringer, 1977). Sediments from pluvial Lake

Table 14-4. Wisconsin Pollen Records from Interior Western North America.

Area	Site	Local Vegetation	Latitude N	Longitude W	Elevation (m)	Approximate Time Range (yr B.P.)	Inferred Vegetation and Age	Reference(s)
				Great Basin and Periphery				
Lake Bonneville, Utah	Crescent Spring	Great Basin desert scrub	41°25'	113°20'	1300	27,700 to present	*Artemisia-Pinus* (woodland?); WM	Mehringer, 1977
Wasatch Mountains, Utah	Snowbird Bog	*Picea-Abies* forest	40°34'	111°45'	2500	12,300 to present	*Picea-Abies* forest; LW	Madsen and Currey, 1979
Lake Bonneville, Idaho	Swan Lake	*Artemisia tridentata* steppe	42°20'	112°00'	1460	12,090 to present	*Picea-Pinus flexilis* forest with *Artemisia*; LW	Bright, 1966
				Southern and Central Rocky Mountains				
Front Range, Colorado	Redrock Lake	*Pinus flexilis-Picea engelmannii-Abies* subalpine forest	40°05'	105°33'	3095	ca. 10,000 to present	Alpine tundra (ca. 300 m above treeline); LW	Maher, 1972
	Devlins Park	*Pinus flexilis, Picea engelmannii-Abies* subalpine forest	40°00'	105°33'	2953	22,400 to 15,230	Alpine tundra (ca. 500 m above treeline); WM	Legg and Baker, 1980
Yellowstone Park, Wyoming	Yellowstone Lake	*Picea-Abies-Pinus contorta* forest	44°17'	110°15'	2380	13,500 to present	*Picea-Abies-Pinus albicaulis* parkland and tundra; LW	Baker, 1976
	Cub Creek Pond	*Pinus contorta* forest with *Picea, Abies*	44°27'	110°15'	2485	14,360 to present	Alpine vegetation, perhaps with groves of *Picea, Abies* and *Pinus*; LW	Waddington and Wright, 1974
	Beaverdam Creek	*Pinus contorta* forest	44°20'	110°09'	2466	ca. 70,000,	Alpine tundra with *Artemisia*; EW	Baker and Richmond, 1978
Crested Butte area, Colorado	Alkali Creek	*Artemisia* steppe with *Picea-Abies-Pseudotsuga* forest	38°45'	107°20'	2800	12,975 to present	*Picea-Abies-Pinus* subalpine forest; LW	Markgraf and Scott, 1981
				Colorado Plateau				
Canyonlands, Utah	Cowboy Cave	Mosaic of *Pinus edulis-Juniperus-Quercus* woodland and *Artemisia* steppe	38°19'	110°12'	1710	13,040-8690	Mosaic of *Artemisia-*Poaceae steppe and *Picea* woodland; LW	Spaulding and Petersen, 1980
Chuska Mountains, New Mexico	Deadman Lake	*Pinus ponderosa* parkland and *Picea-Abies-Populus* forest	36°15'	108°58'	2780	> 28,000 to present	Alpine vegetation and *Picea* parkland; WM	Wright et al., 1973
La Plata Mountains, Colorado	Twin Lakes	*Picea engelmanii-Abies lasiocarpa* subalpine forest	37°30'	108°06'	3290	9800 to present	Same as present (early Holocene)	Petersen and Mehringer, 1976
San Juan Mountains, Colorado	Molas Lake	*Picea engelmanii-Abies lasiocarpa* subalpine forest	37°45'	107°50'	3200	15,450 to present	Alpine tundra; WM	Maher, 1972
Western New Mexico	San Augustin Plains	Chenopod desert scrub and grassland below *Pinus edulis-Juniperus* woodland	33°45'	108°00'	2065	> 27,000 to present	*Pinus-Picea* (woodland?); WM	Clisby and Sears, 1956
				Mojave Desert				
Las Vegas Valley, Nevada	Tule Springs	Mojave desert scrub, *Atriplex* spp., *Larrea*	36°20'	115°10'	703	> 37,000 to 22,600; 12,400 to present	*Juniperus-Artemisia* woodland; LW	Mehringer, 1967

Table 14-4 continued.

Area	Site	Local Vegetation	Latitude N	Longitude W	Elevation (m)	Approximate Time Range (yr B.P.)	Inferred Vegetation and Age	Reference(s)
			Mojave Desert (continued)					
Southeastern California	Searles Lake	Mojave desert scrub, *Atriplex* spp., *Larrea*	35°40'	117°20'	495	32,000 to ca. 5000	*Juniperus* woodland with *Pinus, Artemisia*; WM	Leopold, 1967; Roosma, 1958
	China Lake	Mojave desert scrub, *Atriplex* spp., *Larrea*	35°45'	117°45'	660	Early Wisconsin to 28,000	*Pinus-Artemisia* woodland; EW*	Martin and Mehringer, 1965 and unpublished
			Periphery of the Sonoran Desert					
Sulfur Springs Valley, Arizona	Willcox Playa	Desert scrub and desert grassland	31°30'	109°40'	1260	Sangamon to 23,000	*Pinus-Artemisia* parkland; EW and MW	Martin 1963a; Martin and Mehringer, 1965
San Pedro Valley, Arizona	Lehner Ranch	Desert scrub and desert grassland	31°20'	110°03'	1280	10,940 to present	Grassland; LW	Mehringer and Haynes, 1965
Mogollon Rim, Arizona	Potato Lake	*Pinus ponderosa* parkland	35°00'	11°40'	2210	> 14,400 to present	*Pinus-Picea-Abies-Pseudotsuga* parkland with *Artemisia*; WM*	Whiteside, 1965
			Chihuahuan Desert and Southern High Plains					
Llano Estacado, western Texas	Rich Lake	Grassland	33°15'	102°10'	—	> 26,500 to 17,400	Open boreal woodland; WM	Hafsten, 1961
	Tahoka Lake	Grassland	33°08'	101°40'	—	Wisconsin maximum	Open boreal woodland; WM	Hafsten, 1961
Llano Estacado, New Mexico	Arch Lake	Grassland	34°10'	103°05'	—	Middle Wisconsin (> 23,000) and Late Wisconsin	Open boreal woodland; WM	Hafsten, 1961
Llano Estacado, western Texas	Crane Lake	Grassland	31°25'	102°47'	—	Late Wisconsin to present	Open boreal woodland; LW	Hafsten, 1961
Rio Grande Valley western Texas	Bonfire Shelter	Chihuahuan desert scrub	29°50'	101°15'	427	10,230 to present	*Pinus* (pinyon-type) woodland; LW	Bryant, 1978
Central Texas	Boriack Bog	*Quercus stellata* savanna	30°30'	97°10'	122	15,460 to present	*Alnus* and mixed deciduous tree parkland; with *Picea*; WM	Bryant, 1977; Bryant and Shafer, 1977
	Hershop Bog	*Quercus stellata* savanna	29°35'	97°35'	124	10,574 to present	*Quercus* and mixed deciduous tree parkland; LW	Larson et al., 1972
Guadalupe Mountains, Texas	Williams Cave	Chihuahuan desert scrub	31°55'	104°50'	1500	12,040	*Juniperus-Pinus edulis* woodland; LW	Van Devender et al., 1979
	Upper Sloth Caves	Desert scrub, chaparral	32°00'	104°55'	2000	13,060 to 10,780	Subalpine forest with *Picea*; LW	Van Devender et al., 1979
Central Coahuila, Mexico	Cuatro Cienegas Basin	Chihuahuan desert scrub	27°40'	104°10'	790	> 30,000 to present	Desert scrub in basin and *Pinus-Quercus* woodland on fans; MW and LW	Meyer, 1973; Bryant and Riskind, 1980

Abbreviations: EW, Early Wisconsin; MW, Middle Wisconsin; WM, Wisconsin maximum; LW, latest Wisconsin.
*The age of the inferred vegetation is interpreted in this chapter.

Bonneville show that the lake supported a diverse freshwater diatom flora even though the present Great Salt Lake is too saline for diatoms (Patrick, 1936). During the preceding shallow-lake phase of the Middle Wisconsin interstade (the lowest half-meter of the sequence) (Figure 14-13), relative frequencies of *Pinus* were low and *Artemisia* and Chenopodiineae pollen precentages were as high or higher than they are today, suggesting widespread steppe vegetation.

For the eastern side of Bonneville Basin (Figure 14-4), a pollen diagram from Snowbird Bog, at 2500 m elevation in the Wasatch Mountains, indicates that a climate cooler and drier than that of the present characterized middle elevations during the latest Wisconsin (Madsen and Currey, 1979). Forests supported a somewhat greater

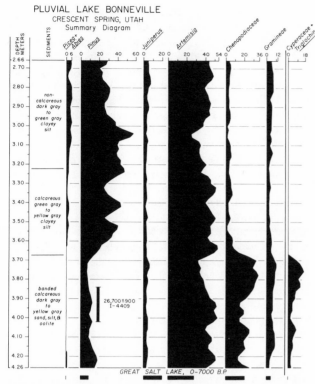

Figure 14-13. A summary pollen diagram of Lake Bonneville sediments and average pollen count for the last 7000 years of a core from the Crescent Spring (Figure 14-4). The diagram shows a transition from shallow water and marshes to an increase in water depth and importance of conifer pollen, including spruce (*Picea*), at about 24,000 yr B.P. The lower values of sagebrush (*Artemisia*) and the higher values of juniper distinguish the Holocene from the Middle Wisconsin; presumably, the latter period was considerably colder. (From Mehringer, 1977.)

abundance of *Picea* (perhaps *P. engelmannii*) than they do now. Higher frequencies of *Artemisia* and other nonarboreal pollen suggest that the forest was more open than it is at present.

The pollen records from Lake Bonneville can be compared with the abundant megafossil records from packrat middens on the southwestern margin of the old pluvial lake. The modern vegetation at packrat midden sites from Smith Creek Canyon, between 1700 and 2150 m elevation in the Snake Range (Figure 14-4) (Thompson, 1979), is pinyon-juniper woodland. The late-Pleistocene vegetation was not strictly analogous to modern subalpine forest, because the subalpine trees *Pinus flexilis, Pinus longaeva, Picea, engelmannii,* and *Juniperus communis* were associated with some high desert shrubs (e.g., *Artiplex confertifolia*) that grow neither in forest nor above high desert today. Both pollen and macrofossil data indicate differences between the Late and Middle Wisconsin vegetation. The Lake Bonneville pollen diagram (Figure 14-13) suggests that, during interstadial time, *Artemisia* and *Chenopodiineae* desert scrub and sedges existed near the lake, perhaps along the fans and marshy shores of a shrinking Lake Bonneville.

SOUTHERN ROCKY MOUNTAINS

Pollen research from Colorado raises some interesting questions about vegetational change at high and middle elevations. Latest-

Wisconsin spectra from Redrock Lake (Figure 14-1) (Maher, 1972), a subalpine site in the Front Range of the Rocky Mountains at 3095 m elevation, indicates that annual pollen accumulation (influx) rates were initially very low (500 to 600 pollen per square centimeter per year), comparable to influx values in low-arctic and forest-tundra regions (Birks and Birks, 1980). The dominant taxon was *Artemisia* (60%), with some *Pinus* and *Picea* pollen. On the basis of influx data, it is evident that *Picea engelmannii-Pinus* subalpine vegetation was established by 10,000 yr B.P. Maher (1972) argues that the occurrence of *Artemisia* pollen (in addition to smaller amounts of pollen from *Quercus, Juniperus, Sarcobatus,* and other Chenopodiineae) probably represents long-distance pollen transport from lower elevations both at present and during the Wisconsin. Also, the tendency for the lake to overflow in spring and early summer may have selectively removed early-blooming pollen types, thereby biasing the frequencies toward late-blooming plants such as *Artemisia* and perhaps reducing observed influx. Such outflow would remove some suspended silts and clays as well. Thus, the cirque-lake sediments have less tree pollen (dispersed in spring and early summer) and more herb and shrub pollen (dispersed throughout the summer and fall) (Maher, 1972).

At Devlins Park, 10 km to the south at 2953 m elevation, a full-glacial pollen sequence dating from 22,400 to 15,000 yr B.P. has been studied by Legg and Baker (1980). The assemblage basically resembles the late-glacial spectra at Redrock Lake, *Artemisia* (40% to 60%) is dominant, and the arboreal pollen (chiefly pine, with spruce and juniper) is a minor element. Lowland forms such as oak and *Sarcobatus,* judged to be products of long-distance transport, are rare (less than 5%). Pollen concentration is fairly high, as is the mean pollen accumulation rate (3000 per square centimeter per year). From these data, Legg and Baker (1980) conclude that *Artemisia* was locally present in the Wisconsin vegetation at Devlins Park. The discovery of an arctic-alpine type (*Koenigia islandica* L.) and high-meadow types (i.e., *Bistorta*) suggests that the local vegetation was probably alpine tundra. Using *Picea/Pinus* ratios (Maher, 1972), Legg and Baker (1980) estimate that, during the full glacial, Devlins Park was about 100 m above the treeline and that the treeline was depressed 500 m below its present altitude. The availability of local climatic stations gave a basis for inferring a mean annual precipitation 15 cm greater than today's, an annual frost-free season 5 days shorter, and a mean July daily maximum temperature about 5 °C cooler.

High percentages of *Artemisia* pollen are present in late-glacial zones at lake sites in the central Rocky Mountains (i.e., Yellowstone Lake, Wyoming [Figure 14-1]) (Baker, 1976), but farther south in Colorado late-Pleistocene sediments contain little or almost no *Artemisia* pollen. The Alkali Creek pollen record from 2800 m elevation in the southern Rocky Mountains (Figure 14-1) records a late-glacial spruce-fir subalpine forest, which was replaced by stands of pine during the early Holocene (Markgraf and Scott, 1981). Thus, the present altitudinal range of subalpine forest may have expanded during the late glacial and covered mountainous areas from the present upper treeline (about 3200 m) to at least 2800 m, some 200 m below the current lower limit of trees. Markgraf and Scott (1981) postulate that the monsoonal winds that bring summer rainfall to the Southwest today were absent from central Colorado prior to about 10,000 yr B.P. and after about 4000 yr B.P. They also suggest that, during the Late Wisconsin, the southern Rocky Mountains received more winter precipitation than they do at present, the result of a southward shift of Pacific frontal storms. Though such a change in seasonality of precipitation is not unlikely, the pollen evidence they present does not by itself appear to demonstrate this assertion.

COLORADO PLATEAU

Cowboy Cave (1710 m elevation), a canyon-side archaeological site that has been studied for both macrofossil and pollen evidence, lies on the northern edge of the Colorado Plateau in Utah (Figure 14-1) (Jen-

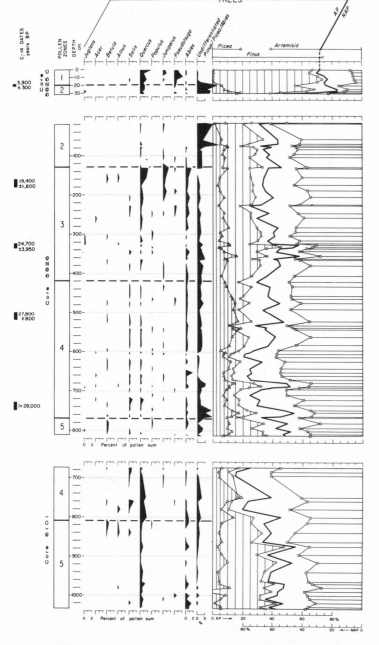

Figure 14-14. Summary pollen diagram from Dead Man Lake, Chuska Mountains, New Mexico. (See Figure 14-1.) Note the contrast between relative frequencies of *Artemisia* during the post-glacial period (zone 1) and the Wisconsin (zones 2 through 5). (From Wright et al., 1973, used with the permission of the Geological Society of America.)

nings, 1980). The present vegetation is a mosaic of pinyon-juniper-oak woodland and *Artemisia tridentata* steppe. The latest-Wisconsin cave sediments studied by Spaulding and Petersen (1980) contain abundant grass pollen, with a rich array of high-desert shrubs and herbs, perhaps representing a well-developed grassy steppe. Low pollen percentages of juniper and montane forest conifers, such as *Picea* and *Pseudotsuga*, are also present. Macrofossil evidence confirms that these trees grew locally along with woodland species (i.e., *Juniperus osteosperma*) and desert-scrub (high-desert) perennials *Artemisia* sec. Tridentatae, *Opuntia* spp., and *Ephedra* spp. *Betula* pollen is assumed to be the stream-side waterbirch (*B. occidentalis* Hook.). Spaulding and Petersen (1980) propose that montane and woodland conifers occupied bedrock exposures and that sagebrush communities occupied the sandy flats, a pattern similar to the present vegetational mosaic. The nearest locality where *Picea*, *Pseudotsuga*, and *Betula* grow together today is 50 km west in the Henry Mountains at elevations above 2200 m. It is evident from Cowboy Cave that montane conifers were more widespread on the Colorado Plateau during the latest Wisconsin and that they coexisted with woodland and high-desert forms. The local vegetation retained some montane elements until at least 8700 yr B.P., albeit with more xerophytic species than during the Late Wisconsin.

Evidence from the eastern edge of the Colorado Plateau also suggests that favorable habitat for sagebrush was greatly expanded during the Wisconsin. One of the most elaborate sequences in the Southwest comes from the Chuska Mountains near the New Mexico-Arizona border (Figure 14-1). Of the four lakes studied, Dead Man Lake at 2780 m elevation has the longest pollen record, dating from more than 28,000 yr B.P. to the present (Figure 14-14) (Wright et al., 1973). The mountain plateau on which the site occurs supports a ponderosa pine parkland. The modern pollen rain (Bent and Wright, 1963) reflects this vegetation, showing a dominance of *Pinus* (60% to 80%).

The late-Pleistocene pollen zones contrast with the Holocene chiefly in the abundance of *Picea* (10% to 20%) and the dominance of *Artemisia* (40% to 60%) (Figure 14-14). *Picea* and *Artemisia* rarely grow together in the Chuska Mountains. Alpine species of *Artemisia* are known in the southern Rocky Mountains to the northeast, but they are characteristically poor pollen producers.

An important aspect of this study is the identification based on pollen morphology of species of pine, now extinct in the Chuska Mountains (Figure 14-15) (Hansen and Cushing, 1973). Pollen assigned to *Pinus flexilis* is common in Wisconsin sediments from Dead Man Lake, further evidence that this species was widespread at higher elevations during Wisconsin time. Two other extralimital pines are *P. aristata* Engelm. (Rocky Mountain bristlecone pine) (Bailey, 1970) and *P. contorta* Doug. ex Loud. (lodgepole pine), which are unknown in the Wisconsin macrofossil record. Ponderosa pine, also identified from pollen, is unknown as well in packrat middens older than 10,000 yr B.P.

Wright and others (1973) interpret the Wisconsin pollen spectra as representing alternating periods of spruce parkland and alpine vegetation at the sites. *Artemisia* pollen, the dominant type, is presumed to be largely blown in from lower elevations. A full-glacial drop in the upper treeline of 900 m (to 2500 m elevation) is postulated; such a decline would place the upper limit of "spruce-fir and limber pine forest" well below the Dead Man Lake site. They further propose a "telescoping" of vegetation zones so that the lower limit of the ponderosa pine zone is depressed by only 400 m while pinyon-juniper-sagebrush woodland covered vast areas on the lower slopes and basin floors. This model of full-glacial vegetation, conceptualized in terms of the present vegetation zonation, is depicted in Figure 14-16.

Packrat middens recently recovered from Chaco Canyon, less than 90 km east of the Chuska Mountains (Figure 14-1), cast doubt on whether a Wisconsin pinyon-juniper-sagebrush woodland was present at lower elevations on the Colorado Plateau. Although only of latest-Wisconsin to earliest-Holocene age (10,600 to 9460 yr B.P.), these middens from 1910 m elevation are dominated by Douglas-fir and Rocky Mountain juniper (*Juniperus scopulorum*), with smaller amounts of limber pine, *Artemisia* sec. Tridentatae, and a trace of spruce. Neither pinyon pine nor the woodland juniper (*Juniperus monosperma* Sarg.) are found in these samples (Betancourt and Van Devender, 1981). (A newly discovered midden locality at 1770 m elevation, immediately west of the Chuska Mountains [latitude 36°09′N, longitude 109°28′W], provides a record of forest apparently dominated by Douglas-fir. Spruce [*Picea pungens* Engelm.], limber pine, dwarf juniper, and Rocky Mountain juniper are also components of this midden assemblage, with a date of 11,900 ± 300 yr B.P. [A-2996].) Ponderosa pine is also absent from these records of montane vegetation at Chaco Canyon, although it is presumed to typify the zone immediately above woodland in pollen-based reconstructions (Figure 14-16).

A major question in the Chuska Mountains and other studies is which pollen types represent local vegetation and which originated in lower-elevation plant communities. Several investigators (i.e., Maher, 1963; Petersen and Mehringer, 1976) show that *Artemis* pollen frequencies are now higher above the timberline than in coniferous forest. This is due to low amounts of arboreal pollen above the treeline and a resulting relative increase in the percentage of pollen blown up from the shrub steppe. Petersen and Mehringer (1976) have made effective use of spruce/sagebrush ratios to reconstruct postglacial timberline fluctuations in the La Plata Mountains (Figure 14-1), where Holocene sediments contain sagebrush-pollen frequencies that are never higher than 20%. But are *Artemisia* frequencies greater than 40% in Wisconsin sediments of mountain lakes solely the product of upslope wind, as many hypothesize? There is now macrofossil evidence from other areas in the Southwest showing mixed associations of steppe species (*Artemisia* sec. Tridentatae, *Atriplex confertifolia*, *Chrysothamnus* spp., etc.) and boreal conifers (e.g., *Pinus flexilis*, *P. longaeva*, *Picea* spp.) at elevations exceeding 2000 m during the full glacial (Tables 14-2 and 14-4) (Cole, 1981; Spaulding, 1981; Thompson and Mead, 1982). Modern vegetation from areas such as the Great Basin also provides examples of the association of sagebrush and montane conifers. Therefore, the dominance of *Artemisia* pollen at Dead Man Lake probably is not the result of long-distance transport, but rather sagebrush was a major element of the local vegetation. If so, then the full-glacial spruce-fir "forest" of Wright and others (1973) may have been a boreal parkland with spruce and an understory of sagebrush. Judging from the relative importance of *Pinus flexilis* in the pollen spectra (Figure 14-15) and its importance in the macrofossil record of the region (Table 14-2), limber pine may also have been a member of the boreal parkland.

The lower range of *Artemisia* at this latitude was at least 970 m, as recorded by a pollen diagram from Stanton's Cave in the upper Grand Canyon of Arizona (Iberall, 1972); macrofossils from nearby packrat middens confirm its presence (Figure 14-16) (Cole, 1981).

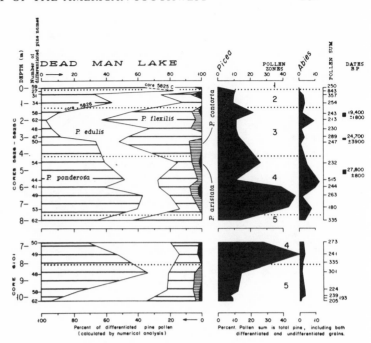

Figure 14-15. Percentages of differentiated pine pollen, spruce, and fir from Dead Man Lake, Chuska Mountains, New Mexico. Both graphs show percentages based on the total pine count. (From Wright et al., 1973, used with the permission of the Geological Society of America.)

Late- and full-glacial pollen spectra from the Stanton's Cave fill show a dominance of *Pinus* and *Artemisia*, with moderate amounts of *Juniperus* and some *Quercus*. Plant cuticles from coprolites at this site record what would be expected in open *Juniperus* woodland with *Artemisia* sec. Tridentatae and other steppe species (Iberall, 1972). Although further analysis is needed, the telescoping model of zone depression (first proposed by Wells [1966: 973]) does not find obvious support in the fossil midden records available thus far. Lower zones appear to have been transformed no less than upper ones.

A long Pleistocene record from the Colorado Plateau region has been recovered from the San Augustin Plains, a former pluvial lake in west-central New Mexico (Figure 14-1). Early work by Clisby and Sears (1956) describes a middle- and late-Pleistocene pollen sequence dominated by pine in association with desert scrub and grassland elements. The Middle and Late Wisconsin segment (the upper 10 m) of the 100-m core (with radiocarbon dates about 19,000 and 27,000 yr B.P.) is unique in the sequence since it records a pronounced rise in spruce. Clisby and Sears (1956) conclude that either the Late Wisconsin was marked by a particularly cold climate or that regional uplift during the late Pleistocene could have occurred. Further study of these deposits may yield a significant paleoclimatic record.

MOJAVE DESERT

Tule Springs in the Las Vegas Valley of southern Nevada at 703 m elevation (Figures 14-2 and 14-4) was the site of an ambitious paleoenvironmental research project (Wormington and Ellis, 1967). The present valley-floor vegetation is saltbush and creosote bush; woodland and forest are restricted to the surrounding mountains at least 10 km away and at elevations exceeding 1800 m. Mehringer (1965, 1967) presents the fossil-pollen record of a complex series of sediments from Tule Springs representing riparian, spring-mound,

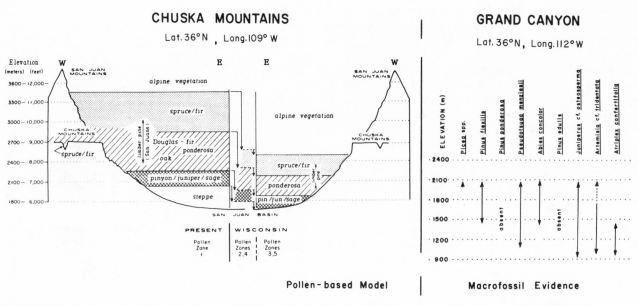

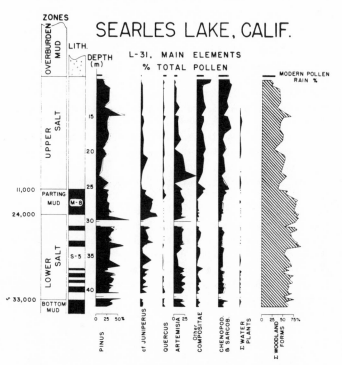

Figure 14-16. Sketch showing the altitudinal zonation of vegetation in the Chuska and San Juan Mountains at the present time (left), and during the Wisconsin (center), as inferred from the Chuska Mountains fossil pollen sequences. (From Wright et al., 1973, used with the permission of the Geological Society of America.) Arrows at the right show the full-glacial elevational range of selected species in the Grand Canyon. (Based on macrofossil evidence presented by Cole, 1981.) No sites were located above 2100 m elevation.

Figure 14-17. Summary pollen diagram from core L-31 from Searles Lake, California (Figure 14-4), showing the relative frequencies of common pollen types. Bars at the top of the section represent the modern pollen rain. Hatched curve on the far right includes those herbs, shrubs, and trees that are known to be associated with woodland and forest communities and therefore have a provenance outside the Searles Basin today. Pinyon-type pine pollen is indicated by the white area to the right of the pine curve (observations available only above S-5). Lithology varies between salt layers (open sections of the column) and mud layers (solid sections). (From Leopold, 1967.)

and lacustrine environments. Relative pollen frequencies from lacustrine mudstone dated between 31,000 to 22,600 yr B.P. are dominated by *Pinus* (30% to 80%), while latest-Wisconsin early-Holocene (14,000 to 9000 yr B.P.) spring sediments contain relatively high percentages of desert scrub taxa (Chenopodiineae and Asteraceae), with less than 20% pine.

Analysis of macrofossils from a series of packrat-midden sites from the Sheep Range to the north of Tule Springs (Figures 14-2 and 14-4) was undertaken, in part, to test pollen-based vegetational reconstructions (Spaulding, 1981). The packrat-midden record confirms only certain aspects of these earlier reconstructions. Mehringer (1967) is correct in stressing the importance of *Juniperus* and *Artemisia* in the late-glacial vegetation of the Las Vegas Valley, in noting that some species occurred more than 1000 m below their present lower limits, and in pointing to the former continuity of woodland between Mojave Desert mountain ranges; but the macrofossils fail to validate the general reconstructions of Wisconsin vegetation based on pollen analysis. Mehringer (1967: 193) postulates that during the Wisconsin maximum "there was a minimal lowering of the present vegetation zones by 1000 m," which suggests the former widespread distribution of mesic pinyon-juniper woodland and fir-pine (*Abies concolor-Pinus ponderosa*) forest or parkland. However, macrofossils document juniper vegetation on the valley flanks and limber pine above (Spaulding, 1981). No ponderosa pine and little pinyon have been found. The arboreal vegetation was of a different composition and was more xeric in character than was previously supposed.

Searles Lake is a playa at 495 m elevation in the western Mojave Desert of California (Figure 14-2 and 14-4). The modern pollen rain (Figure 14-17) reflects largely desert-scrub elements, since the nearest upwind source of conifer pollen in the Sierra Nevada is more than 65 km to the west. The fossil-pollen spectra show relative frequencies of *Artemisia* (20% to 25%) and *Juniperus* (up to 50%) (Figure 14-17) from the Late Wisconsin Parting Mud (Smith, 1977) that are comparable to the latest-Wisconsin spectra from southern Nevada spring mounds (Mehringer, 1967). Pinyon-type pine pollen, identified on the basis of bladder and body-length measurements (Martin, 1963: 20), accounts for only a small percentage of the total pine count at Searles Lake. This agrees with relative scarcity of pinyon in the macrofossil records of woodlands from the eastern Mojave Desert (Spaulding, 1981) and Grand Canyon (Cole, 1981; Phillips, 1977). Pinyon pine may have played a smaller role in the glacial-age woodland of the northern Mojave than it does today (Figure 14-10).

Following the full-glacial juniper maximum at Searles Lake, a marked rise in oak pollen appears in two cores (L-31 [Figure 14-17] and GS-14 [Leopold, 1967]) in the top meter of the Parting Mud. This feature is of special interest because it is of the same age as the latest-Wisconsin (about 13,800 to 12,800 yr B.P.) packrat middens at Robber's Roost in the Scodie Mountains about 60 km west of Searles Lake (Van Devender and Spaulding, 1979). The Robber's Roost record is the only known Wisconsin occurrence of oak megafossils (*Quercus turbinella*) in the Mojave Desert. The presence of *Q. turbinella* in Robber's Roost middens, as well as the appearance of oak pollen at Searles Lake, may well mark the northward migration of oak into the Mojave Desert during latest-Wisconsin time.

Except for the singular prominence of *Juniperus* during the Late Wisconsin, the Searles Lake pollen spectra from the Early, Middle, and Late Wisconsin are rather similar. Pollen and midden evidence suggests that juniper-dominated woodland probably grew on the mountain flanks above the level of pluvial Lake Searles (690 m elevation).

PERIPHERY OF THE SONORAN DESERT

A long and detailed pollen sequence from the Pleistocene of the Southwest comes from Willcox Playa (1260 m), formerly pluvial Lake Cochise, about 100 km east of the eastern boundary of the Sonoran Desert (Figure 14-2). The present vegetation is Chihuahuan desert scrub, and the modern pollen rain is dominated by pollen of Chenopodiineae and Asteraceae. The desert grassland of the flanking bajadas is reflected in the pollen rain by frequencies of Poaceae from 10% to 20%. Woodland with juniper, oak, pinyon (*Pinus discolor* Bailey and Hawksworth), and Chihuahua pine (*P. leiophylla* Schiede and Deppe) begins at 1500 m; and ponderosa pine forest lies above 2100 m. High montane forest with *Abies*, *Pseudotsuga*, and *Picea* is found in the Chiricahua Mountains 40 km west of the playa (Martin, 1963a, 1963b).

The long Pleistocene pollen record from Willcox Playa ends at about 22,000 yr B.P., and so the top zones show the pollen spectra for the Middle Wisconsin and part of the Late Wisconsin. *Pinus* percentages are very high in Late Wisconsin zone II (95% to 99%), contrasting with values of 20% for the modern pollen rain. The Middle Wisconsin (zone III) records somewhat lower percentages of pine (40% to 80%) and a varying array of other types, including *Juniperus*, *Quercus*, Poaceae, and *Artemisia*. Martin and Mosimann (1965) have conducted continuous sampling of isolated core sections and, through statistical techniques, demonstrated short-term vegetational change during the Wisconsin. Martin (1963a) has compared the Middle Wisconsin pollen spectra of Lake Cochise to the present pollen rain at Dead Man Lake in the Chuska Mountains, where a parkland (ponderosa pine) stands at the edge of the lake. Martin (1963a) has also described evidence from species of *Ephedra* indicating that, whereas summer monsoonal rains may have characterized some pre-Wisconsin intervals, summer precipitation was probably lacking during Wisconsin time.

Lehner Ranch Arroyo at 1280 m elevation in southeastern Arizona (Figure 14-2) (Mehringer and Haynes, 1965) provides a latest-Wisconsin and early-Holocene pollen record from alluvial sediments. The current pollen rain at this site in desert scrub along the San Pedro River is dominated by Chenopodiineae, Poaceae, and Asteraceae; arboreal pollen from present woodland and forest, more than 12 km away, accounts for less than 10% of the total. The fossil-pollen spectra directly associated with mammoth (*Mammuthus columbi* Falconer) (Saunders, 1977) and indirectly associated with Clovis-culture artifacts display relative frequencies quite similar to those of today. Mehringer and Haynes (1965) suggest that the vegetation was grassland analogous to that on the upper bajadas and infer that rainfall was no more than 8 to 10 cm higher than now and mean annual temperature no more than 1.7°C to 2.2°C cooler than present. Although macrofossils from the Sonoran Desert, less than 150 km to the northwest, suggest pinyon-juniper-oak woodland on rocky substrate (Figure 14-2) (Van Devender, 1973), the Lehner Ranch pollen record indicates extensive grassland in the San Pedro Valley.

At higher elevations on the Mogollon Rim of central Arizona, Potato Lake (2210 m elevation) yields a pollen sequence from a present-day *Pinus ponderosa* parkland (Figure 14-2) (Whiteside, 1965). Peaks to the east support boreal forest (*Picea engelmannii*, *Pseudotsuga menziesii*, *Abies lasiocarpa*) at higher altitudes. The present pollen rain is dominated by *Pinus* (65%), with some Poaceae and *Quercus*. *Picea*, *Abies*, and *Pseudotsuga* together represent 20% to 40% of the late-glacial and full-glacial pollen count, and *Artemisia* is important (20% to 30%). *Pinus* is relatively unimportant (20%), and Chenopodiineae is more frequent (5% to 15%) in the late-glacial pollen count than in the present-day count. Minor differences exist between the late-glacial and full-glacial assemblages; sagebrush pollen was somewhat more important in the full glacial, whereas pollen of boreal conifers was more abundant in the late glacial.

The Potato Lake record can be compared with the record from packrat middens for the same elevation in the Grand Canyon, about 200 km to the north (Cole, 1981). The abundant full-glacial macrofossils from the Nankoweap locality (Table 14-2) are *Picea engelmannii*, *Pinus flexilis*, *Pseudotsuga menziesii*, and *Abies concolor*. The abundance of *Artemisia* pollen suggests that it was growing nearby (Cole, 1981). Midden assemblages from lower elevations (1450 to 1100 m) (Table 14-1) indicate a *Juniperus* woodland with desert-scrub associates (e.g., *Opuntia erinacea* Engelm. & Bigel., *Agave utahensis* Engelm., and *Atriplex confertifolia*). Thus, we interpret the Potato Lake pollen assemblage as indicating high montane conifers and sagebrush in an open parkland. The small amounts of other desert-scrub and woodland pollen types were probably blown in from elevations below 1600 m south of the Mogollon Rim.

CHIHUAHUAN DESERT AND SOUTHERN HIGH PLAINS

Pollen records from the Llano Estacado (southern High Plains) (Figure 14-3) of western Texas and eastern New Mexico are of special interest, for the area is transitional between the desert West and the more humid Great Plains of North America (Martin, 1975b; Wendorf and Hester, 1975); (see the summary by Bryant, 1977, 1978).

Hafsten (1961) has established some of the first evidence that trees covered extensive areas of the Llano Estacado that are today grassland or desert scrub. The pollen sites lie between 900 and 1000 m elevation and are 80 to 160 km from the nearest stands of conifers. Using pollen data from cores and exposures at Rich, Tahoka, Arch, and Crane Lakes (Figure 14-1), Hafsten has described a composite sequence of vegetational stages for the region. The oldest interval (zone E) is thought to represent a Middle Wisconsin pluvial, a cool-moist period with open woodlands of *Pinus* and *Picea* that ended about 33,500 yr B.P. A Middle Wisconsin interpluvial (zone D; about 33,500 to 22,500 yr B.P.) was dominated by desert scrub and grassland elements. Conifer pollen was a minor component. This assemblage is similar to those of modern sites lying at least 125 km from the nearest conifer stands. In the full- and late-glacial sediments (zone C; 22,500 to 14,000 yr B.P.), proportions of pollen from pine and spruce steadily increase at the expense of steppe elements. By the end of this period, values of *Pinus* (60% to 90%) and *Picea* (5% to 10%) suggest "open boreal woodlands." Hafsten (1961) does not make a distinction between pine-pollen types, but he states that the majority of the full-glacial pine pollen at Crane Lake is "probably from pinyon pines." In contrast, the latest-glacial (zone B; about 14,000 to 10,000 yr B.P.) shows increasing Poaceae, Asteraceae, and Chenopodiineae; this suggests a transitional period in which grassland and steppe species dominate but boreal woodlands persist in favorable habitats. Tentative support for Hafsten's (1961) inferred vegetational sequence is provided by pollen from cave fill at Bonfire Shelter, some 185 km southwest of Crane Lake (Figure 14-1). Bryant and Shafer (1977) describe a shift from a full-glacial pinyon-pine woodland to a late-glacial grassland at this site that probably included isolated clumps of trees. These records agree with macrofossil evidence from middens in southwestern Texas showing the widespread occurrence of Texas pinyon (*Pinus remota*) during the Late Wisconsin (Lanner and Van Devender, 1981) (Table 14-1).

Hafsten's (1961) work provides evidence for the environment of Clovis culture sites found nearby. Remains of extinct megafauna also occur in sediments at several of the sites he studied. In a recent study, pollen analysts Schoenwetter and Oldfield have more intensively sampled Hafsten's key sections and added new pollen sites (Wendorf and Hester, 1975), in general substantiating Hafsten's earlier conclusions and correlations.

The Guadalupe Mountains of western Texas lie about 150 km west-northwest of Crane Lake (Figure 14-3). Van Devender and others (1979) present macrofossil and pollen records from cave sites there that confirm the presence of *Picea* in Texas during the late glacial. Pollen of *Pinus* and *Ostrya* are dominant in Late Wisconsin sediments from the High Sloth Caves at 2000 m elevation. Macrofossils from both middens and cave fill contain a mixture of subalpine, woodland, and even desert-grassland elements. Van Devender and others (1979) note that such a mixture cannot be found in the existing vegetation of the area. This macrofossil evidence also points to major southward extensions of the range of several species during the late Pleistocene. The nearest modern populations of *Juniperus communis* are more than 400 km to the north in central and eastern New Mexico, and *Picea engelmannii* and *Rubus strigosus* Michx. grow together in southeastern New Mexico some 150 km to the north.

The record of forest elements in Texas during the full glacial is not restricted to western Texas. Pollen data from Boriack Bog (Figure 14-1) and nearby Gause Bog in central Texas (Bryant, 1969, 1977) suggest that *Picea* was a minor element (1% to 3%), with eastern deciduous hardwoods and *Pinus* in deposits radiocarbon dated at 15,600 to 12,000 yr B.P. Bryant describes the full-glacial vegetation as deciduous woodland with some conifers present. Undated sediments from two other bogs in central Texas contain similar associations with *Picea* (Potzger and Tharp, 1954); a report of *Abies* pollen at these sites could not be substantiated by later investigations (Bryant, 1977; Graham and Heimsch, 1960). The absence of *Picea* pollen in late-glacial sediments at Hershop Bog 100 km farther south (Figure 14-1) led Bryant (1977) to suppose that the southern limit of spruce in central Texas lay between Boriack Bog and Hershop Bog. Bryant (1977) points out that, on the basis of the present distribution of *Picea* species and their close relation to July temperature (Leopold, 1957; Wolfe and Leopold, 1967), summer temperature was probably 22 °C, or at least 5.6 °C lower than at present during the full glacial.

Thus, full-glacial fossils from western and central Texas record open arboreal vegetation at every site where observations are available. The highlands of the Guadalupe Mountains appear to have been characterized by a queer mixed conifer woodland or forest with some xerophytic elements. The lowlands to the east and south appear to have supported pinyon-juniper-oak woodland, perhaps in a mosaic with scrub grassland. Central Texas localities appear to show an open oak parkland with some conifers and an understory of grasses mixed with shrubby vegetation.

Much farther south (latitude 27 °N) spring sediments in the Cuatro Cienegas Basin of central Coahuila provide a pollen record spanning more than 30,000 years (Figure 14-1). In his analysis of modern and fossil pollen, Meyer (1973) provides evidence from proportions of aquatic-plant pollen that there were probably only minor fluctuations in the local spring habitats during the Wisconsin. Increased relative frequencies of arboreal pollen, however, indicate the expansion of "pine forests and woodland on mountains surrounding the basin," particularly during the Late Wisconsin. Scrub oak-pinyon-acacia vegetation may have occupied the bajadas surrounding the basin. Meyer (1973: 994) suggests that, as a consequence, desert-scrub vegetation was restricted to the valley floor. However, latest-Wisconsin middens from eastern Coahuila and Durango contain juniper and pinyon (*Pinus remota*) associated with succulent desert species (Lanner and Van Devender, 1981; Van Devender, 1978). From these it appears that some desert-scrub plants may have persisted in altered woodland associations rather than being restricted to low-lying refugia. These records have particular biogeographic significance because of the endemic species of snails, isopods, scorpions, and herpetofauna found at Cuatro Cienegas (Meyer, 1973), as well as the rich, endemic succulent flora of the central Chihuahuan Desert.

SUMMARY

The fossil-pollen records of the Southwest provide complementary evidence for several trends documented by macrofossils from packrat middens. The general downward shift of major woodland and forest species is reflected in the Wisconsin-age pollen spectra from the deserts and surrounding highlands. The association of steppe plants and subalpine conifers in the high-elevation midden records is also evident in high-elevation pollen profiles from several areas. Abundant *Artemisia* pollen in the full-glacial spectra from the uplands is explicable in the light of individualistic models of vegetational change based on the macrofossil record. Analogues with current plant-community zonation are less enlightening.

Desert Biota During the Last Ice Age

The few Late Wisconsin packrat-midden records from Mexico as far south as latitude 26°N in the Chihuahuan Desert (Van) Devender, 1978) and latitude 29°N in the Sonoran Desert (Wells and Hunziker, 1976) are dominated by woodland trees. There is no evidence for desert refugia, that is, of desert-scrub vegetation similar to today's "waiting out" the glacial age in southerly latitudes. Instead, many desert plants existed in altered associations dominated by xerophytic conifers (Phillips, 1977; Van Devender, 1978; Van Devender and Spaulding, 1979). Some desert species such as four-wing saltbush, desert almond, Joshua-tree, sagebrush, and shadscale were particularly common in these woodlands (Van Devender, 1977b; Van Devender and Spaulding, 1979).

Small vertebrates also appear to have existed in altered associations. Today, ectotherms such as chuckwalla (*Sauromalus obesus* Baird), desert tortoise (*Gopherus agassizi* Cooper), and banded gecko (*Coleonyx variegatus* Baird) are restricted almost exclusively to deserts. They ranged widely in glacial-age woodlands south of latitude 36°N (Mead, 1981; Van Devender and Mead, 1978; Van Devender et al., 1977).

It appears unnecessary to invoke substantial southward migration to account for the survival of many desert species during glacial ages. These altered species associations were, in a sense, biotic refugia under Wisconsin climatic conditions. Global glacial-interglacial cycles are of high frequency over increments of evolutionary time (10^4 years or more) (Berggren et al., 1980; Imbrie and Imbrie, 1980), and this plasticity of species associations may be evidence of adaptation to repeated environmental changes.

A mystery is posed by the lack of Pleistocene records for some plants that are presently important in warm deserts. These plants include the columnar cacti (*Cereus* spp. *s.l.*), paloverdes (*Cercidium* spp., *Parkinsonia* spp.), ocotillos (*Fouquieria* spp.), and shrubby Asteraceae (such as *Flourensia cernua*, *Parthenium incanum*, *Ambrosia dumosa*, and *A. deltoidea*). The northern part of their ranges was attenuated during the last glacial age, and postglacial climates presumably allowed their expansion. Wells and Hunziker (1976) argue for the Late Wisconsin or early-Holocene dispersal of creosote bush from South America to the Chihuahuan Desert and from there to the Sonoran and Mojave Deserts. As evidence for its recent migration, they cite its importance in many warm-desert communities (giving it the appearance of a robust exotic), its differentiation into three chromosomal races, and its absence in Wisconsin packrat middens. Biogeography and cytology indicate that *Larrea* is indeed a South American plant (Hunziker et al., 1972; Porter, 1974). The hexaploid Mojave Desert creosote bush is farthest from its presumed point of introduction. It is derived from the tetraploid Sonoran race, which in turn comes from an ancestral diploid stock found today in the Chihuahuan Desert (Barbour, 1969; Yang, 1970). However the pace of chromosomal differentiation, even if known, would shed little light on the timing of the introduction. And the apparent vigor of a species in different vegetational types is no evidence for recent introduction *per se*. *Larrea* has been found in Wisconsin packrat middens as well as in the dung of the extinct Shasta ground sloth (*Nothrotheriops shastense* Hofstetter) from Gypsum and Rampart Caves (Figure 14-2) (Hansen, 1978; Laudermilk and Munz, 1934). Moreover, *Larrea* has been present in the Sonoran and Mojave Deserts since at least 10,000 yr B.P. (Spaulding, 1980; Van Devender, 1973); but it did not reach the northern and western limits

of its distribution in the Chihuahuan Desert until the late Holocene, less than 4000 years ago. Thus, an early-Holocene dispersal of creosote bush across the Continental Divide from the Chihuahuan Desert into the Sonoran Desert (Figure 14-1) (Wells and Hunziker, 1976) is quite unlikely. Creosote bush has probably been present in North America on both sides of the Continental Divide since pre-Wisconsin times, as have many other species that are still missing in the Wisconsin macrofossil record.

Faunal Extinction

The end of the Wisconsin was marked by the elimination of more than 35 genera of large mammals in North America, and ecosystems were left with impoverished second and third trophic levels. It was, in a sense, a recent ecologic catastrophe unparalleled before the advent of European humans. Although some workers have taken exception to the idea (Grayson, 1977b; Martin and Neuner, 1978), terminal-Wisconsin extinctions appear to have been restricted almost exclusively to large herbivores, their presumed carnivores, and commensals (Martin, 1967, 1973; Mosimann and Martin, 1975). Martin argues that this was a unique event in terms of the taxa affected, its occurrence without phyletic replacement, and its abruptness. Two types of alternate hypothesis have been offered in explanation. One attributes the demise of late-Pleistocene megafauna to intense hunting pressure by Late Paleolithic (Clovis) people on animal populations totally unused to human predation (Martin, 1967, 1973, 1975a). The other assumes that contemporaneous changes of climate associated with the end of the Wisconsin brought about habitat changes leading to extinctions, which selected against animals of large body size. In his review of the development of explanations for Pleistocene extinctions, Grayson (1980) notes that climatic hypotheses can be separated into those invoking increased continentality (hotter summers and colder winters; e.g., Axelrod, 1967; Guilday et al., 1978; Slaughter, 1967) and those citing desiccation (e.g., Antevs, 1959; Guilday, 1967) as the most likely mechanisms.

The paleoecology of the Shasta ground sloth (*Nothrotheriops shastense*) is the best understood of any extinct vertebrate of the North American Pleistocene. An occasional cave dweller, it ranged from sea level to near 2000 m elevation and frequented vegetation ranging from xeric juniper woodland along the Colorado River valley of Nevada and Arizona and in southern New Mexico (Long and Martin, 1974; Martin et al., 1961; Phillips, 1977; Thompson et al., 1980) to vegetation with montane conifers (spruce, Douglas-fir) in western Texas and southeastern Utah (Hansen, 1980; Spaulding and Petersen, 1980; Van Devender et al., 1979).

Plant cuticles in herbivore dung provide an effective means of reconstructing diet (Dearden et al., 1975), and such methods have been applied to mummified sloth feces from dry caves. The principal dietary components in southern Nevada included Joshua-tree and Utah agave (*Agave utahensis*) (Laudermilk and Munz, 1934), globemallow (*Sphaeralcea ambigua* Gray), and Mormon tea (*Ephedra nevadensis*) (Hansen, 1978) in the lower Grand Canyon and Mormon tea (*Ephedra* sp.), century plant (*Agave* sp.), and shrubs of Rosaceae in southern New Mexico (Eames, 1930; Thompson et al., 1980). From the abundance of desert plants in sloth dung many workers have inferred that desert-scrub vegetation occurred below woodland during the Late Wisconsin (Laudermilk and Munz, 1934; Martin et al., 1961; Mehringer, 1965; Wells and Berger, 1967). However, packrat middens contemporary with the time of the

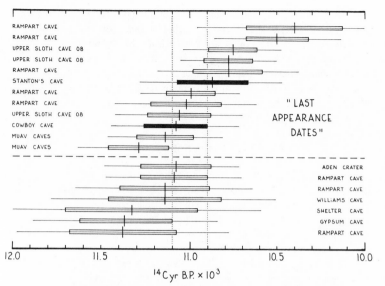

Figure 14-18. Radiocarbon dates younger than 11,400 yr B.P. on dung of extinct herbivores from dry caves in the Southwest. (Modified from Thompson et al., 1980.) Rectangles indicate one standard deviation from the mean (vertical line), and horizontal lines indicate two standard deviations. The solid rectangles indicate dates on herbivores other than ground sloth. Dates plotted above the horizontal dashed line are the highest from a section or unit containing extinct megafauna. Dashed vertical lines represent the 11,000 ± 100 yr B.P. age for Clovis culture in southern Arizona (Haynes, 1980). (Data are from Table 14-4.)

ground sloths indicate that they selected succulents and shrubs growing in juniper woodlands (Phillips, 1977; Thompson et al., 1980) without ingesting much juniper.

Organic deposits in dry caves can provide tight chronologies of megafaunal extinction, and sloth-dung deposits have received the most intense dating tests (Long and Martin, 1974; Long et al., 1974; Thompson et al., 1980). Radiocarbon assays on animal remains from the top of such stratified deposits represent actual terminal or last-appearance dates, not merely the youngest age determination in a given area (Alford, 1974). Only 2 of the 10 terminal dates for ground sloth from Rampart Cave (untrampled dung balls) and Guadalupe Mountains upper sloth Cave 08 (dung balls at the top of the 30- to 40-cm stratum) are significantly younger than 11,000 yr B.P. at one standard deviation (Table 14-5 and Figure 14-18).

These chronologies are important when alternative hypotheses for the cause of megafaunal extinction in the Western Hemisphere are being weighed. The dynamics of the "overkill" model (Mosimann and Martin, 1975) require that megafaunal extinction was synchronous (within the limits of the radiocarbon method) with the advent of Clovis people at about 11,000 yr B.P. (Haynes, 1970, 1980). The model does not exclude the possibility that a less conspicuous human population existed before then (Martin, 1967: 101-2). Although many claim earlier records (e.g., Adovasio et al., 1978; Bryan, 1978; MacNeish, 1976), Clovis is the oldest well-defined cultural complex south of the continental glaciers (West, 1983). Thirty-one radiocarbon dates from the Murray Springs and Lehner mammoth-kill sites allow Haynes (1980) to place Clovis in southern Arizona at 11,000 ± 100 yr B.P. The dry-cave dates indicate a similar age for the final demise of ground sloth in the Grand Canyon and Guadalupe Mountains, as well as a large artiodactyl from Stan-

ton's Cave and *Bison* sp. from Cowboy Cave (Figure 14-18).

It appears that megafaunal extinction took place in the Southwest about 11,000 yr B.P. Such a pattern cannot entirely be explained by the relatively gradual environmental change that characterized the latest-Wisconsin-to-Holocene transition. This was, however, a period of major flux in biotic systems. The disappearance of extensive coniferous woodland and the shrinking of water supplies may somehow be related to the widespread extinction of megafauna. One circumstantial argument against a climatic cause of terminal Pleistocene extinctions is the documented frequency of Quaternary climatic oscillations. At least nine glacial-to-interglacial cycles have occurred in the last 700,000 years (Berggren et al., 1980), and we assume that, by the close of the Wisconsin, the Rancholabrean fauna was well adapted to such change. In this vein, there is little reason to suppose that higher vertebrates would not have been able to adapt to environmental change by simply shifting ranges. Hypotheses attributing megafaunal extinction to environment thus far fail to explain how a single climatic change could have the same impact throughout the Western Hemisphere, from Canada to Patagonia.

Wisconsin Climates: Some Inferences

Moist climatic episodes in the deserts of North America seem out of phase with those of North Africa and Australia (Kutzbach, 1980; Rognon and Williams, 1977; Street and Grove, 1979). In the American West, between 24,000 and 11,000 yr B.P., there was maximum development of alpine glaciers, filling of playa lakes, and expansion of upland vegetation (Benson, 1978; Mehringer, 1977; Morrison, 1965; Smith, 1977; Van Devender and Spaulding, 1979). But the chronology of climate in the western United States is better understood than its nature. Paleoclimatic reconstructions for the Late Wisconsin based on physical evidence vary from a modest 2 °C to 3 °C temperature decline and a 100% relative increase in annual precipitation to a 7 °C to 11 °C temperature drop and even less precipitation than today (Table 14-6).

The fossil record has great potential for reconstructing Wisconsin climate, although many questions remain. Key data on the climate characterizing the modern habitat of many taxa are incomplete or missing. Meteorologic stations are usually placed so that the environment of a plant is not measured (Beatley, 1974), and mean annual values may have little effect on the range of a given plant or animal. Critical variables also include maximum and minimum values, seasonality, evaporation rates, and latitudinal and altitudinal lapse rates. The occasional intense frost or prolonged drought, perhaps occurring only once a decade, may play a significant part in determining species distributions. It is premature to offer specific paleoclimatic reconstructions based on the macrofossil and pollen records, but the fossil biota provide complementary data for some of the models already proposed (Table 14-6). Also, a comparison of the fossil data over this broad area reveals geographic variation that can be attributed to regional variations in the paleoclimate.

TEMPERATURE REGIMES

Paleontologists have pointed to the sympatry of heat-intolerant "northern" animals and cold-sensitive "southern" animals in Wisconsin faunal assemblages from the southern Great Plains and the Southwest. Analogous associations of desert-scrub and woodland plant and animal fossils occur in middens from the warm deserts.

Table 14-5. Radiocarbon Dates Younger Than 11,400 yr B.P. on Extinct
Large Herbivores from Dry Caves in the Southwest

Area	Site	Material Dated	From Top of Section or Unit?	Radiocarbon Age (yr B.P.)	Laboratory Number	Reference
Grand Canyon, Arizona	Rampart Cave	*Nothrotheriops* dung ball	Yes	10,400 ± 275	I-442	Long et al., 1974
Grand Canyon, Arizona	Rampart Cave	*Nothrotheriops* dung ball	Yes	10,500 ± 180	A-2174	Thompson et al., 1980
Grand Canyon, Arizona	Rampart Cave	*Nothrotheriops* dung ball	Yes	10,780 ± 200	A-1067	Long et al., 1974
Grand Canyon, Arizona	Rampart Cave	*Nothrotheriops* dung ball	Yes	11,000 ± 140	A-1066	Long et al., 1974
Grand Canyon, Arizona	Rampart Cave	*Nothrotheriops* dung ball	Yes	11,020 ± 200	A-1068	Long et al., 1974
Grand Canyon, Arizona	Rampart Cave	*Nothrotheriops* dung ball	?	11,090 ± 190	A-1602	Thompson et al., 1980
Grand Canyon, Arizona	Rampart Cave	*Nothrotheriops* dung ball	No	11,140 ± 250	A-1453	Phillips, 1977
Grand Canyon, Arizona	Rampart Cave	*Nothrotheriops* dung ball	No	11,370 ± 300	A-1392	Long et al., 1974
Grand Canyon, Arizona	Muav Caves	*Nothrotheriops* dung ball	Yes	11,140 ± 160	A-1212	Long et al., 1974
Grand Canyon, Arizona	Muav Caves	*Nothrotheriops* dung ball	Yes	11,290 ± 170	A-1213	Long et al., 1974
Grand Canyon, Arizona	Stanton's Cave	Large artiodactyl feces (*Navahoceros*[?], *Oreamnos*[?])	Yes	10,870 ± 200	A-1155	Iberall, 1972
Clark County, Nevada	Gypsum Cave	*Nothrotheriops* dung	No	11,360 ± 260	A-1202	Long et al., 1974
Dona Ana County, New Mexico	Aden Crater	*Nothrotheriops* dung	No	11,080 ± 200	Y-1163B	Long et al., 1974
Dona Ana County, New Mexico	Shelter Cave	*Nothrotheriops* dung	No	11,330 ± 370	A-1878	Long et al., 1974
Guadalupe Mountains, Texas	Cave 08	*Nothrotheriops* dung ball	Yes	10,750 ± 140	A-1583	Spaulding and Martin, 1979
Guadalupe Mountains, Texas	Cave 08	*Nothrotheriops* dung ball	Yes	10,780 ± 140	A-1534	Spaulding and Martin, 1979
Guadalupe Mountains, Texas	Cave 08	*Nothrotheriops* dung ball	Yes	11,060 ± 180	A-1584	Spaulding and Martin, 1979
Guadalupe Mountains, Texas	Williams Cave	*Nothrotheriops* dung ball	Yes	11,140 ± 320	A-1509	Spaulding and Martin, 1979
Canyonlands, Utah	Cowboy Cave*	*Bison* dung	Yes	11,020 ± 180	A-1660	Spaulding and Petersen, 1979

Source: After Thompson et al., 1980.

*Local extinction only.

Such mixing has been attributed to a combination of cool summers and mild winters (Hibbard, 1960; Martin and Neuner, 1978; Slaughter, 1967; Van Devender and Mead, 1978; Van Devender and Spaulding, 1979; Van Devender et al., 1979). This "equable" model holds that glacial winters were free from drastic temperature declines. The hypothesis that arctic air masses may have been trapped north of the edge of the Laurentide ice sheet (Bryson and Wendland, 1967) lends a climatologic basis to such a thesis (Van Devender and Wiseman, 1977).

The question of how cold ice-age winters were depends on elevation and latitude. The warm deserts, apparently characterized by equable glacial winters, lie to the south (below latitude 36°N), with most fossil sites below 1450 m elevation. Full-glacial middens from the north are dominated by xeric woodland or subalpine trees and by Great Basin Desert shrubs (Cole, 1981; Spaulding, 1981; Thompson and Mead, 1982; Wells and Berger, 1967). Steppe species (*Artemisia* sec. Tridentatae, *Atriplex confertifolia*, *Chrysothamnus* spp.) were

common, and mesophytic thermophilous taxa were rare. The Mojave and Great Basin midden records clearly indicate colder winter temperatures compared to sites in the Sonoran and Chihuahuan Deserts, but it is unclear whether the relative decrease in full-glacial winter temperatures was any greater north of about latitude 36°N than farther south.

SUMMER PRECIPITATION

There is a strong theoretical basis for a full-glacial reduction of summer precipitation, relative to that of today, in the Southwest. Summer rainfall in the southwestern deserts depends on the development of a strong Bermuda high (Bryson and Lowry, 1955; Van Devender and Wiseman, 1977; (see also Hales, 1974). Circulation around this system causes a flow of oceanic air from the Gulf of Mexico northwest to the arid West, with the resultant proportions of summer rainfall declining with distance from the Gulf (Hastings and Turner, 1965). Undoubtedly, global cooling would have weakened such sub-

Table 14-6. Paleoclimatic Reconstructions for the Full Glacial of the American Southwest

Reference	Data Base	Location	ΔT_a	ΔT_s	ΔT_w	ΔP	%P	%E
Antevs, 1952	Hydrologic budgets	Lake Lahontan, Nevada	−2.5 to −3.0	—	—	+8 to +16	50 to 100	−30
Broecker and Orr, 1958	Hydrologic budgets	Lake Lahontan, Nevada	−5	—	—	+20	80	−30
Snyder and Langbein, 1962	Hydrologic budgets	Lake Spring, Nevada	−5*	−7.2**	—	+20	67	−30
Mifflin and Wheat, 1979	Hydrologic budgets	Nevada, statewide	−2.8	—	—	+8.4 to −24	68***	−10
Antevs, 1954	Relict snowlines	North-central New Mexico	—	−5.6	—	+23	—	—
Leopold, 1951	Hydrologic budgets and snowline changes	Lake Estancia, New Mexico	−6.6	−9	−2.8	+18 to +25	50 to 70	−23 to −50
Galloway, 1970	Solifluction (?) deposits	Sacramento Mountains, New Mexico	−10.5	—	—	−4.6	—	−51
Bachhuber and McClellan, 1977	Foraminifera distributions	Lake Estancia, New Mexico	—	−9.7	—	—	—	—
Brackenridge, 1978	Relict cirques and cryogenic deposits	Montana to Arizona (lat. 45°40′N to 33°20′N)	−7****	—	—	0	0	—
Reeves, 1966	Hydrologic budgets	Llano Estacado, western Texas	−5	−8	—	+39	89	−27

Note: ΔT_a, ΔT_s, and ΔT_w are changes in °C in annual, summer, and winter temperatures, respectively. ΔP is change in annual precipitation in cm; %P is ΔP/modern P; %E (evaporation) is ΔE/modern E.

*Extrapolated by Morrison (1965), Schumm (1965), and Mifflin and Wheat (1979).

**Extrapolated by Schumm (1965) and Brackenridge (1978).

***Statewide average.

****Minimum estimate.

tropical high-pressure systems, and the increased latitudinal gradient would have restricted their influence to more southerly latitudes. Moreover, the frequency of warm-season precipitation in the deserts depends on local high temperatures and convective uplift. Fewer convective thunderstorms would occur if summer temperatures were lower.

Van Devender (1973 and subsequent publications) has pointed to a distinct lack in Wisconsin middens of fossils of both summer annuals and summer rain-dependent perennials. Ocotillo (*Fouquieria splendens*) is an example of a frost-hardy plant that ranges to elevations exceeding 1525 m, but it is mainly limited to the monsoonal Sonoran and Chihuahuan Deserts. It is absent from the Wisconsin and early-Holocene midden record, which yields many plants characteristic of the relatively summer-dry Great Basin and northern Mojave regions (e.g., Joshua-tree, single-needle pinyon, blackbrush, shadscale, sagebrush). Evidence suggesting a lack of summer precipitation during the Late Wisconsin is also provided by Markgraf and Scott (1981). Thus, we find little substance to Wells's (1979) claim that the existing gradient of summer precipitation was "enhanced" during the Wisconsin.

WINTER PRECIPITATION

The scheme of winter-mild, summer-cool climates for the warm deserts incorporates increased winter precipitation as well. Continental ice sheets and resultant global cooling might lead to an intensification of the Aleutian low and associated Pacific-type storms that today are responsible for much of the winter precipitation in the Southwest (Houghton et al., 1975; Sellers and Hill, 1974). A steepened Arctic-equator temperature gradient might also result in the southerly displacement of frontal storms sweeping in from the Pacific Ocean.

Enhanced zonal flow and an increased frequency in Pacific-type frontal storms are consistent with the woodland midden records, which suggest more winter rainfall than presently occurs in the Sonoran and Chihuahuan Deserts (Van Devender, 1973 and subsequent publications). They are also consistent with evidence of arid conditions in the Pacific Northwest during the full glacial (Barnosky, 1981). Enhanced winter precipitation in the warm deserts may promote relatively mild temperatures through increased cloud cover and relative humidity, reducing the heat lost by nocturnal radiation.

Important questions remain concerning Wisconsin precipitation regimes in the current Great Basin Desert and northern Mojave Desert. Winter seasonality of precipitation is likely, but, in contrast to the warm deserts, there is little biotic evidence for a marked increase in average annual precipitation (Cole, 1982; Spaulding, 1981). The discovery that most playas of southern and south-central Nevada were, at best, ephemeral lakes or marshes during the full glacial (Mifflin and Wheat, 1979) also points to relatively dry conditions north of latitude 36°N (Van Devender and Spaulding, 1979). Surface low-pressure systems (Great Basin lows) draw maritime air up from the southwest and account for much of today's winter precipitation in the eastern and central Great Basin (Houghton et al., 1975; Sellers and Hill, 1974). If, as suggested earlier, there was enhanced zonal flow during the Late Wisconsin, less precipitation would have been drawn into the Great Basin in comparison to the Sonoran and Chihuahuan Deserts (Figure 14-1). The mechanisms causing a significant increase in the annual rainfall in the southern deserts can account for the relatively dry conditions farther north.

CLIMATIC RECONSTRUCTIONS

Table 14-6 shows Late Wisconsin climatic reconstructions, which are based primarily on hydrologic budgets of pluvial lakes and snow-line depression. Mifflin and Wheat (1979) present excellent data on the relative development of pluvial lakes in Nevada but then adopt a quantitative treatment that assumes a corresponding increase in average annual precipitation for each incremental decrease in temperature. The most extreme estimate of temperature reduction (Galloway, 1970) (Table 14-6) is based on the least amount of data (Van Devender and Spaulding, 1979). Brakenridge (1978) presents physical evidence for a 7°C mean annual temperature reduction on the basis of relict cirques, but his discussion of the paleoecologic record is inadequate.

Leopold's (1951) reconstruction of the Late Wisconsin climate in the Estancia Basin (Table 14-6) is in general accord with the packrat-midden data from the northern Chihuahuan Desert. A postulated decline of about 9°C in summer temperature is also supported by the modern affinities of fossil foraminifers from pluvial Lake Estancia (Bachhuber and McClellan, 1977). This is coupled with a suggested modest decline in winter temperatures of less than 3°C. It is difficult, however, to reconcile paleoclimatic reconstructions calling for a greater than 25% increase in average annual precipitation (Table 14-6) with the midden assemblages from the Mojave and Great Basin Deserts. Many paleohydrologic reconstructions are based on the preconception that glacial conditions in temperate regions led to high rainfall in the southwestern deserts. Macrofossil assemblages indicate that this may not have been the case in some parts of the southwestern United States.

Summary and Conclusions

There are many distinctive aspects of both the present-day and Late Wisconsin ecology of the Southwest. This arid region was characterized by increased effective moisture throughout the last glacial age, and widespread woodland occurred in lowlands that today support only desert scrub. Ancient packrat middens from the present Mojave, Sonoran, and Chihuahuan Deserts yield macrofossil records of juniper (*Juniperus* spp.) and juniper-pinyon (*Pinus monophylla, P. edulis, P. remota*) communities from near the valley floors to elevations between 1600 and 2000 m. Species that were common in the glacial-age woodlands south of about latitude 36°N included evergreen oaks (e.g., *Quercus turbinella, Q. dunnii*); succulents such as yuccas, beargrass, and agaves (*Yucca* spp., *Nolina* spp. *Agave* spp.); and certain warm-desert elements such as barrel cactus (*Echinocactus polycephalus*), brittle-bush (*Encelia* spp.), gopher tortoise (*Gopherus agassizi*), and chuckwalla (*Sauromalus obesus*). Above this woodland zone and throughout the present Great Basin Desert north of latitude 37°N, macrofossil and pollen records suggest a full-glacial vegetation dominated by montane and subalpine conifers. In this region bristlecone and limber pine (*Pinus longaeva* and *P. flexilis*) were widely distributed above about 1800 m elevation.

Pollen-stratigraphic data with continuous sequences provide a temporal framework of Wisconsin vegetational changes in the Southwest. Such studies also yield information on regional vegetation that complements packrat-midden data representing local plant associations. Percentages of nonarboreal pollen can give a good indication of general vegetational structure, for example, a basis for discriminating between closed forest and parkland. Both pollen and packrat-midden data suggest that open-structured arboreal vegetation prevailed over much of the Southwest during the full glacial. Previous uncertainties raised by pollen data alone are clarified by positive evidence from macrofossils demonstrating that some xerophytic shrubs had wide ranges and associations during the Wisconsin.

Packrat-midden data are site specific and of high taxonomic resolution. They provide confirmation of the individualistic nature of plant-community change that has been suggested on the basis of some pollen studies in other regions (e.g., Davis, 1981; Watts, 1980). It has long been known that, in the desert West, mesophytic plant species were displaced to lower, presently arid habitats during the Wisconsin. Macrofossil assemblages demonstrate that the relative displacement of different species may have varied by more than 600 m. In the northern Mojave Desert, shadscale (*Atriplex confertifolia*) even occurred more than 300 m above its current upper elevational limit. No modern analogues are known for some of the resultant plant associations found in packrat middens. What is more, certain modern community types such as desert scrub, grassland, Mexican pine-oak woodland, and ponderosa pine parkland or forest have no counterpart in the macrofossil record. There seems to have been a general restructuring of vegetation zones in the Southwest (Cole, 1982). The present combined macrofossil and pollen evidence permits us to define only three widespread vegetational types from the full glacial. These are: (1) a lower woodland zone of pygmy conifers and xerophytic shrubs and succulents restricted to elevations below about 1800 m and latitudes south of about 37°N, (2) a zone of open subalpine or mixed montane conifer vegetation with sagebrush and other steppe shrubs, and (3) a high-elevation alpine tundra zone. This provisional model invites further study from the perspective of modern vegetational distribution.

The high mountains of the desert West served as interglacial refugia for mesophytic plants and animals. These biotic islands were larger and probably more numerous during the Late Wisconsin. Their reduced isolation led to increased migration of boreal mammals and a greater diversity of those species on mountaintops. Postglacial desertification led to a reduction in habitat size and a termination of immigration. Local extinctions of some boreal species followed (Grayson, 1981; Thompson and Mead, 1982). In contrast, the packrat-midden data offer no evidence for an increase in plant species diversity as a result of larger montane islands during the Wisconsin. The macrofossil floras of the large mountains (maximum elevations exceeding 2300 m) studied thus far contain no plant species that do not occur somewhere in those ranges today. Fossil records of extralimital plants are restricted to mountains with low relief and reflect Holocene extinctions only.

Although the Late Wisconsin ecosystems of the Southwest supported a diverse large-mammal fauna, the paleoecology of only one species is relatively well known. The Shasta ground sloth (*Nothrotheriops shastense*) frequented woodlands and montane forests from near sea level to at least 2000 m elevation. It subsisted on a diet of shrubs and succulents that are common in modern desert scrub and woodland vegetation. The radiocarbon dating of ground sloth dung from stratifed cave deposits reveals that its disappearance was contemporaneous with the dates from lowland Clovis-culture archaeological sites. This is circumstantial evidence supporting the ''overkill'' model of megafaunal extinction (Martin, 1973). However, the latest Wisconsin was a time of considerable biotic flux, which may have adversely affected the large mammal populations of the Southwest in some as-yet undefined way.

The full-glacial climate of the Southwest was effectively moist compared to that of the Holocene. Precipitation occurred primarily during the winter, even in the areas that presently receive mostly summer rain. There is evidence for a Late Wisconsin gradient of increasing summer rainfall from northwest to southeast, but it was less pronounced than today's. The relative increase in annual precipitation during the full glacial may have been somewhat greater in the Sonoran and Chihuahuan Deserts to the south than in the Mojave and Great Basin Deserts. Summer temperatures were probably much lower than today's throughout the Southwest. Winters were probably severe north of about latitude 36°N. The relative decline in cold-season temperatures may have been less south of this latitude, even though many of the thermophilous plants that typify the present Sonoran and Chihuahuan Deserts are rare or missing from the fossil record. The ever-increasing data base provided by both packrat-midden and pollen studies offers a rare opportunity for further detailed reconstructions of the paleoenvironments of this unique region.

References

Adovasio, J. M., Gunn, J. D., Donahue, J., and Stukenrath, R. (1978). Meadowcroft Rockshelter, 1977: An overview. *American Antiquity* 43, 632-51.

Alford, J. J. (1974). The geography of mastodon extinction. *The Professional Geographer* 26, 425-29.

Antevs, E. (1952). Cenozoic climates of the Great Basin. *Geologischen Rundschau* 40, 94-108.

Antevs, E. (1954). Climate of New Mexico during the last glaciopluvial. *Journal of Geology* 62, 182-91.

Antevs, E. (1959). Geological age of the Lehner mammoth site. *American Antiquity* 25, 31-34.

Axelrod, D. I. (1967). "Quaternary Extinctions of Large Mammals." University of California Publications in the Geological Sciences 74.

Bachhuber, F. W., and McClellan, W. A. (1977). Paleoecology of marine foraminifera in pluvial Estancia Valley, central New Mexico. *Quaternary Research* 7, 254-67.

Bailey, D. K. (1970). Phytogeography and taxonomy of *Pinus* subsection Balfourianae. *Annals of the Missouri Botanical Garden* 57, 210-49.

Baker, R. G. (1976). "Late Quaternary Vegetation History of the Yellowstone Basin, Wyoming." U.S. Geological Survey Professional Paper 729-E.

Baker, R. G., and Richmond, G. M. (1978). Geology, palynology, and climatic significance of two pre-Pinedale Lake sediment sequences in and near Yellowstone National Park. *Quaternary Research* 10, 226-40.

Barbour, M. G. (1969). Patterns of genetic similarity between *Larrea divaricata* of North and South America. *American Midland Naturalist* 81, 54-67.

Barnosky, C. W. (1981). A record of late Quaternary vegetation from Davis Lake, southern Puget lowland, Washington. *Quaternary Research* 16, 221-39.

Beatley, J. C. (1974). Effects of rainfall and temperature on the distribution and behavior of *Larrea tridentata* (creosote-bush) in the Mojave Desert of Nevada. *Ecology* 55, 245-61.

Beatley, J. C. (1975). Climates and vegetation pattern across the Mojave/Great Basin transition of southern Nevada. *American Midland Naturalist* 93, 53-70.

Beetle, A. A. (1960). "A Study of Sagebrush, the Section Tridentatae of *Artemisia*." Bulletin of the University of Wyoming Experimental Station 368.

Behle, W. H. (1978). Avian biogeography of the Great Basin and Intermountain Region. *In* "Intermountain Biogeography: A Symposium" (K. T. Harper and J. L. Reveal, eds.), pp. 55-80. Great Basin Naturalist Memoirs 2.

Benson, L. V. (1978). Fluctuation in the level of pluvial Lake Lahontan during the last 40,000 years. *Quaternary Research* 9, 300-318.

Bent, A. M., and Wright, H. E., Jr. (1963). Pollen analyses of surface materials and lake sediments from the Chuska Mountains, New Mexico. *Geological Society of America Bulletin* 74, 497-500.

Berggren, W. A., Burckle, L. H., Cita, M. B., Cooke, H. B. S., Funnel, B. M., Gartner, S., Hays, J. D., Kennett, J. P., Opdyke, N. D., Pastouret, L., Shackleton, N. J., and Takayanagi, Y. (1980). Towards a Quaternary time scale. *Quaternary Research* 13, 277-302.

Betancourt, J. L., and Van Devender, T. R. (1981). Holocene vegetation in Chaco Canyon, New Mexico. *Science* 214, 656-58.

Birks, H. J. B., and Birks, H. H. (1980). "Quaternary Paleoecology." Edward Arnold, London.

Bleich, V. C., and Schwartz, O. A. (1975). Observations on the home range of the desert woodrat, *Neotoma lepida intermedia*. *Journal of Mammalogy* 56, 518-19.

Blinman, E., Mehringer, P. J., Jr., and Sheppard, J. C. (1979). Pollen influx and the deposition of Mazama and Glacier Peak tephra. *In* "Volcanic Activity and Human Ecology" (P. D. Sheets and D. K. Grayson, eds.), pp. 393-425. Academic Press, New York.

Brakenridge, G. R. (1978). Evidence for a cold, dry full-glacial climate in the American Southwest. *Quaternary Research* 9, 22-40.

Bright, R. C. (1966). Pollen and seed stratigraphy of Swan Lake, southeastern Idaho. *Tebiwa* 9, 1-47.

Broecker, W. S., and Orr, P. C. (1958). Radiocarbon chronology of Lake Lahontan and Lake Bonneville. *Geological Society of America Bulletin* 69, 1009-32.

Brown, D. E., and Lowe, C. H. (1980). Biotic communities of the Southwest. U.S. Department of Agriculture, Forest Service, General Technical Report RM-78 (map).

Brown, D. E., Lowe, C. H., and Pase, C. P. (1979). A digitized classification system for the biotic communities of North America, with community (series) and association examples for the Southwest. *Journal of the Arizona-Nevada Academy of Science* 14 (supplement 1).

Brown, J. (1971). Mammals on mountain tops: Non-equilibrium insular biography. *American Naturalist* 105, 467-78.

Brown, J. (1978). The theory of insular biogeography and the distribution of boreal birds and mammals. *In* "Intermountain Biogeography: A Symposium" (K. T. Harper and J. L. Reveal, eds.), pp. 209-27. Great Basin Naturalist Memoirs 2.

Bryan, A. L. (ed.) (1978). "Early Man in the New World from a Circum-Pacific Perspective." Occasional Papers of the Department of Anthropology, University of Alberta.

Bryant, V. M., Jr. (1969). Late full-glacial and post-glacial pollen analysis of Texas sediments. Ph.D. dissertation, University of Texas at Austin, Austin.

Bryant, V. M., Jr. (1977). A 16,000 year pollen record of vegetational change in central Texas. *Palynology* 1, 143-56.

Bryant, V.M., Jr. (1978). Late Quaternary pollen records from the east-central periphery of the Chihuahuan Desert. *In* "Transactions of the Symposium on the Biological Resources of the Chihuahuan Desert Region" (R. H. Wauer and D. H. Riskind, eds.), pp. 3-21. U.S. Department of the Interior Transactions and Proceedings Series 3.

Bryant, V. M., Jr., and Riskind, D. H. (1980). The paleoenvironmental record of northeastern Mexico: A review of pollen evidence. *In* "Papers on the Prehistory of Northeastern Mexico and Adjacent Texas" (J. F. Epstein, T. R. Hester, and C. Graves, eds.), pp. 7-31. Center for Archaeological Research, University of Texas at San Antonio, Special Report 9.

Bryant, V. M., Jr., and Shafer, H. J. (1977). "The Late Quaternary Paleoenvironment of Texas: A model for the Archaeologist." Bulletin of the Texas Archaeological Society 48.

Bryson, R. A., and Lowry, W. P. (1955). The synoptic climatology of the Arizona summer precipitation singularity. *American Meteorological Society Bulletin* 36, 329-39.

Bryson, R. A., and Wendland, W. M. (1967). Tentative climatic patterns for some late glacial and post-glacial episodes in central North America. *In* "Life, Land and Water" (S. J. Mayer-Oakes, ed.), pp. 279-98. University of Manitoba Press, Winnipeg, Canada.

CLIMAP Project Members (1976). The surface of the ice-age Earth. *Science* 191, 1131-37.

Clisby, K. H., and Sears, P. B. (1956). San Augustin Plains: Pleistocene climatic changes. *Science* 124, 537-39.

Clokey, I. W. (1951). "Flora of the Charleston Mountains, Clark County, Nevada." University of California at Berkeley Publications in Botany 24.

Cole, K. L. (1981). Late Quaternary environments in the eastern Grand Canyon: Vegetational gradients over the last 25,000 years. Ph.D. dissertation, University of Arizona, Tucson.

Cole, K. L. (1982). Late Quaternary zonation of vegetation in the eastern Grand Canyon. *Science* 217, 1142-45.

Cranford, J. A. (1977). Home range and habitat utilization by *Neotoma fuscipes* as determined by radiotelemetry. *Journal of Mammalogy* 58, 165-72.

Cronquist, A., Holmgren, A. H., Holmgren, N. H., and Reveal, J. L. (1972). "Intermountain Flora." Vol. 1. Hafner Publishing Company, New York.

Daubenmire, R. (1970). "Steppe Vegetation of Washington." Washington State University, Washington Agricultural Experiment Station Technical Bulletin 62.

Davis, M. B. (1981). Quaternary history and the stability of forest communities. *In* "Forest Succession: Concepts and Applications" (D. C. West and H. H. Shugart, eds.), pp. 132-51. Springer-Verlag, New York.

Dearden, B. L., Pegau, R. E., and Hansen, R. M. (1975). Precision of microhistological estimates of ruminant food habits. *Journal of Wildlife Management* 38, 402-7.

Eames, A. J. (1930). Report on ground sloth coprolite from Dona Ana County, New Mexico. *American Journal of Science* 20, 353-56.

Finley, R. B., Jr. (1958). "The Woodrats of Colorado: Distribution and Ecology." University of Kansan Publications in Natural History 10, pp. 213-552.

Galloway, R. W. (1970). The full-glacial in the southwestern United States. *Annals of the Association of American Geographers* 60, 245-56.

Graham, A., and Heimsch, C. (1960). Pollen studies of some Texas peat deposits. *Ecology* 41, 751-63.

Grayson, D. K. (1977a). On the Holocene history of some northern Great Basin lagomorphs. *Journal of Mammalogy* 58, 507-13.

Grayson, D. K. (1977b). Pleistocene avifaunas and the overkill hypothesis. *Science* 195, 691-92.

Grayson, D. K. (1979). Mt. Mazama, climatic change, and Fort Rock Basin archaeofaunas. *In* "Volcanic Activity and Human Ecology" (P. D. Sheets and D. K. Grayson, eds.), pp. 427-58. Academic Press, New York.

Grayson, D. K. (1980). Vicissitudes and overkill: The development of explanations of Pleistocene extinctions. *Advances in Archaeological Method and Theory* 3, 357-403.

Grayson, D. K. (1981). A mid-Holocene record for the heather vole *Phenacomys* cf. *intermedius*, in the central Great Basin and its biogeographic significance. *Journal of Mammalogy* 62, 115-21.

Grayson, D. K. (1982). Toward a history of Great Basin mammals during the past 15,000 years. *In* "Man and Environment in the Great Basin" (D. B. Madsen and J. F. O'Connell, eds.), pp. 82-101. Society for American Archaeology Papers 2.

Guilday, J. E. (1967). Differential extinction during late Pleistocene and recent times. *In* "Pleistocene Extinctions: The Search for a Cause" (P. S. Martin and H. E. Wright, Jr., eds.), pp. 121-40. Yale University Press, New Haven, Conn.

Guilday, J. E., Hamilton, H. W., Anderson, E., and Parmalee, P. W. (1978). "The Baker Bluff Cave Deposit, and the Late Pleistocene Faunal Gradient." Carnegie Museum of Natural History Bulletin 11.

Hafsten, U. (1961). Pleistocene development of vegetation and climate in the southern High Plains as evidenced by pollen analysis. *In* "Paleoecology of the Llano Estacado" (F. Wendorf, ed.), pp. 59-91. Museum of New Mexico Press, Santa Fe.

Hales, J. E., Jr. (1974). Southwestern United States summer monsoon source—Gulf of Mexico or Pacific Ocean? *Journal of Applied Meteorology* 13, 331-42.

Hansen, B. S., and Cushing, E. J. (1973). Identification of pine pollen of late Quaternary age from the Chuska Mountains, New Mexico. *Geological Society of America Bulletin* 84, 1181-99.

Hansen, R. M. (1980). Late Pleistocene plant fragments in the dungs of herbivores at Cowboy Cave. *In* "Cowboy Cave" (J. D. Jennings, ed.), pp. 179-90. University of Utah Anthropological Papers 104.

Harper, K. T., Freeman, D. C., Ostler, W. K., and Klikoff, L. G. (1978). The flora of Great Basin mountain ranges: Diversity, sources, and dispersal ecology. *In* "Intermountain Biogeography: A Symposium" (K. T. Harper and J. L. Reveal, eds.), pp. 81-103. Great Basin Naturalist Memoirs 2.

Hastings, J. R., and Turner, R. M. (1965). "The Changing Mile." University of Arizona Press, Tucson.

Hastings, J. R., Turner, R. M., and Warren, D. K. (1972). "An Atlas of Some Plant Distributions in the Sonoran Desert." University of Arizona Institute of Atmospheric Physics Technical Report 21.

Haynes, C. V., Jr. (1970). Geochronology of man-mammoth sites and their bearing on the origin of the Llano complex. *In* "Pleistocene and Recent Environments of the Central Plains" (W. E. Dort and A. E. Johnson, eds.), pp. 77-92. University Press of Kansas, Lawrence.

Haynes, C. V., Jr. (1980). The Clovis culture. *Canadian Journal of Anthropology* 1, 115-21.

Hibbard, C. W. (1960). An interpretation of Pliocene and Pleistocene climates in North America. *Michigan Academy of Science, Arts, and Letters Annual Report* 62, 5-30.

Houghton, J. G., Sakamoto, C. M., and Gifford, R. O. (1975). "Nevada's Weather and Climate." Nevada Bureau of Mines Special Publication 2.

Hunziker, J. H., Palacios, R. A., deVales, A. G., and Poggio, L. (1972). Species disjunctions in *Larrea*: Evidence from morphology cytogenetics, phenolic compounds, and seed albumins. *Annals of the Missouri Botanical Garden* 59, 224-33.

Iberall, E. R. (1972). Paleoecological studies from fecal pellets: Stanton's Cave, Grand Canyon, Arizona. M.S. thesis, University of Arizona, Tucson.

Imbrie, J., and Imbrie, J. Z. (1980). Modeling the climatic response to orbital variations. *Science* 207, 943-53.

Jennings, J. D. (1980). "Cowboy Cave." University of Utah Anthropological Papers 104.

Johnson, N. K. (1975). Controls of number of species on montane islands in the Great Basin. *Evolution* 29, 545-67.

Johnson, N. K. (1978). Patterns of avian geography and speciation in the Intermountain Region. *In* "Intermountain Biogeography: A Symposium" (K. T. Harper and J. L. Reveal, eds.), pp. 137-60. Great Basin Naturalist Memoirs 2.

Johnston, M. J. (1977). Brief resumé of botanical, including vegetational, features of the Chihuahuan Desert region with special emphasis on their uniqueness. *In* "Transactions of the Symposium on the Biological Resources of Chihuahuan Desert Region" (R. H. Wauer and D. H. Riskind, eds.), pp. 335-62. U.S. Department of the Interior, National Park Service Transactions and Proceedings Series 3.

King, F. B. (1977). An evaluation of the pollen contents of coprolites as environmental indicators. *Journal of the Arizona Academy of Science* 12, 47-52.

King, J. E., and Van Devender, T. R. (1977). Pollen analysis of fossil packrat middens from the Sonoran Desert. *Quaternary Research* 8, 191-204.

Kutzbach, J. E. (1980). Estimates of past climate at Paleolake Chad, North Africa, based on a hydrological and energy-balance model. *Quaternary Research* 14, 210-23.

Lanner, R. M. (1974). Natural hybridization between *Pinus edulis* and *Pinus monophylla* in the American Southwest. *Silvae Genetica* 23, 99-134.

Lanner, R. M., and Van Devender, T. R. (1974). Morphology of pinyon pine needles from fossil packrat middens in Arizona. *Forest Science* 20, 207-11.

Lanner, R. M., and Van Devender, T. R. (1981). Late Pleistocene pinyon pines in the Chihuahuan Desert. *Quaternary Research* 15, 278-90.

Larson, D. A., Bryant, V. M., and Patty, T. S. (1972). Pollen analysis of a central Texas bog. *American Midland Naturalist* 88, 358-67.

Laudermilk, J. D., and Munz, P. A. (1934). "Plants in the Dung of *Nothrotherium* from Gypsum Cave, Nevada." Carnegie Institute of Washington Publication 453, pp. 29-37.

Legg, T. E., and Baker, R. G. (1980). Palynology of Pinedale sediments, Devlins Park, Boulder County, Colorado. *Arctic and Alpine Research* 12, 319-33.

Leopold, E. B. (1957). Some aspects of late-glacial climate in eastern United States. *Verhandlungen der vierten Internationalen Tagung der Quaterbotaniker* 34, 80-85.

Leopold, E. B. (1967). Summary of palynological data from Searles Lake. *In* "Pleistocene Geology and Palynology, Searles Valley, California," pp. 52-66. Unpublished Friends of the Pleistocene Guidebook.

Leopold, L. B. (1951). Pleistocene climate in New Mexico. *American Journal of Science* 249, 152-68.

Leskinen, P. H. (1975). Occurrence of oaks in late Pleistocene vegetation in the Mojave Desert of Nevada, *Madroño* 23, 234-35.

Little, E. L., Jr. (1968). Two new pinyon varieties from Arizona. *Phytologia* 17, 329-42.

Little, E. L., Jr. (1976). "Atlas of United States trees," vol. 3, "Minor Western Hardwoods." U.S. Department of Agriculture, Forest Service, Miscellaneous Publication 1314.

Long, A., Hansen, R. M., and Martin, P. S. (1974). Extinction of the Shasta ground sloth. *Geological Society of America Bulletin* 85, 1843-48.

Long, A., and Martin, P. S. (1974). Death of the American ground sloth. *Science* 638-40.

Mabry, T. J., Hunziker, J. H., and DiFeo, D. R., Jr. (eds.) (1977). "The Creosote Bush: The Biology and Chemistry of *Larrea* in New World Deserts." Dowden, Hutchinson, and Ross, Stroudsburg, Pa.

MacArthur, E. D., and Plummer, A. P. (1978). Sagebrush systematics and evolution. *In* "The Sagebrush Ecosystem", pp. 14-22. Utah State University, College of Natural Resources, Logan.

MacArthur, R. H., and Wilson, E. O. (1967). "The Theory of Island Biogeography." Princeton University Press, Princeton, N.J.

MacNeish, R. S. (1976). Early man in the New World. *American Scientist* 64, 316-27.

Madsen, D. B. (1973). Late Quaternary paleoecology in the southeastern Great Basin. Ph.D. dissertation, University of Missouri, Columbia.

Madsen, D. B., and Currey, D. R. (1979). Late Quaternary glacial and vegetation changes, Little Cottonwood Canyon area, Wasatch Mountains, Utah. *Quaternary Research* 12, 254-70.

Maher, L. J., Jr. (1963). Pollen analyses of surface materials from the southern San Juan Mountains, Colorado. *Geological Society of America Bulletin* 74, 1485-1504.

Maher, L. J., Jr. (1972). Absolute pollen diagram from Redrock Lake, Boulder County, Colorado. *Quaternary Research* 2, 531-53.

Markgraf, V., and Scott, L. (1981). Lower timberline in central Colorado during the past 15,000 yr. *Geology* 9, 231-34.

Martin, L. D., and Neuner, A. M. (1978). The end of the Pleistocene in North America. *Transactions of the Nebraska Academy of Sciences* 6, 117-26.

Martin, P. S. (1963a). Geochronology of pluvial Lake Cochise, southern Arizona: II. Pollen analysis of a 42-meter core. *Ecology* 446, 436-44.

Martin, P. S. (1963b). "The Last 10,000 years: A Fossil Pollen Record of the American Southwest." University of Arizona Press, Tucson.

Martin, P. S. (1967). Prehistoric overkill. *In* "Pleistocene Extinctions: The Search for a Cause" (P. S. Martin and H. E. Wright, Jr., eds.), pp. 75-120. Yale University Press, New Haven, Conn.

Martin, P. S. (1973). The discovery of America. *Science* 179, 969-74.

Martin, P. S. (1975a). Palaeolithic players on the American stage: Man's impact on the late Pleistocene megafauna. *In* "Arctic and Alpine Environments" (J. D. Ives and R. Barry, eds.), pp. 669-700. Methuen, London.

Martin, P. S. (1975b). Vanishings, and future, of the prairie. *Geoscience and Man* 10, 39-49.

Martin, P. S., and Mehringer, P. J., Jr. (1965). Pleistocene pollen analysis and biogeography of the Southwest. *In* "The Quaternary of the United States" (H. E. Wright, Jr., and D. G. Frey, eds.), pp. 433-51. Princeton University Press, Princeton, N.J.

Martin, P. S., and Mosimann, J. E. (1965). Geochronology of pluvial Lake Cochise, southern Arizona: III. Pollen statistics and Pleistocene metastability. *American Journal of Science* 263, 313-58.

Martin, P. S., Sabels, B. E., and Shutter, R., Jr. (1961). Rampart Cave coprolite and paleoecology of the Shasta ground sloth. *American Journal of Science* 259, 102-27.

Mead, J. I. (1981). The last 30,000 years of faunal history within the Grand Canyon, Arizona. *Quaternary Research* 15, 311-44.

Mead, J. I., Thompson, R. S., and Long, A. (1978). Arizona radiocarbon dates IX: Carbon isotope dating of packrat middens. *Radiocarbon* 20, 171-91.

Mehringer, P. J., Jr. (1965). Late Pleistocene vegetation in the Mojave Desert of southern Nevada. *Journal of the Arizona Academy of Science* 3, 172-88.

Mehringer, P. J., Jr. (1966). "Some Notes on the Late Quaternary Biogeography of the Mohave Desert." University of Arizona, Tucson, Geochronology Laboratories Interim Research Report 11.

Mehringer, P. J., Jr. (1967). Pollen analysis of the Tule Springs site, Nevada. *In* "Pleistocene Studies in Southern Nevada" (H. M. Wormington and D. Ellis, eds.), pp. 129-200. Nevada State Museum Anthropological Papers 13.

Mehringer, P. J., Jr. (1977). Great Basin late Quaternary environments and chronology. *In* "Models and Great Basin Prehistory: A Symposium" (D. D. Fowler, ed.), pp. 113-67. Desert Research Institute Publications in the Social Sciences 12.

Mehringer, P. J., Jr., and Haynes, C. V., Jr. (1965). The pollen evidence for the en-

vironment of early man and extinct mammals at the Lehner mammoth site, southeastern Arizona. *American Antiquity* 32, 17-23.

Merriam, C. H. (1890). "Results of a Biological Survey of the San Francisco Mountain Region and Desert of the Little Colorado of Arizona." U.S. Department of Agriculture, North American Fauna 3, pp. 1-136.

Meyer, E. R. (1973). Late-Quaternary paleoecology of the Cuatro Cienegas Basin, Coahuila, Mexico. *Ecology* 54, 982-95.

Mifflin, M. D., and Wheat, M. M. (1979). "Pluvial Lakes and Estimated Pluvial Climates of Nevada." Nevada Bureau of Mines and Geology Bulletin 94.

Morrison, R. B. (1965). Quaternary geology of the Great Basin. *In* "The Quaternary of the United States" (H. E. Wright, Jr., and D. G. Frey, eds.), pp. 265-85. Princeton University Press, Princeton, N.J.

Morrison, R. B., and Frye, J. C. (1965). "Correlations of Middle and Late Quaternary Successions of the Lake Lahontan, Lake Bonneville, Rocky Mountain, Southern Great Plains and Eastern Midwest Areas." Nevada Bureau of Mines Report 9.

Mosimann, J. E., and Martin, P. S. (1975). Simulating overkill by Paleoindians. *American Scientist* 63, 304-13.

Patrick, R. (1936). Some diatoms of Great Salt Lake. *Torrey Botanical Club Bulletin* 63, 157-66.

Petersen, K. L., and Mehringer, P. J., Jr. (1976). Postglacial timberline fluctuations, La Plata Mountains, southwestern Colorado. *Arctic and Alpine Research* 8, 275-88.

Phillips, A. M., III. (1977). Packrats, plants, and the Pleistocene in the lower Grand Canyon. Ph.D. dissertation, University of Arizona, Tucson.

Porter, D. M. (1974). Disjunct distributions in the New World Zygophyllaceae. *Taxon* 23, 239-346.

Potzger, J. E., and Tharp, B. C. (1954). Pollen study of two bogs in Texas. *Ecology* 35, 462-66.

Raun, G. G. (1966). "A population of Woodrats (*Neotoma micropus*) in Southern Texas." Texas Memorial Museum Bulletin 11.

Reeves, C. C., Jr. (1966). Pleistocene climate of the Llano Estacado, II. *Journal of Geology* 74, 642-47.

Rognon, P., and Williams, M. A. J. (1977). Late Quaternary climatic changes in Australia and North Africa: A preliminary interpretation. *Palaeoecology* 21, 285-327.

Roosma, A. (1958). A climatic record from Searles Lake, California. *Science* 128, 716.

Saunders, J. J. (1977). Lehner Ranch revisited. *In* "Paleoindian Lifeways" (E. Johnson, ed.), pp. 48-64. Museum Journal 17, West Texas Museum Association.

Schumm, S. A. (1965). Quaternary paleohydrology. *In* "The Quaternary of the United States" (H. E. Wright, Jr., and D. G. Frey, eds.), pp. 783-94. Princeton University Press, Princeton, N.J.

Sellers, W. D., and Hill, R. H. (eds.) (1974). "Arizona Climate, 1931-1972." University of Arizona Press, Tucson.

Shelford, V. E. (1978). "The Ecology of North America." University of Illinois Press, Urbana.

Shmida, A., and Whittaker, R. H. (1979). Convergent evolution of deserts in the Old and New Worlds. *In* "Werden und vergehen von Pflanzengesellschaften" (O. Wilmanns and R. Tuxen, eds.), pp. 437-49. T. Cramer, Vaduz.

Shreve, F. (1942a). The desert vegetation of North America. *Botanical Review* 8, 195-246.

Shreve, F. (1942b). Grassland and related vegetation in northern Mexico. *Marono* 6, 190-98.

Shreve, F. (1964). Vegetation of the Sonoran Desert. *In* "Vegetation and Flora of the Sonoran Desert" (F. Shreve and I. L. Wiggins), pp. 9-187. Stanford University Press, Stanford, Calif.

Slaughter, B. H. (1967). Animal ranges as a clue to late Pleistocene extinction. *In* "Pleistocene Extinctions: The Search for a Cause" (P. S. Martin and H. E. Wright, Jr., eds.), pp. 155-67. Yale University Press, New Haven, Conn.

Smith, G. I. (1968). Late Quaternary geologic and climatic history of Searles Lake, southeastern California. *In* "Means of Correlation of Quaternary Successions" (R. B. Morrison and H. E. Wright, Jr., eds.), pp. 293-310. University of Utah Press, Salt Lake City.

Smith, G. I. (1977). Paleoclimatic record in the upper Quaternary sediments, Searles Lake, California, U.S.A. *Paleolimnology of Lake Biwa and the Japanese Pleistocene* 4, 577-604.

Smith, G. I., and Street-Perrott, F. A. (1983). Pluvial lakes of the western United States. *In* "Late-Quaternary Environments of the United States," vol. 1, "The Late Pleistocene" (S. C. Porter, ed.), pp. 190-212. University of Minnesota Press, Minneapolis.

Snyder, C. T., Hardman, G., and Zdenek, F. F. (1964). Pleistocene lakes in the Great Basin. U.S. Geological Survey Miscellaneous Geological Investigations Map I-416.

Snyder, C. T., and Langbein, W. B. (1962). The Pleistocene lake in Spring Valley, Nevada, and its climatic implications. *Journal of Geophysical Research* 67, 2385-95.

Spalding, V. M. (1909). "Distribution and Movements of Desert Plants." Carnegie Institute of Washington Publication 113.

Spaulding, W. G. (1980). The presettlement vegetation of the California desert. Unpublished manuscript on file, California Bureau of Land Management, Riverside.

Spaulding, W. G. (1981). The late Quaternary vegetation of a southern Nevada mountain range. Ph.D. dissertation, University of Arizona, Tucson.

Spaulding, W. G., and Martin, P. S. (1979). Ground sloth dung of the Guadalupe Mountains. *In* "Biological Investigations in the Guadalupe Mountains National Park, Texas" (H. H. Genoways and R. J. Baker, eds.), pp. 259-69. U.S. Department of the Interior, National Park Service Proceedings and Transactions Series 4.

Spaulding, W. G., and Petersen, K. L. (1980). Late Pleistocene and early Holocene paleoecology of Cowboy Cave. *In* "Cowboy Cave" (J. D. Jennings, ed.), pp. 163-77. University of Utah Anthropological Papers 104.

Stones, R. C., and Hayward, C. L. (1968). Natural history of the desert woodrat, *Neotoma lepida*. *American Midland Naturalist* 80, 458-76.

Street, F. A., and Grove, A. T. (1979). Global maps of lake-level fluctuations since 30,000 yr B.P. *Quaternary Research* 12, 83-118.

Thompson, R. S. (1979). "Late Pleistocene and Holocene Packrat Middens from Smith Creek Canyon, White Pine County, Nevada." Nevada State Museum Anthropology Papers 17, pp. 363-80.

Thompson, R. S., and Mead, J. I. (1982). Late Quaternary environments and island biogeography in the Great Basin. *Quaternary Research* 17, 39-55.

Thompson, R. S., Van Devender, T. R., Martin, P. S., Foppe, T., and Long, A. (1980). Shasta ground sloth (*Nothrotheriops shastense* Hoffstetter) at Shelter Cave, New Mexico: Environment, diet, and extinction. *Quaternary Research* 14, 360-76.

Tueller, P. T., Beeson, C. D., Tausch, R. J., West, N. E., and Rea, K. H. (1979). "Pinyon-Juniper Woodlands of the Great Basin: Distribution, Flora, Vegetal Cover." U.S. Department of Agriculture, Forest Service Research Paper INT-229.

Van Devender, T. R. (1973). Late Pleistocene plants and animals of the Sonoran Desert: A survey of ancient packrat middens in southwestern Arizona. Ph.D. dissertation, University of Arizona, Tucson.

Van Devender, T. R. (1977a). Comment on "Macrofossil analysis of wood rat (*Neotoma*) middens as a key to the Quaternary vegetational history of arid North America by P. V. Wells." *Quaternary Research* 8, 236-36.

Van Devender, T. R. (1977b). Holocene woodlands in the southwestern deserts. *Science* 198, 189-92.

Van Devender, T. R. (1978). Glaciopluvial woodlands in the Bolson de Mapimi, Durango and Coahuila, Mexico. *In* "American Quaternary Association Fifth Biennial Meeting, Abstracts," p. 234. University of Alberta, Edmonton.

Van Devender, T. R., Freeman, C. E., and Worthington, R. D. (1978). Full glacial and recent vegetation of Livingston Hills, Presidio County, Texas. *Southwestern Naturalist* 23, 289-302.

Van Devender, T. R., and Mead, J. I. (1978). Early Holocene and late Pleistocene amphibians and reptiles in Sonoran Desert packrat middens. *Copeia* 1978 (3), 464-75.

Van Devender, T. R., Phillips, A. M., III, and Mead, J. I. (1977). Late Pleistocene reptiles and small mammals from the lower Grand Canyon of Arizona. *Southwestern Naturalist* 22, 39-66.

Van Devender, T. R., and Riskind, D. H. (1979). Late Pleistocene and early Holocene plant remains from Hueco Tanks State Historical Park: The development of a refugium. *Southwestern Naturalist* 24, 127-40.

Van Devender, T. R., and Spaulding, W. G. (1979). Development of vegetation and climate in the southwestern United States. *Science* 204, 701-10.

Van Devender, T. R., Spaulding, W. G., and Phillips, A. M., III. (1979). Late Pleistocene plant communities in the Guadalupe Mountains, Culberson County,

Texas. *In* "Biological Investigations in the Guadalupe Mountains National Park, Texas" (H. H. Genoways and R. J. Baker, eds.), pp. 13-30. U.S. Department of the Interior, National Park Service Proceedings and Transactions Series 4.

Van Devender, T. R., and Wiseman, F. M. (1977). A preliminary chronology of bioenvironmental changes during the Paleoindian period in the monsoonal Southwest. *In* "Paleoindian Lifeways" (E. Johnson, ed.), pp. 13-27. Museum Journal 17, West Texas Museum Association.

Waddington, J. C. B., and Wright, H. E., Jr. (1974). Late Quaternary vegetational changes on the east side of Yellowstone Park, Wyoming. *Quaternary Research* 4, 175-84.

Wallace, A. R. (1880). "Island Life." Macmillan and Company, London.

Watts, W. A. (1980). Late Quaternary vegetation of central Appalachia and the New Jersey Coastal Plain. *Ecological Monographs* 49, 427-69.

Wells, P. V. (1966). Late Pleistocene vegetation and degree of pluvial climatic change in the Chihuahuan Desert. *Science* 153, 970-75.

Wells, P. V. (1976). Macrofossil analysis of woodrat (*Neotoma*) middens as a key to the Quaternary vegetational history of arid America. *Quaternary Research* 6, 223-48.

Wells, P. V. (1977). Reply to comment by Van Devender. *Quaternary Research* 8, 238-39.

Wells, P. V. (1979). An equable glaciopluvial in the West: Pleniglacial evidence of increased precipitation on a gradient from the Great Basin to the Sonoran and Chihuahuan Deserts. *Quaternary Research* 12, 311-25.

Wells, P. V., and Berger, R. (1967). Late Pleistocene history of coniferous woodland in the Mohave Desert. *Science* 155, 1640-47.

Wells, P. V., and Hunziker, J. H. (1976). Origin of the creosote bush (*Larrea*) deserts of southwestern North America. *Annals of the Missouri Botanical Garden* 63, 843-61.

Wendorf, F., and Hester, J. J. (1975). "Late Pleistocene Environments of the Southern High Plains." Fort Burgwin Research Center Publication 9.

West, F. H. (1983). The antiquity of man in America. *In* "Late-Quaternary Environments of the United States," vol. 1, "The Late Pleistocene" (S. C. Porter, ed.), pp. 364-82. University of Minnesota Press, Minneapolis.

West, N. E., Tausch, R. J., Rea, K. H., and Tueller, P. T. (1978). Phytogeographical variation within juniper-pinyon woodlands of the Great Basin. *In* "Intermountain Biogeography: A Symposium" (K. T. Harper and J. L. Reveal, eds.), pp. 119-36. Great Basin Naturalist Memoirs 2.

Whiteside, M. C. (1965). Paleoecological studies of Potato Lake and its environs. *Ecology* 46, 807-16.

Whittaker, R. H. (1975). "Communities and Ecosystems." Macmillan, New York.

Whittaker, R. H., Buol, S. W., Niering, W. A., and Havens, Y. H. (1968). A soil and vegetation pattern in the Santa Catalina Mountains, Arizona. *Soil Science* 105, 440-50.

Whittaker, R. H., and Niering, W. A. (1965). Vegetation of the Santa Catalina Mountains, Arizona: A gradient analysis of the south slope. *Ecology* 46, 429-52.

Wolfe, J. A., and Leopold, E. B. (1967). Neogene and early Quaternary vegetation of northwestern North America and northeastern Alaska. *In* "The Bering Land Bridge" (D. M. Hopkins, ed.), pp. 193-206. Stanford University Press, Stanford, Calif.

Wormington, H. M., and Ellis, D. (eds.) (1967). "Pleistocene Studies in Southern Nevada." Nevada State Museum Anthropological Paper 13.

Wright, H. E., Jr., Bent, A. M., Hansen, B. S., and Maher, L. J., Jr. (1973). Present past vegetation of the Chuska Mountains, northwestern New Mexico. *Geological Society of America Bulletin* 84, 1155-80.

Yang, T.-W. (1970). Major chromosome races of *Larrea* in North America. *Journal of the Arizona Academy of Science* 6, 41-45.

York, J. C., and Dick-Peddie, W. A. (1969). Vegetation changes in southern New Mexico in the past hundred years. *In* "Arid Lands in Perspective" (W. G. McGinnies and B. J. Goldman, eds.), pp. 157-66. University of Arizona Press, Tucson.

Yurtsev, B. A. (1963). On the floristic relations between steppes and prairies. *Botanica Notiser* 116 (3), 396-408.

Yurtsev, B. A. (1972). Phytogeography of northeastern Asia and the problem of transberingian floristic interrelations. *In* "Floristics and Paleofloristics of Asia and Eastern North America" (A. Graham, ed.), pp. 19-54. Elsevier, Amsterdam.

CHAPTER 15

Vegetational History of the Eastern United States 25,000 to 10,000 Years Ago

W. A. Watts

Introduction

The vegetational history of the unglaciated southeastern United States has been reviewed recently by Whitehead (1973), Delcourt and Delcourt (1979), and Watts (1980b). Wright (1981) discusses the nature of the vegetation for the entire area east of the Rocky Mountains 18,000 years ago, a date considered to represent the time of most extreme cold within the Wisconsin Glaciation. This chapter reviews the vegetational history of the United States from the Great Plains to the East Coast between 25,000 and 10,000 years ago. It describes the vegetation that was present when the Laurentide ice sheet expanded during the Late Wisconsin and the recovery of the vegetation as the ice retreated.

Climate

The most useful parameters for a brief characterization of the climate are the mean annual precipitation (Figure 15-1) and the average number of frost-free days, which defines the length of the growing period (Figure 15-2). Representative climatic data for stations close to some significant fossil sites are listed in Table 15-1. The locations of the sites can be identified on Figure 15-3, which shows all the localities mentioned in the text.

The amounts of precipitation (Figure 15-1) increase from west to east. West of longitude 95°, grassland (prairie) generally prevails and precipitation is inadequate for forest growth. The areas of highest precipitation are at high elevations in the Great Smoky Mountains, where the annual average exceeds 200 cm, along the Gulf Coast, and in tropical Florida. Precipitation is generally greater in the summer than in the winter, especially in southern Florida. Mean July temperatures (Table 15-1) are generally very adequate for tree growth, and the length of the growing season seems to be the most critical factor in determining the species that occur. The length of the growing season decreases from south to north. In tropical Florida, frost occurs very infrequently, and sometimes several winters in succession pass without frost. The southern edge of the boreal forest is roughly defined by a growing season of 120 days. The various lengths of the growing seasons at different stations are recorded in Table 15-1.

Modern Vegetation

Before the area was settled by farmers of European origin, the eastern United States was covered by forest. Most of the trees were of the broad-leaved species, but pine species were abundant, especially in the Southeast. Almost all of the original forests have been cut for timber or cleared for farmland, but stands of primary forest still exist, and there are very large areas of modified secondary forest. The classic study of Braun (1950) still characterizes the eastern forests best. The map of major forest types (Figure 15-4) is a simplified form of the map published in that study. Küchler (1964) provides a generalized map of potential natural vegetation of the United States. Range maps of tree species have been published by Little (1971, 1977, 1978) and Fowells (1965), who also supplies silvicultural data for many tree species.

In many of the areas into which settlement expanded during the 19th century, including large areas of the Midwest and the South, the General Land Office Survey (GLOS) recorded the forest trees present immediately before settlement through the use of a systematic grid scheme (Bourdo, 1956). Both ecologists working with the living flora and paleoecologists working with the fossil records have used these data to reconstruct the presettlement vegetation of the area. The very original study of McAndrews (1966) compares pollen spectra from sediments formed immediately before settlement with the contemporary forest composition reconstructed from GLOS records. Where pollen assemblages of greater age are closely comparable to presettlement pollen spectra, McAndrews also infers in considerable detail the vegetation that produced them. The availability of GLOS records makes studies possible in North America that are of great interest to paleoecologists; no comparable data can exist in Europe, which had a long period of prehistoric agriculture.

The origin of the eastern forests of the United States was the subject of a classical controversy in biogeography. Braun (1950, 1955) maintained that the ''mixed mesophytic forest'' had survived intact in its modern localities since the Tertiary and that Quaternary glaciation had had little effect on the forests. Braun's thesis has been proven incorrect, and Deevey's idea (1949) that very large migratory movements of plant populations must have taken place has been sus-

Contribution 246, Limnological Research Center, University of Minnesota.

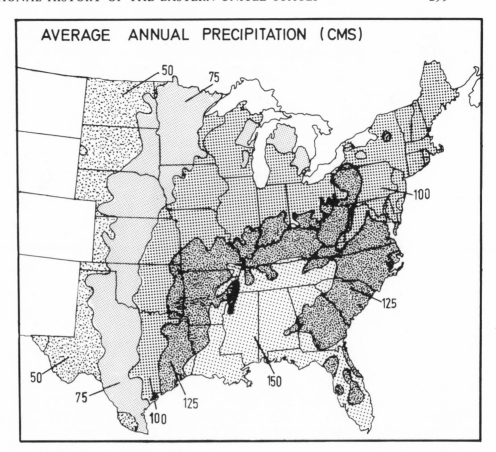

Figure 15-1. Mean annual precipitation of the eastern United States.

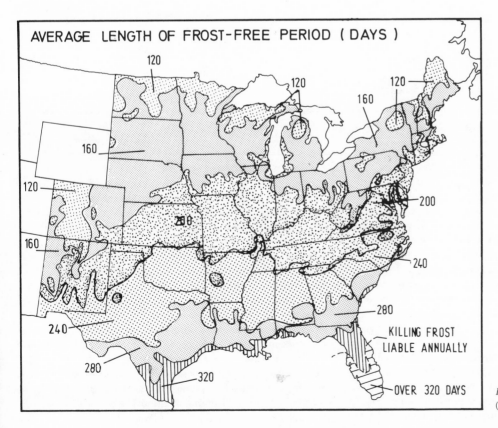

Figure 15-2. Mean duration of frost-free period (growing season) of the eastern United States.

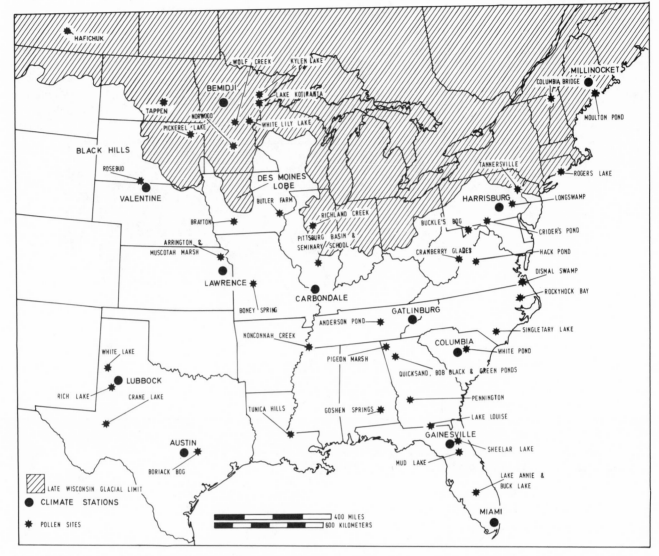

Figure 15-3. Location map for pollen sites and climate stations referred to in the text.

tained. However, Braun's enthusiasm for the "mixed mesophytic forest" has resulted in its recognition as an especially noteworthy plant formation, and it is appropriate to give it special attention.

Mixed mesophytic forest is a species-rich forest of tall, mainly broad-leaved deciduous trees with diverse small trees and shrubs in the understory or in light patches. It reaches its finest development in the Cumberland Plateau and Cumberland Mountains of eastern Tennessee and Kentucky and in the "coves" (sheltered valleys) of the Great Smoky Mountains in the same region. Characteristic trees are *Fagus, Liriodendron, Tilia, Acer saccharum* and other *Acer* species, *Castanea dentata, Aesculus octandra, Quercus rubra, Q. alba, Betula lutea, B. lenta,* and *Tsuga canadensis,* the most abundant conifer. *Magnolia* (four species), *Fraxinus, Halesia* species, *Nyssa sylvatica, Juglans* species, and *Carya* species also occur. (See Table 15-2 for English names.) Ericaceous shrubs are abundant, especially *Rhododendron. Castanea* has been destroyed by disease in the 20th century and survives only as root sprouts. The highest elevations of the Great Smokies (1500 to 2000 m) have forest of *Picea*

rubens and *Abies fraseri.*

The "cove" forests provided a partial analogue for European forests in the latest Pliocene and earliest Pleistocene at such sites as Reuver and Tegelen in the Netherlands (van der Hammen et al., 1971; Zagwijn, 1960, 1963). A detailed comparison, however, cannot be sustained, for many of the genera found in the European sites are now confined to Asia and some may have required warmer climates than that in the Great Smokies, which are unambiguously temperate in climate (Table 15-1). Physiognomically, however, and to some extent floristically, the vegetation of Tegelen somewhat resembles the "mixed mesophytic forest."

The Coastal Plain from Virginia to the Mississippi Delta has predominantly coarse, sandy, nutrient-poor soils, often of Quaternary marine or eolian origin. Pines and sclerophyllous oaks are abundant. Before logging, forests of *Pinus palustris* characterized the region. Similar vegetation occupies much of the Florida peninsula, where there is a rather high frequency of rare and local endemics, especially herbs of sandy soils and fossil dunes (Ward, 1979).

Table 15-1.
Representative Climatic Data for the
Eastern United States

Location	January Mean (°C)	July Mean (°C)	Mean Annual Precipitation (mm)	Number of Frost-Free Days
Bemidji, Minnesota	− 15.7	20.4	595	118
Millinocket, Maine	− 10.1	19.8	1057	114
Valentine, Nebraska	− 6.2	23.5	472	152
Lawrence, Kansas	− 1.3	26.0	900	195
Harrisburg, Pennsylvania	− 0.8	24.1	946	204
Carbondale, Illinois	1.8	26.7	1099	190
Lubbock, Texas	4.3	26.1	478	205
Gatlinburg, Tennessee (Great Smoky Mountains)	4.8	24.1	1345	188
Columbia, South Carolina	7.8	27.1	1066	248
Austin, Texas	10.4	28.8	874	264
Gainesville, Florida	14.2	27.0	1247	285
Miami, Florida	20.0	27.6	1503	365

Note: The climate stations are shown in Figure 15.3.

Much of the Coastal Plain is covered by huge wooded swamps, especially along the major rivers, and shrub-dominated peat bogs ("pocosins" or "bayheads"). *Taxodium* and *Nyssa sylvatica* var. *biflora* dominate in shallow water, and *Magnolia virginiana, Gordonia lasianthus, Persea palustris, Myrica* species, *Clethra, Cyrilla* species, and ericaceous shrubs are abundant in the species-rich shrub bogs. Okefenokee Swamp has a very large stand of *Taxodium*, with local *Nyssa* in shallow water. This vegetation is analogous to certain plant associations of the European Neogene (Teichmüller, 1958; Zagwijn, 1960).

It should be emphasized that the Coastal Plain and most of the Florida peninsula are climatically warm temperate (Table 15-1, Columbia, Gainesville). Tropical vegetation with a diversity of new species occurs in only extreme southern Florida (Figure 15-4), which is essentially frost free. Woodland with predominantly tropical species is confined to the Florida Keys and limited upland areas on the southern Florida mainland. The Everglades, which consist of islands of tropical hardwoods in freshwater swamps dominated by the sedge *Mariscus jamaicense*, grades into coastal mangrove swamps, (*Rhizophora, Laguncularia, Avicennia*) in southwestern Florida. The very great climatic and vegetational differences between the tropical swamps of the Everglades and the temperate *Taxodium* swamps such as the Okefenokee were not fully appreciated in the European literature (Teichmüller, 1958) whenever analogues for European Miocene vegetation were sought.

Elsewhere in the eastern United States, less species-rich deciduous forest occurs. *Quercus* and *Carya* forests predominate in the dry country approaching the prairie-forest border. Gallery forests penetrate far in to the prairie along the floodplains of the major rivers. Before disease destroyed it, *Castanea* was a very important genus in Appalachia. *Fagus* and *Acer saccharum* predominate in mesic sites in the southern Great Lakes region. In the northern and western Great Lakes region and extending to the East Coast is forest in which *Pinus strobus, Tsuga*, and diverse northern hardwoods and conifers such as *Alnus* species, *Betula* species, *Populus* species, *Abies* species, *Picea*

Table 15-2.
Latin and English Equivalents of Plant Names
Mentioned in Text

Latin Name of Tree	English Equivalent	Latin Name of Shrub or Herb	English Equivalent
Abies balsamea	Balsam fir	*Ambrosia*	Ragweed
Abies fraseri	Fraser fir	*Andropogon*	Bluestem grass
Acer saccharum	Sugar maple	*Artemisia*	Sagebrush
Aesculus octandra	Buckeye	*Arundinaria gigantea*	Cane
Alnus crispa	Green alder	*Ceratiola ericoides*	Rosemary
Alnus rugosa	Speckled alder	*Chenopodiaceae*	Pigweed family
Avicennia nitida	Black mangrove	*Clethra*	Sweet pepperbush
Betula glandulosa	Dwarf birch	*Compositae*	Daisy family
Betula lenta	Black birch	*Cyperaceae*	Sedge family
Betula papyrifera	White birch	*Cyrilla*	Titi
Betula populifolia	Gray birch	*Elaegnus commutata*	Silverberry
Carya	Hickory	*Empetrum*	Crowberry
Castanea dentata	Chestnut	*Ephedra*	Mormon tea
Celtis	Hackberry	*Gramineae*	Grass family
Chamaecyparis	White cedar	*Ilex*	Holly
Chrysobalanus	Coco plum	*Mariscus jamaicense*	Sawgrass
Corylus cornuta	Beaked hazel	*Polygonella*	Jointweed
Fagus grandifolia	American beech	*Rhododendron*	Rhododendron
Fraxinus nigra	Black ash	*Sanguisorba canadensis*	Burnet
Gordonia lasianthus	Loblolly bay	*Shepherdia canadensis*	Soapberry
Halesia	Silver bells	*Vitis*	Vine
Juglans cinerea	Butternut		
Juglans nigra	Black walnut		
Juniperus	Juniper		
Lagunculria racemosa	White mangrove		
Larix laricina	Tamarack		
Liquidambar	Sweet gum		
Liriodendron	Tulip tree		
Magnolia acuminata	Cucumber tree		
Magnolia virginiana	Sweet bay		
Myrica	Wax myrtle		
Nyssa sylvatica	Sour gum		
Ostrya	Hophornbeam		
Persea palustris	Red bay		
Picea glauca	White spruce		
Picea mariana	Black spruce		
Picea rubens	Red spruce		
Pinus banksiana	Jack pine		
Pinus clausa	Sand pine		
Pinus palustris	Longleaf pine		
Pinus resinosa	Red pine		
Pinus rigida	Pitch pine		
Pinus strobus	White pine		
Populus balsamifera	Balsam poplar		
Populus tremuloides	Aspen		
Quercus alba	White oak		
Quercus rubra	Red oak		
Quercus stellata	Post oak		
Rhizophora mangle	Red mangrove		
Taxodium	Cypress		
Tilia	Basswood		
Tsuga canadensis	Hemlock		
Ulmus	Elm		

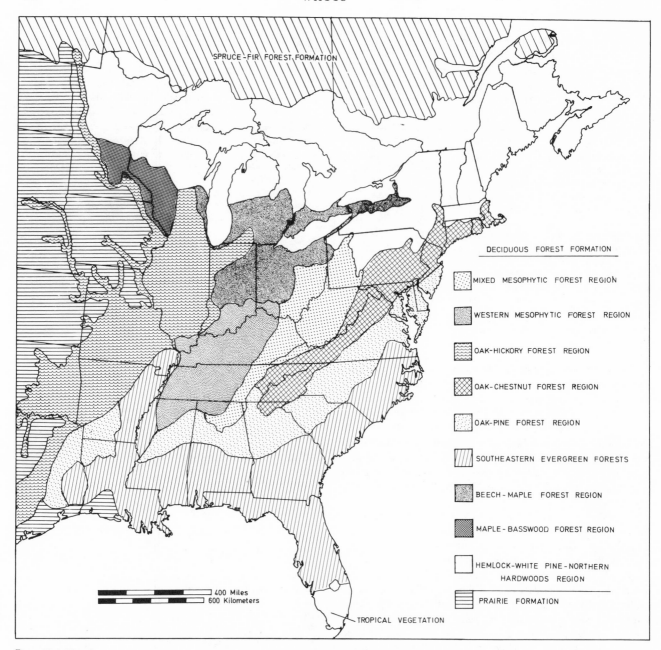

Figure 15-4. Map of main vegetational types of the eastern United States. (Modified from Braun, 1950.)

species, and *Larix laricina* are abundant elements. True boreal forest with a more restricted tree flora, with abundant *Abies* and *Picea*, begins in Canada north of the Great Lakes at about latitude 50°N.

Many tree species of the eastern United States have very wide distributions and occur in many associations, and so different forest types are not necessarily sharply distinguished. Figure 15-5 shows some representative distribution types. *Fagus grandifolia* is a widespread species. *Pinus banksiana* is primarily a boreal forest species but extends into the *P. strobus /Tsuga/*northern hardwoods formation. *Pinus strobus* occurs throughout the Great Lakes region and in Appalachia. *Gordonia lasianthus* is a "bayhead" species of the Coastal Plain and the Florida peninsula but is not tropical. *Chrysobalanus icaco* is a strictly tropical hardwood, abundant on tree

islands ("hammocks") in the Everglades, and widespread in the Caribbean. With few exceptions, the tree distributions of temperate species are centered on Appalachia or the Coastal Plain, which appear to have been the main centers from which tree migrations took place in to the central United States and the Great Lakes region at the end of Wisconsin glaciation. A tabulation of tree and shrub species by state, following maps prepared by Little (1971, 1977, 1978), shows that the Coastal Plain states from Alabama to Virginia have high species diversity (Figure 15-6). In tropical Florida, the number of tree and shrub species (145) is appreciably lower. This emphasizes the fact that the tropical flora of Florida is species poor and consists largely of widespread species.

It is common to speak of the eastern United States flora as being

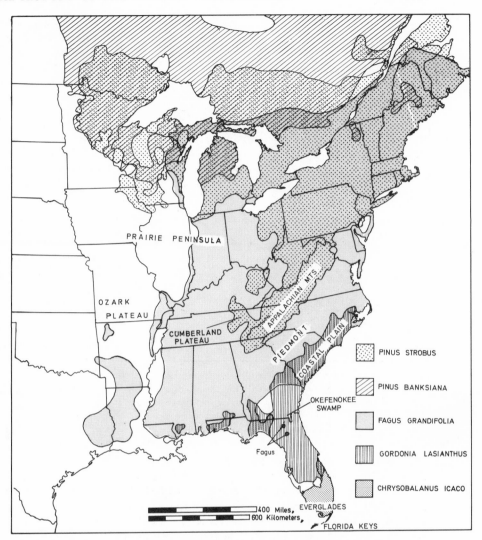

Figure 15-5. Typical distributions of trees: *Chrysobalanus icaco* (tropical), *Gordonia lasianthus* (Coastal Plain), *Pinus strobus* (Appalachia and Great Lakes), *Pinus banksiana* (boreal), and *Fagus grandifolia* (widespread temperate).

rich in species. Small (1933) lists 5100 native vascular plants, excluding ferns, for the southeastern United States south of the northern boundaries of North Carolina and Tennessee and east of the Mississippi River, including Florida's tropical flora. In comparison, Italy has about 4900 native vascular species (Webb, 1978) and California about 4700 (Munz, 1963), both substantially richer floras by unit area. The diversity in tree and shrub species in the southeastern United States (Figure 15-6) appears to be high in comparison with that in Europe, but there may be fewer herbaceous species. The richest woody flora, in which small trees and large shrubs of wetland and acid peatland are an important component, is found in the southern Coastal Plain. Southern Appalachia is also species rich. Although Florida (considered as a unit) has the largest number of woody species (259), the northern part of the state is the southern limit for many temperate and Coastal Plain plants; nearly 100 species, most of which are widespread in the West Indies, are confined to the tropical tip of the Florida peninsula. Central Florida is poor in woody species, a no-man's-land between the truly temperate and tropical floras. Its limited woody flora suggests that this was not a refuge area for temperate trees during the Late Wisconsin. Endemism, in the sense that species have very restricted ranges, reaches its greatest expression in herbaceous and shrubby communities of sandy soils in temperate peninsular Florida and the Florida panhandle (Ward, 1979).

Stratigraphic Concepts and Nomenclature

After a long period in the Wisconsin during which ice withdrew from the Great Lakes region, a major expansion of the Laurentide ice sheet took place, beginning at about 25,000 years ago. The ice sheet was in full retreat once more by 10,000 years ago. As it melted back, interglacial climatic conditions returned. The difficulty in defining the stratigraphic units between 25,000 and 10,000 years ago is discussed by Watts (1980b). Reconciling regional land stratigraphies in current use in the United States with a worldwide scheme of Quaternary stratigraphy based on the record in ocean cores is the problem.

Generalizations about world stratigraphy and chronology in the Quaternary have come to rely on the $^{18}O/^{16}O$ ratio in fossil organisms from ocean cores (Shackleton, 1977). During glacial periods, isotopically light ice accumulated as a result of snowfall on the continents, leaving the oceans slightly enriched in ^{18}O. During interglacial periods, oxygen-isotope ratios and ice-sheet cover approximated present values. Intermediate conditions of inferred ice volume

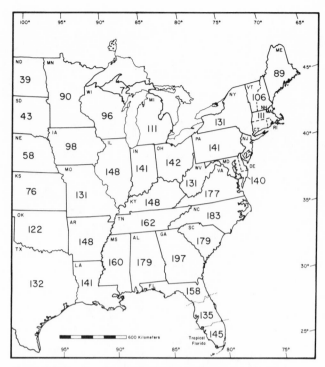

Figure 15-6. Numbers of tree and shrub species by state east of longitude 100°W. Vermont and New Hampshire; Massachusetts, Connecticut, and Rhode Island; and Maryland and Delaware are treated as units. Upper and lower Michigan are scored separately. Florida is divided into northern, central, and tropical regions. (Data from Little, 1971, 1977, 1978.)

1975). Ice was present briefly between 36,000 to 33,000 yr B.P. (Cherrytree stade), and interstadial conditions (Plum Point interstade) existed from 33,000 to 22,500 yr B.P., when the major ice advance of the Late Wisconsin began. An older Port Talbot interstade is also known. The Farmdalian and Plum Point phases may be correlative, although the latter is assigned an earlier starting date. It is difficult to define interstadial limits satisfactorily because of the shortage of datable material in critical stratigraphic positions.

In the Great Lakes region, Late Wisconsin glaciation took the form of the expansion of ice lobes along the basins of the Great Lakes, and the history of each lobe is a complex of advances and retreats not necessarily closely correlated with events in adjoining lobes (Wright, 1976). Each lobe may ultimately reveal a distinctive local chronology. At some distance beyond the glacial front, the combined climatic effect of the fluctuations results in a simplification of the stratigraphy. For the purposes of this chapter, it is assumed that interstadial environments, called for convenience "Farmdalian," were present for some time before 22,000 yr B.P. and that they found their strongest and possibly only expression close to the glacial front in the Midwest. After 22,000 yr B.P., the advance of the Great Lakes ice lobes was general, marking the main Late Wisconsin ice expansion.

The end of the Wisconsin cannot be dated with certainty because of the lag between the beginning of the climatic warming and the expression of the warming in the stratigraphic record at each site. The

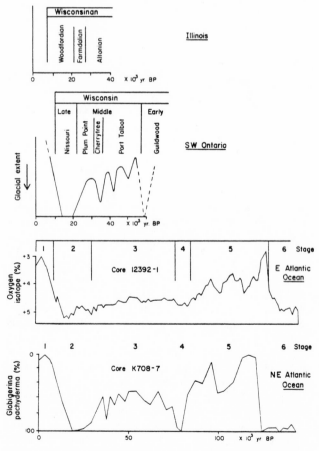

Figure 15-7. ¹⁸O/¹⁶O and Foraminifera profiles from two cores in the eastern North Atlantic (Ruddiman et al., 1977; Shackleton, 1977) compared with land stratigraphy from eastern North America (Berti, 1975; Willman, 1971). (Stage numbers after Emiliani, 1966.)

were "interstadial." On this basis (Figure 15-7), glacial conditions existed briefly from about 84,000 to 73,000 yr B.P., followed by a long, undifferentiated interstade from 73,000 to about 30,000 yr B.P. At 30,000 yr B.P., a climatic deterioration began, culminating in the glacial maximum at about 18,000 yr B.P. Interglacial conditions were restored by a long, slow transition lasting from about 15,000 to about 10,000 yr B.P. (in Watts, 1980b).

For the eastern United States, the classification of the Illinois Geological Survey (Willman, 1971) has been used extensively. During the Woodfordian Glaciation between 22,000 and 12,500 yr B.P., ice of the Lake Michigan lobe extended to southern Illinois. Ice was absent from Illinois from 28,000 to 22,000 yr B.P. (Farmdalian interstade), when dark-colored, slightly organic silt, sometimes described as "peat," accumulated in hollows and when weathering on the upland formed the Farmdale soil. Before 28,000 yr B.P. and extending back to approximately 60,000 yr B.P., an older Altonian glacier occupied northern Illinois. Thus, the Farmdale soil is evidence for a period of warmer climate and ice recession. There is little evidence for a Farmdalian interstadial phase in marine cores. It may not have been observed because of a sampling interval that was too wide or because of core disturbance by bioturbation. Alternatively, it may be too small and local an event to be recorded in the oceanic record. Equivalents of the Farmdale soil have not been identified extensively outside Illinois (Willman, 1971) and Iowa (van Zant et al., 1980). At the moment, caution suggests that "Farmdalian" should be restricted to the margin of the Lake Michigan ice lobe and that its wider application should be subject to further investigation.

A more complex stratigraphy is recorded from the Lake Erie and Lake Ontario regions in southwestern Ontario (Figure 15-7) (Berti,

melting of great ice sheets to expose new land surfaces may occupy thousands of years. Watson and Wright (1980) draw attention to the time-transgressive nature of geologic and biotic responses to climatic change. The ocean record emphasizes the correctness of this view (Figure 15-7), for nearly 5000 years pass between full glacial (about 25,000 to 15,000 yr B.P.) and interglacial equilibria. In this chapter, a transitional phase between 15,000 and 10,000 yr B.P. is proposed. This period is named the late glacial. Thus, the Late Wisconsin (25,000 to 10,000 yr B.P.) includes the full glacial and the late glacial, useful terms that lack formal status.

The Late Wisconsin Full Glacial

THE GREAT PLAINS AND THE
PRAIRIE-FOREST BOUNDARY

The vegetational history of the Great Plains and the prairie-forest transition is poorly understood because of the small number of sites. The localities already studied show that the region was forested with *Picea* during the Late Wisconsin, including the late glacial, and that treeless prairie was not established until shortly before the beginning of the Holocene. At two prairie sites in northeastern Kansas close to the forest margin, Arrington and Muscotah marshes (Figure 15-8), *Picea* provided up to 70% of the pollen rain from about 24,000 yr B.P. until some time between 15,800 and 11,340 yr B.P. (Grüger, 1973). There was then a transitional vegetation with some *Picea*, *Quercus*, *Ulmus*, *Fraxinus*, and *Ostrya* type, with increasing quantities of grass and other herbs until after 9930 yr B.P., when prairie was established. At that time, tree pollen constituted less than 20% of the total. At both sites, the vegetation before 24,000 yr B.P. was relatively diverse. *Picea* was less frequent, and *Pinus*, *Betula*, *Salix*, and *Juniperus* were more abundant than at any other time. Herbs were also more common. Open vegetation with some relatively species-rich woodland of northern type was apparently replaced by *Picea* forest during the Late Wisconsin.

A number of spring deposits have been investigated in the mixed forest and prairie of the Ozark Highlands of western Missouri (King, 1973). At Boney Spring, *Pinus* with diverse herbs was present from before 27,480 yr B.P. to shortly before 21,000 yr B.P., when *Picea* became very abundant. The pine species was *Pinus banksiana*, as is

demonstrated by the presence of cone fragments. The pre-21,000 *P. banksiana*/herb vegetation supported a fauna of mastodon (*Mammut americanum*), horse (*Equus*), and musk-ox (*Symbos*). Jack pine (*Pinus banksiana*) seems to have become extinct in this region after 21,000 yr B.P.; it was not present in the forest that invaded Minnesota as the Late Wisconsin ice sheets melted (Wright, 1968). After 16,500 yr B.P., the *Picea* forest in the Ozarks became more diverse with the addition of broad-leaved tree species. It is believed, on the basis of dated organic debris in mastodon tusk cavities, that *Picea* was still present as late as 13,500 yr B.P. (King, 1973). Abundant macrofossils of *Picea* and *Larix* occur in the full-glacial sediments, which reveal a rich fauna that includes mastodon, *Castoroides* (giant beaver), *Paramylodon* (ground sloth), and *Tapirus* (tapir). After 13,500 yr B.P., pollen preservation is poor and sedimentation discontinuous, so no younger record is available.

Recent work at several sites in Iowa has shown that forest of *Pinus banksiana* and *Picea* characterized the Farmdalian interstade. At the beginning of the Late Wisconsin about 22,000 yr B.P., loess was deposited and *Picea* began to replace *Pinus* at the Butler Farm site (van Zant et al., 1980). At the end of the Late Wisconsin glaciation, *Picea* was the major forest tree available to invade the new landscape of the upper-midwestern and the northern prairie states.

In south-central Illinois outside the Late Wisconsin glacial limit, the presettlement vegetation was a mosaic of *Andropogon* prairie and *Quercus/Carya* woodland (Küchler, 1964). E. Grüger (1972a, 1972b) shows that *Picea* forest was present at Seminary School Basin from before 24,200 yr B.P. to after 21,370 yr B.P. Unfortunately, the radiocarbon dates available from the several sites studied by Grüger are not adequate enough to delimit vegetational events with precision. Seminary School Basin, with its very high *Picea* percentages, contrasts with the nearby Pittsburg Basin (Figure 15-9), where *Picea* is less frequent and *Quercus* is important. If one assumes that the *Picea* maximum at Seminary School Basin corresponds with the advance of Late Wisconsin ice into northern Illinois, the preceding section of the pollen diagram may be dated from before 21,000 yr B.P. back some thousands of years. The pollen spectra are herb dominated, but there is as much as 30% *Pinus* at some levels and 5% to 10% each of *Picea* and *Quercus*. It is possible that the vegetation before 21,000 yr B.P. was entirely herbaceous and that there was long-distance transport of tree pollen. Alternatively, some woodland, including *Pinus banksiana*, as in the Ozarks, may have been present. The Holocene is marked by the disappearance of both *Pinus* and *Picea* and their replacement by *Quercus* and herbs. New cores have been obtained from Pittsburg Basin for use in detailed studies of pollen and paleomagnetic stratigraphy, and more radiocarbon dates would enable estimates of pollen influx to be made. In this way, the hypothesis that the Farmdalian of Illinois was treeless could be tested.

A 16,000-year record from Boriack Bog (Figure 15-10) in east central Texas (Bryant, 1977) completes the list of sites near the present prairie/forest border. This site, near Austin (Table 15-1), lies in an open woodland with *Quercus stellata*, other *Quercus* and *Carya* species, and an understory of prairie grasses and composites. The site has a basal date of 15,460 yr B.P. *Picea* is present only as traces in a few samples. The basal flora is *Alnus* dominated. Apart from the disappearance of *Alnus* at about 10,000 yr B.P. and the steep subsequent rise in grass-pollen percentages, the flora showed little change from Late Wisconsin to the present, and it must be suspected that the local vegetation has been relatively stable. *Alnus* no longer occurs in Texas but is found in southeastern Oklahoma. Its withdrawal during

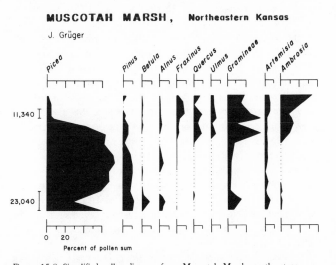

MUSCOTAH MARSH, Northeastern Kansas

J. Grüger

Figure 15-8. Simplified pollen diagram from Muscotah Marsh, northeastern Kansas. (Modified from Grüger, 1973.)

PITTSBURG BASIN, Southern Illinois

E. Grüger

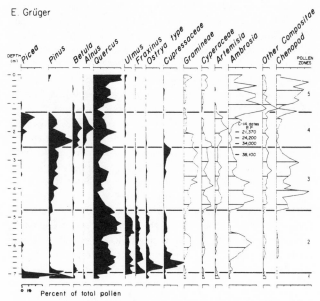

Figure 15-9. Simplified pollen diagram from Pittsburg Basin, southern Illinois. (Modified from Grüger, 1972a.)

BORIACK BOG, East-central Texas

V. Bryant

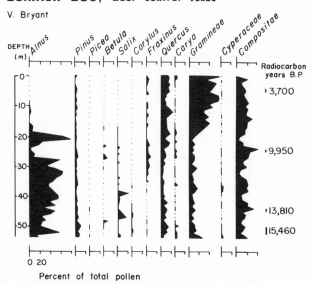

Figure 15-10. Simplified pollen diagram from Boriack Bog, east-central Texas. (Modified from Bryant, 1977.)

the Holocene was hardly for major climatic reasons, for the genus still occurs in the eastern Gulf Coast and northern Florida. The location of Boriack Bog makes it a site of great potential interest, but the rather uninformative pollen diagram suggests that plant-macrofossil studies and some more-detailed pollen analysis (for example, the recognition of taxa within the Compositae), would be of value in attempts to differentiate Holocene and pre-Holocene vegetation.

For western Texas, Hafsten (1961, 1964) shows by a combination of pollen diagrams from Rich Lake and Crane Lake that herbaceous vegetation was present from around 33,500 to 22,500 yr B.P. From

22,000 to 14,000 yr B.P., open boreal woodland was present in an inferred cold, wet climate. The pollen diagrams are dominated by *Pinus*, but *Picea* is present in small percentages. Between 14,000 and 10,000 yr B.P., grassland and boreal woodland were present in a transitional environment. After 10,000 yr B.P., the modern flora appeared, with grasses, chenopods, composites, *Artemisia, Ephedra, Quercus,* and *Juniperus* dominating. In this region, very arid prairie is now characteristic. These western sites are close to the high mountain ranges of western Texas and eastern New Mexico, and so it is probable that the *Pinus* and *Picea* that covered the area during the Late Wisconsin belonged to western species. This is another region in which the modern prairie succeeded forest at the beginning of the Holocene. The early investigations by Hafsten (1961) have recently been confirmed by Oldfield (1975) at White Lake near Lubbock.

THE LOWER MISSISSIPPI VALLEY

At Nonconnah Creek near Memphis, *Picea* supplies nearly 80% of the pollen in alluvial sediments dated between 20,215 and 17,195 yr B.P. (Delcourt et al., 1980). It is accompanied by *Quercus,* in excess of 10% at many levels, and traces of other deciduous and coniferous trees. The macrofossil diagram from a measured profile shows abundant *Picea* and northern aquatics, with some upland herbs. *Vitis* at the top of the profile and *Ilex* at the base are the only possibly mesophytic genera recorded. *Fagus* pollen is absent from most samples. However, the investigators report macrofossils of *Fagus, Liriodendron,* and *Carya* from the Late Wisconsin sediments. In a complex profile of varied alluvial sediments, it seems possible that the macrofossils of *Fagus, Liriodendron,* and *Carya* may be Middle Wisconsin or late glacial in age, or they may even be in secondary position. Mesophytic genera may well have occurred in a *Picea*-dominated environment during the Late Wisconsin full glacial in the Memphis area, but the evidence for such an important fact is not yet securely based on specimens stratified in dated profiles.

THE SOUTHEAST: APPALACHIA AND
THE COASTAL PLAIN

In the Southeast, the Late Wisconsin full-glacial was *Pinus* dominated. A study of fossil pine needles and pollen sizes at Bob Black Pond in northern Georgia (Watts, 1970), Anderson Pond in east-central Tennessee (Figure 15-11) (Delcourt, 1979), and White Pond in South Carolina (Watts, 1980a) shows that *Pinus banksiana* was the species involved. Whitehead (1973) records either *P. banksiana* or *P. resinosa* on the basis of pollen size from Singletary Lake and Rockyhock Bay, North Carolina. *Picea* is present at all these sites. The species is uncertain, but pollen-morphological considerations (Birks and Peglar, 1980) suggest that *P. rubens, P. marinana,* and *P. glauca* all may have been present (Watts, 1980a). Pollen of *Quercus* also occurs in small quantity, with traces of other deciduous trees, including *Ulmus, Carya, Ostrya*-type, and *Fagus.* In addition to pollen of trees, pollen of herbs—especially *Artemisia, Ambrosia,* Compositae, Gramineae, and Cyperaceae—may make up 50% of the pollen. Plants of sand dunes and sandy soils, especially *Polygonella* species, were characteristic of the southern Coastal Plain during the Late Wisconsin. The flora is rather homogeneous at all sites, but the impression of low diversity may be spurious because of the high pollen productivity of *Pinus.*

The nature of the Late Wisconsin full-glacial vegetation is not clear. *Pinus banksiana* must have been very abundant, but its high pollen production may lead to overestimates of its cover. *Picea* species, some broad-leaved trees, and herbs may all have been important. The vegetation is interpreted as *Pinus/Picea* forest or woodland, with some broad-leaved trees either scattered throughout the forest in small populations or concentrated in limited favored habitats such as south-facing slopes or alluvium in river valleys. The presence of herbs suggests that the forest was open or that there were local prairies and sand-hill communities. There is good evidence (Thom, 1970) that sand dunes were being formed on the Coastal Plain by southwesterly winds during the Late Wisconsin.

The climate indicated by the presence of *Pinus banksiana* and *Picea* is that of the boreal forest (Table 15-1, Bemidji) or at least that of Maine, the nearest coastal region where the two kinds of trees cooccur (Table 15-1, Millinocket). These climates indicate a very short growing season and a severe winter. Precipitation is difficult to estimate. Lakes retained water, but this fact may be explained by low temperatures and long winters. The dune movement recorded by Thom suggests deflation in a dry climate, for which there is good evidence from Florida.

Plant macrofossils from nearby Bob Black and Quicksand Ponds (Watts, 1970) have a species-rich flora of northern aquatics, many of them as much as 900 km south of their present range in the Great Lakes region. Similar data are available from Anderson Pond (Delcourt, 1979) and White Pond (Watts, 1980a).

Late Wisconsin *Picea* is also found in several localities in the Tunica Hills of eastern Louisiana (Delcourt and Delcourt, 1977). Macrofossils of *Picea* occur in terrace deposits of streams that dissect the loess-mantled blufflands close to the Mississippi River. A date of 12,740 yr B.P. is available for the base of Fluviatile Terrace I in Tunica Bayou. A basal date is not available from Percy Bluff on Little Bayou Sara, but it is assumed by the investigators to have the same age as the base of Tunica Bayou. The basal sample from Tunica Bayou contains needle fragments of *Picea*, seeds and fruits of *Fagus*, and wood referred to *Magnolia acuminata* (pollen is very sparse). The basal sample from Percy Bluff contains 40% to 50% *Picea*

pollen and frequent *Quercus* and *Ostrya* type, with smaller percentages of *Fagus, Ulmus,* and other deciduous trees, as well as high percentages of Gramineae, Cyperaceae, and Compositae. One seed of *Liriodendron* and fragments of *Arundinaria gigantea* were found at this level. If the two sites are accepted as being of the same age, *Picea* forest accompanied by *Fagus, Quercus, Ostrya* type, *Liriodendron,* and other trees of the mixed mesophytic forest was present in the Tunica Hills at about 12,740 yr B.P. This is consistent with evidence from Anderson Pond in eastern Tennessee (Delcourt, 1979) and from Quicksand Pond and Pigeon Marsh in northern Georgia (Watts, 1975b), which shows that *Picea* briefly increased during the late-glacial transition, accompanied by an increase in *Quercus* and *Ostrya* type. At Pigeon Marsh, Georgia, and Sheelar Lake, Florida, an early expansion of *Fagus* is also notable. The Tunica Hills sites provide evidence, comparable to that at the other sites, that *Picea* was still present at the time deciduous trees were expanding. The richest Tunica Hills macrofossil collections are Holocene, as is demonstrated by the date of 5295 yr B.P. from a log adjacent to the Percy Bluff sample that yielded the richest list of species. The Tunica Hills assemblage is younger than the Late Wisconsin full-glacial flora, which may have been very different in species composition and relative population sizes from the vegetation of 12,740 yr B.P. Delcourt and Delcourt (1979) speculate that the eastern escarpment of the Mississippi River (the blufflands) and ravines cut into the escarpment by stream erosion were a critical refuge area for mesophytic forest trees during the late glacial and possibly earlier and that they were an important migratory pathway for tree populations during times of changing climate. These ideas cannot be evaluated without comparative evidence from other sites within the region.

The Nonconnah Creek site and the Tunica Hills sites illustrate the rewards and difficulties of working with alluvial deposits. Alluvial materials may be rich in macrofossils, especially those of large-fruited deciduous trees not found elsewhere, and rich in pollen, which may be poorly preserved although countable. The sediments, however, are subject to erosion and redeposition, and so the stratigraphy may be discontinuous and must be treated with great caution. The stratigraphy of the Tunica Hills sites and their radiocarbon dating have proved controversial (Delcourt and Delcourt, 1978; Otvos, 1978, 1980). Otvos (1980) claims that the published dates are too young and that the Little Bayou Sara organic alluvial deposits are of Farmdalian age. He has published a new series of dates in support of this view. The controversy is still in progress. Provisionally, the view is accepted here that the Little Bayou Sara material is of late-glacial and Holocene age (Delcourt and Delcourt, 1977).

The abundance of *Pinus banksiana* in the Southeast is one of the most remarkable aspects of the Late Wisconsin flora. Today, in the southern part of its range (Figure 15-5), it occurs mostly on sandy soils and sand plains of glacial origin as well as on rock outcrops. It must have occurred in much more diverse habitats in the past. The sites in northern Georgia, for example, which have abundant needles to prove the local presence of the species, are in rather impermeable silty loams.

The southern limit of the *Pinus banksiana/Picea* flora is not exactly known. *Picea glauca* cones dated at 21,300 yr B.P. (Watts, 1980a) are known from Pennington, Georgia. Together with the findings from the Tunica Hills sites, this finding suggests that *Picea* may have reached the Gulf Coast. At all the sites, the cones are remarkably large. The same is true for cones found at the Nonconnah Creek site (Delcourt et al., 1980). It seems possible that a *P. glauca* population

ANDERSON POND, East-central Tennessee

H. Delcourt

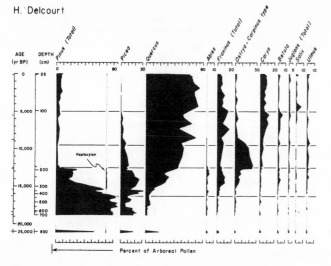

Figure 15-11. Simplified pollen diagram from Anderson Pond, east-central Tennessee, with scale corrected to uniform time scale. (Modified from Delcourt, 1979.)

with distinctive cones was present in the Southeast during the Late Wisconsin and became extinct as spruce populations disappeared at the transition to the Holocene (Delcourt and Delcourt, 1979). It is uncertain whether or not the population merits taxonomic recognition, but it may represent one of the few documented cases of Pleistocene plant extinction.

The site at Goshen Springs, Alabama (Delcourt, 1980), appears to demonstrate a hiatus corresponding with all of the Late Wisconsin. A new date of 8330 ± 90 yr B.P. (Wis-1186) at 205 to 212 cm from the published core (Core 76A) is separated by only 32 cm of sediment from a date of 26,000 yr B.P. The sediment is described as a dark reddish brown, sandy, silty clay. This is interpreted a a result of seasonal exposure to subaerial oxidation of desiccated wet meadows and shallow pools (Delcourt, 1980). Since the flora and sediment are rather homogeneous between 200 and 244 cm, it seems possible that there was a hiatus from immediately after 26,000 yr B.P. until Holocene sedimentation began a little before 8330 yr B.P. Such a hiatus would be consistent with the record at Lake Louise in southern Georgia (Watts, 1971) and Mud Lake in north-central Florida (Watts, 1969), which also show a hiatus before sedimentation resumed in the early Holocene.

The Late Wisconsin full-glacial flora is preserved at Sheelar Lake, north-central Florida (Figure 15-12) (Watts and Stuiver, 1980), between 24,000 and 18,500 yr B.P. *Picea* is absent, so the Tunica Hills and Pennington sites may represent its southern limit. The pollen flora is as much as 80% *Pinus*, with *Quercus* and *Carya* present in significant quantity. Pollen of mesic trees also occurs. Herbs, especially herbs of prairies and sand hills, make up 10% or more of the pollen. The *Pinus* is a two-needle species as yet undetermined. *P. banksiana* is not excluded, but *P. clausa*, a Florida endemic, must also be considered. The interpretation is for an environment dominated by *Pinus* forest and open herbaceous communities with small populations of broad-leaved trees, especially *Quercus* and *Craya*. The silty sediments and the presence of fossil sand dunes suggest that the Late Wisconsin may have had a very dry, windy climate with mobile dunes and deflation of silt.

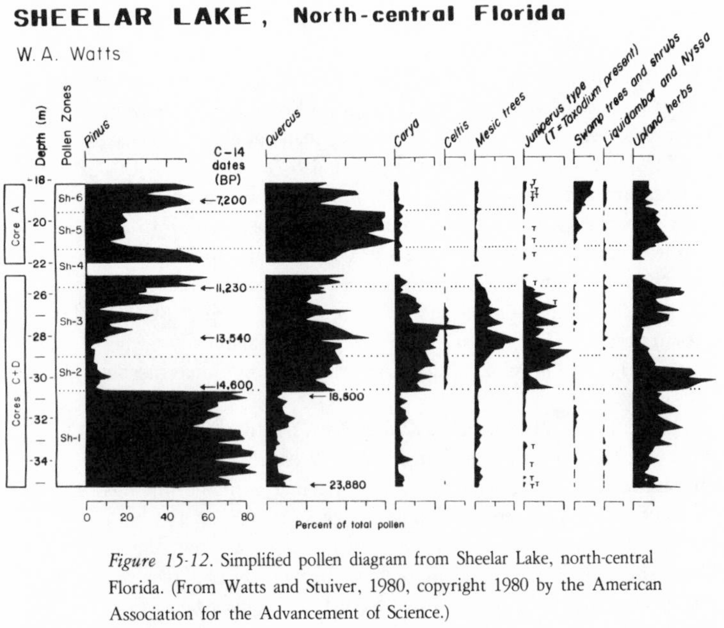

A hiatus at Sheelar Lake exists between 18,500 and 14,800 yr B.P. Elsewhere in Florida and Georgia, lakes were dry during the entire Late Wisconsin, probably because precipitation was not adequate to sustain a high water table in the permeable limestone that forms the region's bedrock. New studies and radiocarbon dates from Lake Annie in southern Florida (Watts, 1975a, 1980b) show that a hiatus from 30,000 to 13,600 yr B.P. is also apparent at this site, which was previously believed to contain a Late Wisconsin full-glacial record.

The Late Wisconsin Full-Glacial Vegetation Near the Glacial Front

MINNESOTA

In Minnesota, the Late Wisconsin glaciation was characterized by complex expansions and contractions of ice lobes at the margin of the main Laurentide ice sheet. The expansions and contractions appear to be a reflection of the dynamics of the main ice sheet and are not clearly correlated with climatic change. Ice sheets from the basin of Lake Superior (Superior lobe) and farther west (Rainy lobe) occupied northeastern and northern Minnesota before 20,000 yr B.P., when they retreated, leaving drumlin fields that were not overridden by later ice (Wright, 1976). Interdrumlin hollows in this region may have a sedimentary record extending back 20,000 years. Wolf Creek (Birks, 1976) is such a site. There, a tundra flora of dwarf shrubs and herbs that included *Dryas*, *Arenaria rubella*, *Silene acaulis*, and *Vaccinium uliginosum* var. *alpinum* was present from 20,500 to 14,700 yr B.P., when a small increase can be observed in the diversity of shrubs and aquatic plants, especially in the macrofossil record. At 13,600 yr B.P., *Picea* arrived locally, and there was simultaneously a great increase in the total pollen influx. Tundra floras identified by pollen and macrofossils are known from Lake Kotiranta, which has a basal date of 16,150 yr B.P. (Wright and Watts, 1969), and Kylen Lake (Birks, 1981) and by pollen at White Lily Lake (Cushing, 1967), which has a basal date of 15,520 yr B.P. (Birks, 1976). The southern limit of the tundra is not known, and it is not certain that the treeline would have maintained a constant position during the Late Wisconsin. The late advance of the Des Moines lobe (Wright, 1976) means that an ice sheet with marginal tundra was present in western, central, or southern Minnesota and northern Iowa until the ice retreated after 13,000 yr B.P. The nearest forest of which there is certain evidence was in southern Illinois (Grüger, 1972a, 1972b) and Missouri (King, 1973). The treeline lay in an unknown and probably fluctuating position between these areas and the Minnesota tundra.

THE EASTERN UNITED STATES

In the East, tundra was present about 30 km outside the glacial limit at Longswamp in eastern Pennsylvania (Watts, 1979). The pollen flora was dominated by Gramineae, with abundant Cyperaceae, *Picea*, *Pinus* and *Betula*. Macrofossils show that *Dryas*, *Empetrum*, and *Vaccinium* were present, together with *Betula glandulosa* (dwarf birch). Rare *Picea* needles suggest that the treeline, or isolated stands of *Picea*, was not far away. The *Pinus* pollen is attributed to long-distance transport. Unfortunately, the Longswamp site has not been securely dated, for all the dates from the very inorganic sediments of the tundra phase lie between 12,400 and 12,000 yr B.P. The dates may appear too young because of the infiltration of organic material from above. A date of 12,500 yr B.P. is considered minimal for the end of the tundra phase.

SHEELAR LAKE, North-central Florida

W. A. Watts

Figure 15-12. Simplified pollen diagram from Sheelar Lake, north-central Florida. (From Watts and Stuiver, 1980, copyright 1980 by the American Association for the Advancement of Science.)

At Crider's Pond (Figure 15-13), about 150 km from the ice margin (Watts, 1979), a peak of *Betula* pollen, probably *B. glandulosa*, suggests that the treeline may have been south of that location before 15,210 yr B.P. It occurs in a pollen assemblage that characteristically includes *Sanguisorba canadensis*. It seems likely that tundra bordered the ice in Pennsylvania toward the end of the full glacial at a time when open *Picea* forest or woodland was less than 100 km to the south. At the same time, areas at higher elevation in the Appalachian Mountains had vegetation dominated by Cyperaceae. The treeline was lower than 800 m at Buckle's Bog (Maxwell and Davis, 1972) and lower than 1000 m at Cranberry Glades (Watts, 1979). West of the Appalachians in Ohio, little information is available about the full-glacial flora, but there is evidence from transported tree trunks that glaciers may have advanced to the northern edge of the *Picea* woodland (Dreimanis and Goldthwait, 1973).

The Late-Glacial Transition

THE SOUTHEAST

In the Florida peninsula, sedimentation began at 14,600 yr B.P. at Sheelar Lake after a short hiatus attributed to a very dry full-glacial climate (Watts and Stuiver, 1980). *Pinus* forest present at 18,500 yr B.P. had been replaced by broad-leaved trees and herbs by 14,600 yr B.P. (Figure 15-12). The beginning of local change in climate and vegetation lay in the hiatus between these dates.

Sheelar Lake is important because it records the details of vegetation changing to the Holocene. From 14,600 to about 14,000 yr B.P., the pollen flora was dominated by *Quercus*, *Carya* (to 20%), *Celtis*, *Juniperus* or *Chamaecyparis*, and herbs, including up to 10% of *Ambrosia*. The inferred vegetation is drought-adapted forest or woodland with some prairie. In the late glacial, the climate apparently became much warmer, deflation of silt ceased, and precipitation increased, leading to a rise in the regional water table sufficient to reflood the basin. By 13,540 yr B.P., mesic trees had displaced the prairie and *Fagus* provided 7% of the pollen. Sheelar Lake lies close to the most southerly stand of *Fagus* in the United States today (Figure 15-5), which may be a relict population from the late glacial. The presence of mesic trees implies that there was a further increase in precipitation, but the climate was probably not yet as warm as today's, nor was there such a long frost-free season (Table 15-1). The present seasonal pattern of wet summers and dry winters may not have been established at that time.

Between 13,500 and 11,200 yr B.P., the mesic trees declined and were replaced by *Pinus*, *Quercus*, and herbs. This change may record the establishment of a climate similar to today's. The sawtooth form of the *Pinus* pollen curve may indicate that *Pinus* was favored competitively by periodic natural fires. *Pinus* and *Quercus* continued to dominate the vegetation at the transition to the Holocene about 10,000 yr B.P.

At Lake Annie, farther south (Figure 15-14), sedimentation began at 13,600 yr B.P. after a hiatus extending back to 30,100 yr B.P. The flora had abundant *Pinus* (probably *P. clausa*), *Ambrosia*, other Compositae, and Gramineae, indicating a dry climate. It differed from the Middle Wisconsin flora, which also contained sand-dune plants such as the empetraceous shrub *Ceratiola* and *Polygonella* species. Lake Annie contains no record of the presence of either temperate or tropical broad-leaved trees during the late glacial, and it is believed that the climate of southern Florida remained dry until the middle

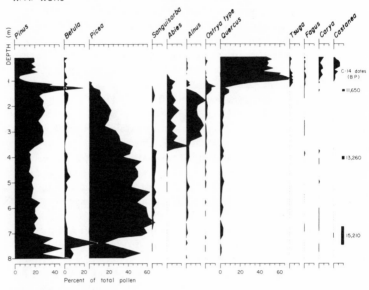

CRIDER'S POND, Southeastern Pennsylvania
W. A. Watts

Figure 15-13. Simplified pollen diagram from Crider's Pond, southeastern Pennsylvania. (From Watts, 1980b, reproduced with the permission of *Annual Reviews of Ecology and Systematics*, volume 11, copyright 1980 by Annual Review, Inc.)

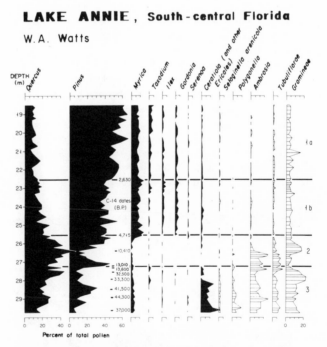

LAKE ANNIE, South-central Florida
W. A. Watts

Figure 15-14. Simplified pollen diagram from Lake Annie, south-central Florida. (Modified from Watts, 1975a.)

Holocene (Watts, 1975a).

At Sheelar Lake, several genera of broad-leaved trees, especially *Carya*, were as common in the Late Wisconsin pollen assemblages as they were in the later Holocene. It is likely that small populations of many broad-leaved tree species survived the dry and cold climate of the glacial period in northern Florida and quickly built large local stands as the climate became wetter and warmer.

Mesic forest species migrated northward time transgressively in response to the climatic change. Deciduous forest rich in *Carya* and *Fagus* had become established at White Pond in South Carolina (Watts, 1980a) by 12,800 yr B.P., at Singletary Lake in North Carolina before 11,000 yr B.P. (Whitehead, 1967), and slightly later at Rockyhock Bay farther north (Whitehead, 1973). *Tsuga* and *Betula* occur at Singletary Lake and Rockyhock Bay but not in significant quantity farther south, and *Carya* and *Fagus* decrease in percentage representation northward. At Dismal Swamp, Virginia (Whitehead and Oaks, 1979), *Fagus* and *Tsuga* were establishing themselves at 11,000 yr B.P., *Betula* was already frequent, but *Carya* did not expand until 9000 yr B.P. Thus, the broad-leaved forest trees formed a loose and changing assemblage of plant populations as they migrated northward, with varying speeds, starting points, and success in maintaining their numbers. *Tsuga* is specially interesting because it can be inferred (Davis, 1976) that its Late Wisconsin refuge cannot have been farther south than North Carolina in the Coastal Plain or the Appalachian Mountains. This may also be true for *Pinus strobus*, known from about 11,000 yr B.P. at Anderson Pond, Rockyhock Bay, and Hack Pond (Craig, 1969). Both species are absent from the Coastal Plain today, except for isolated localities. At White Pond, mesic forest disappeared at about 9500 yr B.P. and gave way to *Quercus*-dominated vegetation that included *Pinus*. This flora lasted until the middle Holocene, a pattern true for all Coastal Plain sites in the Carolinas and Virginia.

West of the Appalachian Ridge at Anderson Pond (Figure 15-11) (Delcourt, 1979), the *Pinus* pollen percentages declined after 15,000 yr B.P. and fell to very low levels by 12,750 yr B.P. During this interval, *Picea* was still frequent, *Abies* was present, *Quercus* had begun to expand, and traces of pollen of *Pinus strobus* occurred. As with *Tsuga*, there is no fossil record of *Pinus strobus* farther south. The increase in the pollen percentages of some of the deciduous trees, especially *Fraxinus* and *Ostrya* type, was beginning as *Pinus banksiana* declined. The change can first be detected as early as 16,300 yr B.P., and it may record the beginning of the climatic warming near the end of the glacial period. Between about 12,500 and 10,000 yr B.P., *Quercus* was abundant, with low *Pinus* and declining *Picea*. *Ostrya* type was also frequent, and *Acer saccharum* was present in significant quantity. The flora had a distinctly mesic aspect, although

Fagus was infrequent and *Tsuga* expanded relatively late, as did *Carya* in this region. The Holocene may be considered to have begun with the completion of the expansion of *Quercus* and *Carya* and the appearance of *Liquidambar* and *Castanea* shortly after 10,000 B.P.

In northwest Georgia at Quicksand Pond, *Quercus* began to increase at about 13,500 yr B.P., at least 2000 years after the Anderson Pond date. The pollen diagrams from the two sites are very similar, and it can be assumed that they record essentially contemporary events, given that they are close geographically and at similar elevations. The apparently younger date for *Quercus* expansion at Quicksand Pond may result from a defect in radiocarbon dating at one of the sites.

Fagus with *Picea* and *Ostrya*-type pollen occur at Pigeon Marsh (Figure 15-15) (Watts, 1975b), a northwestern Georgia site dated to 10,820 yr B.P. *Juglans cinerea* is notably abundant in the *Fagus*/*Picea* assemblage. As at Anderson Pond, the beginning of the Holocene is marked by an abundance of pollen of *Liquidambar*, *Castanea*, and *Nyssa*. Thus, the replacement of the transitional, often strongly mesic forest of the late glacial by distinctly Holocene assemblages had taken place universally in the Southeast by about 10,000 yr B.P.

THE NORTHEAST

At about 13,000 yr B.P., *Pinus banksiana*, *Abies*, tree *Betula*, and *Alnus rugosa* type, abundantly represented by pollen and macrofossils, appeared at Crider's Pond along with a variety of other trees and shrubs. This flora replaced the species-poor spruce woodland recorded for the full glacial. A similar assemblage invaded the *Picea*/*Betula glandulosa* association at Longswamp (Watts, 1979), where *B. populifolia* was also present. The increased productivity of the environment was marked by a change at both sites from silt to organic sedimentation and the arrival of a rich flora of aquatic plants. An environment and flora comparable to that of the boreal forest was replaced essentially by assemblages more like the modern flora of the northern Great Lakes region.

By 11,500 yr B.P., *Picea rubens* had arrived at Crider's Pond, closely followed by *Pinus strobus*. *Quercus* and *Tsuga* increased at about 10,000 yr B.P. at the expense of other conifers and northern hardwoods; this marked the stabilization of the flora in something like

PIGEON MARSH, Northwestern Georgia

W. A. Watts

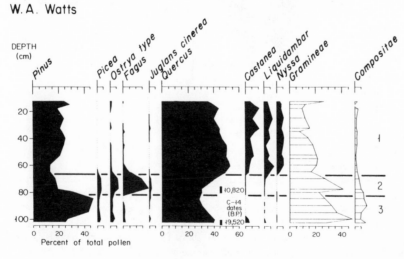

Figure 15-15. Simplified pollen diagram from Pigeon Marsh, northwestern Georgia. (Modified from Watts, 1975b.)

its modern form and the decreasing importance of tree invasions in the region. The rise of *Quercus* and *Tsuga* in Pennsylvania can be used to mark the beginning of the Holocene.

Interesting differences exist between the floral history of the periglacial region and the floral history of the area that was covered by ice. The grass-dominated tundra flora at unglaciated Longswamp is paralleled by sedge-dominated pollen floras at the Tannersville (Watts, 1979) and Rogers Lake (Davis, 1969) sites north of the glacial limit. *Sanguisorba* was never significant in the flora of the glaciated area, whereas *Populus tremuloides* and *Alnus crispa* were insignificant in the stable *Picea* woodland in the unglaciated area but became prominent as invaders of newly exposed ground. *Picea* was not nearly as abundant at Rogers Lake, Tannersville, or Moulton Pond (Davis et al., 1975) as it was at Crider's Pond and Longswamp. There may never have been anything describable as *Picea* forest during the invasion of the deglaciated region in the Northeast. Clearly, the soils and competitive conditions outside and inside the glacial limit were different and favored different species. At Tannersville, in the glaciated area but very close to the ice limit, an invasion of trees in the sequence *Picea—Abies—Pinus banksiana—Betula populifolia—Pinus rigida* can be demonstrated to have occurred between 13,000 and 9000 yr B.P.

The theme of tree invasion has been discussed extensively in recent literature (Davis, 1976; Watts 1973; 1979; 1980b). It has been demonstrated that the population of each tree species behaved independently of all others in its response to climatic change. Species migrated at characteristic speeds for each time and place, accelerating or slowing down with expanding or contracting populations as environmental circumstances and competition dictated. This resulted in plant associations that were very unstable because they were constantly subject to change by new invasions or losses. The time between 15,000 yr B.P., when the first response of trees to the warming climate can be detected in the southeastern United States, to 10,000 or even 9000 yr B.P., when the large-scale migrations had been accomplished, is a time of transition from glacial to interglacial stability. It should be noted, however, that the process of migration continued through the Holocene. *Castanea* was one of the slowest migrants, taking nearly 5000 years to reach southern New England from central Appalachia (Davis, 1976). *Pinus strobus* continued to invade areas westward and reached the prairie-forest border in Minnesota within the last 1000 years (Jacobson, 1979).

The invasion of the deglaciated Northeast by forest was a relatively slow process. In a classic study of pollen-deposition rates at Rogers Lake, Davis (1969) shows that ice had disappeared and tundra was present for over 2000 years before the invasion of *Picea* took place, although the genus was already present in unglaciated Pennsylvania as the ice withdrew. Tundra did not finally disappear from the areas around Moulton Pond until 10,000 yr B.P. (Davis et al., 1975). A species-rich tundra macroflora dated at about 11,500 yr B.P. from Columbia Bridge, Vermont (Miller and Thompson, 1979), confirms that pollen spectra dominated by *Picea*, *Pinus*, and Cyperaceae with low pollen-deposition rates may be consistent with tundra vegetation as identified from macrofossils. The Columbia Bridge macroflora, in addition to herbs and shrubs, contained abundant macrofossils of *Populus balsamifera* and some needles of *Picea*, suggesting that some tree stands were close by.

A comparison of Crider's Pond with Moulton Pond and Columbia Bridge shows that the invasion of the deglaciated landscape of northern New England by tree species from unglaciated Pennsylvania oc-

cupied as much as 3000 years, during which time tundra was present. A crude calculation suggests that *Picea* invaded at a rate of something like 30 km per century. This apparently slow rate can be explained by a slow rate of climatic warming and perhaps also by a need for soil development.

MINNESOTA

In Minnesota, tundra was present at Wolf Creek (Birks, 1976), Kylen Lake (Figure 15-16) (Birks, 1981), Kotiranta Lake (Wright and Watts, 1969), White Lily Lake (Cushing, 1967), and several other sites. Tundra occupied sites between the Des Moines lobe ice and a diminished Lake Superior lobe. It was succeeded at Kylen Lake by a pioneer shrub flora in which *Betula glandulosa*, *Rhododendron lapponicum*, *Salix*, and *Shepherdia canadensis* were present (Birks, 1981). The *Betula glandulosa* expansion at the end of the glaciation was recorded by macrofossils as well as by pollen at several sites in northeastern Minnesota (Watts, 1967). The establishment of forest tundra at Kylen Lake was completed by the invasion of *Picea mariana*, *P. glauca*, and *Larix laricina* at about 12,000 yr B.P. *Betula glandulosa* persisted in open woodland until 10,700 yr B.P., when an invasion by *Pinus banksiana* and northern hardwoods established forest similar to that of the southern boreal forest today.

Cushing (1967) and Wright (1977) show that the northward movement of the treeline, identified by the arrival of *Picea* in Minnesota, was time transgressive. There is no evidence that it was anything but a continuous process once it had begun, and no halts or retreats in its movement have been detected that would suggest climatic fluctuation within the late glacial. The search for equivalents of the climatic events of the European late glacial in Minnesota or elsewhere in North America appears to be futile. Europe's distinctive climatic history (Watts, 1980c) apparently was related to changes in the surface temperatures in the eastern North Atlantic; there were no comparable events in North America.

As the Des Moines lobe retreated, pioneer vegetation was already available for colonization in its immediate neighborhood. At Nor-

KYLEN LAKE, Northeastern Minnesota

H.J.B. Birks

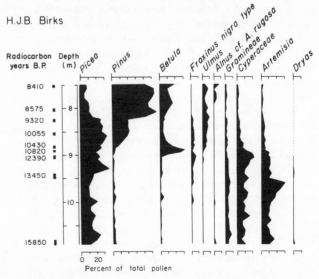

Figure 15-16. Simplified pollen diagram from Kylen Lake, northeastern Minnesota. (Modified from Birks, 1981.)

wood (Ashworth et al., 1981), evidence from pollen, plant macrofossils, and beetles shows that a pioneer vegetation of *Dryas, Shepherdia canadensis, Elaegnus commutata, Populus, Juniperus*, and *Salix* existed in a climate that was no longer tundralike. The pioneer plants and animals were precursors of *Picea* forest, which arrived at 12,000 yr B.P. after a brief lag during which soil development took place. Thus, tundra and tundra climates are known only from northern and eastern Minnesota. Boreal forest was an early invader in southern Minnesota and around the Des Moines lobe, and there was no long lag period like the one that characterized New England.

The presence of high pollen frequencies of *Quercus, Ostrya* type, and *Fraxinus nigra* in the late-glacial *Picea*-dominated floras of Minnesota and elsewhere in the Midwest has long puzzled observers. Was the pollen to be explained by long-distance transport from deciduous forest to the south? Could it, although apparently well preserved, have been secondarily deposited? Or did it represent trees actually present in a *Picea/Larix* forest? Most investigators have hesitantly come to accept the last viewpoint. Fortunately, the issue has been resolved to some extent by the discovery of an acorn cup of *Quercus rubra* and fruits of *Corylus cornuta* in a *Picea* assemblage dated to 12,420 B.P. at Brayton, Iowa (Baker et al., 1980). It now seems to have been demonstrated that some hardwood species were lesser components of the late-glacial *Picea* forest.

The interesting history of *Pinus banksiana* shows that it was either eliminated or became very rare in the Late Wisconsin *Picea* forest that bordered the ice at its maximum. *Larix* was able to expand with *Picea* as soon as the climatic warming began, but *Pinus banksiana* required several thousand years to reestablish its populations in the Midwest. It seems possible that it had to reinvade from the southeastern United States, where it was the most abundant tree of the Late Wisconsin.

Late-glacial *Picea* forest was present everywhere in what is now the northern prairie during the late-glacial phase of the Late Wisconsin. Several sites record the appearance of prairie vegetation as *Picea* forest or woodland disappeared. In the heart of the modern prairie, grassland succeeded *Picea* directly, but near the modern prairie-forest boundary there was a transitional phase during which deciduous trees were present briefly. At Rosebud (near Valentine, Nebraska), a small marsh in South Dakota between prairie-covered fossil dunes, pollen and macrofossils of *P. glauca* predominate in sediments dated to 12,630 yr B.P. (Watts and Wright, 1966). Apart from *Salix*, no other tree species occurs in significant quantity. *Artemisia*, Gramineae, and other herbs make up about 20% of the pollen. At an unknown date not long after 12,580 yr B.P., *Picea* disappeared and was replaced by herbs, especially Chenopodiaceae, *Ambrosia*, and other Compositae. *Pinus* percentages increased slightly. The interpretation is that *Picea* woodland disappeared about 12,000 yr B.P. and was replaced by prairie. *Picea glauca* now occurs in the Black Hills of western South Dakota, 200 km northwest of Rosebud. This population may be relict from continuous Late Wisconsin forest in the western Great Plains. *P. glauca* was still present in central North Dakota at Tappen at 11,480 yr B.P. (Wright, 1970) and at 10,270 yr B.P. at the Hafichuk site in Saskatchewan (Ritchie and de Vries, 1964). This suggests a time-transgressive withdrawal of *Picea* northward in the present prairie region during the late-glacial transition. The *Picea* forest gave way to prairie shortly after 10,000 yr B.P. at Pickerel Lake (Watts and Bright, 1968), after a brief intermediate phase of deciduous woodland. Elsewhere, as the southern margin of

the *Picea* forest withdrew slowly northward, herb-dominated pollen assemblages appeared generally in the modern prairie region by 9000 yr B.P.

In eastern Minnesota, the new land exposed by the melting ice sheet was colonized, after a brief phase of pioneer herbs and shrubs (Ashworth et al., 1981; Cushing, 1967), by homogeneous forest of *Picea* species with *Larix. Populus*, especially *P. balsamifera*, and *Juniperus* may have been frequent (Cushing, 1967). Other deciduous trees, including *Quercus rubra, Fraxinus nigra*, and *Acer negundo*, may also have occurred. The vegetation was similar physiognomically to the modern boreal forest, but it differed in detail in the species present (Amundson and Wright, 1979). Late-glacial spruce forest of this type occurred generally throughout the Midwest. It was replaced by invading forest with *Pinus (probably P. banksiana), Abies balsamea, Alnus rugosa, Betula papyrifera*, and other northern hardwoods and conifers characteristic of the forests of the northern Great Lakes today. The decline of *Picea* at about 10,000 yr B.P. was precipitous, amounting to a population collapse. It seems more likely (Amundson and Wright, 1979; Wright, 1968) that spruce declined because of its inability to prosper in a warming climate with a longer growing season than that it lost in competition to invading tree species. At some sites, the collapse of spruce seems to have occupied as little as 50 years, and there is a brief hiatus before pine rises. This suggests that pine, fir, and northern hardwoods invaded an already dying forest. The time-transgressive northward withdrawal of spruce marks the beginning of the Holocene throughout the Midwest.

In contrast to the diverse climatic events of the late glacial in western Europe, the history of Minnesota vegetation shows that the late-glacial period, after an initial invasion by pioneer herbs and shrubs, was succeeded by spruce-tamarack forest and was stable climatically with little change in a relatively homogeneous vegetation. If the spruce collapse was caused by climatic change, it must be seen as a rather sudden, stepwise increase in warmth, perhaps one to be interpreted as an increase in the length of the growing season. The extent to which a sudden transition from late glacial to Holocene in North America is a general phenomenon has not yet been established. The evidence is rather striking in the Midwest and the Southeast, but the transition in the Northeast is much more gradual. The phenomenon itself and its climatic explanation deserves wider discussion.

References

Amundson, D. C., and Wright, H. E., Jr. (1979). Forest changes in Minnesota at the end of the Pleistocene. *Ecological Monographs* 49, 1-16.

Ashworth, A. C., Schwert, D. P., Watts, W. A., and Wright, H. E., Jr. (1981). Plant and insect fossils at Norwood in south-central Minnesota: A record of late-glacial succession. *Quaternary Research* 16, 66-79.

Baker, R. G., van Zant, K. R., and Dulian, J. J. (1980). Three late- glacial pollen and plant-macrofossil assemblages from Iowa. *Palynology* 4, 197-203.

Berti, A. A. (1975). Paleobotany of Wisconsinan interstadials, eastern Great Lakes region, North America. *Quaternary Research* 5, 591-619.

Birks, H. J. B. (1976). Late-Wisconsin vegetational history at Wolf Creek, central Minnesota. *Ecological Monographs* 46, 395-429.

Birks, H. J. B. (1981). Late Wisconsin vegetational and climatic history at Kylen Lake, northeastern Minnesota. *Quaternary Research* 16, 322-55.

Birks, H. J. B., and Peglar, S. M. (1980). Identification of *Picea* pollen of late Quaternary age in eastern North America: A numerical approach. *Canadian Journal of Botany* 58, 2043-58.

Bourdo, E. A., Jr. (1956). A review of the General Land Office Survey and of its use in the quantitative studies of former forests. *Ecology* 37, 754-68.

Braun, E. L. (1950). "Deciduous Forests of Eastern North America." Blakeston Company, Philadelphia.

Braun, E. L. (1955). The phytogeography of unglaciated eastern United States and its interpretation. *Botanical Review* 21, 297-375.

Brown, C. A. (1938). The flora of Pleistocene deposits in the western Florida parishes, western Feliciana Parish and East Baton Rouge Parish, Louisiana. *Louisiana Department of Conservation, Geological Survey Bulletin* 21, 59-96, 121-29.

Bryant, V. M., Jr. (1977). A 16,000 year pollen record of vegetation change in central Texas. *Palynology* 1, 143-56.

Craig, A. J. (1969). "Vegetational History of the Shenandoah Valley, Virginia." Geological Society of America Special Paper 123, pp. 283-96.

Cushing, E. J. (1967). Late-Wisconsin pollen stratigraphy and the glacial sequence in Minnesota. *In* "Quaternary Paleoecology" (E. J. Cushing and H. E. Wright, Jr., eds.), pp. 59-88. Yale University Press, New Haven, Conn.

Davis, M. B. (1969). Climatic changes in southern Connecticut recorded by pollen deposition at Rogers Lake. *Ecology* 50, 409-22.

Davis, M. B. (1976). Pleistocene biogeography of temperate deciduous forests. *Geoscience and Man* 13, 13-26.

Davis, R. B., Bradstreet, J. E., Stuckenrath, R., Jr., and Borns, H. W., Jr. (1975). Vegetation and associated environments during the past 14,000 years near Moulton Pond, Maine. *Quaternary Research* 5, 435-65.

Deevey, E. S., Jr. (1949). Biogeography of the Pleistocene. *Geological Society of America Bulletin* 60, 1315-1416.

Delcourt, H. R. (1979). Late Quaternary vegetation history of the eastern Highland Rim and adjacent Cumberland Plateau of Tennessee. *Ecological Monographs* 49, 255-80.

Delcourt, P. A. (1978). Quaternary vegetation history of the Gulf Coastal Plain. Ph.D. dissertation, University of Minnesota, Minneapolis.

Delcourt, P. A. (1980). Goshen Springs: Late Quaternary vegetation record for southern Alabama. *Ecology* 61, 371-86.

Delcourt, P. A., and Delcourt, H. R. (1977). The Tunica Hills, Louisiana-Mississippi: Late-glacial locality for spruce and dediduous forest species. *Quaternary Research* 7, 218-37.

Delcourt, P. A., and Delcourt, H. R. (1978). Reply to comments by Ervin G. Otvos, Jr. *Quaternary Research* 9, 253-59.

Delcourt, P. A., and Delcourt, H. R. (1979). Late Pleistocene and Holocene distributional history of the deciduous forest in the southeastern United States. *Veröffentlichungen des geobotanischen Institutes ETH, Stiftung Rübel, Zürich,* 68, 79-107.

Delcourt, P. A., Delcourt, H. R., Brister, R. C., and Lackey, L. E. (1980). Quaternary vegetation history of the Mississippi Embayment. *Quaternary Research* 13, 111-32.

Dreimanis, A., and Goldthwait, R. P. (1973). "Wisconsin Glaciation in the Huron, Erie, and Ontario Lobes." Geological Society of America Memoir 136, pp. 71-106.

Emiliani, C. (1966). Palaeotemperature analysis of Caribbean cores P 6304-8 and P 6304-9 and a generalized temperature curve for the last 425,000 years. *Journal of Geology* 74, 109-26.

Fowells, H. A. (1965). "Silvics of Forest Trees of the United States." U.S. Department of Agriculture, Agriculture Handbook 271.

Grüger, E. (1972a). Late Quaternary vegetation development in south-central Illinois. *Quaternary Research* 2, 217-31.

Grüger, E. (1972b). Pollen and seed studies of Wisconsinan vegetation in Illinois, U.S.A. *Geological Society of America Bulletin* 83, 2715-34.

Grüger, J. (1973). Studies on the late-Quaternary vegetation history of northeastern Kansas. *Geological Society of America Bulletin* 84, 239-50.

Hafsten, U. (1961). Pleistocene development of vegetation and climate in the southern High Plains as evidenced by pollen-analysis. *In* "Paleoecology of the Llano Estacado" (F. Wendorf, ed.), pp. 59-91. Fort Burgwin Research Center, Museum of New Mexico Press, Sante Fe.

Hafsten, U. (1964). A standard pollen diagram for the southern High Plains, U.S.A., covering the period back to the Early Wisconsin glaciation. *In* "Report of the Seventh International Congress on the Quaternary," Warsaw, 1961, Palaeobotanical Section, 2, pp. 407-20.

Jacobson, G. L., Jr. (1979). The palaeoecology of white pine (*Pinus strobus*) in Minnesota. *Journal of Ecology* 67, 697-726.

King, J. E. (1973). Late Pleistocene palynology and biogeography of the western Ozarks. *Ecological Monographs* 43, 539-65.

King, J. E. (1981). Late Quaternary vegetational history of Illinois. *Ecology* 51, 43-62.

Küchler, A. W. (1964). "Potential Natural Vegetation of the Conterminous United States (Map and Manual)." American Geographical Society Special Publication 36.

Lichti-Federovich, S., and Ritchie, J. C. (1968). Recent pollen assemblages from the western interior of Canada. *Review of Paleobotany and Palynology* 7, 297-344.

Little, E. L., Jr. (1971). "Atlas of United States Trees," vol. 1, "Conifers and Important Hardwoods." U.S. Department of Agriculture Miscellaneous Publication 1146.

Little, E. L., Jr. (1977). "Atlas of United States Trees," vol. 4, "Minor Eastern Hardwoods." U.S. Department of Agriculture Miscellaneous Publication 1342.

Little, E. L., Jr. (1978). "Atlas of United States Trees," vol. 5, "Florida." U.S. Department of Agriculture Miscellaneous Publication 1361.

McAndrews, J. H. (1966). Postglacial history of prairie, savanna, and forest in northwestern Minnesota. *Torrey Botanical Club Memoir* 22, 1-72.

Maxwell, J. A., and Davis, M. B. (1972). Pollen evidence of Pleistocene and Holocene vegetation on the Allegheny Plateau, Maryland. *Quaternary Research* 2, 506-30.

Miller, N. G., and Thompson, G. G. (1979). Boreal and western north American plants in the late Pleistocene of Vermont. *Journal of the Arnold Arboretum* 60, 167-218.

Munz, P. A. (1963). "A California Flora." University of California Press, Berkeley.

Oldfield, F. (1975). Pollen analytical results. *In* "Late Pleistocene Environments of the Southern High Plains" (F. Wendorf and J. J. Hester, eds.), pp. 121-47. Fort Burgwin Research Center Publications 9.

Otvos, E. G., Jr. (1978). Comment on "The Tunica Hills Louisiana-Mississippi: Lateglacial locality for spruce and deciduous forest species" by Paul A. Delcourt and Hazel R. Delcourt. *Quaternary Research* 9, 250-52,

Otvos, E. G., Jr. (1980). Age of Tunica Hills (Louisiana-Mississippi): Quaternary fossiliferous creek deposits; problems of radiocarbon dates and intermediate valley terraces in coastal plains. *Quaternary Research* 13, 80-92.

Ritchie, J. C., and de Vries, B. (1964). Contributions to the Holocene paleoecology of westcentral Canada: A late-glacial deposit from the Missouri Coteau. *Canadian Journal of Botany* 42, 677-92.

Ruddiman, W. F., Sancetta, C. D., and McIntyre, A. (1977). Glacial/interglacial response rate of subpolar North Atlantic waters to climatic change: The record in oceanic sediments. *Philosophical Transactions of the Royal Society* series B, 280, 119-42.

Shackleton, N. J. (1977). Oxygen isotope stratigraphy of the middle Pleistocene. *In* "British Quaternary Studies: Recent Advances" (F. W. Shotton, ed.), pp. 1-16. Clarendon Press, Oxford.

Small, J. K. (1933). "Manual of the Southeastern Flora." University of North Carolina Press, Chapel Hill.

Teichmüller, M. (1958). Rekonstruktionen verschiedener Moortypen des Hauptflözes der niederrheinischen Braunkohle. *In* "Die niederrheinische Braunkohlenformation," Fortschritte in der Geologie von Rheinland u. Westfalen 2, pp. 599-612. Nordrhein-Westfalen Geological Survey, Krefeld.

Thom, B. G. (1970). Carolina Bays in Horry and Marion Counties, South Carolina. *Geological Society of America Bulletin* 81, 783-814.

van der Hammen, T., Wijmstra, T. A., and Zagwijn, W. H. (1971). The floral record of the late Cenozoic of Europe. *In* "The Late Cenozoic Glacial Ages" (K. K. Turekian, ed.), pp. 391-424. Yale University Press, New Haven, Conn.

van Zant, K. L., Hallberg, G. R., and Baker, R. G. (1980). A Farmdalian pollen diagram from east-central Iowa. *Proceedings of the Iowa Academy of Science* 87, 52-55.

Ward, D. B. (1967). Southeastern limit of *Fagus grandifolia. Rhodora* 69, 51-54.

Ward, D. B. (1979). "Rare and Endangered Biota of Florida: V. Plants." University Presses of Florida, Gainesville.

Watson, R. A., and Wright, H. E., Jr. (1980). The end of the Pleistocene: A General critique of chronostratigraphic classification. *Boreas* 9, 153-63.

Watts, W. A. (1967). Late-glacial plant macrofossils in Minnesota. *In* "Quaternary

Paleoecology'' (E. J. Cushing and H. E. Wright, Jr., eds.), pp. 89-97. Yale University Press, New Haven, Conn.

Watts, W. A. (1969). A pollen diagram from Mud Lake, Marion County, north-central Florida. *Geological Society of America Bulletin* 80, 631-32.

Watts, W. A. (1970). The full-glacial vegetation of northwestern Georgia. *Ecology* 51, 17-33.

Watts, W. A. (1971). Postglacial and interglacial vegetation history of southern Georgia and central Florida. *Ecology* 52, 676-90.

Watts, W. A. (1973). Rates of change and stability of vegetation in the perspective of long periods of time. *In* "Quaternary Plant Ecology" (H. J. B. Birks and R. G. West, eds.), pp. 195-206. Blackwell Scientific Publications, Oxford.

Watts, W. A. (1975a). A late Quaternary record of vegetation from Lake Annie, south-central Florida. *Geology* 3, 344-46.

Watts, W. A. (1975b). Vegetation record for the last 20,000 years from a small marsh on Lookout Mountain, northwestern Georgia. *Geological Society of America Bulletin* 86, 287-91.

Watts, W. A. (1979). Late Quaternary vegetation of central Appalachia and the New Jersey Coastal Plain *Ecological Monographs* 49, 427-69.

Watts, W. A. (1980a). Late Quaternary vegetation history at White Pond on the inner Coastal Plain of South Carolina. *Quaternary Research* 13, 187-99.

Watts, W. A. (1980b). The late Quaternary vegetation history of southeastern United States. *Annual Review of Ecology and Systematics* 11, 387-409.

Watts, W. A. (1980c). Regional variation in the response of vegetation to late-glacial climatic events in Europe. *In* "Studies in the Late-Glacial of Northwest Europe" (J. J. Lowe, J. M. Gray, and J. E. Johnson, eds.), pp. 1-21. Pergamon Press, New York.

Watts, W. A., and Bright, R. C. (1968). Pollen, seed, and mollusk analysis of a sediment core from Pickerel Lake, northeastern South Dakota. *Geological Society of America Bulletin* 79, 855-76.

Watts, W. A., and Stuiver, M. (1980). Late Wisconsin climate of northern Florida and the origin of species-rich deciduous forest. *Science* 210, 325-27.

Watts, W. A., and Wright, H. E., Jr. (1966). Late-Wisconsin pollen and seed analysis from the Nebraska sandhills. *Ecology* 47, 202-10.

Webb, D. A. (1978). Flora Europaea: A retrospect. *Taxon* 27, 3-14.

Whitehead, D. R. (1967). Full-glacial vegetation in southeastern United States. *In* "Quaternary Paleoecology" (E. J. Cushing and H. E. Wright, Jr., eds.), pp. 273-48. Yale University Press, New Haven, Conn.

Whitehead, D. R. (1973). Late-Wisconsin vegetational changes in unglaciated eastern North America. *Quaternary Research* 3, 621-31.

Whitehead, D. R., and Oaks, R. Q., Jr. (1979). Development history of the Dismal Swamp. *In* "The Great Dismal Swamp" (P. W. Kirk, Jr., ed.), pp. 25-43. University Press of Virginia, Charlottesville.

Willman, H. B. (1971). "Summary of the Geology of the Chicago Area." Illinois State Geological Survey Circular 460, pp. 1-77.

Wright, H. E., Jr. (1968). The roles of pine and spruce in the forest history of Minnesota and adjacent areas. *Ecology* 49, 937-55.

Wright, H. E., Jr. (1970). Vegetational history of the central Plains. *In* "Pleistocene and Recent Environments of the Central Plains" (W. Dort, Jr., and J. K. Jones, Jr., eds.), pp. 157-72. University of Kansas Press, Lawrence.

Wright, H. E., Jr. (1976). Ice retreat and revegetation in the western Great Lakes area. *In* "Quaternary Stratigraphy of North America" (W. C. Mahaney, ed.), pp. 119-32. Dowden, Hutchinson, and Ross, Stroudsburg, Pa.

Wright, H. E., Jr. (1977). Quaternary vegetation history: Some comparisons between Europe and America. *Annual Review of Earth and Planetary Sciences* 5, 123-58.

Wright, H. E., Jr. (1981). Vegetation east of the Rocky Mountains 18,000 years ago. *Quaternary Research* 15, 113-25.

Wright, H. E., Jr., and Watts, W. A. (1969). "Glacial and Vegetational History of Northeastern Minnesota." Minnesota Geological Survey Special Paper 11.

Zagwijn, W. H. (1960). Aspects of the Pliocene and early Pleistocene vegetation in the Netherlands. *Mededelingen van de Geologische Stichting*, C-III-I (5), 1-78.

Zagwijn, W. H. (1963). Pollen-analytic investigations in the Tiglian of the Netherlands. *Mededelingen van de Geologische Stichting* new series 16, 49-71.

CHAPTER 16

Terrestrial Vertebrate Faunas

Ernest L. Lundelius, Jr., Russell W. Graham,
Elaine Anderson, John Guilday, J. Alan Holman,
David W. Steadman, and S. David Webb

Introduction

The vertebrate faunas of the latest Pleistocene (25,000 to 10,000 yr B.P.) of North America are better known than those of any other period in the past. The large number of localities, their broad geographic coverage, and the fact that many of the taxa are still extant allow analyses that are impossible with earlier faunas.

The varying degrees in our understanding of the evolution of the Late Wisconsin fauna is a function of the geographic and chronologic density of the fossil localities. The known faunas are not evenly distributed spatially (Figure 16-1) or temporally (Tables 16-1 through 16-5) but are clustered in those areas and time periods on which intensive research programs have been focused. Thus, the late-Pleistocene faunas of some areas and times are better known than those of others.

Wisconsin vertebrate local faunas from North America are found in many different geographic, geologic, ecologic, and taphonomic settings, which introduce a variety of biases. The fossil records of amphibians, reptiles, and birds of the late Pleistocene are not as well known as is the record of the mammals. In addition, remains of amphibians and reptiles are not as morphologically diagnostic for specific identifications as are those of mammals and birds. Various taxonomic groups of vertebrates from a given site frequently suggest different and sometimes contradictory paleoenvironmental reconstructions; for example, in the Appalachian area mammalian faunas and some birds indicate ''boreal'' conditions but the herpetofaunas and other birds from the same locality indicate a climate similar to that of today or even milder. This situation also occurs in many localities outside the Appalachian area. A taphonomic bias is found in many cave faunas, which usually contain a good sample of the indigenous small animals but frequently yield a poor sample of the large animals, whose remains usually occur as individual specimens in widely scattered localities. This hinders the reconstruction of complete faunas in many areas. Studies of the processes involved in bone accumulations may allow adjustments to be made for these biases in the future, and they may contribute to more accurate reconstructions of past faunas.

A rigorous chronologic framework is mandatory for comprehensive evolutionary studies of faunas. A number of absolute dating techniques can be applied to late-Quaternary faunas, but radiocarbon dating has been the most widely used. Many independent variables can influence the accuracy of radiocarbon dates. Because bone dates on both collagen and apatite fractions appear to be the most vulnerable to contamination (Hassan et al., 1977; Land et al., 1980; Taylor, 1980; Tuross and Hare, 1978), radiocarbon dates on bones are one of the least reliable means of establishing absolute chronologies, but they are one of the most frequently reported types of dates for vertebrate faunas.

The utility of biostratigraphic dating for the late Quaternary is limited by the degree of resolution achievable. Amphibians, reptiles, and small mammals had very low rates of extinction during the late Pleistocene, and so they are of little value for dating. However, the presence or absence of extinct mammalian or avian taxa can be used to differentiate Pleistocene from Holocene faunas. Evolutionary stages within chronoclines for specific mammalian taxa (Nelson and Semken, 1970; Purdue, 1980; Wilson, 1980) may be useful for making finer subdivisions of the Quaternary. Late-Quaternary cultural groups have been narrowly defined and often precisely dated; therefore, cultural remains from archaeological sites can be used as an independent means of dating associated vertebrate faunas.

Many late-Pleistocene local faunas can be divided stratigraphically into faunules; frequently, however, not all of the faunules are associated with absolute dates. Examples of this are the sequences from New Paris No. 4 in Pennsylvania (Guilday et al., 1964) and Baker Bluff Cave in Tennessee (Guilday et al., 1978). Both of these sequences show faunal changes through the later part of the late Pleistocene and into the Holocene. The chronology of the Baker's Bluff Cave sequence is controlled by radiocarbon dates at several levels. Such stratified local faunas can provide useful information about relative chronologic changes in the faunas of a local area.

The authors thank the following individuals for their assistance: Viola Rawn-Schatzinger for permission to use the faunal list from the Jones-Miller site; Judith Lundelius, Melissa Winans, and Betty Clayton for preparing and editing the manuscript; Boyce Cabaniss and Dale Winkler for preparing the map and tables. Financial assistance was provided by the Geology Foundation of the University of Texas at Austin.

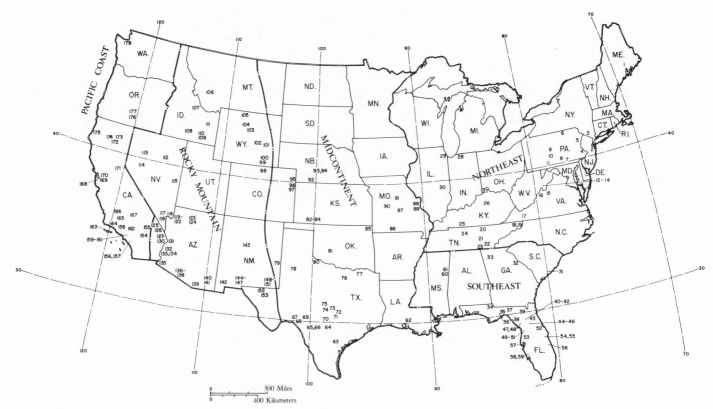

Figure 16-1. Map showing locations of the major late-Pleistocene faunas in the United States. (The numbers correspond to those in Tables 16-1 through 16-5.)

However, correlation of the undated faunules is difficult, and regional reconstructions are more tenuous than they would be if all faunules could be dated (Graham, in press; Logan and Black, 1979).

Slow and uniform rates of deposition like those that occurred at the Chimney Rock Animal Trap (Hager, 1972) may cause temporal compression, which can mask sequential changes in fauna and environment. Faunal mixing by biologic, pedogenic, and geologic processes also may make it difficult to decipher the paleoecologic and paleoenvironmental implications of vertebrate faunas from multifaunule sites. It is often difficult to assess the degree of alteration of faunal composition by these processes, as in the case of the Selby-Dutton fauna (Bannan, 1980; Graham, in press). In the case of other local faunas such as Afton Springs, Oklahoma (Hay, 1920), and Trollinger Springs, Missouri (Saunders, 1977), it is apparent that geologic processes have caused mixing of remains of different ages. Meyer Cave, Illinois (Parmalee, 1967), and Animal Trap, Wyoming (Hager, 1972), show evidence of stratigraphic mixing by the action of animals. Hoyer (1980) has demonstrated that pedogenic processes have been responsible for the downward movement of artifacts at the Cherokee Sewer site in Iowa.

Many supposedly extinct taxa of fossil birds need to be restudied, because many early workers in avian paleontology described new genera and species of birds on the basis of very fragmentary specimens and inadequate modern comparative material. In combination with the systematics problems of living birds, this shortcoming makes it necessary to regard the lists of fossil avian taxa as tentative, particularly in the cases of hawks and eagles (Accipitridae) and grouse (Phasianidae:Tetraoninae). Until these groups are restudied, many identifications of fossils in these families must be regarded as uncertain.

The taxonomy of some groups of mammals also is uncertain at present. The status of many species of *Equus, Bison,* +*Mammuthus, Peromyscus* and many other cricetid rodents, among others, is uncertain, and until restudy of these genera has been completed the correctness of some specific names as used here is in some doubt. (Throughout the chapter, a plus sign indicates an extinct genus and an asterisk an extinct species or subspecies.) In most cases, when the name is in doubt at this time, we have listed it as it was identified by the person who reported it.

Zoogeography

Because the nature of the late-Quaternary vertebrate record makes the demarcation of natural zoogeographic regions for the late Pleistocene difficult to assess, we have arbitrarily divided the United States into five major regions for convenience of discussion (Figure 16-1 and Tables 16-1 through 16-5). A taxonomic list of vertebrate species whose remains have been dated between 25,000 and 10,000 yr B.P. is given in Table 16-6; only those local faunas that we have selected as demonstrating certain phenomena are discussed in the text. The recent book by Kurtén and Anderson (1980) contains a more extensive catalogue of local faunas both for this period and for earlier parts of the Pleistocene.

Many Pleistocene taxa were widely distributed; some were present in virtually every part of the United States. The most conspicuous of these are the large carnivores (e.g., +*Smilodon floridanus,* +*Panthera atrox,* +*Arctodus simus,* +*Homotherium serum, Canis* *dirus, Canis latrans, Ursus americanus, Felix concolor,* and *Felis rufus*) and a number of herbivores (e.g., *Bison,* +*Mammut americanum,* +*Mammuthus* sp., +*Platygonus compressus,* +*Glosso-*

Table 16-1.
List of Localities Producing Major Faunas of Pleistocene Age for
the Northeast Region of the United States.

	Fauna	Location	Depositional Environment	Agents of Accumulation	Age and Method of Dating	Reference(s)	Notes
1.	Presumpscot Formation	Penobscot County, Maine	Shallow marine	Natural accumulation	Late Pleistocene, 13,000-12,000 yr B.P., from faunal assemblage	Marsh, 1872; Stuiver and Borns, 1975	—
2.	Manalapan	Monmouth County, New Jersey	—	—	Rancholabrean? from faunal assemblage	Wetmore, 1958; Steadman, 1980	—
3.	Dutchess Quarry, lower level	Orange County, New York	Cave	Humans and natural accumulation	12,530 ± 370 yr B.P. uncharred bone (*Rangifer*)	Funk et al., 1970	Probable paleo-Indian association
4.	Durham Cave	Bucks County, Pennsylvania	Cave	Natural accumulation	Early Holocene? from faunal assemblage	Leidy, 1889; Mercer, 1897a	Temperate mammalian fauna; but *Rangifer* and *Alces* suggest cooler climate
5.	Hartman's Cave	Monroe County, Pennsylvania	Cave	—	Late Pleistocene/ early Holocene from faunal assemblage	Leidy, 1889	Mammals suggest cooler climate, but fauna may be mixed; bone and stone artifacts are of unknown provenience
6.	Mosherville	Bradford County, Pennsylvania	Glacial kame; Olean drift	—	11,900 ± 750 yr B.P. on uncharred bone (*Platygonus*)	Ray et al., 1970	Youngest date for *Platygonus compressus* in the Northeast
7.	Bootlegger Sink	York County, Pennsylvania	Cave	—	Late Pleistocene from faunal assemblage	Guilday et al., 1966	Mammals indicate boreal woodland
8.	Carlisle Cave	Cumberland County, Pennsylvania	Cave	—	Late Pleistocene? from faunal assemblage	Leidy, 1889; Steadman, 1980	—
9.	Hosterman's Pit	Centre County, Pennsylvania	Cave	—	9240 ± 1000 yr B.P. on charcoal	Guilday, 1967a	Earliest dated Holocene fauna in mid-Appalachians
10.	Frankstown Cave	Blair County, Pennsylvania	Cave	—	Late Pleistocene from faunal assemblage	Peterson, 1926	Extensive megafauna and sparse microfauna suggest woodland
11.	New Paris No. 4	Bedford County, Pennsylvania	Cave; fissure fill	—	11,300 ± 1000 yr B.P. on charcoal	Guilday et al., 1964; Wetmore, 1959	Extensive boreal fauna with *Picea/Pinus* pollen; charcoal drifted in from forest fires
12.	Eagle Cave	Pendleton County, West Virginia	Cave	Owls	Late Pleistocene/ early Holocene from faunal assemblage	Guilday and Hamilton, 1973	Mammalian microfauna suggests boreal woodland
13.	Hoffman School Cave	Pendleton County, West Virginia	Cave	Owls	Late Pleistocene/ early Holocene from faunal assemblage	Guilday and Hamilton, 1978	Mammalian microfauna suggests boreal woodland
14.	Mandy Walters Cave	Pendleton County, West Virginia	Cave	Owls	Late Pleistocene/ early Holocene from faunal assemblage	Guilday and Hamilton, 1978	Mammalian microfauna suggests boreal woodland
15.	Natural Chimneys	Augusta County, Virginia	Cave	—	Late Pleistocene from faunal assemblage	Guilday, 1962; Wetmore, 1962	Fragmentary extinct megafauna; mammalian microfauna suggests boreal woodland
16.	Clarks Cave	Bath County, Virginia	Cave	Owls	Late Pleistocene from faunal assemblage	Guilday et al., 1977	Extensive microfauna indicates boreal woodland and grassland
17.	Saltville	Smyth County, Virginia	Salt spring	Natural accumulation	13,460 ± 420 yr B.P. on Proboscidean tusk	Ray et al., 1967	*Picea/Pinus* pollen indicates boreal woodland
18.	Baker Bluff Cave	Sullivan County, Tennessee	Cave	Owls and packrats (*Neotoma* sp.)	a. 10,560 ± 220 yr B.P. b. 11,640 ± 250 yr B.P. c. 19,100 ± 850 yr B.P.	Guilday et al., 1978	Extensive fauna indicates boreal woodland steppe; Holocene Indian material in disturbed surface levels
19.	Guy Wilson Cave	Sullivan County, Tennessee	Cave	—	19,700 ± 600 yr B.P. on bone (*Platygonus*)	Guilday et al., 1975	Associated boreal fauna; southeasternmost record of *Rangifer tarandus* (36°30′N latitude)

Table 16-1 (continued)

	Fauna	Location	Depositional Environment	Agents of Accumulation	Age and Method of Dating	Reference(s)	Notes
20.	Robinson Cave	Overton County, Tennessee	Cave	—	Late Pleistocene from faunal assemblage	Guilday et al., 1969; Corgan, 1974	Microfauna suggests cooler climate
21.	Big Bone Cave	Van Buren County, Tennessee	Cave	—	Late Pleistocene from faunal assemblage	Mercer, 1897b; Guilday and McGinnis, 1972; Corgan, 1974	Partial skeleton of *Megalonyx jeffersoni* with ligaments and claw sheath
22.	Lookout Cave	Hamilton County, Tennessee	Cave	—	Late Wisconsin from faunal assemblage	Kurtén and Anderson, 1980; Corgan, 1974	—
23.	Nikajack Cave	Marion County, Tennessee	Cave	—	Holocene from cultural association	Shufeldt, 1897; Corgan, 1974	—
24.	First American Bank site, lower level	Davidson County, Tennessee	Cave	Natural accumulation	9410 ± 155 and 10,035 ± 650 yr B.P. on collagen and bone apatite (*Smilodon*)	Guilday, 1977	Youngest published date for *Smilodon*
25.	Savage Cave	Logan County, Kentucky	Cave	—	Late Pleistocene/ early Holocene from faunal assemblage	Guilday and Parmalee, 1979	Paleo-Indian and Indian artifacts with extant and extinct mammals; stratigraphy disturbed by looting
26.	Welsh Cave	Woodford County, Kentucky	Cave	—	12,950 ± 550 yr B.P. on collagen (*Platygonus*)	Guilday et al., 1971	At least 31 *Platygonus*; *Canis dirus* and associated fauna; boreal forest/steppe environment
27.	Big Bone Lick	Boone County, Kentucky	Salt spring	Natural accumulation	17,200 ± 600 yr B.P.	Schultz et al., 1963, 1967; Ives et al., 1967	First North American locality; type locality of *Bootherium bombifrons*, *Cervalces scotti*, *Bison antiquus*, and *Glossotherium harlani*
28.	Potato Creek	St. Joseph County, Indiana	Alluvial	Natural accumulation	Wisconsin from faunal assemblage	Wetmore, 1945	—
29.	Aurora (Wabash Formation)	Cook County, Illinois	Alluvial	Natural accumulation	Late Wisconsin from stratigraphy	Wetmore, 1948	—
30.	Ashmore	Coles County, Illinois	Alluvial	Natural accumulation	Late Wisconsin from stratigraphy	Galbreath, 1944	—

therium harlani, *Castor canadensis*, *Peromyscus maniculatus/ leucopus*, *Ondatra zibethica*, and *Odocoileus*). Although these taxa were widespread, they can still serve as indicators of local environmental conditions.

Reconstructions of faunal provinces for the late Pleistocene have been proposed by Martin and Neuner (1978) and Graham (1979). Martin and Neuner recognize four major faunal complexes, each of which was characterized by the association of a number of taxa over a significant geographic area. The *Ovibos-Dicrostonyx* Faunal Province, based on the association of *Ovibos*, *Dicrostonyx*, and *Rangifer*, occupied a narrow band next to the ice front. The *Symbos-Cervalces* Faunal Province, based on the association of +*Symbos*, +*Cervalces*, +*Castoroides*, and +*Sangamona*, occupied the northeastern United States. The *Chlamytherium (Holmesina)*-Glyptodont Faunal Province, based on the association of +*Holmesina*, glyptodonts, capybaras, and *Tremarctos* sp., covered the southeastern United States. The *Camelops-Navahoceras* Faunal Province, based on the association of +*Camelops* sp., +*Navahoceras*, +*Nothrotheriops*, +*Arctodus simus*, and *Felis *atrox*, took in the central and western areas of the United States. As might be expected, the distributions of some of the taxa that characterize

each of these faunal provinces overlap, but the areas of association do not overlap, and the provinces correspond closely to the major floral provinces of the late Pleistocene as proposed by P. V. Wells (unpublished data).

Graham (1979) recognizes four major Pleistocene faunal provinces in North America, which he likens to the modern superprovinces as defined by Hagmeier (1966). The Northern Mammalian Region was characterized by the association of *Microsorex hoyi*, *Phenacomys* sp., *Dicrostonyx* sp., *Synaptomys borealis*, *Microtus xanthognathus*, *Rangifer* sp., *Ovibos*, +*Mammuthus primigenius* (the range of this species is controversial; in the United States, it may actually be +*M. jeffersoni*). The Eastern Mammalian Region was characterized in part by the association of *Blarina* sp., *Tamias striatus*, *Synaptomys cooperi*, +*Castoroides ohioensis*, +*Symbos* sp., +*Bootherium* sp., and +*Mylohyus* sp. The Southern Mammalian Region was characterized by the association of *Dasypus *bellus*, +*Holmesina* sp., +*Glyptotherium* sp., +*Neochoerus* sp., *Hydrochoerus* sp., and *Neofiber* sp. The Rocky Mountain and Western Mammalian Region was characterized by the association of +*Nothrotheriops shastense*, *Lagurus* sp., *Marmota flaviventris*, +*Euceratherium* sp., +*Preptoceras*

Table 16-2.
List of Localities Producing Major Faunas of Pleistocene Age
for the Southeast Region of the United States

	Fauna	Location	Depositional Environment	Agents of Accumulation	Age and Method of Dating	Reference(s)	Notes
31.	Edisto Beach	Charleston County, South Carolina	Ocean beach	Natural accumulation	12,000-10,000 yr B.P. from faunal assemblage	Roth and Laerm, 1980	Mammals and reptiles suggest woodland/savannah with temperate, equable climate
32.	Little Kettle Creek	Wilkes County, Georgia	Fluvial	Natural accumulation	Late Pleistocene from faunal assemblage	Voorhies, 1974	Small mammals suggest cooler climate
33.	Ladds Quarry	Bartow County, Georgia	Cave; fissure fill	Natural accumulation	Late Pleistocene from faunal assemblage	Ray, 1967; Ray and Lipps, 1968; Holman, 1976; Wetmore, 1967	Extensive fauna, mixed temperate and southern
34.	Chipola River	Jackson County, Florida	Fluvial	Natural accumulation	Wisconsin from faunal assemblage	Webb, 1974a	—
35.	St. Marks River	Leon and Wakulla Counties, Florida	Fluvial	Natural accumulation	Rancholabrean from faunal assemblage	Steadman, 1980	—
36.	Wakulla Springs	Wakulla County, Florida	Spring	Natural accumulation	Late Rancholabrean from faunal assemblage	Brodkorb, 1963	—
37.	Aucilla River	Jefferson County, Florida	Fluvial	Natural accumulation	Wisconsin from faunal assemblage	Webb, 1974a; Olson, 1972; Gillette, 1976	—
38.	Steinhatchie River	Taylor and Dixie Counties, Florida	Fluvial	Natural accumulation	Rancholabrean from faunal assemblage	Steadman, 1980	—
39.	Ichetucknee River	Columbia County, Florida	Fluvial	Natural accumulation	Late Wisconsin/ early Holocene from faunal assemblage	Webb, 1974a; Auffenberg, 1963a, 1963b; Campbell, 1980; Gillette, 1976	Fauna may be mixed
40.	Arredondo a. IA b. IB c. IIA	Alachua County, Florida	Sinkhole; fissure fill	Natural accumulation and owls	Wisconsin to Sangamon from faunal assemblage	Webb, 1974a; Bader, 1957; Brodkorb, 1959	—
41.	Haile XIVA	Alachua County, Florida	Sinkhole; fissure fill	Natural accumulation	Early Wisconsin from faunal assemblage	Martin, 1974	May be older than 25,000 yr B.P.
42.	Hornsby Springs	Alachua County, Florida	Sinkhole	Natural accumulation	9800 ± 270 yr B.P. on shell marl	Brodkorb, 1963; Webb, 1974; Dolan and Allen, 1961	—
43.	Bowman IA	Putnam County, Florida	—	Owls	Rancholabrean from faunal assemblage	Steadman, 1980	—
44.	Mefford Cave I	Marion County, Florida	Cave	Natural accumulation	Late Wisconsin faunal assemblage	Steadman, 1980	—
45.	Oakhurst Quarry	Marion County, Florida	—	—	Rancholabrean from faunal assemblage	Steadman, 1980	—
46.	Eichelberger Cave	Marion County, Florida	Cave	Natural accumulation	Late Wisconsin? from faunal assemblage	Brodkorb, 1956b	—
47.	Devil's Den	Levy County Florida	Sinkhole	Natural accumulation	8000-7000 yr B.P. on bone	Martin and Webb, 1974	—
48.	Waccassassa River, IIB and III	Levy County, Florida	Fluvial	Natural accumulation	Wisconsin from faunal assemblage	Webb, 1974a	—
49.	Davis Quarry	Citrus County, Florida	—	—	Rancholabrean from faunal assemblage	Steadman, 1980	—
50.	Sabertooth Cave	Citrus County, Florida	Sinkhole; fissure fill	Natural accumulation	Wisconsin from faunal assemblage	Webb, 1974a; Holman, 1958	—
51.	Crystal River Power Plant	Citrus County, Florida	Fissure fill	Natural accumulation	14,320 ± 170 yr B.P.	Webb, unpublished data	—

Table 16-2 (continued)

	Fauna	Location	Depositional Environment	Agents of Accumulation	Age and Method of Dating	Reference(s)	Notes
52.	Lake Monroe	Seminole County, Florida	—	—	Rancholabrean? from faunal assemblage	Brodkorb, 1963	—
53.	Crystal Springs	Pasco County, Florida	Fluvial	Natural accumulation	Late Wisconsin? from faunal assemblage	Brodkorb, 1956b	—
54.	Melbourne	Brevard County, Florida	Coastal marsh	Natural accumulation	Wisconsin from faunal assemblage	Webb, 1974a; Gillette, 1976	—
55.	Sebastian Canal	Brevard County, Florida	Coastal marsh	Natural accumulation	Wisconsin from faunal assemblage	Webb, 1974a	—
56.	Vero 2 and 3	Indian River County, Florida	Coastal marsh	Natural accumulation	Wisconsin from faunal assemblage	Webb, 1974a; Weigel, 1962	—
57.	Seminole Field	Pinellas County, Florida	Coastal	Natural accumulation	Wisconsin from faunal assemblage	Webb, 1974a; Simpson, 1929	—
58.	Hog Creek	Sarasota County, Florida	Fluvial	Natural accumulation	Rancholabrean from faunal assemblage	Wetmore, 1931	—
59.	Little Salt Spring	Sarasota County, Florida	Sinkhole	Humans?	12,030 ± 200 yr B.P. on wood spear	Clausen et al., 1979	The dated spear was found run through a specimen of *Geochelone crassiscutata*
60.	Columbus site	Lowndes County, Mississippi		—	Late Pleistocene from faunal assemblage	Jackson and Kaye, 1974b	Fauna includes *Terrapene carolina putnami* and *T. c. triunguis*
61.	Catalpa Creek site	Clay and Lowndes Counties, Mississippi		—	Late Wisconsin from faunal assemblage	Jackson and Kaye, 1974a, 1975	Fauna includes *Geochelone crassiscutata* and the northern *Emydoidea blandingi*

sp., and *Oreamnos* sp. These regions correspond in general to those proposed by Martin and Neuner (1978), but they differ in detail. The Northern Mammalian Region (Graham, 1979) is much broader than the *Ovibos-Dicrostonyx* province (Martin and Neuner, 1978), and it encompasses the western part of the *Symbos-Cervalces* province. The Rocky Mountain and Western Mammalian Region (Graham, 1979) includes only the western part of the *Camelops-Navahoceras* province. Graham points out that these provinces probably can be subdivided into smaller geographic units when more detailed information about faunal distributions becomes available. The differences between the provinces of Martin and Neuner and those of Graham are due partially to different assignments to provinces of subprovinces that are not well defined at present.

NORTHEASTERN REGION

Studies of the Appalachian area are critical to the determination of climatic events between 25,000 and 10,000 yr B.P. Sites with absolute dates have been found in Pennsylvania, Tennessee, and Georgia (Table 16-1); sites with indefinite dates have been found in these states and in Virginia. In the more northern sites, paleoecologic and climatic interpretations based on herpetologic evidence disagree with those based on mammalian evidence.

The unit B faunule of the New Paris No. 4 local fauna in Pennsylvania (Guilday et al., 1964) dates between 10,000 and 11,000 yr B.P. On the basis of the mammalian remains, the investigators postulate a "boreal" climate for the area in the Pleistocene, but the herpetofauna contains some "Carolinian" taxa (Table 16-6), including the salamanders *Desmognathus, Eurycea, Plethodon glutinosus,* and *Notophthalmus viridescens* and the snakes *Agkistro-*

don controtrix and *Carphophis amoenus,* none of which presently are found in "boreal" climates. However, it should be noted that, because of possible stratigraphic mixing, it is not certain that these southern herpetologic taxa were contemporaneous with the boreal forms. The avifauna includes the sharp-tailed grouse (*Tympanuchus phasianellus*), a northern though not strictly "boreal" form (Wetmore, 1959). All other birds from unit B still occur in the area today. The Frankstown Cave local fauna, which is thought to be late Pleistocene (Richmond, 1964), also contains a typical "Carolinian" herpetofauna. The birds from this site still live in the region.

The Natural Chimneys local fauna in Virginia (Guilday, 1962) is considered to be late Pleistocene or early Holocene. The climate is considered to be "boreal" on the basis of its mammalian taxa, which include *Sorex arcticus, Microtus xanthognathus,* and *Synaptomys borealis.* However, the herpetofauna, which includes the coachwhip snake (*Masticophis flagellum*) and the diamondback rattlesnake (*Crotalus adamanteus*) (Guilday, 1962), indicates milder winters than those of today. As in the case of New Paris No. 4, it is not certain that the reptiles and mammals were contemporaneous. The avifauna (Guilday, 1962; Wetmore, 1962) is a mixture of "boreal" species such as the spruce grouse (*Dendragapus canadensis*), the gray jay (*Perisoreus canadensis*), and the magpie (*Pica pica*), as well as more southerly species such as the bobwhite (*Colinus virginianus*), the wild turkey (*Meleagris gallopavo*), the red-bellied woodpecker (*Melanerpes carolinus*), and the red-headed woodpecker (*Melanerpes erythrocephalus*) (Table 16-6). The Clark's Cave fauna in Virginia is thought to be about 11,000 years old because its mammal fauna is similar to that of the New Paris No. 4 (Guilday et al., 1977), but the herpetofauna is similar to that of the area today, with

Table 16-3.
List of Localities Producing Major Faunas of Pleistocene Age
for the Central Region of the United States

Fauna	Location	Depositional Environment	Agents of Accumulation	Age and Method of Dating	Reference(s)	Notes
62. Avery Island	Iberia Parish, Louisiana	Marsh	Natural accumulation	Wisconsin from faunal assemblage	Domning, 1969; Arata, 1964; Gagliano, 1967	—
63. Berclair Terrace site I	Bee County, Texas	Fluvial	Natural accumulation	—	Sellards, 1940	—
64. Friesenhahn Cave a. Upper unit b. Lower unit	Bexar County, Texas	Cave	a. Owls b. Owls and carnivores	a. 9200 yr B.P. on bone b. 20,000-15,000 yr B.P. on bone	Graham, 1976b	—
65. Montell Shelter	Uvalde County, Texas	Cave	—	—	Lundelius, 1967	—
66. Kincaid Shelter	Uvalde County, Texas	Cave	—	—	Lundelius, 1967	—
67. Cueva Quebrado	Val Verde County, Texas	Cave	—	14,000-12,000 yr B.P. on charcoal	Lundelius, unpublished data; Alexander, unpublished data	—
68. Bonfire Shelter	Val Verde County, Texas	Cave	—	9920 ± 150, 10,100 ± 300, and 10,230 ± 160 yr B.P. on charcoal	Dibble and Lorrain, 1968; Dibble, 1970	—
69. Schulze Cave a. Stratum C1 b. Stratum C2	Edwards County, Texas	Cave	a. Natural trap b. Owls	a. 9680 ± 700 yr B.P. on bone b. 9310 ± 310 yr B.P. on bone	Dalquest et al., 1969	—
70. Cave without a Name	Kendall County, Texas	Cave	Owls	10,900 ± 190 yr B.P. on bone	Lundelius, 1967	—
71. Levi Shelter a. Zone I b. Zone II	Travis County, Texas	Cave	a. Humans? b. Humans	a. > 10,000 yr B.P. on shell b. 10,000 ± 175 yr B.P. on shell	Alexander, 1963; Lundelius, 1963	—
72. Laubach Cave a. Site I b. Site II c. Site III	Williamson County, Texas	Cave	—	a. 15,850 ± 500 yr B.P. on bone b. 13,973 ± 310 yr B.P. on bone c. 23,230 ± 490 yr B.P. on bone	Slaughter, 1966; Choate and Hall, 1967	—
73. Longhorn Cavern, Longhorn Breccia	Burnet County, Texas	Cave	Owls and carnivores	Late Wisconsin from faunal assemblage	Semken, 1961; Lundelius, 1967	—
74. Miller's Cave, Travertine unit	Llano County, Texas	Cave	Owls and carnivores	7000 yr B.P. on bone	Patton, 1963; Lundelius, 1967; Weigel, 1967	—
75. Clamp's Cave	San Saba County, Texas	Cave	Carnivore's den?	—	Kurtén, 1963; Lundelius, 1967	—
76. Lewisville	Denton County, Texas	Fluvial	Humans?	> 37,000 yr B.P. on charcoal	Slaughter et al., 1962	Associated with a Clovis point and several hearths; dated charcoal may be burned lignite so that date may be too old
77. Ben Franklin	Fannin and Delta Counties, Texas	Fluvial	Natural accumulation	9550 ± 377 and 11,135 ± 450 yr B.P. on shell	Slaughter and Hoover, 1963	—
78. Lubbock Lake a. Clovis level b. Folsom level	Lubbock County, Texas	Fluvial	Natural accumulation and humans	a. 12,060 ± 100-11,140 ± 80 yr B.P. b. 10,880 ± 90-10,060 ± 70 yr B.P.	Johnson, 1974; Rea, unpublished data	Differences between the Clovis and Folsom faunules indicate a fairly rapid major faunal shift

Table 16-3 (continued)

Fauna	Location	Depositional Environment	Agents of Accumulation	Age and Method of Dating	Reference(s)	Notes
79. Blackwater Draw a. Brown sand wedge b. Gray sand	Roosevelt County, New Mexico	Lacustrine	Natural accumulation and humans	a. ? b. 12,000-11,000 yr B.P. on projectile points	Lundelius, 1972; Slaughter, 1975	—
80. Howard Ranch	Hardeman County, Texas	Fluvial	—	16,805 ± 565 yr B.P. on shell	Dalquest, 1965	—
81. Domebo	Caddo County, Oklahoma	Fluvial	Natural accumulation and humans	11,220 ± 500 yr B.P. on collagen, 9400 ± 300 yr B.P. on organic earth, and 11,045 ± 647 yr B.P. on wood	Slaughter, 1966	—
82. Jones	Meade County, Kansas	Lacustrine	—	29,000-26,700 yr B.P. on shell	Schultz, 1969; Davis, 1975	—
83. Robert	Meade County, Kansas	Marsh	—	11,100 ± 390 yr B.P. on shell	Schultz, 1967	—
84. Shorts Creek	Meade County, Kansas	Fluvial	Natural accumulation	Late Rancholabrean from faunal assemblage and stratigraphy	Stettenheim, 1958	—
85. Afton Springs	Ottawa County, Oklahoma	Sulphur spring	Natural accumulation and humans	Wisconsin? from faunal assemblage	Hay, 1924	Possibly associated with cultural material
86. Zoo Cave	Taney County, Missouri	Cave	Predators?	Late Pleistocene/ early Holocene from faunal assemblage	Hood and Hawksley, 1975	Fauna may be mixed
87. Bat Cave	Pulaski County, Missouri	Cave	Natural accumulation	Late Wisconsin from faunal assemblage	Hawksley et al., 1973	—
88. Crankshaft Cave	Jefferson County, Missouri	Natural trap	Natural accumulation	Late Wisconsin/ early Holocene from faunal assemblage	Parmalee et al., 1969	—
89. Herculaneum Crevice	Jefferson County, Missouri	Cave	—	Late Wisconsin from faunal assemblage	Mehl, 1962	—
90. Boney Spring	Benton County, Missouri	Spring	Natural accumulation	13,700 ± 600 and 16,580 ± 220 yr B.P. on organic matter	Mehringer et al., 1970; Saunders, 1977	—
91. Brynjulfson Caves a. Cave I b. Cave II	Boone County, Missouri	Cave	Natural accumulation	a. 9440 ± 760 yr B.P. on bone b. 2460 ± 230 yr B.P. on bone	Parmalee and Oesch, 1972	Fauna probably is mixed Pleistocene/Holocene
92. Red Willow	Red Willow County, Nebraska	Fluvial	Natural accumulation	Wisconsin/early Holocene from faunal assemblage	Corner, 1977	—
93. Wiggins (Smith) Canyon	Dawson County, Nebraska	Loess	Natural accumulation	Wisconsin from stratigraphy	Tate and Martin, 1968	—
94. Midway Canyon	Dawson County, Nebraska	Loess	Natural accumulation	Wisconsin from stratigraphy	Tate and Martin, 1968	—
95. Selby a. Peorian Loess b. Lacustrine unit	Yuma County, Colorado	a. Aeolian b. Lacustrine	Natural accumulation and humans (?)	Late Pleistocene from stratigraphy	Stanford, 1979; Graham, in press	Possible pre-Clovis kill site
96. Dutton a. Peorian Loess b. Lacustrine unit	Yuma County, Colorado	a. Aeolian b. Lacustrine	Natural accumulation and humans	a. 29,000-14,000 yr B.P. on stratigraphy b. 11,700 yr B.P. on bone	Stanford, 1979; Graham, in press	Possible pre-Clovis and Clovis site
97. Jones-Miller	Yuma County, Colorado	Fluvial	Natural accumulation and humans	10,020 yr B.P. on charcoal	Stanford, no date	Fauna associated with Hell Gap cultural remains

Table 16-4.
List of Localities Producing Major Faunas of Pleistocene Age
for the Rocky Mountain Region of the United States

	Fauna	Location	Depositional Environment	Agents of Accumulation	Age and Method of Dating	Reference(s)	Notes
98.	Chimney Rock Animal Trap	Larimer County, Colorado	Circular depression in the Casper Sandstone	Natural trap	Late Pleistocene from faunal assemblage	Hager, 1972	—
99.	Bell Cave	Albany County, Wyoming	Cave	Raptorial birds, rodents, and and humans	Late Pleistocene/ early Holocene from faunal assemblage	Zeimens and Walker, 1974	—
100.	Horned Owl Cave	Albany County, Wyoming	Cave	Raptorial birds and woodrats	Late Pleistocene/ early Holocene from faunal assemblage	Guilday et al., 1967	—
101.	Little Box Elder Cave	Converse County, Wyoming	Cave	Owls, carnivores, rodents, and humans	Late Pleistocene/ early Holocene from faunal assemblage	Anderson, 1968; Kurtén and Anderson, 1974	—
102.	Casper	Natrona County, Wyoming	Blowout	Natural accumulation and humans	9830 ± 350 yr B.P. on charcoal and 10,080 ± 170 yr B.P. on bone (*Bison*)	Wilson, 1974; Frison, 1974; Frison et al., 1978	Hell Gap kill site; *Camelops* shows butchering marks
103.	Little Canyon Creek Cave	Washakie County, Wyoming	Cave	Owls, rodents, and humans	10,170 ± 250 yr B.P. on bone	D. Walker, unpublished data	—
104.	Colby	Washakie County, Wyoming	Fluvial	Natural accumulation and humans	11,200 ± 220 yr B.P. on bone	Walker and Frison, 1980	Clovis mammoth kill
105.	Natural Trap Cave	Big Horn County, Wyoming	Sinkhole	Natural trap	12,770 ± 990 yr B.P. on bone (*Equus*) 14,670 ± 670 and 17,620 ± 1490 yr B.P. on collagen	Martin et al., 1979	Plains fauna, highly cursorial forms such as *Acinonyx *trumani*
106.	Warm Springs I	Silver Bow County, Montana	Alluvial fan	Natural accumulation	Late Wisconsin/ early Holocene from faunal assemblage	Rasmussen, 1974	—
107.	Jaguar Cave	Lemhi County, Idaho	Cave	Carnivores, woodrats, and humans	11,580 ± 250 and 10,370 ± 350 yr B.P. on charcoal	Kurtén and Anderson, 1972; Dort, 1975; Sadek, 1965	Apparently occupied by paleo-Indians before 12,000 yr B.P.
108.	Wilson Butte Cave	Jerome County, Idaho	Lava blister cave	Owls and and humans	15,000 ± 800 and 14,500 ± 500 yr B.P. on bone	Gruhn, 1961	Fauna associated with very early paleo-Indian remains
109.	Dam	Power County, Idaho	Flood plain	Fluvial	26,500 ± 3500 yr B.P. on bone	Barton, 1975; Kurtén and Anderson, 1980	On Snake River floodplain
110.	Rainbow Beach	Power County, Idaho	Delta	Fluvial	31,300 ± 2300- 21,500 ± 700 yr B.P. on collagen	McDonald and Anderson, 1975	Result of spillover of Lake Bonneville
111.	Wasden	Bonneville County, Idaho	Collapsed lava tube	Owls and humans	7100-12,850 yr B.P. on bone and charcoal	Butler, 1969, 1972	Possible paleo-Indian association
112.	Mineral Hill Cave	Elko County, Nevada	Cave	Natural accumulation	Late Wisconsin/ early Holocene from faunal assemblage	McGuire, 1980	—
113.	Crypt Cave	Pershing County, Nevada	Cave	—	19,705 ± 650 yr B.P. on bone (*Acinonyx*)	Howard, 1958; Kurtén and Anderson, 1980	—

Table 16-4 (continued)

	Fauna	Location	Depositional Environment	Agents of Accumulation	Age and Method of Dating	Reference(s)	Notes
114.	Rattlesnake Hill (Lake Lahontan)	Churchill County, Nevada	Lacustrine	Natural accumulation	Late Pleistocene from stratigraphy	Hall, 1940	—
115.	Smith Creek Cave	White Pine County, Nevada	Cave	Carnivores, rodents, and humans	11,680-9940 yr B.P.	Stock, 1936; Howard, 1952; Miller, 1979	—
116.	Glendale	Clark County, Nevada	Beaver ponds	Natural accumulation	Late Wisconsin from faunal assemblage	Kurtén and Anderson, 1980; Tessman, unpublished data	—
117.	Gypsum Cave	Clark County, Nevada	Cave	Den	11,690 ± 250 yr B.P. on + *Nothrotheriops* dung	Mehringer, 1967	Associated with cultural remains
118.	Tule Springs	Clark County, Nevada	Fluvial	Natural accumulation	> 13,000 yr B.P. from radiocarbon and stratigraphy	Shutler, 1968; Mawby, 1967	Some taxa as old as 40,000 yr B.P. ; cultural associations are unclear
119.	Emery Falls Canyon a. Locality 7343 b. Locality 7345 c. Locality 7346 d. Locality 7312	Mohave County, Arizona	Caves	Packrat	a. 9650 ± 360 on twigs b. 10,310 ± 500 yr B.P. on twigs c. 10,910 ± 450 yr B.P. on twigs d. 11,190 ± 150 yr B.P. on twigs	Van Devender et al., 1977	—
120.	Rampart Cave a. Locality 7331 b. Locality 7349 c. Locality 7311 d. Locality 7350 e. Locality 7445 f. Locality 7204	Mohave County, Arizona	Cave	Packrat middens, den, and carnivores	a. 11,140 ± 250-9520 ± 33 yr B.P. b. 12,230 ± 350 yr B.P. c. 12,600 ± 260 yr B.P. d. 13,510 ± 190 yr B.P. e. 16,330 ± 270 yr B.P. f. 18,890 ± 500-14,810 ± 230 yr B.P. all on twigs and sloth dung	Van Devender et al., 1977; Mead, 1981	Herps come from middens; larger mammals come from stratified dung deposits
121.	Vulture Cave a. Locality 7446 b. Locality 7522 c. Locality 7321	Mohave County, Arizona	Cave	Packrat middens	a. 10,250 ± 290 yr B.P. on twigs b. 11,870 ± 190 yr B.P. on twigs c. 12,770 ± 440 yr B.P. on twigs	Van Devender et al., 1977; Mead and Phillips, 1981	—
122.	Peach Springs Wash	Coconino County, Arizona	Cave	Packrat	12,040 ± 400 yr B.P. on twigs	Van Devender et al., 1977	—
123.	Shinumo Creek	Coconino County, Arizona	Cave	Packrat	13,660 ± 160 yr B.P. on twigs	Van Devender et al., 1977	—
124.	Stanton's Cave	Coconino County, Arizona	Cave	Natural accumulation	Late Pleistocene from faunal assemblage	Parmalee, 1969; Rea, 1980; Rea and Hargrave, no dates; Mead, 1981	—
125.	Kokoweef Cave	San Bernardino County, California	Cave	Natural trap rodents	Late Wisconsin from faunal assemblage	Kurtén and Anderson, 1980; R. E. Reynolds, unpublished data	—
126.	Mescal Cave	San Bernardino County, California	Cave	—	Late Pleistocene from faunal assemblage	Brattstrom, 1958b	—
127.	Whipple Mountain a. Locality 7447 b. Locality 75132	San Bernardino County, California	Caves	Packrat middens	a. 9980 ± 180 yr B.P. on twigs b. 10,430 ± 170 yr B.P. on twigs	Van Devender and Mead, 1978	—

Table 16-4 (continued)

	Fauna	Location	Depositional Environment	Agents of Accumulation	Age and Method of Dating	Reference(s)	Notes
128.	Red Tail Peaks a. Locality 75114 b. Locality 75116 c. Locality 76148	San Bernardino County, California	Caves	Packrat middens	a. 10,030 ± 160 yr B.P. on twigs b. 10,930 ± 170 yr B.P. on twigs c. 13,810 ± 270 yr B.P. on twigs	Van Devender and Mead, 1978	—
129.	Tunnel Ridge a. Locality 7343 b. Locality 7525	San Bernardino County, California	Caves	Packrat middens	a. 10,330 ± 300 yr B.P. on twigs b. 12,670 ± 260 yr B.P. on twigs	Van Devender and Mead, 1978	—
130.	Falling Arches	San Bernardino County, California	Cave	Packrat middens	11,650 ± 190 yr B.P. on twigs	Van Devender and Mead, 1978	—
131.	Artillery Mountains	Mohave County, Arizona	Cave	Packrat middens	12,250 ± 200 yr B.P. on twigs	Van Devender and Mead, 1978	—
132.	New Water Mountains a. Locality 7237 b. Locality 7241 c. Locality 7242	Yuma County, Arizona	Cave	Packrat middens	a. 10,880 ± 900 yr B.P. on twigs b. 7870 ± 750, 11,050 ± 390, 12,090 ± 570 yr B.P., c. 11,000 ± 500 yr B.P. all on twigs	Van Devender and Mead, 1978	—
133.	Brasscap Point	Yuma County, Arizona	Cave	Packrat middens	11,450 ± 400 yr B.P. on twigs	Van Devender and Mead, 1978	—
134.	Burro Canyon	Yuma County, Arizona	Cave	Packrat middens	14,400 ± 300-13,400 ± 250 yr B.P. on twigs	Van Devender and Mead, 1978	—
135.	Welton Hills	Yuma County, Arizona	Cave	Packrat middens	10,740 ± 400-10,580 ± 550 yr B.P.	Van Devender and Mead, 1978	—
136.	Ventana Cave	Pima County, Arizona	Rock shelter	Natural accumulation and humans	11,300 ± 1200 yr B.P. on charcoal	Haury, 1950; Mehringer, 1967	Extinct fauna associated with Clovis artifacts
137.	Wolcott a. Locality 7235 b. Locality 7240	Pima County, Arizona	Cave	Packrat	a. 5020 ± 80 yr B.P. on twigs b. 14,550 ± 800 yr B.P. on twigs	Van Devender and Mead, 1978	—
138.	Tucson Mountain	Pima County, Arizona	Cave	Packrat middens	12,430 ± 400 yr B.P.	Van Devender and Mead, 1978	—
139.	Papago Springs Cave	Santa Cruz County, Arizona	Cave	Natural accumulation	Wisconsin from faunal assemblage	Skinner, 1942	—
140.	Wilcox Gravel Pit	Cochise County, Arizona	—	—	> 11,000 yr B.P. from faunal assemblage	Moodie and Van Devender, 1978	—
141.	North Papago Cave	Cochise County, Arizona	Cave	—	Rancholabrean from faunal assemblage	Rea, 1980	—
142.	Howell's Ridge Cave a. Faunule 1 b. Faunule 2	Grant County, New Mexico	Cave	—	a. 10,720 yr B.P. b. 11,541 yr B.P. from radiocarbon and estimated rates of sedimentation	Van Devender and Worthington, 1977; Harris, 1977b; Smartt, 1972; Howard, 1962a	—
143.	Isleta Cave	Bernalillo County, New Mexico	Cave	—	Late Pleistocene/early Holocene from faunal assemblage	Harris and Findley, 1964	Fauna probably temporally mixed
144.	Robledo Cave	Dona Ana County, New Mexico	Cave	—	Late Pleistocene/Holocene from faunal assemblage	Van Devender et al., 1976	—

Table 16-4 (continued)

	Fauna	Location	Depositional Environment	Agents of Accumulation	Age and Method of Dating	Reference(s)	Notes
145.	Shelter Cave	Dona Ana County, New Mexico	Cave	—	Late Pleistocene/ early Holocene from faunal assemblage	Brattstrom, 1964; Van Devender et al., 1976; Harris, 1977b	—
146.	Conkling Cave	Dona Ana County, New Mexico	Cave	—	Late Pleistocene/ early Holocene from faunal assemblage	Brattstrom, 1964; Van Devender et al., 1976; Harris, 1977b	—
147.	Khulo site	Dona Ana County, New Mexico	Cave	—	Late Pleistocene/ Holocene; 1550 ± 60, 1700 ± 70, and 8210 ± 220 yr B.P. on wood and charcoal	Harris, 1977b	—
148.	Burnet Cave	Eddy County, New Mexico	Cave	—	Late Pleistocene from faunal assemblage	Schultz and Howard, 1935; Harris, 1977b	—
149.	Dry Cave a. TT II (Locality 54) b. Stalag 17 (Locality 23) c. Camel Room (Locality 25) d. Early Man Corridor (Locality 31) e. Bison Chamber (Locality 4) f. Harris's Pocket (Locality 6) g. Animal Fair (Locality 22) h. Lost Valley (Localities 1 and 17) i. Room of the Vanishing Floor (Localities 26 and 27) j. Sabertooth Camel Maze (Localities 2 and 5) k. Balcony Room (Locality 12) l. Locality 3	Eddy County, New Mexico	Cave	Natural accumulation, owls, and predators	a. 10,730 ± 150 yr B.P. on collagen b. 11,880 ± 250 yr B.P. on charcoal c. 12,000 yr B.P. from stratigraphy d. 11,880-15,030 yr B.P. from stratigraphy e. 14,470-10,730 yr B.P. from stratigraphy f. 14,470 ± 250 yr B.P. from Neotoma dung g. 15,030 ± 210 yr B.P. on collagen h. 29,290 ± 1060 yr B.P. on bone carbonate i. 33,590 ± 1500 yr B.P. on bone carbonate j. 25,160 ± 1730 yr B.P. on bone carbonate k. 15,000-10,000 yr B.P. from stratigraphy l. 15,000-10,000 yr B.P. from stratigraphy	Harris, 1970 Holman, 1970; Harris, 1977b	—
150.	Dark Canyon Cave	Eddy County, New Mexico	Cave	—	Late Pleistocene from faunal assemblage	Rickart, 1977; Howard, 1971	—
151.	Rocky Arroyo Cave	Eddy County, New Mexico	Cave	—	12,500-10,000 yr B.P. from faunal assemblage	Howard, 1962a; Harris, 1977b	—
152.	Upper Sloth Cave	Culberson County, Texas	Cave	Owls and carnivores	11,760 ± 610 yr B.P. on Artiodactyl dung	Logan and Black, 1979	—
153.	Williams Cave	Culberson County, Texas	Cave	—	12,500-10,000 yr B.P. from faunal and cultural assemblages	Harris, 1977b	—

Table 16-5
List of Localities Producing Major Faunas of Pleistocene Age
for the Pacific Region of the United States

	Fauna	Location	Depositional Environment	Agent(s) of Accumulation	Age and Method of Dating	Reference(s)	Notes
154.	Schuiling Cave	San Bernardino County, California	Cave	Den, predators, flooding	Late Wisconsin from faunal assemblage	Downs et al., 1959	—
155.	Manix Lake	San Bernardino County, California	Lacustrine	Natural accumulation	Rancholabrean from faunal assemblage	Howard, 1955	—
156.	Imperial Highway (La Habra)	Orange County, California	Alluvial	Natural accumulation	Rancholabrean from faunal assemblage	Miller and DeMay, 1942; Steadman, 1980	—
157.	La Mirada Site	Los Angeles and Orange Counties, California	Fluvial	Natural accumulation	10,690 ± 360yr B.P. on wood	Miller, 1971	—
158.	Rancho La Brea a. Locality 81 b. Locality 2051 c. Locality 10 d. Other sites	Los Angeles County, California	Fluvial	Entrapment in asphalt-impregnated streams and ponds	a. circa 11,000 yr B.P. b. 27,000-22,500 yr B.P. c. 9000 ± 80 yr B.P. d. Late Pleistocene from faunal assemblage	Brattstrom, 1953a, 1954b; Stock, 1972; Howard, 1962b; Woodard and Marcus, 1973; Akersten, 1979; Miller and DeMay, 1942	—
159.	Workman Street and Alhambra Street	Los Angeles County, California	Alluvial	Natural accumulation	Rancholabrean from faunal assemblage	Miller and DeMay, 1942; Steadman, 1980	—
160.	Emery Barrow Pit	Los Angeles County, California	Fluvial	Natural accumulation	Late Wisconsin from faunal assemblage	Kurtén and Anderson, 1980; Langenwalter, unpublished data	—
161.	York Valley	Los Angeles County, California	Fluvial	Natural accumulation	Rancholabrean from faunal assemblage	Miller, 1942	—
162.	Zuma Creek	Los Angeles County, California	Deltaic estuary	Natural accumulation	Late Wisconsin from faunal assemblage	Kurtén and Anderson, 1980	—
163.	Channel Islands (San Miguel, Santa Rosa, Santa Cruz, Anacapa)	Santa Barbara County, California	Alluvial	Natural accumulation and humans (?)	15,800 ± 280 yr B.P. on wood and 29,650 ± 2500 yr B.P. on bone	Johnson, 1978	Island fauna; dwarf proboscideans
164.	Carpinteria	Santa Barbara County, California	Waterhole	Entrapment in asphalt	Late Wisconsin from faunal assemblage	Brattstrom, 1955; Stock, 1937; Miller and DeMay, 1942	—
165.	Maricopa	Kern County, California	Waterhole	Entrapment in asphalt	Late Wisconsin from faunal assemblage	MacDonald, 1967; Kurtén and Anderson, 1980	—
166.	McKittrick	Kern County, California	Waterhole	Entrapment in asphalt	Late Pleistocene from faunal assemblage	Brattstrom, 1953b	—
167.	China Lake	Kern County, California	Lacustrine	Natural accumulation	18,600 ± 4500 yr B.P. on mammoth tusk	Fortsch, 1972; Kurtén and Anderson, 1980	—
168.	Mussel Rock	San Mateo County, California	Alluvial	Natural accumulation	Rancholabrean from faunal assemblage	Miller and Peabody, 1941	—
169.	Mission San Jose	Alameda County, California	Alluvial	Natural accumulation	Rancholabrean? from faunal assemblage	Miller and DeMay, 1942	—
170.	Rodeo	Contra Costa County, California	Marine; alluvial	Natural accumulation	Rancholabrean? from faunal assemblage	Miller and DeMay, 1942	—

Table 16-5 (continued)

Fauna	Location	Depositional Environment	Agents of Accumulation	Age and Method of Dating	Reference(s)	Notes
171. Hawver Cave	Eldorado County, California	Cave	Predator	Wisconsin from faunal assemblage	Kurtén and Anderson, 1980; Stock, 1918; Miller and DeMay, 1942	—
172. Samwel Cave	Shasta County, California	Cave	Den, predators	Late Wisconsin from faunal assemblage	Furlong, 1906; Kurtén and Anderson, 1980; Miller and DeMay, 1942; Payen et al., 1978	—
173. Potter Creek Cave	Shasta County, California	Cave	Den	Wisconsin from faunal assemblage	Stock, 1918; Kurtén and Anderson, 1980; Miller and DeMay, 1942	—
174. Stone Man Cave	Shasta County, California	Cave	—	Rancholabrean from faunal assemblage	Furlong, 1941	—
175. Crannell Junction	Humboldt County, California	Marine	Natural accumulation	Late Rancholabrean from faunal assemblage and stratigraphy	Kohl, 1974	—
176. Fossil Lake	Lake County, Oregon	Lacustrine	Natural accumulation	Wisconsin from faunal assemblage	Jehl, 1967; Allison, 1966; Howard, 1946	—
177. Connley Caves	Lake County, Oregon	Cave	Humans	11,200-7200 yr B.P. from cultural assemblage	Grayson, 1977	—
178. Manis site	Clallam County, Washington	Glacial melt pond	Natural accumulation and humans(?)	12,000 ± 310 yr B.P. on seeds and wood; 11,850 ± 60 yr B.P. on microorganic matter	Gustafson et al., 1979	—

several species (Table 16-6) that would have been unable to survive the "boreal" conditions postulated for the area on the basis of the mammalian fossil fauna. The avifauna is clearly a mixture of northern species, including the spruce grouse, the gray jay, the rock ptarmigan (*Lagopus* cf. *mutus*), and the pine grosbeak (*Pinicola* cf. *enucleator*), and southern species, including the bobwhite, the wild turkey, the red-bellied woodpecker, and the brown thrasher (*Toxostoma rufum*).

The herpetofauna from the 11,000-year-old faunule (levels 6 and 7) of Baker Bluff Cave in Tennessee is totally modern; and, in contrast to the Clark's Cave local fauna, its avifauna contains no boreal species, possibly because of its lower latitude (Guilday et al., 1978). The mammalian taxa include such boreal forms as *Microtus xanthognathus* and *Phenacomys intermedius*.

In most Tennessee local faunas (Table 16-6), the herpetofaunas are similar to those of the area today. At Creek Bend Cave, the presence of wood turtle (*Clemmys insculpta*) suggests a cool-temperate environment (Parmalee and Klippel, 1981), as also is indicated by the late-Pleistocene mammalian fauna of the region (Guilday et al., 1969).

The distribution of the late-Pleistocene avifauna of the northeastern region differs somewhat from that of the modern avifauna, but, except for the historically extinct passenger pigeon (+*Ectopistes migratorius*) in Baker Bluff Cave (Guilday et al., 1978), the only known extinct fossil bird is +*Uria affinis*, a murre of uncertain af-

finities found in marine deposits in Maine (Marsh, 1872). The spectacular extinct scavenging and raptorial birds of the South and West (storks, teratorns, condors, and eagles) are conspicuously absent from the Northeast where all sites with a diverse avifauna occur in caves. This absence may be a sampling bias; however, most of the recorded birds are characteristic of forests and woodlands or their edges, and they probably represent prey of carnivorous mammals and birds that lived in the caves.

The late-Pleistocene mammalian faunas of the northeastern United States are characterized by a number of extinct large species (Table 16-6). Although some, such as +*Cervalces scotti*, +*Castoroides ohioensis*, +*Symbos cavifrons*, and +*Mammut americanum*, were not confined to this region, they were widespread in it and generally reflect forested habitats. Other extinct taxa, such as +*Mammuthus* sp., +*Platygonus compressus*, *Bison bison* *antiquus*, and *Equus*, generally indicate more-open environments. This suggests that the late-Pleistocene northeastern forests may have been more open than those of today. These Pleistocene faunas also contain extant species now found much farther north. The presence of *Dicrostonyx hudsonius* in the New Paris No. 4 site (Guilday et al., 1964) and in New Trout Cave, West Virginia, indicates the presence of tundra, at least locally, along the Appalachian ridge and high plateau areas as far south as latitude 38°50′N (F. Grady, personal communication,

1981). The paucity of *Dicrostonyx* records in the eastern United States supports Wright's (1981) contention that tundra was not extensive in this area during the Pleistocene.

The distribution of boreal and northern species in late-Pleistocene local faunas suggests a biologic zonation in the eastern half of the United States. *Dicrostonyx* has been found only in a narrow zone close to the ice margin (Guilday, 1971). This animal, along with *Ovibos*, characterizes one of the faunal provinces of Martin and Neuner (1978), as discussed earlier. Other boreal microtines such as *Microtus xanthognathus*, *Synaptomys borealis*, and *Phenacomys* sp. extended as far south as northwestern Arkansas (Hallberg et al., 1974) and eastern Tennessee (Guilday et al., 1971). *Synaptomys cooperi* and *Microtus pennsylvanicus* existed still farther south, in Louisiana (Domning, 1969) and Florida (Martin, 1968; Martin and Webb, 1974). *Odobenus rosmarus* (Hay, 1923; S. D. Webb, unpublished data) has been found on the Carolina Coastal Plain and on the continental shelf. The southern limit of this taxon in Pleistocene deposits coincides closely with the southern limit of *Rangifer*.

Notably absent from the entire eastern United States was the large late-Pleistocene camel +*Camelops*. To date, it is not known east of Oklahoma (Kurtén and Anderson, 1980; Webb, 1974a). The antilocaprids also are absent from the late-Pleistocene faunas of the East, although they are known from the late Tertiary and early Pleistocene of Florida (Kurtén and Anderson, 1980; Webb, 1974a).

SOUTHEASTERN REGION

Except for the Ladds local fauna in the Piedmont of northwestern Georgia, all of the known late-Pleistocene herpetofaunal and avifaunal sites in this region are from the southeastern Coastal Plain or the continental shelf (Figure 16-1). The Ladds site contains the southernmost Pleistocene record of the northern wood turtle (*Clemmys insculpta*) and the spruce grouse (*Dendragapus canadensis*). Southern species such as the toad (*Bufo terrestris*) and the giant extinct tortoise (+*Geochelone crassiscutata*) are also present at this site.

A few dozen isolated specimens of large mammals have been dredged from the continental shelves of the Atlantic Ocean and the Gulf of Mexico. The vast majority of the specimens are proboscidean teeth, most often +*Mammut americanum* (Whitmore et al., 1967). +*Mammuthus primigenius* is frequently found in sites north of Chesapeake Bay; +*Mammuthus jeffersoni* is frequently found south of Chesapeake Bay (S. D. Webb, unpublished data). +*Symbos cavifrons*, +*Cervalces scotti* (Whitmore et al., 1967), +*Megalonyx* (S. D. Webb, unpublished data), and *Rangifer* sp. (R. E. Eshelman, personal communication, 1981) also have been reported from sites in the area of Chesapeake Bay (Whitmore et al., 1967). The lack of signs of abrasion on any of these specimens and the large size of many of them (mammoth posterior molars can weight as much as 20 kg) makes reworking unlikely; thus, these sites provide direct evidence that terrestrial environments existed on the shelf during some part of the late Pleistocene. Thus far no radiocarbon dates have been obtained on any of these vertebrate fossils; however, the chronology of Pleistocene sea-level changes limits the possible period of habitation by land animals to the interval between 18,000 and 8000 yr B.P. (Blackwelder et al., 1979). This finding agrees well with radiocarbon dates of 21,000 to 8100 yr B.P. obtained from peat samples dredged from the sediments of this part of the continental shelf (Emery, 1967). A number of late-Pleistocene local faunas in Florida indicate a sea

level lower than at present, which apparently resulted in lower water tables in many parts of Florida. For example, the Crystal River Power Plant site and the Little Salt Spring site, which have produced freshwater mollusks and +*Geochelone*, +*Eremotherium*, +*Holmesina*, +*Megalonyx*, and +*Cuvieronius* are now approximately 20 m below sea level, but they were almost certainly well above sea level at the time the animal remains accumulated.

An important aspect of the very-late-Pleistocene sites in Florida is the widespread occurrence of two extinct tortoises, the giant +*Geochelone crassiscutata* and the small +*Geochelone incisa* (Table 16-6). According to Hibbard (1960), Auffenberg and Milstead (1965), and Holman (1976, 1980), the presence of the large tortoises suggests that there was little or no frost and that there were long periods of warm winter temperatures that allowed feeding activities. These large tortoises evidently survived longer in peninsular Florida than anywhere else in the United States, but the date of the last appearances of these forms is controversial. The Devil's Den site in northern Florida, which contains both +*Geochelone crassiscutata* and +*Geochelone incisa*, has a radiocarbon date on bone of 8000 to 7000 yr B.P. (Kurtén and Anderson, 1980). However, bone dates are often too young, and T. R. Van Devender (personal communication, 1981) believes that the Devil's Den site might be 12,000 to 10,000 years old. Moreover, there are no large specimens of +*Geochelone crassiscutata* from the Devil's Den site, and J. A. Holman (this report) is not completely convinced that the small specimens are juveniles. The youngest reliably dated specimen of +*Geochelone crassiscutata* is from the Ledge faunule of the Little Salt Spring fauna. A wooden spear that had penetrated a cooked tortoise has a radiocarbon date of 12,000 yr B.P. (Clausen et al., 1979).

Northern and southern species of amphibians and reptiles occur in three very-late-Pleistocene local faunas of the southeastern Coastal Plain. In the Catalpa Creek site in northern Mississippi, the northern Blanding's turtle (*Emydoidea blandingii*) occurs with +*Geochelone crassiscutata* (Jackson and Kaye, 1974a, 1975). In the Devil's Den site in northern Florida, the northern toad (*Bufo woodhousei*) occurs with both species of extinct tortoises. At the Vero site in southern Florida, the northern salamander (genus *Ambystoma*) occurs with +*Geochelone crassiscutata* (Weigel, 1962). These occurrences appear to be explained most parsimoniously by hypothesizing an equable climate with warmer winters and cooler summers than those of today.

Birds are well represented in many late-Pleistocene sites in Florida (Table 16-6), but, except for the wild turkey (Steadman, 1980), they have not been fully documented. Some of the richest avifaunas, especially Arredondo and Reddick, are considered to be older than 25,000 yr B.P. and so are excluded from this account. Aside from extinct species, all of the Late Wisconsin taxa still live in Florida except for the nearly extinct "California" condor (*Gymnogyps californianus*) and the now geographically very restricted trumpeter swan (*Olor buccinator*). Thus, the avifaunas imply little environmental change; however, they have not been completely studied. Compared to the Pleistocene avifaunas of the West, particularly the tar-deposit faunas of California, very few extinct species of late-Pleistocene birds occur in Florida. Only three, all probably scavengers, are definitely extinct: the large stork *Ciconia* *maltha and the teratorn +*Teratornis merriami*, both of which formerly ranged from Florida to California, and the caracara *Milvago* *readei, which is not known outside of Florida. Two other possibly extinct species, the grebe *Podilymbus*

Table 16-6.
Taxonomic List Showing the Distribution of Species by Locality

Species	Locality Number(s)	Species	Locality Number(s)
Amphibia		**Amphibia**	
Caudata		Anura (continued)	
Cryptobranchus alleganiensis	18a, 18b, 18c	*Rana pipiens* group	11, 16; 33, 47, 56; 64, 78b, 79a, 80, 86; 117, 145, 149
Cryptobranchus sp.	86	*Rana pipiens* group	11, 16, 18a, 18c
Siren lacertina	56	*Rana* sp.	10, 15, 16, 18a, 18c; 33, 41; 64, 70, 90, 91a; 110, 115; 157
Necturus maculosus	18a, 18b		
Notophthalmus viridescens	10, 11, 15	**Reptilia**	
Amphiuma means	56, 57		
Ambystoma maculatum	18a	Testudines	
Ambystoma opacum	18a, 18b; 90	*Kinosternon bauri*	56
Ambystoma texanum	80	*Kinosternon flavescens*	78b
Ambystoma tigrinum	33, 47; 69, 77; 142a, 142b, 149	*Kinosternon subrubum*	47, 56
Ambystoma sp.	11, 15, 16, 18a, 18b, 18c, 20; 56	*Kinosternon* sp.	24; 31
Desmognathus sp.	11, 15, 18a, 18b	*Sternotherus minor*	56
Desmognathus sp. (or *Leurognathus* sp.)	33	*Sternotherus odoratus*	56
Eurycea sp.	11	*Chelydra serpentina*	15; 31, 47, 56; 80, 91a
Gyrinophilus sp.	33	*Macroclemmys temmincki*	39
Plethodon glutinosus	10, 11; 33; 70	*Clemmys insculpta*	10, 15; 33
Pseudotriton ruber	33	*Clemmys marmorata*	158, 164, 166, 173
Anura		*Clemmys* sp.	157
Hylactophryne augusti	64, 69	*Dierochelys reticularis*	47, 56
Scaphiopus bombifrons	149	*Emydoidea blandingi*	61; 91a
Scaphiopus couchi	69; 142a, 142b, 145	+*Geochelone crassiscutata*	33, 39, 41, 47, 56, 59, 61
Scaphiopus hammondi	115, 142a, 142b, 149	+*Geochelone incisa*	39, 41, 47
Scaphiopus holbrooki	15; 47, 56	+*Geochelone wilsoni*	63, 64, 78a, 79a, 81; 145, 146, 149
Scaphiopus sp.	64; 142a, 142b	+*Geochelone* sp.	31
Leptodactylid	70	*Gopherus agassizii*	117, 120b, 120c, 120e, 120f, 126, 127a, 135, 144, 145, 146, 149; 154, 166
Syrrhophus marnocki	69		
Bufo americanus	10, 11, 16, 20, 33		
Bufo boreas	158b, 158, 173	*Gopherus polyphemus*	41, 47, 56
Bufo cognatus	64	*Gopherus* sp.	31
Bufo punctatus	117, 119b, 119c, 128b, 129a, 149	*Pseudemys concinna*	33
Bufo terrestris	33, 47	*Pseudemys floridana*	31, 59
Bufo speciosus or *cognatus*	80	*Pseudemys nelsoni*	59
Bufo woodhousei *bexarensis*	64, 69	*Pseudemys scripta*	47, 56; 78a, 78b
Bufo woodhousei fowleri	47, 86	*Pseudemys scripta petrolei*	31
Bufo woodhousei woodhousei	69, 149	*Pseudemys* sp.	56; 80
Bufo sp.	15, 18a, 18b, 24; 41, 56; 86, 90, 91a; 115, 142a, 142b	*Terrapene carolina*	15
		Terrapene carolina carolina	33, 41
Hyla arenicolor	142a	*Terrapene carolina carolina* near *bauri*	39, 47, 54, 57
Hyla crucifer	11, 16; 33	*Terrapene carolina putnami*	31, 39, 54, 56, 57, 59; 63, 78a, 79a, 81
Hyla femoralis	20		
Hyla versicolor or *chrysoscelis*	69, 80	*Terrapene carolina putnami* near *triunguis*	60
Hyla sp.	20; 158	*Terrapene carolina triunguis*	64, 70
Pseudacris streckeri	80	*Terrapene ornata*	78a, 78b, 91a; 140
Pseudacris triseriata	149	*Terrapene* sp.	80
Rana aurora	158	*Trionyx* sp.	31; 78a
Rana catesbeiana	15, 16; 47, 56; 77, 78b, 90		
Rana clamitans	16		
Rana palustris	15, 16		

Table 16-6 (continued)

Species	Locality Number(s)	Species	Locality Number(s)
Reptilia (continued)		Reptilia (continued)	
Squamata		Squamata (continued)	
Rhineura floridana	41	*Carphophis* sp.	10, 20, 24
Anolis carolinensis	33, 39, 56	*Chionactis occipitalis*	129b, 130, 137b
Cophosaurus texanus	70; 142a, 142b	*Coluber constrictor*	11, 15; 33, 39, 41, 54, 57; 70; 115, 146; 158, 164
Crotaphytus collaris	64, 70, 80, 117, 119b, 119, 137a, 138, 142a, 142b, 145, 147, 148, 149	*Coluber* sp.	24, 89, 158a
Crotaphytus wislizeni	129a; 164, 166	*Coluber* sp. or *Masticophis* sp.	16, 18a, 18b; 47, 56; 70, 77, 80, 86
Crotaphytus sp.	115, 126	*Diadophis punctatus*	11; 33; 70, 90; 142a, 142b
Holbrookia maculata	142a, 142b	*Diadophis* sp.	10, 15
Phrynosoma cornutum	70; 142a, 142b, 145, 147, 148, 149	*Drymarchon corais*	39, 47, 54, 56, 57
Phrynosoma coronatum	158	*Elaphe guttata*	15; 39, 41, 47; 70; 149
Phrynosoma douglassi	142a, 142b, 147, 148, 149	*Elaphe obsoleta*	11, 20; 39, 47, 57
Phrynosoma modestum	142a, 142b	*Elaphe subocularis*	145, 146
Phrynosoma platyrhinos	115, 117	*Elaphe vulpina*	86
Sauromalus obesus	117, 119d, 120f, 130, 137a; 154	*Elaphe* sp.	18a; 31; 70, 86, 89
Sceloporus graciosus	115; 171	*Farancia abacura*	56, 57
Sceloporus jarrovii	142a, 142b	*Farancia* sp.	39, 47
Sceloporus magister	119b, 119c, 120a, 120b, 120e, 120f, 132a, 133, 134, 135, 142a, 142b	*Ficimia cana*	142a, 142b
Sceloporus occidentalis	126; 158a, 164, 166	*Heterodon nasicus*	70, 79a, 80; 142b
Sceloporus poinsetti	70	*Heterodon platyrhinos*	18a, 18b; 33, 39, 41, 47; 70
Sceloporus undulatus	15, 16; 33, 56; 70, 86, 119b, 119c, 121c, 123, 134, 142a, 142b, 149	*Heterodon simus*	47
Sceloporus variabilis	70	*Heterodon* sp.	10
Sceloporus sp.	15, 16; 56; 86, 119b, 122, 137a, 137b, 148	*Hypsiglena torquata*	115, 121b, 127b, 128c, 132b, 132c, 137b, 142a, 142b
Uta stansburianai	115, 119b, 119c, 133, 135; 158a	*Lampropeltis calligaster*	70
Coeleonyx variegatus	119b, 119c	*Lampropeltis getulus*	18b, 18c; 33, 39, 47, 54, 56, 57; 70, 80; 115, 117, 120d, 128a, 129a, 142a, 142b, 145; 158, 166
Heloderma suspectum	117		
Cnemidophorus sexlineatus	41; 86		
Cnemidophorus tigris	117, 119a, 126, 129a, 137a, 142a, 142b; 158, 171	*Lampropeltis pyromelana*	117, 142b
Cnemidophorus sp.	70; 119c, 129b, 137a, 142a, 142b, 147, 148	*Lampropeltis triangulum*	10, 15, 18a, 18b, 20, 24; 33, 41, 47, 57; 63, 79a, 80, 90
Eumeces fasciatus	18a; 90	*Lampropeltis* sp.	15; 70
Eumeces gilberti	145; 166	*Masticophis flagellum*	15; 41; 115, 145
Eumeces laticeps	16	*Masticophis taeniatus*	117
Eumeces obsoletus	70; 142b	*Nerodia cyclopion*	39
Eumeces skiltonianus	158	*Nerodia fasciata*	39, 56
Eumeces tetragrammus	70	*Nerodia fasciata* or *erythrogaster*	47, 70
Eumeces sp.	20; 56	*Nerodia erythrogaster* or *sipedon*	33; 80
Lichanura trivirgata	128a	*Nerodia sipedon*	11; 33; 86
Leptotyphlops humilis	137a	*Nerodia* sp.	15, 24; 54, 56, 57
Leptotyphlops sp.	142a, 142b	*Nerodia taxispilota*	39, 56
Gerrhonotus multicarinatus	158, 164, 171	*Opheodrys* sp.	15; 70
Ophisaurus compressus	56	*Pyllorhynchus decurtatus*	132b
Xantusia vigilis	158	*Pituophis melanoleucus*	47, 56, 57; 64, 70; 115, 117, 120c, 137b, 142a, 142b, 145, 146; 158b, 158, 166
Arizona elegans	70, 80; 132b, 142a, 142b		
Carphophis amoenus	11, 15; 33; 90	*Pituophis* sp.	24
		Regina alleni	39
		Rhinocheilus lecontei	120e, 129a, 132b, 137a, 138, 142a, 142b

Table 16-6 (continued)

Species	Locality Number(s)
Reptilia (continued)	
Squamata (continued)	
Salvadora sp.	142b, 149
Sonora semiannulata	119b, 132b, 142a, 142b
Stilosoma extenuatum	41
Storeria dekayi	33
Storeria sp.	10, 11, 15; 90
Tantilla sp.	70; 142a, 142b
Thamnophis cyrtopsis	142b
Thamnophis proximus	70; 79a, 80; 149
Thamnophis proximus or *T. sauritus*	90
Thamnophis radix or *marcianus*	80
Thamnophis sauritus	18b
Thamnophis sirtalis	10, 11, 15, 18a; 33, 39, 47; 70, 79a
Thamnophis sp.	10, 18b, 18c, 20; 56, 57; 70, 77, 90; 164
Trimorphodon biscutatus	132b, 133, 142a
Virginia sp.	15
Micrurus fulvius	41, 47; 70
Agkistrodon contortrix	11; 70, 80
Agkistrodon piscivorus	47, 54, 57; 70
Agkistrodon sp.	24; 70
Crotalus adamanteus	15; 39, 47, 54, 56, 57, 59
Crotalus atrox	64; 117, 129a, 145, 146, 149
Crotalus horridus	11, 15, 18a, 18b, 18c; 33; 86
Crotalus mitchelli	117
*Crotalus *potterensis*	173
Crotalus viridis	110, 115, 117, 126; 158a, 166, 171
Crotalus sp.	24; 70, 80, 86, 89; 119c, 142b; 158a, 158, 166, 171
Sistrurus miliarius	41
Crocodilia	
Alligator mississippiensis	31, 47, 56; 89
cf. *Gavialosuchus*	31
Aves	
Gaviiformes	
Gavia immer	36, 52; 175
Procellariiformes	
Puffinus puffinus	54
Pelecaniformes	
*Phalacrocorax *macropus*	176
Phalacrocorax auritus	39, 52, 54, 56, 57, 58; 113, 114, 150; 155, 176
Phalacrocorax sp.	158, 175
Anhinga anhinga	39, 52, 57
Pelecanus erythrorhynchos	84; 114; 155, 176
Pelecanus occidentalis	164

Species	Locality Number(s)
Aves (continued)	
Ciconiiformes	
*Ciconia *maltha*	39, 54, 57; 155, 158, 164, 166
Mycteria americana	39, 59
*Mycteria *wetmorei*	158?
Ciconiid	176
+ *Teratornis merriami*	57; 118, 124; 158, 164, 166
+ *Teratornis incredibilis*	115
+ *Cathartornis gracilis*	158
Gymnogyps californianus	39, 57, 58; 115?, 117, 121, 124, 142, 146, 148, 150, 151; 154, 158, 172, 173?, 174
Gymnogyps sp.	149g?
Coragyps atratus	39, 50, 57; 115, 142, 146, 148, 149a, 149b, 149d, 149g, 149h, 149i, 149j, 150, 151; 158, 164, 166, 171?, 173
+ *Breagyps clarki*	115, 145, 149h, 149j; 158
Cathartes aura	39, 50, 54, 56, 57; 120?, 121, 142, 145, 146, 148, 149b, 151; 158, 171, 172, 173
Cathartes sp.	149j
Gruiformes	
Grus canadensis	2, 30; 39, 54, 57; 148; 158, 166
Grus americana	15; 37, 39, 54, 57; 82; 158
Grus sp.	155?
Aramus garauna	53, 57
Laterallus exilis	78
Rallus longirostris	57
Rallus elegans	39, 42, 57; 91b
Rallus limicola	16; 39; 74, 78; 166, 176
Rallus sp.	115
Porzana carolina	16; 39; 78, 91a, 91b; 145
Coturnicops noveboracensis	91b
Porphyrio martinica	39
Gallinula chloropus	16; 39, 42, 57; 78
Fulica americana	39, 57; 78, 91b; 115, 118, 145, 149i; 154, 158, 176
Rallid	16
Gruiformes (?)	
Podiceps auritus	21?, 22?, 23?; 39, 57
Podiceps caspicus	82; 115, 150; 176
Podiceps grisegena	175
*Podiceps *parvus*	176
Podiceps sp.	149g, 158
Aechmophorus occidentalis	148; 155, 170, 176
Podilymbus podiceps	16; 39, 42, 56, 57; 91b, 115; 158, 166, 176
*Podilymbus *wetmorei*	39
Podicipedid	18

Table 16-6 (continued)

Species	Locality Number(s)	Species	Locality Number(s)
Aves (continued)		Aves (continued)	

Gruiformes (?) (continued)

Species	Locality Number(s)
Eudocimus albus	39, 57
Plegadis chihi	158
Plegadis sp.	57
Ajaia ajaja	158?
Ardea herodias	39, 54, 57; 110; 158, 166, 176
Ardea alba	39, 54, 57; 158, 166
Egretta caerulea	39, 57; 158, 166
Egretta tricolor	57
Ardeola virescens	39, 57; 158, 166
Nyctanassa violacea	39, 54
Nycticorax nycticorax	39; 158, 166
Ixobrychus exilis	91b
Botaurus lentiginosus	16; 39, 56, 57, 58; 158, 176
Ardeid	166

Charadriiformes

Species	Locality Number(s)
Burhinus sp.	150
Pluvialis squatarola	158
Pluvialis dominica	16; 91b?
Charadrius vociferus	15; 158, 166
Charadrius montanus	78; 150, 166
Arenaria interpres	18
Limosa fedoa	158
Limosa sp.	16
Tringa melanoleuca	158, 166, 176
Tringa solitaria	16
Bartramia longicauda	15; 82
Numenius phaeopus	158
Numenius americanus	142; 158, 166, 176?
Numenius sp.	149g
Actitis macularia	16, 18a
Catoptrophorus semipalmatus	15; 115; 158
Erolia minutilla	15
Erolia melanotos	176
Erolia alba	158?
Erolia alpina	158, 166
Erolia sp.	82
Limnodromus griseus	158, 166, 176?
Eroliine	82
Capella gallinago	16, 18a; 115; 158
Philohela minor	15, 16, 18a; 46; 91a
Scolopacid	16, 18a, 18b; 82
Phalaropus fulicarius	124; 158?
Phalaropus lobatus	176
Phalaropodid	115; 155
Himantopus mexicanus	115?; 176
Recurvirostra americana	115; 124; 154; 158, 166, 176
*Phoenicopterus *copei*	155?, 176

Charadriiformes (continued)

Species	Locality Number(s)
*Phoenicopterus *minutus*	155
Larus californicus	176
*Larus *robustus*	176
*Larus *oregonus*	176
Larus philadelphia	176
Larus sp.	18a; 145; 158
Rissa tridactyla	158?
Chlidonias nigra	176
Sterna forsteri	176
*Stercorarius *shufeldti*	176
*Uria *affinis*	1
Uria aalge	168
Synthliboramphus antiquum	175
+ *Mancalla diegense*	175
Charadriiform	82; 166

Anseriformes

Species	Locality Number(s)
*Cygnus *paloregonus*	176
Olor buccinator	29; 39; 176
Olor columbianus	39, 57; 158, 166
Olor sp.	8; 91a
Anser hyperborea/A. caerulescens	78; 110; 158, 166, 176
Anser rossii	158?, 176
Anser albifrons	145?; 158, 176
Anser sp.	84; 164?, 176
*Branta *dickeyi*	166
Branta canadensis	36, 39, 54, 57; 78, 91a; 99, 114, 115, 146; 155, 158, 163, 166?, 169, 173, 176
Branta bernicula	158, 176
*Branta *propinqua*	176
Anserine	91a; 176
+ *Anabernicula gracilenta*	115, 142, 145; 158, 166
+ *Anabernicula oregonensis*	150; 176
Aix sponsa	39; 84, 91b; 150
Anas platyrhynchos	39, 42, 52; 78, 91b; 99, 107, 110, 115, 124, 150; 154, 158, 164, 166, 176
Anas rubripes	33, 39, 42, 52
Anas rubripes or *A. platyrhynchos*	16
Anas fulvigula	57
Anas acuta	39; 82; 115, 142?, 145, 150; 166?, 176
Anas acuta or *A. strepera*	78
Anas strepera	39; 158, 166
Anas americana	39; 91b; 115, 142?; 154?, 166, 176
Anas clypeata	39; 78, 82; 115; 158, 166, 176
Anas discors/A. cyanoptera	15, 16; 39; 74, 78, 91b; 115?, 124; 166, 176

Table 16-6 (continued)

Species	Locality Number(s)	Species	Locality Number(s)
Aves (continued)		Aves (continued)	
Anseriformes (continued)		Accipitriformes (continued)	
Anas *itchtucknee	39	Spizaetus *pliogryps	176
Anas crecca	16, 18a, 39, 52, 53, 57; 78; 115, 124, 142?, 145, 150?; 154, 158, 163, 166, 175, 176	Spizaetus *willetti	115, 142
		Spizaetus sp.	57; 171
		Buteogallus *fragilis	145; 158, 164, 166
Anas sp.	18a; 57; 78, 82, 84, 91a; 158, 163, 166	Buteogallus *milleri	171
		Buteogallus sp.	149g
Aythya valisineria	39; 150; 155?, 158?	+Amplibuteo woodwardi	158
Aythya americana	39, 154, 166, 176	+Wetmoregyps daggetti	158, 164
Aythya collaris	39, 53; 176?	Haliaeetus leucocephalus	39, 50, 54, 57; 99, 146; 158, 164, 166, 176
Aythya marila	115		
Aythya affinis	39, 42, 50, 52, 57; 84; 118; 166, 176	Elanus leucurus	158
		+Neophrontops americanus	150; 158, 164, 166
Aythya sp.	57; 82; 142?; 176	+Neogyps errans	98, 115; 158, 164, 166
Aythyine	78, 82; 115	Circus cyaneus	78; 115; 158, 164?, 166, 176
Bucephala clangula	39; 175	Accipitrid	16, 18a; 39; 91b; 176
Bucephala albeola	15; 39, 57; 166, 176	Incertae Sedis (Fam. Pandionidae)	
Bucephala sp. or Histrionicus sp.	175	Pandion haliaetus	39, 57
Clangula hyemalis	176	Incertae Sedis (Fam. Falconidae)	
+Chendytes sp.	175?	Falco *swarthi	150; 166
Melanitta deglandi	176	Falco rusticolus	99
Melanitta perspicillata	115; 175, 176	Falco mexicanus	107, 115, 142, 146, 148, 150, 151; 158, 166
Lophodytes cucullatus	16, 18a, 18c; 39, 42, 54, 56, 57		
Mergus merganser	29; 39; 110; 154, 176	Falco *oregonus	176
Mergus serrator	29; 39; 176	Falco peregrinus	110, 145, 149i; 158, 166, 173, 176
Mergus sp.	16, 18a; 150?	Falco femoralis	120?
Oxyura jamaicensis	15; 39; 154, 155, 166, 176	Falco columbarius	91b; 107; 158, 166
Anatid	16, 21?, 22?, 23?; 82, 91a, 91b; 99, 118; 166, 172, 176	Falco sparverius	16, 18a; 50; 64, 74; 107, 115, 124, 145, 146, 148, 149g, 150; 158, 164, 166, 172
Accipitriformes			
Accipiter gentilis	18b; 158, 164	Falco sp.	149c; 166
Accipiter cooperii	148, 149g, 150, 151; 158, 164, 166	Milvago *readei	39
Accipiter striatus	15, 16; 145; 158, 164, 172	Polyborus plancus	54, 57; 145, 146, 149b, 149c, 149d, 149g, 150; 158, 163, 164, 166
Buteo jamaicensis	15; 39, 56, 57; 107, 115, 120?, 142?, 145, 149f, 149g, 150; 154, 158, 166, 173	Galliformes	
		Colinus virginianus	15, 16, 21?, 22?, 23?; 39, 46, 50, 52, 54; 64, 91a, 91b
Buteo lineatus	15; 39, 56, 57; 164		
Buteo platypterus	15, 16; 57	Colinus squamata	145
Buteo swainsoni	115, 145, 146, 148, 149c, 149g, 151; 158, 166, 172?	Colinus californica	158, 164, 166, 171
		Colinus pictus	142?, 145, 148, 151; 171, 172, 173
Buteo albonotatus	145?	Colinus sp.	145, 146; 157
Buteo lagopus	115, 150; 158	Odontophorine	74; 99, 150
Buteo regalis	158, 166, 171	Meleagris gallopavo	2, 4, 8, 10, 11, 15, 16, 18a, 21?, 22?, 23?; 28, 30; 33, 37, 38, 39, 43, 44, 49, 50, 54, 56, 57; 91a, 91b; 146, 148?, 151
Buteo lagopus or B. regalis	99		
Buteo nitidus	150		
Buteo sp.	91b; 158, 164, 166		
Aquila chrysaetos	64; 99, 107, 115, 142, 145, 146, 149g, 150; 154, 155, 158, 164, 166, 176	Meleagris *californica	157, 158, 161, 164
		Meleagris *crassipes	139, 141, 142, 145, 148, 150
		Meleagris sp.	35, 45, 58; 78; 153; 156, 159, 169, 173
Spizaetus *grinnelli	158, 164		

Table 16-6 (continued)

Species	Locality Number(s)	Species	Locality Number(s)
Aves (continued)		Aves (continued)	

Galliformes (continued)

Dendragapus obscurus	107; 172, 173, 174, 176
Dendragapus canadensis	15, 16; 33
Bonasa umbellus	10, 11, 14, 15, 16, 18a, 20, 21?, 22?, 23?; 91a; 107, 110; 173
Lagopus mutus	16
Tympanuchus phasianellus	11, 15, 16; 115?; 176
Tympanuchus cupido	91a, 91b
Tympanuchus pallidicinctus	148, 150?, 151
Tympanuchus sp.	21?, 22?, 23?; 74, 78; 152
Tympanuchus sp. or *Centrocercus* sp.	78
Centrocercus urophasianus	107, 110, 115, 124, 142, 145, 146; 176, 177
Tetraonine	16, 18a, 18b, 18c, 20, 21?, 22?, 23?; 99, 107, 150

Columbiformes

Columba fasciata	158, 164, 174
+*Ectopistes migratorius*	14, 15, 16, 18a, 18b, 18c, 20, 21?, 22?, 23?; 33, 39; 91a, 91b; 150; 158
Zenaida macroura	39, 57; 74, 91a; 115, 142, 145, 150; 154, 158, 164, 166

Psittaciformes

Psittacidae	150

Cuculiformes

Coccyzus sp.	16; 91a, 91b
Geococcyx californianus	145; 158, 164, 166
Geococcyx *conklingi	145, 146, 147, 150

Strigiformes

Tyto alba	39, 50; 120?, 142, 145, 146, 149d, 150; 158, 164
Otus asio	16, 18a, 20, 21?, 22?, 23?; 39, 50; 74, 91a, 91b; 107, 145; 158, 164, 173
Otus flammeolus	172
Otus sp.	149c, 149i
Bubo virginianus	16, 18a; 91b; 107, 115, 124, 142, 145, 148, 149f, 149g, 149i, 150, 151; 154, 158, 164, 166, 172, 176
Bubo *sinclairi	172, 173
Glaucidium gnoma	158, 164, 172
Athene cunicularia	145, 146, 149b, 150; 158, 166
Strix occidentalis	115?
Strix varia	39, 46, 50, 54, 56, 57; 91a, 91b
Strix *brea	149i; 158
Asio otus	74; 142, 146?, 149c; 158, 164, 166, 172
Asio flammeus	74, 84, 91a; 107, 115, 148? 149a, 149g, 150?, 151; 158
Asio *priscus	163

Strigiformes (continued)

Asio sp.	14, 16
Aegolius funereus	145
Aegolius acadicus	16; 91b; 115, 145; 158
Strigid	91b

Caprimulgiformes

Chordeiles minor	16, 18a; 74, 91a, 91b; 115
Phalaenoptilus nuttallii	150; 158
Aeronautes saxatalis	115, 145
Chaetura pelagica	16, 18c

Alcediniformes

Ceryle alcyon	15, 16, 18a; 74

Piciformes

Colaptes auratus	11, 15, 16, 18a, 18b, 39, 46; 74, 78, 91a, 91b; 107, 145, 146, 148, 151; 154, 158, 164, 166, 171, 172, 173, 176
Dryocopus pileatus	11, 16, 21?, 22?, 23?; 158
Asyndesmus lewis	158
Melanerpes erythrocephalus	15; 91a?
Melanerpes formicivorus	145
Melanerpes carolinus	15, 16, 56; 91a?, 91b
Sphyrapicus varius	16
Sphyrapicus sp.	158
Dendrocopos villosus	16
Dendrocopos pubescens	15, 16
Dendrocopos sp.	158, 164
Picid	16, 18a, 18b; 91a, 91b; 149b, 149g

Passeriformes

Myiarchus crinitus	91b
Sayornis phoebe	15; 91b
Sayornis saya	145
Sayornis sp.	164
Tyrannus sp.	158
Contopus virens	15
Empidonax sp.	16, 18a; 164
Tyrannid	164
Eremophila alpestris	16; 78, 93; 124, 145, 146; 158, 166
Progne subis	14
Hirundo pyrrhonota	15, 16, 18b; 166
Hirundinid	107
Perisoreus canadensis	15, 16
Gymnorhinus cyanocephalus	145, 146
Aphelocoma coerulescens	158, 164, 166
Cyanocitta cristata	15, 16, 18a, 18b; 91a, 91b
Cyanocitta stelleri	158, 164, 171, 172
+*Protocitta dixi*	74
Pica pica	15, 18b; 94; 99, 115, 145; 158, 164

Table 16-6 (continued)

Species	Locality Number(s)	Species	Locality Number(s)
Aves (continued)		Aves (continued)	
Passeriformes (continued)		Passeriformes (continued)	
Pica nuttalli	158, 164	*Pipilo erythrophthalmus*	91a; 145, 146; 158, 164
Pica sp.	166	*Pipilo fuscus*	145; 158, 164
Nucifraga columbiana	107, 115?; 158	*Pipilo *angelensis*	158
Garruline	20; 115	*Calamospiza melanocorys*	82; 145
Corvus ossifragus	57	*Pooecetes gramineus*	16; 78; 158
Corvus caurinus	158, 164	*Chondestes grammacus*	158
Corvus brachyrhynchos	16; 39, 57; 91a, 91b; 115; 158, 173	*Amphispiza bilineata*	158
Corvus corax	78; 107, 115, 124, 142, 145, 146, 149a, 149b, 149c, 149d, 149g, 149h, 149i?, 150; 154, 158, 164, 166, 171, 176	*Amphispiza belli*	158, 166
		Spizella passerina	158
		Spizella sp.	158
Corvus cryptoleucus	158, 166	*Junco hyemalis*	16; 124
*Corvus *neomexicanus*	149h, 149i, 149j	*Junco* sp.	15
Parus bicolor	16	*Zonotrichia leucophrys*	158
Parus sp.	16; 158, 164	*Zonotrichia albicollis*	15
Sitta canadensis	15, 16; 164	*Melospiza melodia*	15; 158
Sitta pygmaea	164	*Melospiza iliaca*	11, 15; 158, 164
Certhia familiaris	16	*Calcarius* sp.	82
Chamea fasciata	164	Emberizine	18b; 82; 115; 164
Salpinctes obsoletus	107, 124, 145	*Spiza americana*	91a
Catherpes mexicanus	145	*Dolichonyx oryzivorus*	16
Campylorhynchus sp.	166	*Sturnella magna*	46
Cistothorus platensis	16	*Sturnella neglecta*	74; 158, 164, 166
Cinclus mexicanus	124	*Sturnella* sp.	16; 91b
Toxostoma rufum	11, 15, 16	*Molothrus ater*	15; 158
Toxostoma redivivum	158	*Molothrus* sp.	82
Toxostoma bendirei	166	+*Pandanaris convexa*	158
Toxostoma sp.	145	+*Pyelorhamphus molothroides*	145
Oreoscoptes montanus	145; 166	*Euphagus *magnirostris*	158
Mimus polyglottos	78	*Euphagus cyanocephalus*	176
Sialia sialis	16	*Euphagus* sp.	91b?
Sialia mexicana	164	*Quiscalus quiscula*	39, 56, 57; 91a, 91b
Sialia sp.	145, 146; 158, 164	*Quiscalus major*	57
Catharus sp.	15, 16; 91b?; 164?	*Xanthocephalus xanthocephalus*	148, 150
Turdus migratorius	15, 16, 18a; 56; 74, 91b; 115, 124, 145, 146; 158, 164	*Xanthocephalus* or *Sturnella*	176
		Agelaius phoeniceus	15, 16, 18a; 39, 57; 78; 110
Turdus grayi	124	*Agelaius* sp.	82; 115, 146; 158
Regulus sp.	164?	*Agelaius* sp. or *Molothrus* sp.	149g
Anthus spinoletta	16	*Agelaius* sp. or *Euphagus* sp.	176
Bombycilla cedrorum	16; 158, 164	*Icterus spurius*	16
Lanius ludovicianus	145; 158, 166	*Icterus* sp.	158
Mniotilta varia	74	Icterine	82
Dendroica coronata	74	*Carduelis pinus*	158, 164
Seiurus sp.	16	*Carduelis tristis*	158
Vermivora sp.	56	*Carpodacus mexicanus*	145, 146; 166
Paruline	15, 16, 18b; 164	*Coccothraustes vespertina*	158
Piranga sp.	16; 91b	*Pinicola enucleator*	16
Pyrrhuloxia cardinalis	91b	*Loxia curvirostra*	124, 148; 164
Pheucticus melanocephalus	158	Fringillid	16; 78, 82, 91a, 91b,

Table 16-6 (continued)

Species	Locality Number(s)

Aves (continued)

Passeriformes (continued)

Species	Locality Number(s)
Passeriform	14, 18a, 18b, 18c, 20; 33; 82, 87, 91b; 150, 152

Mammalia

Marsupialia

Species	Locality Number(s)
Didelphis virginiana	24; 33, 34, 37, 39, 40a, 40b, 44, 46, 47, 48, 50, 54, 55, 56, 57; 72b, 72c, 80, 86, 88, 91a, 91b

Insectivora

Species	Locality Number(s)
Sorex arcticus	7, 11, 12, 15, 16, 18a, 20; 88, 97
Sorex cinereus	7, 10, 11, 15, 16, 18a, 18b, 18c, 20, 26; 33; 69a, 69b, 70, 77, 80, 82, 83, 86, 87, 88, 97; 99, 103, 152
Sorex dispar	11, 16, 18a, 20
Sorex fumeus	11, 12, 15, 16, 18a, 18b, 18c, 20; 33
Sorex merriami	149a, 149b, 149e, 149f, 149g
Sorex ornatus	158, 164
Sorex palustris	11, 15, 16, 20, 26; 80, 83, 88, 97; 99, 101, 103
Sorex trowbridgii	164
Sorex vagrans	69a, 97; 149a, 149f, 149g
Sorex sp.	88; 98, 101, 110
Microsorex hoyi	7, 11, 15, 16, 18a, 20, 26; 88; 101
Blarina brevicauda	7, 10, 11, 12, 13, 14, 15, 16, 18a, 18b, 18c, 20, 26; 33, 39, 40b, 40c, 47, 48, 54, 56; 64a, 64b, 70, 72b, 74; 80, 86, 87, 88, 89, 91a, 91b
Blarina sp.	69a, 69b
Cryptotis parva	7, 12, 15, 18a, 18c, 20; 41, 47, 48, 56; 64a, 64b, 70, 72b, 74, 80, 88, 89, 91a, 91b; 97; 142, 149a, 149g, 152
Cryptotis sp.	101
Notiosorex crawfordi	64a, 64b, 69a, 69b, 78b; 125, 142, 149a, 149e, 152; 158
Soricid	165
Parascalops breweri	7, 10, 11, 12, 13, 14, 15, 16, 18a, 18b, 18c, 20; 33; 88
Scalopus aquaticus	5, 15, 16, 18a, 18b, 18c, 26; 33, 39, 40b, 40c, 44, 47, 48, 50, 54, 56, 57; 64a, 64b, 69a, 69b, 70, 71a, 74, 76, 80, 86, 89, 91a, 91b, 97
Scapanus latimanus	158, 160, 162, 171, 173
Condylura cristata	7, 11, 15, 16, 18a; 88

Chiroptera

Species	Locality Number(s)
Desmodus *stocki	173
Myotis austroriparius	40b, 47, 56
Myotis evotis	69a, 69b; 101, 139
Myotis grisescens	13, 16; 33, 47; 87, 88, 91b
Myotis keenii	7, 9, 11, 15, 16, 20

Mammalia (continued)

Chiroptera (continued)

Species	Locality Number(s)
Myotis leibii	16
Myotis lucifugus	9, 11, 15, 16; 33; 86, 87, 88; 99
Myotis *rectidentis	72a
Myotis sodalis	86, 87, 88
Myotis subulatus	101
Myotis thysanodes	101, 139, 143
Myotis velifer	64a, 64b, 69a, 69b, 70, 72a, 73, 74; 139, 149e, 152
Myotis volans	101
Myotis sp.	5, 7, 10, 12, 13, 14, 18a, 18b, 18c, 20, 21, 26; 87, 88, 91a, 91b; 125, 149a, 149e, 149f, 149g, 152; 172
Pipistrellus subflavus	7, 11, 13, 15, 16, 18a, 18b, 18c, 20, 26; 33, 44, 47; 69b, 86, 87, 88, 91b
Pipistrellus sp.	64a, 64b, 69a, 74
Lasionycteris noctivagans	99, 101, 152
Eptesicus fuscus	3, 5, 7, 11, 13, 15, 16, 18a, 18b, 18c, 20, 21; 33, 69a, 74, 87, 88, 91b; 99, 101, 149e, 149f, 152
Eptesicus sp.	56
Lasiurus borealis	15, 16; 56; 87
Lasiurus cinereus	87; 149f
Lasiurus intermedius	47
Eumops floridanus	54
Nycticeius humeralis	18c
Plecotus rafinesquii	139
Plecotus townsendii	149f, 152
Plecotus sp.	7, 16, 18a, 20; 88; 149h
Antrozous pallidus	115, 139, 152; 173
Tadarida macrotus	149f, 149j
Tadarida brasiliensis	72c; 139, 149h

Edentata

Species	Locality Number(s)
+ Nothrotheriops shastensis	117, 118, 120, 136, 152; 158, 171, 172, 173
+ Megalonyx jeffersonii	10, 17, 19, 20, 21, 27; 31, 33, 47, 50, 54, 57; 62, 72c, 88, 91a; 109, 110; 158, 160, 173
+ Megalonyx sp.	118; 157, 171, 172
+ Eremotherium mirabile	31, 48, 51
+ Glossotherium harlani	27; 31, 37, 39, 42, 44, 51, 54, 56, 57; 62, 66, 89, 95b, 96a, 96b; 109, 110; 158, 165, 171
Dasypus *bellus	18a, 18b, 18c, 20; 31, 33, 40b, 41, 44, 46, 48, 50, 54, 55, 56, 57; 70, 72c, 74, 77, 86, 88, 91a, 91b
+ Holmesina septentrionalis	31, 34, 37, 39, 40a, 42, 44, 48, 50, 51, 54, 56, 57; 63
+ Glyptotherium floridanum	44, 57
+ Glyptotherium sp.	63, 69b, 72c
+ Edentate	149g, 149i

Table 16-6 (continued)

Species	Locality Number(s)	Species	Locality Number(s)
Mammalia (continued)		Mammalia (continued)	

Lagomorpha

Species	Locality Number(s)
Ochotona princeps	98, 99, 100, 101, 103, 107, 108, 112, 115, 125
Sylvilagus audubonii	74; 111, 139, 143, 148, 149j; 158, 160, 172, 173
Sylvilagus audubonii or *S. floridanus*	80; 149j
Sylvilagus bachmani	158, 160, 164
Sylvilagus nuttallii	7, 9; 37, 40a, 40b, 41, 47, 48, 50, 55, 56, 57; 69a, 69b, 70, 74, 76, 77, 86, 87, 89, 91a, 91b; 148
Brachylagus idahoensis	107, 108, 110, 111, 118, 125
Sylvilagus nutallii	107, 108, 111, 149a, 149c, 149e, 149f, 149g
Sylvilagus palustris	40b, 50, 54, 56, 57
Sylvilagus palustrellus	54, 56
Sylvilagus transitionalis	16; 33
Sylvilagus sp.	4, 5, 7, 10, 15, 24, 26; 31, 32, 44, 46, 59; 64a, 64b, 65, 67, 68, 71a, 71b, 72b, 73, 75, 83, 96a, 97; 98, 99, 100, 101, 105, 109, 110, 112, 115, 117, 118, 125, 142, 149b, 149h, 149i, 152; 157, 162, 165
Lepus alleni	148
Lepus americanus	10, 11, 15, 16, 26; 87, 88; 107, 125; 172
Lepus californicus	69a, 69b, 70, 72b, 72c, 75, 78a, 78b; 100, 115, 118, 120, 136, 139, 149h, 149j; 158, 160, 162, 164
Lepus townsendii	69b; 100, 101, 102, 107, 108, 148, 149f
Lepus sp.	63, 67, 68, 76, 79, 88, 97; 99, 104, 105, 109, 110, 111, 112, 115, 117, 125, 142, 149a, 149b, 149e, 149g, 149i, 152; 165, 171

Rodentia

Species	Locality Number(s)
Aplodontia rufa	171, 172, 173
*Tamias *aristus*	33
Tamias striatus	3, 5, 7, 11, 13, 14, 15, 16, 18a, 18b, 18c, 20; 32, 33; 69a, 69b, 87, 88, 89
Eutamias dorsalis	139
Eutamias minimus	14, 16, 18a; 97; 101, 108, 111
Eutamias sp.	100, 105, 107, 112, 115; 164, 173
Marmota flaviventris	98, 99, 100, 101, 103, 105, 107, 108, 111, 112, 115, 120, 125, 139, 148, 149f, 149g, 152; 173
Marmota monax	4, 5, 7, 11, 12, 13, 14, 15, 16, 18a, 18b, 18c, 20; 33; 86, 87, 88, 89, 91a, 91b
Spermophilus armatus	108
Spermophilus beecheyi	160, 171, 172, 173

Rodentia (continued)

Species	Locality Number(s)
Spermophilus beldingi	108
Spermophilus columbianus	111
Spermophilus franklini	77, 91a
Spermophilus lateralis	97; 98, 99, 100, 101, 107, 125
Spermophilus mexicanus	68, 69b
Spermophilus mexicanus or *S. richardsonii*	80
Spermophilus richardsonii	78a, 82, 83, 95a, 96a; 98, 99, 102, 106, 107, 108, 110, 149f
Spermophilus spilosoma	69b, 97
Spermophilus townsendii	107, 108, 111, 115
Spermophilus tridecemlineatus	7, 11, 12, 13, 14, 15, 16, 18a, 18b, 18c, 20, 26; 78a, 80, 82, 83, 88, 91a, 91b, 97; 99, 101, 102, 149e, 149f, 149g
Spermophilus variegatus	68, 97; 101, 139, 148, 152; 158
Spermophilus sp.	41; 64a, 95a, 96a; 100, 103, 109, 112, 115, 142, 149a, 149h; 173
Cynomys gunnisoni	99, 143
Cynomys leucurus	101, 110
Cynomys ludovicianus	64a, 64b, 65, 69a, 69b, 70, 72a, 76, 78a, 80, 82, 83, 95a, 96a, 97; 106, 136, 142, 148
Cynomys sp.	98, 149f, 149h
Sciurus carolinensis	4, 5, 13, 15, 16, 18a, 18b, 18c; 39, 40b, 40c, 47, 56, 57; 91a, 91b
Sciurus griseus	172
Sciurus niger	20; 47; 76, 86, 89, 91a, 91b
Sciurus sp.	7; 48; 70, 73, 88, 91a, 91b; 164
Tamiasciurus douglasi	172, 173
Tamiasciurus hudsonicus	7, 11, 12, 13, 14, 15, 16, 18a, 18b, 18c, 20, 26; 32; 87, 88
Glaucomys sabrinus	7, 11, 12, 13, 14, 15, 16, 18a, 18b, 20; 172, 173
Glaucomys volans	3, 7, 9, 11, 13, 15, 16, 18a, 18c, 20; 39, 47, 48; 88, 89, 91a, 91b
Sciurid	165
Thomomys bottae	78a; 116, 142, 143, 149e, 149f, 149g, 152; 157, 158, 160, 162, 164
Thomomys bottae or *T. umbrinus*	139
Thomomys microdon	172, 173
Thomomys orientalis	50
Thomomys talpoides	80, 83, 91a, 95a, 96a, 97, 99, 102, 106, 107, 108, 111, 115, 149e, 149f, 149g
Thomomys townsendii	110
Thomomys umbrinus	69a, 69b, 148; 172, 173
Thomomys sp.	67, 68; 98, 101, 103, 109, 112, 115, 118, 125, 143, 148, 149a, 149i; 171

Table 16-6 (continued)

Species	Locality Number(s)
Mammalia (continued)	
Rodentia (continued)	
Geomys bursarius	65, 69a, 69b, 70, 72b, 72c, 74, 76, 78a, 78b, 80, 81, 86, 87, 88, 89, 91a, 91b, 96a, 97
Geomys pinetus	39, 40a, 40b, 40c, 41, 44, 46, 47, 50, 54, 56, 57
Geomys sp.	24, 26; 32; 64a, 64b, 68, 71b, 77, 82, 83
Pappogeomys castanops	143, 148, 149b, 149g, 149j, 152
Pappogeomys sp.	67, 68; 148
Geomyid	149h
Perognathus apache	139
Perognathus californicus	158
Perognathus flavus	97; 143
Perognathus hispidus	64a, 64b, 65, 67, 69a, 69b, 70, 74, 78b, 80, 81, 88, 97
Perognathus longimembris	116
Perognathus merriami	69a, 69b
Perognathus parvus	111
Perognathus sp.	64a, 67, 68, 72b, 80, 82, 101, 125, 142, 149e, 149h, 154, 164, 165
Dipodomys agilis	158
Dipodomys merriami	143
Dipodomys ordii	69a, 69b, 80, 97; 111, 143, 148
Dipodomys spectabilis	142, 143, 149a, 149e, 149h
Dipodomys sp.	64a, 72b; 116, 118, 125, 142, 149j; 164, 165
Castor canadensis	4, 5, 12, 15, 18a, 18b, 18c; 31, 33, 34, 37, 39, 48; 80, 86, 87, 91a, 91b; 101, 103, 107, 109, 110, 116; 172
+ *Castoroides ohioensis*	5, 15, 18a; 31, 39, 48, 51, 54; 77, 85, 89
Oryzomys palustris	33, 39, 40a, 40c, 41, 47, 48, 50, 54, 56, 57; 65, 69a, 80, 91b
Reithrodontomys flavescens	69a, 69b
Reithrodontomys humulis	40c, 56
Reithrodontomys megalotis	64a, 69a, 69b, 80, 88; 158
Reithrodontomys montanus	64a, 64b, 69a, 69b, 73, 78a
Reithrodontomys sp.	64a, 64b, 68, 77, 83; 143, 149e
Peromyscus *anyapahensis	163
Peromyscus boylii	64a, 64b, 69a, 69b; 149e; 171
Peromyscus crinitus	149e, 149f
Peromyscus cumberlandensis	33
Peromyscus difficilis	149f
Peromyscus eremicus	78a; 152
Peromyscus floridanus	40a, 40c, 47
Peromyscus gossypinus	40c, 41, 47, 54, 56; 76
Peromyscus leucopus	7, 11, 15, 16, 20; 33; 69a, 69b, 74, 80, 83, 89, 97; 149e
Peromyscus leucopus or *P. maniculatus*	76

Species	Locality Number(s)
Mammalia (continued)	
Rodentia (continued)	
Peromyscus maniculatus	7, 11, 14, 15, 16, 20; 33; 69a, 69b, 80, 83, 97; 99, 106, 108, 139, 143, 148, 149e, 149f, 149g; 172
Peromyscus *nesodytes	163
Peromyscus pectoralis	69a, 69b
Peromyscus polionotus	40a, 40c, 47, 56
Peromyscus truei	108
Peromyscus sp.	5, 10, 12, 13, 14, 18a, 18b, 18c, 26; 39, 40b; 64a, 64b, 65, 68, 70, 72b, 72c, 73, 74, 82, 87, 88, 91a, 91b; 98, 100, 101, 103, 105, 107, 110, 111, 112, 115, 125, 139, 143, 149a, 149h, 149i, 152; 157, 158, 160, 162
Baiomys taylori	67, 69a, 69b
Onychomys leucogaster	64a, 64b, 67, 68, 69a, 69b, 78a, 80, 82, 88, 95a, 97; 108, 111, 139, 143, 149a, 149f, 149h
Onychomys torridus	143; 158
Onychomys sp.	116, 149g; 164
Ochrotomys nuttallii	39, 40c, 47
Sigmodon hispidus	33, 39, 40a, 40c, 44, 46, 47, 50, 54, 55, 56, 57; 63, 64a, 64b, 65, 68, 69a, 69b, 70, 71a, 72b, 72c, 73, 77, 78b, 80, 81
Sigmodon sp.	41; 68; 142, 149h
Neotoma albigula	69a; 143, 149f, 152
Neotoma cinerea	98, 99, 100, 101, 103, 107, 108, 111, 115, 148, 152; 172, 173
Neotoma floridana	3, 4, 5, 10, 11, 12, 13, 14, 15, 16, 18a, 18b, 18c, 20, 21, 32, 33, 39, 40a, 40b, 46, 47, 48, 54, 56; 69a, 69b, 74, 86, 87, 88, 89, 91a, 91b; 149b
Neotoma fuscipes	171
Neotoma lepida	115, 116, 148
Neotoma mexicana	148, 149f, 152
Neotoma mexicana or *N. albigula*	139
Neotoma micropus	69a, 69b, 78a, 80; 152
Neotoma sp.	64a, 64b, 65, 67, 68, 70, 71b, 72b, 72c, 73, 76, 77, 97; 105, 112, 125, 142, 149a, 149e, 149g, 149h, 149i, 149j; 154, 158, 160, 162, 164
Cleithryonomys gapperi	7, 10, 11, 12, 13, 14, 15, 16, 18a, 18b, 18c, 20, 26; 32; 86, 87, 88, 91a, 91b, 97; 101, 108
Phenacomys intermedius	12, 13, 14, 15, 16, 18a, 18b, 18c, 19, 26; 99, 100, 101, 103, 108
Phenacomys sp.	111
Microtus californicus	120, 136; 158, 160, 164, 172, 173

Table 16-6 (continued)

Species	Locality Number(s)	Species	Locality Number(s)
Mammalia (continued)		Mammalia (continued)	
Rodentia (continued)		Rodentia (continued)	
Microtus chrotorrhinus	7, 11, 12, 13, 14, 15, 16, 18a, 18b, 18c	+*Neochoerus pinckneyi*	31, 37, 39, 54
Microtus longicaudus	97; 101, 108, 148, 149e, 149f	*Hydrochoerus *holmesi*	39, 50, 54, 56, 57; 62
Microtus mexicanus	139, 142, 148, 149e, 149f, 152	Cetacea	
Microtus montanus	78b, 82, 83, 96a; 97; 99, 101, 103, 107, 108, 110, 115, 116, 142	*Physeter* sp.	31
		Tursiops truncatus	31
Microtus ochrogaster	78a, 78b, 82, 83, 96a; 97; 99, 101, 142, 149e, 149h	Carnivora	
Microtus ochrogaster or *M. pinetorum*	18a, 18b, 18c; 64b, 69a, 69b, 71a, 72c, 73, 74, 77, 80, 81, 86, 88, 89, 91a, 91b	*Canis *dirus*	10, 16, 19, 20, 26; 31, 37, 39, 40b, 42, 46, 47, 50, 54, 55, 56, 57; 64a, 64b, 71a, 72c, 75, 79, 86, 87, 91a, 91b; 107, 109, 110, 136, 149j; 157, 158, 164, 165, 167, 171, 172, 173
Microtus pennsylvanicus	5, 6, 7, 10, 11, 12, 13, 14, 15, 16, 18a, 18b, 18c, 26; 40a, 47, 48; 70, 78a, 78b, 80, 82, 83, 86, 87, 88, 89, 91a, 91b, 96a; 97; 101, 103, 106, 142	*Canis familiaris*	47, 54, 57; 97; 107
		Canis latrans	10; 39, 56; 64a, 64b, 69a, 69b, 70, 71b, 72c, 73, 74, 75, 76, 77, 78a, 78b, 79, 80, 85, 86, 87, 91a, 92; 97; 98, 99, 101, 102, 105, 107, 109, 110, 115, 118, 136, 139, 143, 148, 149f, 149g, 149h; 157, 158, 165, 167, 171, 172
Microtus pinetorum	6, 7, 9, 11, 12, 13, 14, 15, 16, 20; 33, 39, 40b, 40c, 44, 47, 48, 56		
Microtus xanthognathus	7, 11, 12, 15, 16, 18a, 18b, 18c, 26; 87		
Microtus sp.	18a, 18b, 18c, 26; 64a, 64b, 65, 70, 76, 87, 89, 91a, 91b, 95a, 96a; 98, 105, 106, 109, 111, 115, 118, 149a, 149g; 165, 171	*Canis lupus*	5, 15; 33; 69b, 76, 78a, 78b, 79, 91a, 92; 98, 99, 101, 107, 115, 139, 143, 149g, 149i, 149j; 154, 173
		Canis rufus	47, 51, 54, 56
Lagurus curtatus	99, 100, 101, 103, 106, 107, 108, 111, 149e, 149f	*Canis* sp.	88, 91a, 91b; 105, 112, 117, 125; 164
Neofiber alleni	33, 39, 40b, 40c, 41, 42, 47, 48, 50, 54, 56, 57	*Alopex* sp.	103
		Vulpes macrotis	78a; 136
Ondatra zibethicus	3, 4, 11, 12, 13, 14, 15, 16, 18a, 18b, 20; 33, 37, 39, 47, 48; 63, 65, 74, 77, 78a, 78b, 79, 80, 81, 87, 88, 91a, 91b, 92; 97; 99, 100, 101, 106, 109, 110, 116, 118, 149f; 178	*Vulpes velox*	69b, 79, 82, 88; 98, 99, 115, 117, 148, 149a, 149c, 149d, 149e, 149f, 149g, 149i, 149j
		Vulpes vulpes	7, 15, 18a, 18b, 18c; 56; 69a, 69b, 86, 87, 88, 91a, 92, 97; 98, 99, 100, 101, 102, 105, 107, 109, 115, 148; 172, 173
*Synaptomys *australis*	39, 40a, 40b, 40c, 47, 48, 50, 54, 56, 57; 88		
Synaptomys borealis	7, 11, 12, 13, 14, 15, 16, 18a, 18b, 18c; 19, 20, 87, 88	*Vulpes* sp.	108, 112, 125, 149b
Synaptomys cooperi	3, 9, 10, 11, 12, 13, 14, 15, 16, 18a, 18b, 18c, 20; 32, 33; 64a, 64b, 69a, 69b, 70, 73, 74, 77, 80, 81, 83, 86, 88, 89, 91a, 91b, 97	*Urocyon cinereoargenteus*	4, 5, 14, 20; 31, 33, 39, 47, 50, 54, 56, 57; 64b, 67, 68, 72c, 80, 88, 91a, 91b, 97; 139, 149h, 149i, 149j; 157, 158, 164, 172, 173
		Urocyon sp.	96b; 112, 125, 143; 154, 160
Dicrostonyx hudsonius	11	*Urocyon* sp. or *Vulpes* sp.	142
Dicrostonyx torquatus	99, 101, 103, 105, 107	+*Arctodus pristinus*	48, 55
Microtine	165	+*Arctodus simus*	67, 78a, 87; 101, 105, 109, 110, 148; 158, 165, 173
Zapus hudsonius	10, 11, 12, 15, 16, 18a, 20; 33; 83, 88, 97	*Tremarctos *floridanus*	31, 33, 37, 39, 46, 47, 48, 54, 56, 57; 72c
Zapus princeps	69a, 69b	*Ursus americanus*	4, 10, 15, 16, 18a, 18c, 20, 27; 33, 39, 40a, 40b, 46, 47, 54, 56; 69a, 69b, 70, 73, 74, 75, 76, 86, 87, 88, 91a, 91b; 109, 139, 142, 143, 149d; 157, 158, 165, 172, 173
Napaeozapus insignis	7, 11, 15, 16, 18a, 18b, 20; 89		
Erethizon dorsatum	4, 5, 7, 9, 10, 11, 12, 13, 14, 15, 16, 18a, 19, 20, 21, 26; 32, 39, 48, 57; 75, 86, 88, 91a, 91b; 98, 99, 101, 103, 107, 115, 143, 149f, 149g, 152; 172		
		Ursus arctos	26; 69a; 98, 101, 107; 158
		Ursus sp.	115; 171, 172

Table 16-6 (continued)

Species	Locality Number(s)	Species	Locality Number(s)
Mammalia (continued)		Mammalia (continued)	
Carnivora (continued)		*Carnivora (continued)*	
Bassariscus astutus	115, 120, 125, 148, 152; 158, 160, 162, 173	*Felis concolor*	33, 54, 57; 69b; 98, 101, 107, 115, 118, 120, 143, 148, 152; 158, 165, 171, 173
*Bassariscus *sonoitensis*	139	*Felis yagouaroundi*	69a, 69b
Bassariscus sp.	149g, 149j	*Lynx canadensis*	107
Procyon lotor	3, 4, 5, 12, 15, 16, 18a, 18b, 18c, 20, 24; 31, 33, 34, 39, 47, 51, 54, 56, 57; 70, 73, 76, 80, 86, 87, 88, 89, 91a, 91b; 158, 171, 172	*Lynx rufus*	20, 24; 33, 39, 40c, 47, 48, 50, 51, 54, 56, 57; 69a, 69b, 70, 71b, 72a, 87, 88, 91b, 98, 99, 101, 102, 107, 115, 117, 143, 148, 149d, 149g, 149h; 157, 158, 164, 165, 172, 173
Martes americana	11, 12, 15, 16, 18c, 20; 97; 98, 99, 108	*Lynx* sp.	112, 118, 120, 125, 149g, 149i, 149j
*Martes *nobilis*	98, 99, 101, 107, 115; 172, 173	+*Acinonyx trumani*	103, 105
Martes pennanti	11, 15, 18a, 20; 33; 87, 91a	+*Homotherium serum*	40a; 64a, 64b, 72b; 109
Mustela erminea	5, 16; 69a, 70, 88, 97; 98, 108, 111	+*Smilodon floridanus*	24; 39, 47, 50, 54, 56, 57; 64a, 64b, 75, 79; 110; 158, 165, 167, 171
Mustela frenata	15, 18a, 18c, 20; 33, 47, 48, 57; 69a, 87, 88, 91a, 97; 98, 99, 100, 101, 106, 107, 108, 109, 111, 115, 149a, 149d, 149f, 149g, 149h, 152; 158, 171, 172, 173	Felid	5; 33, 95b; 160, 167, 171, 173
Mustela nigripes	92; 98, 101, 107, 143	Phocid	157
Mustela nivalis	11, 15, 16, 18a, 20, 26; 88	*Odobenus rosmarus*	31
Mustela vison	14, 15, 16; 80, 88, 91a, 91b; 98, 99, 109, 115	*Halichoerus grypus*	31
Mustela sp.	91b; 105, 112, 117, 148; 164	*Monachus tropicalis*	54
Gulo gulo	98, 101, 105, 107	Proboscidea	
Taxidea taxus	18a, 26; 69b, 75, 82, 91b, 96a; 99, 100, 101, 103, 107, 109, 112, 115, 125, 136, 139, 143, 148, 149g; 154, 158, 164, 165, 172, 173	+*Cuvieronius* sp.	51
+*Brachyprotoma obtusata*	10	+*Mammut americanum*	10, 17, 20, 24, 27, 31, 32, 34, 37, 39, 40b, 42, 47, 48, 50, 54, 55, 56, 57; 62, 63, 64b, 70, 77, 85, 88, 89; 109; 157, 158, 165, 171, 173, 178
+*Bryachyprotoma* sp.	88, 91a	+*Mammuthus columbi*	27; 31, 51; 63, 64a, 64b, 69b, 76, 77, 78a, 80, 85, 95b, 96b; 104, 109
Spilogale gracilis	136	+*Mammuthus columbi* or +*M. floridanus*	39, 48, 54, 57
Spilogale putorius	18b, 18c; 33, 41, 47, 48, 54, 56, 57; 64b, 69a, 69b, 70, 71a, 74, 88, 91a, 91b, 97; 101, 107, 108, 111, 142, 149g, 149h; 158, 164, 172, 173	+*Mammuthus imperator*	81, 85
		+*Mammuthus jeffersoni*	158
Spilogale sp.	112, 115; 171	+*Mammuthus primigenius*	17, 27
Mephitis mephitis	4, 5, 7, 10, 14, 15, 16, 18; 33, 39, 44, 46, 47, 50, 54, 57; 64b, 69a, 69b, 70, 72b, 74, 76, 80, 82, 88, 91a, 91b, 97; 98, 99, 101, 107, 139, 143, 149g; 158, 164, 165, 171, 172, 173	+*Mammuthus* sp.	19, 26; 42, 56; 62, 66, 68, 72b, 79, 92; 105, 110, 111, 116, 118; 160, 163, 167, 172, 173
Mephitis sp.	160	+*Mammuthus* sp. or +*Mammut* sp.	59
Conepatus leuconotus	33; 69a, 69b; 148, 149c	Sirenia	
Lutra canadensis	7; 33, 39, 51, 54, 56, 57; 88, 91a, 91b; 172	*Trichechus manatus*	31, 34, 37, 48, 57
*Panthera *atrox*	39; 92, 95b; 98, 101, 105, 107, 109, 115, 118, 136, 149g; 158, 164, 165	Perissodactyla	
Panthera onca	18c, 21; 31, 33, 39, 47, 54, 56, 57; 69a, 72c, 73; 115, 149i	*Equus * complicatus*	27; 39, 57; 62, 63, 85
*Felis *amnicola*	37, 39, 54	*Equus *conversidens*	67, 79, 80, 92; 99, 101, 104, 107, 109, 139, 149e, 149f
		*Equus *excelsus*	148
		*Equus *leidyi*	50, 57
		*Equus *fraternus*	34, 37, 48, 51, 55; 63
		*Equus *giganteus*	48; 63
		*Equus *niobrarensis*	79, 85, 92

Table 16-6 (continued)

Species	Locality Number(s)	Species	Locality Number(s)
Mammalia (continued)		**Mammalia (continued)**	
Perissodactyla (continued)		Artiodactyla (continued)	
Equus *occidentalis*	117, 136; 158, 162, 164, 172, 173	*Odocoileus virginianus*	3, 4, 5, 7, 9, 10, 13, 14, 15, 16, 18a, 20; 31, 33, 34, 37, 39, 40a, 40b, 42, 44, 46, 47, 48, 50, 51, 54, 55, 56, 57, 59; 62, 64b, 65, 69a, 69b, 70, 71a, 71b, 72c, 73, 74, 77, 80, 85, 88, 91a, 91b; 148
Equus *pacificus*	173		
Equus *scotti*	67, 78a, 79, 80; 109		
Equus *tau*	139, 148		
Equus sp.	5, 10, 17, 24, 26; 31, 33, 40a, 40b, 41, 42, 44, 46, 47, 54, 56; 64b, 66, 68, 69a, 69b, 70, 71a, 71b, 72b, 72c, 73, 75, 76, 77, 78a, 80, 82, 88, 89, 92, 95a, 95b, 96a, 96b; 99, 100, 101, 105, 107, 108, 110, 112, 115, 116, 117, 118, 120, 125, 142, 143, 149a, 149b, 149c, 149d, 149g, 149h, 149i; 154, 157, 160, 165, 167, 172	*Odocoileus* sp.	63, 76, 89, 92, 95b, 96b, 97; 98, 103, 107, 109, 110, 115, 118, 125, 136, 149b, 149g, 149i; 158, 164, 167, 171, 172, 173
		+ *Navahoceros fricki*	67; 101, 148
		+ *Sangamona fugitiva*	10, 18a, 20
		+ *Sangamona* sp.	91a, 92
		+ *Cervalces scotti*	10, 17, 27; 85
Tapirus *californicus*	162	*Alces alces*	4; 85; 101
Tapirus *excelsus*	88	*Alces* sp. or + *Cervalces* sp.	91a
Tapirus *haysii*	27	*Rangifer tarandus*	3, 4, 5, 7, 17, 18a, 19, 27; 92; 107
Tapirus *veroensis*	10, 18a; 33, 34, 40b, 42, 46, 48, 50, 51, 54, 55, 56, 57; 64b, 71a	Cervid	105, 136
Tapirus sp.	19; 31, 37, 39, 40a; 65, 89, 91a; 136, 149h; 158	*Antilocapra americana*	77, 91b; 92; 98, 99, 100, 101, 104, 109, 115, 125, 143, 148, 149e; 158, 160, 162, 165
Artiodactyla			
+ *Mylohyus fossilis*	39, 40b, 54, 57	+ *Capromeryx minor*	115, 149d, 149g, 149h, 149i, 149j; 154, 158
+ *Mylohyus nasutus*	5, 10, 11, 15, 24; 31, 32, 33; 64b, 88	+ *Capromeryx* sp.	78b
+ *Mylohyus* sp.	37, 44, 46, 48, 50, 56; 77	+ *Tetrameryx* sp.	118
+ *Platygonus compressus*	6, 18a, 19, 26; 32, 33, 47, 48, 54, 55, 57; 72a, 72c, 78a, 82, 86, 87, 91a, 91b, 92, 95b, 96a, 96b, 139; 158	+ *Stockoceros conklingi*	136
		+ *Stockoceros onusrosagris*	139, 148
		+ *Stockoceros* sp.	67
+ *Platygonus* sp.	71b, 75, 76, 79; 173	Antilocaprid	72c, 73, 79, 96b; 105, 110
Tayassuid	136	*Bison* *alleni*	109
+ *Camelops hesternus*	78a, 80, 95a, 96a, 96b; 99, 100, 101, 102, 107, 109, 110, 115, 116, 118, 149b, 149c; 157, 158, 164, 165	*Bison bison* *antiquus*	27; 31, 37, 42; 78a, 78b, 79, 80, 92, 95b, 96a, 96b, 97; 102, 109, 139, 148, 149a, 149e; 158, 167
+ *Camelops* sp.	63, 64a, 64b, 66, 68, 72a, 73, 75, 76, 79, 82, 85, 92; 104, 105, 108, 115, 117, 125, 148, 149d, 149h, 149j; 160, 167	*Bison bison bison*	68, 85, 92
		Bison *latifrons*	110
		Bison bison *occidentalis*	80
+ *Hemiauchenia macrocephala*	34, 37, 40a, 40b, 48, 51, 54, 56, 57; 79	*Bison bison* *occidentalis* or *B. b.* *antiquus*	81
+ *Hemiauchenia* sp.	73, 75, 78a, 80; 101, 108, 112, 115, 117, 125, 149a, 149c, 149d, 149g; 154, 165, 167	*Bison* sp.	5, 17; 34, 39, 44, 48, 54, 56, 57; 62, 63, 64a, 64b, 67, 68, 69a, 69b, 70, 71a, 71b, 73, 75, 76, 77, 85, 89; 98, 99, 100, 101, 104, 105, 107, 108, 110, 111, 115, 118, 136, 143; 157, 163, 165, 171, 173, 178
+ *Paleolama mirifica*	31, 39, 40a, 42, 55, 57		
+ *Paleolama* sp.	160		
Camelid	108, 115, 139, 142, 149i; 154	*Oreamnos americanus*	99, 100, 101; 173
Cervus elaphus	3, 4, 5, 9, 13, 16, 18a; 85, 86; 99, 101, 107, 115	*Oreamnos* *harringtoni*	115, 120
		Oreamnos sp.	172
Cervus sp.	139	+ *Symbos australis*	91a
Odocoileus hemionus	99, 100, 101, 116, 117, 148; 157, 160, 162	+ *Symbos cavifrons*	17, 27; 85, 92
		+ *Symbos* sp.	101, 103

Table 16-6 (continued)

Species	Locality Number(s)
Mammalia (continued)	
Artiodactyla (continued)	
+*Bootherium bombifrons*	10, 17, 27
+*Euceratherium collinum*	101, 112, 125, 148; 171, 172, 173
Ovibos moschatus	92
Ovibos sp.	103
Ovibovine	70
*Ovis *catclawensis*	92
Ovis canadensis	73; 98, 99, 100, 101, 103, 105, 107, 112, 115, 116, 117, 125, 148, 149g
Ovis sp.	120, 143

Note: Locality numbers correspond to the numbers in Figure 16-1: northeast region, 1-30; southeast region, 31-61; central region, 62-97; Rocky Mountain region, 98-153; Pacific region, 154-178. Some of the multipart site numbers do not include the sublocality designations because the distributions of the taxa within those sites were not given in the original sources.

*wetmori and the duck *Anas* *itchtucknee*, are only questionably distinct from species still living in the area.

The late-Pleistocene mammalian fauna of the Southeast was characterized by most of the wide-ranging large species found in other parts of the country; these taxa include + *Smilodon*, + *Mammuthus, Bison, Equus*, + *Megalonyx*, and + *Glossotherium* (Table 16-6). Others, including the extinct chipmunk *Tamias* *aristus*, the golden mouse (*Ochrotomys nuttalli*), and the round-tailed water rat (*Neofiber alleni*) were regional endemics. The latter two, which are still extant, have been confined to this region since the late Pleistocene (Frazier, 1977), although *Neofiber* is known from earlier Pleistocene faunas farther north and west (Hibbard, 1970). *Canis rufus* also occurred in the southeast from late-Pleistocene into historic times (Nowak, 1979). The range of *Monachus tropicalis* extended northward to Melbourne in central Florida. Among the extinct taxa, the large short-faced bear (+ *Arctodus pristinus*) has been found only in the East, from West Virginia to Florida (Guilday, 1971; Kurtén, 1967; Kurtén and Anderson, 1980; Webb, 1974a).

The capybaras + *Neochoerus pinckneyi* and *Hydrochoerus* *holmesi* are known from Late Wisconsin faunas of Florida (Kurtén and Anderson, 1980; Webb, 1974a). + *Neochoerus pinckneyi* apparently did not occur outside Florida during the last 25,000 years, although it is known from earlier faunas in Texas (Hay, 1926; Kurtén and Anderson, 1980). *Hydrochoerus* occurs in the Avery Island local fauna from Louisiana (Gagliano, 1967), which is latest Pleistocene in age. The two species have been found together in the Melbourne (Florida) fauna (Webb, 1974a).

The most distinctive aspect of the Gulf and southern Atlantic coastal fauna is the diversity of neotropical species, both extinct and extant. The extinct taxa include + *Eremotherium*, + *Glyptotherium*, + *Cuvieronius*, + *Holmesina*, and + *Geochelone*. + *Palaeolama* and + *Tremarctos* are best known from the Southeast but also have records in coastal Texas. Extant taxa of mammals that now have more southerly ranges include *Hydrochoerus* (as discussed above), *Tapirus, Desmodus, Eumops, Jaguarius, Felis weidii*, and *F. jagourundi. Dasypus* *bellus* had a range that included all of the Gulf Coast in the late Pleistocene, and more recently *D. novemcinctus* has reestablished itself in the same region.

Like the late-Pleistocene faunas in other parts of the United States, faunas in the Southeast contain extant mammals that do not occur in these areas today. Examples are *Microtus pennsylvanicus* and *Synaptomys cooperi* (= *S. australis*), which are found in a number of late-Pleistocene faunas in Florida (Kurtén and Anderson, 1980; Webb, 1974a). The former species has recently been reported to be living still in an extremely limited area of Florida (C. A. Woods, unpublished data), but it is doubtful that this finding invalidates the general conclusions based on the disappearance of this species from most of Florida at the end of the Pleistocene.

MIDCONTINENT REGION

This arbitrarily defined region is very large and ecologically diverse. It includes such areas as the Gulf Coastal Plain, Ozark Highlands, Great Plains, and northern prairie-forest boundary. The Pleistocene faunas of each subregion were different; therefore, each subregion is discussed separately.

There are few coherent local faunas from the Gulf Coastal Plain of Texas (Table 16-3), but a limited number of isolated specimens have been collected. Except for the presence of the extinct + *Geochelone wilsoni* and *Terrapene carolina* *putnami* in the Buckner Ranch

faunas (site 1) in Bee County, southern Texas, the herpetofauna does not differ from the modern fauna (Table 16-6).

Much of our knowledge of the late-Pleistocene mammalian fauna of the Coastal Plain of Louisiana is based on work done during the 19th century and the early 20th century (Hay, 1923; Leidy, 1853). A summary of this early work is given by Domning (1969). Two faunas are known from recent work, one from Avery Island (Arata, 1964; Gagliano, 1964; 1967) and one from West Feliciana Parish (Domning, 1969). These faunas are similar to those of both Florida and the Texas Gulf Coastal Plain.

A number of mammalian taxa found on the western Gulf Coastal Plain extended eastward to Florida, and in Texas they either were confined to the Coastal Plain or were not common outside of it (Table 16-6). These taxa include + *Glyptotherium floridanum, Tapirus* *veroensis*, + *Mammut americanum*, + *Mylohyus nasutus*, and + *Holmesina* (Lundelius, 1967; Sellards, 1940). Some northeastern forms such as + *Castoroides ohioensis* invaded as far south and west as the northeasternmost part of Texas (Slaughter and Hoover, 1963). The northern microtines *Microtus pennsylvanicus* and *Synaptomys cooperi* lived throughout the eastern half of the United States during the Late Wisconsin.

The southern part of the Great Plains includes the Edwards Plateau, trans-Pecos Texas, eastern New Mexico, northwestern Texas, and Oklahoma. Cave deposits in this region have produced diverse microvertebrate faunas, especially on the Edwards Plateau, but fluvial and lacustrine deposits elsewhere have yielded a larger spectrum of the vertebrate fauna (Table 16-3). As is true throughout the United States, the late-Pleistocene fauna of this region had three components: extinct forms, extant forms still living in the area, and extant forms that have been extirpated from the area. The extirpated herpetofauna consists primarily of eastern and southern species, whereas the extirpated mammalian fauna generally consists of northern and eastern forms. The single extirpated bird is a neotropical species (Table 16-6).

Most herpetofaunas in New Mexico and Texas have been dated between 15,000 and 10,000 yr B.P. (Table 16-3). A single Oklahoma fauna, the Domebo Mammoth Kill site, also represents this interval. The Brown Sand Wedge faunule of the Blackwater Draw local fauna in New Mexico (Slaughter, 1975) includes eastern species interpreted as requiring more rainfall than is now available, along with the extinct tortoise + *Geochelone wilsoni*, whose occurrence indicates somewhat milder winters than those of today. The Cave without a Name herpetofauna from Texas (Hill, 1971; Holman, 1968) contains southern and eastern forms that may indicate more rainfall and warmer winters than occur today in that area.

Several herpetologic forms became extinct or were extirpated from the west-central area of the United States at the very end of the Pleistocene. The large box turtle *Terrapene carolina* *putnami*, which is similar in size to *T. c. major* from the Coastal Plain, is found in four localities in New Mexico, Texas, and Oklahoma (Moodie and Van Devender, 1978) (Table 16-6). These turtles are thought to have indicated moister conditions in these areas during the interval 15,000 to 10,000 yr B.P. At the Lubbock Lake site in Texas (Johnson, 1974), these large box turtles were present in the Clovis faunules (11,000 yr B.P.) but were absent from the overlying Folsom faunules (10,000 yr B.P.); this suggests that they became extinct in this area between 11,000 and 10,000 yr B.P. The presence of the small extinct land tortoise + *Geochelone wilsoni* and the desert tortoise *Gopherus agassizi* in New Mexico and Texas, and Oklahoma (Moodie and Van

Devender, 1978) indicates a climate of milder winters and cooler summers than occurs in these areas today. Although the exact dates of extinction or extirpation for these taxa are not known, they appear to have been about 10,000 yr B.P. Almost all of the southern Great Plains herpetofaunas indicate that the climate was more mesic than at present during the interval 15,000 to 10,000 yr B.P., with warmer winters and cooler summers (Table 16-6).

The fact that the extinction of + *Geochelone wilsoni* and *Terrapene carolina* *putnami* and the extirpation of *Gopherus agassizi* all appear to have taken place about 10,000 yr B.P. has been interpreted as a consequence of increasingly cold winters in the area; but it also has been suggested that human predation may have contributed (Moodie and Van Devender, 1978). The extinct toad *Bufo woodhousei* *bexarensis* appears to have been replaced by the modern *Bufo woodhousei woodhousei* at about the same time (Holman, 1969; Moodie and Van Devender, 1978); this has been attributed to climatic change because both species are small forms that would have been an unlikely food source for paleo-Indians.

The only extirpated fossil bird known from the southern Great Plains is the gray-breasted crake (*Laterallus* cf. *exilis*) from the Lubbock Lake site (A. M. Rea, unpublished data) (Table 16-6). *L. exilis* is a neotropical rail that does not occur today north of Belize, although Olson (1974) reported *L. exilis* from the somewhat older Pleistocene site of Reddick, Florida. The single extinct bird from this region is the jay + *Protocitta dixi* from Miller's Cave in central Texas (Weigel, 1967); however, this record is tentative because Weigel listed no osteological characteristics to support his identification. The generic status of + *Protocitta* has been questioned by Steadman and Martin (in press).

The large Pleistocene mammals of the southern Great Plains show few differences from those of the central and northern Plains. *Bison*, + *Camelops*, + *Hemiauchenia*, + *Mammuthus* and *Equus* were widespread. The giant short-faced bear (+ *Arctodus simus*) took the place of the eastern bear + *Arctodus pristinus*.

Eastern species such as *Microtus pinetorum*, *Microtus pennsylvanicus*, *Zapus princeps*, + *Symbos*, *Synaptomys cooperi*, and *Blarina carolinensis* extended as far west as the Edwards Plateau (Dalquest et al., 1969; Lundelius, 1967) (Table 16-6). Today, species such as *Thomomys talpoides* and *Microtus ochrogaster* are restricted to the northern and central Plains, but during the late Pleistocene they were found on the southern Plains of Texas (Dalquest, 1965). *Mustela erminea* and *Sorex cinereus*, whose present distributions include the boreal zone, ranged into central Texas and northern Mexico, respectively, during the late Pleistocene (Dalquest et al., 1969; Lundelius, 1967, 1974).

The late-Pleistocene mammalian record from the Ozark area (Tables 16-1 and 16-6, sites 86 through 89 and site 91a) is derived from a variety of deposits (including caves, springs, and alluvial/colluvial sequences), each of which yields a very different sort of paleoenvironmental information (Tables 16-3 and 16-6). The cave deposits contain primarily microfauna, the alluvial/colluvial sequences contain both microfauna and megafauna, and the spring deposits of western Missouri contain extensive megafauna in association with vegetation.

In the Ozark Highlands, herpetofaunas and avifaunas of very-late-Pleistocene age are confined to Missouri (Tables 16-3 and 16-6). Brynjulfson Cave locality 1 (Parmalee and Oesch, 1972) is noteworthy because it has a date of at least 10,000 yr B.P. The fauna includes the Blanding's turtle (*Emydoidea blandingii*), a species that today occurs well north of the area, and the large armadillo (*Dasypus*

bellus), a southern form. All of the birds from this site have been reported from Missouri within historic times. The Zoo Cave fauna (Table 16-3) has been estimated on the basis of its fauna to be late Pleistocene (Hood and Hawksley, 1975). It contains a northern snake, *Elaphe vulpina*, and the large southern armadillo (*Dasypus* *bellus*). The best explanation for these disharmonious pairings appears to be that the climate was more equable, with cooler summers allowing the presence of *Emydoidea blandingii* and *Elaphe vulpina* and warmer winters allowing the presence of *Dasypus* *bellus*.

The Pleistocene large-mammal fauna of the Ozark area was dominated by + *Mammut americanum*, + *Glossotherium harlani*, + *Symbos cavifrons*, and + *Platygonus compressus*. Eastern species that appear to have been at or close to the western limits of their distribution are + *Sangamona fugitiva*, + *Cervalces scotti*, + *Mylohyus* sp., and + *Castoroides ohioensis*. Southern forms such as *Tapirus* *veroensis* and *Dasypus* *bellus* were represented. The extinct short-nosed skunk (+ *Brachyprotoma* sp.) was present during the late Pleistocene in the Ozark area and the Appalachians; however, the fact that this genus is not known from any Pleistocene fauna from the area between the Appalachians and the Ozarks suggests that these two populations may have been isolated from each other since early in the Pleistocene. *Equus* was poorly represented in the Ozark faunas, in contrast to the faunas farther west. It appears that this region was the zone of change between the predominantly forested environment of the East and the predominantly grassland and savannah environment of the West.

Many of the late-Pleistocene micromammal taxa of the Ozarks now live to the north of the area. Of these taxa, *Sorex arcticus*, *Microtus xanthognathus* and *Synaptomys borealis* now occupy the boreal forest. Others, such as *Clethrionomys gapperi*, *Tamiasciurus hudsonicus*, *Microsorex hoyi*, *Sorex palustris*, *Phenacomys intermedius*, and *Microtus pennsylvanicus*, have their southern boundaries in the midcontinent region south of the boreal forest but north of the Ozarks. A number of open-grassland species, such as *Onychomys leucogaster*, *Geomys bursarius*, *Spermophilus tridecemlineatus*, and *perognathus hispidus*, are represented. These are now ecologically allopatric with the boreal forest forms mentioned above as well as with eastern forest species, such as *Tamias striatus*, *Marmota monax*, and *Glaucomys volans*. These associations support the hypothesis of a more equable late-Pleistocene climate, as suggested by the amphibians, reptiles, and birds.

Pleistocene amphibian and reptilian faunas from the central and northern Great Plains are virtually unknown, and the known mammalian faunas are few. The mammalian fauna was typified by large herbivores, such as *Bison*, + *Camelops*, + *Mammuthus*, *Equus*, + *Glossotherium harlani*, *Antilocapra* sp., *Rangifer* sp., *Ovis* sp., *Ovibos* sp., *Cervus* sp., and + *Platygonus compressus*, all of which indicate open grassland or savannah environments. The micromammalian record is sparse, but as in other areas it is composed of species that are allopatric today.

Similar distributional adjustments have also been observed for alpine microfauna. Species such as *Ochotona princeps*, *Microtus montanus*, *Sorex palustris*, and *Lepus americanus* have migrated from montane habitats in the Rocky Mountains to similar habitats in isolated mountain chains on the northern Plains. Many of these alpine species, including *Microtus montanus*, *Sorex palustris*, and *Lepus americanus*, have also been found in paleo-Indian-age sites, such as the Hell Gap site on the plains of southeastern Wyoming (Roberts, 1970) and the Jones-Miller site in northeastern Colorado

(Rawn-Schatzinger, no date). The distribution of these taxa on the Great Plains suggests cooler climates (compared to today's) with greater effective moisture and lowered altitudinal gradient. Similar phenomena have been observed in the mountains and plains of the southwestern United States (Harris, 1977a, 1977b; Van Devender and Wiseman, 1977).

During the Late Wisconsin, the eastern woodlands also contributed faunal elements to the communities of the Plains. *Blarina* sp., *Cryptotis* sp., *Sciurus* sp., and *Synaptomys* sp. extended their ranges westward, probably in riparian environments along major tributaries that drain toward the east. Only a few of these eastern species, such as *Cryptotis parva* (Harris et al., 1973), penetrated the mountainous environments of the West. In spite of the ameliorated climatic conditions on the Plains, there apparently were barriers that limited the range extensions of eastern and western species (Graham, 1979). At the same time, western species such as *Spermophilus tridecemlineatus*, *Eutamias minimus*, *Taxidea taxus*, and *Ursus arctos* extended their ranges eastward (Guilday et al., 1964). Both of these range extensions were the result of the more varied environments in both regions, which included both woodland and grassland communities.

Late Wisconsin climates did not obliterate longtitudinal and latitudinal environmental gradients, but the modification of these gradients necessitated distributional readjustments for many taxa. Several species, including + *Geochelone wilsoni*, *Dasypus* *bellus*, *Perognathus hispidus*, and *Notiosorex crawfordi*, were restricted to southern environments on the Great Plains. Many northern taxa, including *Dicrostonyx*, *Sorex articus*, *Sorex palustris*, and *Martes americana*, were restricted to the central and northern Plains. Many northern taxa extended much farther south and to lower levels in the Rocky Mountains than they do today. These adjustments in distribution were directly related to the tolerance limits of the individual taxa responding to environmental changes. They did not involve the displacement of immutable life zones; instead, new community organizations were created by the displacement of individual taxa according to their specific environmental tolerances (Graham, 1976a, 1979; Graham and Semken, 1976; King and Graham, 1981).

ROCKY MOUNTAIN AND BASIN AND RANGE REGION

In this region only Arizona and New Mexico have a significant number of amphibian and reptilian local faunas that have absolute dates between 25,000 and 10,000 yr B.P. (Table 16-4). Idaho and Nevada have a few sites with absolute dates and a number of other sites with indefinite dates (Table 16-4). Studies by Van Devender and Mead (1978) and by Van Devender and others (1977) on packrat middens from Arizona indicate more equable climates in the very late Pleistocene than at present. Van Devender and Mead (1978) have studied sites ranging from about 15,000 to 10,000 yr B.P. from Mohave, Yuma, and Pima Counties, Arizona (Table 16-4). Reptilian and amphibian species from these sites (Table 16-4) are desert-scrub forms today, and yet the associated plant microfossils indicate a woodland community. These authors interpret this as reflecting equable climates with mild winters, cool summers, and increased winter precipitation. They doubt that desert scrub retreated to Mexican refugia during glacial times, as has repeatedly been suggested by neoherpetologists. Van Devender and others (1977) have studied sites ranging from about 16,000 to 10,000 yr B.P. in Mohave and Coconino Counties, Arizona (Table 16-4), and report that herpetofaunas from these sites were a mixture of present-day desert and higher woodland species (Tables 16-4 and 16-6) and that the plant macrofossils also represented both desert and woodland species. From these data, they conclude that these mixed biotic communities reflected a more equable climate than that of today, with mild winters, cool summers, and a slightly increased winter rainfall.

In Howell's Ridge Cave faunule 1, from New Mexico (Van Devender and Worthington, 1977), the abundance of mesic forms such as *Phrynosoma douglassi* and *Ambystoma tigrinum* increases with depth, whereas the abundance of xeric forms such as *Phrynosoma modestum* and *Scaphiopus couchi* decreases, although both forms occur throughout the site. The Dry Cave local fauna from New Mexico (Holman, 1970) yielded a herpetofauna containing both low- and high-altitude species, possibly indicating a more equable climate.

No noteworthy records of amphibians and reptiles are found in sites in Power County, Idaho (McDonald and Anderson, 1975), White Pine County, Nevada (Brattstrom, 1958a, 1976), Yuma County, Arizona, (Van Devender et al., 1977), and Cochise County, Arizona (Moodie and Van Devender, 1978). Gypsum Cave, in Clark County, Nevada, which is thought to be between 10,000 and 8000 years old (Brattstrom, 1954a, 1958a), has evidence of both northern and southern extralimital species of amphibians and reptiles (Table 16-4 and 16-6), possibly indicating milder winters and cooler summers.

Thirty-two late-Pleistocene avifaunas have been found in the Rocky Mountain and Basin and Range region, more than in any other of the regions discussed in this chapter (Table 16-4). The distribution of these sites is uneven, with 18 in southern New Mexico and southwestern Texas alone. Nearly all of these sites are in caves.

Unlike avifaunas in the East and Midwest, avifaunas in the West include a significant number of extinct species (Table 16-6). + *Teratornis merriami*, a huge scavenger with a wingspan of about 3.5 m, is recorded from Stanton's Cave, Arizona (Rea and Hargrave, no date), and Tule Springs, Nevada (Mawby, 1967), as well as from sites in Florida and California. Howard (1952) describes + *Teratornis incredibilis* on the basis of a single ulnare from Smith Creek Cave, Nevada; the species is the largest known flying bird in North America, with an estimated wingspan of about 5 m. A radius and rostrum of + *T. incredibilis* has been reported (Howard, 1963, 1972) from the early Pleistocene (Irvingtonian) Vallecito Creek local fauna in California. The extinct condor + *Breagyps clarki* is known from four sites. Other extinct scavengers include two accipitrid vultures, + *Neophrontops americanus* from Dark Canyon Cave, New Mexico (Howard, 1971), and + *Neogyps errans* from Chimney Rock Animal Trap, Colorado (Hager, 1972), and Smith Creek Cave, Nevada (Howard, 1952). Today, accipitrid vultures are confined to the Old World. Two apparently extinct eagles of uncertain affinities are "*Spizaetus*" *willetti* from Smith Creek Cave (Howard, 1935, 1952) and Howell's Ridge Cave faunule 1, New Mexico (Howard, 1962a), and "*Buteogallus*" *fragilis* from Shelter Cave, New Mexico (Howard and Miller, 1933). Other putative extinct raptors include the falcon *Falco* *swarthi* from Dark Canyon Cave, New Mexico (Howard, 1971), and *Strix* *brea*, tentatively reported from Dry Cave, New Mexico (Harris, 1977b).

Extinct shelducks of the genus *Anabernicula* occur in five sites in New Mexico and Nevada (Tables 16-4 and 16-6). Like accipitrid vultures, shelducks presently live only in the Old World. The turkey *Meleagris* *crassipes* has been found in many cave deposits in southern Arizona and southern New Mexico, and it apparently survived into the Holocene (Rea, 1980). A large extinct roadrunner, *Geococcyx *conklingi*, is known from four cave deposits in southern New Mexico (Table 16-6). The extent of the temporal and geographic overlap between *G. *conklingi* and the living *G. californianus* is uncertain. Extinct passerines include the raven *Corvus *neomexicanus* from Dry Cave, New Mexico (Magish and Harris, 1976), and the cowbird + *Pyeloramphus molothroides* from Shelter Cave, New Mexico (Miller, 1932). *Corvus *neomexicanus* may not be specifically distinct from living ravens; the affinities of + *Pyeloramphus* are uncertain.

Because of the drastic vegetational changes that occurred in this region during late-Quaternary times (Spaulding et al., 1982), it is not surprising that many birds also underwent major changes in distribution. Among the well-documented cases are the "California" condor (*Gymnogyps californianus*), known from many fossil sites in Nevada and New Mexico but known historically only from Arizona and the Pacific region (Rea, 1981). The black vulture (*Coragyps atratus*) also is a common fossil in this region, occurring as far north as Smith Creek Cave, Nevada (Howard, 1952); today, it is found in the West only near the United States-Mexico border. Thick-knees (*Burhinus* sp.), large terrestrial charadriiforms that live in and south of tropical Mexico, are unknown today in the United States except for a single specimen from southern Texas (MacInnes and Chamberlain, 1963). Howard (1971) reports *Burhinus* sp. (perhaps two species) from Dark Canyon Cave, New Mexico. Another tropical bird, the clay-colored robin (*Turdus grayi*), has been reported from Stanton's Cave in northern Arizona (Rea and Hargrave, no date). Today, *T. grayi* does not occur regularly north of Tamaulipas on the Gulf Coast of Mexico or north of Guerrero on the Pacific Coast of Mexico.

All of the late-Pleistocene mammals of this region are members of either the alpine or open-intermontane-basin ecotype (Table 16-6). Species such as + *Nothrotheriops shastensis* ranged throughout both intermontane and alpine environments. Today, certain species, such as *Oreamnos americanus*, are restricted to rugged terrain in alpine environments above the treeline in the northern Rocky Mountains. During the Late Wisconsin, *Oreamnos americanus* extended as far east as south-central Wyoming, and a small extinct mountain goat, *Oreamnos *harringtoni*, has been found as far south as San Josecito Cave in the Sierra Madre Oriental of northern Mexico. The extinct North American mountain deer (+ *Navahoceros fricki*) and the extinct brush ox (+ *Euceratherium collinum*) were confined to the mountainous areas of the western half of the United States during the latest Wisconsin.

Both now and in the past *Ovis canadensis* has had a broader geographic and topographic range than *Oreamnos americanus*. Late-Wisconsin bighorn sheep have been found in Chimney Rock Animal Trap (Hager, 1972), Natural Trap (Martin and Gilbert, 1978), Bell Cave (Walker, 1974; Zeimens and Walker, 1974), Horned Owl Cave (Guilday et al., 1967), Isleta Cave (Harris and Findley, 1964), Jaguar Cave (Kurtén and Anderson, 1972) and Little Box Elder Cave (Anderson, 1968). *Ovis canadensis* extended as far east on the Plains as Red Willow (Corner, 1977), southwestern Nebraska. Compared to the limits of *Oreamnos americanus* and perhaps + *Navahoceros fricki*, the broader tolerance limits of *Ovis canadensis* allowed a larger range extension for this species during the Late Wisconsin.

Similar distributional adjustments have been observed for alpine microfauna. Species such as *Ochotona princeps*, *Microtus montanus*, *Sorex palustris*, and *Lepus americanus* have migrated from montane habitats in the Rocky Mountains to similar habitats in

isolated mountain chains on the northern Plains. Of these alpine species, *Microtus montanus*, *Sorex palustris*, *Marmota flaviventris*, and *Lepus americanus* have also been found in sites of paleo-Indian age (10,000 yr B.P.) on the plains of southeastern Wyoming and northeastern Colorado (Rawn-Schatzinger, no date; Roberts, 1970; Walker, in press). During the Late Wisconsin, these species also existed at lower elevations in the mountains. The distribution of these taxa suggests cooler climates with greater effective moisture and a lowering of the altitudinal gradient. Similar phenomena have been observed in the mountains and plains of the southwestern United States (Harris, 1977a, 1977b; Van Devender and Wiseman, 1977).

Most of the intermontane-basin forms represent an extension of part of the Great Plains fauna into the Rocky Mountain and Basin and Range area. Examples are + *Camelops*, + *Mammuthus*, *Bison*, *Equus*, and *Antilocapra*. In this area, many of the large herbivores may have existed in smaller populations more isolated geographically than those on the Great Plains. The known Late Wisconsin records suggest that the American cheetah (*Acinonyx *trumani*) was restricted to this province. One antilocaprid, + *Capromeryx*, appears to have been confined to the southern part of this region and the adjoining part of the Great Plains.

PACIFIC COAST AREA

Of the states in the Pacific Coast area, only California has a significant number of carbon-dated amphibian and reptilian local faunas 25,000 to 10,000 years old. Except for the record of a toad, *Bufo boreas*, and a snake, *Pituophis melanoleucus*, dated about 25,000 to 20,000 yr B.P., all are dated from the time period 15,000 to 10,000 yr B.P. (Tables 16-5 and 16-6).

A few dated herpetologic records (Table 16-6) listed by Brattstrom (1953b, 1954b) from Rancho La Brea (W. A. Akersten, unpublished data) and La Mirada (Miller, 1971) are forms that occur in the areas today. Eight local faunas reported by Van Devender and Mead (1978) from Sonoran Desert packrat middens in San Bernadino County, which are dated at 10,000 to 13,000 yr B.P., are important paleoecologically. These desert-scrub species of amphibians and reptiles (Table 16-6) are associated with plant macrofossils that indicate woodland conditions in the late Pleistocene. Van Devender and Mead (1978: 464) state that "biogeographical scenarios calling for desert faunas to be restricted to Mexican refugia during glacial periods are not supported. Equable climates with mild winters, cool summers and increased winter precipitation are inferred."

Other late-Pleistocene sites in California (Table 16-5) with no absolute dates or with indefinite dates cannot be interpreted ecologically as precisely as Van Devender and Mead interpreted the Sonoran Desert packrat-midden sites. The presence of *Crotalus viridis* in Mescal Cave in San Bernadino County (Brattstrom, 1958b) suggests a cooler late-Pleistocene climate for the area, for it is a "cool-adapted" species that occurs north of that area today. However, this record could just as well be interpreted as reflecting a more equable Pleistocene climate for the area, as suggested by Van Devender and Mead (1978). *Crotalus *potterensis* from the Potter Creek Cave, a site believed to be late Pleistocene, is the only non-chelonian late-Pleistocene amphibian or reptilian species recognized as extinct in the United States. Holman (1981) suggests that this species may be a synonym of *Crotalus viridis*, but Brattstrom, who originally named *C. *potterensis*, has recently reiterated his belief that *C. *potterensis* is a valid species (Brattstrom, 1953b, recent unpublished data).

Two hundred and five species of birds have been recorded from late-Pleistocene sites in the Pacific region, more than the number from any other region. The combined faunal lists of Rancho La Brea, California, and Fossil Lake, Oregon, by themselves include over 75% of these species, although McKittrick, Carpinteria, and several cave deposits in northern California are also rich localities (Table 16-5). Rancho La Brea has the greatest number of extinct land birds, and Fossil Lake has the greatest number of extinct aquatic birds. The majority of late-Pleistocene marine localities in California (e.g., San Pedro, Del Rey Hills, Newport Bay) are generally regarded as being older than 25,000 yr B.P. Therefore, very few marine birds are listed in Table 16-6.

The avifauna of Rancho La Brea (Howard, 1962b) is best known for its abundant and diverse scavenging and raptorial birds, although most other families of North American nonpasserine land birds also are represented. Extinct scavengers or possible scavengers include the storks *Ciconia* *maltha* and *Mycteria* *wetmorei*; the teratorns + *Teratornis merriami* and + *Cathartornis gracilis*; the condor + *Breagyps clarki*; the eagles *Spizaetus* *grinnelli*, *Buteogallus* *fragilis*, + *Amplibuteo woodwardi*, and + *Wetmoregyps daggetti*; and the accipitrid vultures + *Neophrontops americanus* and + *Neogyps errans*. Associated with these extinct forms are extant scavengers: the "California" condor (*Gymnogyps californianus*), the black vulture (*Coragyps atratus*), the turkey vulture (*Cathartes aura*), the golden eagle (*Aquila chrysaetos*), the crested caracara (*Polyborus plancus*), and the common raven (*Corvus corax*). Apparently, the mammalian megafauna of Rancho La Brea provided diverse scavenging opportunities for birds of all sizes to feed on dead, sick, old, or very young animals. Other extinct birds from Rancho La Brea include the shelduck *Anabernicula* *gracilenta*, the turkey *Meleagris* *californica*, the owl *Strix* *brea*, the towhee *Pipilio* *angelensis*, the cowbird + *Pandanaris convexa*, and the blackbird *Euphagus* *magnirostris*. Except for the shelduck (Howard, 1964a, 1964b) and the turkey (Steadman, 1980), all of these extinct species need systematic revision.

Many of the extinct species from the Fossil Lake aquatic avifauna (Howard, 1946; Shufeldt, 1913) are not known elsewhere; some differ from their living counterparts only in minor ways that suggest possibly synonymy. For now, extinct species of grebes, flamingos, gulls, jaegers, swans, geese, ducks, eagles, and falcons must be recognized from Fossil Lake. The flamingo from Fossil Lake, *Phoenicopterus* *copei*, is also known from Manix Lake, California, where a smaller extinct flamingo, *P.* *minutus*, also occurs (Howard, 1955). Now absent from the United States, flamingos may have been common late-Pleistocene inhabitants of the playa lakes of the western United States. Their demise in North America is probably related to the drying of the lakes (Street and Grove, 1979: Figure 12) in combination with aboriginal hunting (Steadman and Martin, in press).

Almost all of the known late-Pleistocene mammalian record for this region comes from California (Table 16-6). The tar deposits of Rancho La Brea and McKittrick have provided the largest and perhaps most complete samples of this environment. Only one other locality with a limited fauna is known: the Manis Mastodon site (Gustafson et al., 1979) from the Olympic Peninsula in Washington. The late-Pleistocene faunas of this area were similar to those of the Rocky Mountains to the east; the mountainous parts of both regions shared taxa such as + *Euceratherium collinum*, *Oreamnos americanus*, and + *Nothrotheriops shastensis*.

The Channel Islands off the Pacific coast of California are unique in being the only island localities in the United States that have a Pleistocene fauna. The fauna contains few taxa, but it shows the phenomenon of dwarfed mammoths seen in island faunas of the Mediterranean and Southeastern Asia (Johnson, 1978). No large predators are known. Two species of *Peromyscus* are larger than mainland forms.

Faunal Changes

Because of the lack of accurate dating techniques, the limited number of precisely dated samples, and the small number of sites with finely stratified fossiliferous deposits, there are few areas in the United States where late-Pleistocene faunal changes can be reliably documented.

In the eastern United States, New Paris No. 4 in Pennsylvania (Guilday et al., 1964) and Baker Bluff Cave in Tennessee (Guilday et al., 1978) have sequences that show faunal changes during the interval 25,000 to 10,000 yr B.P. The sequence at New Paris No. 4 shows a decrease in microtine rodents from 80% of the total number of individuals in the lowest levels (9m) to 34% in the upper levels (4.9 to 5.5 m), which are dated at 11,300 ± 1000 yr B.P. The lower levels contain remains of one tundra form, *Dicrostonyx hudsonius*, and abundant remains of boreal and grassland microtines, such as *Microtus xanthognathus*, *M. pennsylvanicus*, and *Synaptomys borealis*. These species were gradually replaced in higher levels by forest species, such as *Cleithrionomys gapperi* and *Phenacomys* sp., and by more-southern species, such as *Synaptomys cooperi* and *Microtus pinetorum*. Other grassland forms, such as *Spermophilus tridecemlineatus*, decrease in the upper levels, while the woodland shrew *Blarina* and the flying squirrel *Glaucomys* increase. The reptiles in the lowest levels are predominantly natricines. Crotalids and colubrids appear and then increase in relative abundance in the upper levels. Possibly the association of some of the reptiles, such as the crotalids, with boreal mammals such as *Microtus xanthognathus* is the result of the later intrusion of the reptile remains via fissures in the deposit.

The faunal sequence in the Baker Bluff Cave shows a change from a temperate fauna in the lower levels with a high percentage of *Tamias striatus*, *Sciurus carolinensis*, *Glaucomys volans*, *Microtus pinetorum* and/or *M. ochrogaster*, and a small-sized *Blarina* (*B. kirtlandi*) to a more boreal fauna in the upper levels with a higher percentage of *Tamiasciurus hudsonius*, *Glaucomys sabrinus*, *Cleithriomys gapperi*, *Phenacomys intermedius*, *Microtus xanthognathus*, *Sorex arcticus*, *Microsorex hoyi*, and a large *Blarina* (*B. brevicauda*). A sample from the lower levels is dated 19,100 ± 850 yr B.P.; the upper levels are dated at 11,000 to 10,000 yr B.P. *Spermophilus tridecemlineatus*, a midwestern prairie form, also was found in the upper levels. The taxonomic diversity of the soricids is higher in the upper levels (10 taxa) than in the lower levels (6 taxa). This difference apparently reflects a shift from interstadial to glacial conditions that began in the lower levels (Guilday et al., 1978).

The faunas from the Great Plains and Rocky Mountains can be divided into three temporal categories for discussions of chronologic changes. Faunas have been assigned to the first category, pre-Clovis (25,000 to 12,000 yr B.P.), on the basis of radiocarbon dates and stratigraphic position relative to radiocarbon-dated strata or cultural horizons. The second category, Clovis (12,000 to 11,000 yr B.P.), is defined by radiocarbon dates and faunas associated with artifacts from the Clovis culture (Haynes, 1980). The third category, post-Clovis (11,000 to 10,000 yr B.P.), is defined by carbon dates, absence of numerous species of extinct megafauna, or association with post-Clovis paleo-Indian artifacts that have absolute dates within the prescribed interval.

Three faunas from the Great Plains (Jones [Davis, 1975; Hibbard, 1940]) and the Rocky Mountains (Rainbow Beach [McDonald and Anderson, 1975] and Dam [Barton, 1975]) are associated with radiocarbon dates greater than 25,000 yr B.P. However, these dates are of either shell or bone and, therefore, may not reflect the true age of the fauna. The Jones local fauna from Meade County, Kansas, is divisible into an upper *Ambystoma* zone and a lower Classen zone (Davis, 1975). Snail shells from the upper and lower parts of the upper *Ambystoma* zone have yielded radiocarbon dates of 26,700 ± 1500 yr B.P. (I-3461) and 29,000 ± 1300 yr B.P. (I-3462), respectively (B. Miller, personal communication to L. C. Davis, 1975). This fauna suggests a climate that was cooler (by 3 °C to 8 °C) than the present climate and with less annual precipitation, especially in the winter (Davis, 1975). Both the microfauna and the megafauna indicate a treeless grassland, but enough moisture was present to form lakes.

Bone from the Dam site in Idaho (Barton, 1975; Kurtén and Anderson, 1980) has been radiocarbon dated at 26,500 ± 3500 yr B.P. The megafauna indicates an extensive grassland with riparian forest. Bone collagen from the closely related Rainbow Beach local fauna in Idaho (McDonald and Anderson, 1975) has been radiocarbon dated at between 21,500 ± 700 and 31,300 ± 2300 yr B.P. A diverse microfauna of fishes, amphibians, reptiles, birds, and mammals as well as a megafauna indicate aquatic, open-grassland, and forest-scrub habitats.

Six vertebrate faunas from the Great Plains and Rocky Mountains that are definitely between 20,000 and 12,000 years old reflect an equable climate with cool summers, mild winters, and ample precipitation. All are disharmonious faunas and generally indicate open environments. The Gray Sand faunule below the Clovis horizon at Blackwater Draw locality No. 1 (Hester, 1972), Roosevelt County, New Mexico, contains a rich late-Pleistocene vertebrate fauna (Hester, 1972). Lundelius (1972) interprets the climate as being somewhat cooler and moister than that of today and suggests a regional environment of partially wooded grassland. However, only two species, *Terrapene carolina *putnami* and *Ondatra zibethica*, may represent a wooded habitat, while grassland is represented by *Vulpes velox*, + *Camelops*, + *Hemiauchenia*, + *Platygonus*, *Equus*, *Bison*, and *Antilocapra* (Graham, in press).

Faunas of a similar age from Lubbock Lake, Texas (Johnson, 1974), Selby-Dutton, Colorado (Graham, in press), and Natural Trap Cave, Wyoming (Martin and Gilbert, 1978), also suggest extensive grassland environments for that interval. In contrast, the approximately 14,000-year-old fauna from Dry Cave in Eddy County, New Mexico, reflects a mosaic of sagebrush grassland, Sonoran grassland, and heavy riparian growth along drainageways (Harris, 1970). This difference probably is the result of the greater topographic relief at Dry Cave compared to the other sites. The lower levels of the Wasden site are slightly older than 12,000 yr B.P. as determined by two radiocarbon dates on mammoth-bone collagen (Butler, 1972). The fauna from these sediments indicates a grassland environment with little wooded habitat (Butler, 1972; Guilday, 1967b).

The reconstructions of significant areas of grassland on the Great Plains during this time interval are supported by pedologic,

molluscan, and pollen studies (Graham, in press). These grasslands rapidly graded into extensive parkland or savannah in eastern Kansas, Nebraska, and South Dakota.

Well-dated local faunas from 12,000 to 11,000 yr B.P., essentially coincident with the Clovis culture, all indicate open environments; these include Jaguar Cave, Idaho (Kurtén and Anderson, 1972), Colby, Wyoming (Walker and Frison, 1980), Roberts, Kansas (Schultz, 1967), Domebo, Oklahoma (Slaughter, 1966), Lubbock Lake (unit c), Texas (Johnson, 1974), Blackwater Draw Brown Sand Wedge, New Mexico (Slaughter, 1975), and Upper Sloth Cave, Texas (Logan and Black, 1979). In spite of this uniformity, there were regional differences in vegetation and faunas. Eastern New Mexico and western Oklahoma probably supported savanna grassland and parkland. Eastern Colorado, western Kansas, and northern and western Texas were dominated by prairie vegetation. At higher elevations in the northern Basin and Range area of Wyoming and the Rocky Mountains of Idaho, the open environments included faunal elements with tundra affinities (e.g., *Dicrostonyx*).

The climate on the Great Plains and Rocky Mountains 12,000 to 11,000 years ago was not at all like the modern climate. On the lowlands of the Plains, the southern distribution of species such as *Blarina brevicauda*, *Sorex cinereus*, *S. palustris*, *Zapus hudsonius*, *Spermophilus richardsoni*, *Microtus pennsylvanicus*, and *Ovibos*, which reside at higher latitudes and elevations today, indicates that the summers were cooler and perhaps moister than today. Species such as *Ondatra zibethica*, *Synaptomys cooperi*, and *Blarina carolinesis* demonstrate a definite increase in effective moisture. The persistence of southern forms such as + *Geochelone wilsoni*, *Sigmodon hispidus*, and *Dasypus *bellus* in the Clovis faunas suggests that, although the mean annual temperature was probably lower, the winters were not marked by extreme cold spells. These disharmonious associations of species that are distinctly allopatric today suggest that the late-Pleistocene climate was cooler and moister and lacked great seasonal extremes (Graham, 1976a). Such an interpretation is supported by the invertebrate record (Cheatum and Allen, 1966; Wendorf, 1961) and the pollen record (Schoenwetter, 1975; Wilson, 1966).

The temporal position of many of the 11,000 to 10,000-year-old faunas, including Casper, Agate Basin, Hell Gap, Jones-Miller, Lubbock Lake, and Blackwater Draw, can be closely determined on the basis of well-dated cultural sequences. Therefore, a more precise sequential environmental reconstruction can be made for this period of time. These reconstructions are limited only by the scarcity of sites with good faunal records from the Great Plains and Rocky Mountains.

The disconformity between the Clovis (unit C) and Folsom (unit D) horizons at Blackwater Draw marks a period of erosion. This erosion could have occurred in an arid climate with decreased vegetative cover (Harris, 1977). Vertebrate fossils suggest that at the beginning of unit D (dated at 10,490 ± 200 to 10,170 ± 260 yr B.P.) the vegetation was open grassland very similar to that of Clovis times (unit C). During Folsom time (circa 10,500 yr B.P.), the environment was the most mesic and heavily forested on the High Plains (Harris, 1977). More-arid climates and open landscapes returned again before the end of the unit-B deposition.

The faunas from the Hell Gap (Roberts, 1970), Jones-Miller (Rawn-Schatzinger, no date), and Agate Basin (Walker, in press) sites also represent a time when the central plains apparently were forested. Because these paleo-Indian (circa 10,000 yr B.P.) faunas are younger than the fauna from the Folsom horizon at Blackwater Draw and are also more than 640 km farther north, it is difficult to tell without additional evidence whether there were several episodes of forestation and deforestation or whether these faunas represent a slow, gradual retreat of the boreal forest to the north. In either case, the faunal sequence indicates that forestation of the western part of the central and southern Plains took place after Clovis time. The Casper fauna (Wilson, 1974) appears to represent a more-open environment than do the faunas from contemporaneous paleo-Indian sites.

Morphological Clines

A number of mammalian taxa show geographic variation in morphology and/or size during the late Pleistocene. Guilday and others (1967) report that nine species of late-Pleistocene mammals from the New Paris No. 4 fauna show a positive Bergmann's response in comparison with modern populations of that region; these species are *Condylura cristata, Blarina brevicauda, Microsorex hoyi, Sorex cinereus, Tamias striatus, Glaucomys sabrinus, G. volans, Tamiasciurus hudsonicus,* and *Napaeozapus insignis.* (Bergmann's response is the tendency of homoiothermal animals living in cold climates to be larger and thus to have less surface area per unit volume than do animals of the same taxon that live in warmer climates. This apparently promotes improved heat conservation in the cold-climate forms and heat dissipation in the warm-climate forms [Allee et al., 1949].) Four other species from that fauna show a negative Bergmann's response; these are *Synaptomys cooperi, Microtus chrotorrhinus, M. pennsylvanicus,* and *Lepus americanus.* The negative Bergmann's response seen in *Synaptomys cooperi* in the New Paris No. 4 local fauna is supported by its increase in size to the south during the Late Wisconsin, with the largest specimens occurring in Florida (Simpson, 1928). *Platygonus compressus* also shows a negative Bergmann's response (Ray et al., 1970). A negative Bergmann's response is atypical for endothermic vertebrates, but McNabb (1971) demonstrates that it cannot be directly correlated with body size and surface-volume ratios. *Synaptomys cooperi* shows a morphological cline in which the posterior end of the lower incisor extends beyond the M_3, in southern populations but does not reach M_3 in northern populations. This trait led Simpson (1928) to regard the Pleistocene *Synaptomys* of Florida as a separate species, *S. australis.* Patton (1963) shows that morphological intermediates between these two extremes exist in the midcontinent area; this finding indicates that the trait is better regarded as a morphological cline. The muskrat *Ondatra zibethica* shows a positive Bergmann's response and a decrease from north to south in the length/width ratio of M_1 (Nelson and Semken, 1970).

It is now known that the pattern of variation in the short-tailed shrew (*Blarina*) is more complex. Modern populations of *Blarina* in the United States have been placed in the single species *B. brevicauda,* which has been divided into three subspecies: *B. b. brevicauda,* a large form that inhabits the northeastern Great Plains; *B. b. carolinensis,* a small form that inhabits the South and Southeast; and *B. b. kirtlandi,* an intermediate form that inhabits the Midwest and Northeast (Hall and Kelson, 1959). A number of late-Pleistocene local faunas have two or three nonoverlapping size groups of *Blarina* that correspond to the three modern nominal subspecies (Graham and Semken, 1976). This indicates that these three taxa

were behaving as separate species during the Pleistocene, and so they should be recognized as such. The coexistence of these different taxa has been interpreted as indicating a more equable climate during the late Pleistocene (Graham and Semken, 1976).

Taxonomic diversity of shrews and microtines during the late Pleistocene has been shown by Graham (1976a) to decrease from Northeast to Southwest, with a reversal in the Ozark Highlands. This cline parallels the modern one for these groups, but the diversity of each group was greater in the Pleistocene faunas than it is in the modern populations at each site. This indicates a climatic gradient during the Pleistocene similar to the modern one but with more available moisture and more equablity than at present.

Disharmonious Communities

Disharmonious floras and faunas, composed of species that today are separated geographically and appear to be ecologically incompatible, are characteristic of virtually all of the late-Pleistocene biota of the United States. Only in the far-western United States are the late-Pleistocene vertebrate faunas fairly similar in composition to the modern faunas.

There are no modern analogues for the disharmonious biotas or the equable climates that supported them (Dalquest, 1965; Hibbard, 1960; Guilday et al., 1964; Lundelius, 1974, 1976). Different disharmonious groups of species occurred in different parts of the United States. The disharmonious biotas are the result of the individualistic responses of species to the changing environmental conditions of the late Pleistocene (Graham, 1976a, 1979); therefore, these disharmonious biotas were geographically heterogeneous but probably less so than the modern ones.

The disintegration of the disharmonious biotas and the modernization of the vertebrate fauna appear to be related to major environmental changes at the end of the Pleistocene. The more-continental climate of the Holocene accentuated environmental extremes and caused sequential readjustments in the distribution of individual species. The modernization of the vertebrate fauna of any one geographic area was a dynamic phenomenon. For example, in the southern Great Plains and Texas, warming first caused the removal of cool-adapted (''boreal'') species such as *Sorex cinereus, Lepus townsendi, Mustela erminea,* and *Microtus pennsylvanicus.* Less thermally sensitive ''deciduous'' species such as *Blarina, Cryptotis parva, Synaptomys cooperi,* and *Microtus ochrogaster* were extirpated later, probably as a result of increased aridity. In future studies, it may be possible to use the specific sequential dissolution of disharmonious biotas to predict the sequential alterations of the climate.

Like the continuous faunal change in any one geographic locale, the disappearance of disharmonious biotas and the appearance of the modern biota seem to have been time transgressive throughout the Great Plains and Rocky Mountains. Disharmonious biotas in the southern Plains and southern Rocky Mountains had been substantially reorganized by 10,000 yr B.P., and best estimates suggest that these changes began before 11,000 yr B.P. In the northern Plains, disharmonious faunas have been found in association with paleo-Indian bison kill sites (Hell Gap, Jones-Miller, Agate Basin) having ages of 10,000 yr B.P. or younger (Table 16-1). Time-transgressive lags may reflect latitudinal lapse rates in climatic adjustments to the ablation of the Laurentide and Cordilleran ice sheets.

Some of the younger disharmonious faunas on the northern Plains may have been the result of localized climatic amelioration. Glacial meltwater from the late stage of the Pinedale Glaciation may have flowed from the Rocky Mountains down major drainages on the Plains, such as the Platte River, producing favorable microenvironments for more-boreal communities. Other younger disharmonious faunas cannot be explained in this way. One example is the Jones-Miller local fauna (Rawn-Schatzinger, no date), a Hell Gap bison kill site (10,000 yr B.P.) that contains boreal species such as *Sorex arcticus, S. palustris, S. cinereus,* and *Microtus montanus.* This site is located on the Arikaree River, which drains a relatively small area compared to the area drained by the Platte River. In addition, the volume of ice, and consequently the amount of meltwater, for the last stage of the Pinedale Glaciation was much smaller than the corresponding volumes for the Early Pinedale stage (Madole, 1976). Therefore, localized climatic amelioration by glacial meltwater cannot explain this disharmonious fauna.

Extinction

The most obvious faunal change marking the end of the Pleistocene in the United States was the extinction of a large number of vertebrates. Not all groups were affected in the same way. The North American herpetofauna of 25,000 to 10,000 yr B.P. appears to have consisted mostly of the same species that exist today. Of 144 species of reptiles and amphibians represented in late-Pleistocene deposits, only 4 are extinct: 1 snake (*Crotalus *potterensis,* whose taxonomic status is doubtful) and 3 species of turtles of the genus +*Geochelone.* The number of avian taxa that became extinct is uncertain because of the problems with identification and taxonomy, as discussed above. As many as 28 species of birds may have become extinct at the end of the Pleistocene. The majority of these come from faunas in the southern United States, particularly the Southwest. Approximately a third of the extinct taxa were either predators or scavengers; several of the extinct taxa were large forms. With a few exceptions, such as the skunk +*Brachyprotoma,* the chipmunk *Tamias *aristus,* and the small antilocaprid +*Capromeryx,* all of the mammals that became extinct were large, with body weights greater than 60 kg.

Studies of the radiocarbon chronology of the extinctions were begun by Jelinek (1957) and Martin (1958). A more comprehensive study by Hester (1960) concludes that 17 genera of mammals disappeared between 20,000 and 7000 yr B.P., with the majority becoming extinct about 8000 yr B.P. A study by Martin (1967), who used additional dates and reevaluations of Hester's dates, concludes that the major episodes of extinction occurred between 11,000 and 10,000 yr B.P. Martin's data indicate that some taxa, such as +*Capromeryx* and +*Smilodon,* last occurred as early as 15,000 yr B.P. and that +*Mylohyus* may have survived somewhat beyond 10,000 yr B.P.; however, recently obtained dates on the +*Smilodon* material from the American Bank site in Tennessee suggest a much later date of disappearance for this genus. More recent data show that +*Homotherium* probably disappeared earlier, between 20,000 and 18,000 yr B.P. (Graham, 1976b), and that +*Camelops* survived until 10,000 yr B.P. in the northern Great Plains (Frison et al., 1978).

The date of the major episode of extinction coincides closely with the date of the breakdown of disharmonious assemblages and the date of the northward retreat of most boreal species and with evidence of increased climatic extremes. Late-Pleistocene faunas in all parts of North America that are dated at 10,000 yr B.P. or older contain numerous disharmonious pairs of species. Those faunas with younger dates have few or no disharmonious pairs (Semken, 1982).

Summary and Conclusions

The interval from 25,000 to 10,000 yr B.P. was a period of great faunal and climatic change. Fossil vertebrate faunas from most of this interval seem to indicate a more equable climate than that of today, with smaller seasonal extremes. Taxa that we think of today as "boreal," "temperate," or "subtropical" have been found in associations that do not occur under today's climatic conditions. Fossil vertebrate faunas that date from 11,000 to 10,000 yr B.P. show an apparently sudden shift toward "modern" associations of taxa and the extinction of many species, especially large mammals. These changes coincide closely with the time of disruption of the equable climate pattern, as estimated from other lines of evidence. This relation suggests that climatic disruption may have played a major role in the wave of extinctions and range restrictions that came near the end of this interval.

Although much has been learned in recent decades, many parts of the overall pattern of life during this time remain unclear. There is a serious need for taxonomic revision of many groups. The uneven coverage of various ages, taxonomic groups, and geographic areas, together with numerous depositional biases, create many problems in correlation among sites and faunas. Our incomplete understanding of the factors involved in the environmental tolerances of living animals makes it difficult to interpret the distributions and associations of fossil taxa. Although the overall cause of the extinctions appears to be climatic change, the exact way in which this acted on the various taxa is not well understood. Particularly puzzling is understanding why some taxa became extinct when others with seemingly similar environmental needs were not affected by the climatic changes. Finally, we are prevented from precisely determining the extent and timing of faunal changes in this interval by the lack of completely reliable dating methods.

It may be possible to resolve some of these uncertainties by applying recently developed techniques. Many others can be dealt with by the more thorough study of known faunas or by the wider application of techniques available but never put to their fullest use. Solutions to the remaining problems will have to wait for the discovery of new faunas or new ways of dealing with known faunas.

References

Aikens, C. M. (1970). "Hogup Cave." University of Utah Anthropological Papers 93.

Akersten, W. A. (1979). New mammalian records from the late Pleistocene of Rancho La Brea. *Bulletin of the Southern California Academy of Science* 78, 141-43.

Alexander, H. L., Jr. (1963). The Levi site: A paleo-Indian campsite in central Texas. *American Antiquity* 281, 510-28.

Allee, W. C., Park, O., Park, T., Emerson, A. E., and Schmidt, K. P. (1949). "Principles of Animal Ecology." W. B. Saunders, Philadelphia.

Allison, I. S. (1966). "Fossil Lake, Oregon: Its Geology and Fossil Faunas." Oregon State Monograph Studies in Geology 9.

Anderson, E. (1968). "Fauna of the Little Box Elder Cave, Converse County, Wyoming: The Carnivora." University of Colorado Studies Series, Earth Sciences 6.

Anderson, E. (1974). A survey of the late Pleistocene and Holocene mammalian fauna of Wyoming. *In* "Applied Geology and Archeology: The Holocene History of Wyoming" (M. V. Wilson, ed.), pp. 78-87. Geological Survey of Wyoming Report of Investigations 10.

Arata, A. A. (1964). Fossil vertebrates from Avery Island. *In* "An Archeological Survey of Avery Island" (S. M. Gagliano, ed.), Appendix, pp. 66-72. Avery Island, Inc., and Coastal Studies Institute, Louisiana State University, Baton Rouge.

Auffenberg, W. (1963a). The fossil snakes of Florida. *Tulane Studies in Zoology* 10, 131-216.

Auffenberg, W. (1963b). Fossil Testudinine Turtles of Florida, Genera *Geochelone* and *Floridemys. Bulletin of the Florida State Museum* 7, 53-97.

Auffenberg, W., and Milstead, W. W. (1965). Reptiles in the Quaternary of North America. *In* "The Quaternary of the United States" (H. E. Wright, Jr., and D. G. Frey, eds.), pp. 557-68. Princeton University Press, Princeton, N.J.

Bader, R. S. (1957). Two Pleistocene mammalian faunas from Alachua County, Florida. *Bulletin of the Florida State Museum* 2, 53-75.

Bannan, D. (1980). Stratigraphy and sedimentology of late Quaternary sediments in the High Plains depressions of Yuma and Kit Carson Counties, Colorado. M.S. thesis, University of California, Davis.

Barton, J. B. (1975). A preliminary report of the American Falls Dam local fauna, Idaho (abstract). *Proceedings of the Utah Academy of Sciences, Arts, and Letters* 52, 76.

Blackwelder, D. W., Pilkey, O. H., and Howard, J. D. (1979). Late Wisconsinan sea level curvature for the southeastern United States continental shelf. *Science* 204, 618-20.

Brattstrom, B. H. (1953a). The amphibians and reptiles from Rancho La Brea. *Transactions of the San Diego Society of Natural History* 11, 365-92.

Brattstrom, B. H. (1953b). Records of Pleistocene reptiles from California. *Copeia* 1953, 174-79.

Brattstrom, B. H. (1954a). Amphibians and reptiles from Gypsum Cave, Nevada. *Bulletin of the Southern California Academy of Science* 53, 8-12.

Brattstrom, B. H. (1954b). The fossil pit vipers (Reptilia:Crotalidae) of North America. *Transactions of the San Diego Society of Natural History* 12, 31-46.

Brattstrom, B. H. (1955). Small herpetofauna from the Pleistocene of Carpinteria, California. *Copeia* 1955, 138-39.

Brattstrom, B. H. (1958a). Additions to the Pleistocene herpetofauna of Nevada. *Herpetologica* 14, 36.

Brattstrom, B. H. (1958b). New records of Cenozoic amphibians and reptiles from California. *Bulletin of the Southern California Academy of Science* 57, 5-12.

Brattstrom, B. H. (1961). Some new fossil tortoises from western North America with remarks on the zoogeography and paleoecology of tortoises. *Journal of Paleontology* 35, 343-60.

Brattstrom, B. H. (1964). Amphibians and reptiles from cave deposits in south-central New Mexico. *Bulletin of the Southern California Academy of Science* 63, 93-103.

Brattstrom, B. H. (1976). A Pleistocene herpetofauna from Smith Creek Cave, Nevada. *Bulletin of the Southern California Academy of Science* 75, 283-84.

Brodkorb, P. (1956a). Pleistocene birds from Crystal Springs, Florida. *Wilson Bulletin* 68, 158.

Brodkorb, P. (1956b). Pleistocene birds from Eichelberger Cave, Florida. *Auk* 73, 136.

Brodkorb, P. (1959). The Pleistocene avifauna of Arredondo, Florida. *Bulletin of the Florida State Museum, Biological Sciences* 4, 269-91.

Brodkorb, P. (1963). Catalogue of fossil birds, part 1 (Archaeopterygiformes through Ardeiformes). *Bulletin of the Florida State Museum, Biological Sciences* 7, 179-293.

Brodkorb, P. (1964). Catalogue of fossil birds, part 2 (Anseriformes through Galliformes). *Bulletin of the Florida State Museum, Biological Sciences* 8, 195-335.

Brodkorb, P. (1967). Catalogue of fossil birds, part 3 (Ralliformes, Ichthyorniformes, Charadriiformes). *Bulletin of the Florida State Museum, Biological Sciences* 11, 99-220.

Butler, B. R. (1969). More information on the frozen ground features and further interpretation of the small mammal sequence at the Wasden site (Owl Cave), Bonneville County, Idaho. *Tebiwa* 12, 58-63.

Butler, B. R. (1972). The Holocene or postglacial ecological crisis on the eastern Snake River Plain. *Tebiwa* 15, 49-63.

Campbell, K. E., Jr. (1980). A review of the Rancholabrean avifauna of the Itchtucknee River, Florida. *Contributions in Science, Natural History Museum of Los Angeles County* 330, 119-29.

Cheatum, E. P., and Allen, D. (1966). Ecological significance of the fossil fresh-water and land snails from the Domebo Mammoth Kill site. *In* "Domebo: A Paleo-indian Mammoth Kill in the Prairie Plains" (F. C. Leonhardy, ed.), pp. 36-43. Contributions of the Museum of the Great Plains 1.

Choate, J. R., and Hall, E. R. (1967). Two new species of bats, genus *Myotis*, from a Pleistocene deposit in Texas. *American Midland Naturalist* 78, 531-34.

Clausen, C. J., Cohen, A. D., Emiliani, C., Holman, J. A., and Stipp, J. J. (1979). Little Salt Spring, Florida: A unique underwater site. *Science* 203, 609-14.

Corgan, J. X. (1974). Fossil birds of Tennessee. *Migrant* 45, 81-85.

Corner, R. G. (1977). A late Pleistocene-Holocene vertebrate fauna from Red Willow County, Nebraska. *Transactions of the Nebraska Academy of Sciences* 4, 77-93.

Dalquest, W. W. (1965). New Pleistocene formation and local fauna from Hardeman County, Texas. *Journal of Paleontology* 39, 63-79.

Dalquest, W. W., Roth, E., and Judd, F. (1969). The mammal fauna of Schulze Cave, Edwards County, Texas. *Bulletin of the Florida State Museum* 13, 206-76.

Davis, L. C. (1975). Late Pleistocene geology and paleoecology of the Spring Valley basin, Meade County, Kansas. Ph.D. dissertation, University of Iowa, Iowa City.

Dibble, D. S. (1970). On the significance of additional radiocarbon dates from Bonfire Shelter, Texas. *Plains Anthropologist* 15, 251-54.

Dibble, D. S., and Lorrain, D. (1968). "Bonfire Shelter: A Stratified Bison Kill Site, Val Verde County, Texas." Texas Memorial Museum Miscellaneous Papers 1.

Dolan, E. M., and Allen, G. T. (1961). "An Investigation of the Darby and Hornsby Springs Sites, Alachua Co., Florida." Florida Geological Survey Special Publication 7, pp. 1-24.

Domning, D. P. (1969). A list, bibliography and index of the fossil vertebrates of Louisiana and Mississippi. *Transactions of the Gulf Coast Association of Geological Societies* 19, 385-422.

Dort, W. (1975). Archaeo-geology of Jaguar Cave, upper Birch Creek valley, Idaho. *Tebiwa* 17, 33-57.

Downs, T. (1954). Pleistocene birds from the Jones fauna of Kansas. *Condor* 56, 207-21.

Downs, T., Howard, H., Clements, T., and Smith, G. A. (1959). Quaternary animals from Schuiling Cave in the Mojave Desert, California. *Contributions in Science, Los Angeles County Museum* 29, 1-21.

Emery, K. O. (1967). Freshwater peat on the continental shelf. *Science* 158, 1301-7.

Fortsch, D. E. (1972). A late Pleistocene vertebrate fauna from the northern Mojave Desert of California. M.S. thesis, University of Southern California, Los Angeles.

Frazier, M. F. (1977). New records of *Neofiber leonardi* (Rodentia, Cricetidae) and the paleoecology of the genus. *Journal of Mammalogy* 58, 368-73.

Frison, G. C. (ed.) (1974). "The Casper Site: A Hell Gap Bison Kill on the High Plains." Academic Press, New York.

Frison, G. C., Walker, D. N., Webb, S. D., and Zeimens, G. M. (1978). Paleo-Indian procurement of *Camelops* on the northwestern Plains. *Quaternary Research* 10, 385-400.

Funk, R. E., Fisher, D. W., and Reilly, E. M., Jr. (1970). Caribou and paleo-Indian in New York State: A presumed association. *American Journal of Science* 268, 181-86.

Furlong, E. L. (1906). The exploration of Samwell Cave. *American Journal of Science* 22, 235-47.

Furlong, E. L. (1941). Stone Man Cave, Shasta County, California. *Science* 94, 414-15.

Gagliano, S. M. (ed.) (1964). "An Archaeological Survey of Avery Island." Avery Island, Inc., and Coastal Studies Institute, Louisiana State University, Baton Rouge.

Gagliano, S. M. (1967). "Occupation Sequence at Avery Island." Louisiana State University Coastal Study Series 22.

Galbreath, E. C. (1944). *Grus canadensis* from the Pleistocene of Illinois. *Condor* 46, 35.

Gillette, D. D. (1976). A new species of small cat from the late Quaternary of southeastern United States. *Journal of Mammalogy* 57, 664-76.

Graham, R. W. (1976a). Late Wisconsin mammal faunas and environmental gradients of the eastern United States. *Paleobiology* 2, 343-50.

Graham, R. W. (1976b). Pleistocene and Holocene mammals, taphonomy, and paleoecology of the Friesenhahn Cave local fauna, Bexar County, Texas. Ph.D. dissertation, University of Texas, Austin.

Graham, R. W. (1979). Paleoclimates and late Pleistocene faunal provinces in North America. *In* ''Pre-Llano Cultures of the Americas: Paradoxes and Possibilities'' (R. L. Humphrey and D. J. Stanford, eds.), pp. 46-69. Anthropological Society of Washington, Washington, D.C.

Graham, R. W. (1981). ''Preliminary Report on Late Pleistocene Vertebrates from the Selby and Dutton Archeological/Paleontological Sites, Yuma County, Colorado.'' Contributions in Geology, University of Wyoming 20, pp. 33-56.

Graham, R. W., and Semken, H. A., Jr. (1976). Paleoecological significance of the short-tailed shrew (*Blarina*), with a systematic discussion of *Blarina ozarkensis*. *Journal of Mammalogy* 57, 433-49.

Grayson, D. K. (1977). A review of the evidence for early Holocene turkeys in the northern Great Basin. *American Antiquity* 42, 110-14.

Gruhn, R. (1961). ''The Archeology of Wilson Butte Cave, South-Central Idaho.'' Occasional Papers of the Idaho State College Museum 6.

Guilday, J. E. (1962). The Pleistocene local fauna of the Natural Chimneys, Augusta County, Virginia. *Annals of the Carnegie Museum* 36, 87-122.

Guilday, J. E. (1967a). The climatic significance of the Hosterman's Pit local fauna, Centre County, Pennsylvania. *American Antiquity* 32, 231-32.

Guilday, J. E. (1967b). Differential extinction during late Pleistocene and recent times. *In* ''Pleistocene Extinctions: The Search for a Cause'' (P. S. Martin and H. E. Wright, Jr., eds.), pp. 121-40. Yale University Press, New Haven, Conn.

Guilday, J. E. (1971). The Pleistocene history of the Appalachian mammal fauna. *In* ''The Distributional History of the Biota of the Southern Appalachians: Part III. Vertebrates'' (P. C. Holt, ed.), pp. 233-62. Virginia Polytechnic Institute State University Research Division Monograph 4.

Guilday, J. E. (1977). Sabertooth cat, *Smilodon floridanus* (Leidy), and associated fauna from a Tennessee Cave (40 Dv 40), the First American Bank site. *Journal of the Tennessee Academy of Science* 52, 84-94.

Guilday, J. E., and Adam, E. K. (1967). Small mammal remains from Jaguar Cave, Lemhi County, Idaho. *Tebiwa* 10, 26-36.

Guilday, J. E., and Hamilton, H. W. (1973). The late Pleistocene small mammals of Eagle Cave, Pendleton County, West Virginia. *Annals of the Carnegie Museum* 44, 45-58.

Guilday, J. E., and Hamilton, H. W. (1978). Ecological significance of displaced boreal mammals in West Virginia caves. *Journal of Mammalogy* 59, 176-81.

Guilday, J. E., Hamilton, H. W., and Adam, E. K. (1967). ''Animal Remains from Horned Owl Cave, Albany County, Wyoming.'' Contributions to Geology, University of Wyoming 6, pp. 97-99.

Guilday, J. E., Hamilton, H. W., Anderson, E., and Parmalee, P. W. (1978). The Baker Bluff Cave deposit, Tennessee, and the late Pleistocene faunal gradient. *Bulletin of the Carnegie Museum of Natural History* 11, 1-67.

Guilday, J. E., Hamilton, H. W., and McCrady, A. D. (1966). The bone breccia of Bootlegger Sink, York County, Pennsylvania. *Annals of the Carnegie Museum* 38, 145-63.

Guilday, J. E., Hamilton, H. W., and McCrady, A. D. (1969). The Pleistocene vertebrate fauna of Robinson Cave, Overton County, Tennessee. *Palaeovertebrata* 2, 25-75.

Guilday, J. E., Hamilton, H. W., and McCrady, A. D. (1971). The Welsh Cave peccaries (*Platygonus*) and associated fauna, Kentucky Pleistocene. *Annals of the Carnegie Museum* 43, 249-320.

Guilday, J. E., Hamilton, H. W., and Parmalee, P. W. (1975). Caribou (*Rangifer tarandus* L.) from the Pleistocene of Tennessee. *Journal of the Tennessee Academy of Science* 50, 109-12.

Guilday, J. E., and McGinnis, H. (1972). Jaguar (*Panthera onca*) from Big Bone Cave, Tennessee and east central North America. *Bulletin of the National Speleological Society* 34, 1-14.

Guilday, J. E., Martin, P. S., and McCrady, A. D. (1964). New Paris No. 4: A Pleistocene cave deposit in Bedford County, Pennsylvania. *Bulletin of the National Speleological Society* 26, 121-94.

Guilday, J. E., and Parmalee, P. W. (1979). Pleistocene and recent vertebrate remains from Savage Cave (15L011), Kentucky. *Western Kentucky Speleological Survey Annual Report* 1979, 5-10.

Guilday, J. E., Parmalee, P. W., and Hamilton, H. W. (1977). The Clark's Cave bone deposit and the late Pleistocene paleoecology of the central Appalachian Mountains of Virginia. *Bulletin of the Carnegie Museum of Natural History* 2, 1-87.

Gustafson, C. E., Gilbow, D., and Daughtery, R. D. (1979). The Manis mastodon site: Early man on the Olympic Peninsula. *Canadian Journal of Archeology* 3, 157-64.

Hager, M. (1972). ''A Late Wisconsin-Recent Vertebrate Fauna from the Chimney Rock Animal Trap, Larimer County, Wyoming.'' Contributions in Geology, University of Wyoming 2, pp. 63-71.

Hagmeier, E. M. (1966). A numerical analysis of the distributional patterns of North American mammals: A reevaluation of the provinces. *Systematic Zoology* 15, 279-99.

Hall, E. R. (1940). An ancient nesting site of the white pelican in Nevada. *Condor* 42, 87-88.

Hall, E. R., and Kelson, K. R. (1959). ''The Mammals of North America.'' Ronald Press, New York.

Hallberg, G. R., Semken, H. A., and Carson, L. C. (1974). Quaternary records of *Microtus xanthognathus* (Leach), the yellow-cheeked vole, from northwestern Arkansas and southwestern Iowa. *Journal of Mammalogy* 15, 640-45.

Harris, A. H. (1970). The Dry Cave mammalian fauna and late pluvial conditions in southeastern New Mexico. *Texas Journal of Science* 22, 3-27.

Harris, A. H. (1977a). Biotic environments of the Paleoindian. *In* ''Paleoindian Lifeways'' (E. Johnson, ed.), pp. 1-12. The Museum Journal 17. West Texas Museum Association, Lubbock, Texas.

Harris, A. H. (1977b). Wisconsin age environments in the northern Chihuahuan Desert: Evidence from the higher vertebrates. *In* ''Transactions of the Symposium on the Biological Resources of the Chihuahuan Desert Region, United States and Mexico'' (R. H. Wauer and D. H. Riskind, eds.), pp. 23-52. National Park Service Transactions and Proceedings Series 3. U.S. Government Printing Office, Washington, D.C.

Harris, A. H., and Findley, J. S. (1964). Pleistocene-Recent fauna of the Isleta Caves, Bernalillo County, New Mexico. *American Journal of Science* 262, 114-20.

Harris, A. H., Smartt, R. A., and Smartt, W. A. (1973). *Cryptotis parva* from the Pleistocene of New Mexico. *Journal of Mammalogy* 54, 512-13.

Hassan, A. A., Termine, J. D., and Haynes, C. V. (1977). Mineralogical studies on bone apatite and their implications for radiocarbon dating. *In* ''Radiocarbon Dating and Methods of Low Level Counting,'' pp. 163-68. International Atomic Energy Agency, Vienna.

Haury, E. (1950). ''The Stratigraphy and Archeology of Ventana Cave, Arizona.'' University of New Mexico Press, Albuquerque.

Hawksley, O., Reynolds, J. F., and Foley, R. L. (1973). Pleistocene vertebrate fauna of Bat Cave, Pulaski County, Missouri. *Bulletin of the National Speleological Society* 35, 61-87.

Hay, O. P. (1920). Descriptions of some Pleistocene vertebrates found in the United States. *Proceedings of the United States National Museum* 58, 83-146.

Hay, O. P. (1923). ''The Pleistocene of North America and Its Vertebrated Animals from States East of the Mississippi River and from the Canadian Provinces East of Longitude 95°.'' Carnegie Institution of Washington Publication 322.

Hay, O. P. (1924). ''The Pleistocene of the Middle Region of North America and Its Vertebrated Animals.'' Carnegie Institution of Washington Publication 322a.

Hay, O. P. (1926). A collection of Pleistocene vertebrates from southwestern Texas. *Proceedings of the United States National Museum* 68, 1-18.

Haynes, C. V. (1980). The Clovis culture. *Canadian Journal of Anthropology* 1, 115-21.

Hester, J. J. (1960). Pleistocene extinctions and radiocarbon dating. *American Antiquity* 26, 58-77.

Hester, J. J. (1972). ''Blackwater Locality No. 1: A Stratified Early Man Site in Eastern New Mexico.'' Fort Burgwin Research Center Publication 8.

Hibbard, C. W. (1940). A new Pleistocene fauna from Meade County, Kansas. *Transactions of the Kansas Academy of Science* 43, 417-25.

Hibbard, C. W. (1960). Pliocene and Pleistocene climates in North America. *Annual Report of the Michigan Academy of Science, Arts, and Letters* 62, 5-30.

Hibbard, C. W. (1970). Pleistocene mammal local faunas from the Great Plains and central lowland provinces of the United States. *In* ''Pleistocene and Recent Environments of the Central Great Plains'' (W. Dort, Jr., and J. K. Jones, Jr., eds.), pp. 395-433. Kansas University Department of Geology Special Publication 3.

Hill, H. W. (1971). Pleistocene snakes from a cave in Kendall County, Texas. *Texas Journal of Science* 22, 209-16.

Holman, J. A. (1958). The Pleistocene herpetofauna of Saber-Tooth Cave, Citrus County, Florida. *Copeia* 1958, 276-80.

Holman, J. A. (1968). A Pleistocene herpetofauna from Kendall County, Texas. *Quarterly Journal of the Florida Academy of Science* 31, 165-72.

Holman, J. A. (1969). Pleistocene amphibians from a cave in Edwards County, Texas. *Texas Journal of Science* 21, 63-67.

Holman, J. A. (1970). A Pleistocene herpetofauna from Eddy County, New Mexico. *Texas Journal of Science* 22, 29-39.

Holman, J. A. (1976). Paleoclimatic implications of "ecologically incompatible" herpetological species (late Pleistocene: southeastern United States). *Herpetologica* 32, 290-95.

Holman, J. A. (1980). Paleoclimatic implications of Pleistocene herpetofaunas of eastern and central North America. *Transactions of the Nebraska Academy of Science* 8, 131-40.

Holman, J. A. (1981). A review of North American Pleistocene snakes. *Publications of the Museum, Michigan State University, Paleontology Series* 7, 263-306.

Hood, C. H., and Hawksley, O. (1975). A Pleistocene fauna from Zoo Cave, Taney County, Missouri. *Missouri Speleology* 15, 1-42.

Howard, H. (1946). "A Review of the Pleistocene Birds of Fossil Lake, Oregon." Carnegie Institution of Washington Publication 551, pp. 141-95.

Howard, H. (1952). The prehistoric avifauna of Smith Creek Cave, Nevada, with a description of a new giant raptor. *Southern California Academy of Science Bulletin* 51, 50-54.

Howard, H. (1955). "Fossil Birds from Manix Lake, California." U.S. Geological Survey Professional Paper 264-J, pp. 199-205.

Howard, H. (1958). An ancient cormorant from Nevada. *Condor* 60, 411-13.

Howard, H. (1962a). Bird remains from a prehistoric cave deposit in Grant County, New Mexico. *Condor* 64, 241-42.

Howard, H. (1962b). "A Comparison of Avian Assemblages from Individual Pits at Rancho La Brea, California." Contributions in Science, Los Angeles County Museum 58, pp. 3-24.

Howard, H. (1963). "Fossil Birds from the Anza-Borrego Desert." Contributions in Science, Los Angeles County Museum 73.

Howard, H. (1964a). Fossil Anseriformes. *In* "The Waterfowl of the World" (J. Delacour, ed.), pp. 233-326. Country Life Limited, London.

Howard, H. (1964b). A new species of the "pygmy goose," *Anabernicula*, from the Oregon Pleistocene, with a discussion of the genus. *American Museum Novitates* 200, 1-14.

Howard, H. (1971). Quaternary avian remains from Dark Canyon Cave, New Mexico. *Condor* 73, 237-40.

Howard, H. (1972). The incredible teratorn again. *Condor* 74, 341-43.

Howard, H., and Miller, A. H. (1933). Bird remains from cave deposits in New Mexico. *Condor* 35, 15-18.

Hoyer, B. E. (1980). The geology of the Cherokee sewer site. *In* "The Cherokee Sewer Excavations: Holocene Ecology and Human Adaptations in Northwestern Iowa" (D. C. Anderson and H. A. Semken, Jr., eds.), pp. 21-66. Academic Press, New York.

Ives, P. C., Levin, B., Oman, C. L., and Rubin, M. (1967). United States Geological Survey radiocarbon dates IX. *Radiocarbon* 9, 505-29.

Jackson, C. G., and Kaye, J. M. (1974a). The occurrence of Blanding's turtle, *Emydoidea blandingi*, in the late Pleistocene of Mississippi (Testudines: Testudinidae). *Herpetologica* 30, 417-19.

Jackson, C. G., and Kaye, J. M. (1974b). Occurrence of box turtles (Testudines: Testudinidae) in the Pleistocene of Mississippi. *Herpetologica* 30, 11-13.

Jackson, C. G., and Kaye, J. M. (1975). Giant tortoises in the late Pleistocene of Mississippi. *Herpetologica* 31, 421.

Jehl, J. R., Jr. (1967). Pleistocene birds from Fossil Lake, Oregon. *Condor* 69, 24-27.

Jelinek, A. J. (1957). Pleistocene faunas and early man. *Michigan Academy of Science, Arts, and Letters Papers* 6, 225-37.

Johnson, D. L. (1977). The California ice-age refugium and the Rancholabrean extinction problem. *Quaternary Research* 8, 149-79.

Johnson, D. L. (1978). The origin of island mammoths and the Quaternary land bridge history of the Northern Channel Island, California. *Quaternary Research* 10, 204-5.

Johnson, E. (1974). Zooarcheology and the Lubbock Lake site. *In* "History and Prehistory of the Lubbock Lake Site" (C. C. Black, ed.), pp. 107-22. The Museum Journal 15. West Texas Museum Association, Lubbock, Texas.

King, F. B., and Graham, R. W. (1981). Effects of ecological and paleoecological patterns on subsistence and paleoenvironmental reconstructions. *American Antiquity* 46, 128-42.

Kohl, R. F. (1974). A new late Pleistocene fauna from Humboldt County, California. *Veliger* 17, 211-19.

Kurtén, B. (1963). Fossil bears from Texas. *Pearce-Sellards Series, Texas Memorial Museum* 1, 3-15.

Kurtén, B (1967). Pleistocene bears of North America: II. Genus *Arctodus*, short-faced bears. *Acta Zoologica Fennica* 117, 1-60.

Kurtén, B., and Anderson, E. (1972). The sediments and fauna of Jaguar Cave: II. The fauna. *Tebiwa* 15, 21-45.

Kurtén, B., and Anderson, E. (1974). Association of *Ursus arctos* and *Arctodus simus* (Mammalia:Ursidae) in the late Pleistocene of Wyoming. *Breviora* 426, 1-26.

Kurtén, B., and Anderson, E. (1980). "Pleistocene Mammals of North America." Columbia University Press, New York.

Land, L. S., Lundelius, E. L., Jr., and Valastro, S., Jr. (1980). Isotopic ecology of deer bones. *Palaeogeography, Palaeoclimatology, Palaeoecology* 32, 143-52.

Leidy, J. (1853). Description of an extinct species of American lion: *Felis atrox*. *Transactions of the American Philosophical Society New Series* 10, 319-21.

Leidy, J. (1889). Notice and description of fossils in caves and crevices of the limestone rocks of Pennsylvania. *Geological Survey of Pennsylvania Annual Report* 1887, 1-20.

Logan, L. E., and Black, C. C. (1979). The Quaternary vertebrate fauna of Upper Sloth Cave, Guadalupe Mountains National Park, Texas. *In* "Biological Investigations in the Guadalupe National Park, Texas" (H. H. Genoways and R. J. Baker, eds.), pp. 141-58. National Park Service, Proceedings and Transactions Series, No. 4. U.S. Government Printing Office, Washington, D.C.

Lundelius, E. L., Jr. (1967). Late Pleistocene and Holocene faunal history of central Texas. *In* "Pleistocene Extinctions: The Search for a Cause" (P. S. Martin and H. E. Wright, Jr., eds.), pp. 287-319. Yale University Press, New Haven, Conn.

Lundelius, E. L., Jr. (1972). Vertebrate remains from the Gray Sand. *In* "Blackwater Locality No. 1: A Stratified Early Man Site in Eastern New Mexico" (J. J. Hester, ed.), pp. 148-63. Fort Burgwin Research Center Publication 8.

Lundelius, E. L., Jr. (1974). The last fifteen thousand years of faunal change in North America. *In* "History and Prehistory of the Lubbock Lake Site" (C. C. Black, ed.), pp. 141-60. The Museum Journal 15. West Texas Museum Association, Lubbock, Texas.

Lundelius, E. L., Jr. (1976). Vertebrate paleontology of the Pleistocene: An overview. *In* "Geoscience and Man' (R. C. West, ed.), vol. 13, pp. 49-59. Louisiana State University, Baton Rouge.

McDonald, H. G., and Anderson, E. (1975). A late Pleistocene fauna from southeastern Idaho. *Tebiwa* 18, 19-37.

MacDonald, J. R. (1967). The Maricopa Brea. *Museum Alliance Quarterly, Los Angeles County Museum of Natural History* 6, 21-24.

McGuire, K. R. (1980). Cave sites, faunal analysis, and big game hunters of the Great Basin: A caution. *Quaternary Research* 14, 263-68.

MacInnes, C. D., and Chamberlain, E. B. (1963). The first record of the double-striped thick-knee in the United States. *Auk* 80, 79.

McNabb, B. K. (1971). On the ecological significance of Bergmann's Rule. *Ecology* 52, 845-54.

Madole, R. F. (1976). Glacial geology of the Front Range, Colorado. *In* "Quaternary Stratigraphy of North America" (W. C. Mahoney, ed.), pp. 297-318. Dowden, Hutchinson, and Ross, Stroudsburg, Pa.

Magish, D. P., and Harris, A. H. (1976). Fossil ravens from the Pleistocene of Dry Cave, Eddy County, New Mexico. *Condor* 78, 399-404.

Marsh, O. C. (1872). Notice of some new Tertiary and post-Tertiary birds. *American Journal of Science, Series* 3 (4), 256-62.

Martin, L. D. (1977). A whooping crane from the late Pleistocene of Kansas. *Bulletin of the Kansas Ornithological Society* 28, 22-23.

Martin, L. D., and Gilbert, B. M. (1978). "An American Lion, *Panthera atrox*, from Natural Trap Cave, North Central Wyoming." Contributions to Geology, University of Wyoming 16, pp. 95-101.

Martin, L. D., and Gilbert, B. M. (1979). Excavations at Natural Trap. *Transactions of the Nebraska Academy of Science* 6, 107-16.

Martin, L. D., Gilbert, B. M., and Chomko, S. A. (1979). *Dicrostonyx* (Rodentia) from the late Pleistocene of northern Wyoming. *Journal of Mammalogy* 60, 193-95.

Martin, L. D., and Neuner, A. M. (1978). The end of the Pleistocene in North America. *Transactions of the Nebraska Academy of Science* 6, 117-26.

Martin, P. S. (1958). Pleistocene ecology and biogeography of North America. In "Zoogeography" (C. L. Hubbs, ed.), pp. 375-420. American Association for the Advancement of Science Publication 51.

Martin, P. S. (1967). Prehistoric overkill. In "Pleistocene Extinctions: The Search for a Cause" (P. S. Martin and H. E. Wright, Jr., eds.), pp. 75-120. Yale University Press, New Haven, Conn.

Martin, R. A. (1968). Late Pleistocene distribution of *Microtus pennsylvanicus*. *Journal of Mammalogy* 49, 265-71.

Martin, R. A. (1974). Fossil vertebrates from the Haile XIVA Fauna, Alachua County. In "Pleistocene Mammals of Florida" (S. D. Webb, ed.), pp. 100-113. University of Florida Presses, Gainesville.

Martin, R. A., and Webb, S. D. (1974). Late Pleistocene mammals from the Devil's Den Fauna, Levy County. In "Pleistocene Mammals of Florida" (S. D. Webb, ed.), pp. 114-45. University of Florida Presses, Gainesville.

Mawby, J. E. (1967). "Fossil Vertebrates of the Tule Springs Site, Nevada." Nevada State Museum Anthropological Papers 13, pp. 105-28.

Mead, J. I. (1981). The last 30,000 years of faunal history within the Grand Canyon, Arizona. *Quaternary Research* 15, 311-26.

Mead, J. I., and Phillips, A. M., III (1981). The late Pleistocene and Holocene fauna and flora of Vulture Cave, Grand Canyon, Arizona. *Southwestern Naturalist* 26, 257-88.

Mehl, M. G. (1962). "Missouri's Ice Age Animals." Missouri Geological Survey Water Resources Series 1, pp. 1-104.

Mehringer, P. J., Jr. (1967). The environment of extinction of the late Pleistocene megafauna in the arid southwestern United States. In "Pleistocene Extinctions: The Search for a Cause" (P. S. Martin and H. E. Wright, Jr., eds.), pp. 247-66. Yale University Press, New Haven, Conn.

Mehringer, P. J., Jr., King, J. E., and Lindsay, E. H. (1970). A record of Wisconsin age vegetation and fauna from the Ozarks of western Missouri. In "Pleistocene and Recent Environments of the Central Great Plains" (W. Dort, Jr., and J. K. Jones, Jr., eds.), pp. 173-83. University of Kansas Department of Geology Special Publication 3.

Mercer, H. C. (1897a). An exploration of Durham Cave, Bucks County, Pennsylvania, in 1893. In "Researches upon the Antiquity of Man," pp. 149-78. University of Pennsylvania Monographs in Philology, Literature, and Archaeology 6.

Mercer, H. C. (1897b). The finding of the remains of the fossil sloth at Big Bone Cave, Tennessee, in 1896. *Proceedings of the American Philosophical Society* 36, 36-70.

Miller, A. H. (1932). An extinct icterid from Shelter Cave, New Mexico. *Auk* 49, 38-41.

Miller, A. H., and Peabody, F. E. (1941). An additional Pleistocene occurrence of the murre, *Uria aalge*. *Condor* 43, 78.

Miller, L. H. (1942). A new fossil bird locality. *Condor* 44, 283-83.

Miller, L. H., and DeMay, L. (1942). The fossil birds of California: An avifauna and bibliography with annotations. *University of California Publications in Zoology* 47, 47-142.

Miller, S. J. (1979). The archeological fauna of four sites in Smith Creek Canyon. In "The Archeology of Smith Creek Canyon, Eastern Nevada" (D. R. Tuohy and D. L. Randall, eds.), pp. 273-329. Nevada State Museum Anthropology Papers 17.

Miller, W. E. (1971). "Pleistocene Vertebrates of the Los Angeles Basin and Vicinity (Exclusive of Rancho La Brea)." Los Angeles County Museum of Natural History Science Bulletin 10.

Moodie, K. B., and Van Devender, T. R. (1978). Fossil box turtles (Genus *Terrapene*) from southern Arizona. *Herpetologica* 34, 172-74.

Nelson, R. S., and Semken, H. A., Jr. (1970). Paleoecological and stratigraphic significance of the muskrat in Pleistocene deposits. *Bulletin of the Geological Society of America* 81, 3733-38.

Nowak, R. M. (1979). "North American Quaternary *Canis*." Monograph of the University of Kansas Museum of Natural History 6.

Olson, S. L. (1972). A whooping crane from the Pleistocene of north Florida. *Condor* 74, 341.

Olson, S. L. (1974). The Pleistocene rails of North America. *Condor* 76, 169-75.

Orr, P. C. (1956a). "Dwarf Mammoths and Man on Santa Rosa Island." University of Utah Anthropological Papers Department of Anthropology 26, pp. 74-86.

Orr, P. C. (1956b). Radiocarbon dates from Santa Rosa Island: I *Bulletin of the Santa Barbara Museum of Natural History Department of Anthropology* 2, 1-10.

Parmalee, P. W. (1967). A recent cave bone deposit in southwestern Illinois. *Bulletin of the National Speleological Society* 29, 119-47.

Parmalee, P. W. (1969). California condor and other birds from Stanton's Cave, Arizona. *Journal of Arizona Academy of Sciences* 5, 204-6.

Parmalee, P. W., and Klippel, W. E. (1981). Remains of the wood turtle *Clemmys insculpta* (LeConte) from a late Pleistocene deposit in middle Tennessee. *American Midland Naturalist* 105, 413-16.

Parmalee, P. W., and Oesch, R. D. (1972). "Pleistocene and Recent Faunas from the Brynjulfson Caves, Missouri." Illinois State Museum Report of Investigations 25.

Parmalee, P. W., Oesch, R. D., and Guilday, J. E. (1969). "Pleistocene and Recent Vertebrate Faunas from Crankshaft Cave, Missouri." Illinois State Museum Report of Investigations 14.

Patton, T. H. (1963). Fossil vertebrates from Miller's Cave, Llano County, Texas. *Bulletin of the Texas Memorial Museum* 7, 1-41.

Payen, L. A., Hall, M. C., and Kelley, M. D. (1978). Radiocarbon and obsidian hydration studies of Samwell Cave. In "American Quaternary Association Fifth Biennial Meeting, Abstracts," pp. 231. University of Alberta, Edmonton.

Peterson, O. A. (1926). The fossils of the Frankstown Cave, Blair County, Pennsylvania. *Annals of the Carnegie Museum* 16, 249-315.

Preston, R. E. (1979). "Late Pleistocene Cold-Blooded Vertebrate Faunas from the Midcontinental United States: I. Reptilia:Testudines, Crocodilia." University of Michigan Museum of Paleontology Papers in Paleontology 19.

Purdue, J. R. (1980). Clinal variation of some mammals during the Holocene in Missouri. *Quaternary Research* 13, 242-58.

Rasmussen, D. L. (1974). New Quaternary mammal localities in the upper Clark Fork River valley, western Montana. *Northwestern Geology* 3, 62-70.

Rawn-Schatzinger, V. (no date). The Jones-Miller local fauna, Yuma County, Colorado. In The Jones-Miller Site: A Hell Gap Bison Kill in Yuma County, Colorado (D. J. Stanford, ed.). Unpublished manuscript on file at the Smithsonian Institute, Washington, D.C.

Ray, C. E. (1967). Pleistocene mammals from Ladds, Bartow County, Georgia. *Bulletin of the Georgia Academy of Science* 25, 120-50.

Ray, C. E., Cooper, B. N., and Benninghoff, W. S. (1967). Fossil mammals and pollen in a late Pleistocene deposit at Saltville, Virginia. *Journal of Paleontology* 41, 608-22.

Ray, C. E., Denny, C. S., and Rubin, M. (1970). A peccary, *Platygonus compressus* LeConte, from drift of Wisconsin age in northern Pennsylvania. *American Journal of Science* 268, 78-94.

Ray, C. E., and Lipps, L. (1968). Additional notes on the Pleistocene mammals from Ladds, Georgia. *Bulletin of the Georgia Academy of Science* 26, 63.

Rea, A. M. (1980). Late Pleistocene and Holocene turkeys in the Southwest. *Contributions in Science, Los Angeles County Museum* 330, 209-24.

Rea, A. M. (1981). California condor captive breeding: A recovery proposal. *Environment Southwest* 492, 8-12.

Rea, A. M., and Hargrave, L. L. (no date). Bird bones from Stanton's Cave, Arizona. Unpublished manuscript.

Richmond, N. D. (1964). Fossil amphibians from Frankstown Cave, Pennsylvania. *Annals of the Carnegie Museum* 36, 225-28.

Rickart, E. A. (1977). Pleistocene lizards from Burnet and Dark Canyon Caves, Guadalupe Mountains, New Mexico. *Southwestern Naturalist* 21, 519-22.

Roberts, M. C. (1970). Late glacial and postglacial environments in southeastern Wyoming. *Palaeogeography, Palaeoclimatology, Palaeoecology* 8, 5-17.

Ross, R. C. (1935). A new genus and species of pygmy goose from the McKittrick Pleistocene. *Transactions of the San Diego Society of Natural History* 8, 107-14.

Roth, J. A., and Laerm, J. (1980). A late Pleistocene vertebrate assemblage from Edisto Island, South Carolina. *Brimleyana* 3, 1-29.

Sadek, H. (1965). Appendix: Distribution of the bird remains at Jaguar Cave. *Tebiwa* 8, 20-28.

Sadek-Kooros, H. (1966). Jaguar Cave: An early man site in the Beaverhead Mountains of Idaho. Ph.D. dissertation, Harvard University, Cambridge, Mass.

Saunders, J. J. (1977). ''Late Pleistocene Vertebrates of the Western Ozark Highland, Missouri.'' Illinois State Museum Report of Investigations 33.

Schoenwetter, J. (1975). Pollen analytical results: Part I. *In* ''Late Pleistocene Environments of the Southern High Plains'' (F. Wendorf and J. J. Hester, Eds.), pp. 103-20. Fort Burgwyn Research Center Publication 9.

Schultz, C. B., and Howard, E. B. (1935). The fauna of Burnet Cave, Guadalupe Mountains, New Mexico. *Proceedings of the Academy of Natural Sciences of Philadelphia* 87, 273-98.

Schultz, C. B., Tanner, L. G., Whitmore, F. C., and Crawford, E. C. (1963). Paleontologic investigations at Big Bone Lick State Park, Kentucky: A preliminary report. *Science* 142, 1167-69.

Schultz, C. B., Tanner, L. G., Whitmore, F. C., Ray, L. L., and Crawford, E. C. (1967). Big Bone Lick, Kentucky. *Museum Notes of the University of Nebraska State Museum* 33, 1-12.

Schultz, G. R. (1967). Four superimposed late Pleistocene vertebrate faunas from southwest Kansas. *In* ''Pleistocene Extinctions: The Search for a Cause'' (P. S. Martin and H. E. Wright, Jr., eds.), pp. 321-36. Yale University Press, New Haven, Conn.

Schultz, G. R. (1969). ''Geology and Paleontology of a Late Pleistocene Basin in Southwest Kansas.'' Geological Society of America Special Paper 105.

Sellards, E. H. (1940). New Fossil localities in Texas. *Bulletin of the Geological Society of America* 51, 1977-78.

Semken, H. A., Jr. (1961). Fossil vertebrates from Longhorn Cavern, Burnet County, Texas. *Texas Journal of Science* 13, 290-310.

Semken, H. A., Jr. (1983). The Holocene mammalian biography and climatic change in the eastern and central United States. *In* ''Late-Quaternary Environments of the United States,'' vol. 2, ''The Holocene'' (H. E. Wright, Jr., ed.), pp. 182-207. University of Minnesota Press, Minneapolis.

Shufeldt, R. W. (1897). On fossil bird-bones obtained by expeditions of the University of Pennsylvania from the bone caves of Tennessee. *American Naturalist* 31, 645-60.

Shufeldt, R. W. (1913). Review of the fossil fauna of the desert region of Oregon, with a description of additional material collected there. *Bulletin of the American Museum of Natural History* 32, 123-78.

Shutler, R. (1968). Tule Springs: Its implications to early man studies in North America. *Contributions to Anthropology, Eastern New Mexico University* 1, 19-26.

Sibley, C. (1939). Fossil fringillids from Rancho La Brea. *Condor* 41, 126-27.

Simpson, G. G. (1928). Pleistocene mammals from a cave in Citrus County, Florida. *American Museum Novitates* 328, 1-16.

Simpson, G. G. (1929). Pleistocene mammalian fauna of the Seminole Field, Pinellas County, Florida. *American Museum of Natural History Bulletin* 56, 561-99.

Sinclair, W. J. (1903). A preliminary account of the exploration of Potter Creek Cave, Shasta County, California. *Science* 17, 708-12.

Skinner, M. S. (1942). The fauna of Papago Springs Cave, Arizona, and a study of *Stockoceras. American Museum of Natural History Bulletin* 80, 143-220.

Slaughter, B. H. (1966). The vertebrates of the Domebo local fauna, Pleistocene of Oklahoma. *In* ''Domebo: A Paleoindian Mammoth Kill in the Prairie Plains'' (F. C. Leonhardy, ed.), pp. 31-35. Contributions of the Museum of the Great Plains 1.

Slaughter, B. H. (1975). Ecological interpretation of the Brown Sand Wedge local fauna. *In* ''Late Pleistocene Environments of the Southern High Plains'' (F. Wendorf and J. J. Hester, eds.), pp. 179-92. Fort Burgwin Research Center Publication 9.

Slaughter, B. H., Crook, W. W., Jr., Harris, R. K., Allen, D. C., and Seifert, M. (1962). ''The Hill-Shuler Local Faunas of the Upper Trinity River, Dallas and Denton Counties, Texas.'' University of Texas Bureau of Economic Geology Report of Investigations 48.

Slaughter, B. H., and Hoover, R. (1963). Sulphur River Formation and the Pleistocene mammals of the Ben Franklin local fauna. *Journal of the Graduate Research Center, Southern Methodist University* 31 (3), 132-48.

Smartt, R. A. (1977). The ecology of late Pleistocene and recent *Microtus* from south-central and south-western New Mexico. *Southwestern Naturalist* 22, 1-19.

Spaulding, W. G., Leopold, E. B., and Van Devender, T. R. (1983). Late Wisconsin paleoecology of the American Southwest. *In* ''Late-Quaternary Environments of the United States,'' vol. 1, ''The Late Pleistocene'' (S. C. Porter, ed.), pp. 259-93.

University of Minnesota Press, Minneapolis.

Stanford, D. J. (1979). The Selby and Dutton sites: Evidence for a possible pre-Clovis occupation of the High Plains. *In* ''Pre-Llano Cultures of the Americas: Paradoxes and Possibilities'' (R. L. Humphrey and D. J. Stanford, eds.), pp. 101-23. Anthropological Society of Washington, Washington, D.C.

Stanford, D. J. (ed.) (no date). The Jones-Miller Site: A Hell Gap Bison Kill in Yuma County, Colorado. Unpublished manuscript on file at the Smithsonian Institution, Washington, D.C.

Steadman, D. W. (1980). A review of the osteology and paleontology of turkeys (Aves:Meleagridinae). *Contributions in Science, Los Angeles County Museum* 330, 131-207.

Steadman, D. W., and Martin, P. S. (in press). Late Pleistocene extinction of birds in North America. *In* ''Quaternary Extinctions'' (P. S. Martin, ed.). University of Arizona Press, Tucson.

Stettenheim, P. (1958). Bird fossils from the late Pleistocene of Kansas. *Wilson Bulletin* 70, 197-99.

Stock, C. (1918). The Pleistocene fauna from Hawver Cave. *University of California Publications, Bulletin of the Department of Geology* 10, 461-515.

Stock, C. (1931). Problems of antiquity presented in Gypsum Cave, Nevada. *Scientific Monthly* 32, 22-32.

Stock, C. (1935). Exiled elephants of the Channel Islands, California. *Scientific Monthly* 41, 205-14.

Stock, C. (1936). A new mountain goat from the Quaternary of Smith Creek Cave, Nevada. *Southern California Academy of Science Bulletin* 35, 149-53.

Stock, C. (1937). Tar babies. *Westways* 29, 26-27.

Stock, C. (1943). Foxes and elephants of the Channel Islands. *Museum Alliance Quarterly, Los Angeles County Museum of Natural History* 3, 6-9.

Stock, C. (1972). ''Rancho La Brea: A Record of Pleistocene Life in California.'' Los Angeles County Museum Science Series, Paleontology 11.

Stock, C., and Furlong, E. L. (1928). The Pleistocene elephants of Santa Rosa Island, California. *Science* 68, 140-41.

Storer, R. W. (1976). The Pleistocene pied-billed grebes (Aves:Podicipedidae). *Smithsonian Contributions to Paleobiology* 27, 147-53.

Street, F. A., and Grove, A. T. (1979). Global maps of lake-level fluctuations since 30,000 years B.P. *Quaternary Research* 12, 83-118.

Stuiver, M., and Borns, H. W. (1975). Late Quaternary marine invasion in Maine: Chronology and associated crustal movement. *Geological Society of America Bulletin* 86, 99-104.

Tate, J., Jr., and Martin, L. D. (1968). Horned lark and black-billed magpie from the Pleistocene of Nebraska. *Condor* 70, 183.

Taylor, R. E. (1980). Radiocarbon dating of Pleistocene bone: Toward criteria for the selection of samples. *Radiocarbon* 22, 969-79.

Tihen, J. A. (1958). Comments on the osteology and phylogeny of ambystomatid salamanders. *Bulletin of the Florida State Museum* 3, 1-50.

Tihen, J. A. (1962). A review of New World fossil bufonids. *American Midland Naturalist* 68, 1-50.

Tuross, N., and Hare, P. E. (1978). Collagen in fossil bone. *Carnegie Institution of Washington Yearbook* 77, 891-95.

Van Devender, T. R., and Mead, J. I. (1978). Early Holocene and late Pleistocene amphibians and reptiles in Sonoran Desert packrat middens. *Copeia* 1978, 464-75.

Van Devender, T. R., Moodie, K. B., and Harris, A. H. (1976). The desert tortoise (*Gopherus agassizi*) in the Pleistocene of the northern Chihuahuan Desert. *Herpetologica* 32, 298-304.

Van Devender, T. R., Phillips, A. M., III, and Mead, J. I. (1977). Late Pleistocene reptiles and small mammals from the lower Grand Canyon of Arizona. *Southwestern Naturalist* 22, 49-66.

Van Devender, T. R., and Wiseman, F. M. (1977). A preliminary chronology of bioenvironmental changes during the Paleoindian period in the monsoonal Southwest. *In* ''Paleoindian Lifeways'' (E. Johnson, ed.), pp. 13-27. The Museum Journal 17. West Texas Museum Association, Lubbock, Texas.

Van Devender, T. R., and Worthington, R. D. (1977). The herpetofauna of Howell's Ridge Cave and the Paleoecology of the northwestern Chihuahuan Desert. *In* ''Transactions of the Symposium on Biological Resources of the Chihuahuan Desert'' (R. H. Wauer and D. H. Riskind, eds.), pp. 85-106. United States and Mexico National Park Service Transactions and Proceedings 13.

Voorhies, M. R. (1974). Pleistocene vertebrates with boreal affinities in the Georgia Piedmont. *Quaternary Research* 4, 85-93.

Walker, D. (1974). A Pleistocene gyrfalcon. *Auk* 91, 820-21.

Walker, D. N. (in press). The early Holocene vertebrate fauna from the Agate Basin site, Niobrara County, Wyoming. *In* "The Agate Basin Site: A Record of the Paleoindian Occupation of the Northwestern High Plains" (G. C. Frison and D. J. Stanford, eds.) Academic Press, New York.

Walker, D. N., and Frison, G. C. (1980). The late Pleistocene mammalian fauna from the Colby Mammoth Kill site, Wyoming. *Contributions to Geology, University of Wyoming* 19, 69-79.

Webb, S. D. (1974a). Chronology of Florida Pleistocene mammals. *In* "Pleistocene Mammals of Florida" (S. D. Webb, ed.), pp. 5-31. University Presses of Florida, Gainesville.

Webb, S. D. (1974b). Pleistocene llamas of Florida, with a brief review of the Llamini. *In* "Pleistocene Mammals of Florida" (S. D. Webb, ed.), pp. 170-213. University Presses of Florida, Gainesville.

Weigel, R. D. (1962). "Fossil Vertebrates of Vero, Florida." State of Florida Board of Conservation Division of Geology Special Publication 10.

Weigel, R. D. (1967). Fossil birds from Miller's Cave, Llano Co., Texas. *Texas Journal of Science* 19, 107-9.

Wendorf, F. (1961). An interpretation of late Pleistocene environments of the Llano Estacado. *In* "Paleoecology of the Llano Estacado" (F. Wendorf, ed.), pp. 115-33. Fort Burgwin Research Center Publication 1.

Wetmore, A. (1931). The fossil birds of North America. *In* "Check-list of North American Birds," 4th ed., pp. 407-42. American Ornithological Union, Lancaster, Pa.

Wetmore, A. (1945). Record of the turkey from the Pleistocene of Indiana. *Wilson Bulletin* 47, 204.

Wetmore, A. (1948). A Pleistocene record for *Mergus merganser* in Illinois. *Wilson Bulletin* 60, 240.

Wetmore, A. (1958). Miscellaneous notes on fossil birds. *Smithsonian Miscellaneous Collections* 135, 1-11.

Wetmore, A. (1959). Notes on certain grouse of the Pleistocene. *Wilson Bulletin* 71, 178-82.

Wetmore, A. (1962). Birds. *In* The Pleistocene local fauna of the Natural Chimneys, Augusta County, Virginia (J. E. Guilday, ed.). *Annals of the Carnegie Museum* 36, 87-122.

Wetmore, A. (1967). Pleistocene Aves from Ladds, Georgia. *Bulletin of the Georgia Academy of Sciences* 25, 151-53.

Whitmore, F. C., Jr., Emery, K. O., Cooke, H. B. S., and Swift, D. P. J. (1967). Elephant teeth from the Atlantic continental shelf. *Science* 156, 1477-81.

Wilson, L. R. (1966). Palynology of the Domebo site. *In* "Domebo: A Paleoindian Mammoth Kill in the Prairie Plains" (F. C. Leonhardy, ed.), pp. 44-50. Contributions to the Museum of the Great Plains 1.

Wilson, M. (1974). The Casper local fauna and its fossil bison. *In* "The Casper Site: A Hell Gap Bison Kill on the High Plains" (G. C. Frison, ed.), pp. 113-24. Academic Press, New York.

Wilson, M. (1980). Morphological dating of late Quaternary bison on the northern Plains. *Canadian Journal of Anthropology* 1, 81-85.

Woodard, G. D., and Marcus, L. F. (1973). Rancho La Brea fossil deposits: A re-evaluation from stratigraphic and geological evidence. *Journal of Paleontology* 47, 54-69.

Wright, H. E., Jr. (1981). Vegetation east of the Rocky Mountains 18,000 years ago. *Quaternary Research* 18, 113-25.

Zeimens, G., and Walker, D. N. (1974). Bell Cave, Wyoming: Preliminary archeological and paleontological investigations. *In* "Applied Geology and Archeology: the Holocene History of Wyoming" (M. Wilson, ed.), pp. 88-90. Geological Survey of Wyoming Report of Investigations 10.

CHAPTER 17

Late Wisconsin Fossil Beetles in North America

*Alan V. Morgan, Anne Morgan, Allan C. Ashworth,
and John V. Matthews, Jr.*

Introduction

Paleoentomological research in North America has been the subject of a number of papers produced over the last century. Many of these papers have outlined theories that today are unacceptable. For example, the theory that extinction was caused by Pleistocene glaciations today is usually regarded as invalid, and more recent findings are proving that many so-called extinct Coleoptera species found in Pleistocene assemblages are, in fact, still extant. Indeed, studies by Matthews (1970, 1976) show the amazing longevity of our present beetle fauna, with some species remaining unchanged for more than 5 million years.

In 1965, at the Seventh Congress of the International Association for Quaternary Research in Boulder, Colorado, two papers were presented on fossil insects (Coope, 1965; Shotton, 1965). Although they reflected work then in progress in Britain, they served to stimulate interest among North American workers in Pleistocene insect studies. Until that time, most of the theories regarding Pleistocene insect movements relied only on the knowledge of their present ranges.

This chapter is an attempt to synthesize the results of North American research during the period from 1965 until the present, and it concentrates primarily on the Coleoptera. Only brief mention is made of papers that lie outside the geographic and chronologic framework of this volume or that refer indirectly to groups other than Coleoptera. Reviews by Ashworth (1979), Coope (1970, 1979), Matthews (1977, 1980), and Morgan and Morgan (1980a, 1980b) summarize the work of the last decade, and earlier work (around 1890 to 1940) is reviewed by Ashworth (1979) and Morgan and Morgan (1980a).

The number of researchers currently working on North American coleopterous faunas is still relatively small; nevertheless, substantial progress has been made in analyzing fossil assemblages from different parts of the continent. We have chosen to divide North America into three principal sectors (Figure 17-1): (1) the northwestern area of the continent, including Alaska and adjacent parts of the Northwest Territories and the Yukon Territory; (2) the southwestern part of the continent, primarily California, Arizona, New Mexico, and Texas; and (3) the north-central and northeastern part of the continent, en-compassing the area from North Dakota to Vermont and the adjacent Canadian provinces of Ontario and Quebec.

Northwestern North America

Studies of North American fossil insects, in the tradition of those carried out in Britain by G. R. Coope and his colleagues, began with Matthew's preliminary study of some Alaskan samples in 1964. Various geologists in Alaska and the Yukon had noted the presence of "beetle carapaces" in the highly fossiliferous silts exposed at placer gold mines.

The early studies, except for one (Matthews, 1968), remain unpublished, and the focus of the research on Alaskan fossil insects has now shifted to the west and the north (Matthews, 1974; Morgan et al., 1979). Investigations have also commenced in the northern Yukon Territory, which, along with unglaciated Alaska, constitutes the region of eastern Beringia. One area in the northern Yukon Territory near the settlement of Old Crow is the scene of current research on Pleistocene insects and environments (Matthews, 1975). More than most, the Yukon studies point out some of the problems inherent in dealing with northern fossil insects.

In general, the Late Wisconsin Alaskan assemblages have a lower diversity than those farther south, and the taxa represented by fossils seldom display the provocative range disjunctions evidenced by fossils from more-southern sites. Another problem is that many of the Alaskan fossils represent taxonomic groups that are still in need of revision or for which there is little more than anecdotal accounts on distribution and habitat requirements. An attempt has been made to circumvent these problems by comparing the relative abundance of

We wish to thank the coleopterists at the Biosystematics Research Institute, Ottawa, and the Smithsonian Institution, Washington, D.C., for their valuable assistance on the taxonomy and ecology of modern and fossil beetles. We also thank George Ball, University of Alberta, Edmonton, and Russell Coope, University of Birmingham, England, for their constant encouragement and critical reading of the manuscript. Financial support was provided by the Natural Sciences and Engineering Research Council of Canada and the National Science Foundation of the United States. Finally, we would like to pay tribute to the late Carl Lindroth, whose work has been an inspiration to us all.

groups of fossil taxa having similar environmental requirements (Matthews, 1982). Unfortunately, this procedure involves percentage comparisons, and it requires large fossil assemblages—usually larger than the number that can be isolated at any one site. These factors reduce the paleoenvironmental resolving power of Alaska-Yukon insect assemblages, and this means that the fossil data must be evaluated in conjunction with other types of fossil and stratigraphic evidence.

Most of the alternate evidence comes from palynology. Pollen studies at several sites show that between 25,000 and approximately 13,000 years ago virtually all of eastern Beringia was treeless. Even shrub birch and alder, which are so common in the modern vegetation, were rare or absent. The regional vegetation was tundra, but there is disagreement about whether it was primarily a depauperate sedge-moss and fell-field tundra, like that of some present-day montane and high-Arctic lowlands, or whether it included significant tracts of steppelike vegetation. It is certain, however, that the vegetation of much of Alaska and the Yukon was unlike the poorly drained *Eriophorum-Carex-Arctophila*-type tundra of the lowlands (e.g., the northern Coastal Plain of Alaska) or the mesic ericaceous-shrub and birch-dominated tundra of better-drained sites. Several unique aspects of the late-Pleistocene physical environment may be responsible for such departures. Among them are the heightened eolian activity (Carter, 1981; Hopkins, 1982; Péwé, 1975) and the enhancement of climatic continentality when the Bering Sea was replaced by a wide land bridge (Barry, 1982).

Plant macrofossils, although not as well studied as pollen, also show that the Late Wisconsin environment of eastern Beringia differed from that of the present tundra. For example, according to examinations of macroscopic plant remains, parts of eastern Beringia 13,000 years ago contained more areas suitable for plants adapted to well-drained or slightly alkaline conditions than they do today. To some extent, this is also the type of environment indicated by the numerous vertebrate fossils, for many of them refer to taxa that either depended on grazing or were otherwise maladapted for an existence in boggy terrain (Matthews, 1982).

Starting about 13,000 years ago, shrub birch increased its impor-

tance as a component of the eastern Beringia "tundra." For a brief time, isolated stands of poplars comprised the only forest-type vegetation (in fact, they grew beyond their present distributional limit), but by 10,000 years ago spruce forests apparently were reoccupying large tracts of interior Alaska and the Yukon. The advent of spruce taiga and poorly drained peatlands similar to those of the present was delayed until later in the Holocene, perhaps even until as late as 4000 years ago (T. Ager, personal communication, 1980).

The sequence just described is a brief account of the environmental succession of the Late Wisconsin in Alaska and the Yukon. Though highly generalized, it provides a backdrop for the discussion of the insects from that area. One of the most important faunas is from Eva Creek, near Fairbanks in central Alaska (Figure 17-1). Five Eva Creek faunas have been studied so far; accounts of three of them have been published (Matthews, 1968). Those three come from a part of the Eva Creek section that we now believe (contrary to Matthews, 1968, 1970) ranges in age from early Middle Wisconsin to Late Wisconsin. The lowest sample (3-1A) of the Eva series is probably no older than approximately 70,000 years; the uppermost one (3-3C) has been dated at 24,400 ± 600 yr B.P. (I-2116).

The Eva Creek samples (Matthews, 1968) appear to represent a sequence of progressively drier and more treeless communities. Eva 3-1A contains spruce needles, a few scolytid beetles, and fragments of other insects normally associated with trees. Despite the tundra aspect of the pollen sample (Matthews, 1970), the vegetation of the lowland probably contained scattered groves of *Picea*, tree birch, and *Populus*. Local environments were poorly drained, but there must have been nearby open grassy areas that account for the fossil carabid, *Harpalus amputatus* Say. By 24,400 yr B.P., the Eva Creek Valley was drier and possibly more steppelike. The weevil *Lepidophorus lineaticollis* Kirby dominates the 3-3C assemblage (53% of all beetles), and the most abundant species of the subgenus *Cryobius* is *Pterostichus kotzebuei* Ball, rather than *P. nivalis* Sahlberg as in sample 3-1A. The former species seems to prefer more xerophilous tundra sites, whereas the latter is frequently collected in mesic tundra areas where mosses are abundant (Ball, 1966). *Cymindis*, a carabid of very dry substrates, is also present in Eva 3-3C. Unlike the Eva 3-1A sample, none of the insects in Eva 3-3C suggests that trees were present, either locally or regionally. This finding is substantiated by the pollen spectrum for level 3-3C (Matthews, 1970), which also shows higher percentages of *Artemisia* than the lower samples show.

The Eva Creek samples differ in several ways from most others in Alaska and the Yukon. The taxonomic diversity is low; Byrrhidae are rare, water beetles are not represented, and the ground beetle *Cryobius* is dominant. Holocene assemblages from the same area (Matthews, 1982) are more diverse, contain an abundance of taxa that ordinarily are associated with forested habitats (Scolytidae, Formicidae), along with many fossils of both aquatic and hygrophilous species.

The Cape Deceit exposure near the town of Deering, on the northern shore of the Seward Peninsula (Figure 17-1), contains sediments ranging in age from early to late Pleistocene (Guthrie and Matthews, 1971; Matthews, 1974; Repenning, 1980). The sediments are rich in fossils, particularly insects (Matthews, 1974), but the Late Wisconsin interval is represented by only a single 12,420-year-old sample (S-6).

In the S-6 sample, fossils of the weevil *Lepidophorus lineaticollis* and the byrrhid *Morychus* are predominant (40% and 16%, respectively), whereas carabid fossils of *Cryobius* type account for only 26%. This represents a reduction from 65% in an Early Wisconsin

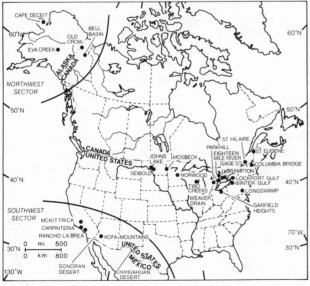

Figure 17-1. Locations of Late Wisconsin fossil sites mentioned in the text.

Cape Deceit sample (S-5), which is thought to imply mesic tundra like that of present times. When both the plant macrofossils and insects are considered, sediments from S-6 seem to have been deposited on the shore of an ephemeral pond in a tundra region with a lot of bare ground and well-drained substrates. Assemblages of insects similar to that of S-6 occurred in the Yukon 31,300, 15,900 and 13,500 years ago (J. V. Matthews, Jr., unpublished data).

The Yukon is an area of eastern Beringia with a rich and diverse Quaternary history (Hughes, 1972; Lichti-Federovich, 1973, 1974; Matthews, 1975; Morlan, 1980). A great deal of information on Quaternary invertebrates from exposures near Old Crow is now in the final stages of compilation. However, because of their age, most are not directly relevant to this chapter. Two exceptions are small assemblages dated at 15,600 and 13,500 yr B.P. from the Bell Basin (Figure 17-1) (Matthews, 1982).

Southwestern United States

The traditional repositories of the northern insect faunas in bog and lacustrine sediments are largely absent in the more arid areas of the American Southwest. The early-Pleistocene Rita Blanca deposits of western Texas are the only lake sediments from which insects have been described (Sleeper, 1969; Sublette, 1969; Werner, 1969). Reports of Late Wisconsin fossils within the region are from unusual sources. The most significant of these are the southern Californian asphalt deposits and packrat (*Neotoma*) middens, which occur in dry caves and rock shelters throughout the region. In addition, Waage (1976) described some dipterous larvae and puparia from the dung of the extinct Shasta ground sloth.

THE ASPHALT DEPOSITS OF SOUTHERN CALIFORNIA

A variety of insects has been reported from the asphalt deposits at Rancho La Brea in Los Angeles County, McKittrick in Kern County, and Carpinteria in Santa Barbara County (Figure 17-1). Rancho La Brea is by far the most thoroughly investigated of these and the only one with an extensive Late Wisconsin record. The popular image of lakes of liquid asphalt trapping unsuspecting animals has been dispelled by Woodard and Marcus (1973), who envision instead the postdepositional permeation of fluvial-channel and floodplain sediments by upward-moving asphalt. Convective stirring is no longer believed to have occurred on a large scale, and the deposits have a demonstrable, though complex, stratigraphy.

Disarticulated beetle fragments are common fossils readily extracted and cleaned from the asphalt matrix with an ultrasonic cleaner and xylene as a solvent. Pierce (1947) reported remains of water beetles (Haliplidae, Dytiscidae, Hydrophilidae), carrion beetles (Silphidae, Dermestidae, Histeridae), predaceous and scavenging beetles (Cicindelidae, Carabidae, Tenebrionidae), dung beetles (Scarabaeidae), and plant-feeding beetles (Cerambycidae, Curculionidae).

Beetle studies have concentrated on taxonomic rather than ecologic or stratigraphic aspects. In the earliest report, Grinnell (1908) listed several species consisting mostly of described California taxa but including new species of tenebrionids, which were later reassigned by Blaisdell (1909) to existing taxa. Of the specimens he studied, Pierce (1944, 1946a, 1946b, 1948, 1949, 1954a, 1954b, 1954c) identified a few of the carabids, tenebrionids, silphids, and scarabs as extant species, but most he considered new and he described 10 species and 15 subspecies. Unfortunately, he described

some species from solitary fragments of exoskeleton, notable examples being the silphid *Nicrophorus obtusiscutellum* from a scutellum and the scarab *Serica kanakoffi* from a head. The only justification he offered for designating subspecific rank to numerous specimens was the fact that they were fossils, which is not in accord with modern practice in systematics.

Recent revisionary studies of Pierce's tenebrionids and silphids (Doyen and Miller, 1980; Miller and Peck, 1979) provide new synonymies and demonstrate, with the possible exception of the tenebrionid *Coniontis remnans*, that the La Brea species are extant members of the southern California fauna. Almost certainly the dung-feeding scarab *Copris pristinus* described by Pierce (1946b) is extinct. Matthews and Halffter (1968) confirm that designation and comment that the closest living species are in the *C. rebouchei* complex. They consider the closest living relative is *C. lecontei* Matthews, whose northern range barely extends into Arizona.

Miller and others (1981) have recently reexamined all of Pierce's La Brea scarab fossils. They accept *C. pristinus* as a valid species and further find *Onthophagus everestae* Pierce to be a valid and probably extinct species. They have not attempted a specific reevaluation of *Paleocopris labreae* Pierce because of the broken and incomplete nature of the holotype, but they have reassigned the genus *Paleocopris* as a junior synonym of *Phaneus*. Among the extant La Brea species, *Canthon praticola* LeConte no longer lives in California. Miller and others (1981) attribute the extinctions and changes in the range of the scarabs to changes in climate, extinctions of mammals, or a combination of both. Tempting as it is, there is no paleontologic evidence actually linking the extinction of dung-feeding beetles to the well-known La Brea extinctions of mammals.

NEOTOMA MIDDENS OF THE AMERICAN SOUTHWEST

The other major sources of Wisconsin insects in the Southwest are the indurated middens of packrats or woodrats (*Neotoma*) that commonly occur in dry caves and rock shelters throughout the region (Spaulding et al., 1982). The middens consist of dung, bones, and plant and insect remains cemented by urine. The plant fossils have been intensively studied and have been applied in a detailed regional analysis of vegetational dynamics and climatic change (Van Devender and Spaulding, 1979).

The abundance and quality of insect remains vary considerably among middens. The fossils are mostly disarticulated skeletal parts, but occasionally the preservation of individual fossils is spectacular. A 14,400-year-old specimen of the tenebrionid *Stibia tuckeri* Casey from a Kofa Mountain midden includes an entire exoskeleton missing only the terminal antennal and leg segments. Even in well-preserved specimens, the genitalia are rarely preserved. Telltale openings in the abdominal segments suggest the postmortem activity of dermestid larvae. Occasionally, specimens are badly corroded, possibly as a result of microbial activity. Almost all specimens are uniformly light yellowish brown; the cause of this loss of pigmentation has not been investigated, but the urine cement is a possible solvent.

There are few published studies of midden insects, modern or fossil. The only Late Wisconsin records are reports of the ptinid *Niptus abstrusus* Spilman (Ashworth, 1973) and a few weevil and scarab fragments from the Chihuahuan Desert of western Texas, together with a description of a new ptinid species, *Ptinus priminidi* Spilman, from the Sonoran Desert of Arizona and California (Spilman, 1976). There is no way of telling whether the latter is an extinct species or a living member of the desert fauna.

Ptinids are common fossils, but several other insects are represented in the midden assemblages from the Sonoran desert (A. C. Ashworth, unpublished data). As would be expected, many of the taxa are intimately associated with the pack rat and its waste products, scarabs and tenebrionids being expecially common. Some carabids and tenebrionids were scavengers of the immediate area of the midden; some plant-feedings scarabs, weevils, and bostrichids may have been brought in with nest-building materials. There are no records of packrats feeding on beetles. With the possible exception of *P. priminidi*, the species identified from Sonoran Desert middens are inhabitants of the region at present. Some of the species, however, live at higher elevations than those at which their fossils occurred.

North-Central and Northeastern United States and Adjacent Areas of Canada

Perhaps no other portion of the North American continent has been so extensively studied as the region running from North Dakota to Quebec and south to Pennsylvania. The stratigraphy of the region is relatively well known, particularly when compared to the northwestern and southwestern portions of the continent; there is an abundance of material for radiometric dating, and the taxonomy of the insect fauna is moderately well known. This area is of great interest to investigators studying the Pleistocene, because it marks the zone that saw the fluctuating Late Wisconsin ice front finally reach its maximum extent. In this region, insect assemblages that both predate and postdate the Late Wisconsin advance are known.

A number of earlier faunas are described that are, respectively, of Sangamon age (Hammond et al., 1979; Prest et al., 1976), Early Wisconsin age (Ashworth, 1980; Morgan, 1972; Morgan and Morgan, 1980a), and Middle Wisconsin age (Morgan and Morgan, 1980b). The only site that predates the Late Wisconsin advance and is younger than 25,000 yr B.P. is the Garfield Heights site (Figure 17-1) in Cleveland, Ohio, which was described by Coope (1968). There is good evidence from the insect faunas described so far, however, that a rich and varied assemblage of Coleoptera inhabited the area just before it was overrun by Late Wisconsin ice. Presumably, the insects were forced south to positions along the ice margin in the Dakotas, Minnesota, Wisconsin, Illinois, Indiana, Ohio, Pennsylvania, and New York. They probably survived in confined climatic zones until the retreat of the ice, which began approximately 17,000 yr B.P. The insect fauna colonized the open country left bare by the receding ice, and there are a large number of sites where well-preserved beetle remains indicate the presence of these pioneer species. An examination of these faunas is only now revealing both the transient geographic nature and the temporary associations of many of the species as they migrated northward to the zones they presently occupy.

The earliest Late Wisconsin site from which insect remains have been recovered is at Garfield Heights in Ohio (Coope, 1968; Morgan, Morgan, and Miller, 1982). The insects were recovered from the base of a varved silt-and-clay unit that was covered by two Woodfordian tills. A tree trunk recovered nearby from the same horizon (and located about 2 m away from the organic debris examined for beetles) provided radiocarbon dates of 24,600 ± 800 yr B.P. (W-71) and 23,310 ± 390 yr B.P. (K361-4). The known modern distribution of the few identified species is close to the treeline in Canada and Alaska, but they do extend south into the northern portion of the United States (particularly to montane refugia in the Northeast) and are present in high-altitude areas of the Rocky Mountains and the Sierra

Nevada. The fauna is not typical of true tundra but occurs in open, barren country with scattered trees and low vegetation, and it provides a glimpse of conditions outside the advancing Wisconsin ice margin.

Unfortunately, little is known of conditions along the ice front at the maximum advance. Insect faunas that were approximately contemporaneous with the maximum ice advance are being studied from a number of sites in Indiana, Ohio, and Tennessee. These faunas contain predominantly boreal and treeline insects that apparently survived in these regions during the glacial maximum (A. V. and A. Morgan, unpublished data). Preliminary reports by Schwert and others (1981) indicate that the climate of east-central Iowa 17,200 years ago was cold enough to support an insect assemblage with tundra and treeline affinities. Ground beetles are represented by several species of the subgenus *Cryobius*, together with *Diacheila polita* Faldermann, *Elaphrus lapponicus* Gyllenhal, *Agonum quinquepunctatum* Motschulsky, *Pterostichus vermiculosis* Ménétries, and *Amara alpina* Paykull. Many of these species are also present in sites about 20,000 years old located close to the terminus of the Laurentide ice sheet in south-central Illinois (Morgan and Morgan, 1982).

The oldest known insect fauna (Morgan, Elias, and Morgan, 1982) postdating the maximum ice advance comes from Longswamp, Pennsylvania (Figure 17-1). It has been suggested that the Longswamp site is more than 15,000 years old (Watts, 1979). The sequence lies in a small sag pond in dolostones along the eastern margin of the Appalachian Mountains. Watts (1979: 433) points to "clear evidence for tundra vegetation" at the base of this site, but the associated insects indicate open-ground conditions rather than true tundra. For example, the staphylinid, *Tachinus elongatus* Gyllenhall, is transcontinental and northern, living in wet organic soils (Campbell, 1973), but it is not a true tundra inhabitant. Similarly, the carabids *Notiophilus semistriatus* Say (transcontinental) and *Trechus crassiscapus* Lindroth (northeastern) are typical residents of the boreal zone, the former in open-ground situations and the latter in swampy areas with *Betula* and *Alnus* (Lindroth, 1961). The presence of species such as the scolytids *Scierus annectans* LeConte and *Polygraphus rufipennis* (Kirby) indicates the proximity of coniferous trees, while other fossils, believed to be more than 12,500 years old, represent species that are today largely confined to southern Canada. Hence, the earliest fauna at Longswamp (contemporaneous with "tundra plants") seems to represent open-ground areas adjacent to stands of coniferous trees.

The Weaver Drain site of southern Michigan (Figure 17-1), dated at about 13,800 yr B.P. (Morgan et al., 1981), consists of a sequence of gray silts containing abundant insect and plant fossils. The plants, which include dwarf willow, dwarf birch, and *Dryas*, indicate a "tundra plant" assemblage. The beetle fauna substantiates a similar ecology and includes open-ground species (e.g., *Amara glacialis* Mannerheim) found in either treeless country or on the tundra. However, a solitary scolytid elytron belonging to *Polygraphus rufipennis* (Kirby), which feeds on conifers, was also recovered from the sequence. This suggests that isolated stands of trees were close to the site.

Examples of other insect faunas that crossed the transition between open-ground and forested regions have been recorded from sites in Minnesota, southern Ontario, northern New York, and Quebec. The Norwood fossil assemblage in east-central Minnesota (Figure 17-1) accumulated in a hollow formed by the melting of buried ice on the eastern margin of the Des Moines lobe (Ashworth et al., 1981). At the Norwood site, an initially open-ground fauna was replaced by a forest fauna at 12,400 ± 60 yr B.P. (QL-1083). Since no shifts in latitudinal distribution patterns could be detected between the faunas, the change has been interpreted as a local successional change rather than a climatic event. Two of the open-ground taxa, the hydrophilid *Helophorus arcticus* Brown and the carabid *Cymindis unicolor* Kirby, are species occurring close to the treeline in northern Canada today. *C. unicolor* is frequently found in dry, open clearings within the coniferous forest of the Yukon.

A fauna believed to be of similar age (around 12,500 yr B.P.) and indicative of open-ground conditions, has been recovered from the base of a kettle deposit at Brampton, near Toronto, Ontario (Figure 17-1). *C. unicolor* Kirby, *Helphorus arcticus* Brown, and *Asaphidion yukonense* Wickham have also recovered from the open-ground portion of this kettle deposit. Both the Norwood and Brampton faunas contain species that seem to indicate warmer conditions than those implied from the aforementioned species. For example, *Opisthius richardsoni* Kirby in the Norwood fauna is not known to be present above the treeline or on the tundra (Lindroth, 1961), but it lives on open ground close to water. Similarly, the basal deposits of the Brampton site contain an elytron of a tiger beetle, *Cicindela limbalis* Klug, a species usually found in the boreal-forest zone; no tiger beetles are known from tundra areas of North America (Morgan and Freitag, 1982). Many other species in both faunas indicate open ground but not true tundra.

The Brampton site has other interesting implications with respect to the migration rate of coniferous trees into the Lake Erie/Lake Ontario region. South of the former Lake Ontario ice lobe near Buffalo, New York, the Winter Gulf site (Figure 17-1), dated at about 12,800 to 12,500 yr B.P., possesses an insect fauna that indicates a relatively rapid transition from an open marsh habitat to a spruce-covered environment (Schwert and Morgan, 1980). The ice front was retreating from the limit of the Port Huron readvance, dated at approximately 13,000 yr B.P., and probably was about 50 km north of Winter Gulf at that time (Calkin and McAndrews, 1980). About 150 km northwest of Winter Gulf, on the northern side of the Lake Ontario lobe, the Brampton site was accumulating the first organic debris, with "tundra plant" assemblages including *Dryas integrifolia*, *Vaccinium uliginosum*, dwarf willow, and dwarf birch. It was not until around 12,000 yr B.P. that the first scolytid beetles appeared simultaneously with plant macrofossils and cones of *Picea glauca*.

Yet another site containing open-ground plant communities is at Columbia Bridge in northern Vermont (Figure 17-1). Two small insect assemblages are currently being studied by J. V. Matthews, Jr. Most taxa at Columbia Bridge, including the beetles *Pterostichus* cf. *adstrictus* Eschscholtz, *Bembidion matatum* Gemminger and Harold, and *Asaphidion yukonense* Wickham, are relatively temperate in distribution. None of the Columbia Bridge fossils represents an obligate tundra species; bark beetles (scolytids) are conspicuously absent from the site, as they are from the lower levels of Brampton and Norwood. Findings from the insect fossils are compatible with the conclusion made by Miller and Thompson (1979) that the Columbia Bridge sediments accumulated in an essentially nonforested landscape but one containing a rich mixture of arctic-alpine and rather temperate shrubs and herbs. The insect fauna indicates that the climate there was significantly warmer than it was 300 to 400 years later and 150 km farther north at St. Eugene, Quebec (Figure 17-1).

Interesting reconstructions of the Late Wisconsin environment of southern Quebec can be made by comparing the St. Eugene fauna with one from St. Hilaire near Montreal (Figure 17-1). Both sites were formerly at the shoreline of the Late Wisconsin Champlain Sea. The St. Eugene assemblage has been dated at 11,050 ± 130 yr B.P. (QU-488), whereas St. Hilaire dates to 10,100 ± 150 yr B.P. (GSC-2200). The differences between the two sites testify to rapid changes in the insect fauna in this portion of eastern North America between 11,000 and 10,000 yr B.P. Because the sites were located near a marine coast, they contain many taxa that frequent shoreline seaweed or live near beaches. There are no insects indicative of forested conditions in the St. Eugene fauna (Mott et al., 1981), but the assemblage includes a number of obligate and facultative tundra species such as *Amara alpina* Paykull, *A. glacialis* Mannerheim, and *Elaphrus parviceps* Van Dyke. In order to account for such species, one must assume that the summer climate at St. Eugene was considerably colder than it was at St. Hilaire, possibly approaching present conditions in northern Quebec.

While most of the insects in the St. Eugene assemblage can be found in northern Quebec today or can be expected to occur there, some (such as the ground beetle *Bembidion mckinleyi* Fall) are currently restricted to northwestern North America and extend westward to Fennoscandia. Such marked changes in distribution are shown by several other fossil species from northeastern North America.

The St. Hilaire locality, dating to 10,000 yr B.P., is at the base of a steep-sided island (up to 350 m altitude) within the Champlain Sea. The richness and diversity of the fossil assemblage is probably due in part to the fact that the deltaic detritus from which the insect fragments come includes samples of all vegetational zones on the island. Even so, it is obvious from the insect and plant fossils that the St. Hilaire assemblage represents a boreal environment with a regional climate only slightly cooler than today's. Although the pollen, plant, and insect fossils collectively point to an island that was mostly covered with coniferous forest, oaks were also present. This conclusion is based on the rather high percentage of oak pollen and two insects (one a tree hopper, the other a scarabaeid beetle) that feed on oak trees or are otherwise associated with them.

Fossil assemblages at Two Creeks, Wisconsin (Figure 17-1) (Morgan and Morgan, 1979), and at Gage Street in Kitchener, Ontario (Schwert, 1978), indicate that at approximately 11,800 yr B.P. both areas were covered by relatively open *Picea* woodland. These conditions prevailed until at least 10,600 yr B.P. on the coast of Lake Huron in southern Ontario, as illustrated by faunas recovered from the Parkhill site, northwest of London, Ontario (A. V. Morgan, unpublished data), dated at 10,870 yr B.P. (WAT-376), and Eighteen Mile River (Ashworth, 1977), dated at 10,600 yr B.P. (GSC-1127) (Figure 17-1). Work by Miller (1980) shows that a boreal/spruce forest/beetle fauna characterized Lockport Gulf (Figure 17-1) east of Niagara Falls at approximately the same time before being replaced about 9600 yr B.P. by a fauna indicative of mixed coniferous and deciduous woodland. Both the Lockport fauna and the Eighteen Mile River fauna show the presence of northern boreal species in what might seem to be local cooler microclimates. The Eighteen Mile River site may have been modified by cool winds blowing eastward off glacial Lake Algonquin, whereas the Lockport Gulf site may have been cold because of its position in the lee of the north-facing Niagara Escarpment (Morgan, Miller, and Morgan, 1982).

Farther west, the stagnating ice margin left behind kettle depressions that have been examined for insects. These include the Seibold and Johns Lake assemblages (Figure 17-1) from the Missouri Coteau, with radiocarbon dates of 9800 and 10,800 yr B.P., respectively (Ashworth and Brophy, 1972; A. C. Ashworth, unpublished data). The Mosbeck assemblage (Figure 17-1) (Ashworth et al., 1972), dated at approximately 10,000 yr B.P., was obtained from the eastern shoreline of Lake Agassiz in Minnesota. The faunas associated with these sites, together with the aforementioned Norwood locality, have their greatest affinity with the modern fauna of the boreal forest and contrast sharply with the prairie fauna of the midcontinent today. Between the oldest assemblage at Norwood and the youngest at Seibold, there is a small but perceptible shift from northern to southern boreal-forest elements, possibly implying a slight increase in summer temperatures.

Discussion

The Wisconsin ice that advanced from the Canadian Shield forced most of the northern insect fauna into ice-free areas, or refugia. One recognized refugium is Beringia, which extended from the extreme western Northwest Territories and most of the Yukon westward into Alaska and across the then-dry floor of the Bering-Chukchi Seas into Siberia. The second main refugium is believed to have existed south of the Wisconsin ice limit running from Washington State through Montana and the Dakotas and southeast into Wisconsin, Illinois, northeastern Pennsylvania, and New York State. The existence of a refugium off the eastern coast of North America has long been debated. With a lowered sea level, extensive areas of nonglaciated, dry land probably existed on the continental shelf of both the United States and Canada. Bones of terrestrial mammals recovered from the shelf probably indicate that this was dry ground at the time of maximum ice advance, but insect evidence is absent.

NORTHWESTERN REFUGIA

The late-Pleistocene invertebrate assemblages from Alaska and the Yukon suggest tundra environments with dry, well-drained substrates. From the periglacial structures at the sites it seems evident that the climate was cold enough to form permafrost but that it may not have fostered the moss-dominated wetland of modern permafrost terrain. Eolian activity was apparently greater then than it is now throughout all of Beringia. Whether the regional environment was steppe-tundra or arctic fell-field cannot be determined from insect fossils alone, but much circumstantial evidence argues for steppe-tundra or at least for a mosaic consisting of steppe, fell-field, and moss-sedge tundra.

Fragments of weevils numerically dominate most Alaskan and Yukon fossil insect faunas. Most belong to a single distinctive species, *Lepidophorus lineaticollis* Kirby, which now inhabits dry sites within forest and tundra areas of northwestern North America, probably occurring no farther south than northern Alberta. Although *Lepidophorus* fossils by themselves do not signify any particular regional environment, they are often most abundant in samples that also contain numerous obligate tundra species such as *Amara alpina* Paykull and some species of *Cryobius*.

During the late Pleistocene, *Lepidophorus* should have had easy access to Siberia across the Bering Land Bridge. Other weevils, particularly those with riparian affinities, apparently entered Alaska from Siberia via that route (Berman et al., 1979). Curiously enough, *Lepidophorus* is only a rare element in the present fauna of Chukotka

and is absent from Siberian fossil assemblages (Kiselyov, *in* Sher et al., 1979; S. V. Kiselyov, personal communication, 1979).

The subfamily Cleoninae is another group of weevils that also may someday provide important evidence concerning Pleistocene environments in eastern Beringia. Several species of Cleoninae appear in assemblages from exposures in the Kolyma lowland in eastern Siberia (Kiselyov, *in* Sher et al., 1979), and they provide the best evidence for the movement of true Mongolian steppe species into that area during the Pleistocene. Fossils of *Cleonus plumbeus* (LeConte), a North American species, have been reported from both the early- and late-Pleistocene assemblages at Cape Deceit (Matthews, 1974), but there is now good reason to question those identifications. Some of the specimens actually appear to be more similar to one of the Asian steppe species of Cleoninae than to *C. plumbeus* (LeConte), but a definite opinion on this matter must await the revision of the group.

A few byrrhid fossils (pill beetles) occur in almost all Pleistocene insect assemblages, but rarely are they as abundant as in some of the Late Wisconsin samples from Beringia (e.g., S-6 at Cape Deceit). In eastern Siberian sites, byrrhids similar to *Morychus*, but referred instead to the species *Chrysobyrrhulus rutilans* Motschulsky, are spectacularly abundant (Kiselyov, *in* Sher et al., 1979; Matthews, 1974). Kiselyov considers *Chrysobyrrhulus* to be an indicator of "cryoxerotic" conditions. We are not certain that the abundant *Morychus* fossils in Alaska-Yukon samples have a similar meaning. However, *Morychus* and *Lepidophorus* fossils are often codominants in eastern Beringian samples; this suggests dry substrates. *Morychus* is rarely collected; hence, we still know little about the habitat requirements of the various beetle species. Studies on the systematics and ecology of all northern byrrhids and weevils are desperately needed in order to understand better the environmental significance of these fossils.

SOUTHERN REFUGIA

The insect faunas that inhabited the southern part of Canada and the northern part of the United States during Sangamon and Early and Middle Wisconsin times were displaced southward at the time of the maximum Late Wisconsin ice advance. Insufficient data analysis of sites lying close to the maximum ice limit, in terms of time and space, preclude a detailed discussion of these faunal assemblages. However, the abundant open-ground and boreal species that are present in late-glacial sites in formerly glaciated areas attest to the successful survival of these species south of the ice front. The colonizing wave of beetle species following the retreating ice front reflects diachronous faunal assemblages migrating northward. Although many species found fossilized in sites in the northern United States are transcontinental in distribution, there are some interesting exceptions with restricted modern ranges far from their fossil occurrences. Extreme examples of such species are *Diacheila polita* Faldermann, *Anotylus gibbulus* (Eppelsheim), and *Holoboreaphilus nordenskioldi* (Mäklin), which so far have been found in Sangamon and Wisconsin deposits (Morgan and Morgan, 1980b, 1981; Schwert et al., 1981).

Examples of Late Wisconsin species of this type include the carabids *Opisthius richardsoni* Kirby and *Asaphidion yukonense* Wickham, the weevil *Vitavitus thulius* Kissinger, and the scolytid *Carphoborus andersoni* Swaine.

Opisthius richardsoni *Kirby*

Lindroth (1961) describes this beetle as pronouncedly western, with a distribution extending from Alaska south to California and east as far as Montana and Wyoming (Figure 17-2). *O. richardsoni* Kirby has not been found above the treeline or on the true tundra (Lindroth, 1961), and it is believed to be a pioneer species restricted to unstable riparian environments. The only known fossil occurrence of *O. richardsoni* Kirby is from late-glacial deposits at Norwood, Minnesota (Ashworth et al., 1981), some 1200 km east of the closest known modern record. This species must have survived both in the Alaska-Yukon refugium and in the southern refugium during Wisconsin time, and it must have been able to colonize the open disturbed ground following ice retreat. The study of more fossil sites should provide further clues to the postglacial migrational routes in the central and western United States.

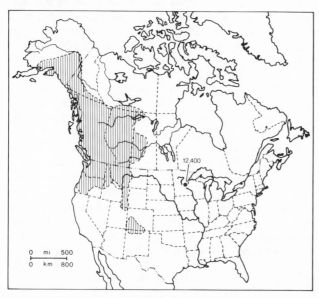

Figure 17-2. Approximate modern distribution (lined) and Late Wisconsin fossil locality (black circle) of the carabid *Opisthius richardsoni* Kirby.

Asaphidion yukonense *Wickham*

This carabid beetle is also confined to the western part of North America today, but it has been found as a fossil in both eastern and central North America (Figure 17-3). *A. yukonense* Wickham is usually found in open areas where the vegetation is restricted to patches of tiny mosses (Lindroth, 1963); such habitats must have been widely available in the recently deglaciated regions during late-glacial time. The greatest distance between known fossil and modern records is more than 3000 km; the minimum distance is more than 2000 km.

Vitavitus thulius *Kissinger*

Until recently, this species was known only from a single modern specimen (Kissinger, 1973), but in 1976 several individuals were taken in pitfall traps beside the Thelon River (Figure 17-4) in the central Northwest Territories (M. Barlow, in private collection of R. E. Morlan). Numerous attempts have been made to locate this weevil in riparian sites in the Yukon and Alaska without success. However, this species has been found on an upland dolomitic fell-field area, in association with *Morychus* and *Lepidophorus*, in the northern Yukon.

V. thulius Kissinger is relatively abundant in the early-Pleistocene deposits of the Kolyma Basin and at Cape Deceit, Alaska, but it is missing in samples of Late Wisconsin age (Matthews, 1974). *V.*

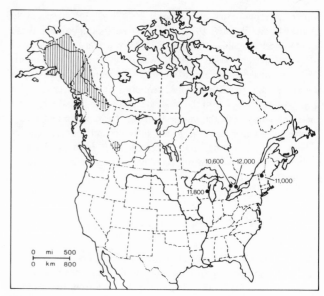

Figure 17-3. Approximate modern distribution (lined) and Late Wisconsin fossil localities (black circles) of the carabid *Asaphidion yukonense* Wickham.

thulius Kissinger occurs with *Lepidophorus* in samples from the Bell Basin in the northern Yukon, and both of these weevil species have been found in Holocene samples from the Mackenzie Valley. Fossils of *Vivavitus* have not yet been recorded from interior Alaska, but the species does occur in Early Wisconsin samples from the midcontinent (Ashworth, 1980), possibly at Columbia Bridge in Vermont (J. V. Matthews, Jr., unpublished data) and in the lower levels of the Brampton site near Toronto (A. V. Morgan, unpublished data).

V. thulius Kissinger, therefore, seems to have been abundant in some areas of Alaska, the Yukon, and eastern Siberia during the early Pleistocene. It may have been displaced to the south during the coldest phases of the late Pleistocene, and it has only managed a partial reoccupation of its former range during the Holocene. Because the modern occurrences are in glaciated areas, this species undoubtedly

Figure 17-4. Modern (black circles) and Late Wisconsin fossil localities (open circles) of the weevil *Vitavitus thulius* Kissinger.

Figure 17-5. Modern (black circles) and Late Wisconsin fossil localities (open circles) of the scolytid *Carphoborus andersoni* Swaine.

survived the Late Wisconsin maxima elsewhere, perhaps in the Beringian region or in areas south of the ice sheet.

Carphoborus andersoni *Swaine*

This bark beetle today is known from very few localities in extreme northwestern North America, but it has a long fossil history in the southern and central parts of Canada and the neighboring United States (Figure 17-5). *C. andersoni* Swaine has been recorded from Sangamon and Early and Middle Wisconsin sites, as well as sites that postdate ice retreat (Morgan and Morgan, 1980b). Undoubtedly, this beetle is well suited to the boreal habitat (it has been recorded from *Picea glauca*), and it seems peculiar that the species has not been found more recently in central and eastern North America.

The presence of beetles with contrasting modern and fossil occurrences indicates that considerable range changes took place as recently as 12,000 to 10,000 yr B.P. Such major distributional shifts are also recorded for fossil species in northwestern Europe (Coope, 1970) and show that the insects in large portions of the Northern Hemisphere reacted in this manner to the ecological and climatic changes that occurred during latest Wisconsin time.

Conclusions

Research on Pleistocene insects is providing a wealth of exciting information. Some of the more important aspects include the evolutionary changes in species, the longevity of species, the paleoclimate, the Pleistocene refugia, and the real evidence that fossil distributions provide for determining zoogeographic shifts through time.

Over the last decade, such studies in North America have helped to provide a general framework on which future studies can be based. Many of the problems are due to a lack of knowledge concerning the taxonomy and range of the present fauna, although the Biological Survey of Canada, a current project of the National Museums of Canada, promises to alleviate some of these inadequacies. Studies of the Beringian refugium, as well as the zoogeography of the Holarctic fauna, will benefit from joint Soviet-North American studies through the collaboration of all paleoecologists working on Pleistocene assemblages.

References

Ashworth, A. C. (1973). Fossil beetles from a fossil wood rat midden in western Texas. *The Coleopterists Bulletin* 27, 139-40.

Ashworth, A. C. (1977). A Late Wisconsin coleopterous assemblage from southern Ontario and its environmental significance. *Canadian Journal of Earth Sciences* 14, 1625-34.

Ashworth, A. C. (1979). Quaternary Coleoptera studies in North America: Past and present. *In* "Carabid Beetles: Their Evolution, Natural History and Classification" (T. L. Erwin, G. E. Ball, and D. R. Whitehead, eds.), pp. 395-406. Proceedings of the First International Symposium of Carabidology, Dr. W. Junk bv Publishers, The Hague.

Ashworth, A. C. (1980). Environmental implications of a beetle assemblage from the Gervais Formation (Early Wisconsinan?), Minnesota. *Quaternary Research* 13, 200-212.

Ashworth, A. C., and Brophy, J. A. (1972). A late Quaternary fossil beetle assemblage from the Missouri Coteau, North Dakota. *Geological Society of America Bulletin* 83, 2981-88.

Ashworth, A. C., Clayton, L., and Bickley, W. B. (1972). The Mosbeck site: A paleoenvironmental interpretation of the late Quaternary history of Lake Agassiz based on fossil insect and mollusc remains. *Quaternary Research* 2, 176-88.

Ashworth, A. C., Schwert, D. P., Watts, W. A., and Wright, H. E., Jr. (1981). Plant and insect fossils at Norwood in south-central Minnesota: A Record of late-glacial succession. *Quaternary Research* 16, 66-79.

Ball, G. E. (1966). A revision of the North American species of the subgenus *Cryobius* Chaudoir. *Opuscula Entomologica* 28, 1-166.

Barry, R. G. (1982). Approaches to reconstructing the climate of the steppe-tundra biome. *In* "Paleoecology of Beringia" (D. M. Hopkins, J. V. Matthews, Jr., C. E. Schweger, and S. B. Young, eds.), 195-204. Academic Press, New York.

Berman, D. I., Vinokurov, N. N., and Korotjaev, B. A. (1979). On the Beringian connections of Curculionidae, Carabidae and Heteroptera faunas. *In* "14th Pacific Science Congress, Khabarovsk, Abstracts, Committee CD," pp. 206-7.

Blaisdell, F. E. (1909). "A Monographic Revision of the Coleoptera Belonging to the Tenebrionid Tribe Eleodiini Inhabiting the United States, Lower California, and Adjacent Islands. United States National Museum Bulletin 63.

Calkin P. E., and McAndrews, J. H. (1980). Geology and paleontology of two late Wisconsin sites in western New York State. *Geological Society of America Bulletin* 91, 295-306.

Campbell, J. M. (1973). "A Revision of the Genus *Tachinus* Coleoptera: Staphylinidae) of North and Central America." Entomological Society of Canada Memoir 90.

Carter, L. D. (1981). A Pleistocene sand sea on the Alaskan Coastal Plain. *Science* 211, 381-83.

Coope, G. R. (1965). The value of fossil insect faunas in the understanding of Quaternary ecologies and climates. *In* "International Association for Quaternary Research Seventh Congress Abstracts," p. 74. Boulder, Colo.

Coope, G. R. (1968). Insect remains from silts below till at Garfield Heights, Ohio. *Geological Society of America Bulletin* 79, 753-56.

Coope, G. R. (1970). Interpretations of Quaternary insect fossils. *Annual Review of Entomology* 15, 97-120.

Coope, G. R. (1979). Late Cenozoic fossil Coleoptera: Evolution, biogeography, and ecology. *Annual Review of Ecology and Systematics* 10, 247-67.

Doyen, J. T., and Miller, S. E. (1980). Review of Pleistocene darkling ground beetles of the California asphalt deposits (Coleoptera:Tenebrionidae, Zopheridae). *Pan-Pacific Entomologist* 56, 1-10.

Grinnell, F. (1908). Quaternary myriapods and insects of California. *University of California Department of Geology Bulletin* 5, 207-15.

Guthrie, R. D., and Matthews, J. V., Jr. (1971). The Cape Deceit fauna: Early Pleistocene mammalian assemblage from the Alaskan Arctic. *Quaternary Research* 1, 474-510.

Hammond, P., Morgan, A., and Morgan, A. V. (1979). On the *gibbulus* group of *Anotylus*, and fossil occurrences of *Anotylus gibbulus* (Staphylinidae). *Systematic Entomology* 4, 215-21.

Hopkins, D. M. (1982). Aspects of the paleogeography of Beringia during the late Pleistocene. *In* "Paleoecology of Beringia" (D. M. Hopkins, J. V. Matthews, Jr., C. E. Schweger, and S. B. Young, eds.), 3-28. Academic Press, New York.

Hughes, O. L. (1972). "Surficial Geology of Northern Yukon Territory and Northwestern District of Mackenzie, Northwest Territories." Geological Survey of Canada Paper 69-36, pp. 1-11.

Kissinger, D. G. (1973). A new weevil genus from America north of the Arctic Circle and notes of fossils from Pliocene and Pleistocene sediments (Coleoptera:Curculionidae). *The Coleopterists Bulletin* 27, 193-200.

Lichti-Federovich, S. (1973). Palynology of six sections of late Quaternary sediments from the Old Crow River, Yukon Territory. *Canadian Journal of Botany* 51, 533-64.

Lichti-Federovich, S. (1974). "Palynology of Two Sections of Late Quaternary Sediments from the Porcupine River, Yukon Territory." Geological Survey of Canada Paper 74-23, pp. 1-6.

Lindroth, C. H. (1961). The ground beetles of Canada and Alaska: Part 2. *Opuscula Entomologica* 20, 1-200.

Lindroth, C. H. (1963). The ground beetles of Canada and Alaska: Part 3. *Opuscula Entomologica* 24, 201-408.

Lindroth, C. H. (1966). The ground beetles of Canada and Alaska: Part 4. *Opuscula Entomologica* 29, 409-648.

Matthews, E. G., and Halffter, G. (1968). New Data on American *Copris* with discussion of a fossil species (Coleopt., Scarab.). *Ciencia* (Mexico) 26, 147-62.

Matthews, J. V., Jr. (1968). A paleoenvironmental analysis of three late Pleistocene assemblages from Fairbanks, Alaska. *Quaestiones Entomologicae* 4, 202-24.

Matthews, J. V., Jr. (1970). Quaternary environmental history of interior Alaska: Pollen samples from organic colluvium and peats. *Arctic and Alpine Research* 2, 241-51.

Matthews, J. V., Jr. (1974). Quaternary environments at Cape Deceit (Seward Peninsula, Alaska): Evolution of a tundra ecosystem. *Geological Society of America Bulletin* 85, 1353-84.

Matthews, J. V., Jr. (1975). "Incongruence of Macrofossil and Pollen Evidence: A Case from the Late Pleistocene of the Northern Yukon Coast." Geological Survey of Canada Paper 75-1B, pp. 139-46.

Matthews, J. V., Jr. (1976). "Insect Fossils from the Beaufort Formation: Geological and Biological Significance." Geological Survey of Canada Paper 76-1B, pp. 217-27.

Matthews, J. V., Jr. (1977). Coleoptera fossils: Their potential for dating and correlation of late Cenozoic sediments. *Canadian Journal of Earth Sciences*, 14, 2339-47.

Matthews, J. V., Jr. (1980). Tertiary land bridges and their climate: Backdrop for development of the present Canadian fauna. *Canadian Entomologist* 112, 1089-1103.

Matthews, J. V., Jr. (1982). East Beringia during Late Wisconsin time: arctic-steppe or fell-field? A review of the biotic evidence. *In* "Paleoecology of Beringia" (D. M. Hopkins, J. V. Matthews, Jr., C. E. Schweger, and S. B. Young, eds.), pp. 127-50. Academic Press, New York.

Miller, N. J., and Thompson, G. G. (1979). Boreal and western North American plants in the late Pleistocene of Vermont. *Journal of the Arnold Arboretum* 60, 167-218.

Miller, R. F. (1980). Palaeoentomological analysis of a postglacial site in northwestern New York State. M.S. thesis, University of Waterloo, Waterloo, Ontario.

Miller, S. E., Gordon, R. D., and Howden, H. F. (1981). Reevaluation of Pleistocene scarab beetles from Rancho La Brea, California (Coleoptera:Scarabaeidae). *Proceedings of the Entomological Society of Washington* 83, 625-30.

Miller, S. E., and Peck, S. B. (1979). Fossil carrion beetles of Pleistocene California asphalt deposits, with a synopsis of Holocene California Silphidae (Insecta:Coleoptera:Silphidae). *Transactions of the San Diego Society of Natural History* 19, 85-106.

Morgan, A. (1972). The fossil occurrence of *Helophorus arcticus* Brown (Coleoptera:Hydrophilidae) in Pleistocene deposits of the Scarborough Bluffs, Ontario. *Canadian Journal of Zoology* 50, 555-58.

Morgan, A. V., Elias, S. A., and Morgan, A. (1981). Paleoenvironmental implications of a late glacial insect assemblage from south-east Michigan. *In* "Abstracts of the Geological Association of Canada Annual Meeting," 6, p. A-41. Calgary, Alta.

Morgan, A. V., Elias, S. A., and Morgan, A. (1982). Insect fossils from a late glacial site at Longswamp, Pennsylvania. *Geological Society of America, Program with Abstracts (North Central Section)*, p. 266.

Morgan, A. V., and Freitag, R. (1982). The occurrence of *Cicindela limbalis* Klug, (Coleoptera:Cicindelidae) in a late glacial site at Brampton, Ontario. *The Coleopterists Bulletin* 36, 105-8.

Morgan, A. V., and Morgan, A. (1979). The fossil Coleoptera of the Two Creeks Forest Bed, Wisconsin. *Quaternary Research* 12, 226-40.

Morgan, A. V., and Morgan, A. (1980a). Beetle bits: The science of paleoentomology. *Geoscience Canada* 7, 22-29.

Morgan, A. V., and Morgan, A. (1980b). Faunal assemblages and distributional shifts of Coleoptera during the late Pleistocene in Canada and the northern United States. *Canadian Entomologist* 112, 1105-28.

Morgan, A. V., and Morgan, A. (1981). Paleoentomological methods of reconstructing paleoclimate with reference to interglacial and interstadial insect faunas of southern Ontario. *In* "Quaternary Paleoclimate" (W. C. Mahaney, ed.), pp. 173-92. Geo Abstracts Limited, Norwich, England.

Morgan, A. V., and Morgan, A. (1982). Fossil insect faunas associated with the maximum Late Wisconsinan ice advance in the northern United States. *Geological Society of America, Program with Abstracts* 14, 570.

Morgan, A. V., Miller, R. F., and Morgan, A. (1982). Paleoenvironmental reconstruction of southwestern Ontario between 11,000 and 10,000 yr B.P. using fossil insects as indicators. *In* "Third North American Paleontological Convention Proceedings," pp. 381-86. Montreal, Quebec.

Morgan, A. V., Morgan, A., and Carter, L. D. (1979). Paleoenvironmental interpretation of a fossil insect fauna from bluffs along the lower Colville River, Alaska. *In* "The United States Geological Survey in Alaska: Accomplishments during 1978" (K. M. Johnson and J. R. Williams, eds.), pp. 41-43. U.S. Geological Survey Circular 804-B.

Morgan, A. V., Morgan, A., and Miller, R. F. (1982). Late Farmdalian and Early Woodfordian insect assemblages from Garfield Heights, Ohio. *Geological Society of America, Program with Abstracts (North-Central Section)* p. 267.

Morlan, R. E. (1980). "Taphonomy and Archaeology in the Upper Pleistocene of the Northern Yukon Territory: A Glimpse of the Peopling of the New World." National Museum of Man Mercury Series, Ottawa.

Mott, R. J., Anderson, T. W., and Matthews, J. V., Jr. (1981). Late glacial paleoenvironments of sites bordering the Champlain Sea based on pollen and macrofossil evidence. *In* "Quaternary Paleoclimate" (W. C. Mahaney, ed.), pp. 129-71, Geo Abstracts Limited, Norwich, England.

Péwé, T. L. (1975). "Quaternary Geology of Alaska." U.S. Geological Survey Professional Paper 835.

Pierce, W. D. (1944). Fossil arthropods of California: 1. Introductory statement. *Bulletin of the Southern California Academy of Sciences* 43, 1-3.

Pierce, W. D. (1946a). Fossil arthropods of California: 10. Exploring the minute world of the California asphalt deposits. *Bulletin of the Southern California Academy of Sciences* 45, 113-18.

Pierce, W. D. (1946b). Fossil arthropods of California: 11. Descriptions of the dung beetles (Scarabaeidae) of the tar pits. *Bulletin of the Southern California Academy of Sciences* 45, 119-32.

Pierce, W. D. (1947). Fossil arthropods of California: 13. A progress report on the Rancho La Brea asphaltum studies. *Bulletin of the Southern California Academy of Sciences* 46, 136-38.

Pierce, W. D. (1948). Fossil arthropods of California: 16. The carabid genus *Elaphrus* in the asphalt deposits. *Bulletin of the Southern California Academy of Sciences* 16, 53-54.

Pierce, W. D. (1949). Fossil arthropods of California: 17. The silphid burying beetles in the asphalt deposits. *Bulletin of the Southern California Academy of Sciences* 48, 55-70.

Pierce, W. D. (1954a). Fossil arthropods of California: 18. The Tenebrionidae-Tentyriinae of the asphalt deposits. *Bulletin of the Southern California Academy of Sciences* 53, 35-45.

Pierce, W. D. (1954b). Fossil arthropods of California: 19. The Tenebrionidae-

Scaurinae of the asphalt deposits. *Bulletin of the Southern California Academy of Sciences* 53, 93-98.

Pierce, W. D. (1954c). Fossil arthropods of California: 20. The Tenebrionidae-Coniontinae of the asphalt deposits. *Bulletin of the Southern California Academy of Sciences* 53, 142-56.

Prest, V. K., Terasmae, J., Matthews, J. V., Jr., and Lichti-Federovich, S. (1976). Late Quaternary history of Magdalen Islands, Quebec. *Maritime Sediments* 12, 39-59.

Repenning, C. A. (1980). Faunal exchanges between Siberia and North America. *In* "The Ice-Free Corridor and Peopling of the New World" (N. W. Rutter and C. E. Schweger, eds.), pp. 37-44. Canadian Journal of Anthropology 1.

Schwert, D. P. (1978). Paleoentomological analyses of two postglacial sites in eastern North America. Ph.D. dissertation, University of Waterloo, Waterloo, Ontario.

Schwert, D. P., Ashworth, A. C., and Baker, R. G. (1981). An Arctic insect assemblage from the Late Wisconsinan of midcontinental North America. *Geological Society of America, Program with Abstracts* 13, 550.

Schwert, D. P., and Morgan, A. V. (1980). Paleoenvironmental implications of a late glacial insect assemblage from northwestern New York. *Quaternary Research* 13, 93-110.

Sher, A. V., Kaplina, T. N., Giterman, R. E., Lozhkin, A. V., Arkhangelov, A. A., Kiselyov, S. V., Kouznetsov, Y. V., Virina, E. I., and Zazhigin, V. S. (1979). Late Cenozoic of the Kolyma lowland. *In* "Guidebook to Tour XI, Navka," p. 115. 14th Pacific Science Congress, Khabarovsk, U.S.S.R.

Shotton, F. W. (1965). The movements of insect populations in the British Pleistocene. *In* "International Association for Quaternary Research Seventh Congress Abstracts," p. 427. Boulder, Colo.

Sleeper, E. L. (1969). Two new Cleoninae from the Rita Blanca lake deposits. *In* "Paleoecology of an Early Pleistocene Lake on the High Plains of Texas" (R. Y.

Anderson and D. W. Kirkland, eds.), pp. 131-33. Geological Society of America Memoir 113.

Spaulding, W. G., Leopold, E. B., and Van Devender, T. R. (1983). Late Wisconsin paleoecology of the American Southwest. *In* "Late-Quaternary Environments of the United States," vol. 1, "The Late Pleistocene" (S. C. Porter, ed.), pp. 259-93. University of Minnesota Press, Minneapolis.

Spilman, T. J. (1976). A new species of the fossil *Ptinus* from fossil wood rat nests in California and Arizona (Coleoptera, Ptinidae), with a postscript on the definition of a fossil. *The Coleopterists Bulletin* 30, 239-44.

Sublette, J. E. (1969). Aquatic insects of the Rita Blanca lake deposits. *In* "Paleoecology of an Early Pleistocene Lake on the High Plains of Texas." (R. Y. Anderson and D. W. Kirkland, eds.), pp. 117-22. Geological Society of America Memoir 113.

Van Devender, T. R., and Spaulding, W. G. (1979). Development of vegetation and climate in the southwestern United States. *Science* 204, 701-10.

Waage, J. K. (1976). Insect remains from ground sloth dung. *Journal of Paleontology* 50, 991.

Watts, W. A. (1979). Late Quaternary vegetation of central Appalachia and the New Jersey Coastal Plain. *Ecological Monographs*, 49, 427-69.

Werner, F. G. (1969). Terrestrial insects of the Rita Blanca lake deposits. *In* "Paleoecology of an Early Pleistocene Lake on the High Plains of Texas" (R. Y. Anderson and D. W. Kirkland, eds.), pp. 123-30. Geological Society of America Memoir 113.

Woodard, G. D., and Marcus, L. F. (1973). Rancho La Brea fossil deposits: A re-evaluation from stratigraphic and geological evidence. *Journal of Paleontology* 47, 54-69.

CHAPTER 18

The Antiquity of Man in America

Frederick Hadleigh West

Introduction

Any attempt to summarize in a few pages a subject as intricate, as large, and as contentious as this one is bound to amount to something of a fiction. Some idea of the variety of interpretation may be gained from summaries such as those by MacNeish (1976, 1978) and Bryan (1978a) and from more-encyclopedic regional surveys by Griffin (e.g., 1965, 1978) and Williams and Stoltman (1965). In addition, practically every paper ever written on any aspect of the subject is partially interpretive in character. Extended treatments of the materials are to be found in Wormington (1957), MacGowan and Hester (1962), and Sellards (1952). These remain essential works for the general student of early American prehistory.

For this chapter, a somewhat more explicitly synthetic approach is taken. This is done partially in deference to the limitations of the author's competency—freely acknowledged—and to the limitations of space but more particularly in the conviction that the evidence now available makes this course preferable. Nevertheless, it is clear that this or any other account is effective to the degree to which it marshals good evidence in support of the theses advanced. To fail in this regard is to invite the charge of purveying just-so stories, that is, pure fiction.

Certainly the most basic question in the study of early man in the New World is when did the earliest immigrants appear? Involved here is a great deal more than simple antiquarianism, for no proper understanding of the building of the large, complexly structured edifice that was pre-Columbian America can even begin without a knowledge of its basic building blocks.

Among students of the problem it may be said that there exist basically two positions on the questions of time of entry. Although there is an almost universal consensus in the fundamental importance of Clovis culture, there are archaeologists who also see Clovis as representing the earliest movement. Other archaeologists maintain just as stoutly that other cultures had flourished in the Americas long before the Clovis people made their appearance. This dichotomy of opinion forms a convenient format for discussion of the problem. Clearly, this difference in viewpoint stems more from the equivocal nature of much of the evidence than from the preconceptions of the investigators—a point to bear in mind on those occasions when discussions of the subject overheat.

The Question of Pre-Clovis Remains in the Americas

There have been a number of recent statements on the several facets of this problem. (See especially the series of papers edited by Bryan [1978a], the reviews of MacNeish [1976, 1978], and the volume by Carter [1980].) At present, the question turns principally on the interpretation of bone material found in the Old Crow Flats of the northern Yukon Territory (Bonnichsen, 1978, 1979; Irving, 1978; Irving and Harington, 1973; Irving et al., 1977; Morlan, 1978, 1979, 1980); similar material at the Dutton, Selby, and Lamb Spring sites in Colorado (Stanford, 1979b; Stanford et al., no date); the Calico Mountains site in California (Simpson, 1980); the Timlin site in New York State (Bryan et al., 1980; Stagg et al., 1980); the Taber child site in Alberta (Stalker, 1969, 1977); the Lively complex in Alabama (Josselyn, 1965; Lively, 1965); China Lake in California (Davis, 1975, 1978); the Texas Street site in California (Carter, 1980); and several others that have been matters of debate for many years. (See Figure 18-1). These include such localities as Sheguiandah in Ontario (Lee, 1955), Santa Rosa Island off the California coast (Orr, 1968), and Del Mar and others in southern California (Carter, 1980). Farther afield, the recent discoveries at El Bosque in Nicaragua (Gruhn, 1977, 1978; Page, 1978), Taima-taima in Venezuela (Bryan et al., 1978), and Pikimachay in Peru (MacNeish, 1979) must also be considered.

The recently discovered Meadowcroft Rockshelter site in Pennsylvania is a special case in this context, but its radiocarbon dating

It is a pleasure to record my gratitude to several people whose aid much facilitated the preparation of this chapter: Robert E. Funk, state archaeologist of New York, for allowing me to examine portions of the West Athens Hill collection; Peter J. Fetchko, director, and John R. Grimes, assistant curator of ethnology, Peabody Museum of Salem, for allowing me to study the collection from Bull Brook; and Nancy J. Schmidt, director of the Tozzer Library, Harvard University, for helping me use that facility. In addition, several scholars gave freely of their expert knowledge: George C. Frison, Robert E. Funk, Robert Gal, John R. Grimes, R. Dale Guthrie, C. Vance Haynes, Jr., Dennis Stanford, and Robert Stuckenrath. It was a singular pleasure, as well, to spend some time with William A. Eldridge, Nicola Vaccaro, and Frank Vaccaro, whose dedication (with that of the other Vaccaro brothers) to their excavation of the Bull Brook site is the only reason that that remarkable collection survives. Finally, as always, my gratitude goes to my wife Constance for her patience and encouragement.

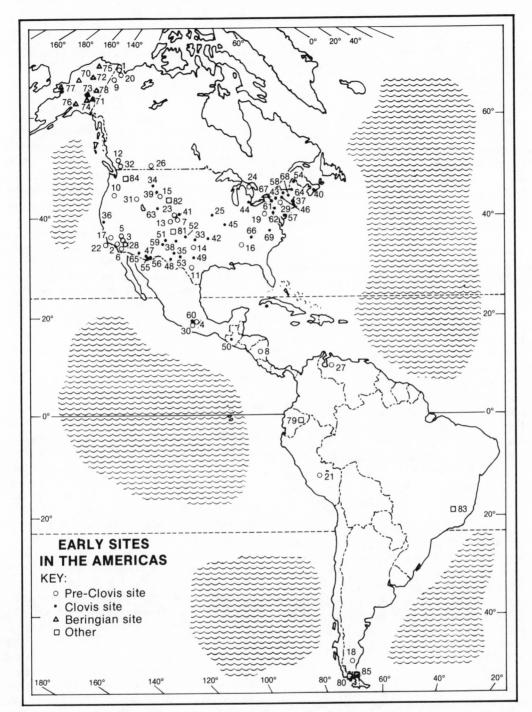

**EARLY SITES
IN THE AMERICAS**
KEY:
○ Pre-Clovis site
• Clovis site
△ Beringian site
□ Other

Figure 18-1. Early sites in the Americas. Shown on the map are the locations of sites discussed in the text under one of the following four headings: Pre-Clovis, Clovis, Beringian, and unclassified. **Pre-Clovis:** 1, British Mountain, Yukon Territory; 2, Buchanan Canyon, California; 3, Calico Mountains, California; 4, Caulapan, El Horno, Hueyatlaco, Mirador, Tecaxco, Mexico (Valsequillo vicinity); 5, China Lake, California; 6, Del Mar, California; 7, Dutton, Colorado; 8, El Bosque, Nicaragua; 9, Flint Creek, Yukon Territory; 10, Fort Rock Cave, Oregon; 11, 41 ME 3, Texas; 12, Frazer Canyon, British Columbia; 13, Lamb Spring, Colorado; 14, Lewisville, Texas; 15, Little Canyon Creek, Wyoming; 16, Lively complex, Alabama; 17, Los Angeles, California; 18, Los Toldos, Argentina; 19, Meadowcroft Rockshelter, Pennsylvania; 20, Old Crow Flats (vicinity), Yukon Territory; 21, Pikimachay, Peru; 22, Santa Rosa Island, California; 23, Selby, Colorado; 24, Sheguiandah, Ontario; 25, Shriver, Missouri; 26, Taber, Alberta; 27, Taima-taima, Venezuela; 28, Texas Street, California; 29, Timlin, New York; 30, Tlapacoya, Mexico; 31, Wilson Butte Cave, Idaho; 32, Yale, British Columbia. **Clovis:** 33, Anadarko, Oklahoma; 34, Anzik, Montana; 35, Blackwater Draw, New Mexico; 36, Borax Lake, California; 37, Bull Brook, Massachusetts; 38, Burnet Cave, New Mexico; 39, Colby, Wyoming; 40, Debert, Nova Scotia; 41, Dent, Colorado; 42, Domebo, Oklahoma; 43, Dutchess Quarry Cave, New York; 44, Holcombe, Michigan; 45, Kimmswick, Missouri; 46, Kings Road, New York; 47, Lehner, Arizona; 48, Leon Creek, Texas; 49, Levi Rockshelter, Texas; 50, Los Tapiales, Guatemala; 51, Lucy, New Mexico; 52, Miami, Texas; 53, Midland (Scharbauer), Texas; 54, Munsungun, Maine; 55, Murray Springs, Arizona; 56, Naco, Arizona; 57, Plenge, New Jersey; 58, Reagan, Vermont; 59, Sandia, New Mexico; 60, Santa Isabel Iztapan, Mexico; 61, Shawnee-Minisink, Pennsylvania; 62, Shoop, Pennsylvania; 63, Union Pacific, Wyoming; 64, Vail, Maine; 65, Ventana Cave, Arizona; 66, Wells Creek, Tennessee; 67, West Athens Hill, New York; 68, Whipple, New Hampshire; 69, Williamson, Virginia. **Beringian:** 70, Akmak, Alaska; 71, Dry Creek, Alaska; 72, Girl's Hill, Alaska; 73, Mount Hayes, Alaska; 74, Phipps (Mount Hayes 111), Alaska; 75, Putu, Alaska; 76, Ravine Lake, Alaska; 77, Trail Creek, Alaska; 78, Village (Healy Lake), Alaska. **Unclassified:** 79, El Inga, Ecuador; 80, Fells Cave, Chile; 81, Folsom, New Mexico; 82, Hell Gap, Wyoming; 83, Lagoa Santa Cave, Brazil; 84, Marmes, Washington; 85, Palli Aike, Chile.

certainly places it within the general category of pre-Clovis remains (Adovasio, Gunn, Donahue, and Stuckenrath, 1977, 1978; Adovasio, Gunn, Donahue, Stuckenrath, et al., 1978).

The most usual position taken in works presenting arguments for the presence of man in North America prior to Clovis times is that the first entry took place at some point in the Early to Middle Wisconsin. Gruhn (1977: 163) puts the matter succinctly: "Evidence from a number of archaeological sites distributed in the western part of the hemisphere from the Yukon into South America now indicate a *minimum* possible date of 40,000 years ago for the earliest entry of man into the North American continent." MacNeish (1978: 479) sees evidence of man in South America 20,000 years ago and finds "related remains . . . in North America from at least 34,000 to 60,000 years ago." Dragoo (1980) affirms a continuing conviction that the Americas were occupied at least by 30,000 years ago. The Old Crow evidence is seen by Morlan (1980: 277) also as extending back 60,000 years. Krieger (1979), discussing the Calico Mountains site, suggests that the movements responsible for these materials could have taken place any time between 150,000 and 45,000 years ago. Carter (1980: 3) estimates the period of human occupancy of the Americas at about 100,000 years. Meadowcroft, by virtue of its radiocarbon chronology is considered as giving evidence of occupation dating back to 16,000 or perhaps 19,000 years ago (Adovasio, Gunn, Donahue, Stuckenrath, et al., 1978: 158).

In discussing these posited earliest occupations, the several authors tend to characterize them minimally as being pre-Clovis or as having been dominated by pebble-tool or chopping-tool industries of a not very advanced sort. Carter (1978) sees, in addition, the presence of milling stones in certain contexts (for example, the Lajollan, which he places in the area of 50,000 years ago). Krieger (1979) equates an as-yet-unnamed American assemblage (consisting of Calico, Buchanan Canyon, California, and three sites in Texas) with Middle Palaeolithic Mousterian.

Probably the most ambitious organizational scheme proposed for this long period is that of MacNeish (1976, 1978). He describes a sequence of four stages, namely, Big Game Collectors (I), Unspecialized Hunters (II), Specialized Hunters (III), and Ecozonal Specialized Hunters (IV). Stage I evidence is seen in the early phases at Pikimachay, at the Lewisville site in Texas, at the Yale site in British Columbia, in the Calico Mountains, with the "skeletons from southern California," and at the Taber child site. Stage II is evidenced in North America by Meadowcroft Rockshelter, Levi Rockshelter of Texas ("older than 13,750 ± 410 yr B.P.?"), Old Crow, Santa Rosa Island, and British Mountain (Yukon Territory) and, to the south, Valsequillo and Tlapacoya in Mexico (both dating beyond 20,000 yr B.P.) and El Bosque in Central America (placed in the same time range). The Stage III hunters are regarded as having left their North American evidence at Fort Rock Cave in Oregon, Flint Creek in the Yukon Territory, the Midland Man site in Texas, early Frazer Canyon in British Columbia, early Hell Gap in Wyoming, Wilson Butte Cave in Idaho, Meadowcroft Rockshelter in Pennsylvania, and Taima-taima in northern South America. Stage IV is represented in North America by the various Clovis and later paleo-Indian complexes and in South America by several sites and complexes of greater or lesser familiarity to northerners (for example, Fell's Cave in Chile, El Inga in Ecuador, Lagoa Santa Cave in Brazil, and Tlapacoya Il in Mexico, among others).

In his "Overview of Paleo-American Prehistory," Bryan (1978b: 306) maintains that the case for pre-Clovis has been rendered firm by

evidence "from widely separated localities such as Pikimachay, Tlapacoya, Meadowcroft, and Old Crow."

REPRESENTATIVE SITES AND EVIDENCE OF THE PRE-CLOVIS OCCUPATION

Bryan's broad characterization of the evidence would presumably meet no serious argument from most other researchers of the pre-Clovis time period: "The basic Paleo-American tool kit contained few, if any, standardized stone or bone tool types" (Bryan, 1978b: 311). MacNeish (1978) acknowledges that his stages I and III are poorly defined but maintains that stage II is "less ethereal" primarily due to the relative abundance of material from four sites and localities: (1) layers 11 and 12 of Los Toldos Cave in southern Argentina, (2) the Ayacucho complex of Pikimachay (zone h and H1), (3) several sites in the Valsequillo vicinity of Puebla (Tecaxco, Mirador, El Horno, and Hueyatlaco) and the nearby site of Caulapan (which has a radiocarbon date in the appropriate range of around 20,000 years), and (4) the lower levels of the Meadowcroft site (with radiocarbon dates ranging back more than 28,000 years).

It is essential at this point to consider briefly some of these critical pieces of evidence. For this purpose they are reduced to a single group—evidence of pre-Clovis man.

At Pikimachay the two lower components (Pacaicasa and Ayacucho) leave some doubt as to whether all or any of the illustrated specimens are true artifacts (MacNeish, 1979: Figures 21-23; see also the comment of Lynch, 1974). This, combined with a series of bone (presumably collagen) dates, strongly urges caution, at the least, until much fuller and more convincing evidence has been presented.

The acceptance of the El Bosque materials, evidently likewise crude to the degree that serious questions about their human manufacture can be raised, depends in part upon an assumption that the chert in question is exotic to the site. C. V. Haynes (personal communication, 1977) reports that this chert does, in fact, occur in veins close to the site (note Gruhn's [1978: 261 and Figure 1] reference to "tabular flaked chert" and the illustrations of two such specimens). Perhaps the final word on El Bosque should be Gruhn's (1978: 262): "Once again a major problem in dealing with a very early New World site is the inability of archaeologists to distinguish conclusively the handiwork of man from natural events in an unfamiliar context."

The Taima-taima site in Venezuela appears to provide firm evidence of El Jobo culture dating to 13,000 yr B.P. (Bryan et al., 1978). The association of El Jobo hunting apparatus with mastodon remains seems clear; the matter of dating, less so. Rouse and Cruxent (1963) report dates of around 16,000 yr B.P. and 14,000 yr B.P. for the site of Muaco, like Taima-taima, lying close to a spring or water hole. These authors acknowledge the cooccurrence at Muaco of El Jobo artifacts, bones of extinct fauna, and modern bottle glass. Nevertheless, the cited bone dates are said to pertain to a paleo-Indian occupation and possibly the antiquity of El Jobo. Lynch (1974) in assessing this problem observes that Cruxent attributes the disturbed associations to the upwelling of spring water. Although Taima-taima should, according to Lynch (1974: 373), be free of "gross modern disturbance," he notes that Bryan in 1973 admitted that there, as well, the activity of a spring and the trampling of animals may have disturbed the associations. It is difficult to see why the El Jobo point, found lying "within the cavity of the right pubis of the juvenile mastodon" (Bryan et al., 1978: Figure 2), should be, by implication, exempted from such movement. Whereas earlier reported dates for Taima-taima were run on "non-carbonate carbon content from the grey (fossiliferous) sand'' (Bryan, 1973: 247) as well as on bone, the assays reported in 1978 were on twig fragments and yielded results within the same range as the previous series (12,980 ± 85 yr B.P. and 14,200 ± 300 yr B.P. [Bryan et al., 1978: Figure 4]). Haynes's (1974) discussion of this site is still pertinent. The possibility appears to remain that older carbonates suspended in the spring water contaminated the bone, the gray soil (both per Haynes's suggestion), *and* the twigs. The situation is less than clear-cut.

At Tlapacoya, near Mexico City, Mirambell (1978) reports two hearths associated with an extensive extinct fauna on a Pleistocene beach of Lake Chalco. There appear to be critical problems of artifact identification and provenience, however, suggesting that a proper assessment of this postulated 20,000-year-old occupation must await further clarification.

The collections from the Calico Mountains (or Hills) site in the Mojave Desert of California (Carter, 1980; Simpson, 1978, 1980) associated with alluvial-fan deposits have been estimated to date to Sangamon times (Simpson, 1978) or earlier (Haynes, 1973). Haynes's question—artifacts or geofacts?—remains: the objects found by Simpson (1980: Plates 4-7) simply do not appear to be artifacts. This impression is supported by Duvall and Venner's (1979) study, which concluded that all were natural. So saying, it must be admitted that those objects illustrated by Bryan (1978b: Figures 1-6) are not easily dismissed. More studies like those of Duvall and Venner (1979) and Reeves (1980) are obviously needed, here as elsewhere.

The Taber child site (Stalker, 1969, 1977) of southeastern Alberta presents a different set of problems. The poorly preserved elements of an infant's skeleton were discovered under circumstances that Stalker interprets as positively indicative of Middle to Early Wisconsin age. The several radiocarbon dates cited by Stalker (1977: 125) in the 30,000 yr B.P. range are from a nearby bluff that is presumed to correlate with that of the Taber find. No artifacts were found with the original 1961 discovery and no further material has turned up since. Although it is averred that intrusion (i.e., intentional burial) can be ruled out, the possibility that the original find occurred in slumped soil appears to remain. At least, it would seem that, inasmuch as the bed from which the Taber child is presumed to have come lacks direct dating and its position in situ was not observed, the possibility of either intrusion or slumping from a higher position in the section cannot be positively ruled out. Thus, a definitive conclusion cannot yet be drawn. (Since portions of the infant's cranium are present, one could suppose that, if the Taber child were indeed in the 60,000-year time range, the investigators who worked with it might have found some anatomical features commensurate with that high antiquity.)

On Santa Rosa Island in California, the occurrence of the charred bones of an insular pygmy form of mammoth is held by some to be evidence of the presence of human hunters (Carter, 1980; MacNeish, 1976; Orr, cited *in* Wormington, 1957). Artifactual material is apparently limited to what Wormington (1957: 198) refers to as "one chipped stone." Radiocarbon dates of 29,700 ± 3000 yr B.P. (L-290R) and 30,400 ± 2500 yr B.P. (UCLA-1898) (Berger, 1975) have been obtained on burned bone, but there appears to be no clear proof of the agency of man here.

Putatively ancient human skeletal remains from several sites in California have recently been reinvestigated in the hope of establishing their ages and demonstrating the validity of aspartic acid racemization dating (Bada and Helfman, 1975; Hare, 1974). The development of the particular protocols employed by Bada and Helfman and their associates has been combined with refinements of

bone-collagen extraction for radiocarbon dating (Berger, 1975) applied to the same specimens. The results, in round numbers, have been: 17,000 yr B.P. for the Laguna skull, 23,600 yr B.P. for Los Angeles Man, and 48,000 to 41,000 yr B.P. for Del Mar Man (Bada and Helfman, 1975; Masters and Bada, 1978). Since the application of the technique in these instances depends upon radiocarbon assays on bone collagen, the precautions pointed out in a number of recent studies (Hassan and Hare, 1978; Hedges and Wallace, 1980; Kessels and Dungworth, 1980; Tuross et al., 1980) must be observed with particular rigor in this research. If the indexes upon which the calibrations are based are incorrect, then, obviously, so too will be any results derived from them. If anything, the California skeletons are more problematical now than they were before these studies were undertaken.

Although workers in that region generally agree that Fort Rock Cave in Oregon provides evidence of early occupation (Aikens, 1982; Bedwell, 1973), the mixed associations in the lowest stratum leave no question but that the assay, 13,200 ± 750 yr B.P. (Gak-1738), is best disregarded.

Wilson Butte Cave in Idaho (Gruhn, 1961, 1965) has produced radiocarbon dates of 14,500 ± 500 yr B.P. (M-1409) and 15,000 ± 800 yr B.P. (M-1410) in its lower strata. The former date pertains to the lowest occupation level; the latter, to the underlying stratum. The cultural material (Wilson Butte I) consists of a straight-based biface and a retouched blade but little else (Gruhn, 1961). Fragmentary bones of horse and camel are assumed to represent quarry, although Gruhn (1961) notes, there are no projectile points in this assemblage. As Haynes (1974) observes, these radiocarbon dates are at unity. The material dated (rodent bone) leaves questions in one's mind because of the substance itself and because (in the case of the younger date) the bones were collected "over several horizontal excavation units in order to get enough to analyze" (Haynes, 1974: 379). Moreover, Haynes's implication that the higher sample(s) may actually have been reworked from below would leave open the question of whether these dates were even pertinent to an artifact-bearing stratum.

The Sheguiandah site on Manitoulin Island, Ontario, has been a matter of some controversy virtually since it was first reported (Lee, 1954, 1955). The essential question revolves around the age of the lowest levels of the site. Lee holds that this material is interbedded with a thin till (Lee, 1955: 70-71) but rejects the geologist Sanford's tentative suggestion that a minor glacial readvance (for which there is no evidence elsewhere in the region) might account for this deposit (*in* Lee, 1955: 70; actually, Sanford offered two other possible causes but clearly favored that of ice deposition). Thus, Lee (1979) sees the lowest stratum at Sheguiandah as either a pre-Late Wisconsin interstade or a Sangamon interstade. One radiocarbon date of 9130 ± 250 yr B.P. (W-345) on peat just overlying the occupation in question is considered by Lee (1956) as supporting his interpretation and by Griffin (1965) as supporting his view that Sheguiandah is completely Holocene in age.

In Texas, the Midland Man (Scharbauer site) skeletal material (Wendorf and Krieger, 1959) is associated with extinct fauna (*Capromeryx* and *Equus*) and a projectile-point base identified as possibly either Midland or Clovis in type. Radiocarbon dates of more than 20,000 and less than 9000 yr B.P. were all viewed as spurious by the investigators. In the best estimate of Wendorf and Krieger (1959), Midland Man should date in the range of 12,000 to 11,000 yr B.P.

Various recent finds in the Yuha Desert of southern California are reported to have produced evidence of great age. These include the discovery of what are termed ridge-back tools in stratigraphic sections that would place them as older than 50,000 yr B.P. (Childers, 1977; Childers and Minshall, 1980) and a human burial for which a radiocarbon age of about 22,000 yr B.P. is claimed (Bischoff et al., 1976). A critique of the latter finding (Payen et al., 1978) casts serious doubt upon the validity of these caliche (calcium carbonate) dates. The ridge-back tools likewise appear to be problematical; Childers (1977: 248) admits that "a few could be fortuitous."

The Timlin site in New York State was originally reported as giving evidence of pre-Wisconsin occupation, the conclusion being based upon an interpretation of its stratigraphic position "in direct association with ice-contact gravels" (Raemsch and Vernon, 1977; Stagg et al., 1980 [paper written in 1973]). The material illustrated appears to include both evident artifacts and possible geofacts (Stagg et al., 1980). Raemsch and Vernon's (1977) more explicit statement assigning an early age to the Timlin site drew vigorous criticism from Cole and others (1977). A reinvestigation of the site in 1979 and 1980 by Bryan and others (1980) led to the conclusion that the gravels in question are fluvial in origin and postglacial in age. Bryan and others (1980: 52), nevertheless, conclude that there is evidence here of a pre-Clovis unifacial stone tool industry and cite a radiocarbon date of 16,040 ± 170 yr B.P. (SI-4128) as indicating the "earliest human occupation of the deglaciated Northeast." However, inasmuch as this material is mostly rolled and clearly redeposited, it is difficult to imagine how any associated radiocarbon-datable material could actually refer to the former (and unknown) place of occupation. A side-notched projectile point in the deposit is said to have occurred in a somewhat higher position (A. L. Bryan and R. Gruhn, personal communication, 1980), but even so, on the basis of present evidence, it cannot be ruled out as a constituent of the total assemblage. The total assemblage has the appearance of quarry-workshop material and is, therefore, difficult to characterize. Subsequent to the reporting cited here, additional and much younger radiocarbon dates have been obtained (R. E. Funk, personal communication, 1981).

The Lewisville site in Texas has been enigmatic since it was first described (Crook and Harris, 1957; Krieger, 1962; Wormington, 1957). The report of a Clovis biface associated with hearths dating to more than 37,000 yr B.P. resulted in the discrediting of both. Since the site was flooded shortly after it was uncovered, Lewisville could have rested forever in limbo by providing evidence either for very early human occupation in Texas or for being a nonsite, depending upon the biases of the reviewer. The recent draining of the reservoir and its reinvestigation by D. Stanford (personal communication, 1980) has resulted in a new, different, and credible account. Briefly, Stanford concludes that this is a Clovis site, that the point was in situ, that the hearths (from which the early dates were obtained) are hearths, but that the substance dated contained lignite (used evidently by the Clovis hunters).

The series of sites in the Old Crow Flats in the northern Yukon Territory, from which there have been recovered artifacts exclusively of bone, has aroused a great deal of interest and discussion (Bonnichsen, 1978, 1979, 1981; Guthrie, 1980, 1981; Irving, 1978; Irving and Harington, 1973; Irving et al., 1977; Jopling et al., 1981; Morlan, 1978, 1979, 1980; Myers et al., 1980). Apart from the original find (a fleshing tool made from a caribou tibia), there is some controversy over the identification of subsequent finds as artifacts. No stone tools have been discovered with them; in fact, stone tools of any age are sparse in the region (Morlan, 1980). All or most of this

material is redeposited. The original series of radiocarbon dates (Irving and Harington, 1973), averaging about 25,000 yr B.P., were run on bone apatite rather than on collagen and are now discounted. For these several reasons, the status of these finds is still most unclear.

Similar materials are reported by Stanford (1979b) from the Dutton, Selby, and Lamb Spring sites in Colorado (Stanford et al., no date). The bone horizons at Dutton underlie a Clovis level; at Lamb Spring they are separated from a unit of the Cody complex. As with the Old Crow collections, these findings are regarded by Stanford (1979a) as possible representations of the work of pre-Clovis peoples, but a convincing demonstration of that status has yet to be made.

Meadowcroft Rockshelter in Pennsylvania is a deeply stratified site that has been the scene of an intensive multidisciplinary investigation (Adovasio, Gunn, Donahue, and Stuckenrath, 1977, 1978; Adovasio, Gunn, Donahue, Stuckenrath, et al., 1978, 1980). Of concern in the present context is lower stratum IIa, which contains a small lithic assemblage. This stratum has produced radiocarbon dates ranging from 19,100 ± 810 yr B.P. (SI-2062) to 12,800 ± 870 yr B.P. (SI-2489). As both Stoltman (1978: 713) and Dincauze (1981: 3) observe, the assemblage is difficult to assign culturally since it consists of one simple straight-based biface, a number of retouched flakes ("Mungai knives"), small blades (most unretouched), and various flakes (some of which may have served as tools of expediency). The suite of early dates is consistent. There is, however, no extinct fauna as might be expected of a site that dates well into the Late Wisconsin. Instead, the large forms found throughout all occupational levels consist of white-tailed deer and wapiti. Guilday (1967) discusses local fauna from two sinkholes in western Pennsylvania. Hosterman's Pit with modern fauna (including the two animals just named) is dated at 9240 ± 1000 yr B.P. (M-1291). The previously reported New Paris No. 4 sinkhole (105 km southwest of Hosterman's Pit) was dated at 11,250 ± 1000 yr B.P. (Y-727) but yielded an assemblage of "cold weather species characteristic of central Canada today" (Guilday, 1967: 232). These two sites are regarded by Guilday as bracketing the major environmental transition from the late-glacial period to the Holocene. The implications for the interpretation of Meadowcroft are clear. (See also Mead, 1980.) It is because of these apparent anomalies, as well as others, that Haynes (1980) raises the question of there having been a possible systematic contamination of the dated charcoal. Full reporting of Meadowcroft may dispel some of the present controversy.

Several works (e.g., Bryan, 1978b; MacNeish, 1976, 1978; Wormington, 1957) list other pre-Clovis localities. In addition, many comprehensive interpretive accounts of American archaeology, of greater or lesser scope (e.g., Chard, 1975; Jennings, 1974; Willey, 1966), incorporate portions of this material. An attempt has been made here to single out the sites that are at present acknowledged to be the most important of the pre-Clovis sites in the sense of supporting the general proposition of the existence of these earlier populations.

EVALUATION OF PRE-CLOVIS EVIDENCE

As one reviews the evidence for pre-Clovis occupations, it is difficult not to be struck by what is, or should be, an amazingly anomalous condition: with the possible exception of those containing only bone, no two sites appear to show the same kind of artifactual material, even when they are relatively close both temporally and geographically. In fact, the only obvious unifying principle is that all are held to date from the Late Wisconsin or earlier.

This lack of standardization could be attributed to the exiguousness of the American record for the pre-Clovis period both in the number of sites and in the amount of material present at any given site. If, however, it is maintained that what is seen results from the activities of small, isolated pockets of humanity, then there logically is generated another series of vexatious questions: *why* were populations apparently so low or, more accurately, *how* were they kept so low? Simply considered alone, such a conception seems totally at odds with the greater part of the record of human fecundity. It is perhaps in recognition of this difficulty that pre-Clovis peoples are sometimes referred to as a category of "big-game collectors" (MacNeish, 1978); other investigators emphasize (on unstated evidence) that the pre-Clovis peoples were at least equally concerned with small-game hunting, fishing, and gathering plants (Gruhn and Bryan, 1977). But, if there was an abundant and varied late-Pleistocene fauna present in America and if its counterparts were being hunted by man in the Old World, what would account for such variance here?

Cultural standardization comparable to that which characterizes, for example, Mousterian-stage complexes of the Old World (Bordes, 1968) is absent. Such a lack of regularization or patterning of culture generally at this late stage in human evolution is difficult to understand and to interpret.

A more disturbing corollary of this problem is the low level of technology generally held to characterize pre-Clovis people. The period of concern is, after all, *late* in the human record—late Sangamon to Late Wisconsin. In the Old World, this same period of time finds humans with quite advanced Palaeolithic cultures (and for our purposes here that term specifically includes Mousterian and Mousterian-related cultures). Whether Mousterian (or Mousteroid) or any of the several major Upper Palaeolithic traditions is considered, all are characterized by *diversified* and *specialized* lithic and bone-tool kits. By 30,000 years ago, Upper Palaeolithic people of western Europe were painting on cave walls accurate depictions of the large animals that were their prey. By perhaps 20,000 years ago, the Upper Palaeolithic folk at Malta in central Siberia were evidencing the same level of advanced technologic—and artistic—ability in a portable art that rivals the European Magdalenian. The essential point here is that, virtually without exception and wherever the archaeological record may be read unambiguously, the past 125,000 years are marked by highly evolved, sophisticated lithic technology. On this basis alone, pre-Clovis as presently portrayed is difficult to comprehend and difficult to accept. The interpretive problems that seem to adhere to every one of the several pre-Clovis sites render still more nebulous the interpretation of this body of material as a whole. To set America apart from what appears to be the normal course of evolution on the basis of such tenuous evidence seems unwarranted.

The Clovis Hunters

If the picture of ancient American hunters becomes clearer with the Clovis period, it also becomes vastly more complicated by reason of the enormous mass of detail available in the archaeological record. The first discovery of distinctive fluted points near Folsom, New Mexico, in 1926 (Figgins, 1927) clearly established the association of early hunters with an extinct form of bison (*Bison occidentalis*). The immediate surge of enthusiasm quickly resulted in further discoveries of these associations within the same general region of the southwestern Great Plains. The first positive association of man and mammoths was found at the Dent site in Colorado in 1932 (Figgins, 1933).

In the same year, a Folsom point was found at Blackwater Draw near Clovis, New Mexico. This finding precipitated the intense scientific concern with that site which continues to the present day. The early history of Folsom and Clovis research is summarized by Wormington (1957) and Hester (1972).

In 1936 and 1937, the first discoveries of fluted points and mammoth remains were made at Blackwater Draw (Cotter, 1937, 1938). From these findings the Clovis point as a type was named and described. Through the intervening years, a great deal of research was carried out in the Southwest and to a lesser extent in the East. Where organic preservation was favorable, Clovis was consistently associated with elements of late-Rancholabrean fauna (Table 18-1). Still, that most important of questions, the age of Clovis, remained a matter of estimation until, with the discovery of the Lehner site in southeastern Arizona, it became possible to date by radiocarbon the association of Clovis points with the remains of nine mammoths, as well as those of horse, tapir, and extinct bison (Haury et al., 1959). As with the previously excavated Naco mammoth site (Haury et al., 1953), there was at Lehner no question but that the Clovis people had been successful mammoth hunters. The four oldest (and most acceptable) radiocarbon dates for Lehner average 11,340 yr B.P. (Haury et al., 1959: 25).

Where radiocarbon dates are available from Clovis sites, there is a significant clustering within an approximately 600-year-long time span. Moreover, there is enough similarity in cultural content to support the arguments of Mason (1962), Green (1963), Hester (1966), and others that we are dealing here with a single large tradition. It is for this reason that the name Clovis is here applied throughout its enormous range, as suggested by Hester (1966). At present, it is not possible even to suggest whether—let alone how many—subtraditions of Clovis exist. Clear it is, however, that Clovis—whether found in Nova Scotia or Guatemala—is a highly distinctive, readily recognizable entity.

In addition to the distinctive fluted bifaces that constitute their "index fossil," Clovis assemblages show a frequent occurrence of simple lenticular and ovoid bifaces (the former usually in about the same size range as the points), of thumbnail scrapers (sometimes spurred) made on blades and flakes, of gravers, of large sidescrapers frequently on blades, and of grinding stones. Occasionally found are bone or ivory implements, such as the widely distributed beveled bone points and the shaft wrench found at Murray Springs, Arizona (Haynes and Hemmings, 1968). Perhaps somewhat less well known is the fact that common Clovis property includes prepared blade cores, blades struck from them, burins (sometimes with notches), lithic wedges (*pièces esquillées*), limaces, and other perhaps more-localized tools (e.g., the bifacial twist drills of Bull Brook and several other eastern sites). (As a result of the recent reassembling and study of the entire Bull Brook collection [Grimes, 1979], it appears that, contrary to Hester's findings [1966: Table 1], both blades and burins are present in that assemblage [author's observation].)

As it has been observed more than once, if the elements just enumerated were given as much attention as are Clovis points, there could be little discussion about the fundamentally Upper Palaeolithic character of these assemblages. Clovis points reveal a good deal of variation both within individual sites (e.g., Lehner, Bull Brook) and across their range of distribution. The former observation should be kept in mind since occasional attempts are made to read major chronologic and other significance into the intersite variability of this one form. Cotter (1962) has repeated his earlier logical but still-

Table 18-1.
Representative Clovis-Period and Clovis-Related Sites

Site	Classification	Radiocarbon Dates in yr B.P. (Laboratory Number)	Faunal Associations	Reference(s)	Comments
Clovis (Blackwater No. 1), New Mexico	Clovis	11,170 ± 360 (A-481), 11,040 ± 500 (A-490), 11,630 ± 400 (A-491)	Mammoth, bison, camel	Howard, 1935; Cotter, 1937; Hester, 1972	Type locality; habitation and kill site
Burnet Cave, New Mexico	Clovis	—	Musk ox, caribou, horse, camel, four-horned antelope	Howard, 1935	Associations questioned
Sandia, New Mexico	Proto-Clovis?	—	Mammoth, horse, bison, camel	Hibben and Bryan, 1941	Sandia fluted and unfluted lanceolate bifaces; related to Clovis; underlies Folsom
Lucy, New Mexico	Proto-Clovis?	—	Mammoth	Roosa, 1956	Sandia kill site; as with Sandia, single ^{14}C date discounted
Ventana Cave, Arizona	Clovis?	11,200 ± 500 (A-203)	Horse, sloth, tapir	Haury, 1950	—
Naco, Arizona	Clovis	—	Mammoth	Haury et al., 1953	Kill site
Lehner, Arizona	Clovis	11,180 ± 140 (K-554), 11,290 ± 500 (M-811)	Mammoth, bison, horse, tapir	Haury et al., 1959	Kill site; several dates in range shown
Murray Springs, Arizona	Clovis	11,230 ± 340 (A-805)	Mammoth, bison, horse	Hemmings, 1970; Haynes, 1976	Kill site
Borax Lake, California	Clovis	—	—	Harrington, 1948; Meighan and Haynes, 1970	Stratigraphically precedes other projectile-point forms here
Santa Isabel Iztapan, Mexico	Clovis?	—	Mammoth	Aveleyra Arroyo de Anda, 1956	Lenticular and straight-based simple bifaces; blade tools
Los Tapiales, Guatemala	Clovis	—	—	Gruhn and Bryan, 1977	—
Dent, Colorado	Clovis	11,200 ± 500 (I-622)	Mammoth	Figgins, 1933	Kill site
Union Pacific, Wyoming	Clovis?	11,280 ± 350 (I-449)	Mammoth	Irwin et al., 1962	Lenticular and ovate bifaces
Colby, Wyoming	Clovis	11,200 ± 220 (RL-397), 10,548 ± 141 (SMU-254)	Mammoth, camel	Frison, 1976	Processing and storage site
Anzik, Montana	Clovis	—	—	Lahren and Bonnichsen, 1971	Burial cache?
Domebo, Oklahoma	Clovis	11,220 ± 500 (SI-172), 11,200 ± 600 (SI-175)	Mammoth	Leonhardy, 1966	Kill site
Anadarko, Oklahoma	Clovis?	—	—	Hammatt, 1970	Cache of blades, blade cores, and burins (?) attributed to Clovis
Miami, Texas	Clovis	—	Mammoth	Sellards, 1952	Kill site
Lewisville, Texas	Clovis	—	Mammoth, horse, bison (?)	Crook and Harris, 1958; D. A. Stanford, personal communication, 1979	Habitation; spuriously old ^{14}C dates run on lignite
Leon Creek (41 BX 52), Texas	Clovis	—	Bison	Henderson, 1980; personal communication, 1981	Habitation?; collections not analyzed; some provenience problems with Folsom components

Table 18-1 (continued).

Site	Classification	Radiocarbon Dates in yr B.P. (Laboratory Number)	Faunal Association	Reference(s)	Comments
Levi, Texas	Clovis?	10,000 ± 175 (O-1106)	Horse, bison	Alexander, 1963, 1973	O-1106 on shell considered too young
Kimmswick, Missouri	Clovis	—	Mastodon	Graham et al., 1981	Kill site; two superimposed Clovis levels
Holcombe, Michigan	Clovis?	—	Caribou	Fitting et al., 1966	Habitation
Williamson, Virginia	Clovis	—	—	McCary, 1951; Benthall and McCary, 1973	Habitation
Shoop, Pennsylvania	Clovis	—	—	Witthoft, 1952; Cox, 1972	Habitation
Shawnee-Minisink, Pennsylvania	Clovis	10,590 ± 300 (W-2994), 10,750 ± 600 (W-3134)	—	McNett et al., 1977	Habitation
Plenge, New Jersey	Clovis	—	—	Kraft, 1973	Habitation
West Athens Hill, New York	Clovis	—	—	Funk, 1973	Quarry workshop
Kings Road, New York	Clovis	—	—	Funk, Weinman, and Weinman, 1969	Habitation; surface collection
Dutchess Quarry Cave, New York	(Cumberland fluted point)	12,530 ± 370 (I-4137)	Caribou	Funk et al., 1969; Funk, Walters, and Ehlers, 1969; Guilday, 1969	Some question on associations
Bull Brook, Massachusetts	Clovis	9340 ± 400 (M-807), 8940 ± 400 (M-810)	Caribou?	Byers, 1954, 1959; Jordan, 1960; Grimes, 1979	Habitation; these dates and others considered too young
Whipple, New Hampshire	Clovis	—	Caribou	Curran, personal communications, 1981	Habitation
Reagan, Vermont	Clovis	—	—	Ritchie, 1953	Habitation
Munsungun (site 154-14), Maine	Clovis	—	—	Bonnichsen et al., 1980	—
Vail, Maine	Clovis	11,120 ± 180 (Beta-1833), 10,300 ± 90 (SI-4617)	—	Gramley, 1981	Habitation; "killing ground" nearby
Debert, Nova Scotia	Clovis	10,604 ± 45 (average of 11 samples, University of Pennsylvania dates)	—	MacDonald, 1968	Habitation

overlooked suggestion that size variation probably is attributable to different "calibers" for different game species. Perhaps the diagnosis of one who attests to having handled "thousands of fluted points" should also be cited. In his discussion of the Wells Creek (Tennessee) assemblage, Dragoo (1973: 47) states, "Although early Clovis points vary some in size and individual configuration, I can find no significant difference between those found in the West or East in basic typology or technology." One can but echo Bushnell and McBurney (1959: 101), who addressed a "plea to the scholarly 'splitters' that they shift their gaze from the minute variations of a single tool type to take a wider view of their material."

Although the evidence from the East is still elusive, that from the West indicates that Clovis people were indeed nomadic big-game hunters dependent upon the surviving elements of late-Rancholabrean fauna. The Kimmswick site findings encourage the speculation that other eastern Clovis people were at least part-time proboscidean hunters as well. Whether the Clovis hunters are to be considered primary causes of the extirpation of this fauna is still a matter of controversy. Table 18-1 lists a number of Clovis or Clovis-period sites selected rather arbitrarily on the basis of their providing either data on faunal associations, stratigraphic evidence, evidence of the geographic spread of Clovis, or some combination of these.

Thousands of Clovis points have been found in North America. Were all these distributions on record, doubtless some reasonable interpretive assessment could be made on the basis of locations or densities (after the fashion of Williams and Stoltman, 1965, for example).

Minimally, one may note that surface finds of Clovis points have not been reported within the limits of late advances of the Laurentide ice sheet. In other respects, too, Clovis localities tend to correlate with late-glacial land and water features, examples of which may be seen in such studies as Quimby (1958) on the Lake Michigan basin, Curran and Dincauze (1977) on the Connecticut Valley, and Loring (1980) on the Champlain Sea. In addition, of course, many of the well-known mammoth and mastodon kill sites record a perfect correlation with local water features that developed under climatic regimes of that period.

Examination of the available radiocarbon dates clearly shows that the clustering noted by Haynes (1967), Martin (1973), and others continues to hold. The recently obtained date for the Vail site (Gramley, 1981) is particularly interesting in that it brings into question (along with a similar date from Debert) the supposed temporal priority of western Clovis. If there is a lag, it begins to appear small to the point of being uninterpretable. This is not to suggest that there is no chronology within Clovis, and MacDonald's (1968) subdivisions may be on the whole valid. On the other hand, it seems clear that some of the dates on which these assessments are based are themselves suspect (e.g., those of Bull Brook and Naco).

If the earlier eastern dates are valid, and there is every reason to believe they are, and if the depictions of Late Wisconsin fauna for the East are correct, then the likelihood is nil that eastern Clovis hunters pursued any but those now-extinct animals. Once more, the great significance of the Kimmswick mastodon association comes to mind. The several finds of caribou in the extreme Northeast from this time period suggest a landscape that would have supported mastodon equally well. Whatever the prey, the size of the eastern sites suggests an abundant food supply, and, although there is some contrary opinion, there seems to be little in these assemblages to suggest that the easterners were anything other than hunters.

Origins of Clovis Culture: Evidence from the North

Clovis would seem to be, by anyone's definition, the earliest well-delineated North American archaeological tradition. Its origins have been a matter of intense interest and speculation since the time of its first discovery. With some variation, these theories have seen it as either homegrown from a preexisting base or as exogenous. Bryan (1978b: 311) holds the opinion that Clovis was derived from a "basic Paleo-American tool kit [containing] few, if any, standardized stone or bone tool types." MacNeish (1976) appears to take the same stand. But it is not necessary to suppose an endogenous origin for Clovis in order, likewise, to propose the existence of pre-Clovis populations. Green (1963) derives Clovis ("Llano complex") from the Siberian Upper Palaeolithic but construes it as representing the later of two basic migrations into the New World. Much the same position is taken by MacDonald (1968). Wormington (1957, 1971) simply leaves the question open. Mason (1962) follows essentially the same course, indicating that the state of knowledge at the time of his writing required the presence of an antecedent population. Chard (1959) favors an Early Wisconsin movement of chopping-tool cultures into America that forms the basis for all later major developments. Bushnell and McBurney, (1959: 101), in a rejoinder to Chard, trace American origins to "the great events of Upper Palaeolithic spread between the 25th and 20th millennia B.C." To judge from Bushnell's (1962) later statement, though the matrix may have been more

than 20,000 years old, the first American representatives were taken by him to be Clovis of around 11,000 years ago.

In his discussion of burin-faceted projectile points of the Clovis period, Epstein (1963) suggests that this trait's origin (and, by implication, Clovis's) is in the Old World. Alexander (1973) concurs on similar grounds. Witthoft's (1952) assessment of the fluted-point people (based on his analysis of the Shoop site in Pennsylvania) is essentially the same as that subsequently embraced, elaborated, and modified by Haynes (e.g., 1966), Hester (1966), Jelinek (1971), Martin (1973), and West (1981). These people were, according to Witthoft (1952: 493-94), Upper Palaeolithic migrants; they were "newcomers . . . not . . . many generations away from Bering Strait." His later statement (1954) emphasizing this line of thought has been disputed by Krieger (1954), who points out that the assumed Old World origin of Clovis projectile points is unsupported by any direct evidence. Hester (1966) suggests that the Clovis tool kit developed in isolation from a slightly earlier Upper Palaeolithic base. Haynes (1966) would derive people and tool kit in one migration, employing the same reasoning later used by Martin (1973) and by Mosimann and Martin (1975) in their computer simulation, and have this population rapidly overspread North America. In Müller-Beck's (1966) scheme, Clovis is seen to have developed from an earlier Mousteroid base, which entered sub-Laurentide America some time after 26,000 years ago. Mochanov (1977) derives Middle Wisconsin pre-Clovis as well as Clovis itself from Dyuktai. In Dikov's (1979) view, the stemmed points at the base of his Ushki sequence suggest an ancestral relationship with the Marmes site in Washington State; this was followed by a later wave of (Dyuktai-like) "proto Eskimo-Aleut" people, evidence for whom comes from the succeeding two levels of the Ushki sites. Dragoo (1973) sees in Wells Creek evidence supporting the Müller-Beck hypothesis. Rouse (1976) assigns Clovis to a "Middle Lithic Age" and apparently sees it as developing in place. Stanford (1979a: 148) finds no evidence to indicate how or from where Clovis ("Llano") derives. It is now possible to make some rather specific statements on just these points.

THE BERINGIAN TRADITION

Table 18-2 lists a short series of sites from Alaska and Siberia. The area they represent is large, extending from the upper Aldan River system of present-day Yakutia through central Kamchatka to south-central Alaska, a distance of some 4000 km west to east. Although the distances that separate the two groupings and the fact that they occur on different continents might suggest an equal cultural distance between them, all are, in fact, contiguous culturally and temporally. Those listed in Table 18-2 are part of a much longer series assigned by West (1981) to a "Beringian tradition"—a distinctive Upper Palaeolithic tradition that occupied the refugium of Beringia during the Late Wisconsin and lingered in diminished condition into the early Holocene of Alaska.

Assemblages of the Beringian tradition are characterized by lenticular to ovoid to straight-based simple bifaces, prepared blade and microblade cores, endscrapers on flakes and blades, multiple burins often on notched flakes, blade and microblade implements, pebble choppers (or flake cores), sandstone abraders, small implements on burin spalls, and various scrapers sometimes made on large flakes and sometimes on large blades. The abundance of burins, typically with facet-edge wear, attests to the common use of antler, ivory, and bone. In only a few cases are there found actual specimens made from organic materials. Nevertheless, it is certain that projectile points like

Table 18-2. Representative Beringian-Tradition Sites

Site	Classification	Radiocarbon Dates in yr B.P. (Laboratory Number)	Faunal Associations (Extinct Forms)	Reference(s)	Comments
Ust'-Mil II, Yakutia	Dyuktai	35,400 ± 600 (Le-654) to 12,200 ± 170 (Le-953)	Mammoth, wooly rhinoceros, horse	Mochanov, 1975	Three horizons identified
Ikhine I (horizon B), Yakutia	Dyuktai	30,200 ± 500 (GIN-1019)	Mammoth, bison, horse	Mochanov and Fedoseeva, 1968; Mochanov, 1975	—
Upper Troitskaya, Yakutia	Dyuktai	18,300 ± 180 (Le-905), 17,680 ± 250 (Le-906)	Mammoth, wooly rhinoceros, bison, horse	Mochanov, 1975	—
Dyuktai Cave, Yakutia	Dyuktai	14,100 ± 100 (GIN-404) to 12,000 ± 120 (Le-907)	Mammoth, bison, sheep, horse	Mochanov, 1970, 1975	—
Berelekh, Yakutia	Dyuktai	12,930 ± 80 (GIN-1021)	Mammoth	Mochanov, 1972, 1975	—
Ushki I, Kamchatka	Ushkovskaye	14,300 ± 200 (GIN-167) to 10,760 ± 110 (Mag-219)	Horse, bison, moose, salmon	Dikov, 1977, 1979	Lowest level: stemmed points, blades, burins, simple bifaces
Putu, Alaska	?	11,470 ± 540 (SI-2382), 8450 ± 130 (WSU-318), 6090 ± 430 (Gak-4939)	—	Alexander, 1974; personal communication, 1980	Fluted points, burins, blades, other bifaces
Dry Creek, Alaska	Denali	11,120 ± 85 (SI-2880), 10,690 ± 250 (SI-1561)	Bison, elk	Powers and Hamilton, 1978	Habitation
Village (Healy Lake), Alaska	Denali?	10,150 ± 250 (SI-737)	—	Cook, 1969	Chindadn assemblage; habitations; may be Denali
Phipps (Mount Hayes 111), Alaska	Denali	10,150 ± 280 (UGa-572)	—	West, 1981	Habitation; 1 of 16 Denali sites in Tangle Lakes district
Akmak, Alaska	American paleo-Arctic tradition	9857 ± 155 (K-1583)	Caribou	Anderson, 1970	Habitation; Denali related
Trail Creek, Alaska	—	9070 ± 150 (K-980)	—	Larsen, 1968	Perhaps related to Denali

the slotted forms found in some Baikalian sites were made and armed with microblade cutting edges. Several slotted antler points from Trail Creek (Larsen, 1968) may well provide direct evidence on this Beringian form. An extensive bone industry, perhaps including portable art, may well have been present in some Beringian cultures. Bifaces probably functioned as knives; such wear studies as have been conducted tend to bear this out.

Where evidence survives (principally in Siberia), it clearly shows that at least some of the Beringians were successful hunters of big game (Table 18-2). It is instructive, in view of the purported superiority of the Clovis projectile point, to observe in the Dyuktai culture sites the presence not only of proboscideans but rhinoceroses as well. The Dry Creek site is the only Alaskan representative in which unusual faunal remains (bison and wapiti) occur. (Both animals are regionally extinct; earlier reports of further extinct forms there were incorrect.) Whether any of the species sought by the Dyuktai hunters was utilized by the Beringian-tradition Denali people of Alaska is problematical. (Denali is taken to be the earliest of the

Alaskan Beringian cultures.) The few sites of this culture that have produced faunal remains have shown, apart from the qualified exception of Dry Creek, completely modern forms, mainly caribou and mountain sheep.

Dyuktai and Denali are so closely related as to be best considered parts of a cultural continuum (West, 1981). The only significant differences between them are those of place and time of occurrence. Some of the most characteristic artifacts of Denali (e.g., the technologically complex wedge-shaped "Gobi" blade core form) occur throughout the Dyuktai time span. Although earliest Ushki has yielded stemmed points and perhaps other divergent forms, Ushki quickly assumes the character of a Dyuktai-related sequence. It is not now clear how its earliest form should be interpreted.

There appears to be no readily available reason to question the validity of the numerous radiocarbon dates available for the Dyuktai-Denali series (West, 1981). Their acceptance, however, requires an interpretation that, though supported by other lines of evidence, cannot ignore the peculiar environmental conditions that existed. The

Beringian refugium in Wisconsin time provided habitats for a group of large mammals in every way compatible to that of sub-Laurentide America. Kurtén and Anderson (1980), in fact, refer these animals to the Rancholabrean. These include mammoths, mastodons, ground sloths, lions, saber-toothed cats, short-faced bears, two forms of horses, camels, wapiti, stag moose, two species of musk oxen, bison, yaks, and saiga antelope. These occurrences are summarized by West (1981) on the basis of information from Guthrie (1968, 1972), Péwé (1975), and Harington (1971) (the Kurtén and Anderson [1980] compilation was not then available). Those data, together with reconstructions of vegetation, suggest the extensive development of grasslands in Wisconsin times. Details of the "steppe-tundra" biome construct are controversial (Colinvaux, 1980; Cwyner and Ritchie, 1980; Matthews, 1976); Guthrie (1968) suggests that the large-mammal evidence virtually demands the presence of extensive herbaceous landscapes. Central Beringia—that great region now under the Bering Sea and much of the Arctic Ocean—for reasons of topography and location, probably came closest to supporting a fully developed arctic steppe. If this supposition is correct, then there the Beringian fauna and, of course, the Beringian hunters were likely to have had the densest concentrations. This central Beringian region was submerged once during Middle Wisconsin time between around 34,000 and 24,000 years ago (Hopkins, 1973) and again definitively (for archaeological purposes) about 13,000 years ago. The Alaskan Beringian sites are taken to represent relict populations dating from the time of the dissolution of the central Beringian landscape and the final severing of a land connection between the Asian and North American continents (West, 1981).

It is essential to emphasize that *there are no credible archaeological remains in northeast Asia that predate Dyuktai.* The Beringian tradition, therefore, may be suggested as the matrix from which Clovis culture arose far to the south. Two lines of evidence argue for the correctness of this hypothesis.

Continuities in Artifact Forms

The general resemblance between Clovis- and Beringian-tradition artifact categories has already been brought out. The primary difference between them is the distinctive fluted biface characteristic of Clovis. Witthoft's (1952) suggestion based on findings at the Shoop site can be broadened to encompass the full range of Clovis (and Folsom) fluted bifaces: in a technologic sense, these are blade-core analogues with the blade-removal technique adapted to the requirements of basal (and medial) thinning. In various Beringian assemblages, there occur some blade cores having the same platform angles and on which, evidently, were used the same removal techniques. Included in this category are those varieties informally referred to as "Tuktu" in Alaska and "epi-Levallois" in Siberia (West, 1981: Figure 10). The writer has examined blade cores from West Athens Hill that easily fit into the latter category. Bonnichsen and others (1980: Figure 12) illustrate a rather nondescript specimen from Munsungun Lake. Several specimens shown by Dragoo (1973: Figures 25-27) from Wells Creek could be similarly characterized. Some of the *pièces esquillées* could be microblade cores (as per Witthoft's [1952] suggestion), but most are probably not intentional blade cores (MacDonald, 1968).

Worked or unworked blades are virtually ubiquitous in Clovis collections. From the recently discovered site at Leon Creek in Texas, J. Henderson (personal communication, 1981) describes blades in an apparent Clovis context similar to those from the type site, a resemblance noted as well for the Anadarko (Oklahoma) blades (Hammatt, 1970).

Although Patterson (1977) appears to regard it as pre-Clovis, the surface site 41 ME 3 in Texas—which contains burins, blades, and one subconical blade core—could be placed in the Clovis tradition.

Notched burins on flakes occur at Murray Springs in Arizona (Hemmings, 1970), and unnotched burins at Los Tapiales in Guatemala. Alexander (1963) reports notched-flake burins in what is identified as a Clovis zone at the Levi site. Hammatt (1970: Figure 4) illustrates apparent burins from the Anadarko cache. Objects that appear to be notched-flake burins occur in Bull Brook and burin-faceted flakes can be found in the West Athens Hill collection (F. H. West, unpublished data). Epstein (1963) demonstrates the existence of Clovis points with burin facets, but the writer has no information on the frequency of this trait in excavated Clovis components.

Limaces are known from several eastern sites (Plenge, Bull Brook, and Vail, for example) and occur as well in some Alaskan Beringian collections (such as Girl's Hill [R. Gal, personal communication, 1979] and collections made by the writer at Ravine Lake and Mount Hayes 72). In addition, "proto-limaces" (Bordes, 1961) occur in several Denali collections. (Although Bordes's use of the term "proto-limace" presumably is intended to point up an evolution in this tool type, there is no basis whatever in the material under consideration here for such an assumption.) Thus far, the limace appears to be unknown in the West, and the eastern specimens conform more nearly to European forms than do the northern ones.

Bifaces of variable size and simple form (lenticular, ovoid, and straight based and with random bifacial thinning) are found in many Clovis collections, among them Wells Creek, Plenge, Kings Road, West Athens Hill, Vail, Murray Springs, and, of course, the Union Pacific (Wyoming) and Santa Isabel Iztapan (Mexico) mammoth kills. At Union Pacific and Santa Isabel Iztapan, there was a distinct paucity of material. Although the reporting is uneven, the distribution of these traits is wide enough to suggest that all may properly be considered parts of Clovis culture.

Despite the fact that core-and-blade technology appears to dominate Beringian assemblages, blades were seldom used as blanks for further manufactures. Most blades are miniaturized (microblades) and their dominant purpose, it seems, was to provide cutting edges on antler projectile points. Through the range of Beringian cultures, full-sized blades and implements made on them do occur. They are few, however, and the correlative preferential use of flakes finds a parallel in Clovis. In turn, that preference may have had historical significance, for, even though the Beringian tradition is clearly Upper Palaeolithic, one of the significant ways in which the Siberian Upper Palaeolithic cultures differ from the classical French series, is in their *conservatism*: there are pronounced Mousterian elements throughout, and this is especially true in the Beringian series. The limace in western Europe is almost exclusively Mousterian; Levallois flake cores (and products) occur in certain Beringian collections from both sides of the Bering Strait, and, of course, there is the just-mentioned preferential use of flakes in small manufactures. In view of the exceedingly wide distribution of Mousterian (or Mousteroid) industries, however, there is no need to read anything more than simple conservatism into the Beringian (and Clovis) occurrences—the consequences, presumably, of their artisans being at the uttermost end of the range of the species. Müller-Beck's (1966) perception of Mousteroid elements in Clovis is certainly correct; how they came to be there, in the light of findings made since 1966, is, of course, differently interpreted.

It is not being maintained that there is assemblage identity between any particular Beringian culture and Clovis. It *is* suggested, however, that the level of resemblance between Clovis and almost any given Beringian culture is of the same order as that to be found between any two Beringian complexes. To a northern specialist, in other words, Clovis assemblages have a very familiar cast; inevitably, Witthoft's prescient comment that the Clovis materials at Shoop look as though they had not long been removed from the Bering Strait is recalled.

Continuities in Adaptation

If the general typological similarity between Beringian tradition and Clovis culture is accepted as essentially correct, then the next question is how shall this circumstance be interpreted? Does it suggest only approximate contemporaneity? Obviously, even though temporal proximity is a factor, it may be subsidiary to another. Simply stated, the Beringians were hunters of elephants, horses, bison, and other large animals. The Clovis people hunted those same genera in the regions south of the Laurentide ice sheet. Although assertions are frequently made that early man must surely have been also a gatherer of plant foods, beyond the fragile logic of the assertion itself, there is no evidence to support this assumption anywhere in the American record of this era. Plant foods likely were used, but that they may have been a significant part of the Clovis dietary seems neither provable nor logical. The Clovis people appear to have lived in very much the same way as their postulated northern ancestors. Therefore, it is reasonable to assume that the resemblances in artifact inventories must, in a measure, reflect this continuity in adaptation.

Immediate Antecedents of Clovis Culture

What may be said of the *immediate* antecedents of Clovis if its ultimate antecedents are to be found within the matrix of Beringian tradition? Fluted points have been found in northern Alaska. The only serious suggestion that these points are ancient and are possibly the precursors of Clovis points to the south, however, is Alexander's (1973) (which was endorsed by Morlan, 1977) for the Putu site. The situation there is far from clear. There is a wide range of radiocarbon dates; and, moreover, a similar assemblage, also from the Brooks Range (Girl's Hill [R. Gal, personal communication, 1981]) is estimated to date to about 7000 yr B.P. Several other surface finds of fluted points have been made in Alaska in the same general region of the Brooks Range (Clark and Clark, 1975; Solecki, 1951; Thompson, 1948), and one generally similar point was found in situ with the Denbigh Flint assemblage at the type site (Giddings, 1964). A tight distribution and an almost invariant style suggest that a short period of time is represented and that the Girl's Hill interpretation will turn out to apply to all. The sites appear to be late enough and may, indeed, represent a "backwash" from regions to the south, as suggested years ago by Krieger (1954). It is also possible that they may represent a local late outgrowth in late-Beringian technology. All are classified as sites of the Beringian tradition (West, 1981). As is well known, nothing resembling that point form is recorded from Siberia. At any rate, it does not appear possible, on present evidence, to derive the Clovis point from the Far North.

If the last assertion is correct, and if Clovis can be conceived to be essentially a Beringian variant onto which is grafted that distinctive biface, then what would be the appearance of Clovis minus the point? Are there sites on record that could be interpreted as such an entity? Several possible candidates exist and need to be examined with this possibility in mind.

Materials from Sandia at the type site in New Mexico, the Shriver site in Missouri (Reagan et al., 1978), and possibly Little Canyon Creek Cave in Wyoming (Frison and Walker, 1978) underlie Folsom remains. The peculiar Sandia bifaces have been found in situ only at Sandia Cave and at the Lucy site (about 80 km distant). That there is a close relationship to Clovis is clear; these two assemblages could, in fact, represent a peculiar local and temporal variant of Clovis. They could, however, also represent a slightly earlier and antecedent form of Clovis in which there is evident *one* of the possible technologic pathways from a previous reliance on bone points toward the development of the Clovis fluted form. The asymmetry of the Sandia bifaces inevitably leads one to the supposition that they may have functioned more as knives than as projectile heads. According to Roosa (1962), the type 1 specimens from the Lucy site in New Mexico show edge wear indicating that they, at least, were used as knives. On the other hand, there is no evidence known to this writer to confute the equal possibility that some of the large (and bilaterally symmetrical) Clovis bifaces may not have served basically as knives. In other words, if the Sandia bifaces are knives and the Clovis specimens are projectile points, then they are not comparable and both Sandia and Lucy may represent aberrant or skewed Clovis assemblages. The presence of bone artifacts at Sandia similar to the Sandia 1 point type (Hibben and Bryan, 1941: Plate 13, No. 2) could be used to support such an evolutionary argument. At neither site are the radiocarbon dates of any real help.

The bottom (pre-Folsom) component of the Shriver site is small and nondiagnostic. There is nothing evident that would distinguish it from Clovis material. The assignment of it to pre-Clovis (Reagan et al., 1978) is based upon a thermoluminescence determination on chert from the overlying Folsom level. It seems obvious that the estimated age of nearly 13,000 yr B.P. for a Folsom assemblage should be discarded, thus leaving only the stratigraphic and typological relationships to be taken into account. There is no reason to doubt that the bottommost component antedates Folsom here, but equally there is no reason to suppose that the assemblage is anything other than Clovis.

Little Canyon Creek presents a similar situation, except that the exact ascription of the overlying stratum is undeterminable, consisting solely of a hearth dated at 10,170 ± 250 yr B.P. (RL-641) (Frison and Walker, 1978). Because the component in question underlies an unconformity that separates the two strata and because musk oxen, lemmings, and other northern forms are present along with a few nondiagnostic artifacts, the investigators have tentatively proposed that this site might represent a pre-Clovis occupation, too. From reasoning similar to that applied in the Shriver case, the component could be designated a Clovis occupation except that in Wyoming this faunal assemblage generally dates to around 12,000 yr B.P. Thus, Little Canyon Creek may be one of the most likely antecedent Clovis prospects. In turn, that speculation gives rise to another: the singular form of the Clovis bifaces from the Colby site of Wyoming (Frison, 1976) may represent, like Sandia, another Clovis prototype.

The site of Santa Isabel Iztapan in Mexico could easily be interpreted as Clovis or immediately antecedent Clovis. As with most of these situations, the quantity of material is very limited, discouraging protracted speculation. Similarly, Wilson Butte Cave in Idaho could fit into the antecedent category, but, as noted above, it is difficult to interpret.

Finally, the lowest cultural stratum of the Meadowcroft site in Pennsylvania could conceivably be Clovis or antecedent Clovis. There is nothing apparent in the small assemblage that rules out such an assignment, nor, for that matter, an assignment as Archaic. Clearly, this question will be resolved on the basis of evidence other than that of a purely archaeological nature.

Let us revert once more to a reading of the northern evidence. The writer has argued elsewhere that it seems most likely that only one small band (the view favored also by Haynes [1966] and Martin [1973]) of Beringians traversed an inhospitable Mackenzie Corridor and that their projectile weapons carried antler points armed with laterally mounted microblades. For reasons that *may* have had to do with non-availability of the type of antler preferred, the Clovis point was developed and came to be the primary implement of the chase. Alternatively, the new Clovis mounting system (foreshaft and point) may simply have proved to be superior to the old methods. Judge (1973) argues convincingly that the Folsom biface was easily made and easily broken while the bone foreshafts, or sockets, being relatively durable, tended to be retained and thus actually dictated the basal dimensions of subsequently manufactured points.

It seems likely that the movement down the Mackenzie Corridor was quickly made. This consideration, along with their supposed origin point on the Bering Land Bridge, renders it extremely unlikely that these earliest antecedent (proto-Clovis) people would differ in their cultural baggage in any significant way from their northern kin. As stressed above, even with the addition of a new and dominant weapon, the Clovis hunters continued to bear the clear stamp of their Beringian origins.

Reflections on Man and Late-Quaternary Environments

There would likely be no argument with the proposition that food production could not have arisen anywhere in the Middle Palaeolithic world. The reasons are evident: that revolutionary way of coping with nature was the product both of the sapiency that overtook the species during the Wisconsin glacial age and, bound in with it, of the cultural advances marked by the vastly more sophisticated exploitation of nature.

Mousterian hunters of Middle Palaeolithic France were not the same kind of predators as the Lower Perigordians who immediately followed them. The evidence of this difference range from the *relatively* simple and unspecialized cultural equipment they carried, to the indications of low population densities, and finally to evidence of a much more restricted areal distribution. J.-Ph. Rigaud (personal communication, 1979) puts the ratio of known Middle Palaeolithic sites to those of the Upper Palaeolithic in the Dordogne region of France at about 1.0:1.92. In their imperfect way, these statistics do record population numbers. Analogously, the size of a given assemblage (or, better, a series of the same type) may record something of both population numbers and densities. Mousterian-stage culture, although much more widespread than its predecessors, generally appears to have been a middle-latitude phenomenon. More-northerly regions were reached only later. The presence of Mousterian Neanderthalers in Europe through the first third of the Würm Glaciation may constitute a rather important exception, or it may mean that a great deal more needs to be learned about climatic oscillations through

that period, about the severity of the environments in which these sites occur, and about whether those occupations actually were continuous through that period.

Present archaeological evidence suggests that the first occupations of boreal latitudes came late in the last glacial age and are connected with the rapid spread of Upper Palaeolithic peoples. Even if there is some small question as to the extent of Mousterian adaptation to the cold of Würmian middle-latitude western Europe, there assuredly is none concerning the various Upper Palaeolithic cultures that followed. There is abundant evidence of the animals hunted, of the probable use of skin tents, and, perhaps most significantly, of the use of tailored fur clothing. There is even evidence that some Magdalenian peoples during the waning of the Pleistocene chose to follow to the north and mountainous east their particular subarctic ecologic niche.

It was asserted earlier in this chapter that in northeastern Siberia there is no credible evidence of man's presence prior to the appearance of the Upper Palaeolithic Dyuktai culture. Similarly, for Siberia at large, there are no verifiable remains of the Middle Palaeolithic time period or cultural form. Given these constraints, it is difficult to envision a movement of people across the Bering Strait and thence down into the American continents at times when there simply were no people in the staging area from which these putative movements were to have taken place. It will not do to turn the problem on its head and to argue that, since there *is* evidence for man in America during Sangamon or Early Wisconsin times, the few highly dubious sites known in Siberia must therefore be accepted (Chard, 1974).

On a preindustrial level it is extremely unlikely that any but the most advanced hunters could have entered, mastered, and then flourished in this severe arctic environment (West, 1981). There is no suggestion known to this writer that Sangamon interglacial conditions there were any less rigorous than those of today. Included here, after all, are the coldest regularly inhabited places in the world. A January average of about −50°C at Verkhoyansk is vastly more stressful than, say, −4.7°C at Peking or 2.5°C at Paris.

The early Beringians were Arctic peoples; they perfected the requisite adaptation over a period of more than 20,000 years in an environment drastically different from that which later came to characterize the region. (The efficiency of Beringian adaptation can be measured by the fact that in its late phase it was able to survive in an early-Holocene environment that must have been impoverished in the extreme.) All substantive evidence indicates that the Beringians were highly efficient inland Arctic hunters who, when it became habitable about 20,000 years ago, moved into the central region of Beringia and only much later into Alaska.

THE CONSEQUENCES OF ENVIRONMENTAL CHANGE

The climatic warming that got underway about 14,000 years ago set the stage for fundamental change in Beringia. The period from 14,000 to 11,000 yr B.P. must have witnessed rapid degradation and diminution of the steppe tundra biome in central Beringia. By 11,000 years ago, the region was breached by the rising waters of the Bering Sea and the Arctic Ocean. Much more than simple environmental change was involved: a vast land area ceased to exist. Probably it was these events that triggered the emigration of that band of Beringians who became the immediate Clovis ancestors. Perhaps as much as 500 to 1000 years later, the last survivors of central Beringia began to establish themselves in very specific habitat regions of Alaska (West, 1981). During this 3000-year period, most of the Pleistocene Beringian fauna became extinct, but the exact role of man is presently unclear. Further communication between the surviving sectors of Beringia was cut off, and at this point the prehistory of northeastern Siberia and that of Alaska diverge; no significant contact was reestablished until the advent at Bering Strait of Eskimo culture.

Wright (1981) summarizes a number of recent studies of Late Wisconsin vegetation east of the Rocky Mountains. Generally, zonation is found to have been weakly developed, but boreal forest (without strict modern analogues) locally extended well into the south, and tundra was found discontinuously across the northern Midwest into Pennsylvania, New York, and New England. At some of the studied localities, these conditions persisted until 12,000 to 10,000 yr B.P. The contrast with modern environments is striking and the implications for archaelogical interpretation are profound.

The course of Late Wisconsin environmental change was sharply different in the Far Northwest and the region south of the Laurentide ice sheet, and those differences fundamentally affected the course of American prehistory. Alaska, as a cul-de-sac occupied by relatively few species, appears never to have held out that possibility of the development of a foraging existence like that developed elsewhere under Holocene conditions. Consequently, Alaska appears to have been very nearly depopulated in this period (West, 1981). Such was not the case in sub-Laurentide America, where there is recorded a fairly orderly progression paralleling that of the Old World.

Pleistocene Extinctions

In general, Martin's (1973) overkill hypothesis appears to have merit, but, as many investigators have suggested, it may go a bit beyond the mark. (See the variety of views expressed in Martin and Wright, 1967.) Major environmental changes were in process that would surely have had a profound effect upon the distributions, if not the survival, of the species that did become extinct. Perhaps, as many have suggested, the essential role of the Clovis hunters was that of tipping an already precarious balance. In any event, though the archaeological data assuredly do suggest a Clovis population crash (but perhaps only among certain segments) that would correlate with the destruction of large portions of the Rancholabrean fauna, in contrast to the northern case, a fairly rapid recovery ensued, but in new cultural forms. The reason for the difference in result surely lies in the far greater population base number here combined, of course, with the factor just mentioned—a varied and infinitely richer series of habitat types.

Crossing Environmental Frontiers

Passing from one major environmental zone, for which adaptation had been achieved, to another is sometimes held to have required long periods of time. Thus, if verifiable remains of man dating to 11,000 years ago are found in Patagonia, beyond an intervening series of widely different biomes, then one must infer the lapse of long periods as the peculiar problems of each new zone were mastered. There is certainly logic in this premise, and something of this nature must have been involved as the populations of early man moved southward from the Bering Strait. However, this view may suffer from an agrarian bias. The basic concern of the hunter (at least for early man) is with one class of animal and the phenomenology pertaining to that one class. Moreover, within the class of mammals, he narrowed his preference to a relatively few species. The record from Siberia to South America suggests a particular emphasis upon

elephant hunting. Surely in this context there cannot have been a great deal of difference between mammoths and mastodons. The environment of the hunter is mobile. He may be able to operate from a home base, but his food source has no fixed locus and so neither have his economic activities. Hunters are especially mobile when their quarry are subject to obligatory seasonal migrations.

It is not known at present how many different faunal assemblages or habitats the mammoth might have occupied during Clovis time. But, even with the partial evidence now available, it is known that mammoths were being hunted in Siberia, the American Plains, the American Southwest, and the Valley of Mexico. The evidence for mastodons is less clear, but the Kimmswick discovery should provide support for several disputed associations (e.g., Quimby, 1956; Williams, 1957). If mixed coniferous and broad-leaved forest rather than mature coniferous forest could have provided optimum mastodon browse (Skeels, cited in Hester, 1966), then Wright's (1981) characterization of the eastern boreal forest of 18,000 years ago (and later) assumes particular significance. The supposition that the Clovis people throughout their range were elephant hunters seems a very much stronger case.

Practically all sites that have yielded faunal remains have shown the presence of several—invariably large—species. This, too, is important in assessing the possible negative role of environmental frontiers. If one of the preferred quarry species was elephant, and elephants could have occurred almost universally in the Americas at the time, then the taking of other species in varied settings would have been fairly easily accomplished. But more to the point is this: hunters of large mammals have skills that transfer readily and widely.

The present reading of the evidence favors a rapid spread of Clovis and its immediate descendants as far as the tip of South America, where the Fell's Cave (Chile) occupation appears to date to about 11,000 yr B.P. Projectile heads similar to those found at Fell's Cave are reported from Central America (Bird and Cooke, 1978), and a recent study of human dentition carried out on skeletal material from the culturally related Chilean cave of Palli Aike (Turner and Bird, 1981) has led the investigators to conclusions that parallel those presented here.

POST-CLOVIS HUNTERS

In attempting to understand the pattern of later Beringian settlement and cultural differentiation in Alaska, it has been proposed that there came into existence a number of insular situations that fostered varying degrees of cultural change and then operated to fix those changes (West, 1981). It seems a valuable model, and, because the differentiation that later occurred in Clovis territory appears to be generally parallel, it is tempting to suggest that an analogous condition prevailed in sub-Laurentide America and that an analogous model would aid in the explication of these differences.

In Clovis time, there existed essentially one culture type spread over an immense area. The portrayal of the Clovis peoples as recently arrived hunters rapidly spreading into unoccupied lands is overwhelmingly persuasive. The short period of time during which this remarkable expansion was accomplished, followed discontinuously by different and more local cultures, suggests the rapid depletion of the resource on which Clovis grew and a resultant crash in the Clovis population. From this greatly reduced base of discontinuously distributed populations, there developed variant hunting cultures more particularly adapted to game species with more restricted ranges. The latter condition, in turn, may be presumed to represent a response on the part of those species to the new environmental zonations quickly becoming established in the early Holocene. The enforced areal specialization would mean diminished mobility of people and ideas. With poorer conditions for hunters, this period (11,000 to 8000 yr B.P.) probably witnessed an increasing reliance on plant foods, presaging the heavy dependence that marked the fully developed Archaic cultures. That there appear to be wide differences in the various beginnings of the Archaic (Curran and Dincauze, 1977; Griffin, 1978) may be taken to reflect, at least in part, the regionally differential transition into full Holocene conditions.

Summary: Equifacts and Earliest Man in North America

It would be folly to dismiss the possibility that firm evidence for pre-Clovis man may one day be found. At the same time, the question is simply not as open as it once was. Because the preponderance of evidence is against the likelihood, any such presentation will have to be clear and totally unambiguous. As a means of coping with one of the recurrent problems, perhaps one simple device could be profitably applied to the immediate evaluation of ambiguous field data. This entails acknowledging first that it is almost always possible to find evidence in the field that defies ready explanation, be it archaeological, geologic, or biologic. Recognizing that there exists this class of object—*equifacts*—will go far toward clarifying our thinking and purging our literature. (An equifact may be defined as an object of stone, bone, wood, or some other natural substance that has been modified by simple, usually random operations. Equifacts may be man-made or natural.) The only immediate way by which an equifact can be assigned with some certainty as either of human or natural origin is by context, and, of course, even that identification can only be based in probability.

At its best, the archaeological record consists of fragments and filaments (in D. W. Clark's phrase) allowing partial reconstructions and always permitting degrees of dissent and variant interpretation. Nevertheless, it is possible to discriminate between good evidence and unsound evidence. A chain of weak evidence does not gain strength by being made longer. The emerging reconstruction of the Clovis hunters and their expansion—completely apart from the theory presented here as to their origins—is becoming one of the best-documented and most coherent bodies of evidence in the North American archaeological record. No matter what position one takes on the question of the earliest human immigrations, the uniquely dominant organism that was Clovis must be given full account.

At present, all *unequivocal* evidence favors placing man's first entry into the Americas at about 13,000 to 12,000 yr B.P. The earliest precursors of the earliest Americans were the Arctic-adapted eastern Siberian Upper Palaeolithic folk who flourished in the Late Wisconsin refugium of Beringia. The Clovis hunters were direct and immediate descendants of these northern hunters and retained much of the original Beringian culture. The details of their spread and later differentiation will continue to form some of the most engrossing chapters of American archaeology.

References

Adovasio, J. M., Gunn, J. D., Donahue, J., and Stuckenrath, R. (1977). Meadowcroft Rockshelter: Retrospect 1976. *Pennsylvania Archaeologist* 47, 1-93.

Adovasio, J. M., Gunn, J. D., Donahue, J., and Stuckenrath, R. (1978). Meadowcroft Rockshelter, 1977: An overview. *American Antiquity* 43, 632-51.

Adovasio, J. M., Gunn, J. D., Donahue, J., Stuckenrath, R., Guilday, J. E., and Lord, K. (1978). Meadowcroft Rockshelter. *In* "Early Man in America from a Circum-Pacific Perspective" (A. L. Bryan, ed.), pp. 140-80. Occasional Papers 1, Department of Anthropology, University of Alberta, Edmonton.

Adovasio, J. M., Gunn, J. D., Donahue, J., Stuckenrath, R., Guilday, J. E., and Volman, K. (1980). Yes, Virginia, it really is that old: A reply to Haynes and Mead. *American Antiquity* 45, 588-95.

Aikens, C. M. (1983). Environmental archaeology of the western United States. *In* "Late-Quaternary Environments of the United States," vol. 2, "The Holocene" (H. E. Wright, Jr., ed.), pp. 239-51. University of Minnesota Press, Minneapolis.

Alexander, H. L., Jr. (1963). The Levi site: A paleo-Indian campsite in central Texas. *American Antiquity* 28, 510-28.

Alexander, H. L., Jr. (1974). The association of Aurignacoid elements with fluted point complexes in North America. *In* "International Conference on the Prehistory and Paleoecology of Western North American Arctic and Subarctic" (S. Raymond and P. Schledermann, eds.), pp. 21-32. University of Calgary Archaeological Association, Calgary, Alta.

Anderson, D. D. (1970). Microblade traditions in northwestern Alaska. *Arctic Anthropology* 7, 2-16.

Aveleyra Arroyo de Anda, L. (1956). The second mammoth and associated artifacts at Santa Isabel Iztapan, Mexico. *American Antiquity* 22, 12-28.

Bada, J. L., and Helfman, P. M. (1975). Amino acid racemization dating of fossil bones. *World Archaeology* 7, 160-73.

Bedwell, S. F. (1973). "Fort Rock Basin Prehistory and Environment." University of Oregon Books, Eugene.

Benthall, J. L., and McCary, B. C. (1973). The Williamson site: A new approach. *Archaeology of Eastern North America* 1, 127-32.

Berger, R. (1975). Advances and results in radiocarbon dating: Early man in America. *World Archaeology* 7, 174-84.

Bird, J. B., and Cooke, R. (1978). The occurrence in Panama of two types of paleo-Indian projectile points. *In* "Early Man in America from a Circum-Pacific Perspective" (A. L. Bryan, ed.), pp. 263-72. Occasional Papers 1, Department of Anthropology, University of Alberta, Edmonton.

Bischoff, J. L., Merriam, R., Childers, W. M., and Protsch, R. (1976). Antiquity of man in America indicated by radiometric dates on the Yuha burial site. *Nature* 261, 128-29.

Bonnichsen, R. (1978). Critical arguments for Pleistocene artifacts from the Old Crow Basin, Yukon: A preliminary statement. *In* "Early Man in America from a Circum-Pacific Perspective" (A. L. Bryan, ed.), pp. 102-18. Occasional Papers 1, Department of Anthropology, University of Alberta, Edmonton.

Bonnichsen, R. (1979). "Pleistocene Bone Technology in the Bering Refugium." National Museum of Man, Mercury Series, Archaeological Survey of Canada Paper 89.

Bonnichsen, R. (1981). Response to Guthrie 1980. *Quarterly Review of Archaeology* 2 (2), 17.

Bonnichsen, R., Konrad, V., Clary, V., Gibson, T., and Schnurrenberger, D. (1980). "Archaeological Research at Munsungun Lake: 1980 Preliminary Technical Report of Activities." Munsungun Lake Papers 1, Institute for Quaternary Studies, University of Maine, Orono.

Bordes, F. (1961). "Typologie du Paléolithique Ancien et Moyen." 2nd ed. Delmas, Bordeaux.

Bordes, F. (1968). "The Old Stone Age." McGraw-Hill, New York.

Bryan, A. L. (1973). Palaeoenvironments and cultural diversity in late Pleistocene South America. *Quaternary Research* 3, 237-56.

Bryan, A. L. (ed.). (1978a). "Early Man in America from a Circum-Pacific Perspective." Occasional Papers 1, Department of Anthropology, University of Alberta, Edmonton.

Bryan, A. L. (1978b). An overview of paleo-American prehistory from a circum-Pacific perspective. *In* "Early Man in America from a Circum-Pacific Perspective" (A. L. Bryan, ed.), pp. 306-27. Occasional Papers 1, Department of Anthropology, University of Alberta, Edmonton.

Bryan, A. L., Casamiguela, R. M., Cruxent, J. M., Gruhn, R., and Ochsenius, C. (1978). An El Jobo mastodon kill at Taima-taima, Venezuela. *Science* 200, 1275-77.

Bryan, A. L., Schnurrenberger, D., and Gruhn, R. (1980). An early postglacial lithic industry in east-central New York State. "Abstracts, American Quaternary Association," p. 52. University of Maine, Orono.

Bushnell, G. H. S. (1962). "The First Americans." McGraw-Hill, New York.

Bushnell, G., and McBurney, C. (1959). New World origins seen from the Old World. *Antiquity* 33, 93-101.

Byers, D. S., (1954). Bull Brook: A fluted point site in Ipswich, Massachusetts. *American Antiquity* 19, 343-51.

Byers, D. S. (1959). Radiocarbon dates for the Bull Brook site, Massachusetts. *American Antiquity* 24, 427-29.

Carter, G. F. (1978). The American Paleolithic. In "Early Man in America from a Circum-Pacific Perspective" (A. L. Bryan, ed.), pp. 10-19. Occasional Papers 1, Department of Anthropology, University of Alberta, Edmonton.

Carter, G. F. (1980). "Earlier Than You Think." Texas A&M University Press, College Station.

Chard, C. S. (1959). New World origins: A reappraisal. *Antiquity* 33, 44-49.

Chard, C. S. (1974). "Northeast Asia in Prehistory." University of Wisconsin Press, Madison.

Chard, C. S. (1975). "Man in Prehistory." 2nd ed. McGraw-Hill, New York.

Childers, W. M. (1977). Ridge-back tools of the Colorado Desert. *American Antiquity* 42, 242-48.

Childers, W. M., and Minshall, H. L. (1980). Evidence of early man exposed at Yuha Pinto Wash. *American Antiquity* 45, 297-308.

Clark, D. W., and Clark, A. M. (1975). Fluted points from the Batza Tena obsidian source of the Koyukuk River region, Alaska. *Anthropological Papers of the University of Alaska* 17, 31-38.

Cole, J. R., Godfrey, L. R., Funk, R. E., Kirkland, J. T., and Starna, W. A. (1977). On some Paleolithic tools from northeast North America. *Current Anthropology* 18, 541-46.

Colinvaux, P. A. (1980). Vegetation of the Bering Land Bridge revisited. *Quarterly Review of Archaeology* 1, 2.

Cook, J. P. (1969). The early prehistory of Healy Lake, Alaska. Ph.D. dissertation, University of Wisconsin, Madison.

Cotter, J. L. (1937). The occurrence of flints and extinct animals in pluvial deposits near Clovis, New Mexico: Part IV. Report on excavations at the gravel pit in 1936. *Proceedings, Philadelphia Academy of Natural Sciences* 89, 2-16.

Cotter, J. L. (1938). The occurrence of flints and extinct animals in pluvial deposits near Clovis, New Mexico: Part VI. Report on field season of 1937. *Proceedings, Philadelphia Academy of Natural Sciences* 90, 113-17.

Cotter, J. L. (1962). Comment on Mason's "The paleo-Indian tradition in eastern North America." *Current Anthropology* 3, 250-52.

Cox, S. L. (1972). "A Re-Analysis of the Shoop Site." Smithsonian Institution, Washington, D.C.

Crook, W. W., Jr., and Harris, R. K. (1957). Hearths and artifacts of early man near Lewisville, Texas, and associated faunal material. *Bulletin of the Texas Archaeological Society* 28, 7-97.

Crook, W. W., Jr., and Harris, R. K. (1958). A Pleistocene campsite near Lewisville, Texas. *American Antiquity* 23, 233-46.

Curran, M. L., and Dincauze, D. F. (1977). Paleoindians and paleo-lakes: New data from the Connecticut drainage. In Amerinds and Their Palaeoenvironments in Northeastern North America (W. S. Newman and B. Salwen, eds.). *Annals of the New York Academy of Sciences* 288, 333-48.

Cwyner, L. C., and Ritchie, J. C. (1980). Arctic steppe-tundra: A Yukon perspective. *Science* 208, 1375-77.

Davis, E. L. (1975). The "exposed archaeology" of China Lake, California. *American Antiquity* 40, 39-53.

Davis, E. L. (1978). Associations of people and a Rancholabrean fauna at China Lake, California. In "Early Man in America from a Circum-Pacific Perspective" (A. L. Bryan, ed.), pp. 183-217. Occasional Papers 1, Department of Anthropology, University of Alberta, Edmonton.

Dikov, N. N. (1977). "Archaeologic Monuments in Kamchatka, Chukotka and the Upper Reaches of the Kolyma." ("Arkheologicheskie Pamaitniki Kamchatki, Chukotki i Verknei Kol'im'i.") Iztadel'stvo, Nauka, Moscow. (In Russian.)

Dikov, N. N. (1979). "Ancient Cultures of Northeastern Asia." ("Drevnie Kul'tur'i Severo-Vostochnoi Azii.") Iztadel'stvo, Nauka, Moscow. (In Russian.)

Dincauze, D. F. (1981). The Meadowcroft papers. *Quarterly Review of Archaeology* 2, 3.

Dragoo, D. W. (1973). Wells Creek: An early man site in Stewart County, Tennessee. *Archaeology of Eastern North America* 1, 1-55.

Dragoo, D. W. (1980). The trimmed-core tradition in Asiatic-American contacts. In "Early Native Americans" (D. L. Browman, ed.), pp. 69-82. World Anthropology series, Mouton, The Hague.

Duvall, J. D., and Venner, W. T. (1979). A statistical analysis of the lithics from the Calico site (5BCM 1500A), California. *Journal of Field Archaeology* 6, 455-62.

Epstein, J. F. (1963). The burin-faceted projectile point. *American Antiquity* 29, 187-201.

Figgins, J. D. (1927). The antiquity of man in America. *Natural History* 27, 229-39.

Figgins, J. D. (1933). A further contribution to the antiquity of man in America. *Proceedings, Colorado Museum of Natural History* 12 (2).

Fitting, J. E., DeVisscher, J., and Wahla, E. J. (1966). "The Paleo-Indian Occupation of the Holcombe Beach." Anthropological Paper 27, Museum of Anthropology, University of Michigan, Ann Arbor.

Frison, G. C. (1976). Cultural activity associated with prehistoric mammoth butchering and processing. *Science* 194, 728-30.

Frison, G. C., and Bradley, B. A. (1980). "Folsom Tools and Technology at the Hanson Site, Wyoming." University of New Mexico Press, Albuquerque.

Frison, G. C., and Walker, D. M. (1978). The archaeology of Little Canyon Creek and its associated late Pleistocene fauna. In "American Quaternary Association, Fifth Biennial Meeting Abstracts," p. 200.

Funk, R. E. (1973). The West Athens Hill site (Cox 7). In "Aboriginal Settlement Patterns in the Northeast" (W. A. Ritchie and R. E. Funk, eds.), pp. 9-36. New York State Museum and Sciences Service Memoir 20.

Funk, R. E., Walters, G. R., and Ehlers, W. F., Jr. (1969). A radiocarbon date for early man from the Dutchess Quarry Cave, Orange County, New York. *Bulletin of the New York State Archaeological Association* 46, 19-21.

Funk, R. E., Walters, G. R., Ehlers, W. F., Jr., Guilday, J. E., and Connally, G. G. (1969). The archaeology of Dutchess Quarry Cave, Orange County, New York. *Pennsylvania Archaeologist* 39, 7-22.

Funk, R. E., Weinman, T. P., and Weinman, P. L. (1969). The King's Road site: A recently discovered paleo-Indian manifestation in Greene County, New York. *Bulletin of the New York State Archaeological Association* 45, 1-23.

Giddings, J. L. (1964). "The Archaeology of Cape Denbigh." Brown University Press, Providence, R.I.

Graham, R. W., Haynes, C. V., Jr., Johnson, D. L., and Kay, M. (1981). Kimmswick: A Clovis-mastodon association in eastern Missouri. *Science* 213, 1115-17.

Gramley, R. M. (1981). A new paleo-Indian site in the state of Maine. *American Antiquity* 46, 354-61.

Green, F. E. (1963). The Clovis blades: An important addition to the Llano complex. *American Antiquity* 29, 145-65.

Griffin, J. B. (1965). Late Quaternary prehistory in the northeastern woodlands. In "The Quaternary of the United States" (H. E. Wright, Jr., and D. J. Frey, eds.), pp. 655-67. Princeton University Press, Princeton, N.J.

Griffin, J. B. (1978). Eastern United States. In "Chronologies in New World Archaeology" (R. E. Taylor and C. W. Meighan, eds.), pp. 51-70. Academic Press, New York.

Grimes, J. R. (1979). A new look at Bull Brook. *Anthropology* 3, 109-30.

Gruhn, R. (1961). "The Archaeology of Wilson Butte Cave, South-Central Idaho." Occasional Papers 6, Idaho State College Museum, Pocatello.

Gruhn, R. (1965). Two early radiocarbon dates from the lower levels of Wilson Butte Cave, south-central Idaho. *Tebiwa* 8, 57.

Gruhn, R. (1977). Earliest man in the Northeast: A hemisphere-wide perspective. In Amerinds and Their Palaeoenvironments in Northeastern North America (W. S. Newman and B. Salwen, eds.). *Annals of the New York Academy of Sciences* 288, 163-64.

Gruhn, R. (1978) A note on the excavations at El Bosque, Nicaragua in 1975. In "Early Man in America from a Circum-Pacific Perspective" (A. L. Bryan, ed.), pp. 261-62. Occasional Papers 1, Department of Anthropology, University of Alberta, Edmonton.

Gruhn, R., and Bryan, A. L. (1977). Los Tapiales: A paleo-Indian campsite in the Guatemalan Highlands. *Proceedings of the American Philosophical Society* 121, 235-73.

Guilday, J. E. (1967). The climatic significance of the Hosterman's Pit local fauna, Centre County, Pennsylvania. *American Antiquity* 32, 231-32.

Guilday, J. E. (1969). A possible caribou-paleo-Indian association from Dutchess Quarry Cave, Orange County, New York. *Bulletin of the New York State Archaeological Association* 45, 24-29.

Guthrie, R. D. (1968). Paleoecology of the large mammal community in interior Alaska during the last Pleistocene. *The American Midland Naturalist* 79, 346-63.

Guthrie, R. D. (1972). Re-creating a vanished world. *National Geographic* 141, 294-301.

Guthrie, R. D. (1980). The first Americans? The elusive Arctic bone culture. *Quarterly Review of Archaeology* 1, 2.

Guthrie, R. D. (1981). Rejoinder to Bonnichsen 1981. *Quarterly Review of Archaeology* 2, 18.

Hammatt, H. H. (1970). A paleo-Indian butchering kit. *American Antiquity* 35, 141-52.

Hare, P. E. (1974). Amino acid dating: A history and an evaluation. *MASCA Newsletter* 10, 4-7.

Harington, C. R. (1971). Ice age mammal research in the Yukon Territory and Alaska. *In* "Early Man and Environments in Northwest North America" (R. A. Smith and J. W. Smith, eds.), pp. 35-51. University of Calgary Archaeological Association, Calgary, Alta.

Harrington, M. R. (1948). "An Ancient Site at Borax Lake." Southwest Museum Papers 16.

Hassan, A., and Hare, P. E. (1978). Amino acid analysis in radiocarbon dating of bone collagen. *In* "Archaeological Chemistry: II" (G. F. Carter, ed.), pp. 109-16. American Chemical Society, Washington, D.C.

Haury, E. W. (1950). "The Stratigraphy and Archaeology of Ventana Cave, Arizona." University of Arizona Press and University of New Mexico Press, Tucson and Albuquerque.

Haury, E. W., Antevs, E., and Lance, J. F. (1953). Artifacts with mammoth remains, Naco, Arizona. *American Antiquity* 19, 1-24.

Haury, E. W., Sayles, E. B., and Wasley, W. W. (1959). The Lehner mammoth site, southeastern Arizona. *American Antiquity* 25, 2-30.

Haynes, C. V., Jr. (1966). Elephant-hunting in North America. *Scientific American* 214, 104-12.

Haynes, C. V., Jr. (1967). Carbon-14 dates and early man in the New World. *In* "Pleistocene Extinctions: The Search for a Cause" (P. S. Martin and H. E. Wright, Jr., eds.), pp. 267-86. Yale University Press, New Haven, Conn.

Haynes, C. V., Jr. (1973). The Calico site: Artifacts or geofacts. *Science* 181, 305-10.

Haynes, C. V., Jr. (1974). Palaeoenvironments and cultural diversity in late Pleistocene South America: A reply to A. L. Bryan. *Quaternary Research* 4, 378-82.

Haynes, C. V., Jr. (1976). Mammoth hunters of the USA and USSR. *In* "Beringia in Cenozoic." Far Eastern Branch, Academy of Sciences, Vladivostok. (In Russian.)

Haynes, C. V., Jr. (1980). Palaeoindian charcoal from Meadowcroft Rockshelter: Is contamination a problem? *American Antiquity* 45, 582-87.

Haynes, C. V., Jr., and Hemmings, E. T. (1968). Mammoth-bone arrowshaft wrench from Murray Springs, Arizona. *Science* 159, 186-87.

Hedges, R. E. M., and Wallace, C. J. A. (1980). The survival of protein in bone. *In* "Biogeochemistry of Amino Acids" (P. E. Hare, T. C. Hoering, and K. King, Jr., eds.), pp. 35-40. Wiley-Interscience, New York.

Hemmings, E. T. (1970). Early man in the San Pedro Valley, Arizona. Ph.D. dissertation, University of Arizona, Tucson.

Henderson, J. (1980). A preliminary report of Texas Highway Department excavations at site 41Bx52: The paleo component. (Mimeographed).

Hester, J. J. (1966). Origins of the Clovis culture. *In* "Proceedings of the 36th International Congress of Americanists," pp. 128-38.

Hester, J. J. (1972). "Blackwater Locality No. 1: A Stratified Early Man Site in Eastern New Mexico." Publication of the Fort Burgwin Research Center 8.

Hibben, F. C., and Bryan, K. (1941). "Evidences of Early Occupation of Sandia Cave, New Mexico, and Other Sites in the Sandia-Manzano Region: Correlation of the Deposits of Sandia Cave, New Mexico, with the Glacial Chronology." Smithsonian Miscellaneous Collections 99 (23).

Hopkins, D. M. (1973). Sea level history of Beringia during the past 250,000 years. *Quaternary Research* 3, 520-40.

Howard, E. B. (1935). Evidence of early man in North America. *University of Pennsylvania Museum Journal* 24 (2-3).

Irving, W. N. (1978). Pleistocene archaeology in Beringia. *In* "Early Man in America from a Circum-Pacific Perspective" (A. L. Bryan, ed.), pp. 96-101. Occasional Papers 1, Department of Anthropology, University of Alberta, Edmonton.

Irving, W. N., and Harington, C. R. (1973). Upper Pleistocene radiocarbon-dated artifacts from the northern Yukon. *Science* 179, 335-40.

Irving, W. N., Mayhall, J. T., Melbye, F. J., and Beebe, B. F. (1977). A human mandible in probable association with a Pleistocene faunal assemblage in eastern Beringia: A preliminary report. *Canadian Journal of Archaeology* 1, 81-93.

Irwin, C., Irwin, H., and Agogino, G. (1962). Wyoming muck tells of battle: Ice age man vs. mammoth. *National Geographic* 121, 828-37.

Irwin, H. T. (1971). Developments in early man studies in western North America, 1960-1970. *In* Symposium on Early Man in North America, New Developments: 1960-1970 (R. Shutler, Jr., ed.). *Arctic Anthropology* 8, 42-67.

Jelinek, A. J. (1971). Early man in the New World: A technological perspective. *Arctic Anthropology* 8, 15-21.

Jennings, J. D. (1974). "Prehistory of North America." 2nd ed. McGraw-Hill, New York.

Jopling, A. V., Irving, W. N., and Beebe, B. F. (1981). Stratigraphic sedimentological and faunal evidence for the occurrence of pre-Sangamonian artefacts in northern Yukon. *Arctic* 34, 3-33.

Jordan, D. F. (1960). The Bull Brook site in relation to "fluted point" manifestations in eastern North America. Ph.D. dissertation, Harvard University, Cambridge, Mass.

Josselyn, D. W. (1965). "The Lively Complex." (Mimeo, privately issued. 408 Broadway, Edgewood, Birmingham, Alabama 35209.)

Judge, W. J. (1973). "Paleoindian Occupation of the Central Rio Grande Valley." University of New Mexico Press, Albuquerque.

Kessels, H. J., and Dungworth, G. (1980). Necessity of reporting amino acid compositions of fossil bones where racemization analyses are used for geochronological applications: Inhomogeneities of D/L amino acids in fossil bones. *In* "Biogeochemistry of Amino Acids" (P. E. Hare, T. C. Hoering, and K. King, Jr., eds.), pp. 527-41. Wiley-Interscience, New York.

Kraft, H. C. (1973). The Plenge site: A paleo-Indian occupation site in New Jersey. *Archaeology of Eastern North America* 1, 56-117.

Krieger, A. D. (1954). A comment on "fluted point relationships" by John Witthoft. *American Antiquity* 19, 273-75.

Krieger, A. D. (1962). The earliest cultures in the western United States. *American Antiquity* 28, 138-43.

Krieger, A. D. (1979). The Calico site and Old World Paleolithic industries. *In* "Pleistocene Man at Calico: A Report on the Calico Mountain Excavations, San Bernardino County, California" (W. C. Schuiling, ed.), pp. 69-74. San Bernardino County Museum Association, Redlands, Calif.

Kurtén, B., and Anderson, E. (1980). "Pleistocene Mammals of North America." Columbia University Press, New York.

Lahren, L. A., and Bonnichsen, R. (1971). The Anzick site: A Clovis complex site in southwestern Montana. Paper presented to the 36th Annual Meeting of the Society for American Archaeology, Norman, Okla.

Larsen, H. (1968). Trail Creek: Final report on the excavation of two caves on Seward Peninsula, Alaska. *Acta Arctica*, Fasc. 15.

Lee, T. E. (1954). The first Sheguiandah expedition, Manitoulin Island, Ontario. *American Antiquity* 20, 101-11.

Lee, T. E. (1955). The second Sheguiandah expedition, Manitoulin Island, Ontario. *American Antiquity* 21, 63-71.

Lee, T. E. (1956). Position and meaning of a radiocarbon sample from the Sheguiandah site, Ontario. *American Antiquity* 22, 79.

Lee, T. E. (1979). Commentary on the Calico conference. *In* "Pleistocene Man at Calico: A Report on the Calico Mountains Excavations, San Bernardino County, California" (W. C. Schuiling, ed.), pp. 103-4. San Bernardino County Museum Association, Redlands, Calif.

Leonhardy, F. C. (1966). "Domebo: A Paleo-Indian Mammoth Kill in the Prairie-Plains." Contribution of the Museum of the Great Plains 1.

Lively, M. (1965). "Preliminary Report on a Pebble Tool Complex in Alabama." (Mimeo, privately issued. 1912 St. Charles Avenue, Birmingham, Alabama 35211.)

Loring, S. (1980). Paleo-Indian hunters and the Champlain Sea: A presumed association. *Man in the Northeast* 19, 15-41.

Lynch, T. F. (1974). Early man in South America. *Quaternary Research* 4, 356-77.

McCary, B. C. (1951). A workshop site of early man, Dinwiddie County, Virginia. *American Antiquity* 17, 9-17.

MacDonald, G. F. (1968). "Debert: A Palaeo-Indian Site in Central Nova Scotia." Anthropology Paper of the National Museums of Canada 16.

MacGowan, K., and Hester, J. A., Jr. (1962). "Early Man in the New World." Doubleday, New York.

MacNeish, R. S. (1976). Early man in the New World. *American Scientist* 64, 316-27.

MacNeish, R. S. (1978). Late Pleistocene adaptations: A new look at early peopling of the New World as of 1976. *Journal of Anthropological Research* 34, 475-96.

MacNeish, R. S. (1979). The early man remains from Pikimachay Cave, Ayacucho Basin, Highland, Peru. *In* "Pre-Llano Cultures of the Americas: Paradoxes and Possibilities" (R. L. Humphrey and D. Stanford, eds.), pp. 1-48. Anthropological Society of Washington, Washington, D.C.

McNett, C. W., Jr., McMillan, B. A., and Marshall, S. B. (1977). The Shawnee-Minisink site. *In* Amerinds and Their Palaeoenvironments in Northeastern North America (W. S. Newman and B. Salwen, eds.). *Annals of the New York Academy of Sciences* 288, 282-96.

Martin, P. S. (1973). The discovery of America. *Science* 179, 969-74.

Martin, P. S., and Wright, H. E., Jr. (1967). "Pleistocene Extinctions: The Search for a Cause." Yale University Press, New Haven, Conn.

Mason, R. J. (1962). The paleo-Indian tradition in eastern North America. *Current Anthropology* 3, 227-83.

Masters, P. M., and Bada, J. L. (1978). Amino acid racemization dating of bone and shell. *In* "Archaeological Chemistry: II" (G. F. Carter, ed.), pp. 117-38. American Chemical Society, Washington, D.C.

Matthews, J. V., Jr. (1976). Arctic steppe: An extinct biome. *In* "American Quaternary Association, Fourth Biennial Meeting, Abstracts," pp. 73-77. Arizona State University, Tempe.

Mead, J. I. (1980). Is it really that old? A comment about the Meadowcroft Rockshelter "overview." *American Antiquity* 45, 579-82.

Meighan, C. W., and Haynes, C. V., Jr. (1970) The Borax Lake site revisited. *Science* 162, 1121-23.

Mirambell, L. (1978). Tlapacoya: A late Pleistocene site in central Mexico. *In* "Early Man in America from a Circum-Pacific Perspective" (A. L. Bryan, ed.), pp. 221-30. Occasional Papers 1, Department of Anthropology, University of Alberta, Edmonton.

Mochanov, Yu. A. (1970). Dyuktai Cave: A new Palaeolithic site in northeast Asia (the results of 1967). (Dyuktaiskaia peschera: Novyi paleoliticheskii pamaitnik Severo-Vostochnoi Azii.) *In* "Po Sledam Drevnikh Kul'tur Yakutiia." Yakutskoe Knizhnoe Izdatel-stvo, Yakutsk. (In Russian.)

Mochanov, Yu. A. (1972). New data on the Bering Sea route of the peopling of America. (Novye dannye o Beringomorskom puti zaselenii Ameriki). *Sovetskaia Etnografiia* 2, 98-102. (In Russian.)

Mochanov, Yu. A. (1975). The stratigraphy and absolute chronology of the Palaeolithic of northern East Asia. (Straigrafiya i absoliutnaya khronologiya paleolita Severo-Vostochoi Azii.) *In* "Yakutiia and Its Neighbors in Antiquity." (Works of the Prilensk Archaeological Expedition.) (Yu. A. Mochanov, ed.) Yakutskii filial, SO AN S.S.S.R., Yakutsk. (In Russian.)

Mochanov, Yu. A. (1977). "The Most Ancient Stages in the Settlement by Man of Northeast Asia." Academy of Science U.S.S.R., Siberian Branch, Novosibirsk.

Mochanov, Yu. A., and Fedoseeva, S. A. (1968). The Palaeolithic site of Ikhine in Yakutiia. (Paleoliticheskaia stoianka Ikhine v Yakutii.) *Sovetskaia Arkheologiia* 4, 244-48. (In Russian.)

Morlan, R. E. (1977). Fluted point makers and the extinction of the arctic-steppe biome in eastern Beringia. *Canadian Journal of Archaeology* 1, 95-108.

Morlan, R. E. (1978). Early man in northern Yukon Territory: Perspectives as of 1977. *In* "Early Man in America from a Circum-Pacific Perspective" (A. L. Bryan, ed.), pp. 78-95. Occasional Papers 1, Department of Anthropology, University of Alberta, Edmonton.

Morlan, R. E. (1979). A stratigraphic framework for Pleistocene artifacts from Old Crow River, northern Yukon Territory. *In* "Pre-Llano Cultures of the Americas: Paradoxes and Possibilities" (R. L. Humphrey and D. Stanford, eds.), pp. 125-46.

Anthropological Society of Washington, Washington, D.C.

Morlan, R. E. (1980). "Taphonomy and Archaeology in the Upper Pleistocene of the Northern Yukon Territory: A Glimpse of the Peopling of the New World." National Museum of Man, Mercury Series, Archaeological Survey of Canada Paper 94.

Mosimann, J. E., and Martin, P. S. (1975). Simulating overkill by paleoindians. *American Scientist* 63, 304-13.

Müller-Beck, H. (1966). Paleohunters in America: Origins and diffusion. *Science* 152, 1191-1210.

Myers, T. P., Voorhies, M. R., and Corner, R. G. (1980). Spiral fractures and bone pseudotools at palaeontological sites. *American Antiquity* 45, 483-90.

Orr, P. C. (1968). "Prehistory of Santa Rosa Island." Santa Barbara Museum of Natural History, Santa Barbara, Calif.

Page, W. D. (1978). The geology of the El Bosque archaeological site, Nicaragua. *In* "Early Man in America from a Circum-Pacific Perspective" (A. L. Bryan, ed.), pp. 231-60. Occasional Papers 1, Department of Anthropology, University of Alberta, Edmonton.

Patterson, L. W. (1977). A discussion of possible Asiatic influences on Texas Pleistocene lithic technology. *Bulletin of the Texas Archaeological Society* 48, 27-45.

Payen, L. A., Rector, C. H., Ritter, E., Taylor, R. E., and Ericson, J. E. (1978). Comments on the Pleistocene age assignment and associations of a human burial from the Yuha Desert, California. *American Antiquity* 43, 448-53.

Péwé, T. L. (1975). "Quaternary Geology of Alaska." U.S. Geological Survey Professional Paper 835.

Powers, W. R., and Hamilton, T. D. (1978). Dry Creek: A late Pleistocene human occupation in central Alaska. *In* "Early Man in America from a Circum-Pacific Perspective" (A. L. Bryan, ed.), pp. 72-77. Occasional Papers 1, Department of Anthropology, University of Alberta, Edmonton.

Quimby, G. I. (1956). The locus of the Natchez pelvis find. *American Antiquity* 22, 77-79.

Quimby, G. I. (1958). Fluted points and geochronology of the Lake Michigan basin. *American Antiquity* 23, 247-54.

Raemsch, B. E., and Vernon, W. W. (1977). Some Paleolithic tools from northeast North America. *Current Anthropology* 18, 97-99.

Reagan, M. J., Rowlett, R. M., Garrison, E. G., Dort, W., Jr., Bryant, V. M., Jr., and Johannsen, C. J. (1978). Flake tools stratified below paleo-Indian artifacts. *Science* 200, 1272-75.

Reeves, B. O. K. (1980). Fractured cherts from Pleistocene fossiliferous beds at Medicine Hat, Alberta. *In* "Early Native Americans" (D. L. Browman, ed.), pp. 83-98. World Anthropology series, Mouton, The Hague.

Ritchie, W. A. (1953). A probable paleo-Indian site in Vermont. *American Antiquity* 18, 249-58.

Roosa, W. B. (1956). The Lucy site in central New Mexico. *American Antiquity* 21, 310.

Roosa, W. B. (1962). Comment on Mason's "The paleo-Indian tradition in eastern North America." *Current Anthropology* 3, 263.

Rouse, I. (1976). Peopling of the Americas. *Quaternary Research* 6, 597-612.

Rouse, I., and Cruxent, J. M. (1963). Some recent radiocarbon dates for western Venezuela. *American Antiquity* 28, 537-40.

Schuiling, W. C. (1979). "Pleistocene Man at Calico: A Report on the Calico Mountains Excavations." San Bernardino County Museum Association, Redlands, Calif.

Sellards, E. H. (1952). "Early Man in America." University of Texas Press, Austin.

Shutler, R., Jr. (ed.) (1971). Symposium on early man in North America, new developments: 1960-1970. *Arctic Anthropology* 8, 1-91.

Simpson, R. D. (1978). The Calico Mountains archaeological site. *In* "Early Man in America from a Circum-Pacific Perspective" (A. L. Bryan, ed.), pp. 218-20. Occasional Papers 1, Department of Anthropology, University of Alberta, Edmonton.

Simpson, R. D. (1980). The Calico Mountains site: Pleistocene archaeology in the Mojave Desert, California. *In* "Early Native Americans" (D. L. Browman, ed.), 7-20. World Anthropology series, Mouton, The Hague.

Solecki, R. S. (1951). Notes on two archaeological discoveries in northern Alaska, 1950. *American Antiquity* 17, 55-57.

Stagg, R. M., Vernon, W. W., and Raemsch, B. E. (1980). Wisconsin and pre-

Wisconsin stone industries of New York State and related tools from a shop site near Tula, Mexico. *In* "Early Native Americans" (D. L. Browman, ed.), pp. 41-68. World Anthropology series, Mouton, The Hague.

Stalker, A. M. (1969). Geology and age of the early man site at Taber, Alberta. *American Antiquity* 34, 425-28.

Stalker, A. M. (1977). Indications of Wisconsin and earlier man from the southwest Canadian prairies. *In* "Amerinds and Their Paleoenvironments in Northeastern North America" (W. S. Newman and B. Salwen, eds.). *Annals New York Academy of Science* 288, 119-36.

Stanford, D. (1979a). Afterword: Resolving the question of New World origins. *In* "Pre-Llano Cultures of the Americas: Paradoxes and Possibilities" (R. L. Humphrey and D. Stanford, eds.), pp. 147-50. Anthropological Society of Washington, Washington, D.C.

Stanford, D. (1979b). The Selby and Dutton sites: Evidence for a possible pre-Clovis occupation of the High Plains. *In* "Pre-Llano Cultures of the Americas: Paradoxes and Possibilities" (R. L. Humphrey and D. Stanford, eds.), pp. 102-24. Anthropological Society of Washington, Washington, D.C.

Stanford, D., Wedel, W. R., and Scott, G. R. (no date). Archaeological investigations of Lamb Spring site. Unpublished manuscript.

Stoltman, J. B. (1978). Temporal models in prehistory: An example from eastern North America. *Current Anthropology* 19, 703-46.

Thompson, R. M. (1948). Notes on the archaeology of the Utukok River, northwestern Alaska. *American Antiquity* 14, 62-65.

Turner, C. G., II, and Bird, J. B. (1981). Dentition in Chilean paleo-Indians and peopling of the Americas. *Science* 212, 1053-55.

Tuross, N., Eyre, D. R., Holtrop, M. E., Glimcher, M. J., and Hare, P. E. (1980). Collagen in fossil bones. *In* "Biogeochemistry of Amino Acids" (P. E. Hare, T. C. Hoering, and K. King, Jr., eds.), pp. 53-64. Wiley-Interscience, New York.

Wendorf, F., and Krieger, A. D. (1959). New light on the Midland discovery. *American Antiquity* 25, 66-78.

West, F. H. (1981). "The Archaeology of Beringia." Columbia University Press, New York.

Willey, G. R. (1966). "An Introduction to American Archaeology: North and Middle America." Vol. 1. Prentice-Hall, Englewood Cliffs, N.J.

Williams, S. (1957). The Island 35 mastodon: Its bearing on the age of Archaic cultures in the East. *American Antiquity* 22, 359-72.

Williams, S., and Stoltman, J. B. (1965). An outline of southeastern United States prehistory with particular emphasis on the paleo-Indian era. *In* "The Quaternary of the United States" (H. E. Wright, Jr., and D.J. Frey, eds.), pp. 669-83. Princeton University Press, Princeton, N.J.

Witthoft, J. (1952). A paleo-Indian site in eastern Pennsylvania: An early hunting culture. *Proceedings of the American Philosophical Society* 96, 464-95.

Witthoft, J. (1954). A note on fluted point relationships. *American Antiquity* 19, 271-73.

Wormington, H. M. (1957). "Ancient Man in North America." Denver Museum of Natural History, Popular Series, 4.

Wormington, H. M. (1971). Comments on early man in North America, 1960-1970. *Arctic Anthropology* 8, 83-91.

Wright, H. E., Jr. (1981). Vegetation east of the Rocky Mountains 18,000 years ago. *Quaternary Research* 15, 113-25.

Climatology

CHAPTER 19

Paleoclimatic Evidence from Stable Isotopes

Irving Friedman

Introduction

The thermodynamic properties of molecules that contain different isotopic species differ slightly. This fact suggests that during various chemical and physical equilibrium reactions these molecules react at differing rates and that separation of isotopically differing molecules occurs. For example, during the precipitation of calcium carbonate from water (under equilibrium conditions), the stable isotopes of oxygen of mass 16 and 18 distribute themselves in such a manner that the calcium carbonate becomes enriched in oxygen-18 relative to the water. In other words, the ratio of ^{18}O to ^{16}O is higher in the calcium carbonate than in the water. The degree of such enrichment is a function of the temperature at which the reaction occurs. This fact has led Urey (1947) to suggest that the measurement of $^{18}O/^{16}O$ ratios in fossil shells could result in the determination of the temperatures at which the organisms lived. Because this isotopic separation is a mass-dependent function, the separation is greater for the light elements in which the percentage difference in mass among the isotopic molecules is greatest. This is illustrated by a comparison of the percentage difference in mass between the isotopic molecules H_2 (mass 2) and HD (mass 3). The molecule containing heavy hydrogen (deuterium) is 50% heavier than the molecule composed of two light (protium) hydrogens. When one examines the molecules $H^{79}Br$ and $H^{81}Br$, one finds that the mass difference is only 2.5%. One would then expect that isotopic separation processes would be more efficient and would result in greater separations in reactions involving H_2-HD than for reactions involving $H^{79}Br$-$H^{81}Br$. In general, natural processes that result in measurable isotopic separation are restricted to molecules composed principally of elements having atomic weights of less than 40.

Natural water is composed of a mixture of the isotopic molecules $H_2^{16}O$, $H_2^{18}O$, $HD^{16}O$, $H_2^{17}O$, $HD^{17}O$, and $HD^{18}O$. The first three molecules are relatively abundant compared to the last three, and they are the ones whose abundance is usually measured. The distribution of these three species in any water sample is a function of the history of the water, including its source and trajectory, and so they can be used to monitor many of the physical and chemical processes that the water has undergone. The water source and trajectory, and

the processes of evaporation and condensation, all affect the isotopic composition of the water, and all are elements of climate. Therefore, studies of the D/H and $^{18}O/^{16}O$ ratios of natural waters can give information on present and past climates. Other stable-isotopic ratios, such as carbon ($^{13}C/^{12}C$) have also been useful in this regard. For a detailed discussion of the changes in O^{18} and deuterium in precipitation, the reader is referred to Dansgaard (1964) and Friedman and others (1964).

The climatic record, as deduced from stable-isotopic evidence, can be divided into two categories—the oceanic and the continental.

Marine Carbonates

The development of the carbonate paleotemperature method by H. C. Urey and his co-workers (Urey, 1947; Epstein et al., 1951, 1953) in the early 1950s was the first use of stable isotopes in paleoclimate research. When calcium carbonate is precipitated in equilibrium with dissolved bicarbonate, ^{18}O distributes itself between the solid phase (calcate or argonite) and the liquid phases (water, bicarbonate) in such a way that the ratio of ^{18}O to ^{16}O in the solid is higher than that in the other phases by an amount that is determined by the temperature. The measured ratio of ^{18}O to ^{16}O in the solid carbonate is then determined by the $^{18}O/^{16}O$ ratio in the water and by the temperature.

When water evaporates from the oceans, the water vapor is depleted in ^{18}O relative to the ocean. This is due to the fact that $H_2^{18}O$ has a lower vapor pressure than $H_2^{16}O$. As this depleted but warm water vapor cools and loses water by precipitation as rain, the remaining water vapor is further depleted in ^{18}O because the $H_2^{18}O$ tends to be enriched in the liquid (rain) compared to the vapor, again because its vapor pressure is lower than that of $H_2^{16}O$. With cooling and loss of water, the vapor becomes progressively depleted in ^{18}O. By the time the vapor is cooled to below freezing, the snow that is precipitated is highly depleted compared to the ocean water, the original source of the water vapor. Thus, Antarctic ice has $\delta^{18}O$ values of $-50‰$ to $-60‰$ compared to ocean water with a $\delta^{18}O$ of 0.

The buildup of glacier ice not only removes water from the oceans, it also enriches the ocean in ^{18}O by locking up water (ice) depleted in

[16]O. The change in the [18]O content of the oceans from full-glacial to interglacial conditions depends upon the amount of water tied up as glacier ice and the average $\delta^{18}O$ of this ice.

Emiliani (1955), after applying the Urey technique to Foraminifera tests from deep-sea cores, reported an abrupt change in the isotopic composition of the carbonate tests with depth in the cores and, since the change in [18]O content is too large to be accounted for only on the basis of oceanic temperature changes, concluded that the isotopic composition of the ocean changed between 20,000 and 11,000 yr B.P. This change implies a very rapid melting of the Laurentide ice sheet.

Examinations of the [18]O of Foramifera separated from deep-sea cores collected in the Gulf of Mexico by Kennett and Shackleton (1975) and by Emiliani and others (1978) have shown that the last melting of the Laurentide ice sheet began about 17,000 yr B.P., reached a peak about 13,500 yr B.P., and was essentially complete by about 11,500 yr B.P. This conclusion assumes that meltwater drained via the Mississippi River valley. Minor melting may have gone on past 11,500 years, but, because the meltwater would then have drained mainly to the Atlantic Ocean via the Gulf of St. Lawrence, the evidence is not to be found in Foraminifera samples from the Gulf of Mexico.

Isotopic Evidence from Tree Rings

Within the last 7 years, a number of investigators (Schiegl and Vogel, 1970; Libby and Pandolfi, 1974; Schiegl, 1974; Wilson and Grinstead, 1975; Epstein et al., 1976) have suggested that the isotopic ($\delta^{18}O$, δD, $\delta^{13}C$) composition of a sample of wood or of its components may reflect the climate prevailing during the tree's period of growth. In particular, the δD of treated wood and treated cellulose appears to represent the δD of the water utilized by the tree during its growth. Yapp and Epstein (1977) have examined a number of trees from North America that had ^{14}C dates ranging from 9500 to 22,000 yr B.P. and deduced the δD of the water used by these trees. They conclude that meteoric waters that precipitated over ice-free regions of North America during the interval between 22,000 and 14,000 yr B.P. were enriched in δD by an average of 19‰ compared to present-day precipitation in the same regions (Figure 19-1). They suggest that this enrichment possibly was caused by lower ocean temperatures and smaller temperature gradients between the ocean and the precipitation sites that exist at present. They further suggest that warmer winters and cooler summers characterized this glacial age. The isotopic studies show that the transition from glacial to interglacial conditions in North America occurred between 12,900 and 10,000 yr B.P. but may not have been synchronous over North America. In particular, samples from Seattle, Washington, and from the Guadalupe Mountains of New Mexico appear to lag the change to interglacial conditions, as compared to the remainder of the tree samples (New Jersey, Ohio, Indiana, Wisconsin, Minnesota, Michigan, South Dakota, Utah, and California). This apparent increase in δD of precipitation in the ice-free areas occurred at the same time that δD decreased at high latitudes, as shown from the analysis of Greenland and Antarctic ice cores (Dansgaard et al., 1969; Epstein et al., 1970). Yapp and Epstein have found that the δD of precipitation during the interval between 10,000 and 9500 yr B.P. was similar to that of present-day precipitation.

Burk and Stuiver (1981) present evidence that shows that the $\delta^{18}O$ in tree-ring cellulose correlates well with that in the water contained in the leaves. The $\delta^{18}O$ in leaf water, in turn, is determined by that in

the precipitation as modified by evapotranspiration. They suggest that, if the $\delta^{18}O$ of the precipitation can be estimated either by δD measurements of cellulose or by some other method, then paleoclimatic relative humidity can be determined by comparing the [18]O of cellulose with the [18]O content of precipitation.

Precipitation Regimes

The enrichment in deuterium of precipitation over the ice-free areas of central North America during the glacial maximum can be accounted for by a change in the air-mass trajectories that carry moisture from the oceans. At present, the moisture sources for a large part of North America are the Gulf of Mexico and the Pacific Ocean. The δD of present-day precipitation over the middle of the continent varies greatly, depending upon which of these two trajectories is the moisture path. For example, the δD of the annual snowpack over parts of the Rocky Mountains of Colorado sampled at different years at the same site and time of year (April 1) can vary by as much as 30‰ (I. Friedman, K. Hardcastle, and J. Gleason, un-

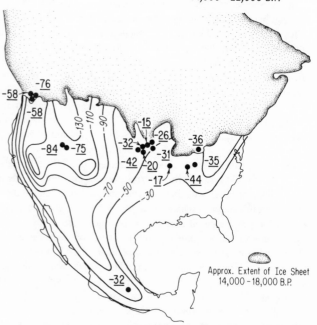

Inferred δD Values (Underlined) of Meteoric Water in the Interval 14,000–22,000 B.P.

Iso–δD Contours are for Modern Meteoric Water (Taylor,1974)

Figure 19-1. Map of North America showing the distribution of 16 δD values of meteoric waters (underlined) from 15 different sites during the Late Wisconsin glacial maximum (about 22,000 to 14,000 yr B.P.) as inferred from δD values of tree cellulose C-H hydrogen. The glacial-age meteoric waters of the 15 sites on the map have, on the average, δD values that are 19‰ more positive than the corresponding modern meteoric waters at those sites as deduced from the modern iso-δD contours. The glacial-age distributional pattern of δD values is similar to the modern distribution but is systematically shifted by the positive bias of the ancient waters. The North American coastline is that of modern sea level and does not reflect the lowest level of the glacial maximum. The numbers designating contour line δD values should be divided by 1.04. (From Yapp and Epstein, 1977; used with the permission of the authors.)

published data). This variation can be ascribed to the variation in the air-mass trajectory bringing moisture to the sampling site. Air from the Pacific must pass over the Sierra Nevadas and the high mountains of the Basin and Range as well as the western Rockies before reaching Colorado. The deuterium is preferentially lost from the air due to the lower vapor pressure of HDO as compared to H_2O during the passage over these high mountains. In contrast, air from the Gulf of Mexico traverses little high country before arriving in Colorado and therefore has a higher δD compared to air masses from the Pacific.

If the air masses that contribute to precipitation over the middle of the continent were to shift so that more air masses derived from the Gulf of Mexico were involved, not only would a higher δD for the precipitation result but also an increased amount of precipitation. This effect is particularly pertinent to areas west of the 100th meridian. To the east of this datum, the Gulf of Mexico and the Atlantic Ocean are the dominant moisture sources at present.

The increased δD of water in the midwestern United States (Ohio to Wisconsin) postulated by Yapp and Epstein may be explained by an increase of summer (high δD) versus winter (low δD) precipitation (Friedman et al., 1964). This change in precipitation regime may have occurred by a diminution of winter precipitation caused by the presence of high pressure over the Laurentide ice sheet deflecting storm tracks far to the south.

The results of Yapp and Epstein's (1977) work on trees dating from early in the present interglaciation (10,000 to 9500 yr B.P.) show that the δD of water was about the same at that time as it is at present; this implies that the immediately postglacial climate was similar to the present climate in the ice-free areas that Yapp and Epstein sampled. As they point out, the δD of tree rings can be a powerful tool in understanding past climate when coupled with investigations of the present regime of δD in precipitation.

Cave Deposits

In contrast to the data of Yapp and Epstein (1977) showing increased a δD in precipitation in the ice-free areas during the glacial maximum, Harmon and others (1979) have found that the δD of water contained in fluid inclusions trapped in calcite deposited in caves in Iowa, West Virginia, Kentucky, and Missouri during this same period have a δD lower than that of modern precipitation. They claim an average shift of about 12‰ during the Wisconsin Glaciation for precipitation over the ice-free areas of North America.

Speleothem data from Coldwater Cave in Iowa, in an area that was close to the southern edge of the Laurentide ice sheet, cover the time period of 25,000 ± 4000 to 6000 ± 1000 yr B.P. All the fluid inclusions in this case have a δD lower than the δD of modern precipitation.

The apparent contradiction between the results of Yapp and Epstein (1977) and those of Harmon and others (1979) may reflect the fact that trees utilize shallow groundwater representing precipitation that fell just before or during the growing season whereas the cave deposits result from deeper groundwater that may be quite old—perhaps many hundreds or even thousands of years old. In addition, the deep groundwater is more likely to be "contaminated" with glacial meltwater even during a glacial period, when subglacial melting may contribute to the deep groundwater. Harmon and others (1979) argue against such effects in their Iowa cave; the local topography suggests to them that the water source for this cave is seepage derived from meteoric precipitation on the land surface im-

mediately above the cave. They suggest that the discrepancy may result from large differences in the δD of summer and winter precipitation, with the trees utilizing summer (higher δD) precipitation but the cave deposits accumulating winter-derived meltwater (lower δD) during the spring. Although the reasons for the apparent discrepancy are not clear, data from both types of samples (trees and cave deposits) must reflect something about climate, and the elucidation of this problem should result in a better understanding of the climate prevalent during and after the Wisconsin glacial maximum.

Two other lines of isotopic evidence are in apparent conflict with inferences from speleothem data regarding the isotopic composition of precipitation in the ice-free areas during the Late Wisconsin. One piece of evidence supports the hypothesis.

Samples of water from underground aquifers with ^{14}C dates or estimated ages of as much as 30,000 yr B.P. have the same δD as present-day precipitation. Although most of the samples are from the southern United States (Texas, Florida, California), some are from areas somewhat farther north. Hanshaw and others (1980) conclude from this similarity in δD that the mean continental winter and spring temperatures were not much different from today, even during the glacial maximum, and that winter and spring storm tracks also were similar. Although the ^{14}C dating of the Florida aquifers is quite firm, the dating of the other aquifers leaves much to be desired. Conclusions based on this evidence, therefore, should be viewed as tenuous.

Hanshaw and others (1980) have analyzed one sample of subglacially precipitated calcite collected from central Ohio that was overlain by till estimated to be older than 13,500 yr B.P. They state that the $\delta^{18}O$ of this sample indicates that it formed from water having an isotopic composition similar to that of *local* modern mean annual precipitation. The authors state that this conclusion is highly speculative and is based on only one sample.

In a study of the $\delta^{18}O$ and δD ratios of formation waters from the Michigan and Illinois basins, Clayton and others (1966) observe that a few of their samples had δD and $\delta^{18}O$ ratios that were very much lighter than present-day precipitation in the recharge areas for these basins. Because of the absence of any process that would change these ratios after precipitation, they conclude that these samples originated during a climate cooler than the present's and that these samples represent water that precipitated during Pleistocene glaciation.

Snowpack Data

The isotopic composition of precipitation in the Sierra Nevada of California also has been used as an indicator of climate. Friedman and Smith (1972) compare the low deuterium content of the seasonal snowpack for the unusually wet winter of 1968-1969 (two to three times normal) to that of the dry 1969-1970 (about three-fourths of normal). The average temperature for October 1 to March 31 for the wet year was about 1.5° lower than that for the dry year in the climatic district under consideration. During the Pleistocene pluvials, the storms are likely to have had the same character and trajectory as those of the wet and cool 1968-1969 winter, when the δD was 20‰ less than that of the subsequent two "normal" years.

In a paper describing the δD of samples of precipitation collected in individual storms during the 1968-1969 season in eastern California, Smith and others (1979) conclude that Pleistocene precipitation had δD values of as much as 50‰ to 85‰ more negative than normal values for present-day precipitation and that the net 1968-1969 season was isotopically intermediate between the present and full-glacial climates of the Pleistocene.

The unpublished snow-core data of I. Friedman and K. Hardcastle throw some light on the influence of air-mass trajectory on the δD of winter precipitation in the Rocky Mountains of Montana, Wyoming, Colorado, and New Mexico. The cores were collected at 170 stations on April 1, 1971, at 83 stations on April 1, 1972, and at 30 stations on April 1, 1973. From data on snow collections made in different areas but at the same elevation, stations in the San Juan Mountains have been shown to have a higher δD by 50‰ than stations farther northeast in the Front Range. In the north-south-trending Front Range, the samples (all collected near the crest of the range at elevations between 3000 and 3500 m) show an abrupt change in δD within a distance of about 50 km. The transition point at which this change in δD occurs shifts north and south from snow season to snow season. The change in δD is ascribed to the influence of Gulf of Mexico moisture that penetrates northward (mainly during the late fall and early spring) and contributes "heavy" δD to the precipitation. The depth of this northward penetration of Gulf of Mexico moisture varies from year to year.

From the above discussion, it is apparent that the δD and the $\delta^{18}O$ of precipitation during the glacial maximum of the Pleistocene in the ice-free areas south of the Laurentide ice sheet are still in question. Some of the obvious discrepancies in the conclusions reached by various investigators might be explained by their use of incorrect present-day precipitation data that are based only on broad estimates and ignore variations in δD and $\delta^{18}O$, both in time (including season) and in space. These isotopic differences, as can be seen from the data quoted previously and from the data of Smith and others (1979) and Friedman and Smith (1972), can be quite large, at least as large as the inferred differences found by others between the present and glacial maximum δD and $\delta^{18}O$ values in precipitation.

Lake Sediments

The $\delta^{13}C$ in the organic components of lake sediments has been used by Stuiver (1975) as a climatic indicator. The organic component is largely a mixture of terrestrial and aquatic plant material. The terrestrial organic carbon could vary as a function of climate because land plants fall in two groups having different values of $\delta^{13}C$. The first group consists of most trees and shrubs and has a $\delta^{13}C$ of from $-24‰$ to $-34‰$. The second group is composed mainly of desert and salt-marsh plants and tropical grasses and has a $\delta^{13}C$ of from $-6‰$ to $-19‰$. Changes in climate could shift the mix of plant types and result in a change in the $\delta^{13}C$ of debris entering the lakes.

Floating aquatic plants utilize atmospheric carbon dioxide as their carbon source. Because it is believed that the $\delta^{13}C$ of atmospheric carbon dioxide changed but little during the Pleistocene and Holocene, this source of carbon should have remained rather constant during a climatic change.

The submerged aquatic plants have been shown by Oana and Deevey (1960) to utilize either dissolved bicarbonate or dissolved carbon dioxide. The relative amount of each used by the submerged plants is a function of water hardness. The bicarbonate $\delta^{13}C$ is in equilibrium with atmospheric carbon dioxide and has a $\delta^{13}C$ of from 0‰ to 2‰. The dissolved carbon dioxide in equilibrium with this bicarbonate is depleted in ^{13}C by about 10‰. However, in the lakes studied by Oana and Deevey, the submerged aquatic plants con-

tributed little to the organic matter in recent sediments. The phytoplankton and zooplankton also fractionate ^{13}C, and this fractionation appears to depend upon the availability of dissolved carbon dioxide. In freshwater lakes, water hardness influences this fractionation; in hard-water lakes, the plankton show the maximum depletion of ^{13}C.

Stuiver (1975) has found that, in 4 of 12 lakes studied, $\delta^{13}C$ changes correlate with climatic changes. He believes that, in spite of the complex interplay among variable ^{13}C sources in the sediments and the climate, $\delta^{13}C$ measurements of lake sediments can often yield information about climate.

Stuiver (1970) has also investigated the $\delta^{18}O$ of mollusk shells and marls from freshwater lake sediments. The $\delta^{18}O$ of these samples is determined by both the temperature of formation and the $\delta^{18}O$ of the water from which they formed. The $\delta^{18}O$ of the water, in turn, is a function of the oxygen-isotopic composition of the oceans and ocean-water temperature, as well as the difference in temperature between the source of the precipitation (the ocean surface) and the temperature of the precipitation. In addition, evaporation of the water from a lake surface can change its ^{18}O content.

Stuiver (1975) has been able to measure increases in ^{18}O in four lakes (in Maine, New York, Indiana, and South Dakota) that occurred during the Hypsithermal; this finding suggests a higher temperature during that interval. The post-Hypsithermal $\delta^{18}O$ values are constant and indicate an extremely stable climate (stable storm tracks and temperatures). This conclusion is consistent with the findings derived from underground-aquifer and speleothem data.

In principle, a climatic record should be evident in the deuterium content of hydrous minerals deposited in lake sediments. If one could experimentally determine at various temperatures the isotopic fractionation between a saline brine and a growing solid phase found in a lake deposit, it should be possible to use this information for climatic reconstruction. With this in mind, Matsuo and others (1972) have investigated the isotopic fractionation between brine and borax and between gaylussite and trona, all phases found in deposits from Searles Lake, California. They also have measured the deuterium concentration in brine, borax, gaylussite, and trona separated from Searles Lake cores. Although many consistent relations are found in cores, several conclusions are obvious. (1) The borax, although originally primary, dehydrated pseudomorphously to tincalconite during the altithermal and was then reconstructed (also pseudomorphously) back to borax. These transformations erased the original isotopic record in the borax. (2) The gaylussite is secondary and was formed by the reaction of calcium carbonate (aragonite) with a sodium-bearing brine. The isotopic record in the mineral is not easy to interpret. (3) The trona is probably secondary after natron but formed within months of burial. (4) The brines now present in the lake sediments are not the brines that were originally trapped with the sediment; the originals were replaced by brines that formed later. Fluid inclusions present in halite that is found in most of the salt layers in Searles Lake may provide a sample of the original brine, and future research will investigate this possibility.

Conclusions

The use of stable-isotope ratios of hydrogen, oxygen, and carbon for paleoclimatic reconstructions has just begun. Although some contradictions are evident, the techniques show great promise for the future.

To date, the rapid melting of the Laurentide ice sheet between 17,000 and 11,500 years ago has been confirmed by isotopic evidence. The relative stability of climate from the Hypsithermal to the present also seems to have been demonstrated. The isotopic composition and, therefore, the history of precipitation during glacial buildup prior to 14,000 yr B.P. is still in doubt.

Experimental techniques are now fairly well developed, and the emphasis in the future will be on finding suitable samples that retain a paleoclimatic fingerprint. In addition, more emphasis will have to be placed on achieving an understanding of how various climatic regimes affect the distribution of these isotopes. To this end, a closer rapport between stable-isotope geochemists and meteorologists than has existed in the past will be needed.

References

Burk, R. L., and Stuiver, M. (1981). Oxygen isotope ratios in trees reflect mean annual temperature and humidity. *Science* 211, 1417-19.

Clayton, R. N., Friedman, I., Graf, D. L., Mayeda, T. K., Meets, W. F., and Shimp, N. F. (1966). Origin of saline formation waters. I: Isotopic composition. *Journal of Geophysical Research* 71, 3869-82.

Dansgaard, W. (1964). Stable isotopes in precipitation. *Tellus* 16, 436-68.

Dansgaard, W., Johnson, S. J., and Moller, J. (1969). One thousand centuries of climatic record from Camp Century on the Greenland ice sheet. *Science* 166, 377-81.

Emiliani, C. (1955). Pleistocene temperatures. *Journal of Geology* 63, 538-78.

Emiliani, C., Rooth, C., and Stipp, J. J. (1978). The Late Wisconsin flood into the Gulf of Mexico. *Earth and Planetary Science Letters* 41, 159-62.

Epstein, S., Buchsbaum, R., Lowenstam, H. A., and Urey, H. C. (1951). Carbonate-water isotopic temperature scale. *Geological Society of America Bulletin* 62, 417-26.

Epstein, S., Buchsbaum, R., Lowenstein, H. A., and Urey, H. C. (1953). Revised carbonate-water isotopic temperature scale. *Geological Society of America Bulletin* 64, 1315-26.

Epstein, S., Sharp, R. P., and Gow, A. J. (1970). Antarctic ice sheet: Stable isotope analysis of Byrd Station cores and interhemispheric climatic implications. *Science* 168, 1570-72.

Epstein, S., Yapp, C. J., and Hall, J. H. (1976). The determination of the B/H ratio of nonexchangeable hydrogen in cellulose extracted from aquatic and land plants. *Earth and Planetary Science Letters* 30, 241-51.

Friedman, I., Redfield, A. C., Schoen, B., and Harris, J. (1964). The variations of the deuterium content of natural waters in the hydrologic cycle. *Reviews of Geophysics* 2, 177-224.

Friedman, I., and Smith, G. I. (1972). Deuterium content of snow as an index to winter climate in the Sierra Nevada area. *Science* 176, 790-93.

Hanshaw, B. B., Winograd, I. J., and Pearson, F. J., Jr. (1980). Stable isotope studies of subglacially precipitated carbonates and of ancient ground water: Paleoclimatic implications. *In* "Proceedings of the International Meeting on Stable Isotopes in Tree-Ring Research" (G. C. Jacoby, ed.), pp. 102-4. U.S. Department of Energy CONF-7907180, New Platz, N.Y.

Harmon, R. S., Schwarcz, H. P., and O'Neil, J. R. (1979). D/H ratios in speleothem fluid inclusions: A guide to variations in the isotopic composition of meteoritic precipitation. *Earth and Planetary Science Letters* 42, 254-66.

Kennett, J. P., and Shackleton, N. J. (1975). Laurentide ice sheet meltwater recorded in Gulf of Mexico deep-sea cores. *Science* 188, 147-50.

Libby, L. M., and Pandolfi, L. J. (1974). Temperature dependence of isotope ratios in tree rings. *Proceedings of the National Academy of Sciences (U.S.A.)* 71, 2482.

Matsuo, S., Friedman, I., and Smith, G. I. (1972). Studies of Quaternary saline lakes: I. Hydrogen isotope fractionation in saline minerals. *Geochemica et Cosmochimica Acta* 36, 427-35.

Oano, S., and Deevey, E. S. (1960). Carbon 13 in lake waters and its possible bearing on paleolimnology. *American Journal of Science* 258A, 253-72.

Schiegl, W. E. (1974). Climatic significance of deuterium abundance in tree rings of Picea. *Nature* 251, 582.

Schiegl, W. E., and Vogel, J. C. (1970). Deuterium content of organic matter. *Earth and Planetary Science Letters* 7, 307-13.

Smith, G. I., Friedman, I., Kleiforth, H., and Hardcastle, K. (1979). Areal distribution of deuterium in eastern California precipitation, 1968-1969. *Journal of Applied Meteorology* 18, 172-88.

Stuiver, M. (1970). Oxygen and carbon isotope ratios of fresh-water carbonates as climatic indicators. *Journal of Geophysical Research* 75, 5247-57.

Stuiver, M. (1975). Climate versus changes in ^{13}C content of the organic component of lake sediments during the late Quaternary. *Quaternary Research* 5, 251-62.

Urey, H. C. (1947). The thermodynamic properties of isotopic substances. *Journal of Chemical Society* 1947, 562-81.

Wilson, A. T., and Grinstead, M. J. (1975). Paleotemperatures from tree rings and the D/H ratio of cellulose as a biochemical thermometer. *Nature* 257, 287-88.

Yapp, C. J., and Epstein, S. (1977). Climatic implications of D/H ratios of meteoric water over North America (9500-22,000 B.P.) as inferred from ancient wood cellulose C-H hydrogen. *Earth and Planetary Science Letters* 34, 333-50.

CHAPTER 20

Late-Pleistocene Climatology

R. G. Barry

Introduction

The last decade has witnessed a remarkable upsurge of interest in ice-age climates and their causes. At the popular level, this has given rise to alarmist speculation about an impending cooling and "snow-blitz" inception of renewed glacial conditions or, alternatively, to a "super-interglacial" period resulting from anthropogenic impacts on world climate (Calder, 1974; Ponte, 1976). At the same time, dramatic scientific progress has been made in our basic understanding of the mechanisms of climate. Several factors have contributed to this progress. First, systematic efforts have been made to collate and interpret proxy data on paleoenvironmental conditions worldwide. Second, new or improved analytical and interpretive techniques have become available, backed by radiometric dating and computer methods. Third, meteorologists have become deeply interested in the mechanisms and variability of global climate.

The wealth of information available across a wide range of environmental disciplines makes the presentation of a complete and definitive synthesis of paleoclimatic conditions an almost impossible task. The reader should be prepared to follow up specific details in other chapters or sources. Recent reference works that are pertinent to the topics covered in this chapter include Cline and Hays (1976), Denton and Hughes (1981b), Flint (1971), Imbrie and Imbrie (1979), Lamb (1977), and Pittock and others (1978).

Methods of Paleoclimatic Reconstruction

A brief review of the principal methods of reconstructing paleoclimatic parameters is appropriate. One approach involves the treatment of proxy data sources (i.e., environmental or biologic indexes of climate), and a second involves strictly meteorological considerations.

PROXY SOURCES

Physical-Environmental Sources

The most direct paleoclimatic inferences are obtained from proxy sources that involve only the physical environment. These include evidence of former snow or ice extent, frozen-ground features, lake levels, oxygen isotopes in ice cores and speleothems, and sand dunes. Only phenomena that relate to proxy records in the United States are considered here.

One of the criteria most widely used for assessing glacial climatic severity is the degree of lowering of the snow line in mountains. Snow-line gradient serves as a direct indicator of climatic gradients and an indirect indicator of atmospheric circulation (Barry, 1981a). The calculation of snow line from existing and past determinations of ice masses is described by Flint (1971: 63), Østrem (1974), and Heuberger (1974). Several different definitions and methods of determination appear in the literature (Andrews and Miller, 1972). In an analysis of the southwestern United States, Brakenridge (1978) draws attention to the possible errors arising from the limitations of both the orographic snow line, determined from the altitude of relict cirque floors, and the climatic snow line, estimated from modern climatic data. Because cirques can form during glacial onset, valley glaciers can simply extend to lower elevations as climates become more severe, so that new cirques are not excavated. Hence, orographic snow-line depression based on cirques may underestimate glacial climate in extensively glaciated terrain. In the other case, the modern climatic snow line is above many mountain summits, and so the uncertain extrapolation of climatic parameters is necessary. Moreover, cirques may have been occupied only during earlier glacial cycles.

Differences between modern and (presumed) late-Pleistocene snow lines have been determined by numerous investigators for the western United States and elsewhere. Some of the major sources of information for all the main western mountain ranges are Kaiser (1966) and Flint (1971: 475); these are supplemented or amended for individual ranges in the West by Crandell (1965), Porter (1964, 1977), Richmond (1965), Wahrhaftig and Birman (1965), Brakenridge

I wish to thank J. E. Kutzbach for reading this chapter in its entirety and W. D. McCoy, L. D. Carter, H. R. Delcourt, P. A. Delcourt, T. D. Hamilton, D. M. Hopkins, and K. L. Pierce for their comments and suggestions on individual sections. I am indebted to Mrs. M. Strauch for patiently typing the manuscript and to R. MacDonald for drafting the figures.

(1978), and Zwick (1980); for Alaska by Péwé (1975); and for Hawaii by Porter (1979). A summary of this information is provided in Table 20-1. (See also Hamilton and Thorson, 1982, and Porter et al., 1982.)

The climatic interpretation of snow-line lowering is conventionally made in terms of an estimate of reduced mean summer temperature based on mean atmospheric lapse rates. Free-air lapse rates are not necessarily identical with station lapse rates, although they are usually close (Barry, 1981b). Moreover, ice bodies cause some local atmospheric cooling, which appears to increase at altitudes above 3 km (Hess, 1973). The relationship between climatic variables and snow line (equilibrium-line altitude or glaciation threshold) has received insufficient scrutiny, however. Brakenridge (1978) shows that in the southern Rocky Mountains (latitude 40°N to 46°N) the present orographic snow line matches the slope and elevation of the $-6°C$ ground-level annual isotherm much better than the 0°C July free-air isotherm. For the Cascade Range in Washington State, Porter (1977) demonstrates that the inland rise of the glaciation threshold correlates best with the mean precipitation of the accumulation season ($r^2 = 0.86$), compared with a much poorer fit with July freezing level ($r^2 = 0.40$). Pierce reports a similar relationship with accumulation for a transect from the Pacific coast to Montana along latitude 44°30′N (Porter et al., 1982). Consequently, the climatic changes inferred from the snow-line lowerings shown in Table 20-1 must be treated with considerable caution.

Isotopic studies of ice cores, from which atmospheric temperatures have been estimated, are not discussed here, since they do not relate specifically to the United States. However, it is worth noting the potential for making paleotemperature estimates from isotopic analysis of speleothems (cave deposits of calcite). Thompson and others (1974) identify the timing of minimum temperatures during

the Wisconsin in West Virginia as 70,000 to 60,000 yr B.P. and 30,000 to 6000 yr B.P. on the basis of uranium-thorium dating of intervals when there was an absence of cave calcite deposition. Harmon and others (1979a) place the major temperature minimum in northeastern Iowa at 17,000 ± 2000 yr B.P., also on the basis of uranium-thorium dating. Changes in meteoric precipitation inferred from deuterium ratios in speleothem fluid inclusions are at present too uncertain to be of much value. Harmon and others (1979b) note the effects of coastal-inland temperature gradients, annual temperature range, storm-track location, and isotopic composition of ocean waters on the glacial-interglacial range of values. (See Friedman, 1982.)

Another widely used indicator of paleotemperature regime is former periglacial phenomena (Péwé, 1982), although Black (1976a: 93) states that "of all those fossilized phenomena, perhaps the only truly diagnostic forms reflecting the temperature at time of formation are ice wedges." Washburn (1980b) proposes a mean annual temperature isotherm of $-5°$ for ice-wedge casts. Péwé (1973) summarizes periglacial evidence adjacent to the Late Wisconsin Laurentide ice margin, but Black (1976b) is critical of many of the purported sites. Ice wedges could be expected close to the ice margin in areas that had light snowfall and strong winds, and locations in the High Plains east to North Dakota obviously satisfy these requirements. Mears's (1981) findings in Wyoming support this idea.

The most extensively used indicator of precipitation is the occurrence of high lake shorelines or lake-sediment evidence of higher lake levels. Past "pluvial" conditions have been inferred from such evidence in many tropical and subtropical regions. More-reliable information is now becoming available as a result of firmer radiometric dating of high lake stands (Smith and Street-Perrott, 1982).

A preliminary global synthesis for the last 30,000 years by Street

Table 20-1.
Differences (in Meters) between Modern and Late-Pleistocene Snow-Line Altitudes

Location	Latitude	Basis	Δh (m)	Source
Alaska:				
Brooks Range	67°-68°	—	300 (W)-100 (E)	Porter et al., 1982
Alaska-Chugach Range[a]	62°-63°	Orographic snow line[b]	200	Péwé, 1975
Kenai Peninsula[a]	60°	Orographic snow line[b]	800	Péwé, 1975
Western Cordillera:				
Olympic Mountains[a]	48°	Orographic snow line	1300	Kaiser, 1966
North Cascades[a]	47°-48°	Orographic snow line	1200	Kaiser, 1966
North Cascades	47°-48°	Glaciation level	900 ± 100	Porter, 1977
Klamath Mountains[a]	40°-41°	Climatic snow line	600	Wahrhaftig and Birman, 1965
Sierra Nevada[a]	37°-38°	Climatic snow line	750	Wahrhaftig and Birman, 1965
Sierra Nevada[a]	37°-38°	Orographic snow line	1000	Kaiser, 1966
Rocky Mountains:				
Yellowstone	44°45′	ELA[c]	900-1200	Porter et al., 1982
Wind River Mountains[a]	43°	Orographic snow line	1100	Kaiser, 1966
Front Range, Colorado[a]	40°	Orographic snow line	800	Kaiser, 1966
Front Range, Colorado[a]	40°	ELA	1000	Haefner, 1976
Sangre de Cristo Range[a]	38°	Orographic snow line	900	Kaiser, 1966
Hawaii:				
Mauna Kea.	20°	ELA	935 ± 190	Porter, 1979

[a]Age uncertain.

[b]Termed "climatic snow line" but estimated from the lowest north-facing cirque floors.

[c]ELA means equilibrium-line altitude.

and Grove (1979) shows that, in Australia and intertropical Africa, the last glacial maximum was a time of low lake levels and the early Holocene was a time of high stands. Lake levels in the western United States were generally low from about 40,000 to 25,000 yr B.P. and high from about 20,000 to 11,000 yr B.P., and then, after a sharp decline, they remained low until the present (Smith, 1976; Mehringer, 1977; Benson, 1978; Scott et al., 1980).

The hydrologic conditions required for the existence of lakes in the Great Basin and southern High Plains are widely interpreted as implying increased precipitation and runoff as a result of lower temperatures that caused evaporation to decrease (Schumm, 1965; Reeves, 1966, 1973). The temperature reduction during glacial times has been *assumed* to be much greater in summer than in winter. In a test of hydrologic budget equations formulated by Leopold (1951) and Snyder and Langbein (1962), Weide and Weide (1977) show that both models give similar results for the Warner Lakes of southern Oregon. Lakes could exist when a 5 °C reduction in mean annual temperature and a 33% increase in precipitation are assumed. By assuming a 10 °C to 11 °C reduction, Galloway (1970) concludes that Lake Estancia, New Mexico, could be maintained even with a 10% to 20% decrease in precipitation. Careful reassessment of snow-line lowering, relict cryogenic deposits, and timberline changes in the Southwest leads Brakenridge (1978) to infer a minimum cooling of 7 °C to 8 °C in summer *and* winter. A recalculation of Leopold's and Galloway's budgets that uses this value leads to the conclusion that maximum lake levels can be sustained with precipitation close to present amounts.

Kutzbach (1980) demonstrates that possibility of modeling the energy and hydrologic balances of past and present lakes. In simulations for the lake basin of paleo-Lake Chad in West Africa, linear and nonlinear models are developed, the latter based on the climatonomic approach of Lettau (1969). Kutzbach avoids the direct estimation of lake evaporation from temperature estimates and uses radiation parameters instead. P. A. Kay (personal communication, 1981) has made a perliminary attempt to apply this formulation to Lake Bonneville. With modifications to include seasonal effects and moisture recycling across North America, the methods offer considerable promise for future studies of paleohydrologic problems.

BIOLOGICAL SOURCES

The major source of existing paleoclimatic evidence from the continental United States for the late Quaternary is palynologic material in lake sediments or peat deposits. Current work is adding alternative evidence from insect assemblages (Morgan et al., 1982), pack-rat middens (Wells, 1976), and mollusks that confirms or modifies the more-traditional biologic methods. In addition, objective multivariate computational approaches to the paleoclimatic interpretation of biologic data have become more common since the development of transfer function techniques (Webb and Bryson, 1972; Bryson and Kutzbach, 1974). A review of the various methods and their relative advantages is given by Sachs and others (1977). The basic model utilizes linear algebraic equations that can be solved by a variety of methods, including linear multiple regression, stepwise regression, principal-components analysis plus multiple regression (Imbrie and Kipp, 1971), and canonical correlation (Webb and Bryson, 1972). Sachs and others suggest that, given adequate data on modern environmental conditions and modern representation of the fossil assemblage (i.e.,

without substantial evolutionary change), useful transfer functions can be developed. Most of the algorithms yield comparable results, but the use of several on the same data facilitates the discovery of "no-analogue" situations. These can arise when the calibration data from the modern environment are unable to represent some element of the paleoenvironment. The same type of problem could also occur if some biologic evolutionary changes had taken place, particularly if a species had become extinct. Both problems are more likely to be involved in studying Pleistocene environments.

SYNOPTIC-DYNAMIC CLIMATOLOGICAL STUDIES

Empirical Approaches

Until recently, reconstructions of past climate were qualitative and based largely on analogue techniques. Attempts to map hypothetical patterns of atmospheric circulation under glacial conditions were made as early as the turn of the century by Harmer (1901), and several meteorologists have endeavored to update this type of approach (Willett, 1950; Lamb, 1964). Such reconstructions often utilize the idea of selecting present extremes—snowy winters, cool summers—and analyzing the associated circulation anomalies as an analogue of past regimes. This assumes that the range of modern climatic conditions contains extreme events that were more frequent in the past and gave rise to a significantly different mean regime. For radically different external and internal boundary conditions in the climate system, such as occur during glacial intervals, this assumption is invalid, at least on a global basis. Regionally, however, analogue methods may still be helpful in ascertaining some of the possible anomaly patterns.

The idea that changes in the frequency of the atmospheric circulation pattern are the primary cause of climatic fluctuations is well established, but it merits critical examination. Recent investigations suggest that even short-term regional anomalies of temperature and precipitation cannot always be closely tied to changes in circulation pattern (Barry, 1981a). There appear to be significant changes in the climatic characteristics of individual circulation types themselves; this fact suggests that multiple causes may be involved. Changes in surface characteristics may occur, and/or larger-scale controls may override the regional effects. For example, an analysis of recent climatic fluctuations in the western United States in relation to synoptic circulation types shows that during extremely warm or cold seasons there is a tendency for all types to be warmer or colder than their average values (Barry et al., 1981).

An alternative method of circulation reconstruction has been pioneered by Bryson (1966; Bryson et al., 1970). First, he demonstrated close spatial correspondence between vegetation boundaries and air-mass boundaries. The boreal forest-tundra ecotone, for example, is closely followed by the median location of the boundary of arctic air in summer, a relationship that also holds in Eurasia (Krebs and Barry, 1970). By extension, Bryson and others have traced the movement of air-mass boundaries during the late Pleistocene and Holocene on the basis of mapping paleobotanical evidence of vegetational distribution.

In a temporal context, changes in several climatic parameters including air-mass dominance have been reconstructed from a palynologic record at Kirchner Marsh, Minnesota (Webb and Bryson, 1972). Figure 20-1 illustrates some features of this analysis.

Given a set of such point reconstructions throughout North America, the earlier qualitative assessments could be much refined to provide an alternative approach to numerical circulation models of the type described next.

Numerical-Model Studies

A semiempirical approach to paleoclimatic reconstruction using thermodynamic principles was first developed by Lamb and his colleagues (Lamb et al., 1966; Lamb and Woodroffe, 1970). The procedure requires paleoenvironmental estimates of surface air temperature for the warm and cold seasons at specified times in the past over most of the Northern Hemisphere. From the isotherms so obtained, a map of 1000-500 millibar thickness is constructed (the thickness of this lower tropospheric air layer is proportional to its mean temperature). In turn, the thickness map is used to infer areas of cyclogenesis and anticyclogenesis on the basis of dynamic theory. Finally, a schematic mean surface-pressure distribution is prepared from the preceding information. By this means, maps of January and July circulation have been prepared for 20,000 yr B.P., 11,500 yr B.P., and five intervals during the Holocene. More recently, Chappell (1978) has followed this procedure to examine possible circulation modes for glacial inception in Canada. By examining the January circulation simulated with estimated land and sea ice at 10,500 yr B.P. compared with the same land conditions but with modern sea-ice boundaries, he shows that the latter sets up a strong Newfoundland low center with moisture influx towards Labrador-Ungava. Although the results from such analyses are much less precise and detailed than those obtainable by numerical models, the approach allows the use of information as input that is quite heterogeneous when compared with what is currently required for experiments using general circulation models (GCMs).

There is a wide variety of modeling strategies for climatic-change studies, and many of the results have only limited applicability to specific reconstructions of late-Pleistocene conditions. A general introduction to GCM experiments is provided by Barry (1975, 1981a).

There have been three principal simulations of global climate at the last glacial maximum: those of the National Center for Atmospheric Research (NCAR), (Williams et al., 1974), the Rand-Oregon State University (Rand-OSU) (Gates, 1976a, 1976b), and the Geophysical Fluid Dynamics Laboratory (GFDL) (Manabe and Hahn, 1977). The boundary conditions used for the NCAR-model experiment were derived from the available literature; those for the other two experiments used the first set of CLIMAP data (CLIMAP Project Members, 1976). All models assumed fixed ocean-surface conditions, thereby eliminating potentially crucial atmosphere-ocean feedbacks.

Comparisons among the three experiments are not readily made because of important differences in the input data and in the model properties. For example, the NCAR experiment underestimated the intensity of sea-surface temperature gradients in the middle latitudes at 18,000 yr B.P.; in the GFDL experiment a coding error resulted in a deep depression in the ice over Hudson Bay, although, ironically, the erroneous pattern might be close to some recent interpretations of ice-sheet topography in the area. The various results have been examined by J. Williams (1979) and Heath (1979), and some general points have emerged. For July in the Northern Hemisphere, the models agree that generally cooler, drier conditions existed over the continents. The zonally averaged jet-stream system moved equator-

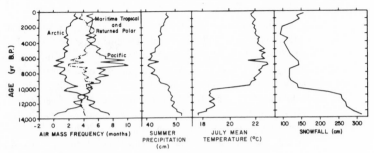

Figure 20-1. Reconstructions of late-glacial and Holocene July mean temperature, summer precipitation, and annual snowfall at Kirchner Marsh, Minnesota, based on pollen-climate transfer functions. (After Webb and Bryson, 1972, and Bryson and Kutzbach, 1974.)

ward and was strengthened in association with the major middle-latitude cyclone tracks. More detailed intermodel comparisons will necessitate an experiment in which several models use the same (revised) CLIMAP data base.

Pleistocene Climatic History

THE LATE-CENOZOIC SETTING

It is now usual to refer to the late-Cenozoic glacial ages rather than simply Quaternary glacial ages, and it is apparent that these events reflect the culmination of a shift in paleogeographic conditions spanning about 50 million years. Donn and Shaw (1977) identify Northern Hemisphere climatic trends as being linked with the northward displacement of North America and Eurasia during the Mesozoic and Cenozoic. However, although this displacement may have set the stage by steepening the poleward temperature gradient, events in the Southern Hemisphere also seem to have been involved. Temperatures in high-southern-latitude oceans fell sharply during the Eocene, apparently as part of the global cooling trend. This high-latitude cooling resulted in the formation of Antarctic bottom water with near-freezing temperatures about 38 million years ago (Savin et al., 1975; Shackleton and Kennett, 1975b). The separation of Australia from Antarctica and the later opening of the Drake Passage between South America and Antarctica finally allowed a circum-Antarctic deepwater circulation to be established about 23 million years ago (Kennett, 1980). The significant buildup of Antarctic ice began in the middle-Miocene (14 to 10 million years ago), and by the late Miocene (6 to 5 million years ago) the Antarctic ice volume may have been up to 50% greater than it is now (Shackleton and Kennett, 1975a). Schnitker (1980) argues that this growth and the subsequent decrease in the amount of Antarctic ice by early-Pliocene time were associated with changes in global oceanic circulations, but it seems likely that the Antarctic ice sheet underwent progressive "starvation" due to its size (including sea-ice cover) and its distance from sources of moisture. After the decrease in the amount of Antarctic ice, full-scale global glaciation was triggered by Pliocene tectonic activity. The complete closure of the Isthmus of Panama about 3.5 million years ago strengthened the Gulf Stream system, and a renewed link with the Mediterranean Sea made the North Atlantic more saline. The warmth of the North Atlantic seems to be a key to continental glaciation in adjacent land areas.

The earliest land evidence for glaciation in North America comes from the Wrangell Mountains in Alaska, where tillites are interbedded with volcanic rocks dated radiometrically as 10 to 8 million years old, with another series dated as 3.6 million years old (Denton and Armstrong, 1969). In California, the Deadman Pass till has also been dated by potassium-argon methods as about 3 million years old (Curry, 1966). Strong glacial cycles and the establishment of the modern oceanic circulation are dated from 3.2 million years ago (Shackleton and Opdyke, 1977). The continental ice sheets reached two-thirds the size of the Late Wisconsin ice by about 2.4 million years ago, according to Herman and Hopkins (1980). The Arctic Ocean's pack-ice cover is thought by Clark (1971) to have formed at this time, although Hunkins and others (1971) and Herman and Hopkins (1980) date its inception at 0.7 million years ago. Moisture from this source does not appear to be involved in the growth of the North American or Fennoscandian ice sheets during the late Pleistocene.

THE QUATERNARY PERIOD

The lower stratigraphic boundary of the Pleistocene, conventionally identified by the first occurrence of cold marine (Calabrian) fauna in southern Italy, is arbitrary in a global climatic context (Berggren and van Couvering, 1974). Any cooling event is likely to be geographically metachronous, and pronounced cold intervals occurred earlier elsewhere. The Pliocene/Pleistocene boundary is commonly dated at about 1.8 to 1.6 million years ago, the age of the Olduvai normal-polarity subzone (Bandy, 1972); coincidentally, this definition accords quite closely with the present dating of the first major glaciation (Nebraskan) in North America (Haq et al., 1977).

In a global context, the Quaternary is marked first by widespread middle-latitude continental glaciation and second by pronounced and recurrent glacial/interglacial oscillations. During at least the last 500,000 years, these oscillations have a time scale of approximately 100,000 years. Oxygen-isotope records from deep-sea sediments, which reflect global ice volume, show 21 cycles in the last 2.3 million years, according to van Donk (1976). In an independent study, Shackleton and Opdyke (1976) identify 9 cycles within the Brunhes paleomagnetic epoch spanning the last 0.73 million years. The European loess record indicates 17 "interglacial" intervals since the Olduvai epoch (1.7 million years ago) (Fink and Kukla, 1977). Kukla (1978) contends that the classical 4 (or 5) glacial stages in Europe and North America represent misinterpretations of the glacial-geomorphological evidence.

An impressive element of the late-Pleistocene record is the fact that conditions as warm as those of the present (Holocene) interglacial interval have typically lasted for only about 10% of each glacial/interglacial cycle (Emiliani, 1972). Also striking is the apparent rapidity of the glacial terminations in the deep-sea isotopic record (Broecker and van Donk, 1970) and of interglacial/glacial transitions in other oceanic and land records. Several rapid cooling events spanning only a few hundred years with an intensity of up to 5 °C per 50 years have been identified (Flohn, 1979), although such events are by no means confined to the interglacial/glacial transitions.

Because problems in stratigraphic interpretation and dating make interregional and land-sea stratigraphic correlations uncertain, discussions of possible climatic regimes occurring before the last glaciation cannot be undertaken. On the basis of the supposed similarities in the extent and pattern of earlier ice limits in the United States (Flint, 1971: Figure 18.11), we can infer broadly similar climatic controls, although there may have been some differences in the case of mountain glaciation of the American West.

The inception of ice sheets in North America after the last in-

terglaciation (isotope stage 5e, about 125,000 to 118,000 yr B.P.) has been described recently by Andrews and Barry (1978), who postulate on the basis of inferred global ice volumes that the initial ice-sheet buildup began about 115,00 yr B.P. However, there is no unequivocal field evidence of ice in North America until immediately before the St. Pierre interstade, 85,000 yr B.P. From climatological and glaciologic considerations (Barry et al., 1975; Andrews and Mahaffy, 1976; L. D. Williams, 1978, 1979), ice caps are thought to have formed first on upland plateaus in the eastern Canadian Arctic, although Denton and Hughes (1981a) argue for an ice-shelf origin of ice domes. Ice in the Queen Elizabeth Islands probably was never connected with ice farther south and, because of its distance from sources of moisture in the Atlantic, must have remained a modest size (England, 1976). Presumably, the Greenland ice sheet originated earlier and remained essentially stable, although no estimate has yet been made of the timing of this event.

As ice caps built up in Baffin Island-northern Keewatin and northern Labrador-Ungava (Andrews and Barry, 1978: Figure 3), ice on the more-northern Arctic islands must have virtually ceased to grow because of the decreased ability of cyclonic systems and moisture to move northward. After a recession during the St. Pierre interstade, major ice-sheet growth occurred rapidly during the oxygen-isotope stage 5/4 transition (about 75,000 years ago). Ruddiman and others (1980) and Ruddiman and McIntyre (1981) suggest that the surface of the North Atlantic was relatively warm and thereby provided the necessary moisture, although it is not yet clear whether this warmth extended into the Labrador Sea-Davis Strait region (T. Kellogg, personal communication, 1980). However, Fillon and Duplessy (1980) indicate that the eastern Labrador Sea was ice free at least in summer during episodes of ice growth about 75,000 years ago and parts of the late Pleistocene (perhaps 25,000 to 18,000 and 15,000 to 13,000 years ago). The general model proposed by Ruddiman and others lends support to the earlier suggestions that moisture was associated with a southeasterly air flow (cf. Chappell, 1978) and cyclones moving along the Labrador coast into Davis Strait (Barry, 1966; Lamb and Woodroffe, 1970; Brinkmann and Barry, 1972). Ives and others (1975) propose that upland plateaus in the eastern Canadian Arctic and sub-Arctic probably underwent "instantaneous glacierization" owing to snow-line lowering. Koerner (1980) shows that such a lowering does not cause a step change in regional albedo so that an augmented feedback does not necessarily result. However, the general argument remains unchanged. More problematic is the apparently persistent nature of the large-scale circulation regime required to account for rapid but prolonged ice accumulation (Barry, 1981a).

The Laurentide ice sheet seems to have expanded rapidly westward and southward coincidently with the growth of the Cordilleran ice sheet (about 75,000 years ago). It is possible that surge mechanisms were involved in this process (Andrews and Barry, 1978), because on climatic grounds substantial ice accumulation in the Prairie region is not easily explained. During the long Port Talbot interstade (70,000 to 45,000 years ago), the Laurentide ice sheet may have retreated to a "core area" (similar to the one that existed about 90,000 years ago) and then advanced rapidly again after 25,000 years ago.

Recent field studies of till trains around Hudson Bay have led Shilts (1980) to conclude that, during the Late Wisconsin maximum, Keewatin ice formed the earliest major center, followed by the Laurentide center, and that there may have been no later Hudson Bay dome as had been suggested by earlier investigators. Mayewski and others (1981: 118) point out that dispersal centers for erratics need not be locations of ice domes when the ice sheet has zones where the ice is frozen to the bedrock. However, evidence that Hudson Bay was ice free along its southern shore approximately 105,000, 75,000, and 35,000 years ago (Shilts et al., 1981) points to the need to consider seriously a multidome model of the Laurentide ice sheet. The possible climatic implications of this recent evidence have not yet been evaluated.

The most completely documented paleoclimatic event during the Pleistocene is the last glacial maximum about 18,000 yr B.P. World maps of paleoenvironmental variables for February and August 18,000 yr B.P. have been compiled by the CLIMAP project members (1976). These include sea-surface temperatures and sea-ice extent, based on assemblages of foraminiferal and other marine faunal records, land-surface albedos from palynologic evidence of vegetational changes, and extent and thickness of ice masses (Denton and Hughes, 1981b). These data have been used as primary inputs for simulations (using numerical models) of the atmospheric circulation and meteorological parameters.

The land evidence for the last glacial maximum that is reliably dated within 5000 years is remarkably meager; even in North America there are only 31 well-dated proxy sites (Peterson et al., 1979). For mean annual values, the temperature depression ranges between about 3°C and 8°C below present in the United States. The largest depressions probably occurred in the northeastern and Great Lakes regions near the Laurentide ice sheet's margin. Washburn (1980b) suggests a mean annual temperature decrease of 12°C to 15°C in these areas; but, on the basis of recalculations of snow-line lowering, Pierce (in Porter et al., 1982) now proposes decreases up to 14°C to 17°C in the Rocky Mountains. It appears to have been moister in the deserts of the Southwest at 18,000 yr B.P. and drier in the Southeast and Alaska; in the Midwest, different sites record a variable pattern of changes.

In the global context, the CLIMAP reconstructions show an average reduction of sea-surface temperature of only 2.3°C, compared with 5.3°C for the average reduction of ground temperature in the Northern Hemisphere. Maximum ocean cooling (5°C to 15°C) occurred in northern high latitudes, contrasting with little or no cooling in the subtropical gyres and creating a strong middle-latitude thermal gradient.

In several respects, 18,000 yr B.P. is a less-than-ideal reference point. From a climatological standpoint, solar radiation values for the caloric northern summer at latitudes 60° to 70°N were then relatively close to modern levels, having reached a minimum some 5000 to 7000 years earlier (Berger, 1980). Also, it must be presumed that the forces that led to deglaciation were already beginning to be felt at 18,000 yr B.P., so that the climate was not long in equilibrium with the ice extent at the glacial maximum.

The brevity of any purported equilibrium is apparent from glacial and geomorphic evidence. Various North American records suggest that interstadial conditions occurred quite widely in the time interval 22,000 to 19,500 yr B.P. Moreover, deglaciation appears to have begun about 17,000 yr B.P. in the Great Lakes region (Dreimanis, 1977) and New England, although this trend was briefly reversed by several subsequent ice readvances. In western Ohio, the ice margin retreated about 320 km between 18,000 and 14,000 yr B.P., and,

when minor readvances are considered, this represents an average of perhaps 25 years per kilometer (Forsyth and Goldthwait, 1980). The earliest expansion of deciduous forest marking a vegetational adjustment to climatic amelioration took place about 16,500 yr B.P. in the southeastern United States (Delcourt and Delcourt, 1979).

Regional Features of Late Wisconsin Climate in the United States

In this section, interpretations of the inferred regional climatic characteristics of the United States during the Late Wisconsin are discussed and compared, where possible, with expected patterns based on meteorological considerations.

ALASKA

Environmental conditions at 18,000 yr B.P. ranged from mountain glaciation, which was extensive in the Alaska and Chugach Ranges but less pronounced in the Brooks Range, to a problematic arctic steppe or steppe tundra on the Beringian lowlands, which was made more extensive by the lowered sea level. A detailed examination of conditions in Beringia was recently carried out through a Wenner-Gren symposium on the paleoecology of the arctic steppe-mammoth biome (Hopkins et al., 1982).

Pollen spectra indicate a unique flora including herbaceous elements with steppe affinities and tundra species (Matthews, 1974a). At present, such a flora occurs in vestigial form only in a few small areas of northeastern Siberia. There has been considerable debate about the climatic implications of this flora. The palynologic evidence may be misleading, for several of the genera represented include species with diverse environmental tolerances. Matthews (1974b) identifies the occurrence of prairie insects at Deering, Alaska, about 13,000 to 12,000 yr B.P., but these may reflect a treeless environment rather than one analogous to the modern steppes. The vegetational associations also probably comprised a mosaic related to local topoclimatic and soil factors; such a landscape would provide the diverse ecologic niches necessary for the varied megafauna.

Hopkins (1979) postulates the occurrence of short but warmer and drier summers than at present and long, cold winters for the Late Wisconsin in Beringia. However, the evidence for warmth, other than in locations where föhn winds were involved or on south-facing slopes, does not seem compelling. Bowling (1979) suggests that the high-amplitude, long-wave pattern of the 1976-1977 winter season, which caused a southerly flow at 700 mb and heavy snowfall on the coastal ranges and the Alaska Range and dry conditions in the lowlands, provides a synoptic model for the last glacial maximum.

At present, dry, mild summers in interior Alaska are also associated with a southerly to southeasterly flow accompanying a tilted northwest-southeast 700-mb ridge over Alaska (Streten, 1974). This general type of pattern is proposed as a model of late-glacial summer circulation (Barry, 1982) because it can account for *relatively* warm, dry conditions in the interior lowlands and conditions that facilitate glacier decay, at least in the Brooks Range. Dune fields in the lower Tanana Valley dating from the late glaciation, as well as dunes near Cape Espenberg on the Chukchi Sea coast, indicate northeasterly prevailing winds, as at present during winter, although the season of sand movement is uncertain (D. M. Hopkins, personal communication, 1981). Broadly contemporaneous dunes on the northern Coastal Plain suggest east-northeasterly winds (Carter, 1981). These may reflect summer and winter sand movement, because the snow cover appears to have been very thin (L. D. Carter, personal communication, 1981). These surface-wind directions are broadly compatible with the suggested upper-level mean southeasterly flow over southern Alaska.

The snow-line lowering inferred by Péwé (1975) is now considered to represent *Early* Wisconsin glaciation, at least in the Brooks Range (Porter et al., 1982); for 18,000 yr B.P., a minimum lowering of 300 m at the western end and only 100 m at the eastern end of the Brooks Range is suggested. This implies a minimum cooling of 1 °C to 2 °C, although the aridity suggested by the vegetational evidence (Cwynar and Ritchie, 1980) would explain the limited snow-line depression if the temperature reduction were greater. The similarity of modern and Late Wisconsin snow-line gradients also implies that there was no substantial change in the source of moisture apart from a reduced contribution from the Bering-Chukchi Seas resulting from the exposure of the continental shelf. Hollin and Schilling (1981) note that the suggested southerly source of moisture for the Itkillik II Glaciation (around 24,000 to 17,000 yr B.P.) in the Brooks Range (Hamilton and Porter, 1975) may only represent a topographically induced asymmetry in the ice distribution; however, the reconstructed glaciation threshold for this glaciation parallels the modern one (Porter et al., 1982). Today, the Gulf of Alaska and the North Pacific Ocean are the only likely sources of major moisture input to the Alaskan glaciers. Presumably, accumulation-season snowfall occurred in the Brooks Range, with occasional systems advecting moisture from a southwesterly direction and perhaps also in summer associated with local land evaporation during the melt season. (Compare the modern evaporation data of Weller and Holmgren, 1974.) At present, orographic effects yield total annual amounts of precipitation in the Brooks Range of about 50 cm, or almost double the estimated true amounts received on the Arctic Coastal Plain (Wendler et al., 1974; C. S. Benson, personal communication, 1981).

Climatic amelioration in Alaska was pronounced by about 14,700 yr B.P., when birch-shrub tundra replaced steppe tundra at Birch Lake (latitude 64°N, longitude 147°W) (Ager, 1975), although late-glacial ice readvances did occur in the Brooks Range about 12,800 to 12,500 yr B.P. (Nelson and Creager, 1977; Porter et al., 1982). This fluctuation may reflect the availability of additional moisture. Dunes on St. Lawrence Island indicate a southeasterly airflow between about 12,000 and 10,000 yr B.P., according to D. M. Hopkins (personal communication, 1981); conditions were also relatively wet, despite the movement of sand.

HAWAII

The Pleistocene glaciations of Hawaii provide an interesting contrast with Alaska in view of the large decrease in the snow line necessary to form and sustain an ice cap that extended down to 3010 m. Porter (1979; Porter et al., 1982) infers a 935 ± 190 m drop in the snow line for the last glacial maximum (Late Makanaka) on Mauna Kea, which implies a decrease of approximately 5 °C in summer temperature on the mountains. This figure contrasts with the estimated 1 °C to 2 °C reduction of August sea-surface temperature at 18,000 yr B.P. (CLIMAP Project Members, 1976). Kraus (1973) argues that a small decrease in tropical sea-surface temperature causes an approximately threefold reduction in equivalent potential temperature (θ_e), which is the quantity conserved when air rises *without dilution by entrainment* in tropical cumulonimbi. Hence, a slight cooling of tropical surface waters theoretically can be amplifed at higher levels.

However, model results obtained by Rowntree (1976: 588) show no such amplification. In particular, cooling seems to be limited to a shallow surface layer. (Potential temperature [θ] is the absolute temperature attained by an air parcel brought to 1000-mb pressure by a dry-adiabatic process [i.e., warmed 9.8 °C per kilometer of descent]. Equivalent potential temperature [θ_e] is the maximum potential temperature of an air parcel brought first to its dew point temperature by a saturated adiabatic process. Air at 1000-mb pressure, in thermal equilibrium with the sea surface and having $\theta = 301$ K, has a value of $\theta_e = 374$ K, compared with $\theta_e = 358$ K for $\theta = 298$ K.)

Vertical contrasts in inferred cooling are also reported by Webster and Streten (1978) in the analogous case of New Guinea. Apparently, they discount Kraus's argument and suggest that either there must be errors in the CLIMAP reconstructions (since the land evidence from the high mountains is drawn from several independent sources) or steeper mean lapse rates reflect an increased frequency of synoptic circulations permitting cold-air incursions associated with meridional upper-air trough patterns.

At present, snow falls on the highest mountain summits in the Hawaiian islands three or four times per month in January-February (C. S. Ramage, personal communication, 1981), with a tendency for occasional falls 30 cm deep in October and May. Patches may extend to 2200 m but rarely persist below 3500 m. The annual total precipitation at the Mauna Loa Observatory is only 475 mm, with a seasonal maximum in winter caused by cold-front storms or upper-level disturbances associated with cutoff lows in the upper westerlies. The dominant trade-wind inversion, which is located at about 1800 m elevation in summer, suppresses orographic-convective cloud development and gives rise to a mean annual potential moisture deficit of 690 mm at the Mauna Loa Observatory (Juvik et al., 1978), although there is a secondary rainfall maximum in August associated with disturbances in the easterlies.

Porter (1979) notes that the ice cap descended to lower limits on the southeastern side of Mauna Kea, which implies a wind pattern and moisture source similar to those of the present but with a weaker and/or less-persistent trade-wind inversion and more frequent easterly disturbance activity. This would permit enhanced orographic-convective cloud development and precipitation. Presumably, also, frequencies of winter systems were more like extreme seasons in the present period, with six to eight cold fronts, several subtropical cyclones, and other upper-level lows.

THE PACIFIC NORTHWEST AND THE WESTERN CORDILLERA

Until recently, most paleoenvironmental information for this region was obtained from the glacial-geomorphologic record. Late Wisconsin glaciation involved Cordilleran ice incursions from British Columbia, local mountain ice caps, and piedmont and alpine glaciers (Porter et al., 1982; Waitt and Thorson, 1982). The record is unusual in that the main phase of the last (Fraser) glaciation (about 15,000 to 14,000 yr B.P.) postdates the conventional 18,000 yr B.P. date for the Late Wisconsin maximum. It is climatically significant that the Puget lobe was in near equilibrium according to Thorson (1980). Consequently, glacier surges were probably not involved in the advance of this lobe.

For the southern part of the North Cascade Range in Washington State, Porter (1977) calculates that a 900-m lowering of the glaciation threshold during the Fraser Glaciation would have required a reduction of 5.5 °C in mean ablation-season temperature, with precipitation

20% to 30% less than it is now. This assessment is also supported by Heusser's (1977) palynologic evidence from the Olympics-Cascades area and his proposal that, during the Fraser Glaciation, this region must have resembled the Mount St. Elias region today. This implies that there was less precipitation then than there is now in western Washington. Heusser originally proposed an average July cooling of at least 7 °C during the last glaciation in the Olympic Peninsula, but this hypothesis has been revised to about 3 °C to 4 °C on the basis of a transfer-function analysis of palynologic data (Heusser et al., 1980). One of the main conclusions from this analysis is that conditions were cold and dry between 28,000 and 13,000 yr B.P. except for a short-lived moist episode about 18,000 yr B.P. (Figure 20-2A).

The extent and timing of glaciation in the Sierra Nevada are not well known, although valley glaciers in the higher southern and central parts of the mountains were larger and reached lower elevations on the western slopes than on the eastern side (Wahrhaftig and Birman, 1965; Porter et al., 1982). For the Cascade Range of Oregon, where ice fields covered the higher sections during the Fraser Glaciation (Crandell, 1965), Scott (1977) estimates a 950-m lowering of equilibrium-line altitudes. The Wisconsin (?) "orographic" snow line determined by Kaiser (1966) for these ranges shows a steady rise with deceasing latitude south of 42°N, but it is interesting to note that his depiction of the modern snow line shows an even steeper rise in the southern Sierra Nevada, which implies a greater snow-line lowering in the latter area.

The climatic requirements for ice accumulation at Donner Pass (altitude 2200 m, latitude 39°N) have been examined in detail by Curry (1969), who shows that, given cool and cloudy summers, a winter snowfall of 12 m would allow snow cover to persist through the summer. The present average is 10.2 m, and the difference represents perhaps two additional storms per season. Curry suggests that such a difference, persisting for 20 to 60 years, could account for the late-Neoglacial ice advances. About 20% of the annual precipitation is not accounted for in the April-first snowpack figures used by Curry and with cooler summers some of this would fall as snow; however, the overall evidence favors less precipitation than now and, therefore, a greater cooling effect.

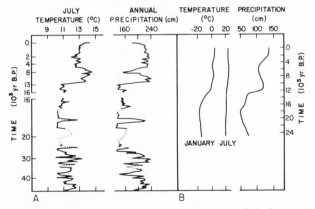

Figure 20-2. Estimated temperature and precipitation changes during late-Pleistocene and Holocene time. (A) Olympic Peninsula, Washington, based on pollen-climate transfer functions. (After Heusser et al., 1980.) (B) West of the Cumberland Plateau in Tennessee, inferred from the pollen record. (After Delcourt, 1979.)

THE SOUTHWEST

The southwestern part of the United States is considered unique in its having experienced a "pluvial-lacustral regime" with precipitation exceeding present values during the full glacial (Morrison, 1965). Alternative interpretations have suggested a cold, dry glacial climate with increases only in effective moisture (Galloway, 1970; Brakenridge, 1978).

This large area is highly diverse climatically. The southwestern deserts proper (Sonoran, Mohave, and Chihuahuan) lie in the zone of summer rainfall; California experiences a warm, dry summer and a winter precipitation maximum associated with Pacific cyclonic activity (the "Mediterranean regime"); and the Great Basin has a hot, dry summer with spring and early-winter precipitation. The Llano Estacado of western Texas and eastern New Mexico is different again, with spring and autumn maxima (Trewartha, 1961). Important advances have recently been made in improving the chronologic control of the stratigraphic record, especially for the Great Basin, and analyses of vegetational evidence from pack-rat middens have greatly added to our knowledge of Late Wisconsin-Holocene paleoenvironments.

Studies confirm a picture of high lake levels during the last glacial maximum (Smith and Street-Perrott, 1982). Benson (1978) proposes very high stands of Lake Lahontan, Nevada, between 25,000 and 22,000 yr B.P. and again between 13,500 and 11,000 yr B.P., with moderately high levels from 20,000 to 15,000 yr B.P. In contrast, levels were low from 40,000 to 25,000 yr B.P. and very low from 9000 to 5000 yr B.P. Currey (1980), Scott and others (1980), and McCoy (1981) postulate maximum levels for Lake Bonneville about 17,000 to 15,000 yr B.P.; such a situation would represent a 7000 to 10,000-year lag behind full-glacial conditions in the northern Rocky Mountains, according to Scott and others. However, Madsen and Currey (1979) note that tills were laid down in the Wasatch Range, Utah, about 20,000 to 19,000 yr B.P., with later recessional moraines built about 14,000 to 13,000 yr B.P. The recession of Lake Bonneville occurred rapidly about 12,500 yr B.P., and during one or more phases of maximum Holocene aridity (about 7500 to 5500 yr B.P.) the lake was reduced to a playa (Currey and James, 1982). A parallel history is identified in the Searles Valley of California, where perennial and mostly deep lakes existed between about 24,000 and 10,000 yr B.P. (Smith, 1976). Searles Lake itself became a dry salt flat thereafter until 6000 yr B.P. (Smith and Litticoat, 1980). In the Wasatch Range, deglaciation is dated at about 12,500 yr B.P., a timing comparable to recent findings in Colorado (Porter et al., 1982); but cool, dry conditions appear to have persisted until about 8000 yr B.P. (Madsen and Currey, 1979).

Macrofossils from desert pack-rat middens have been extensively used by Wells (1979), Van Devender and Spaulding (1979), Cole (1980), and Spaulding and others (1982) to reconstruct the late-glacial vegetation. In a strong critique of Brakenridge's analysis, Wells (1979) argues for a substantial lowering of juniper woodland into the northern Sonoran Desert and all of the Mohave and Chihuahuan Deserts. The absence of vegetational zonation during the full glacial in Chihuahua, together with the absence of the thermophyllous species, indicate a moist equable climate with mild winters. The modern lower limit of woodland declines *southward* with latitude in this area in response to the southeastward increase in summer "monsoonal" rainfall. Wells proposes that this vegetational gradient was considerably steeper between latitudes 36°N and 34°N during the last glaciation and, accordingly, that there must have been at least more effective total moisture available. Moreover, because the temperature reduction is not likely to have been greater at lower elevations in the south than at more-northern higher elevations, Wells concludes that summer rains were heavier.

Today, the Gulf of California and the eastern Pacific are the principal sources of moisture for July-August rainfall in southern Arizona and westward. As Wells (1979: 323) notes, minimal vegetational change occurred in southern Sonora, so the Gulf of Mexico was probably the source of much of the enhanced precipitation in southern Arizona and New Mexico during the Late Wisconsin.

An alternative interpretation is offered by Van Devender and Spaulding (1979). Their reconstructions of vegetation from midden evidence also indicate the presence of woodland in most of the southwestern deserts during the last glacial maximum. They note that the interpretations of Galloway and Brakenridge imply a 1750- to 1250-m lowering (respectively) of vegetational zones, and they conclude that, if this situation did occur, it could only have operated above 2000 m elevation. In contrast to Galloway, Brakenridge, and Wells, they infer mild, wet winters and cool summers. The plant associations suggest that glacial-maximum winters were mild, whereas the fauna indicates winters similar to the present. Possibly, cold "northers" were less frequent or less severe than they are now. Van Devender and Spaulding argue that, even if summers were cooler, the hydrologic calculations would still imply increased precipitation. However, the stability of the pinyon-juniper woodland from 22,000 to 11,000 yr B.P. suggests that temperature changes were less significant than moisture changes during the Late Wisconsin. About 11,000 yr B.P., the montane vegetation changed toward modern associations, and mesophytic elements decreased at lower elevations. For the Grand Canyon (where potential problems of anomalous topoclimate and ecologic niches may compound interpretation problems), on the basis of macrofossils from middens, Cole (1980) infers a cold full glacial with dry summers and annual precipitation close to modern values but with some increase probable in winter amounts. These suggestions of augmented Pacific cyclonic activity in winter are rejected by Wells (1979) on the grounds that, as today, the mountain ranges must have given rise to large-scale and localized rain-shadow effects.

Relevant to these inland studies is evidence from coastal California. Johnson (1977) reports that late-Quaternary dunes of widely varying age point to little difference in the dominant wind direction from the present direction and that buried soils in the dunes are rich in carbonate; the latter finding suggests that precipitation amounts at that time were not sufficient to cause leaching, as they are not at present. Nevertheless, Muhs (1980) identifies a major phase of dune building about 21,000 yr B.P. on San Clemente Island. He proposes that dust originating in the Mohave Desert was associated with frequent Santa Ana conditions; there northeasterly winds occur in the cool half of the year when a Great Basin anticyclone develops after the passage of a trough eastward. The Santa Anas' modern frequency is about 35 to 40 days per year in coastal southern California.

Johnson (1977) also notes that fossil floras of coastal coniferous forest elements and sclerophyllous vegetation were widely distributed during the Late Wisconsin, a situation that would suggest some increase in effective moisture. Gorsline and Prensky (1975) document an order-of-magnitude increase in offshore terrigenous sedimentation from about 17,000 to 12,000 yr B.P. that could indicate increased runoff resulting from a general increase in precipitation; however, they favor a regime with more-frequent intense rains and floods

within an essentially semiarid climate. This is compatible with Johnson's suggestion that the core of the Mediterranean climatic regime remained essentially intact, although it may have been displaced a few degrees southward during the Late Wisconsin. Johnson proposes that, if temperatures in the offshore California Current were lower due to enhanced upwelling (as Gorsline and Prensky propose), then coastal fogs might have been more frequent. This would enhance moisture effectiveness directly by fog drip and indirectly through lower summer temperatures along the coast.

The question of precipitation and temperature changes in the Great Basin in relation to the inferred and redated paleohydrologic record has been reexamined by McCoy (1981), who concludes that the mean annual temperature for the northern Bonneville Basin during the maximum of the last lake cycle (about 16,000 to 15,000 yr B.P.) was lowered at least 7 °C but no more than 16 °C. This would have required a regional precipitation of 190% or 50% of present amounts, respectively. However, temperatures might have been as much as 12 °C to 14 °C lower during the glacial maximum (regionally about 20,000 years ago), accompanied by a reduction of about 25% in precipitation. The Great Basin lakes at that time stood at moderate levels. The late-glacial rise in lake levels appears to have been associated with higher rainfall (almost present amounts) and some warming. A recent modeling assessment of evaporation for Lake Lahontan (Benson, 1981) suggests that an increase in total cloudiness of 10%, associated with a 10 °C temperature lowering, could account for past high stands. Possible effects of a seasonal cover of lake ice are not evaluated, although this could further suppress evaporation.

The late-Pleistocene mountain glaciation in the intermontane basin area has received little attention. Moran (1974) notes a parallelism across Nevada and Utah, at about latitude 40 °N, between the west southwestern/east-northeastern orientation of the Pleistocene snow line (on the basis of minimum cirque-floor altitudes), on the one hand, and the modern winter location of a zone of air-mass confluence— and gradient of potential temperature indicative of a frontal zone— on the other hand. He suggests that the frontal boundary was anchored by the mountain ranges and that winter snowfall must have been the main control of the mountain glaciers. However, this hypothesis neglects the necessity for a significant reduction in summer ablation compared with the present period. For the Colorado Front Range, for example, Johnson (1979) finds a 2.5-m loss of mass in summer on the Arikaree Glacier. Moreover, the accumulation on this glacier is nearly three times the snowfall amount, a result of the localized wind eddies set up by airflow crossing the Continental Divide, which forms the cirque headwall. The effect of the eddies does not extend down-valley. Hence, in these latitudes (about 40 °N), reduced summer ablation rather than increased snowfall is the key to late-Pleistocene alpine glaciation.

THE ROCKY MOUNTAINS

The chronology of late-Pleistocene glaciation in the Rocky Mountains has undergone substantial revision since the review prepared by Richmond (1965) for the Seventh Congress of the International Association for Quaternary Research was published (Clark, 1980). Dated sites are still scarce, but the extent of the *maximum* Pinedale ice of the last glaciation has been mapped from literature sources and climatic estimates (Hollin and Schilling, 1981).

Pierce's (1979) detailed studies in northern Yellowstone Park, an area of major interconnected ice caps, show a glaciation threshold during the Pinedale maximum of about 2800 m in the north, near

Gardiner, declining southwestward to 2600 m. The snow-line depression is estimated to have been at least 800 m in Montana and Wyoming and probably between 950 and 1200 m, according to K. L. Pierce (unpublished data) and in agreement with the earlier estimates of Richmond (1965) and Kaiser (1966). The extensive glaciation in Montana was apparently favored by Pacific storms reaching the area via the Snake River Plain. For example, the modern snowpack at 2700 m altitude contains eight or nine times more water in western Yellowstone than in the Bighorn Mountains and the Wind River Range (Porter et al., 1982).

For the Colorado Front Range, Haefner (1976) estimates equilibrium-line altitudes of about 3100 m for Pinedale glaciers east of the divide and altitudes perhaps 50 m lower to the west (compared with a modern climatic snow line above 4100 m). For the time around 14,000 to 12,000 yr B.P., Legg and Baker (1980) suggest a 500-m depression of the treeline. A similar 600-m lowering of late-glacial life zones is proposed from mammalian evidence in southeastern Wyoming (Roberts, 1971). Given a westerly wind regime like that of the present day, late-Pinedale July mean *maximum* temperatures may have been reduced about 5 °C (assuming dry-adiabatic air descent across the divide). Mean minimum temperatures in modern summers show little variation with altitude between 3050 m and 3750 m (Barry, 1973), although greater katabatic effects would have occurred during late-Pinedale time. Modern analogies indicate cooling of a few degrees immediately adjacent to glacier termini with off-glacier winds, but this effect does not extend far from the ice (Rannie, 1977). On the basis of the present altitudinal gradient on the eastern slope up to Niwot Ridge at 3750 m, Legg and Baker also propose increased precipitation. However, this station has a winter-season maximum reflecting a western-slope regime, whereas up to at least 3100 m (only 6 km to the east) there is a distinct spring maximum indicating up-slope effects from the east (Barry, 1973). Hence, the validity of Legg and Baker's argument is questionable. Precipitation is not likely to have been heavier than at present (Porter et al., 1982) when Pleistocene atmospheric circulation and climate are considered.

Data from the San Juan Mountains and the Sangre de Cristo Range of southern Colorado are mostly Holocene or late glacial. However, for the Chuska Mountains of northwestern New Mexico (latitude 36 °N, longitude 109 °W) at 2780 m, Wright and others (1973) identify an alpine vegetation or spruce parkland during the last glacial maximum. They suggest a treeline depression of 800 to 1000 m, which would point to a summer cooling of about 5 °C to 7 °C, less than half that in the northern Rocky Mountains.

THE GREAT PLAINS

In the broad sense, the Great Plains region of subhumid climate extends eastward from the foot of the Rocky Mountains as the Prairie Peninsula, which extends into Iowa, Missouri, and Illinois, and southward from north of the Canadian border to the southern High Plains of western Texas. The location and shape of the Prairie Peninsula reflect the predominance of dry westerly airflow across the continent (Trewartha, 1961: 262; Bryson, 1966). The Great Plains have a strongly continental climate, with a wide annual temperature range that increases to the north. Most of the area experiences a summer rainfall maximum, but the rains occur mainly as intense thundershowers. Summer droughts are common and often protracted, while winter snowfall is meager.

The late-Quaternary climatic record from the Plains is still poorly known. Quaternary studies have concentrated on the glacial stratigraphy and periglacial features in the North (Moran et al., 1976), and few suitable sites have been located for pollen studies within the core of the area (Grüger, 1973), although several useful records exist along the eastern margins.

In the Upper Midwest, Late Wisconsin ice lobes affected northwestern Minnesota and the eastern Dakotas, advancing first from the northeast and later from the northwest. Moran and others (1976) suggest that the second lobe developed as a result of "orographic" precipitation associated with the margin of the first tongue to the east, but no meteorological evidence exists at present for such effects upwind of mountain barriers. The widespread occurrence of periglacial phenomena and loess deposition in the Upper Midwest is indicative of a cold, dry Late Wisconsin climate with a thin snow cover (Washburn, 1980a; Mears, 1981). The Nebraska Sandhills were formerly interpreted in this context, also, although the major dune activity is now dated to Holocene time (Ahlbrandt and Fryberger, 1980). Evidence of a cold, dry regime seems difficult to accommodate within the same time interval as that which permitted the growth of a major ice sheet to the north, although recent analysis of glacial landforms in the plains (Moran et al., 1980) suggests that the basal ice was frozen to the underlying surface in North Dakota during the Laurentide maximum. A numerical calculation indicates a mean annual temperature between −7 °C and −10 °C with low accumulation (about 30 to 40 cm per year), comparable to the modern polar ice sheets. In northeastern Iowa, 50 km west of the driftless area, Harmon, Schwarcz, Ford, and Koch (1979) estimate from speleothem data a possible minimum mean annual temperature of 2 °C during the Late Wisconsin maximum of the Lake Michigan lobe about 17,000 yr B.P. This would have been 8 °C less than the modern mean annual temperature. The implied difference in mean annual temperature of 9 °C to 12 °C between North Dakota and Iowa is double the present gradient (5 °C) and must reflect the cooling influence of the ice in North Dakota, assuming that both estimates are reliable. Harmon and others also infer cooling of between 0.6 °C and 1.0 °C per 1000 years following a relatively mild and moist interval about 23,000 yr B.P.

The modern grassland vegetation of the Great Plains has been a feature of the area for only the last 8000 years or so (King, 1981). During the Late Wisconsin, the central and northern Great Plains supported a widespread boreal forest predominantly composed of spruce. This has been documented for the period between about 14,000 and 10,000 yr B.P. from localities that were within the limits of Late Wisconsin ice in eastern North Dakota and South Dakota, Minnesota, and central Iowa (Birks, 1976; Wright, 1976, 1981; Van Zant, 1979; Szabo, 1980). Longer palynologic records from northeastern Kansas (Grüger, 1973) and the Ozark Highlands, Missouri (King, 1973), show that spruce forest was also dominant during the Late Wisconsin maximum. In Kansas and western Missouri, spruce forests were present from about 23,000 to at least 15,000 yr B.P. A record of deciduous woodland in central Illinois until 21,000 yr B.P., with spruce occurring during the glacial maximum only 60 km from the edge of the ice, remains problematical (Grüger, 1972). In central Texas, the vegetation toward the end of the glacial maximum (16,000 yr B.P.) was open deciduous woodland with some conifers, including spruce (Bryant, 1977). In western Texas, Oldfield and Schoenwetter (1975) suggest continuous ponderosa pine and spruce woodland during the Tahoka "pluvial" (21,000 to 14,000 yr B.P.).

Boreal spruce forest was replaced directly by prairie in the western

Great Plains, but farther east it was succeeded first by other boreal elements and then by deciduous trees. According to Wright (1976), these transformations began in the South about 12,500 yr B.P. The change to prairie took place about 10,000 yr B.P. in Kansas but did not affect the vegetation in Minnesota for another 2000 years.

The problem of vegetational succession makes climatic inferences from the palynologic record especially difficult in this area, as in the southeastern and Atlantic coastal sectors. Independent interpretations based, for instance, on insect assemblages could aid in determining whether or not the climatic trend during late-glacial time was indeed unidirectional. The climatic contrasts in both temperature and moisture between the northern and southern Great Plains must have been much greater at 18,000 yr B.P. than they are now, although these difference cannot yet be quantified.

THE GREAT LAKES REGION AND THE NORTHEAST

The Late Wisconsin record in this area is dominated by studies of the Laurentide ice sheet's margins. Dreimanis and Goldthwait (1973) identify three distinct pulses in the eastern Great Lakes area about 21,000, 19,500 and 18,000 years ago. For Illinois, Frye and Willman (1973) propose a fluctuating trend toward climatic amelioration after about 19,000 years ago; however, the fluctuations of the Lake Michigan lobe may have involved processes of glacial mechanics as well as purely climatically controlled oscillations. Retreat intervals affecting most of the major lobes in the Great Lakes region are identified by Dreimanis (1977) at about 21,000 to 20,000, 17,500 (more pronounced to the west), and about 15,500 years ago (the Erie interstade of Mörner and Driemanis [1973] to the east). There was a widespread readvance (Cary) about 14,500 years ago southwest of the Great Lakes, which was followed by the fastest and most extensive Late Wisconsin recession that lasted until about 13,300 years ago. The final readvance (Port Huron) occurred in the central and eastern Great Lakes area about 13,000 years ago, and this was followed by more-or-less continuous glacial retreat with minor interruptions just before the Holocene transition, by which time the Laurentide ice sheet had retreated from most of the Great Lakes region.

In contrast to the three recorded advances of the Lake Superior lobe between 17,000 and 12,000 yr B.P., the vegetational record from central and northeastern Minnesota within 100 km of these lobes implies a unidirectional climatic amelioration (Birks, 1976, 1981). A tundra-like flora, with no modern analogue, was present between 20,500 and 14,700 yr B.P., with spruce woodland arriving by 13,600 yr B.P. and some deciduous elements arriving around 12,250 yr B.P. Wright (1976) suggests that ice-margin fluctuations may have been controlled by snow accumulation over the ice domes rather than by ablation at the margin. There is no clear evidence for surging, although there is a strong possibility that the soft, deformable sediments underlying most of this area would have enabled the formation of low-profile ice lobes (Boulton and Jones, 1979). This factor would have greatly facilitated rapid changes in the extent and configuration of the ice margin in the Great Lakes area.

The interpretation of fluctuations of the ice sheet's margin has received considerable scrutiny by Mayewski and others (1981), who conclude that the events during the period between 21,000 and 17,000 yr B.P., when the ice south of the Great Lakes remained near its maximum, and during the widespread recession at 14,000 yr B.P., were climatically controlled. On the basis of glaciologic concepts, they suggest that the maximum stage was associated with a Laurentide ice sheet dominated by a single major ice divide whereas at 14,000

yr B.P. marine ice streams were draining much of the eastern part of the ice sheet as a result of the rising sea level and the global warming trend. These events, according to Mayewski and others (1981), gave rise to a stage with several ice domes. However, as noted earlier, work around Hudson Bay indicates a predominance of a multidomed ice sheet throughout the Wisconsin.

A less precise source of evidence about paleoenvironmental conditions in the area south and west of the Great Lakes is the distribution of Late Wisconsin loess. Ruhe (1976) notes that the loess is time transgressive since it spans the interval 25,000 to 10,500 yr B.P. in Iowa and Illinois. The deposits thin east of the Wabash River valley, Illinois, demonstrating westerly airflow during some part of the year, probably summer and autumn.

Deglaciation began about 17,000 yr B.P. in western Long Island, New York (Conally and Sirkin, 1973; Sirkin, 1977), with the ice margin receding slowly and taking almost 4000 years to retreat 280 km to the northern Hudson Valley. Then, between 13,200 and 12,600 yr B.P., it receded northward another 225 km. A tundra fringe was present at least in late-glacial time, until about 12,500 to 12,000 yr B.P. in southern Connecticut, western Massachusetts, and northeastern Pennsylvania (Whitehead, 1979). It also probably extended along the Appalachians to the crest of the Great Smoky Mountains (2000 m altitude) (Wright, 1981). The occurrence of *Pinus banksiana* (jack pine) in the late-glacial boreal forest of the Appalachian Highlands and the Atlantic Coastal Plain, in contrast to its absence in the Midwest, suggests that conditions may have been milder in the former regions, according to Wright. Sirkin (1977) also postulates the occurrence of grassland on the exposed continental shelf off the east coast. The ice-marginal areas during late-glacial times may have been considerably warmer than modern arctic tundra environments, according to recent work by Schwert and Morgan (1980) on an insect fauna dated at about 12,700 yr B.P. near Lake Erie in upper New York State.

THE SOUTHEAST

The literature reveals considerable disagreement over the extent of vegetational and climatic changes during the last glaciation in the eastern and southeastern United States. Early studies hypothesize either negligible vegetational change since the Tertiary period or marked southward displacement of forests due to severe cooling. In a critical reevaluation based on recent research, Delcourt and Delcourt (1979, 1981) establish a more moderate viewpoint. They propose that, north of about latitude 34°N, boreal taxa displaced deciduous forest taxa, which shifted to the Gulf of Mexico and the southern Atlantic Coastal Plain.

Important refuges for endemic mesic species probably existed along the loess blufflands to the east of the Mississippi River and on bottomlands with favorable microclimates. A vegetational reconstruction for Nonconnah Creek on blufflands near Memphis, Tennessee (latitude 35°N, longitude 90°W), for example, indicates spruce dominance from 23,000 to 16,500 yr B.P.; however, the area also remained a continuous glacial refuge for many deciduous forest taxa (Delcourt et al., 1980). In Georgia, the Carolinas, and central Tennessee, *Pinus banksiana* (jack pine)-*Picea* forests were widespread during the last glaciation (Watts, 1975b; Delcourt, 1979). These have no exact modern analogue, but they resemble the central and eastern part of the boreal forest. For the Coastal Plain of South Carolina (latitude 34°N), Watts (1980b) estimates a late Wisconsin temperature difference (compared to the present) of 7°C (to 20°C) less in

July and of 18 °C (to −10 °C) less in January, with perhaps only slightly less precipitation than now. At Anderson Pond, just west of the Cumberland Plateau in Tennessee (latitude 36 °N, longitude 85 °30′W), however, on the basis of modern pollen/climate analogues of fossil pollen spectra, Delcourt (1979) estimates that mean annual precipitation may have been reduced by more than half. Mean annual temperatures were as much as 15 °C lower than they are today during the full glacial, with most of the change occurring in winter (Figure 20-2B). In support of such cooling, Maxwell and Davis (1972) draw attention to the occurrence of tundra 300 km south of the ice border at only 800 m elevation in the Appalachians. For eastern Pennsylvania, 60 km from the ice, Watts (1979) also identifies a cold, dry, windy tundra environment; however, for central Delaware and Maryland, Sirkin and others (1977) record pine for 17,800 yr B.P.

The marked cooling about 23,000 yr B.P. appears to have led to pond formation at several sites in eastern North America between latitudes 34 °N and 37 °N, apparently as a result of the reduced evaporation. In southern Alabama (latitude 32 °N), however, evidence of xeric oak-hickory forest suggests summer drought stress (Delcourt, 1980). Watts (1975a, 1980a; Watts and Stuiver, 1980) also identifies very dry conditions in Florida between 37,000 and 13,000 yr B.P., with xeric shrubs and only a few trees present; in this case, the drawdown of water tables due to the lower sea level may have been an additional factor.

The full-glacial ''boreal''/deciduous forest ecotone, approximately along latitude 34 °N east of the Mississippi Valley, is interpreted as a sharp climatic gradient (Delcourt, 1980). In terms of the air-mass scheme of Bryson and others (1970), it probably paralleled the winter limit of the mean arctic frontal zone, although this far south the arctic-air-mass properties would have been muted.

The vegetation, presumably after some significant time lag, indicates climatic amelioration beginning about 16,500 years ago in Tennessee (Delcourt and Delcourt, 1979). Plant fossils from the Tunica Hills in Louisiana and Mississippi (Delcourt and Delcourt, 1977) and from southwestern Tennessee (Delcourt et al., 1980) suggest locally cool conditions during the late glacial in the lower

Mississippi Valley. This may be attributable to fog and clouds associated with the mixing of warm Gulf air with cold air over the glacial meltwater in the Mississippi Valley (Emiliani et al., 1978; Delcourt and Delcourt, 1979). The interpretation of the stratigraphy of the Tunica Hills site, however has been challenged (Otvos, 1980).

A subsequent ''postglacial'' regime of warmth and summer drought was established about 12,500 yr B.P. on the Gulf Coastal Plain (Delcourt, 1980), whereas cool-temperate mixed mesic forest was present from about 12,500 to 8000 yr B.P. between latitude 34 °N and 37 °N (Delcourt and Delcourt, 1981). Climatic amelioration with increased moisture apparently occurred about 13,000 yr B.P. in Florida (Watts, 1975a), South Carolina (Watts, 1980b), and southern Pennsylvania (Watts, 1979) but not until 10,800 yr B.P. in northwestern Georgia (Watts, 1975b). Both precipitation and winter temperatures rose rapidly between 13,000 and 12,000 yr B.P. in central Tennessee, according to Delcourt (1979). For South Carolina, Watts (1980b) proposes that most of the late-glacial temperature rise occurred in winter, with some lengthening of the growing season and 15% less precipitation than now.

Climatic Synthesis

Several climatic reconstructions have been attempted for the last glacial maximum. These range in complexity from semiempirical approaches (e.g., Lamb and Woodroffe, 1970), to numerical simulations using general circulation models (GCMs) (Williams et al., 1974; Gates, 1976a, 1976b; Manabe and Hahn, 1977) to simplified global models (Otto-Bleisner et al., 1982). Before these reconstructions are compared with inferences based on field evidence, however, modern climatic anomaly patterns are discussed.

Figure 20-3 summarizes the anomalies of summer climate around 18,000 yr B.P. inferred from biologic and geologic evidence. Nearly all of the continental United States, with the exception of the Southwest and perhaps the Great Basin, was both colder and drier than it is at present, although the amount of quantitative information arranged on a *seasonal* basis is very limited (Peterson et al., 1979). As

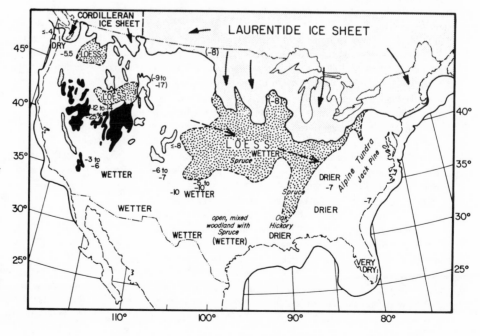

Figure 20-3. Summary of estimated summer temperature and annual precipitation changes for 18,000 yr B.P. shown in relation to other paleoenvironmental data. In a few cases, mean annual temperature estimates are shown in parentheses. (Based on Bachhuber and McClellan, 1977; Bryant, 1977; Delcourt, 1979; Delcourt, 1980; Grüger, 1973; Heusser et al., 1980; King, 1973; McCoy, 1981; Moran, 1976; Oldfield and Schoenwetter, 1975; Peterson et al., 1979: appendix; Porter, 1977; Porter et al., 1982; Smith, 1976; Washburn, 1980a; Watts, 1970, 1980b; Wright et al., 1973.)

Heusser and others (1980) point out for the Pacific Northwest, there is a striking difference between late-Pleistocene and Holocene regimes. During Holocene time, including the present, anomalies of temperature and precipitation in North America have tended to be *inversely* correlated (i.e., cold-wet/warm-dry). It is instructive, therefore, to examine individual seasons in the recent record in terms of anomalies of circulation and weather pattern.

October 1952 was an exceptional month: it was drier than average over the entire contiguous United States (Namias, 1955). Temperatures were above average west of about longitude 100°W, associated with an amplified upper ridge, and below average in the East due to outbreaks of continental polar air from Canada. Subcontinental-scale anomalies are the normal pattern. Diaz (1981) has performed eigenvector analyses of anomalies of seasonal temperature, seasonal precipitation, and annual cyclonic/anticyclonic activity since 1895. He notes that each of the first three precipitation eigenvectors shows regional differences. Moreover, in summer they account for a smaller fraction of the variance than in the other seasons (only 42% in the case of precipitation), and the spatial coherence of the patterns is reduced in scale. Namias (1967) also notes that dry regimes typically affect no more than one-quarter of the contiguous United States in summer, compared with up to one-half in winter. Walsh and Mostek (1980) identify as their first eigenvector mode, for *annual* precipitation anomalies, a pattern that is dry everywhere and is centered in Ohio, but the pattern accounts for only 11% of the variance. Thus, it is impractical to consider individual years. Instead, summer examples of Diaz's first three precipitation modes, for both signs, have been scrutinized through the accounts of circulation regimes published in *Monthly Weather Review*. The same procedure was also followed for the seasonal anomaly types identified by Blasing and Lofgren (1980). In almost all cases, cold-wet and warm-dry associations are predominant. The former typically occurs with strong zonal westerly flow, particularly in the West, and the latter with amplified upper-level ridges. Hence, it does not seems feasible to identify modern anomalies of continental scale bearing any significant relationship to those identified for 18,000 yr B.P.

In the case of the Southwest, identified as being cooler and wetter in Figure 20-3, it may be worth noting that, at the present time, wet summers are commonly caused by a tongue of moist air from the Gulf of Mexico. Wet conditions may also reflect tropical storm movement over Texas into New Mexico. Changes in the Great Basin area are uncertain climatologically and in terms of the field evidence and its dating control. Kay (1982) shows three seasonal precipitation regimes in different parts of the basin that are associated with three distinct meteorological controls: winter-spring cyclones from the Pacific, late-winter cyclogenesis within the Great Basin, and summer convective activity in moist air from the Gulf of California and possibly the Gulf of Mexico. Hence, the area is not a homogeneous climatic unit, and different sections of it need not respond in the same manner to a specific type of climatic shift.

Modeling studies using the University of Wisconsin low-order spectral GCM (Otto-Bliesner et al., 1982) suggest a more winterlike circulation for 18,000 yr B.P., with an augmented westerly jet stream south of the ice margin in North America (J. E. Kutzbach, personal communication, 1981). Kutzbach points out that this pattern would favor upper-level divergence, rising air motion, increased cloudiness, and increased precipitation in the right-entrance sector of the jet (in the Southwest) and the left-exit sector (in eastern Canada and the northeastern United States). Subsidence and reduced precipitation

would tend to predominate in the other two sectors. Except for the Northeast, where the evidence for increased precipitation is inferential (i.e., snowfall over the ice sheet may be required on glaciologic grounds), this simple scheme fits the sign of the anomalies shown in Figure 20-3.

Of the high-resolution models, the NCAR GCM July experiment (Williams et al., 1974) shows no evidence of a southward displacement of the westerlies by the Laurentide ice sheet, although the zonal winds are significantly increased in strength and cyclonic activity is displaced equatorward. Precipitation is also much reduced, except in the Southwest (Williams and Barry, 1975). Gates (1976b), using the Rand-OSU GCM, finds a major displacement of the westerly jet but a reduction in precipitation in the West (California) and an increase in the northern Great Plains. Simulated temperature decreases range from 4°C in Texas, to 15°C in the northern Rocky Mountains, to as much as 18°C in southern California (Gates, 1976a). With the GFDL model, Manabe and Hahn (1977) report reduced precipitation over the ice sheet and the Southwest but increases in the Southeast. Although dryness and major temperature reductions over the continents are featured in several model experiments and the July circulation for 18,000 yr B.P. is shown to be more winterlike, many of the details differ (Heath, 1979; Williams, 1979).

Other empirical reconstructions of atmospheric circulation for the last glacial maximum or late-glacial time should also be mentioned. Lamb and Woodroffe (1970), updated by Wright and Lamb (1973), have derived upper-air and surface-pressure patterns for January and July from inferred distributions of 1000-500-mb thickness, which corresponds approximately to the mean temperature of the lowest 5.5 km of the atmosphere. Their main conclusion is that there was less seasonal variation in circulation intensity 18,000 years ago than there has been during the Holocene. However, the July map, which shows a low centered over the Great Lakes area, implies cloudy and wet conditions in most of the central United States. For January 18,000 yr B.P., the mean isobars are much more zonal over the country than they are at present, although blocking is indicated in the North Atlantic. The case for enhanced air circulation in summer has been argued by Van Heuklon (1977) to account for the wide distribution of loess in the midwestern United States. A modern analogue suggests the transport of dust from a source in North Dakota and/or southern Saskatchewan to western Illinois. Van Heuklon notes that Alberta lows, which generally are dry systems in the Midwest, would create synoptic airflow patterns suitable for such dust transport. For late-glacial and Holocene intervals, Bryson and others (1970) postulate mean frontal locations over North America based on the modern relationships among air masses, frontal boundaries, and biotic distributions. They propose that the Prairie Peninsula was restricted to eastern Colorado and western Kansas during the late glacial (13,000 to 10,000 years ago). In winter, arctic air still dominated the central United States to latitude 35°N.

Conclusions

Although some strands of a consensus regarding glacial and late-glacial climatic conditions are beginning to emerge, the global circulation patterns giving rise to them are still open to considerable research and discussion. Indeed, more progress has been made recently in developing a general framework for understanding the primary forcings that influence the climate system and their time scales, given the major changes in external boundary conditions that existed in late-Pleistocene time, than in formulating definitive regional models of climatic anomalies in relation to circulation patterns. There is now general agreement that the orbital elements of obliquity (about 41,000-year oscillation) and precession (19,000- and 23,000-year oscillations) are present in climatic records, as exemplified by ^{18}O in deep-ocean cores (Hayes et al., 1976; Kominz et al., 1979), but there is also recognition of the important nonlinearities in the climate system. This is especially apparent for the 100,000-year oscillations. Glacier buildup and deglaciation are closely modulated by ocean/land-ice exchanges of heat and moisture (Emiliani, 1978; Ruddiman and McIntyre, 1981). Moreover, the apparent problem of the inadequate atmospheric energy available for removing the ice sheets, once formed, can be resolved when allowance is made for isostatic processes (Budd and Smith, 1981) and the calving of marine-based ice streams (Denton and Hughes, 1981a). On a shorter time scale, Karlén and Denton (1976) note the apparent repetition of "Little Ice Age" episodes during late-glacial as well as Holocene time. They also suggest that this may reflect a forcing due to solar activity, although accepted physical models for evaluating this idea do not yet exist.

One area where progress should shortly be possible can be noted. Synoptic studies clearly show that short-term (10- to 20-year) climatic fluctuations are strongly influenced by shifts in the tropospheric long-wave pattern (van Loon and Williams, 1976), although this is less pronounced in summer than in other seasons. The contribution of changes in hemispheric wave number, amplitude, and phase angle (or longitudinal location of the wave troughs) to changes in late-Pleistocene climate is not yet known. An improvement in the quantitative description of continental climatic conditions at 18,000 yr B.P. and during subsequent intervals would greatly facilitate the resolution of this question.

References

Ager, T. A. (1975). "Late Quaternary Environmental History of the Tanana Valley, Alaska." Ohio State University Institute for Polar Studies Report 54.

Ahlbrandt, T. S., and Fryberger, S. G. (1980). Eolian deposits in the Nebraska Sand Hills. In "Geologic and Paleoecologic Studies of the Nebraska Sand Hills," pp. 1-24. U.S. Geological Survey Professional Papers 1120-A, B, and 1120-C.

Andrews, J. T., and Barry, R. G. (1978). Glacial inception and disintegration during the last glaciation. Annual Reviews of Earth and Planetary Science 6, 205-28.

Andrews, J. T., and Mahaffy, M. A. W. (1976). Growth rate of the Laurentide ice sheet and sea level lowering (with emphasis on the 115,000 yr B.P. sea level low). Quaternary Research 6, 167-84.

Andrews, J. T., and Miller, G. H. (1972). Quaternary history of northern Cumberland Peninsula, Baffin Island, N.W.T., Canada: Part IV. Maps of the present glaciation limits and lowest equilibrium line altitude for north and south Baffin Island. Arctic and Alpine Research 4, 45-59.

Bachhuber, F. W., and McClellan, W. A. (1977). Paleoecology of marine foraminifera in pluvial Estancia Valley, central New Mexico. Quaternary Research 7, 254-67.

Bandy, O. L. (1972). Neogene planktonic foraminiferal zones, California and some geologic implications. Palaeogeography, Palaeoclimatology, Palaeoecology 12, 131-50.

Barry, R. G. (1966). Meteorological aspects of the glacial history of Labrador-Ungava with special reference to atmospheric vapour transport. Geographical Bulletin 8, 319-40.

Barry, R. G. (1973). A climatological transect on the east slope of the Front Range, Colorado. Arctic and Alpine Research 5, 89-110.

Barry, R. G. (1975). Climate models in paleoclimatic reconstruction. Palaeogeography, Palaeoclimatology, Palaeoecology 17, 123-37.

Barry, R. G. (1981a). Atmospheric circulation and climatic change: I. Approaches to paleoclimatic reconstructions; II. Case studies. In "Climatic Variations and

Variability: Facts and Theories" (A. Berger, ed.), pp. 323-44. D. Reidel, Dordrecht.

Barry, R. G. (1981b). "Mountain Weather and Climate." Methuen and Company, London.

Barry, R. G. (1982). Approaches to reconstructing the climate of the steppe-tundra biome. In "Paleoecology of Beringia" (D. M. Hopkins, J. V. Matthews, Jr., C. E. Schweger, and S. B. Young, eds.), pp. 195-204. Academic Press, New York.

Barry, R. G., Andrews, J. T., and Mahaffy, M. A. (1975). Continental ice sheets: Conditions for growth. Science 190, 979-81.

Barry, R. G., Kiladis, G., and Bradley, R. S. (1981). Synoptic climatology of the western United States in relation to climatic fluctuations during the twentieth century. Journal of Climatology 1, 97-113.

Benson, L. V. (1978). Fluctuation in the level of pluvial Lake Lahontan during the last 40,000 years. Quaternary Research 9, 300-318.

Benson, L. V. (1981). Paleoclimatic significance of lake level fluctuations in the Lahontan Basin. Quaternary Research 16, 390-403.

Berger, A. (1980). A critical review of modeling the astronomical theory of paleoclimates and the future of our climate. In "Proceedings, Symposium on Sun and Climate," pp. 325-56. Centre National Recherches Scientifiques, Toulouse.

Berggren, W. A., and van Couvering, J. A. (1974). The late Neogene. Palaeogeography, Palaeoclimatology, Palaeoecology 16, 1-215.

Birks, H. J. B. (1976). Late-Wisconsin vegetational history at Wolf Creek, central Minnesota. Ecological Monographs 46, 395-429.

Birks, H. J. B. (1981). Late Wisconsin vegetational and climatic history at Kylen Lake, northeastern Minnesota. Quaternary Research 16, 322-55.

Black, R. F. (1976a). Features indicative of permafrost. Annual Review of Earth and Planetary Sciences 4, 75-94.

Black, R. F. (1976b). Periglacial features indicative of permafrost: Ice and soil wedges. Quaternary Research 6, 3-26.

Blasing, T. J., and Lofgren, G. R. (1980). Seasonal climatic anomaly types for the North Pacific sector and western North America. Monthly Weather Review 108, 700-719.

Boulton, G. S., and Jones, A. S. (1979). Stability of temperate ice caps and ice sheets resting on beds of deformable sediment. Journal of Glaciology 24, 29-43.

Bowling, S. A. (1979). Meteorological interpretations of ice age climate in Alaska (abstract). Bulletin of the American Meteorological Society 60, 845.

Brakenridge, G. R. (1978). Evidence for a cold, dry full-glacial climate in the American Southwest. Quaternary Research 9, 22-40.

Brinkmann, W. A. R., and Barry, R. G. (1972). Palaeoclimatological aspects of the synoptic climatology of Keewatin. Northwest Territories, Canada. Palaeogeography, Palaeoclimatology, Palaeoecology 11, 77-91.

Broecker, W. S., and van Donk, J. (1970). Insolation changes, ice volumes, and the O^18 record in deep ocean cores. Reviews of Geophysics and Space Physics 8, 169-98.

Bryant, V. M. (1977). A 16,000 year pollen record of vegetational change in central Texas. Palynology 1, 143-56.

Bryson, R. A. (1966). Air masses, streamlines, and the boreal forest. Geographical Bulletin 8, 228-69.

Bryson, R. A., Baerreis, D. A., and Wendland, W. M. (1970). The character of late- and postglacial climatic changes. In "Pleistocene and Recent Environments of the Central Great Plains" (W. Dort and J. K. Jones, eds.), pp. 54-74. University of Kansas Press, Lawrence.

Bryson, R. A., and Kutzbach, J. E. (1974). On the analysis of pollen-climate canonical transfer functions. Quaternary Research 4, 162-74.

Budd, W. F., and Smith, I. N. (1981). The growth and retreat of ice sheets in response to orbital radiation changes. In "Symposium on Sea Level, Ice Sheets and Climatic Change" (I. Allison, ed.), pp. 369-409. International Association of Scientific Hydrology Publication 131.

Calder, N. (1974). "The Weather Machine and the Threat of Ice." British Broadcasting Corporation, London.

Carter, L. D. (1981). A Pleistocene sand sea on the Alaskan Arctic Coastal Plain. Science 211, 381-83.

Chappell, J. (1978). Theories of upper Quaternary ice ages. In "Climatic Change and Variability: A Southern Perspective" (A. B. Pittock, L. A. Frakes, D. Jenssen, J. A. Peterson, and J. W. Zillman, eds.), pp. 211-25. Cambridge University Press, Cambridge.

Clark, D. L. (1971). The Arctic Ocean ice cover and its late Cenozoic history. *Bulletin of the American Geological Society* 82, 3313-23.

Clark, M. J. (1980). Glacial geology: Overview. *In* "Geoecology of the Colorado Front Range" (J. D. Ives, ed.), pp. 9-23. Westview Press, Boulder, Colo.

CLIMAP Project Members (1976). The surface of the ice-age Earth. *Science* 1131-44.

Cline, R. M., and Hays, J. D. (eds.) (1976). "Investigation of Late Quaternary Paleoceanography and Paleoclimatology." Geological Society of America Memoir 145.

Cole, K. (1980). Late Pleistocene and Holocene vegetational gradients in the Grand Canyon, Arizona. *In* "American Quaternary Association, Sixth Biennial Meeting, Abstracts and Program, 18-20 August 1980," p. 57. Institute for Quaternary Studies, University of Maine, Orono.

Conally, C. G., and Sirkin, L. A. (1973). Wisconsinan history of the Hudson-Champlain lobe. *In* "The Wisconsin Stage" (R. F. Black, R. P. Goldthwait, and H. B. Willman, eds.), pp. 47-70. Geological Society of America Memoir 136.

Crandell, D. R. (1965). The glacial history of western Washington and Oregon. *In* "The Quaternary of the United States" (H. E. Wright, Jr., and D. G. Frey, eds.), pp. 341-54. Princeton University Press, Princeton, N.J.

Currey, D. R. (1980). Events associated with the last cycle of Lake Bonneville—Idaho, Nevada and Utah. *In* "American Quaternary Association, Sixth Biennial Meeting, Abstracts and Program, 18-20 August 1980," p. 59-60. Institute for Quaternary Studies, University of Maine, Orono.

Currey, D. R., and James, S. R. (1982). Paleoenvironments of the northeastern Great Basin and northeastern basin rim region: A review of geological and biological evidence. *In* "Man and Environments in the Great Basin" (D. B. Madsen and J. F. O'Connell, eds.), pp. 27-52. Society for American Archaeology Papers 2.

Curry, R. R. (1966). Holocene climatic and glacial history of the central Sierra Nevada, California. *In* "United States Contributions to Quaternary Research" (S. A. Schumm and W. C. Bradley, eds), pp. 1-47. Geological Society of America Special Paper 123.

Cwynar, L. C., and Ritchie, J. O. (1980). Arctic steppe-tundra: A Yukon perspective. *Science* 208, 1375-77.

Delcourt, H. R. (1979). Late Quaternary vegetational history of the eastern highland rim and adjacent Cumberland Plateau of Tennessee. *Ecological Monographs* 49, 255-80.

Delcourt, P. A. (1980). Goshen Springs: Late Quaternary vegetation record for southern Alabama. *Ecology* 61, 371-86.

Delcourt, P. A., and Delcourt, H. R. (1977). The Tunica Hills, Louisiana-Mississippi: Late glacial locality for spruce and deciduous forest species. *Quaternary Research* 7, 218-37.

Delcourt, P. A., and Delcourt, H. R. (1979). Late Pleistocene and Holocene distributional history of the deciduous forest in the southeastern United States. *Veroffentlichungen des Geobotanischen Institutes der ETH, Zurich* 68, 79-107.

Delcourt, P. A., and Delcourt, H. R. (1981). Vegetation maps for eastern North America: 40,000 yr B.P. to the present. *In* "Geobotany" (R. C. Romans, ed.), vol. 2, pp. 123-65. Plenum Publications, New York.

Delcourt, P. A., Delcourt, H. R., Brister, R. C., and Lackey, L. E. (1980). Quaternary vegetation history of the Mississippi Embayment. *Quaternary Research* 13, 111-32.

Denton, G. H., and Armstrong, R. L. (1969). Miocene-Pliocene glaciations in southern Alaska. *American Journal of Science* 267, 1121-42.

Denton, G. H., and Hughes, T. J. (1981a). The Arctic ice sheet: An outrageous hypothesis. *In* "The Last Great Ice Sheets" (G. H. Denton and T. J. Hughes, eds.), pp. 437-67. John Wiley and Sons, New York.

Denton, G. H., and Hughes, T. J. (eds.) (1981b). "The Last Great Ice Sheets." John Wiley and Sons, New York.

Diaz, H. F. (1981). Eigenvector analysis of seasonal temperature and precipitation over the contiguous United States and related changes in the climatology of synoptic scale systems: Part II. Spring, summer, fall, and annual. *Monthly Weather Review* 109, 1285-1304.

Donn, W. L., and Shaw, D. M. (1977). Model of climatic evolution based on continental drift and polar wandering. *Bulletin of the Geological Society of America* 88, 390-96.

Dreimanis, A. (1977). Late Wisconsin glacial retreat in the Great Lakes region, North America. *Annals of the New York Academy of Sciences* 288, 70-89.

Dreimanis, A., and Goldthwait, R. P. (1973). Wisconsin glaciation in the Huron, Erie and Ontario lobes. *In* "The Wisconsinan Stage" (R. F. Black, R. P. Goldthwait, and H. B. Willman, eds.), pp. 71-106. Geological Society of America Memoir 136.

Easterbrook, D. J. (1976). Quaternary geology of the Pacific Northwest. *In* "Quaternary Stratigraphy of North America" (W. C. Mahaney, ed.), pp. 441-62. Dowden, Hutchinson, and Ross, Stroudsburg, Pa.

Emiliani, C. (1972). Quaternary Hypsithermals. *Quaternary Research* 2, 270-73.

Emiliani, C. (1978). The causes of ice ages. *Earth and Planetary Science Letters* 37, 349-52.

Emiliani, C., Rooth, C., and Stipp, J. J. (1978). The Late Wisconsin flood into the Gulf of Mexico. *Earth and Planetary Science Letters* 41, 159-63.

England, J. (1976). Late Quaternary glaciation of the eastern Queen Elizabeth Islands, N.W.T., Canada: Alternative models. *Quaternary Research* 6, 185-202.

Fillon, R. H., and Duplessy, J. C. (1980). Labrador Sea bio-, tephro-, oxygen isotope stratigraphy and late Quaternary paleoceanography trends. *Canadian Journal of Earth Science* 17, 831-54.

Fink, J., and Kukla, G. J. (1977). Pleistocene climates in central Europe: At least 17 interglacials after the Olduvai event. *Quaternary Research* 7, 363-71.

Flint, R. F. (1971). "Glacial and Quaternary Geology." John Wiley and Sons, New York.

Flohn, H. (1979). On time scales and causes of abrupt paleoclimatic events. *Quaternary Research* 12, 135-49.

Forsyth, J. L., and Goldthwait, R. P. (1980). Rapid Late Wisconsin deglaciation of western Ohio. *In* "American Quaternary Association, Sixth Biennial Meeting, Abstracts and Program, 18-20 August 1980," p. 80-81. Institute for Quaternary Studies, University of Maine, Orono.

Friedman, I. (1983). Paleoclimatic evidence from stable isotopes. *In* "Late-Quaternary Environments of the United States," vol. 1, "The Late Pleistocene" (S. C. Porter, ed.), pp. 385-89. University of Minnesota Press, Minneapolis.

Frye, J. C., and Willman, H. B. (1973). Wisconsinan climatic history interpreted from Lake Michigan lobe deposits and soils. *In* "The Wisconsinan Stage" (R. F. Black, R. P. Goldthwait, and H. B. Willman, eds.), pp. 135-52. Geological Society of America Memoir 136.

Galloway, R. W. (1970). The full glacial climate in the southwestern United States. *Annals of the Association of American Geographers* 60, 246-56.

Gates, W. L. (1976a). Modeling the ice-age climate. *Science* 191, 1131-44.

Gates, W. L. (1976b). The numerical simulation of ice-age climate with a global general circulation model. *Journal of Atmospheric Sciences* 33, 1844-73.

Gorsline, D. S., and Prensky, S. E. (1975). Paleoclimatic inferences for late Pleistocene and Holocene from California Continental Borderland basin sediments. *In* "Quaternary Studies" (R. P. Suggate and M. M. Cresswell, eds.), pp. 147-54. Royal Society of New Zealand Bulletin 13.

Graf, W. L. (1976). Cirques as glacier locations. *Arctic and Alpine Research* 8, 79-80.

Grüger, E. (1972). Late Quaternary vegetation development in south-central Illinois. *Quaternary Research* 2, 217-31.

Grüger, J. (1973). Studies on the late-Quaternary vegetation history of northeastern Kansas. *Geological Society of America Bulletin* 83, 239-50.

Haefner, B. D. (1976). Glacial valley morphometry in the Front Range of Colorado. M.A. thesis, University of Colorado, Boulder.

Hales, J. E., Jr. (1976). Southwestern United States summer monsoon source: Gulf of Mexico or Pacific Ocean? *Journal of Applied Meteorology* 13, 331-42.

Hamilton, T. D., and Porter, S. C. (1975). Itkillik Glaciation in the Brooks Range, northern Alaska. *Quaternary Research* 5, 471-98.

Hamilton, T. D., and Thorson, R. M. (1983). The Cordilleran ice sheet in Alaska. *In* "Late-Quaternary Environments of the United States," vol. 1, "The Late Pleistocene" (S. C. Porter, ed.), pp. 38-52. University of Minnesota Press, Minneapolis.

Haq, B. U., Berggren, W. A., and van Couvering, J. A. (1977). Corrected age of the Pliocene-Pleistocene boundary. *Nature* 269, 483-88.

Harmer, F. W. (1901). The influence of the winds upon climate during the Pleistocene epoch: A paleometeorological explanation. *Quaternary Journal of the Geological Society of London* 57, 405-78.

Harmon, R. S., Schwarcz, H. P., Ford, D. C., and Koch, D. L. (1979). An isotopic paleotemperature record for Late Wisconsinan time in northeast Iowa. *Geology* 7, 430-33.

Harmon, R. S., Schwarcz, H. P., and O'Neil, J. R. (1979). D/H ratios in speleothem fluid inclusions: A guide to variations in the composition of meteoric precipitation? *Earth and Planetary Science Letters* 42, 254-66.

Hays, J. D., Imbrie, J., and Shackleton, N. J. (1976). Variations in the Earth's orbit: Pacemaker of the ice ages. *Science* 194, 1121-32.

Heath, G. R. (1979). Simulations of a glacial paleoclimate by three different atmospheric general circulation models. *Palaeogeography, Palaeoclimatology, Palaeoecology* 26, 291-303.

Herman, Y., and Hopkins, D. M. (1980). Arctic oceanic climate in late Cenozoic time. *Science* 209, 557-62.

Hess, M. (1973). A method for determining the influence of mountain glaciers on the climate. *Arctic and Alpine Research* 5, A183-86.

Heuberger, H. (1974). Alpine Quaternary glaciation. *In* "Arctic and Alpine Environments" (J. D. Ives and R. G. Barry, eds.), pp. 319-38. Methuen and Company, London.

Heusser, C. J. (1977). Quaternary palynology of the Pacific slope of Washington. *Quaternary Research* 8, 282-306.

Heusser, C. J., Heusser, L. E., and Streeter, S. S. (1980). Quaternary temperatures and precipitation for the north-west coast of North America. *Nature* 286, 702-74.

Hollin, J. T., and Schilling, D. H. (1981). Late Wisconsin-Weichselian mountain glaciers and small ice caps. *In* "The Last Great Ice Sheets" (G. H. Denton and T. J. Hughes, eds.), pp. 179-206. John Wiley and Sons, New York.

Hopkins, D. M. (1979). Landscape and climate of Beringia during late Pleistocene and Holocene time. *In* "The First Americans: Origins, Affinities and Adaptations" (W. S. Laughlin and A. B. Harper, eds.), pp. 15-41. Gustaf Fischer, New York.

Hopkins, D. M., Matthews, J. V., Jr., Schweger, C. E., and Young, S. B. (eds.) (1982). "Paleoecology of Beringia." Academic Press, New York.

Hunkins, K., Be, A. W. H., Opdyke, N. D., and Mathieu, G. (1971). The late Cenozoic history of the Arctic Ocean. *In* "The Late Cenozoic Glacial Ages" (K. K. Turekian, ed.), pp. 215-38. Yale University Press, New Haven, Conn.

Imbrie, J., and Imbrie, K. P. (1979). "Ice Ages: Solving the Mystery." Enslow Publishers, Short Hills, N.J.

Imbrie, J., and Kipp, N. G. (1971). A new micropaleontological method for quantitative paleoclimatology: Applications to a late Pleistocene Caribbean core. *In* "The Late Cenozoic Glacial Ages" (K. K. Turekian, ed.) pp. 71-181. Yale University Press, New Haven, Conn.

Ives, J. D., Andrews, J. T., and Barry, R. G. (1975). Growth and decay of the Laurentide ice sheet and comparisons with Fenno-Scandinavia. *Die Naturwissenschaften* 62, 118-25.

Johnson, D. L. (1977). The late Quaternary climate of coastal California: Evidence for an ice age refugium. *Quaternary Research* 8, 154-79.

Johnson, J. B. (1979). Mass balance studies on the Arikaree Glacier. Ph.D. dissertation, University of Colorado, Boulder.

Juvik, J. O., Singleton, D. C., and Clarke, G. G. (1978). Climate and water balance on the island of Hawaii. *In* "Mauna Loa Observatory: A 20th Anniversary Report" (J. Miller, ed.), pp. 129-40. National Oceanic and Atmospheric Administration, Environmental Research Laboratories, Boulder, Colo.

Kaiser, K. (1966). Probleme und Ergebnisse der Quartarforschung in den Rocky Mountains (i.w.S.) und angrenzenden Gebieten. *Zeitschrift für Geomorphologie* 10, 264-302.

Karlén, W., and Denton, G. H. (1976). Holocene glacial variations in Sarek National Park, northern Sweden. *Boreas* 5, 25-56.

Kay, P. A. (1982). A perspective on Great Basin paleoclimates. *In* "Man and Environments of the Great Basin" (D. B. Madsen and J. F. O'Connell, eds.), pp. 76-81. Society for American Archaeology Papers 2.

Kennett, J. K. (1980). Paleoceanographic and biogeographic evolution of the Southern Ocean during the Cenozoic, and Cenozoic microfossil datums. *Palaeogeography, Palaeoclimatology, Palaeoecology* 31, 123-52.

King, J. E. (1973). Late Pleistocene palynology and biogeography of the western Missouri Ozarks. *Ecological Monographs* 43, 539-65.

King, J. E. (1981). Late Quaternary vegetational history of Illinois. *Ecological Monographs* 51, 43-62.

Koerner, R. M. (1980). Instantaneous glacierization, the rate of albedo change, and feedback effects at the beginning of the ice age. *Quaternary Research* 13, 153-59.

Kominz, M. A., Heath, G. R., Ku, T. L., and Pisias, N. G. (1979). Brunhes time scales

and the interpretation of climatic change. *Earth and Planetary Science Letters* 45, 394-410.

Kraus, E. B. (1973). Comparison between ice age and present general circulations. *Nature* 245, 129-33.

Krebs, J. S., and Barry, R. G. (1970). The arctic front and the tundra-taiga boundary in Eurasia. *Geographical Review* 60, 548-54.

Kukla, G. (1978). The classical European glacial stages: Correlation with deep sea sediments. *Transactions of the Nebraska Academy of Science* 6, 57-93.

Kutzbach, J. E. (1980). Estimates of past climate at paleolake Chad, North Africa, based on hydrological and energy-balance model. *Quaternary Research* 14, 210-33.

Lamb, H. H. (1964). Fundamentals of climate. *In* "Problems of Palaeoclimatology" (A. E. M. Nairn, ed.), pp. 8-44. Interscience, New York.

Lamb, H. H. (1977). "Climate: Present, Past and Future." Vol. 2. Methuen and Company, London.

Lamb, H. H., Lewis, R. P. W., and Woodroffe, A. (1966). Atmospheric circulation and the main climatic variables between 8000 and 0 BC: Meteorological evidence. *In* "World Climate from 8000 to 0 BC" (J. S. Sawyer, ed.), pp. 174-217. Royal Meteorological Society, London.

Lamb, H. H., and Woodroffe, A. (1970). Atmospheric circulation during the last ice age. *Quaternary Research* 1, 29-58.

Legg, T. E., and Baker, R. G. (1980). Palynology of Pinedale sediments, Devlins Park, Boulder County, Colorado. *Arctic and Alpine Research* 12, 319-33.

Leopold, L. B. (1951). Pleistocene climates in New Mexico. *American Journal of Science* 249, 152-68.

Lettau, H. H. (1969). Evapotranspiration climatonomy: I. A new approach to numerical prediction of monthly evapotranspiration, run off, and soil moisture storage. *Monthly Weather Review* 97, 691-99.

Levin, B., Ives, P. C., Oman, C. L., and Rubin, M. (1965). U.S. Geological Survey radiocarbon dates VIII. *Radiocarbon* 7, 372-98.

McCoy, W. D. (1981). Quaternary aminostratigraphy of the Bonneville and Lahontan Basins, western United States, with paleoclimatic implications. Ph.D. dissertation, University of Colorado, Boulder.

Madsen, D. B., and Currey, D. R. (1979). Late Quaternary glacial and vegetation changes, Little Cottonwood Canyon area, Wasatch Mountains, Utah. *Quaternary Research* 12, 254-70.

Manabe, S., and Hahn, D. G. (1977). Simulation of the tropical climate of an ice age. *Journal of Geophysical Research* 82, 38898-3911.

Matthews, J. V., Jr. (1974a) Wisconsin environment of interior Alaska: Pollen and macrofossil analysis of a 27 meter core from the Isabella Basin (Fairbanks, Alaska). *Canadian Journal of Earth Science* 11, 828-44.

Matthews, J. V., Jr. (1974b). Quaternary environments of Cape Deceit (Seward Peninsula, Alaska): Evolution of a tundra ecosystem. *Geological Society of America Bulletin* 85, 1353-84.

Maxwell, J. A., and Davis, M. B. (1972). Pollen evidence of Pleistocene and Holocene vegetation on the Allegheny Plateau, Maryland. *Quaternary Research* 2, 506-30.

Mayewski, P. A., Denton, G. H., and Hughes, T. J. (1981). Late Wisconsin ice sheets in North America. *In* "The Last Great Ice Sheets" (G. H. Denton and T. J. Hughes, eds.), p. 67-178. John Wiley and Sons, New York.

Mears, B., Jr. (1981). Periglacial wedges and the late Pleistocene environment of Wyoming's intermontane basins. *Quaternary Research* 15, 171-98.

Mehringer, P. J. (1977). Great Basin late Quaternary environments and chronology. *In* "Models and Great Basin Prehistory: A Symposium" (D. D. Fowler, ed.), pp. 113-67. Desert Research Institute Publication in the Social Sciences 12.

Moran, J. M. (1974). Possible coincidence of a modern and a glacial-age climatic boundary in the montane West, United States. *Arctic and Alpine Research* 6, 319-21.

Moran, S. R., Arndt, M., Bluemle, J. P., Camara, M., Clayton, L., Fenton, M. M., Harris, K. L., Hobbs, H. C., Keatinge, R., Sackreiter, D. K., Salomon, N. L., and Teller, J. (1976). Quaternary stratigraphy and history of North Dakota, southern Manitoba and northwestern Wisconsin. *In* "Quaternary Stratigraphy of North America" (W. C. Mahaney, ed.), pp. 133-58. Dowden, Hutchinson, and Ross, Stroudsburg, Pa.

Moran, S. R., Clayton, L., Hooke, R. LeB., Fenton, M. M., and Andriashek, L. D. (1980). Glacier-bed landforms of the prairie region of North America. *Journal of Geology* 25, 457-76.

Morgan, A. V., Morgan, A., Ashworth, A. C., and Matthews, J. V., Jr. (1983). Late Wisconsin fossil beetles in North America. In "Late-Quaternary Environments of the United States," vol. 1, "The Late Pleistocene" (S. C. Porter, ed.), pp. 354-63. University of Minnesota Press, Minneapolis.

Mörner, N. A., and Dreimanis, A. (1973). The Erie interstade. In "The Wisconsinan Stage" (R. F. Black, R. P. Goldthwait, and H. B. Willman, eds.), pp. 107-34. Geological Society of America Memoir 136.

Morrison, R. B. (1965). Quaternary geology of the Great Basin. In "The Quaternary of the United States" (H. E. Wright, Jr., and D. G. Frey, eds.), pp. 265-86. Princeton University Press, Princeton, N.J.

Muhs, D. R. (1980). Quaternary stratigraphy and soil development, San Clemente Island, California. Ph.D. dissertation, University of Colorado, Boulder.

Namias, J. (1955). Some meteorological aspects of drought. Monthly Weather Review 83, 199-205.

Namias, J. (1967). Further studies of drought over northeastern United States. Monthly Weather Review 95, 497-508.

Nelson, C. H., and Creager, J. S. (1977). Displacement of Yukon-derived sediment from Bering Sea to Chukchi Sea during Holocene time. Geology 5, 141-46.

Oldfield, F., and Schoenwetter, J. (1975). Discussion of the pollen-analytical evidence. In "Late Pleistocene Environments of the Southern High Plains" (F. Wendorf and J. J. Hester, eds.), pp. 149-78. Fort Burgwin Research Center Publications 9.

Østrem, G. (1974). Present alpine ice cover. In "Arctic and Alpine Environments" (J. D. Ives and R. G. Barry, eds.), pp. 225-50. Methuen and Company, London.

Otto-Bliesner, B. L., Branstator, G. W., and Houghton, D. D. (1982). A global low-order spectral general circulation model. Part I: Formulation and seasonal climatology. Journal of the Atmospheric Sciences 39, 929-48.

Otvos, E. G., Jr. (1980). Age of Tunica Hills (Louisiana-Mississippi) Quaternary fossiliferous creek deposits: Problems of radiocarbon dates and intermediate valley terraces in coastal plains. Quaternary Research 13, 80-92.

Peterson, G. M., Webb, T., III, Kutzbach, J. E., van der Hammen, T., Wijmstra, T. A., and Street, F. A. (1979). The continental record of environmental conditions at 18,000 years B.P.: An initial evaluation. Quaternary Research 19, 47-82.

Péwé, T. L. (1973). Ice wedge casts and past permafrost distribution in North America. Geoforum 15, 15-26.

Péwé, T. L. (1975). "Quaternary Geology of Alaska." U.S. Geological Survey Professional Paper 835.

Péwé, T. L. (1983). The periglacial environment in North America during Wisconsin time. In "Late Quaternary Environments of the United States," vol. 1, "The Late Pleistocene" (S. C. Porter, ed.), pp. 157-89. University of Minnesota Press, Minneapolis.

Pierce, K. L. (1979). "History and Dynamics of Glaciation in the Northern Yellowstone Park Area." U.S. Geological Survey Professional Paper 729-F.

Pittock, A. B., Frakes, L. A., Jenssen, D., Peterson, J. A., and Zillman, J. W. (1978). "Climatic Changes and Variability: A Southern Perspective." Cambridge University Press, Cambridge.

Ponte, L. (1976). "The Cooling." Prentice-Hall, Englewood Cliffs, N.J.

Porter, S. C. (1964). Composite Pleistocene snowline of Olympic Mountains and Cascade Range, Washington. Bulletin of the American Geological Society 75, 477-82.

Porter, S. C. (1977). Present and past glaciation threshold in the Cascade Range, Washington, U.S.A.: Topographic and climatic controls, and paleoclimatic implications. Journal of Glaciology 18, 101-16.

Porter, S. C. (1979). Hawaiian glacial ages. Quaternary Research 12, 161-87.

Porter, S. C., Pierce, K. L., and Hamilton, T. D. (1983). Late Wisconsin mountain glaciation in the western United States. In "Late-Quaternary Environments of the United States," vol. 1, "The Late Pleistocene" (S. C. Porter, ed.), pp. 71-111. University of Minnesota Press, Minneapolis.

Rannie, W. F. (1977). A note on the effect of a glacier on the summer thermal climate of an ice-marginal area. Arctic and Alpine Research 9, 301-4.

Reeves, C. C., Jr. (1966). Pleistocene climate of the Llano Estacado, II. Journal of Geology 74, 642-47.

Reeves, C. C., Jr. (1973). The full-glacial climate of the southern High Plains, west Texas. Journal of Geology 81, 693-704.

Richmond, G. M. (1965). Glaciation of the Rocky Mountains. In "The Quaternary of the United States" (H. E. Wright, Jr., and D. G. Frey, eds.), pp. 217-30. Princeton University Press, Princeton, N.J.

Roberts, M. F. (1971). Late glacial and postglacial environments in southeastern Wyoming. Palaeogeography, Palaeoclimatology, Palaeoecology 8, 5-17.

Rowntree, P. R. (1976). Tropical forcing of atmospheric motions in a numerical model, Quarterly Journal of the Royal Meteorological Society 102, 583-606.

Ruddiman, W. F., and McIntyre, A. (1981). Moisture flux from the North Atlantic: Amplification of the 23,000-year Milankovitch forcing period. Science 212, 617-27.

Ruddiman, W. F., McIntyre, A., Niebler-Hunt, V., and Durazzi, J. T. (1980). Oceanic evidence for the mechanism of rapid Northern Hemisphere glaciation. Quaternary Research 13, 33-64.

Ruhe, R. V. (1976). Stratigraphy of mid-continent loess, U.S.A. In "Quaternary Stratigraphy of North America" (W. C. Mahaney, ed.), pp. 197-212. Dowden, Hutchinson, and Ross, Stroudsburg, Pa.

Sachs, H. M. (1973). Late Pleistocene history of the North Pacific: Evidence from a quantitative study of radiolaria in core V21-173. Quaternary Research 3, 89-98.

Sachs, H. M., Webb, T., III, and Clark, D. R. (1977). Paleoecological transfer functions. Annual Review of Earth and Planetary Sciences 5, 159-78.

Savin, S. M., Douglas, R. G., and Stehli, F. G. (1975). Tertiary marine paleotemperatures. Geological Society of America Bulletin 86, 1499-1510.

Schnitker, D. (1980). Global paleoceanography and its deep water linkage to the Antarctic glaciation. Earth Science Reviews 16, 1-20.

Schumm, S. A. (1965). Quaternary paleohydrology. In "The Quaternary of the United States" (H. E. Wright, Jr., and D. G. Frey, eds.), pp. 783-94. Princeton University Press, Princeton, N.J.

Schwert, D. P., and Morgan, A. V. (1980). Paleoenvironmental implications of a late glacial insect assemblage from northwestern New York. Quaternary Research 13, 93-110.

Scott, W. E. (1977). Quaternary glaciation and volcanism, Metolius River area, Oregon. Geological Society of America Bulletin 88, 113-24.

Scott, W. E., McCoy, W. D., Shroba, R. R., and Miller, R. D. (1980). New interpretation of the late Quaternary history of Lake Bonneville, western United States. In "American Quaternary Association, Sixth Biennial Meeting, Abstracts and Program, 18-20 August 1980," p. 168-69. Institute for Quaternary Studies, University of Maine, Orono.

Shackleton, N. J., and Kennett, J. P. (1975a). Late Cenozoic oxygen and carbon isotopic changes at DSDP site 284: Implications for the glacial history of the Northern Hemisphere and Antarctica. In "Initial Reports of the Deep Sea Drilling Project 29" (J. P. Kennett, R. E. Houtz, and others, eds.), pp. 801-7. U.S. Government Printing Office, Washington, D.C.

Shackleton, N. J., and Kennett, J. P. (1975b). Paleotemperature history of the Cenozoic and the initiation of Antarctic glaciations: Oxygen and carbon isotopes in DSDP sites 277, 279, and 281. In "Initial Reports of the Deep Sea Drilling Project 29" (J. P. Kennett, R. E. Houtz, and others, eds.), pp. 743-55. U.S. Government Printing Office, Washington, D.C.

Shackleton, N. J., and Opdyke, N. D. (1976). Oxygen-isotope and paleomagnetic stratigraphy of Pacific core v28-239 late Pliocene to latest Pleistocene. In "Investigations of Late Quaternary Paleoceanography and Paleoclimatology" (R. M. Cline and J. D. Hays, eds.), pp. 449-64. Geological Society of America Memoir 145.

Shackleton, N. J., and Opdyke, N. G. (1977). Oxygen isotope and palaeomagnetic evidence for early Northern Hemisphere glaciation. Nature 270, 216-19.

Shilts, W. W. (1980). Flow patterns in the central North American ice sheets. Nature 286, 213-18.

Shilts, W. W., Miller, G. H., and Andrews, J. T. (1981). Glacial flow indicators and Wisconsin glacial chronology, Hudson Bay/James Bay lowlands: Evidence against a Hudson Bay ice divide. Geological Society of America, Abstracts with Programs 13, 553.

Sirkin, L. A. (1977). Late Pleistocene vegetation and environments in the middle Atlantic region. Annals of the New York Academy of Sciences 288, 206-17.

Sirkin, L. A., Denny, C. S., and Rubin, M. (1977). Late Pleistocene environment of the central Delmarva Peninsula, Delaware, Maryland. Geological Society of America Bulletin 88, 139-47.

Smith, G. I. (1976). Paleoclimatic record in the upper Quaternary sediments of Searles Lake, California, U.S.A. In "Paleolimnology of Lake Biwa and the Japanese Pleistocene" (S. Horie, ed.), vol. 4, pp. 577-604. Kyoto University, Takashima, Japan.

Smith, G. I., and Liddicoat, J. C. (1980). Late Tertiary-to-present sequence of pluvial events in Searles Valley, California, and their relation to global ice volumes. *In* "American Quaternary Association, Sixth Biennial Meeting, Abstracts and Program, 18-20 August 1980," p. 57. Institute for Quaternary Studies, University of Maine, Orono.

Smith, G. I., and Street-Perrott, A. (1983). Pluvial lakes of the western United States. *In* "Late-Quaternary Environments of the United States," vol. 1, "The Late Pleistocene" (S. C. Porter, ed.), pp. 190-212. University of Minnesota Press, Minneapolis.

Snyder, C. T., and Langbein, W. B. (1962). The Pleistocene lake in Spring Valley, Nevada, and its climatic implications. *Journal of Geophysical Research* 67, 2385-94.

Spaulding, W. G., Leopold, E. B., and Van Devender, T. R. (1983). Late Wisconsin paleoecology of the American Southwest. *In* "Late-Quaternary Environments of the United States," vol. 1, "The Late Pleistocene" (S. C. Porter, ed.), pp. 259-93. University of Minnesota Press, Minneapolis.

Street, A. F., and Grove, A. T. (1979). Global maps of lake-level fluctuations since 30,000 years B.P. *Quaternary Research* 12, 83-118.

Streten, N. A. (1974). Some features of the summer climate of interior Alaska. *Arctic* 27, 273-86.

Szabo, J. P. (1980). Two pollen diagrams from Quaternary deposits in east-central Iowa, U.S.A. *Canadian Journal of Earth Science* 17, 453-58.

Thompson, P., Schwarcz, H. P., and Ford, D. C. (1974). Continental Pleistocene climatic variations from speleothem and isotopic data. *Science* 184, 893-95.

Thorson, R. W. (1980). Ice-sheet glaciation of the Puget lowland, Washington, during the Vashon stade (late Pleistocene). *Quaternary Research* 13, 303-21.

Trewartha, G. T. (1961). "The Earth's Problem Climates." University of Wisconsin Press, Madison.

Van Devender, T. R., and Spaulding, W. G. (1979). Development of vegetation and climate in the southwestern United States. *Science* 204, 701-10.

van Donk, J. (1976). O¹⁸ record of the Atlantic Ocean for the entire Pleistocene period. *In* "Investigation of Late Quaternary Paleoceanography and Paleoclimatology" (R. M. Cline and J. D. Hays, eds.), pp. 147-64. Geological Society of America Memoir 145.

Van Heuklon, T. K. (1977). Distant source of 1976 dustfall in Illinois and Pleistocene weather models. *Geology* 5, 693-95.

van Loon, H., and Williams, J. (1976). The connection between mean trends of temperature and circulation at the surface: Part I, Winter; Part II, Summer. *Monthly Weather Review* 104, 365-80 and 1003-11.

Van Zant, K. (1979). Late-glacial and postglacial pollen and plant macrofossils from Lake West Okoboji, northwestern Iowa. *Quaternary Research* 12, 358-80.

Wahrhaftig, C., and Birman, J. H. (1965). The Quaternary of the Pacific mountain system in California. *In* "The Quaternary of the United States" (H. E. Wright, Jr., and D. G. Frey, eds.), pp. 299-340. Princeton University Press, Princeton, N.J.

Waitt, R. B., Jr., and Thorson, R. M. (1983). The Cordilleran ice sheet in Washington, Idaho, and Montana. *In* "Late-Quaternary Environments of the United States," vol. 1, "The Late Pleistocene" (S. C. Porter, ed.), pp. 53-70. University of Minnesota Press, Minneapolis.

Walsh, J. E., and Mostek, A. (1980). A quantitative analysis of meteorological anomaly patterns over the United States, 1900-1977. *Monthly Weather Review* 108, 615-30.

Washburn, A. L. (1980a). "Geocryology: A Survey of Periglacial Processes and Environments." Halsted Press, John Wiley and Sons, New York.

Washburn, A. L. (1980b). Permafrost features as evidence of climatic change. *Earth Science Reviews* 15, 327-402.

Watts, W. A. (1970). The full-glacial vegetation of northwestern Georgia. *Ecology* 51, 17-33.

Watts, W. A. (1975a). A late Quaternary record of vegetation from Lake Annie, south-central Florida. *Geology* 3, 344-46.

Watts, W. A. (1975b). Vegetation record for the last 20,000 years from a small marsh on Lookout Mountain, northwestern Georgia. *Geological Society of America Bulletin* 86, 287-91.

Watts, W. A. (1979). Late Quaternary vegetation of central Appalachia and the New Jersey Coastal Plain. *Ecological Monographs* 49, 427-69.

Watts, W. A. (1980a). The late Quaternary vegetation history of southeastern United States. *Annual Review of Ecology and Systematics* 11, 387-409.

Watts, W. A. (1980b). Late Quaternary vegetation history of White Pond on the inner Coastal Plain of South Carolina. *Quaternary Research* 13, 187-99.

Watts, W. A., and Stuiver, M. (1980). Late Wisconsin climate of northern Florida and the origin of species-rich deciduous forest. *Science* 210, 325-27.

Webb, T., III, and Bryson, R. A. (1972). Late- and postglacial climatic change in the northern Midwest, U.S.A.: Quantitative estimates derived from fossil pollen spectra by multivariate statistical analysis. *Quaternary Research* 2, 70-115.

Webster, P. J., and Streten, N. A. (1978). Late Quaternary ice age climates of tropical Australia: Interpretation and reconstructions. *Quaternary Research* 10, 279-309.

Weide, D. L., and Weide, M. L. (1977). Time, space, and intensity in Great Basin paleoecological models. *In* "Models and Great Basin Prehistory: A Symposium" (D. D. Fowler, ed.), pp. 79-111. Desert Research Institute Publications in the Social Sciences 12.

Weller, G., and Holmgren, B. (1974). The microclimates of the Arctic tundra. *Journal of Applied Meteorology* 13, 854-62.

Wells, P. V. (1976). Macrofossil analysis of wood rat (*Neotoma*) middens as a key to the Quaternary vegetational history of arid America. *Quaternary Research* 6, 223-48.

Wells, P. V. (1979). An equable glaciopluvial in the West: Pleniglacial evidence of increased precipitation on a gradient from the Great Basin to the Sonoran and Chihuahuan Deserts. *Quaternary Research* 12, 311-25.

Wendler, G., Ishikawa, N., and Streten, N. (1974). The climate of the McCall Glacier, Brooks Range, Alaska, in relation to its geographical setting. *Arctic and Alpine Research* 6, 307-18.

Whitehead, D. R. (1979). Late-glacial and postglacial vegetational history of the Berkshires, western Massachusetts. *Quaternary Research* 12, 333-57.

Willett, H. C. (1950). The general circulation at the last (Würm) glacial. *Geografiska Annaler* 32, 179-87.

Williams, J. (1979). GCM experiments using glacial period and related boundary conditions. *In* "Report of the JOC Study Conference on Climate Models: Performance, Intercomparisons and Sensitivity Studies" (W. L. Gates, ed.), vol. 1, pp. 525-37. Global Atmospheric Research Programme Publications Series 22, World Meteorological Organization/International Council of Scientific Unions.

Williams, J., and Barry, R. G. (1975). Ice age experiments with the NCAR general circulation model: Conditions in the vicinity of the northern continental ice sheets. *In* "Climate of the Arctic" (G. Weller and S. A. Bowling, eds.), pp. 143-49. Geophysical Institute, University of Alaska, Fairbanks.

Williams, J., Barry, R. G., and Washington, W. M. (1974). Simulation of the atmospheric circulation using the NCAR global circulation model with ice age boundary conditions. *Journal of Applied Meteorology* 13, 305-17.

Williams, L. D. (1978). Ice sheet initiation and climatic influences of expanded snow cover in Arctic Canada. *Quaternary Research* 10, 141-49.

Williams, L. D. (1979). An energy-balance model of potential glacierization of northern Canada. *Arctic and Alpine Research* 11, 443-56.

Wright, H. E., Jr. (1976). Vegetational history of the Central Plains. *In* "Pleistocene and Recent Environments of the Central Great Plains" (W. Dort, Jr., and J. K. Jones, Jr., eds.), pp. 159-83. University Press of Kansas, Lawrence.

Wright, H. E., Jr. (1981). Vegetation east of the Rocky Mountains, 18,000 years ago. *Quaternary Research* 15, 13-25.

Wright, H. E., Jr., Bent, A. M., Hansen, B. S., and Maher, L. J., Jr. (1973). Present and past vegetation of the Chuska Mountains, northwestern New Mexico. *Geological Society of America Bulletin* 84, 1155-80.

Wright, H. E., Jr., and Lamb, H. H. (1973). A second approximation to the circulation patterns prevailing at the time of the last glacial maximum. *In* "Mapping the Atmospheric and Ocean Circulation and Other Climatic Parameters at the Time of the Last Glacial Maximum about 17,000 Years Ago," pp. 104-7. Climatic Research Unit, Report CRU-RP2, School of East Anglia. Norwich, England.

Zwick, T. T. (1980). A comparison between the modern and composite Pleistocene snow-lines, Absaroka and Beartooth Mountains, Montana-Wyoming, U.S.A. *Journal of Glaciology* 25, 347-52.